STATICS AND MECHANICS OF MATERIALS

Anthony Bedford • Wallace Fowler • Kenneth M. Liechti

University of Texas at Austin

Prentice
Hall

Pearson Education, Inc.
Upper Saddle River, New Jersey 07458

CIP Data Available.

Vice President and Editorial Director, ECS: *Marcia J. Horton*
Executive Editor: *Eric Svendsen*
Associate Editor: *Dee Bernhard*
Vice President and Director of Production and Manufacturing, ESM: *David W. Riccardi*
Executive Managing Editor: *Vince O'Brien*
Managing Editor: *David A. George*
Production Editor: *Tamar Savir*
Director of Creative Services: *Paul Belfanti*
Creative Director: *Carole Anson*
Managing Editor of Audio/Visual Assets: *Patricia Burns*
Audio/Visual Editor: *Xiaohong Zhu*
Art Director: *Jonathan Boylan*
Interior Designer: *Circa 86*
Cover Designer: *John Christiana*
Manufacturing Manager: *Trudy Pisciotti*
Manufacturing Buyer: *Lisa McDowell*

About the Cover: Photograph of *Fallingwater*, designed by Frank Lloyd Wright, located in Mill Run, PA, courtesy of Super Stock, Inc.
Additional cover images courtesy of Tony Stone Images.

© 2003 Pearson Education, Inc.
Pearson Education, Inc.
Upper Saddle River, New Jersey 07458

MATLAB is a registered trademark of The MathWorks, Inc., 3 Apple Hill Drive, Natick, MA 01760-2098.

Mathcad is a registered trademark of MathSoft Engineering and Education, 101 Main St., Cambridge, MA 02142-1521.

Printed in the United States of America

10   9   8   7   6   5   4   3   2   1

ISBN 0-13-028593-5

Pearson Education Ltd., *London*
Pearson Education Australia Pty, Ltd., *Sydney*
Pearson Education Singapore, Pte. Ltd.
Pearson Education North Asia Ltd., *Hong Kong*
Pearson Education Canada, Ltd., *Toronto*
Pearson Educación de Mexico, S.A. de C.V.
Pearson Education—Japan, *Tokyo*
Pearson Education Malaysia, Pte. Ltd.
Pearson Education, *Upper Saddle River*, *New Jersey*

CONTENTS

3 Forces   81

4 Systems of Forces and Moments   115

Preface

Our original objective in writing this book was to present the foundations and applications of statics and mechanics of materials as we do in the classroom. We used many sequences of figures, emulating the gradual development of a figure by a teacher explaining a concept. We stressed the importance of visual analysis in gaining understanding, especially through the use of free-body diagrams. Because inspiration is so conducive to learning, we based many of our examples and problems on a variety of modern engineering applications.

Key Series Features

This volume completes our series in Engineering Mechanics that includes

1. *Engineering Mechanics—Statics*, Third Edition, by Bedford and Fowler
2. *Engineering Mechanics—Dynamics*, Third Edition, by Bedford and Fowler
3. *Engineering Mechanics—Statics & Dynamics*, Third Edition, by Bedford and Fowler
4. *Engineering Mechanics—Statics Principles*, by Bedford and Fowler
5. *Engineering Mechanics—Dynamics Principles*, by Bedford and Fowler
6. *Engineering Mechanics—Statics & Dynamics Principles*, by Bedford and Fowler
7. *Mechanics of Materials* by Bedford and Liechti
8. *Statics and Mechanics of Materials* by Bedford, Fowler, and Liechti (*this volume*)

Positive responses from users and reviewers have led us to retain the basic organization, content, and features. During our preparation of this volume, we examined how we presented each concept, example, figure, summary statement, and problem. Where necessary, we made changes, additions, or deletions to simplify and clarify the presentation. This volume features

- Problems that contain elements of engineering design have been marked with a new $\mathscr{D}$ icon.

- We have added new examples where users indicated more were needed. Many of the new examples continue our emphasis on realistic and motivational applications and engineering design.

- New sets of **Study Questions** appear after most sections to help students check their retention of key concepts.

- Each example is clearly labeled for its teaching purpose.

- A redesigned text and added photographs throughout help students connect the text real world applications and situations.

Examples that Teach

The Strategy/Solution/Discussion framework employed by most of our examples is designed to emphasize the critical importance of good problem-solving skills. Our objective is to teach students how to approach problems and critically judge the results.

"Strategy" sections show the preliminary planning needed to begin a solution. What principles and equations apply? What must be determined, and in what order?

The solution is then described in detail, using sequences of figures when needed to clarify the steps.

"Discussion" sections point out properties of the solution, or comment on alternative solution methods, or suggest out ways to check answers.

Example 8.3

Analyzing a Friction Brake

The motion of the disk in Fig. 8.11 is controlled by the friction force exerted at C by the brake ABC. The hydraulic actuator BE exerts a horizontal force of magnitude F on the brake at B. The coefficients of friction between the disk and the brake are μ_s and μ_k. What couple M is necessary to rotate the disk at a constant rate in the counterclockwise direction?

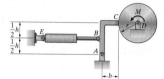

Figure 8.11

Strategy

We can use the free-body diagram of the disk to obtain a relation between M and the reaction exerted on the disk by the brake, then use the free-body diagram of the brake to determine the reaction in terms of F.

Solution

We draw the free-body diagram of the disk in Fig. a, representing the force exerted by the brake by a single force R. The force R opposes the counterclockwise rotation of the disk, and the friction angle is the angle of kinetic friction $\theta_k = \arctan \mu_k$. Summing moments about D, we obtain

$$\Sigma M_{(\text{point } D)} = M - (R \sin\theta_k)r = 0.$$

Then, from the free-body diagram of the brake (Fig. b), we obtain

$$\Sigma M_{(\text{point } A)} = -F\left(\frac{1}{2}h\right) + (R\cos\theta_k)h - (R\sin\theta_k)b = 0.$$

We can solve these two equations for M and R. The solution for the couple M is

$$M = \frac{(1/2)hr\,F\sin\theta_k}{h\cos\theta_k - b\sin\theta_k} = \frac{(1/2)hr\,F\mu_k}{h - b\mu_k}.$$

(a) The free-body diagram of the disk.

(b) The free-body diagram of the brake.

Discussion

If μ_k is sufficiently small, then the denominator of the solution for the couple, $(h\cos\theta_k - b\sin\theta_k)$, is positive. As μ_k becomes larger, the denominator becomes smaller, because $\cos\theta_k$ decreases and $\sin\theta_k$ increases. As the denominator approaches zero, the couple required to rotate the disk approaches infinity. To understand this result, notice that the denominator equals zero when $\tan\theta_k = h/b$, which means that the line of action of R passes through point A (Fig. c). As μ_k becomes larger and the line of action of R approaches point A, the magnitude of R necessary to balance the moment of F about A approaches infinity and, as a result, M approaches infinity.

(c) The line of action of R passing through point A.

Commitment to Students and Instructors

In preparing this volume and the accompanying solutions manual, we have taken precautions to ensure accuracy to the best of our ability. We have each solved the new problems in an effort to be sure that their answers are correct and that they are of an appropriate level of difficulty. Karim Nohra of the University of South Florida also checked the text, examples, problems and solutions manual. Any errors that remain are the responsibility of the authors.

We welcome communication from students and instructors concerning errors or areas for improvement. Our mailing address is Department of Aerospace Engineering and Engineering Mechanics, University of Texas at Austin, Austin, Texas 78712. Our electronic mail address is *abedford@mail.utexas.edu*.

Supplements

Instructor's Solutions Manual This supplement, available to instructors contains complete solutions to all of the problems and several sample syllabi. Each solution comes with problem statement as well as associated artwork.

Study Packs are designed to give students the tools to improve their mechanics study skills. They consists of three study components: a free body-diagram workbook, a Visualization CD based on Working Model Software, and an access code to a website with 500 sample Statics and Dynamics problems and solutions.

- **Free-Body Diagram Workbook** prepared by Peter Schiavone of the University of Alberta. This workbook begins with a tutorial on free body diagrams and then includes 50 practice problems of progressing difficulty with complete solutions. Further "strategies and tips" help students understand how to use the diagrams in solving the accompanying problems.

- **Working Model CD** contains 25 pre-set simulations of Statics examples that include questions for further exploration. Simulations are powered by the Working Model Engine and were created with actual artwork from the text to enhance their correlation with the text.

- **Password-Protected Website** contains 500 sample Statics and Dynamics problems for students to study. All problems are supplemental and do not appear in this volume. Student passwords are printed on the inside cover of the Free-Body Diagram Workbook. To access this site, students should go to http://www.prenhall.com/bedford and follow the on-line directions to register.

The Study Packs are available as a stand-alone item or can be bundled with this volume at additional cost. Order stand-alone Study Packs with the ISBN 0-13-061574-9.

MATLAB® /Mathcad® Tutorials Twenty tutorials showing how to use computational software in engineering mechanics. Each tutorial discusses a basic mechanics concept, and then shows how to solve a specific problem related to this concept using MATLAB/Mathcad. There are twenty tutorials each for MATLAB and Mathcad, and are available in PDF format from the password-protected area of the Bedford website. Passwords appear in each student study pack. Worksheets were developed by Ronald Larsen and Stephen Hunt of Montana State University—Bozeman.

Website—http://www.prenhall.com/bedford contains mechanics multiple-choice and True/False quizzes developed by Karim Nohra of the University of South Florida. Web Assessment, MATLAB/Mathcad tutorials, and Study Pack questions and solutions are all available at the password protected part of this website. Passwords for the protected portion are printed in the Statics Study Pack.

ADAMS Simulations for Dynamics Mechanical Dynamics, Inc. has created over 100 simulations of problems from *Dynamics* using their ADAMS simulation/protyping software. Professors and students can simulate and observe the effects of changing parameters in systems and gain deeper insight into their behavior. Simulations also come with an accompanying avi "movie" file. Files are located at the password-protected part of the website. Students should use their study pack passwords and professors should contact their Prentice Hall representative for professor access. Qualified adopters may also be able to obtain free site licenses. Contact *university@adams.com* for more information.

Web Assessment Software PH Grade Assist lets students solve statics problems with randomized variables so each student solves a slightly different problem. For more information and pricing, contact your local Prentice Hall representative or visit **www.prenhall.com/bedford** and choose Engineering Mechanics.

Acknowledgments

Many students and teachers have given us insightful comments on our texts. The following academic colleagues have made valuable suggestions regarding this series.

Edward E. Adams
Michigan Technological University

Raid S. Al-Akkad
University of Dayton

Jerry L. Anderson
Memphis State University

James G. Andrews
University of Iowa

Robert J. Asaro
University of California, San Diego

Leonard B. Baldwin
University of Wyoming

Gautam Batra
University of Nebraska

Mary Bergs
Marquette University

Spencer Brinkerhoff
Northern Arizona University

L.M. Brock
University of Kentucky

William (Randy) Burkett
Texas Tech University

Donald Carlson
University of Illinois

Major Robert M. Carpenter
U.S. Military Academy

Douglas Carroll
University of Missouri, Rolla

Paul C. Chan
New Jersey Institute of Technology

Namas Chandra
Florida State University

James Cheney
University of California, Davis

Ravinder Chona
Texas A & M University

Anthony DeLuzio
Merrimack College

Mitsunori Denda
Rutgers University

James F. Devine
University of South Florida

Craig Douglas
University of Massachusetts, Lowell

Marijan Dravinski
University of Southern California

S. Olani Durrant
Brigham Young University

Estelle Eke
California State University, Sacramento

William Ferrante
University of Rhode Island

Robert W. Fitzgerald
Worcester Polytechnic Institute

George T. Flowers
Auburn University

Mark Frisina
Wentworth Institute

Robert W. Fuessle
Bradley University

William Gurley
University of Tennessee, Chattanooga

John Hansberry
University of Massachusetts, Dartmouth

W. C. Hauser
California Polytechnic University Pomona

Linda Hayes
University of Texas - Austin

R. Craig Henderson
Tennessee Technological University

James Hill
University of Alabama

Allen Hoffman
Worcester Polytechnic Institute

Edward E. Hornsey
University of Missouri, Rolla

Robert A. Howland
University of Notre Dame

Joe Ianelli
University of Tennessee, Knoxville

Ali Iranmanesh
Gadsden State Community College

David B. Johnson
Southern Methodist University

E. O. Jones, Jr.
Auburn University

Serope Kalpakjian
Illinois Institute of Technology

Kathleen A. Keil
California Polytechnic University San Luis Obispo

Yohannes Ketema
University of Minnesota

Seyyed M. H. Khandani
Diablo Valley College

Charles M. Krousgrill
Purdue University

B. Kent Lall
Portland State University

Kenneth W. Lau
University of Massachusetts, Lowell

Norman Laws
University of Pittsburgh

William M. Lee
U.S. Naval Academy

Donald G. Lemke
University of Illinois, Chicago

Richard J. Leuba
North Carolina State University

Richard Lewis
Louisiana Technological University

Bertram Long
Northeastern University

V. J. Lopardo
U.S. Naval Academy

Frank K. Lu
University of Texas, Arlington

K. Madhaven
Christian Brothers College

Gary H. McDonald
University of Tennessee

James McDonald
Texas Technical University

Jim Meagher
California Polytechnic State University, San Luis Obispo

Lee Minardi
Tufts University

Norman Munroe
Florida International University

Shanti Nair
University of Massachusetts, Amherst

Saeed Niku
California Polytechnic State University, San Luis Obispo

Harinder Singh Oberoi
Western Washington University

James O'Connor
University of Texas, Austin

Samuel P. Owusu-Ofori
North Carolina A& T State University

Venkata Panchakarla
Florida State University

Assimina A Pelegri
Rutgers University

Noel C. Perkins
University of Michigan

David J. Purdy
Rose-Hulman Institute of Technology

Colin E Ratcliffe
U.S. Naval Academy

Daniel Riahi
University of Illinois

Charles Ritz
California Polytechnic State University Pomona

George Rosborough
University of Colorado, Boulder

Robert Schmidt
University of Detroit

Robert J. Schultz
Oregon State University

Patricia M. Shamamy
Lawrence Technological University

Sorin Siegler
Drexel University

L. N. Tao
Illinois Institute of Technology

Craig Thompson
Western Wyoming Community College

John Tomko
Cleveland State University

Kevin Z. Truman
Washington University

John Valasek
Texas A &M University

Dennis VandenBrink
Western Michigan University

Thomas J. Vasko
University of Hartford

Mark R. Virkler
University of Missouri, Columbia

William H. Walston, Jr.
University of Maryland

Reynolds Watkins
Utah State University

Charles White
Northeastern University

Norman Wittels
Worcester Polytechnic Institute

Julius P. Wong
University of Louisville

T. W. Wu
University of Kentucky

Constance Ziemian
Bucknell University

Books of this kind represent a collaborative effort by many individuals. It has been our good fortune to work with extremely talented and agreeable people at Prentice Hall. In large measure we owe the improvements in this edition to our editor and mentor Eric Svendsen. We have not met a more creative, hard-working and conscientious person in publishing. Eric's efforts and ours were overseen, inspired, and supported by Marcia Horton. Tamar Savir, our production editor, helped us through the day-to-day crises of the production process. David George was our Managing Editor. Jon Boylan and John Christiana were responsible for the book's design, and Xiaohong Zhu coordinated the art program. Karim Nohra of the University of South Florida carefully examined the text, examples and problems for accuracy. Among the many other people on whom we relied, we mention particularly Dee Bernhard, Lisa McDowell, Trudy Pisciotti, and Holly Stark.

Increasingly, a textbook is only part of an integrated set of pedagogical tools. Our website was created and is maintained by Daniel Sandin. Peter Schiavone has written a free-body diagram workbook that supplements and expands upon the book's treatment. Ronald Larsen and Steven Hunt have developed MATLAB and Mathcad worksheets based on problems and examples in the book for optional use by instructors and students.

And we thank our wives, for their continued support, patience, and acceptance of years of lost weekends.

Anthony Bedford, Wallace Fowler, and Kenneth M. Liechti
Austin, Texas

Anthony Bedford is Professor of Aerospace Engineering and Engineering Mechanics at the University of Texas at Austin. He received the B.S. degree from the University of Texas at Austin, the M.S. degree from the California Institute of Technology, and the Ph.D. degree from Rice University in 1967. He has industrial experience at Douglas Aircraft Company and at TRW, where he did structural dynamics and trajectory analyses for the Apollo program. He has been on the faculty of the University of Texas at Austin since 1968. He is a member of the University of Texas Academy of Distinguished Teachers and has received several teaching awards over the years.

Dr. Bedford's main professional activity has been education and research in engineering mechanics. He has been principal investigator on grants from the National Science Foundation and the Office of Naval Research, and from 1973 until 1983 was a consultant to Sandia National Laboratories, Albuquerque, New Mexico. His other books include *Hamilton's Principle in Continuum Mechanics*, *Introduction to Elastic Wave Propagation* (with D.S. Drumheller), and *Mechanics of Materials* (with K.M. Liechti).

Wallace T. Fowler holds the Paul D. and Betty Robertson Meek Professorship in Engineering in the Department of Aerospace Engineering and Engineering Mechanics at University of Texas at Austin. Dr. Fowler received the B.A., M.S., and Ph.D. degrees from the University of Texas at Austin, and has been on the faculty there since 1965. During the Fall of 1976, he was on the staff of the United States Air Force Test Pilot School, Edwards Air Force Base, California, and in 1981–1982 he was a visiting professor at the United States Air Force Academy. Since 1991 he has been Associate Director of the Texas Space Grant Consortium.

Dr. Fowler's areas of teaching and research are dynamics, orbital mechanics, and spacecraft mission design. He is author or coauthor of technical papers on trajectory optimization, attitude dynamics, and space mission planning and has also published papers on the theory and practice of engineering teaching. He has received numerous teaching awards including the Chancellor's Council Outstanding Teaching Award, the General Dynamics Teaching Excellence Award, the Halliburton Education Foundation Award of Excellence, the ASEE Fred Merryfield Design Award, and the AIAA-ASEE Distinguished Aerospace Educator Award. He is a member of the Academy of Distinguished Teachers at the University of Texas at Austin. He is a licensed professional engineer, a member of several technical societies, and a Fellow of both the American Institute of Aeronautics and Astronautics and the American Society for Engineering Education. In 2000–2001, he served as president of the American Society for Engineering Education.

Kenneth M. Liechti is a Professor of Aerospace Engineering and Engineering Mechanics at the University of Texas at Austin and holds the E. P. Schoch Professorship in Engineering. He received the B.Sc. degree in aeronautical engineering from Glascow University and the M.S. and Ph.D. degrees in aeronautics from the California Institute of Technology. He gained industrial experience at the Fort Worth Division of General Dynamics prior to joining the faculty of the University of Texas at Austin in 1982.

His primary areas of teaching and research are in the mechanics of materials and fracture mechanics. He is the author or coauthor of papers on interfacial fracture, fracture in adhesively bonded joints, and the nonlinear behavior of polymers. He has consulted on fracture problems with several companies.

Dr. Liechti is a Fellow of the American Society of Mechanical Engineers and a member of the Society for Experimental Mechanics, the American Academy of Mechanics, and the Adhesion Society. He is an associate editor of the journal *Experimental Mechanics*, published by the Society for Experimental Mechanics.

Photo Credits

Chapter 1

Opener, *Tony Stone Images*
Figure 1.5, *NASA Johnson Space Station*

Chapter 2

Opener, *Tony Stone Images*
Figure P2.29, *Corbis*
Figure P2.125, *NASA*
Figure P2.41, *Corbis*
Figure P2.125, *NASA*

Chapter 3

Opener, *Tony Stone Images*
Figure P3.20, *Corbis*

Chapter 4

Opener, *Tony Stone Images*

Chapter 5

Opener, *NASA Headquarters*

Chapter 6

Opener, *Photo Researchers, Inc.*
Figure 6.3, Brownie Harris/*The Tock Market*

Chapter 7

Opener, *Tony Stone Images*

Chapter 8

Opener, *Tony Stone Images*
Figure 8.4, *Courtesy of Uzi Landman*
Figure 8.13a, John Reader/*Photo Researchers, Inc.*
Figure 8.20a, *Courtesy of SKF Industries*
Figure P8.47, *Rainbow*
Figure P8.95, *Omni-Photo Communications, Inc.*

Chapter 9

Figure P9.54, *Beech Aircraft Corporation*

Chapter 10

Opener, Robert Laberge/*Allsport Photography (USA, Inc.)*
Figure 10.32, Prof. Roy E. Olsen
Figure 10.41, provided by the authors

Chapter 11

Openers, *Ride & Drive* magazine
Figure 11.16, authors
Figure P11.19, *NASA Headquarters*

Chapter 12

Opener, Richard Pasley/*Stock Boston*
Figure 12.38, Don Morely/*Tony Stone Images*
Figure P12.8, *Tony Stone Images*
Figure P12.16, *Textron Inc.*
Figure P12.64, Dale Boyer/*Tony Stone Images*
Figure P12.43, *NASA Headquarters*
Figure P12.48, *NASA Headquarters*

Chapter 13

Opener, French Government Tourist Office
Figure P13.11, French Government Tourist Office
Figure 13.10, *Corbis*
Figure P13.40, *NASA/Glenn Research Center*

Chapter 14

Openers, Julius Shulman

Chapter 15

Opener, Brian Parker/*Tom Stack & Associates*
Figure 15.1, authors

Chapter 16

Openers, authors

Chapter 17

Openers, Bob Rowan/*Corbis*
Figure 17.2, authors

STATICS AND MECHANICS OF MATERIALS

The architects and engineers are guided by the principles of statics during each step of the design and construction of a building. Statics is one of the sciences underlying the art of structural design.

Introduction

Engineers are responsible for the design, construction, and testing of the devices we use, from simple things such as chairs and pencil sharpeners to complicated ones such as dams, cars, airplanes, and spacecraft. They must have a deep understanding of the physics underlying these devices and must be familiar with the use of mathematical models to predict system behavior. Students of engineering begin to learn how to analyze and predict the behavior of physical systems by studying mechanics.

1.1 Engineering and Mechanics

How do engineers design complex systems and predict their characteristics before they are constructed? Engineers have always relied on their knowledge of previous designs, experiments, ingenuity, and creativity to develop new designs. Modern engineers add a powerful technique: They develop mathematical equations based on the physical characteristics of the devices they design. With these mathematical models, engineers predict the behavior of their designs, modify them, and test them prior to their actual construction.

At its most basic level, mechanics is the study of forces and their effects. The results obtained in elementary mechanics apply directly to many fields of engineering. Mechanical and civil engineers who design structures use the equilibrium equations derived in statics. Mechanics was the first analytical science; consequently fundamental concepts, analytical methods, and analogies from mechanics are found in virtually every field of engineering. Students of chemical and electrical engineering gain a deeper appreciation for basic concepts in their fields such as equilibrium, energy, and stability by learning them in their original mechanical contexts. By studying mechanics, they retrace the historical development of these ideas.

1.2 Learning Mechanics

Mechanics consists of broad principles that govern the behavior of objects. In this book we describe these principles and provide examples that demonstrate some of their applications. Although it is essential that you practice working problems similar to these examples, and we include many problems of this kind, our objective is to help you understand the principles well enough to apply them to situations that are new to you. Each generation of engineers confronts new problems.

Problem Solving

In the study of mechanics you learn problem-solving procedures you will use in succeeding courses and throughout your career. Although different types of problems require different approaches, the following steps apply to many of them:

- Identify the information that is given and the information, or answer, you must determine. It's often helpful to restate the problem in your own words. When appropriate, make sure you understand the physical system or model involved.
- Develop a *strategy* for the problem. This means identifying the principles and equations that apply and deciding how you will use them to solve the problem. Whenever possible, draw diagrams to help visualize and solve the problem.

- Whenever you can, try to predict the answer. This will develop your intuition and will often help you recognize an incorrect answer.

- Solve the equations and, whenever possible, interpret your results and compare them with your prediction. This last step is a *reality check*. Is your answer reasonable?

Calculators and Computers

Most of the problems in this book are designed to lead to an algebraic expression with which to calculate the answer in terms of given quantities. A calculator with trigonometric and logarithmic functions is sufficient to determine the numerical value of such answers. The use of a programmable calculator or a computer with problem-solving software such as *Mathcad* or MATLAB is convenient, but be careful not to become too reliant on tools you will not have during tests.

Engineering Applications

Although the problems are designed primarily to help you learn mechanics, many of them illustrate uses of mechanics in engineering. We also include problems that emphasize two essential aspects of engineering:

- *Design.* Some problems ask you to choose values of parameters to satisfy stated design criteria.
- *Safety.* Some problems ask you to evaluate the safety of devices and choose values of parameters to satisfy stated safety requirements.

Subsequent Use of This Text

This book contains tables and information you will find useful in subsequent engineering courses and throughout your engineering career. In addition, you will often want to review fundamental engineering subjects, both during the remainder of your formal education and when you are a practicing engineer. The most efficient way to do so is by using the textbooks with which you are familiar. Your engineering textbooks will form the core of your professional library.

1.3 Fundamental Concepts

Some topics in mechanics will be familiar to you from everyday experience or from previous exposure to them in mathematics and physics courses. In this section we briefly review the foundations of elementary mechanics.

Numbers

Engineering measurements, calculations, and results are expressed in numbers. You need to know how we express numbers in the examples and problems and how to express the results of your own calculations.

Significant Digits This term refers to the number of meaningful (that is, accurate) digits in a number, counting to the right starting with the first nonzero digit. The two numbers 7.630 and 0.007630 are each stated to four significant digits. If only the first four digits in the number 7,630,000 are known to be accurate, this can be indicated by writing the number in scientific notation as 7.630×10^6.

If a number is the result of a measurement, the significant digits it contains are limited by the accuracy of the measurement. If the result of a measurement is stated to be 2.43, this means that the actual value is believed to be closer to 2.43 than to 2.42 or 2.44.

Numbers may be rounded off to a certain number of significant digits. For example, we can express the value of π to three significant digits, 3.14, or we can express it to six significant digits, 3.14159. When you use a calculator or computer, the number of significant digits is limited by the number of digits the machine is designed to carry.

Use of Numbers in This Book You should treat numbers given in problems as exact values and not be concerned about how many significant digits they contain. If a problem states that a quantity equals 32.2, you can assume its value is 32.200. ... We express intermediate results and answers in the examples and the answers to the problems to at least three significant digits. If you use a calculator, your results should be that accurate. Be sure to avoid round-off errors that occur if you round off intermediate results when making a series of calculations. Instead, carry through your calculations with as much accuracy as you can by retaining values in your calculator.

Space and Time

Space simply refers to the three-dimensional universe in which we live. Our daily experiences give us an intuitive notion of space and the locations, or positions, of points in space. The distance between two points in space is the length of the straight line joining them.

Measuring the distance between points in space requires a unit of length. We use both the International System of units, or SI units, and U.S. Customary units. In SI units, the unit of length is the meter (m). In U.S. Customary units, the unit of length is the foot (ft).

Time is, of course, familiar—our lives are measured by it. The daily cycles of light and darkness and the hours, minutes, and seconds measured by our clocks and watches give us an intuitive notion of time. Time is measured by the intervals between repeatable events, such as the swings of a clock pendulum or the vibrations of a quartz crystal in a watch. In both SI units and U.S. Customary units, the unit of time is the second (s). The minute (min), hour (hr), and day are also frequently used.

If the position of a point in space relative to some reference point changes with time, the rate of change of its position is called its *velocity*, and the rate of change of its velocity is called its *acceleration*. In SI units, the velocity is expressed in meters per second (m/s) and the acceleration is expressed in meters per second per second, or meters per second squared (m/s^2). In U.S. Customary units, the velocity is expressed in feet per second (ft/s) and the acceleration is expressed in feet per second squared (ft/s^2).

Newton's Laws

Elementary mechanics was established on a firm basis with the publication in 1687 of *Philosophiae naturalis principia mathematica*, by Isaac Newton. Although highly original, it built on fundamental concepts developed by many others during a long and difficult struggle toward understanding (Fig. 1.1).

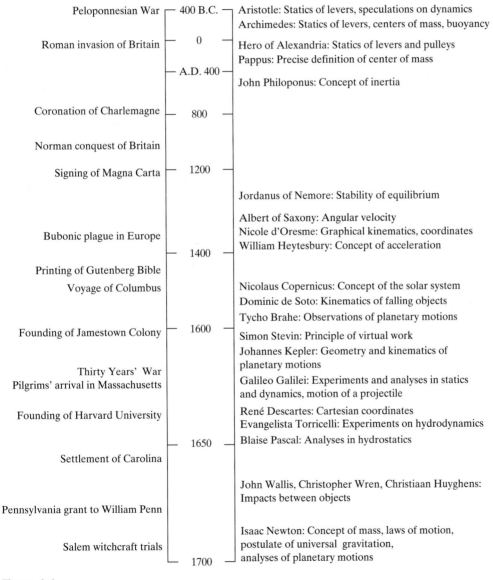

Figure 1.1
Chronology of developments in mechanics up to the publication of Newton's *Principia* in relation to other events in history.

Newton stated three "laws" of motion, which we express in modern terms:

1. *When the sum of the forces acting on a particle is zero, its velocity is constant. In particular, if the particle is initially stationary, it will remain stationary.*

2. *When the sum of the forces acting on a particle is not zero, the sum of the forces is equal to the rate of change of the linear momentum of the particle. If the mass is constant, the sum of the forces is equal to the product of the mass of the particle and its acceleration.*

3. *The forces exerted by two particles on each other are equal in magnitude and opposite in direction.*

Notice that we did not define force and mass before stating Newton's laws. The modern view is that these terms are defined by the second law. To demonstrate, suppose that we choose an arbitrary object and define it to have unit mass. Then we define a unit of force to be the force that gives our unit mass an acceleration of unit magnitude. In principle, we can then determine the mass of any object: We apply a unit force to it, measure the resulting acceleration, and use the second law to determine the mass. We can also determine the magnitude of any force: We apply it to our unit mass, measure the resulting acceleration, and use the second law to determine the force.

Thus Newton's second law gives precise meanings to the terms *mass* and *force*. In SI units, the unit of mass is the kilogram (kg). The unit of force is the newton (N), which is the force required to give a mass of one kilogram an acceleration of one meter per second squared. In U.S. Customary units, the unit of force is the pound (lb). The unit of mass is the slug, which is the amount of mass accelerated at one foot per second squared by a force of one pound.

Although the results we discuss in this book are applicable to many of the problems met in engineering practice, there are limits to the validity of Newton's laws. For example, they don't give accurate results if a problem involves velocities that are not small compared to the velocity of light $(3 \times 10^8 \text{ m/s})$. Einstein's special theory of relativity applies to such problems. Elementary mechanics also fails in problems involving dimensions that are not large compared to atomic dimensions. Quantum mechanics must be used to describe phenomena on the atomic scale.

Study Questions

1. What is the definition of the significant digits of a number?
2. What are the units of length, mass, and force in the SI system?

1.4 Units

The SI system of units has become nearly standard throughout the world. In the United States, U.S. Customary units are also used. In this section we summarize these two systems of units and explain how to convert units from one system to another.

International System of Units

In SI units, length is measured in meters (m) and mass in kilograms (kg). Time is measured in seconds (s), although other familiar measures such as minutes (min), hours (hr), and days are also used when convenient. Meters, kilograms, and seconds are called the *base units* of the SI system. Force is measured in newtons (N). Recall that these units are related by Newton's second law: One newton is the force required to give an object of one kilogram mass an acceleration of one meter per second squared:

$$1\ N = (1\ kg)(1\ m/s^2) = 1\ kg\text{-}m/s^2.$$

Because the newton can be expressed in terms of the base units, it is called a *derived unit*.

To express quantities by numbers of convenient size, multiples of units are indicated by prefixes. The most common prefixes, their abbreviations, and the multiples they represent are shown in Table 1.1. For example, 1 km is 1 kilometer, which is 1000 m, and 1 Mg is 1 megagram, which is 10^6 g, or 1000 kg. We frequently use kilonewtons (kN).

Table 1.1 The common prefixes used in SI units and the multiples they represent.

Prefix	Abbreviation	Multiple
nano-	n	10^{-9}
micro-	μ	10^{-6}
milli-	m	10^{-3}
kilo-	k	10^{3}
mega-	M	10^{6}
giga-	G	10^{9}

U.S. Customary Units

In U.S. Customary units, length is measured in feet (ft) and force is measured in pounds (lb). Time is measured in seconds (s). These are the base units of the U.S. Customary system. In this system of units, mass is a derived unit. The unit of mass is the slug, which is the mass of material accelerated at one foot per second squared by a force of one pound. Newton's second law states that

$$1\ lb = (1\ slug)(1\ ft/s^2).$$

From this expression we obtain

$$1\ slug = 1\ lb\text{-}s^2/ft.$$

We use other U.S. Customary units such as the mile (1 mi = 5280 ft) and the inch (1 ft = 12 in.). We also use the kilopound (kip), which is 1000 lb.

Angular Units

In both SI and U.S. Customary units, angles are normally expressed in radians (rad). We show the value of an angle θ in radians in Fig. 1.2. It is defined to be the ratio of the part of the circumference subtended by θ to the radius of the circle. Angles are also expressed in degrees. Since there are 360 degrees

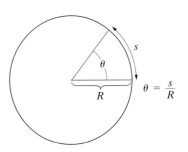

Figure 1.2
Definition of an angle in radians.

(360°) in a complete circle, and the complete circumference of the circle is $2\pi R$, 360° equals 2π rad.

Equations containing angles are nearly always derived under the assumption that angles are expressed in radians. Therefore when you want to substitute the value of an angle expressed in degrees into an equation, you should first convert it into radians. A notable exception to this rule is that many calculators are designed to accept angles expressed in either degrees or radians when you use them to evaluate functions such as $\sin \theta$.

Conversion of Units

Many situations arise in engineering practice that require you to convert values expressed in units of one kind into values in other units. If some data in a problem are given in terms of SI units and some are given in terms of U.S. Customary units, you must express all of the data in terms of one system of units. In problems expressed in terms of SI units, you will occasionally be given data in terms of units other than the base units of seconds, meters, kilograms, and newtons. You should convert these data into the base units before working the problem. Similarly, in problems involving U.S. Customary units, you should convert terms into the base units of seconds, feet, slugs, and pounds. After you gain some experience, you will recognize situations in which these rules can be relaxed, but for now the procedure we propose is the safest.

Converting units is straightforward, although you must do it with care. Suppose that we want to express 1 mi/hr in terms of ft/s. Since one mile equals 5280 ft and one hour equals 3600 seconds, we can treat the expressions

$$\left(\frac{5280 \text{ ft}}{1 \text{ mi}}\right) \quad \text{and} \quad \left(\frac{1 \text{ hr}}{3600 \text{ s}}\right)$$

as ratios whose values are 1. In this way we obtain

$$1 \text{ mi/hr} = 1 \text{ mi/hr} \times \left(\frac{5280 \text{ ft}}{1 \text{ mi}}\right) \times \left(\frac{1 \text{ hr}}{3600 \text{ s}}\right) = 1.47 \text{ ft/s}.$$

We give some useful unit conversions in Table 1.2.

Table 1.2 Unit conversions.

Time	1 minute	=	60 seconds
	1 hour	=	60 minutes
	1 day	=	24 hours
Length	1 foot	=	12 inches
	1 mile	=	5280 feet
	1 inch	=	25.4 millimeters
	1 foot	=	0.3048 meters
Angle	2π radians	=	360 degrees
Mass	1 slug	=	14.59 kilograms
Force	1 pound	=	4.448 newtons

Study Questions

1. What are the base units of the SI and U.S. Customary systems?
2. What is the definition of an angle in radians?

Example 1.1

Converting Units of Pressure

The pressure exerted at a point of the hull of the deep submersible in Fig. 1.3 is 3.00×10^6 Pa (pascals). A pascal is 1 newton per square meter. Determine the pressure in pounds per square foot.

Figure 1.3
Deep Submersible Vehicle.

Strategy

From Table 1.2, 1 pound = 4.448 newtons and 1 foot = 0.3048 meters. With these unit conversions we can calculate the pressure in pounds per square foot.

Solution

The pressure (to three significant digits) is

$$3.00 \times 10^6 \, \text{N/m}^2 = 3.00 \times 10^6 \, \text{N/m}^2 \times \left(\frac{1 \, \text{lb}}{4.448 \, \text{N}} \right) \times \left(\frac{0.3048 \, \text{m}}{1 \, \text{ft}} \right)^2$$

$$= 62{,}700 \, \text{lb/ft}^2.$$

Discussion

From the table of unit conversions in the inside front cover, 1 Pa = 0.0209 lb/ft^2. Therefore an alternative solution is

$$3.00 \times 10^6 \, \text{N/m}^2 = 3.00 \times 10^6 \, \text{N/m}^2 \times \left(\frac{0.0209 \, \text{lb/ft}^2}{1 \, \text{N/m}^2} \right)$$

$$= 62{,}700 \, \text{lb/ft}^2.$$

Example 1.2

Determining Units from an Equation

Suppose that in Einstein's equation

$$E = mc^2,$$

the mass m is in kilograms and the velocity of light c is in meters per second.
(a) What are the SI units of E?
(b) If the value of E in SI units is 20, what is its value in U.S. Customary base units?

Strategy

(a) Since we know the units of the terms m and c, we can deduce the units of E from the given equation.
(b) We can use the unit conversions for mass and length from Table 1.2 to convert E from SI units to U.S. Customary units.

Solution

(a) From the equation for E,

$$E = (m\ \text{kg})(c\ \text{m/s})^2.$$

the SI units of E are kg-m^2/s^2.
(b) From Table 1.2, 1 slug = 14.59 kg and 1 ft = 0.3048 m. Therefore

$$1\ \text{kg-m}^2/\text{s}^2 = 1\ \text{kg-m}^2/\text{s}^2 \times \left(\frac{1\ \text{slug}}{14.59\ \text{kg}}\right) \times \left(\frac{1\ \text{ft}}{0.3048\ \text{m}}\right)^2$$

$$= 0.738\ \text{slug-ft}^2/\text{s}^2.$$

The value of E in U.S. Customary units is

$$E = (20)(0.738) = 14.8\ \text{slug-ft}^2/\text{s}^2.$$

Discussion

In part (a) we determined the units of E by using the fact that an equation must be dimensionally consistent. That is, the dimensions, or units, of each term must be the same.

1.5 Newtonian Gravitation

Newton postulated that the gravitational force between two particles of mass m_1 and m_2 that are separated by a distance r (Fig. 1.4) is

$$F = \frac{Gm_1m_2}{r^2},$$ (1.1)

Figure 1.4
The gravitational forces between two particles are equal in magnitude and directed along the line between them.

where G is called the universal gravitational constant. Based on this postulate, he calculated the gravitational force between a particle of mass m_1 and a homogeneous sphere of mass m_2 and found that it is also given by Eq. (1.1), with r denoting the distance from the particle to the center of the sphere. Although the earth is not a homogeneous sphere, we can use this result to approximate the weight of an object of mass m due to the gravitational attraction of the earth.

$$W = \frac{Gmm_E}{r^2},$$ (1.2)

where m_E is the mass of the earth and r is the distance from the center of the earth to the object. Notice that the weight of an object depends on its location relative to the center of the earth, whereas the mass of the object is a measure of the amount of matter it contains and doesn't depend on its position.

When an object's weight is the only force acting on it, the resulting acceleration is called the acceleration due to gravity. In this case, Newton's second law states that $W = ma$, and from Eq. (1.2) we see that the acceleration due to gravity is

$$a = \frac{Gm_E}{r^2}.$$ (1.3)

The *acceleration due to gravity at sea level* is denoted by g. Denoting the radius of the earth by R_E, we see from Eq. (1.3) that $Gm_E = gR_E^2$. Substituting this result into Eq. (1.3), we obtain an expression for the acceleration due to gravity at a distance r from the center of the earth in terms of the acceleration due to gravity at sea level:

$$a = g\frac{R_E^2}{r^2}.$$ (1.4)

Since the weight of the object $W = ma$, the weight of an object at a distance r from the center of the earth is

$$W = mg\frac{R_E^2}{r^2}.$$ (1.5)

At sea level ($r = R_E$), the weight of an object is given in terms of its mass by the simple relation

$$W = mg.$$ (1.6)

The value of g varies from location to location on the surface of the earth. The values we use in examples and problems are $g = 9.81$ m/s^2 in SI units and $g = 32.2$ ft/s^2 in U.S. Customary units.

Study Questions

1. Does the weight of an object depend on its location?
2. If you know an object's mass, how do you determine its weight at sea level?

Example 1.3

Determining an Object's Weight

In its final configuration, the International Space Station (Fig. 1.5) will have a mass of approximately 450,000 kg.

(a) What would be the weight of the ISS if it were at sea level?

(b) The orbit of the ISS is 354 km above the surface of the earth. The earth's radius is 6370 km. What is the weight of the ISS (the force exerted on it by gravity) when it is in orbit?

Figure 1.5
International Space Station.

Strategy

(a) The weight of an object at sea level is given by Eq. (1.6). Because the mass is given in kilograms, we will express g in SI units: $g = 9.81$ m/s^2.
(b) The weight of an object at a distance r from the center of the earth is given by Eq. (1.5).

Solution

(a) The weight at sea level is

$$W = mg$$
$$= (450{,}000)(9.81)$$
$$= 4.41 \times 10^6 \text{ N}.$$

(b) The weight in orbit is

$$W = mg\,\frac{R_E^2}{r^2}$$
$$= (450{,}000)(9.81)\,\frac{(6{,}370{,}000)^2}{(6{,}370{,}000 + 354{,}000)^2}$$
$$= 3.96 \times 10^6 \text{ N}.$$

Discussion

Notice that the force exerted on the ISS by gravity when it is in orbit is approximately 90% of its weight at sea level.

Problems

1.1 Express the fractions $\frac{1}{3}$ and $\frac{2}{3}$ to three significant digits.

1.2 What is the value of e (the base of natural logarithms) to five significant digits?

1.3 A machinist drills a circular hole in a panel with radius $r = 5$ mm. Determine the circumference C and the area A of the hole to four significant digits.

1.4 The opening in a soccer goal is 24 ft wide and 8 ft high. Use these values to determine its dimensions in meters to three significant digits.

1.5 The central span of the Golden Gate Bridge is 1280 m long. What is its length in miles to three significant digits?

1.6 Suppose that you have just purchased a Ferrari F355 coupe and you want to know whether you can use your set of SAE (U.S. Customary unit) wrenches to work on it. You have wrenches with widths $w = 1/4$ in., $1/2$ in., $3/4$ in., and 1 in., and the car has nuts with dimensions $n = 5$ mm, 10 mm, 15 mm, 20 mm, and 25 mm.

Defining a wrench to fit if w is no more than 2% larger than n, which of your wrenches can you use?

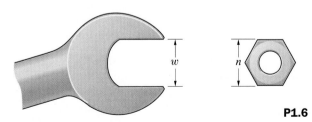

P1.6

1.7 The orbital velocity of the International Space Station is 7690 m/s. Determine its velocity in km/hr and in mi/hr to three significant digits.

1.8 High-speed "bullet trains" began running between Tokyo and Osaka, Japan, in 1964. If a bullet train travels at 240 km/hr, what is its velocity in mi/hr to three significant digits?

1.9 In December 1986, Dick Rutan and Jeana Yeager flew the *Voyager* aircraft around the world nonstop. They flew a distance of 40,212 km in 9 days, 3 minutes, and 44 seconds.
(a) Determine the distance they flew in miles to three significant digits.
(b) Determine their average speed (the distance flown divided by the time required) in kilometers per hour, miles per hour, and knots (nautical miles per hour) to three significant digits.

1.10 Engineers who study shock waves sometimes express velocity in millimeters per microsecond (mm/μs). Suppose the velocity of a wavefront is measured and determined to be 5 mm/μs. Determine its velocity: (a) in m/s; (b) in mi/s.

1.11 The kinetic energy of a particle of mass m is defined to be $\frac{1}{2}mv^2$, where v is the magnitude of the particle's velocity. If the value of the kinetic energy of a particle at a given time is 200 when m is in kilograms and v is in meters per second, what is the value when m is in slugs and v is in feet per second?

1.12 The acceleration due to gravity at sea level in SI units is $g = 9.81$ m/s^2. By converting units, use this value to determine the acceleration due to gravity at sea level in U.S. Customary units.

1.13 A *furlong per fortnight* is a facetious unit of velocity, perhaps made up by a student as a satirical comment on the bewildering variety of units engineers must deal with. A furlong is 660 ft (1/8 mile). A fortnight is 2 weeks (14 nights). If you walk to class at 2 m/s, what is your speed in furlongs per fortnight to three significant digits?

1.14 The cross-sectional area of a beam is 480 in^2. What is its cross-sectional area in m^2?

1.15 At sea level, the weight density (weight per unit volume) of water is approximately 62.4 lb/ft^3. 1 lb = 4.448 N, 1 ft = 0.3048 m, and $g = 9.81$ m/s^2. Using only this information, determine the mass density of water in kg/m^3.

1.16 A pressure transducer measures a value of 300 lb/in^2. Determine the value of the pressure in pascals. A pascal (Pa) is one newton per meter squared.

1.17 A horsepower is 550 ft-lb/s. A watt is 1 N-m/s. Determine the number of watts generated by (a) the Wright brothers' 1903

Boeing 747
UNITED STATES OF AMERICA

Wright Brothers' Flier (shown to scale)

P1.17

airplane, which had a 12-horsepower engine; (b) a modern passenger jet with a power of 100,000 horsepower at cruising speed.

1.18 In SI units, the universal gravitational constant $G = 6.67 \times 10^{-11}$ N-m^2/kg^2. Determine the value of G in U.S. Customary base units.

1.19 If the earth is modeled as a homogeneous sphere, the velocity of a satellite in a circular orbit is

$$v = \sqrt{\frac{gR_E^2}{r}},$$

where R_E is the radius of the earth and r is the radius of the orbit.
(a) If g is in m/s^2 and R_E and r are in meters, what are the units of v?
(b) If $R_E = 6370$ km and $r = 6670$ km, what is the value of v to three significant digits?
(c) For the orbit described in (b), what is the value of v in mi/s to three significant digits?

1.20 In the equation

$$T = \frac{1}{2}I\omega^2,$$

the term I is in kg-m^2 and ω is in s^{-1}.
(a) What are the SI units of T?
(b) If the value of T is 100 when I is in kg-m^2 and ω is in s^{-1}, what is the value of T when it is expressed in terms of U.S. Customary base units?

1.21 The aerodynamic drag force D exerted on a moving object by a gas is given by the expression

$$D = C_D S \frac{1}{2}\rho v^2,$$

where the drag coefficient C_D is dimensionless, S is a reference area, ρ is the mass per unit volume of the gas, and v is the velocity of the object relative to the gas.
(a) Suppose that the value of D is 800 when S, ρ, and v are expressed in SI base units. By converting units, determine the value of D when S, ρ, and v are expressed in U.S. Customary base units.
(b) The drag force D is in newtons when the expression is evaluated using SI base units and is in pounds when the expression is evaluated using U.S. Customary base units. Using your result from (a), determine the conversion factor from newtons to pounds.

1.22 The pressure p at a depth h below the surface of a stationary liquid is given by

$$p = p_s + \gamma h,$$

where p_s is the pressure at the surface and γ is a constant.
(a) If p is in newtons per meter squared and h is in meters, what are the units of γ?
(b) For a particular liquid, the value of γ is 9810 when p is in newtons per meter squared and h is in meters. What is the value of γ when p is in pounds per foot squared and h is in feet?

1.23 The acceleration due to gravity is 1.62 m/s^2 on the surface of the moon and 9.81 m/s^2 on the surface of the earth. A female astronaut's mass is 57 kg. What is the maximum allowable mass of her spacesuit and equipment if the engineers don't want the total weight on the moon of the woman, her spacesuit and equipment to exceed 180 N?

1.24 A person has a mass of 50 kg.
(a) The acceleration due to gravity at sea level is $g = 9.81 \text{ m/s}^2$. What is the person's weight at sea level?
(b) The acceleration due to gravity on the surface of the moon is 1.62 m/s^2. What would the person weigh on the moon?

1.25 The acceleration due to gravity at sea level is $g = 9.81 \text{ m/s}^2$. The radius of the earth is 6370 km. The universal gravitational constant $G = 6.67 \times 10^{-11} \text{ N-m}^2/\text{kg}^2$. Use this information to determine the mass of the earth.

1.26 A person weighs 180 lb at sea level. The radius of the earth is 3960 mi. What force is exerted on the person by the gravitational attraction of the earth if he is in a space station in orbit 200 mi above the surface of the earth?

1.27 The acceleration due to gravity on the surface of the moon is 1.62 m/s^2. The radius of the moon is $R_M = 1738 \text{ km}$.

Determine the acceleration due to gravity of the moon at a point 1738 km above its surface.

Strategy: Write an equation equivalent to Eq. (1.4) for the acceleration due to gravity of the moon.

1.28 If an object is near the surface of the earth, the variation of its weight with distance from the center of the earth can often be neglected. The acceleration due to gravity at sea level is $g = 9.81 \text{ m/s}^2$. The radius of the earth is 6370 km. The weight of an object at sea level is mg, where m is its mass. At what height above the surface of the earth does the weight of the object decrease to $0.99 \, mg$?

1.29 The centers of two oranges are 1 m apart. The mass of each orange is 0.2 kg. What gravitational force do they exert on each other? (The universal gravitational constant $G = 6.67 \times 10^{-11} \text{ N-m}^2/\text{kg}^2$.)

1.30 At a point between the earth and the moon, the magnitude of the earth's gravitational acceleration equals the magnitude of the moon's gravitational acceleration. What is the distance from the center of the earth to that point to three significant digits? The distance from the center of the earth to the center of the moon is 383,000 km, and the radius of the earth is 6370 km. The radius of the moon is 1738 km, and the acceleration due to gravity at its surface is 1.62 m/s^2.

Vectors can specify the positions of points of a structure. Vectors are used to describe and analyze quantities that have magnitude and direction, including positions, forces, moments, velocities, and accelerations.

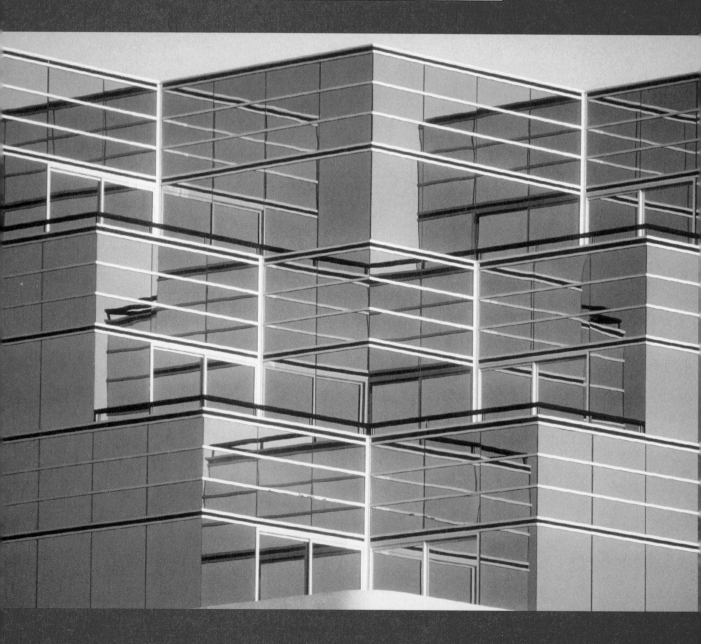

Vectors

To describe a force acting on a structural member, both the magnitude of the force and its direction must be specified. In engineering we deal with many quantities that have both magnitude and direction and can be expressed as vectors. In this chapter we review vector operations, resolve vectors into components, and give examples of engineering applications of vectors.

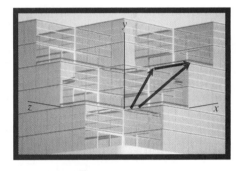

Vector Operations and Definitions

Engineers designing a structure must analyze the positions of its members and the forces acting on them. When designing a machine, they must analyze the velocities and accelerations of its moving parts. These and many other physical quantities important in engineering, can be represented by vectors and analyzed by vector operations. Here we review fundamental vector operations and definitions.

2.1 Scalars and Vectors

A physical quantity that is completely described by a real number is called a *scalar*. Time is a scalar quantity. Mass is also a scalar quantity. For example, you completely describe the mass of a car by saying that its value is 1200 kg.

In contrast, you have to specify both a nonnegative real number, or *magnitude*, and a direction to describe a vector quantity. Two vector quantities are equal only if both their magnitudes and their directions are equal.

The position of a point in space relative to another point is a vector quantity. To describe the location of a city relative to your home, it is not enough to say that it is 100 miles away. You must say that it is 100 miles west of your home. Force is also a vector quantity. When you push a piece of furniture across the floor, you apply a force of magnitude sufficient to move the furniture and you apply it in the direction you want the furniture to move.

We will represent vectors by boldfaced letters, **U**, **V**, **W**, …, and will denote the magnitude of a vector **U** by |**U**|. A vector is represented graphically by an arrow. The direction of the arrow indicates the direction of the vector, and the length of the arrow is defined to be proportional to the magnitude. For example, consider the points A and B of the mechanism in Fig. 2.1a. We can specify the position of point B relative to point A by the vector $\mathbf{r}_{AB}$ in Fig. 2.1b. The direction of $\mathbf{r}_{AB}$ indicates the direction from point A to point B. If the distance between the two points is 200 mm, the magnitude $|\mathbf{r}_{AB}| = 200$ mm.

The cable AB in Fig. 2.2 helps support the television transmission tower. We can represent the force the cable exerts on the tower by a vector **F** as shown. If the cable exerts an 800-N force on the tower, $|\mathbf{F}| = 800$ N.

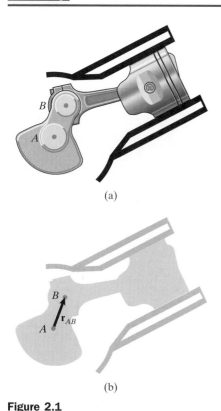

(a)

(b)

Figure 2.1
(a) Two points A and B of a mechanism.
(b) The vector $\mathbf{r}_{AB}$ from A to B.

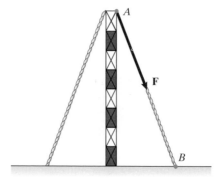

Figure 2.2
Representing the force cable AB exerts on the tower by a vector **F**.

2.2 Rules for Manipulating Vectors

Vectors are a convenient means for representing physical quantities that have magnitude and direction, but that is only the beginning of their usefulness. Just as you manipulate real numbers with the familiar rules for addition, subtraction, multiplication, and so forth, there are rules for manipulating vectors. These rules provide you with powerful tools for engineering analysis.

Vector Addition

When an object moves from one location in space to another, we say it undergoes a *displacement*. If we move a book (or, speaking more precisely, some point of a book) from one location on a table to another, as shown in Fig. 2.3a, we can represent the displacement by the vector **U**. The direction of **U** indicates the direction of the displacement, and |**U**| is the distance the book moves.

Suppose that we give the book a second displacement **V**, as shown in Fig. 2.3b. The two displacements **U** and **V** are equivalent to a single displacement of the book from its initial position to its final position, which we represent by the vector **W** in Fig. 2.3c. Notice that the final position of the book is the same whether we first give it the displacement **U** and then the displacement **V** or we first give it the displacement **V** and then the displacement **U** (Fig. 2.3d). The displacement **W** is defined to be the sum of the displacements **U** and **V**:

$$\mathbf{U} + \mathbf{V} = \mathbf{W}.$$

The definition of vector addition is motivated by the addition of displacements. Consider the two vectors **U** and **V** shown in Fig. 2.4a. If we place them head to tail (Fig. 2.4b), their sum is defined to be the vector from the tail of **U** to the head of **V** (Fig. 2.4c). This is called the *triangle rule* for vector addition. Figure 2.4(d) demonstrates that the sum is independent of the order

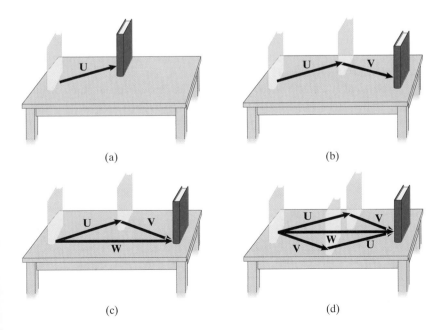

(a)

(b)

(c)

(d)

Figure 2.3
(a) A displacement represented by the vector **U**.
(b) The displacement **U** followed by the displacement **V**.
(c) The displacements **U** and **V** are equivalent to the displacement **W**.
(d) The final position of the book doesn't depend on the order of the displacements.

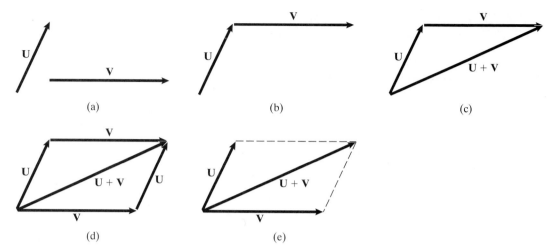

Figure 2.4

(a) Two vectors **U** and **V**.

(b) The head of **U** placed at the tail of **V**.

(c) The triangle rule for obtaining the sum of **U** and **V**.

(d) The sum is independent of the order in which the vectors are added.

(e) The parallelogram rule for obtaining the sum of **U** and **V**.

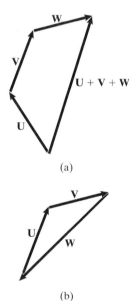

(a)

(b)

Figure 2.5

(a) The sum of three vectors.

(b) Three vectors whose sum is zero.

in which the vectors are placed head to tail. From this figure we obtain the *parallelogram rule* for vector addition (Fig. 2.4e).

The definition of vector addition implies that

$$\mathbf{U} + \mathbf{V} = \mathbf{V} + \mathbf{U} \quad \text{Vector addition is commutative.} \tag{2.1}$$

and

$$(\mathbf{U} + \mathbf{V}) + \mathbf{W} = \mathbf{U} + (\mathbf{V} + \mathbf{W}) \quad \text{Vector addition is associative.} \tag{2.2}$$

for any vectors **U**, **V**, and **W**. These results mean that when you add two or more vectors, you don't need to worry about the order in which you add them. The sum is obtained by placing the vectors head to tail in any order. The vector from the tail of the first vector to the head of the last one is the sum (Fig. 2.5a). If the sum is zero, the vectors form a closed polygon when they are placed head to tail (Fig. 2.5b).

A physical quantity is called a vector if it has magnitude and direction and obeys the definition of vector addition. We have seen that a displacement is a vector. The position of a point in space relative to another point is also a vector quantity. In Fig. 2.6, the vector $\mathbf{r}_{AC}$ from A to C is the sum of $\mathbf{r}_{AB}$ and $\mathbf{r}_{BC}$.

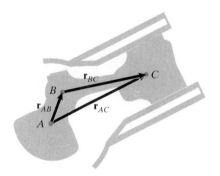

Figure 2.6

Arrows denoting the relative positions of points are vectors.

A force has direction and magnitude, but do forces obey the definition of vector addition? We will assume that they do. In the study of dynamics it is shown that Newton's second law implies that force is a vector.

Product of a Scalar and a Vector

The product of a scalar (real number) a and a vector $\mathbf{U}$ is a vector written as $a\mathbf{U}$. Its magnitude is $|a||\mathbf{U}|$, where $|a|$ is the absolute value of the scalar a. The direction of $a\mathbf{U}$ is the same as the direction of $\mathbf{U}$ when a is positive and is opposite to the direction of $\mathbf{U}$ when a is negative.

The product $(-1)\mathbf{U}$ is written as $-\mathbf{U}$ and is called "the negative of the vector $\mathbf{U}$." It has the same magnitude as $\mathbf{U}$ but the opposite direction. The division of a vector $\mathbf{U}$ by a scalar a is defined to be the product

$$\frac{\mathbf{U}}{a} = \left(\frac{1}{a}\right)\mathbf{U}.$$

Figure 2.7 shows a vector $\mathbf{U}$ and the products of $\mathbf{U}$ with the scalars 2, -1, and $1/2$.

The definitions of vector addition and the product of a scalar and a vector imply that

$$a(b\mathbf{U}) = (ab)\mathbf{U}, \quad \text{The product is associative with} \tag{2.3}$$
$$\text{respect to scalar multiplication.}$$

$$(a + b)\mathbf{U} = a\mathbf{U} + b\mathbf{U} \quad \text{The product is distributive} \tag{2.4}$$
$$\text{with respect to scalar addition.}$$

and

$$a(\mathbf{U} + \mathbf{V}) = a\mathbf{U} + a\mathbf{V} \quad \text{The product is distributive} \tag{2.5}$$
$$\text{with respect to vector addition.}$$

for any scalars a and b and vectors $\mathbf{U}$ and $\mathbf{V}$. We will need these results when we discuss components of vectors.

Vector Subtraction

The difference of two vectors $\mathbf{U}$ and $\mathbf{V}$ is obtained by adding $\mathbf{U}$ to the vector $(-1)\mathbf{V}$:

$$\mathbf{U} - \mathbf{V} = \mathbf{U} + (-1)\mathbf{V}. \tag{2.6}$$

Consider the two vectors $\mathbf{U}$ and $\mathbf{V}$ shown in Fig. 2.8a. The vector $(-1)\mathbf{V}$ has the same magnitude as the vector $\mathbf{V}$ but is in the opposite direction (Fig. 2.8b). In Fig. 2.8c, we add the vector $\mathbf{U}$ to the vector $(-1)\mathbf{V}$ to obtain $\mathbf{U} - \mathbf{V}$.

Unit Vectors

A *unit vector* is simply a vector whose magnitude is 1. A unit vector specifies a direction and also provides a convenient way to express a vector that has a particular direction. If a unit vector $\mathbf{e}$ and a vector $\mathbf{U}$ have the same direction, we can write $\mathbf{U}$ as the product of its magnitude $|\mathbf{U}|$ and the unit vector $\mathbf{e}$ (Fig. 2.9),

$$\mathbf{U} = |\mathbf{U}|\mathbf{e}.$$

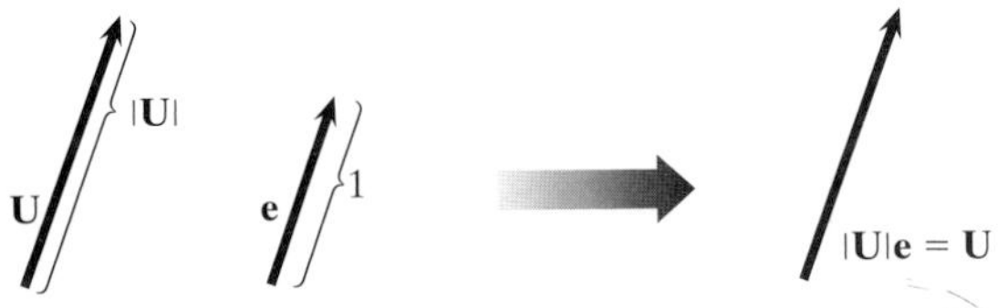

Figure 2.7
(a) A vector $\mathbf{U}$ and some of its scalar multiples.

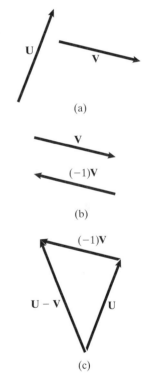

Figure 2.8
(a) Two vectors $\mathbf{U}$ and $\mathbf{V}$.
(b) The vectors $\mathbf{V}$ and $(-1)\,\mathbf{V}$.
(c) The sum of $\mathbf{U}$ and $(-1)\,\mathbf{V}$ is the vector difference $\mathbf{U} - \mathbf{V}$.

Figure 2.9
Since $\mathbf{U}$ and $\mathbf{e}$ have the same direction, the vector $\mathbf{U}$ equals the product of its magnitude with $\mathbf{e}$.

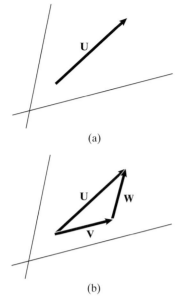

(a)

(b)

Figure 2.10
(a) A vector **U** and two intersecting lines.
(b) The vectors **V** and **W** are vector components of **U**.

Any vector **U** *can be regarded as the product of its magnitude and a unit vector that has the same direction as* **U**. Dividing both sides of this equation by $|\mathbf{U}|$:

$$\frac{\mathbf{U}}{|\mathbf{U}|} = \mathbf{e},$$

we see that dividing any vector by its magnitude yields a unit vector that has the same direction.

Vector Components

When a vector **U** is expressed as the sum of a set of vectors, each vector of the set is called a *vector component* of **U**. Suppose that the vector **U** shown in Fig. 2.10a is parallel to the plane defined by the two intersecting lines. We can express **U** as the sum of vector components **V** and **W** that are parallel to the two lines, as shown in Fig. 2.10b. We say that **U** is *resolved* into the vector components **V** and **W**.

Study Questions

1. What is the triangle rule for vector addition?
2. Vector addition is commutative. What does that mean?
3. If you multiply a vector **U** by a number a, what do you know about the resulting vector $a\mathbf{U}$?
4. What is a unit vector?

Example 2.1

Adding Vectors

Figure 2.11 is an initial design sketch of part of the roof of a sports stadium that is to be supported by the cables *AB* and *AC*. The forces the cables exert on the pylon to which they are attached are represented by the vectors $\mathbf{F}_{AB}$ and $\mathbf{F}_{AC}$. The magnitudes of the forces are $|\mathbf{F}_{AB}| = 100$ kN and $|\mathbf{F}_{AC}| = 60$ kN. Determine the magnitude and direction of the sum of the forces exerted on the pylon by the cables (a) graphically and (b) by using trigonometry.

Strategy

(a) By drawing the parallelogram rule for adding the two forces *with the vectors drawn to scale*, we can measure the magnitude and direction of their sum.
(b) We will calculate the magnitude and direction of the sum of the forces by applying the laws of sines and cosines (Appendix A, Section A.2) to the triangles formed by the parallelogram rule.

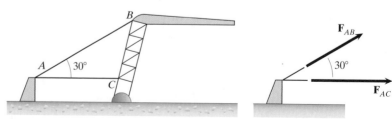

Figure 2.11

Solution

(a) We graphically construct the parallelogram rule for obtaining the sum of the two forces with the lengths of $\mathbf{F}_{AB}$ and $\mathbf{F}_{AC}$ proportional to their magnitudes (Fig. a). By measuring the figure, we estimate the magnitude of the vector $\mathbf{F}_{AB} + \mathbf{F}_{AC}$ to be 155 kN and its direction to be 19° above the horizontal.

(b) Consider the parallelogram rule for obtaining the sum of the two forces (Fig. b). Since $\alpha + 30° = 180°$, the angle $\alpha = 150°$. By applying the law of cosines to the shaded triangle,

$$\left|\mathbf{F}_{AB} + \mathbf{F}_{AC}\right|^2 = \left|\mathbf{F}_{AB}\right|^2 + \left|\mathbf{F}_{AC}\right|^2 - 2\left|\mathbf{F}_{AB}\right|\left|\mathbf{F}_{AC}\right| \cos \alpha$$

$$= (100)^2 + (60)^2 - 2(100)(60) \cos 150°,$$

we determine that the magnitude $\left|\mathbf{F}_{AB} + \mathbf{F}_{AC}\right| = 155$ kN.

To determine the angle β between the vector $\mathbf{F}_{AB} + \mathbf{F}_{AC}$ and the horizontal, we apply the law of sines to the shaded triangle:

$$\frac{\sin \beta}{\left|\mathbf{F}_{AB}\right|} = \frac{\sin \alpha}{\left|\mathbf{F}_{AB} + \mathbf{F}_{AC}\right|}.$$

The solution is

$$\beta = \arcsin\left(\frac{\left|\mathbf{F}_{AB}\right| \sin \alpha}{\left|\mathbf{F}_{AB} + \mathbf{F}_{AC}\right|}\right) = \arcsin\left(\frac{100 \sin 150°}{155}\right) = 18.8°$$

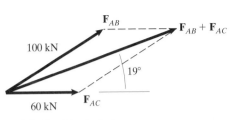

100 kN

60 kN

(a) Graphical solution.

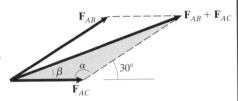

(b) Trigonometric solution.

Discussion

Engineering applications of vectors usually require the precision of analytical solutions, but experience with graphical solutions can help you understand vector operations. Carrying out a graphical solution can also help you formulate an analytical solution.

Example 2.2

Resolving a Vector into Components

The force $\mathbf{F}$ in Fig. 2.12 lies in the plane defined by the intersecting lines L_A and L_B. Its magnitude is 400 lb. Suppose that you want to resolve $\mathbf{F}$ into vector components parallel to L_A and L_B. Determine the magnitudes of the vector components (a) graphically and (b) by using trigonometry.

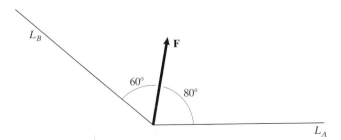

Figure 2.12

Strategy

The parallelogram rule (Fig. 2.4e) clearly indicates how we can resolve **F** into components parallel to L_A and L_B.

Solution

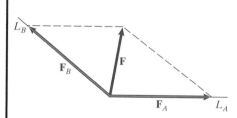

(a) Graphical solution.

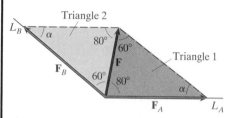

(b) Trigonometric solution.

(a) We draw dashed lines from the head of **F** parallel to L_A and L_B to construct the vector components, which we denote $\mathbf{F}_A$ and $\mathbf{F}_B$ (Fig. a). By measuring the figure, we estimate their magnitudes to be $|\mathbf{F}_A| = 540$ lb and $|\mathbf{F}_B| = 610$ lb.

(b) Consider the force **F** and the vector components $\mathbf{F}_A$ and $\mathbf{F}_B$ (Fig. b). Since $\alpha + 80° + 60° = 180°$, the angle $\alpha = 40°$. By applying the law of sines to triangle 1,

$$\frac{\sin 60°}{|\mathbf{F}_A|} = \frac{\sin \alpha}{|\mathbf{F}|},$$

we obtain the magnitude of $\mathbf{F}_A$:

$$|\mathbf{F}_A| = \frac{|\mathbf{F}| \sin 60°}{\sin \alpha} = \frac{400 \sin 60°}{\sin 40°} = 539 \text{ lb}.$$

Then by applying the law of sines to triangle 2,

$$\frac{\sin 80°}{|\mathbf{F}_B|} = \frac{\sin \alpha}{|\mathbf{F}|},$$

we obtain the magnitude of $\mathbf{F}_B$:

$$|\mathbf{F}_B| = \frac{|\mathbf{F}| \sin 80°}{\sin \alpha} = \frac{400 \sin 80°}{\sin 40°} = 613 \text{ lb}.$$

Problems

Refer to the following diagram when solving Problems 2.1 through 2.5.

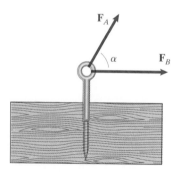

P2.1–2.5

2.1 The magnitudes $|\mathbf{F}_A| = 60$ N and $|\mathbf{F}_B| = 80$ N. The angle $\alpha = 45°$. Graphically determine the magnitude of the sum of the forces $\mathbf{F} = \mathbf{F}_A + \mathbf{F}_B$ and the angle between $\mathbf{F}_B$ and **F**.

Strategy: Construct the parallelogram for determining the sum of the forces, drawing the lengths of $\mathbf{F}_A$ and $\mathbf{F}_B$ proportional to their magnitudes and accurately measuring the angle α, as we did in Example 2.1. Then you can measure the magnitude of their sum and the angle between their sum and $\mathbf{F}_B$.

2.2 The magnitudes $|\mathbf{F}_A| = 40$ N and $|\mathbf{F}_A + \mathbf{F}_B| = 80$ N. The angle $\alpha = 60°$. Graphically determine the magnitude of $\mathbf{F}_B$.

2.3 The magnitudes $|\mathbf{F}_A| = 100$ lb and $|\mathbf{F}_B| = 140$ lb. The angle $\alpha = 40°$. Use trigonometry to determine the magnitude of the sum of the forces $\mathbf{F} = \mathbf{F}_A + \mathbf{F}_B$ and the angle between $\mathbf{F}_B$ and **F**.

Strategy: Use the laws of sines and cosines to analyze the triangles formed by the parallelogram rule for the sum of the forces as we did in Example 2.1. The laws of sines and cosines are given in Section A.2 of Appendix A.

2.4 The magnitudes $|\mathbf{F}_A| = 40$ N and $|\mathbf{F}_A + \mathbf{F}_B| = 80$ N. The angle $\alpha = 60°$. Use trigonometry to determine the magnitude of $\mathbf{F}_B$.

2.5 The magnitudes $|\mathbf{F}_A| = 100$ lb and $|\mathbf{F}_B| = 140$ lb. If α can have any value, what are the minimum and maximum possible values of the magnitude of the sum of the forces $\mathbf{F} = \mathbf{F}_A + \mathbf{F}_B$, and what are the corresponding values of α?

2.6 The angle $\theta = 30°$. What is the magnitude of the vector $\mathbf{r}_{AC}$?

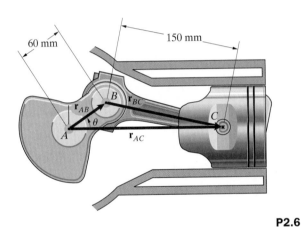

P2.6

2.7 The vectors $\mathbf{F}_A$ and $\mathbf{F}_B$ represent the forces exerted on the pulley by the belt. Their magnitudes are $|\mathbf{F}_A| = 80$ N and $|\mathbf{F}_B| = 60$ N. What is the magnitude $|\mathbf{F}_A + \mathbf{F}_B|$ of the total force the belt exerts on the pulley?

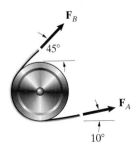

P2.7

2.8 The magnitude of the vertical force $\mathbf{F}$ is 80 kN. If you resolve it into components $\mathbf{F}_{AB}$ and $\mathbf{F}_{AC}$ that are parallel to the bars AB and AC, what are the magnitudes of the components?

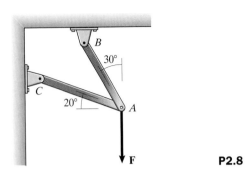

P2.8

2.9 The rocket engine exerts an upward force of 4 MN (meganewtons) magnitude on the test stand. If you resolve the force into vector components parallel to the bars AB and CD, what are the magnitudes of the components?

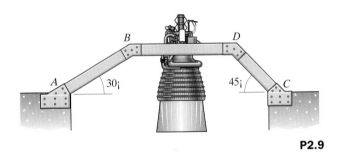

P2.9

2.10 If $\mathbf{F}$ is resolved into components parallel to the bars AB and BC, the magnitude of the component parallel to bar AB is 4 kN. What is the magnitude of $\mathbf{F}$?

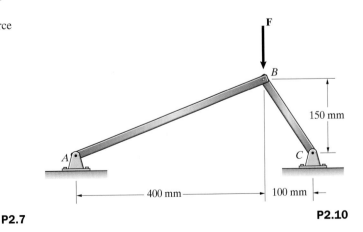

P2.10

2.11 The forces acting on the sailplane are represented by three vectors. The lift **L** and drag **D** are perpendicular, the magnitude of the weight **W** is 3500 N, and **W** + **L** + **D** = **0**. What are the magnitudes of the lift and drag?

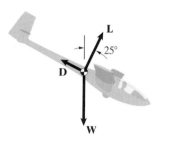

P2.11

2.12 The suspended weight exerts a downward 2000-lb force **F** at *A*. If you resolve **F** into vector components parallel to the wires *AB*, *AC*, and *AD*, the magnitude of the component parallel to *AC* is 600 lb. What are the magnitudes of the components parallel to *AB* and *AD*?

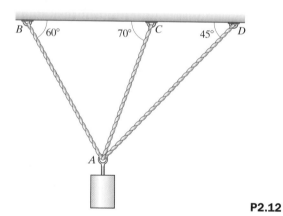

P2.12

2.13 The wires in Problem 2.12 will safely support the weight if the magnitude of the vector component of **F** parallel to each wire does not exceed 2000 lb. Based on this criterion, how large can the magnitude of **F** be? What are the corresponding magnitudes of the vector components of **F** parallel to the three wires?

2.14 Two vectors $\mathbf{r}_A$ and $\mathbf{r}_B$ have magnitudes $|\mathbf{r}_A|$ = 30 m and $|\mathbf{r}_B|$ = 40 m. Determine the magnitude of their sum $\mathbf{r}_A + \mathbf{r}_B$
(a) if $\mathbf{r}_A$ and $\mathbf{r}_B$ have the same direction.
(b) if $\mathbf{r}_A$ and $\mathbf{r}_B$ are perpendicular.

2.15 A spherical storage tank is supported by cables. The tank is subjected to three forces: the forces $\mathbf{F}_A$ and $\mathbf{F}_B$ exerted by the cables and the weight **W**. The weight of the tank $|\mathbf{W}|$ = 600 lb. The vector sum of the forces acting on the tank equals zero. Determine the magnitudes of $\mathbf{F}_A$ and $\mathbf{F}_B$ (a) graphically and (b) by using trigonometry.

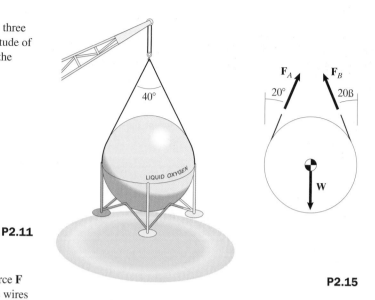

P2.15

2.16 The rope *ABC* exerts forces $\mathbf{F}_{BA}$ and $\mathbf{F}_{BC}$ on the block at *B*. Their magnitudes are $|\mathbf{F}_{BA}|$ = $|\mathbf{F}_{BC}|$ = 800 N. Determine $|\mathbf{F}_{BA} + \mathbf{F}_{BC}|$ (a) graphically and (b) by using trigonometry.

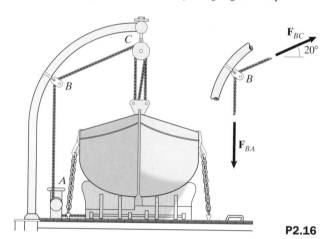

P2.16

2.17 Two snowcats tow a housing unit to a new location at McMurdo Base, Antarctica. (The top view is shown. The cables are horizontal.) The sum of the forces $\mathbf{F}_A$ and $\mathbf{F}_B$ exerted on the unit is parallel to the line *L*, and $|\mathbf{F}_A|$ = 1000 lb. Determine $|\mathbf{F}_B|$ and $|\mathbf{F}_A + \mathbf{F}_B|$ (a) graphically and (b) by using trigonometry.

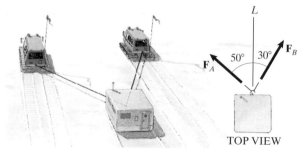

TOP VIEW

P2.17

2.18 A surveyor determines that the horizontal distance from A to B is 400 m and that the horizontal distance from A to C is 600 m. Determine the magnitude of the horizontal vector $\mathbf{r}_{BC}$ from B to C and the angle α (a) graphically and (b) by using trigonometry.

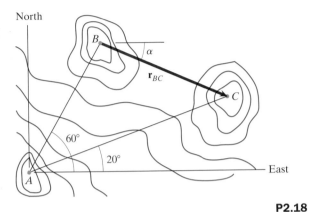

North

B

α

$\mathbf{r}_{BC}$

C

$60°$

$20°$

A

East

P2.18

2.19 The vector $\mathbf{r}$ extends from point A to the midpoint between points B and C. Prove that

$$\mathbf{r} = \tfrac{1}{2}\left(\mathbf{r}_{AB} + \mathbf{r}_{AC}\right).$$

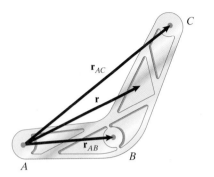

C

$\mathbf{r}_{AC}$

$\mathbf{r}$

$\mathbf{r}_{AB}$

B

A

P2.19

2.20 By drawing sketches of the vectors, explain why

$$\mathbf{U} + (\mathbf{V} + \mathbf{W}) = (\mathbf{U} + \mathbf{V}) + \mathbf{W}.$$

Cartesian Components

Vectors are much easier to work with when they are expressed in terms of mutually perpendicular vector components. Here we explain how to resolve vectors into cartesian components in two and three dimensions and give examples of vector manipulations using components.

2.3 Components in Two Dimensions

Consider the vector $\mathbf{U}$ in Fig. 2.13a. By placing a cartesian coordinate system so that $\mathbf{U}$ is parallel to the x-y plane, we can resolve it into vector components $\mathbf{U}_x$ and $\mathbf{U}_y$ parallel to the x and y axes (Fig. 2.13b),

$$\mathbf{U} = \mathbf{U}_x + \mathbf{U}_y.$$

Then by introducing a unit vector $\mathbf{i}$ defined to point in the direction of the positive x axis and a unit vector $\mathbf{j}$ defined to point in the direction of the positive y axis (Fig. 2.13c), we can express the vector $\mathbf{U}$ in the form

$$\mathbf{U} = U_x\mathbf{i} + U_y\mathbf{j}. \qquad (2.7)$$

The scalars U_x and U_y are called *scalar components of* $\mathbf{U}$. *When we refer simply to the components of a vector, we will mean its scalar components.* We will call U_x and U_y the x and y components of $\mathbf{U}$.

The components of a vector specify both its direction relative to the cartesian coordinate system and its magnitude. From the right triangle formed by the vector $\mathbf{U}$ and its vector components (Fig. 2.13c), we see that

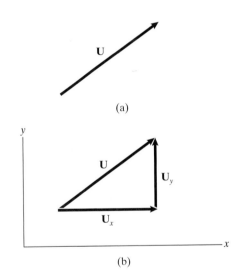

$\mathbf{U}$

(a)

y

$\mathbf{U}$

$\mathbf{U}_y$

$\mathbf{U}_x$

x

(b)

y

$\mathbf{U}$

$\mathbf{U}_y = U_y\mathbf{j}$

$\mathbf{U}_x = U_x\mathbf{i}$

$\mathbf{j}$

$\mathbf{i}$

x

(c)

Figure 2.13
(a) A vector $\mathbf{U}$.
(b) The vector components $\mathbf{U}_x$ and $\mathbf{U}_y$.
(c) The vector components can be expressed in terms of $\mathbf{i}$ and $\mathbf{j}$.

the magnitude of **U** is given in terms of its components by the Pythagorean theorem,

$$|\mathbf{U}| = \sqrt{U_x^2 + U_y^2}. \tag{2.8}$$

With this equation you can determine the magnitude of a vector when you know its components.

Manipulating Vectors in Terms of Components

The sum of two vectors **U** and **V** in terms of their components is

$$\begin{aligned}
\mathbf{U} + \mathbf{V} &= \left(U_x\mathbf{i} + U_y\mathbf{j}\right) + \left(V_x\mathbf{i} + V_y\mathbf{j}\right) \\
&= \left(U_x + V_x\right)\mathbf{i} + \left(U_y + V_y\right)\mathbf{j}.
\end{aligned} \tag{2.9}$$

The components of **U** + **V** are the sums of the components of the vectors **U** and **V**. Notice that in obtaining this result we used Eqs. (2.2), (2.4), and (2.5).

It is instructive to derive Eq. (2.9) graphically. The summation of **U** and **V** is shown in Fig. 2.14a. In Fig. 2.14b we introduce a coordinate system and resolve **U** and **V** into their components. In Fig. 2.14c we add the x and y components, obtaining Eq. (2.9).

The product of a number a and a vector **U** in terms of the components of **U** is

$$a\mathbf{U} = a\left(U_x\mathbf{i} + U_y\mathbf{j}\right) = aU_x\mathbf{i} + aU_y\mathbf{j}.$$

The component of $a\mathbf{U}$ in each coordinate direction equals the product of a and the component of **U** in that direction. We used Eqs. (2.3) and (2.5) to obtain this result.

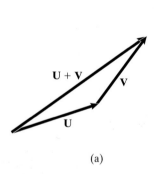

(a)

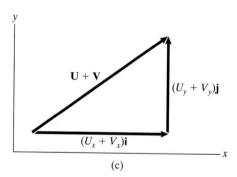

(b)   (c)

Figure 2.14
(a) The sum of **U** and **V**.
(b) The vector components of **U** and **V**.
(c) The sum of the components in each coordinate direction equals the component of **U** + **V** in that direction.

Position Vectors in Terms of Components

We can express the position vector of a point relative to another point in terms of the cartesian coordinates of the points. Consider point A with coordinates $\left(x_A, y_A\right)$ and point B with coordinates $\left(x_B, y_B\right)$. Let $\mathbf{r}_{AB}$ be the vector that specifies the position of B relative to A (Fig. 2.15a). That is, we denote the

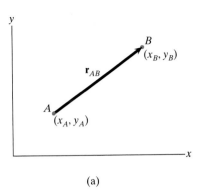

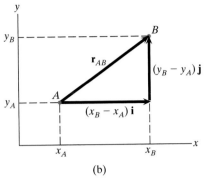

Figure 2.15
(a) Two points A and B and the position vector $\mathbf{r}_{AB}$ from A to B.
(b) The components of $\mathbf{r}_{AB}$ can be determined from the coordinates of points A and B.

vector *from* a point A *to* a point B by $\mathbf{r}_{AB}$. We see from Fig. 2.15b that $\mathbf{r}_{AB}$ is given in terms of the coordinates of points A and B by

$$\mathbf{r}_{AB} = (x_B - x_A)\mathbf{i} + (y_B - y_A)\mathbf{j}. \qquad (2.10)$$

Notice that the x component of the position vector from a point A to a point B is obtained by subtracting the x coordinate of A from the x coordinate of B, and the y component is obtained by subtracting the y coordinate of A from the y coordinate of B.

Study Questions

1. How are the scalar components of a vector defined in terms of a cartesian coordinate system?
2. If you know the scalar components of a vector, how can you determine its magnitude?
3. Suppose that you know the coordinates of two points A and B. How do you determine the scalar components of the position vector of point B relative to point A?

Example 2.3

Adding Vectors in Terms of Components

The forces acting on the sailplane in Fig. 2.16 are its weight $\mathbf{W} = -600\mathbf{j}$ (lb), the drag $\mathbf{D} = -200\mathbf{i} + 100\mathbf{j}$ (lb), and the lift $\mathbf{L}$.
(a) If the sum of the forces on the sailplane is zero, what are the components of $\mathbf{L}$?
(b) If the lift $\mathbf{L}$ has the components determined in (a) and the drag $\mathbf{D}$ increases by a factor of 2, what is the magnitude of the sum of the forces on the sailplane?

Strategy

(a) By setting the sum of the forces equal to zero, we can determine the components of $\mathbf{L}$. (b) Using the value of $\mathbf{L}$ from (a), we can determine the components of the sum of the forces and use Eq. (2.8) to determine its magnitude.

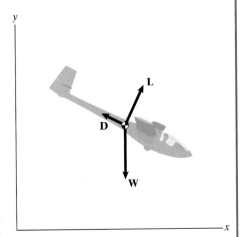

Figure 2.16

Solution

(a) We set the sum of the forces equal to zero:

$$\mathbf{W} + \mathbf{D} + \mathbf{L} = \mathbf{0}.$$

$$(-600\mathbf{j}) + (-200\mathbf{i} + 100\mathbf{j}) + \mathbf{L} = \mathbf{0}.$$

Solving for the lift, we obtain

$$\mathbf{L} = 200\mathbf{i} + 500\mathbf{j} \text{ (lb)}.$$

(b) If the drag increases by a factor of 2, the sum of the forces on the sailplane is

$$\mathbf{W} + 2\mathbf{D} + \mathbf{L} = (-600\mathbf{j}) + 2(-200\mathbf{i} + 100\mathbf{j}) + (200\mathbf{i} + 500\mathbf{j})$$

$$= -200\mathbf{i} + 100\mathbf{j} \text{ (lb)}.$$

From Eq. (2.8), the magnitude of the sum is

$$|\mathbf{W} + 2\mathbf{D} + \mathbf{L}| = \sqrt{(-200)^2 + (100)^2} = 224 \text{ lb}.$$

Example 2.4

Determining Components in Terms of an Angle

Hydraulic cylinders are used to exert forces in many mechanical devices. The force is exerted by pressurized liquid (hydraulic fluid) pushing against a piston within the cylinder. The hydraulic cylinder AB in Fig. 2.17 exerts a 4000-lb force $\mathbf{F}$ on the bed of the dump truck at B. Express $\mathbf{F}$ in terms of components using the coordinate system shown.

Strategy

When the direction of a vector is specified by an angle, as in this example, we can determine the values of the components from the right triangle formed by the vector and its components.

Solution

We draw the vector $\mathbf{F}$ and its vector components in Fig. a. From the resulting right triangle, we see that the magnitude of $\mathbf{F}_x$ is

$$|\mathbf{F}_x| = |\mathbf{F}| \cos 30° = (4000) \cos 30° = 3460 \text{ lb}.$$

$\mathbf{F}_x$ points in the negative x direction, so

$$\mathbf{F}_x = -3460\mathbf{i} \text{ (lb)}.$$

The magnitude of $\mathbf{F}_y$ is

$$|\mathbf{F}_y| = |\mathbf{F}| \sin 30° = (4000) \sin 30° = 2000 \text{ lb}.$$

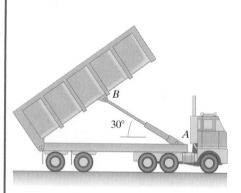

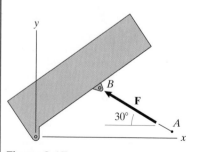

Figure 2.17

The vector component $\mathbf{F}_y$ points in the positive y direction, so

$$\mathbf{F}_y = 2000\mathbf{j} \text{ (lb)}.$$

The vector $\mathbf{F}$ in terms of its components is

$$\mathbf{F} = \mathbf{F}_x + \mathbf{F}_y = -3460\mathbf{i} + 2000\mathbf{j} \text{ (lb)}.$$

The x component of $\mathbf{F}$ is -3460 lb, and the y component is 2000 lb.

Discussion

When you determine the components of a vector, you should check to make sure they give you the correct magnitude. In this example,

$$|\mathbf{F}| = \sqrt{(-3460)^2 + (2000)^2} = 4000 \text{ lb}.$$

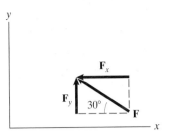

(a) The force $\mathbf{F}$ and its components form a right triangle.

Example 2.5

Determining Vector Components

The cable from point A to point B exerts an 800-N force $\mathbf{F}$ on the top of the television transmission tower in Fig. 2.18. Resolve $\mathbf{F}$ into components using the coordinate system shown.

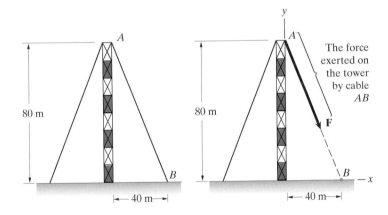

Figure 2.18

Strategy

We determine the components of $\mathbf{F}$ in three ways.

First Method From the given dimensions we can determine the angle α between $\mathbf{F}$ and the y axis (Fig. a), then determine the components from the right triangles formed by the vector $\mathbf{F}$ and its components.

Second Method The right triangles formed by $\mathbf{F}$ and its components are similar to the triangle OAB in Fig. a. We can determine the components of $\mathbf{F}$ by using the ratios of the sides of these similar triangles.

Third Method From the given dimensions we can determine the components of the position vector $\mathbf{r}_{AB}$ from point A to point B [Fig. b]. By dividing this vector by its magnitude, we will obtain a unit vector $\mathbf{e}_{AB}$ with the same direction as $\mathbf{F}$ (Fig. c), then obtain $\mathbf{F}$ in terms of its components by expressing it as the product of its magnitude and $\mathbf{e}_{AB}$.

Solution

First Method Consider the force $\mathbf{F}$ and its vector components (Fig. a). The tangent of the angle α between $\mathbf{F}$ and the y axis is $\tan \alpha = 40/80 = 0.5$, so $\alpha = \arctan(0.5) = 26.6°$. From the right triangles formed by $\mathbf{F}$ and its vector components, the magnitude of $\mathbf{F}_x$ is

$$|\mathbf{F}_x| = |\mathbf{F}| \sin 26.6° = (800) \sin 26.6° = 358 \text{ N}$$

and the magnitude of $\mathbf{F}_y$ is

$$|\mathbf{F}_y| = |\mathbf{F}| \cos 26.6° = (800) \cos 26.6° = 716 \text{ N}.$$

Since $\mathbf{F}_x$ points in the positive x direction and $\mathbf{F}_y$ points in the negative y direction, the force $\mathbf{F}$ is

$$\mathbf{F} = 358\mathbf{i} - 716\mathbf{j} \text{ (N)}$$

Second Method The length of the cable AB is $\sqrt{(80)^2 + (40)^2} = 89.4$ m. Since the triangle OAB in Fig. a is similar to the triangle formed by $\mathbf{F}$ and its vector components,

$$\frac{|\mathbf{F}_x|}{|\mathbf{F}|} = \frac{OB}{AB} = \frac{40}{89.4}.$$

Thus the magnitude of $\mathbf{F}_x$ is

$$|\mathbf{F}_x| = \left(\frac{40}{89.4}\right)|\mathbf{F}| = \left(\frac{40}{89.4}\right)(800) = 358 \text{ N}.$$

We can also see from the similar triangles that

$$\frac{|\mathbf{F}_y|}{|\mathbf{F}|} = \frac{OA}{AB} = \frac{80}{89.4},$$

so the magnitude of $\mathbf{F}_y$ is

$$|\mathbf{F}_y| = \left(\frac{80}{89.4}\right)|\mathbf{F}| = \left(\frac{80}{89.4}\right)(800) = 716 \text{ N}.$$

Thus we again obtain the result

$$\mathbf{F} = 358\mathbf{i} - 716\mathbf{j} \text{ (N)}.$$

Third Method The vector $\mathbf{r}_{AB}$ in Fig. b is

$$\mathbf{r}_{AB} = (x_B - x_A)\mathbf{i} + (y_B - y_A)\mathbf{j} = (40 - 0)\mathbf{i} + (0 - 80)\mathbf{j}$$
$$= 40\mathbf{i} - 80\mathbf{j} \text{ (m)}.$$

We divide this vector by its magnitude to obtain a unit vector $\mathbf{e}_{AB}$ that has the same direction as the force $\mathbf{F}$ (Fig. c):

$$\mathbf{e}_{AB} = \frac{\mathbf{r}_{AB}}{|\mathbf{r}_{AB}|} = \frac{40\mathbf{i} - 80\mathbf{j}}{\sqrt{(40)^2 + (-80)^2}} = 0.447\mathbf{i} - 0.894\mathbf{j}.$$

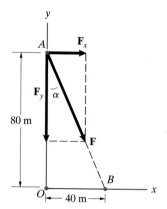

(a) Vector components of $\mathbf{F}$.

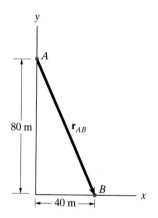

(b) The vector $\mathbf{r}_{AB}$ form A to B.

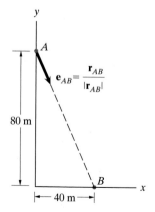

(c) The unit vector $\mathbf{e}_{AB}$ pointing from A toward B.

The force $\mathbf{F}$ is equal to the product of its magnitude $|\mathbf{F}|$ and $\mathbf{e}_{AB}$:

$$\mathbf{F} = |\mathbf{F}|\mathbf{e}_{AB} = (800)(0.447\mathbf{i} - 0.894\mathbf{j}) = 358\mathbf{i} - 716\mathbf{j} \; (\text{N}).$$

Example 2.6

Determining an Unknown Vector Magnitude

The cables A and B in Fig. 2.19 exert forces $\mathbf{F}_A$ and $\mathbf{F}_B$ on the hook. The magnitude of $\mathbf{F}_A$ is 100 lb. The tension in cable B has been adjusted so that the total force $\mathbf{F}_A + \mathbf{F}_B$ is perpendicular to the wall to which the hook is attached.
(a) What is the magnitude of $\mathbf{F}_B$?
(b) What is the magnitude of the total force exerted on the hook by the two cables?

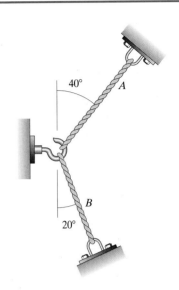

Strategy

The vector sum of the two forces is perpendicular to the wall, so the sum of the components parallel to the wall equals zero. From this condition we can obtain an equation for the magnitude of $\mathbf{F}_B$.

Solution

(a) In terms of the coordinate system shown in Fig. a, the components of $\mathbf{F}_A$ and $\mathbf{F}_B$ are

$$\mathbf{F}_A = |\mathbf{F}_A| \sin 40°\mathbf{i} + |\mathbf{F}_A| \cos 40°\mathbf{j},$$
$$\mathbf{F}_B = |\mathbf{F}_B| \sin 20°\mathbf{i} - |\mathbf{F}_B| \cos 20°\mathbf{j}.$$

The total force is

$$\mathbf{F}_A + \mathbf{F}_B = (|\mathbf{F}_A| \sin 40° + |\mathbf{F}_B| \sin 20°)\mathbf{i}$$
$$+ (|\mathbf{F}_A| \cos 40° - |\mathbf{F}_B| \cos 20°)\mathbf{j}.$$

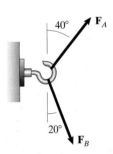

Figure 2.19

By setting the component of the total force parallel to the wall (the y component) equal to zero,

$$|\mathbf{F}_A| \cos 40° - |\mathbf{F}_B| \cos 20° = 0,$$

we obtain an equation for the magnitude of $\mathbf{F}_B$:

$$|\mathbf{F}_B| = \frac{|\mathbf{F}_A| \cos 40°}{\cos 20°} = \frac{(100) \cos 40°}{\cos 20°} = 81.5 \; \text{lb}.$$

(b) Since we now know the magnitude of $\mathbf{F}_B$, we can determine the total force acting on the hook:

$$\mathbf{F}_A + \mathbf{F}_B = (|\mathbf{F}_A| \sin 40° + |\mathbf{F}_B| \sin 20°)\mathbf{i}$$
$$= \big[(100) \sin 40° + (81.5) \sin 20°\big]\mathbf{i} = 92.2\mathbf{i} \; (\text{lb}).$$

The magnitude of the total force is 92.2 lb.

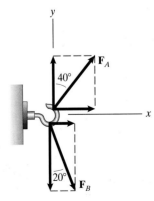

(a) Resolving $\mathbf{F}_A$ and $\mathbf{F}_B$ into components parallel and perpendicular to the wall.

Discussion

We can obtain the solution to (a) in a less formal way. If the component of the total force parallel to the wall is zero, we see in Fig. (a) that the magnitude of the vertical component of $\mathbf{F}_A$ must equal the magnitude of the vertical component of $\mathbf{F}_B$:

$$\left|\mathbf{F}_A\right| \cos 40° = \left|\mathbf{F}_B\right| \cos 20°.$$

Therefore the magnitude of $\mathbf{F}_B$ is

$$\left|\mathbf{F}_B\right| = \frac{\left|\mathbf{F}_A\right| \cos 40°}{\cos 20°} = \frac{(100) \cos 40°}{\cos 20°} = 81.5 \text{ lb}.$$

Problems

2.21 A force $\mathbf{F} = 40\mathbf{i} - 20\mathbf{j}$ (N). What is its magnitude $|\mathbf{F}|$?
   Strategy: The magnitude of a vector in terms of its components is given by Eq. (2.8).

2.22 An engineer estimating the components of a force $\mathbf{F} = F_x\mathbf{i} + F_y\mathbf{j}$ acting on a bridge abutment has determined that $F_x = 130$ MN, $|\mathbf{F}| = 165$ MN, and F_y is negative. What is F_y?

2.23 A support is subjected to a force $\mathbf{F} = F_x\mathbf{i} + 80\mathbf{j}$ (N). If the support will safely support a force of magnitude 100 N, what is the allowable range of values of the component F_x?

2.24 If $\mathbf{F}_A = 600\mathbf{i} - 800\mathbf{j}$ (kip) and $\mathbf{F}_B = 200\mathbf{i} - 200\mathbf{j}$ (kip), what is the magnitude of the force $\mathbf{F} = \mathbf{F}_A - 2\mathbf{F}_B$?

2.25 If $\mathbf{F}_A = \mathbf{i} - 4.5\mathbf{j}$ (kN) and $\mathbf{F}_B = -2\mathbf{i} - 2\mathbf{j}$ (kN), what is the magnitude of the force $\mathbf{F} = 6\mathbf{F}_A + 4\mathbf{F}_B$?

2.26 Two perpendicular vectors $\mathbf{U}$ and $\mathbf{V}$ lie in the x-y plane. The vector $\mathbf{U} = 6\mathbf{i} - 8\mathbf{j}$ and $|\mathbf{V}| = 20$. What are the components of $\mathbf{V}$?

2.27 A fish exerts a 40-N force on the line that is represented by the vector $\mathbf{F}$. Express $\mathbf{F}$ in terms of components using the coordinate system shown.

2.28 A person exerts a 60-lb force $\mathbf{F}$ to push a crate onto a truck. Express $\mathbf{F}$ in terms of components.

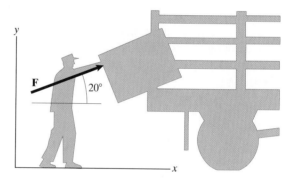

P2.28

2.29 The missile's engine exerts a 260-kN force $\mathbf{F}$. Express $\mathbf{F}$ in terms of components using the coordinate system shown.

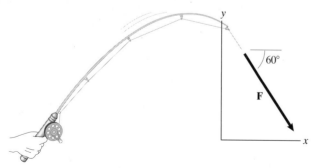

P2.27

P2.29

2.30 The coordinates of two points A and B of a truss are shown. Express the position vector from point A to point B in terms of components.

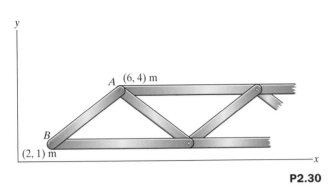

P2.30

2.31 The points $A, B, \ldots$ are the joints of the hexagonal structural element. Let $\mathbf{r}_{AB}$ be the position vector from joint A to joint B, $\mathbf{r}_{AC}$ the position vector from joint A to joint C, and so forth. Determine the components of the vectors $\mathbf{r}_{AC}$ and $\mathbf{r}_{AF}$.

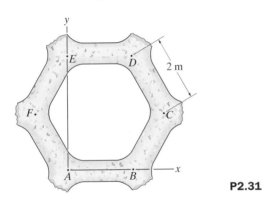

P2.31

2.32 For the hexagonal structural element in Problem 2.31, determine the components of the vector $\mathbf{r}_{AB} - \mathbf{r}_{BC}$.

2.33 The coordinates of point A are $(1.8, 3.0)$ m. The y coordinate of point B is 0.6 m and the magnitude of the vector $\mathbf{r}_{AB}$ is 3.0 m. What are the components of $\mathbf{r}_{AB}$?

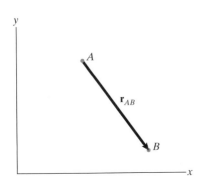

P2.33

2.34 (a) Express the position vector from point A of the front-end loader to point B in terms of components.
(b) Express the position vector from point B to point C in terms of components.
(c) Use the results of (a) and (b) to determine the distance from point A to point C.

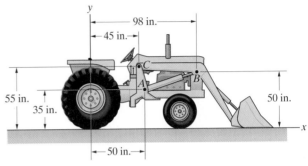

P2.34

2.35 Consider the front-end loader in Problem 2.34. To raise the bucket, the operator increases the length of the hydraulic cylinder AB. The distance between points B and C remains constant. If the length of the cylinder AB is 65 in., what is the position vector from point A to point B?

2.36 Determine the position vector $\mathbf{r}_{AB}$ in terms of its components if (a) $\theta = 30°$; (b) $\theta = 225°$.

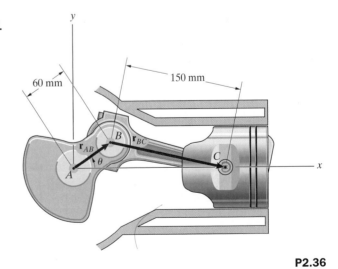

P2.36

2.37 In Problem 2.36 determine the position vector $\mathbf{r}_{BC}$ in terms of its components if (a) $\theta = 30°$; (b) $\theta = 225°$.

2.38 A surveyor measures the location of point A and determines that $\mathbf{r}_{OA} = 400\mathbf{i} + 800\mathbf{j}$ (m). He wants to determine the location of a point B so that $|\mathbf{r}_{AB}| = 400$ m and $|\mathbf{r}_{OA} + \mathbf{r}_{AB}| = 1200$ m. What are the cartesian coordinates of point B?

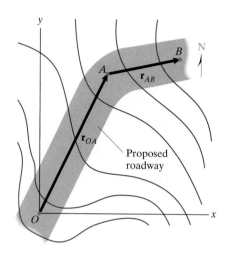

P2.38

2.39 Bar AB is 8.5 m long and bar AC is 6 m long. Determine the components of the position vector $\mathbf{r}_{AB}$ from point A to point B.

P2.39

2.40 For the truss in Problem 2.39, determine the components of a unit vector $\mathbf{e}_{AC}$ that points from point A toward point C.

Strategy: Determine the components of the position vector from point A to point C and divide the position vector by its magnitude.

2.41 The x and y coordinates of points A, B, and C of the sailboat are shown.
(a) Determine the components of a unit vector that is parallel to the forestay AB and points from A toward B.
(b) Determine the components of a unit vector that is parallel to the backstay BC and points from C toward B.

P2.41

2.42 Consider the force vector $\mathbf{F} = 3\mathbf{i} - 4\mathbf{j}$ (kN). Determine the components of a unit vector $\mathbf{e}$ that has the same direction as $\mathbf{F}$.

2.43 Determine the components of a unit vector that is parallel to the hydraulic actuator BC and points from B toward C.

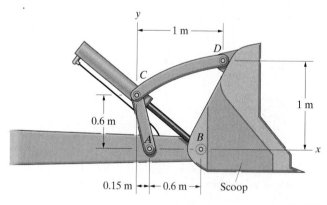

P2.43

2.44 The hydraulic actuator BC in Problem 2.43 exerts a 1.2-kN force $\mathbf{F}$ on the joint at C that is parallel to the actuator and points from B toward C. Determine the components of $\mathbf{F}$.

2.45 A surveyor finds that the length of the line OA is 1500 m and the length of the line OB is 2000 m.
(a) Determine the components of the position vector from point A to point B.
(b) Determine the components of a unit vector that points from point A toward point B.

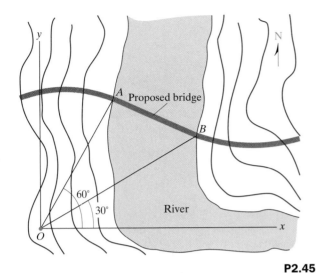

P2.45

nitude of the vector sum of the forces by resolving the forces into components, and compare your answer with that of Problem 2.16.

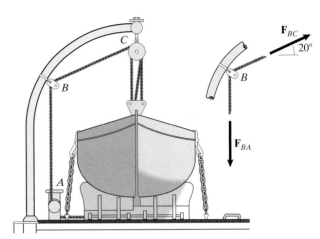

P2.48

2.46 The positions at a given time of the Sun (**S**) and the planets Mercury (**M**), Venus (**V**), and Earth (**E**) are shown. The approximate distance from the Sun to Mercury is 57×10^6 km, the distance from the Sun to Venus is 108×10^6 km, and the distance from the Sun to the Earth is 150×10^6 km. Assume that the Sun and planets lie in the x–y plane. Determine the components of a unit vector that points from the Earth toward Mercury.

2.49 The magnitudes of the forces are $|\mathbf{F}_1| = |\mathbf{F}_2| = |\mathbf{F}_3| = 5$ kN. What is the magnitude of the vector sum of the three forces?

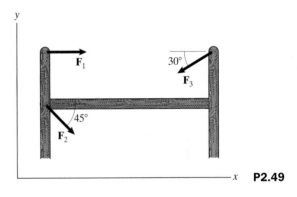

P2.49

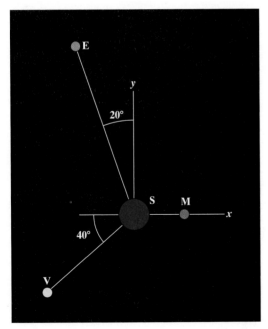

P2.46

2.47 For the positions described in Problem 2.46, determine the components of a unit vector that points from the Earth toward Venus.

2.48 The rope ABC exerts forces $\mathbf{F}_{BA}$ and $\mathbf{F}_{BC}$ on the block at B. Their magnitudes are $|\mathbf{F}_{BA}| = |\mathbf{F}_{BC}| = 800$ N. Determine the mag-

2.50 Four groups engage in a tug-of-war. The magnitudes of the forces exerted by groups B, C, and D are $|\mathbf{F}_B| = 800$ lb, $|\mathbf{F}_C| = 1000$ lb, and $|\mathbf{F}_D| = 900$ lb. If the vector sum of the four forces equals zero, what are the magnitude of $\mathbf{F}_A$ and the angle α?

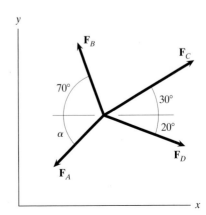

P2.50

2.51 The total thrust exerted on the launch vehicle by its main engines is 200,000 lb parallel to the y axis. Each of the two small vernier engines exerts a thrust of 5000 lb in the directions shown. Determine the magnitude and direction of the total force exerted on the booster by the main and vernier engines.

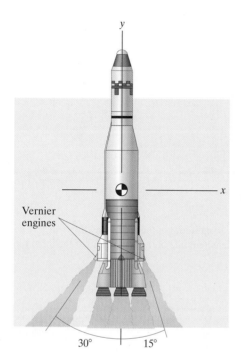

Vernier engines

30°   15°

P2.51

2.52 The magnitudes of the forces acting on the bracket are $|\mathbf{F}_1| = |\mathbf{F}_2| = 2$ kN. If $|\mathbf{F}_1 + \mathbf{F}_2| = 3.8$ kN, what is the angle α? (Assume that $0 \le \alpha \le 90°$.)

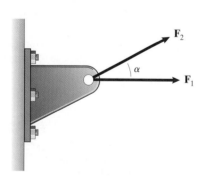

P2.52

2.53 The figure shows three forces acting on a joint of a structure. The magnitude of $\mathbf{F}_C$ is 60 kN , and $\mathbf{F}_A + \mathbf{F}_B + \mathbf{F}_C = \mathbf{0}$. What are the magnitudes of $\mathbf{F}_A$ and $\mathbf{F}_B$?

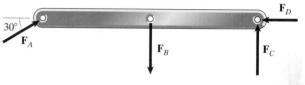

P2.53

2.54 Four forces act on a beam. The vector sum of the forces is zero. The magnitudes $|\mathbf{F}_B| = 10$ kN and $|\mathbf{F}_C| = 5$ kN. Determine the magnitudes of $\mathbf{F}_A$ and $\mathbf{F}_D$.

P2.54

2.55 Six forces act on a beam that forms part of a building's frame. The vector sum of the forces is zero. The magnitudes $|\mathbf{F}_B| = |\mathbf{F}_E| = 20$ kN, $|\mathbf{F}_C| = 16$ kN, and $|\mathbf{F}_D| = 9$ kN. Determine the magnitudes of $\mathbf{F}_A$ and $\mathbf{F}_G$.

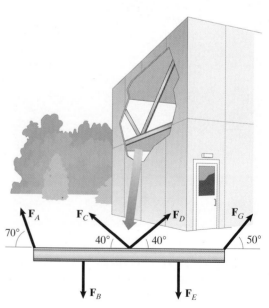

P2.55

2.56 The total weight of the man and parasail is $|\mathbf{W}| = 230$ lb. The drag force $\mathbf{D}$ is perpendicular to the lift force $\mathbf{L}$. If the vector sum of the three forces is zero, what are the magnitudes of $\mathbf{L}$ and $\mathbf{D}$?

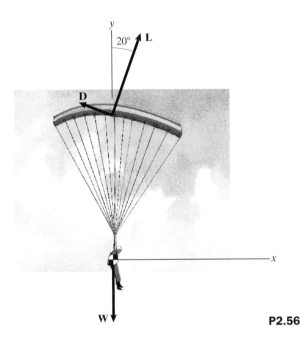

P2.56

2.57 Two cables AB and CD extend from the rocket gantry to the ground. Cable AB exerts a force of magnitude 10,000 lb on the gantry, and cable CD exerts a force of magnitude 5000 lb.
(a) Using the coordinate system shown, express each of the two forces exerted on the gantry by the cables in terms of scalar components.
(b) What is the magnitude of the total force exerted on the gantry by the two cables?

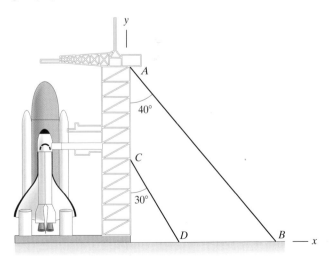

P2.57

2.58 The cables A, B, and C help support a pillar that forms part of the supports of a structure. The magnitudes of the forces exerted by the cables are equal: $|\mathbf{F}_A| = |\mathbf{F}_B| = |\mathbf{F}_C|$. The magnitude of the vector sum of the three forces is 200 kN. What is $|\mathbf{F}_A|$?

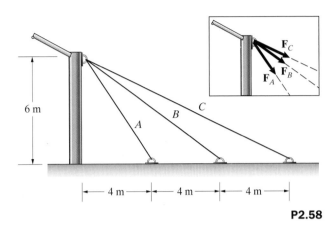

P2.58

2.59 The cable from B to A on the sailboat shown in Problem 2.41 exerts a 230-N force at B. The cable from B to C exerts a 660-N force at B. What is the magnitude of the total force exerted at B by the two cables? What is the magnitude of the downward force (parallel to the y axis) exerted by the two cables on the boat's mast?

2.60 The structure shown forms part of a truss designed by an architectural engineer to support the roof of an orchestra shell. The members AB, AC, and AD exert forces $\mathbf{F}_{AB}$, $\mathbf{F}_{AC}$, and $\mathbf{F}_{AD}$ on the joint A. The magnitude $|\mathbf{F}_{AB}| = 4$ kN. If the vector sum of the three forces equals zero, what are the magnitudes of $\mathbf{F}_{AC}$ and $\mathbf{F}_{AD}$?

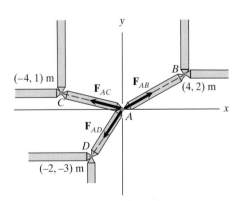

P2.60

2.61 The distance $s = 45$ in.
(a) Determine the unit vector $\mathbf{e}_{BA}$ that points from B toward A.
(b) Use the unit vector you obtained in (a) to determine the coordinates of the collar C.

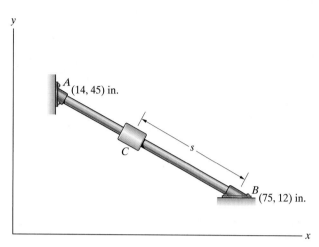

P2.61

2.62 In Problem 2.61, determine the x and y coordinates of the collar C as functions of the distance s.

2.63 The position vector $\mathbf{r}$ goes from point A to a point on the straight line between B and C. Its magnitude is $|\mathbf{r}| = 6$ ft. Express $\mathbf{r}$ in terms of scalar components.

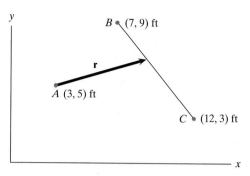

P2.63

2.64 Let $\mathbf{r}$ be the position vector from point C to the point that is a distance s meters from point A along the straight line between A and B. Express $\mathbf{r}$ in terms of scalar components. (Your answer will be in terms of s.)

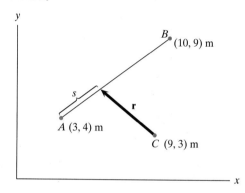

P2.64

2.4 Components in Three Dimensions

Many engineering applications require you to resolve vectors into components in a three-dimensional coordinate system. In this section we explain this technique and demonstrate vector operations in three dimensions.

Let's first review how to draw objects in three dimensions. Consider a three-dimensional object such as a cube. If we draw the cube as it appears when your line of sight is perpendicular to one of its faces, we obtain the diagram shown in Fig. 2.20a. In this view the cube appears two-dimensional; you can't see the dimension perpendicular to the page. To remedy this, we can draw the cube as it appears if you move upward and to the right (Fig. 2.20b). In this oblique view you can see the third dimension. The hidden edges of the cube are shown as dashed lines.

We can use this method to draw three-dimensional coordinate systems. In Fig. 2.20c we align the x, y, and z axes of a three-dimensional cartesian coordinate system with the edges of the cube. The three-dimensional representation of the coordinate system is shown in Fig. 2.20d.

The coordinate system in Fig. 2.20d is *right-handed*. If you point the fingers of your right hand in the direction of the positive x axis and bend

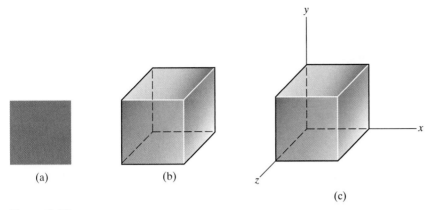

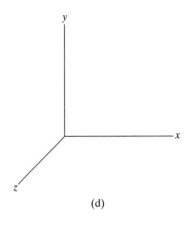

(a) (b) (c) (d)

Figure 2.20
(a) A cube viewed with the line of sight perpendicular to a face.
(b) An oblique view of the cube.
(c) A cartesian coordinate system aligned with the edges of the cube.
(d) Three-dimensional representation of the coordinate system.

them (as in preparing to make a fist) toward the positive y axis, your thumb will point in the direction of the positive z axis (Fig. 2.21). When the positive z axis points in the opposite direction, the coordinate system is left-handed. For some purposes, it doesn't matter which coordinate system you use. However, some equations we will derive do not give correct results with a left-handed coordinate system. For this reason we will use only right-handed coordinate systems.

We can resolve a vector $\mathbf{U}$ into vector components $\mathbf{U}_x$, $\mathbf{U}_y$, and $\mathbf{U}_z$ parallel to the x, y, and z axes (Fig. 2.22):

$$\mathbf{U} = \mathbf{U}_x + \mathbf{U}_y + \mathbf{U}_z. \tag{2.11}$$

(We have drawn a box around the vector to help you visualize the directions of the vector components.) By introducing unit vectors $\mathbf{i}$, $\mathbf{j}$, and $\mathbf{k}$ that point in the positive x, y, and z directions, we can express $\mathbf{U}$ in terms of scalar components as

$$\mathbf{U} = U_x\mathbf{i} + U_y\mathbf{j} + U_z\mathbf{k}. \tag{2.12}$$

We will refer to the scalars U_x, U_y, and U_z as the x, y, and z components of $\mathbf{U}$.

Magnitude of a Vector in Terms of Components

Consider a vector $\mathbf{U}$ and its vector components (Fig. 2.23a). From the right triangle formed by the vectors $\mathbf{U}_y$, $\mathbf{U}_z$, and their sum $\mathbf{U}_y + \mathbf{U}_z$ (Fig. 2.23b), we can see that

$$\left|\mathbf{U}_y + \mathbf{U}_z\right|^2 = \left|\mathbf{U}_y\right|^2 + \left|\mathbf{U}_z\right|^2. \tag{2.13}$$

The vector $\mathbf{U}$ is the sum of the vectors $\mathbf{U}_x$ and $\mathbf{U}_y + \mathbf{U}_z$. These three vectors form a right triangle (Fig. 2.23c), from which we obtain

$$\left|\mathbf{U}\right|^2 = \left|\mathbf{U}_x\right|^2 + \left|\mathbf{U}_y + \mathbf{U}_z\right|^2.$$

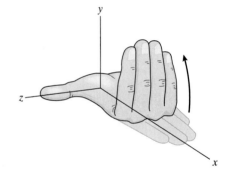

Figure 2.21
Recognizing a right-handed coordinate system.

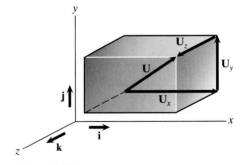

Figure 2.22
A vector $\mathbf{U}$ and its vector components.

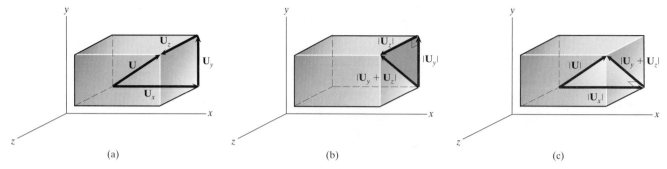

Figure 2.23
(a) A vector **U** and its vector components.
(b) The right triangle formed by the vectors $\mathbf{U}_y$, $\mathbf{U}_z$, and $\mathbf{U}_y + \mathbf{U}_z$.
(c) The right triangle formed by the vectors **U**, $\mathbf{U}_x$, and $\mathbf{U}_y + \mathbf{U}_z$.

Substituting Eq. (2.13) into this result yields the equation

$$|\mathbf{U}|^2 = |\mathbf{U}_x|^2 + |\mathbf{U}_y|^2 + |\mathbf{U}_z|^2 = U_x^2 + U_y^2 + U_z^2.$$

Thus the magnitude of a vector **U** is given in terms of its components in three dimensions by

$$|\mathbf{U}| = \sqrt{U_x^2 + U_y^2 + U_z^2}. \tag{2.14}$$

Direction Cosines

We described the direction of a vector relative to a two-dimensional cartesian coordinate system by specifying the angle between the vector and one of the coordinate axes. One of the ways we can describe the direction of a vector in three dimensions is by specifying the angles θ_x, θ_y, and θ_z between the vector and the positive coordinate axes (Fig. 2.24a).

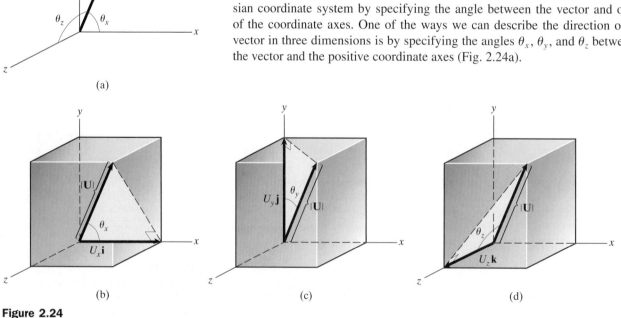

Figure 2.24
(a) A vector **U** and the angles θ_x, θ_y, and θ_z.
(b)–(d) The angles θ_x, θ_y, and θ_z and the vector components of **U**.

In Figs. 2.24b–d, we demonstrate that the components of the vector **U** are given in terms of the angles θ_x, θ_y, and θ_z. by

$$U_x = |\mathbf{U}| \cos\theta_x, \quad U_y = |\mathbf{U}| \cos\theta_y, \quad U_z = |\mathbf{U}| \cos\theta_z. \quad (2.15)$$

The quantities $\cos\theta_x$, $\cos\theta_y$, and $\cos\theta_z$ are called the *direction cosines* of **U**. The direction cosines of a vector are not independent. If we substitute Eqs. (2.15) into Eq. (2.14), we find that the direction cosines satisfy the relation

$$\cos^2\theta_x + \cos^2\theta_y + \cos^2\theta_z = 1. \quad (2.16)$$

Suppose that **e** is a unit vector with the same direction as **U**, so that

$$\mathbf{U} = |\mathbf{U}|\mathbf{e}.$$

In terms of components, this equation is

$$U_x\mathbf{i} + U_y\mathbf{j} + U_z\mathbf{k} = |\mathbf{U}|(e_x\mathbf{i} + e_y\mathbf{j} + e_z\mathbf{k}).$$

Thus the relations between the components of **U** and **e** are

$$U_x = |\mathbf{U}|e_x, \quad U_y = |\mathbf{U}|e_y, \quad U_z = |\mathbf{U}|e_z.$$

By comparing these equations to Eqs. (2.15), we see that

$$\cos\theta_x = e_x, \quad \cos\theta_y = e_y, \quad \cos\theta_z = e_z.$$

The direction cosines of a vector **U** are the components of a unit vector with the same direction as **U**.

Position Vectors in Terms of Components

Generalizing the two-dimensional case, let's consider a point A with coordinates (x_A, y_A, z_A) and a point B with coordinates (x_B, y_B, z_B). The position vector $\mathbf{r}_{AB}$ from A to B, shown in Fig. 2.25a, is given in terms of the coordinates of A and B by

$$\mathbf{r}_{AB} = (x_B - x_A)\mathbf{i} + (y_B - y_A)\mathbf{j} + (z_B - z_A)\mathbf{k}. \quad (2.17)$$

The components are obtained by subtracting the coordinates of point A from the coordinates of point B (Fig. 2.25b).

Components of a Vector Parallel to a Given Line

In three-dimensional applications, the direction of a vector is often defined by specifying the coordinates of two points on a line that is parallel to the vector. You can use this information to determine the components of the vector.

Suppose that we know the coordinates of two points A and B on a line parallel to a vector **U** (Fig. 2.26a). We can use Eq. (2.17) to determine the position vector $\mathbf{r}_{AB}$ from A to B (Fig. 2.26b). We can divide $\mathbf{r}_{AB}$ by its magnitude to obtain a unit vector $\mathbf{e}_{AB}$ that points from A toward B (Fig. 2.26c). Since $\mathbf{e}_{AB}$ has the same direction as **U**, we can determine **U** in terms of its scalar components by expressing it as the product of its magnitude and $\mathbf{e}_{AB}$.

More generally, suppose that we know the magnitude of a vector **U** and the components of any vector **V** that has the same direction as **U**. Then $\mathbf{V}/|\mathbf{V}|$

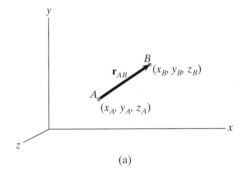

(a)

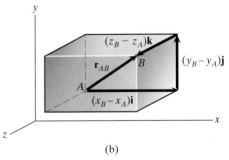

(b)

Figure 2.25
(a) The position vector from point A to point B.
(b) The components of $\mathbf{r}_{AB}$ can be determined from the coordinates of points A and B.

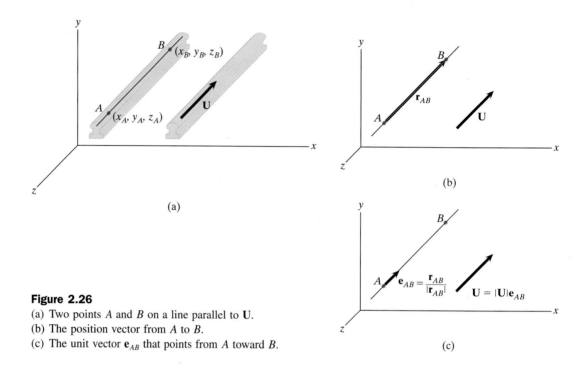

Figure 2.26
(a) Two points A and B on a line parallel to $\mathbf{U}$.
(b) The position vector from A to B.
(c) The unit vector $\mathbf{e}_{AB}$ that points from A toward B.

is a unit vector with the same direction as $\mathbf{U}$, and we can determine the components of $\mathbf{U}$ by expressing it as $\mathbf{U} = |\mathbf{U}|(\mathbf{V}/|\mathbf{V}|)$.

Study Questions

1. How do you identify a right-handed coordinate system?
2. If you know the scalar components of a vector in three dimensions, how can you determine its magnitude?
3. What are the direction cosines of a vector? If you know them, how do you determine the components of the vector?
4. Suppose that you know the coordinates of two points A and B in three dimensions. How do you determine the scalar components of the position vector of point B relative to point A?

Example 2.7

Magnitude and Direction Cosines of a Vector

An engineer designing a threshing machine determines that at a particular time the position vectors of the ends A and B of a shaft are $\mathbf{r}_A = 3\mathbf{i} - 4\mathbf{j} - 12\mathbf{k}$ (ft) and $\mathbf{r}_B = -\mathbf{i} + 7\mathbf{j} + 6\mathbf{k}$ (ft).
(a) What is the magnitude of $\mathbf{r}_A$?
(b) Determine the angles θ_x, θ_y, and θ_z between $\mathbf{r}_A$ and the positive coordinate axes.
(c) Determine the scalar components of the position vector of end B of the shaft relative to end A.

Strategy

(a) Since we know the components of $\mathbf{r}_A$, we can use Eq. (2.14) to determine its magnitude.
(b) We can obtain the angles θ_x, θ_y, and θ_z from Eqs. (2.15).
(c) The position vector of end B of the shaft relative to end A is $\mathbf{r}_B - \mathbf{r}_A$.

Solution

(a) The magnitude of $\mathbf{r}_A$ is

$$|\mathbf{r}_A| = \sqrt{r_{Ax}^2 + r_{Ay}^2 + r_{Az}^2} = \sqrt{(3)^2 + (-4)^2 + (-12)^2} = 13 \text{ ft.}$$

(b) The direction cosines of $\mathbf{r}_A$ are

$$\cos \theta_x = \frac{r_{Ax}}{|\mathbf{r}_A|} = \frac{3}{13},$$

$$\cos \theta_y = \frac{r_{Ay}}{|\mathbf{r}_A|} = \frac{-4}{13},$$

$$\cos \theta_z = \frac{r_{Az}}{|\mathbf{r}_A|} = \frac{-12}{13}.$$

From these equations we find that the angles between $\mathbf{r}_A$ and the positive coordinate axes are $\theta_x = 76.7°$, $\theta_y = 107.9°$, and $\theta_z = 157.4°$.
(c) The position vector of end B of the shaft relative to end A is

$$\mathbf{r}_B - \mathbf{r}_A = (-\mathbf{i} + 7\mathbf{j} + 6\mathbf{k}) - (3\mathbf{i} - 4\mathbf{j} - 12\mathbf{k})$$

$$= -4\mathbf{i} + 11\mathbf{j} + 18\mathbf{k} \text{ (ft).}$$

Example 2.8

Determining Scalar Components

The crane in Fig. 2.27 exerts a 600-lb force $\mathbf{F}$ on the caisson. The angle between $\mathbf{F}$ and the x axis is 54°, and the angle between $\mathbf{F}$ and the y axis is 40°. The z component of $\mathbf{F}$ is positive. Express $\mathbf{F}$ in terms of components.

Figure 2.27

Strategy

Only two of the angles between the vector and the positive coordinate axes are given, but we can use Eq. (2.16) to determine the third angle. Then we can determine the components of **F** by using Eqs. (2.15).

Solution

The angles between **F** and the positive coordinate axes are related by

$$\cos^2\theta_x + \cos^2\theta_y + \cos^2\theta_z = (\cos 54°)^2 + (\cos 40°)^2 + \cos^2\theta_z = 1.$$

Solving this equation for $\cos\theta_z$, we obtain the two solutions $\cos\theta_z = 0.260$ and $\cos\theta_z = -0.260$, which tells us that $\theta_z = 74.9°$ or $\theta_z = 105.1°$. The z component of the vector **F** is positive, so the angle between **F** and the positive z axis is less than 90°. Therefore $\theta_z = 74.9°$.

The components of **F** are

$$F_x = |\mathbf{F}|\cos\theta_x = 600\cos 54° \quad = 353 \text{ lb},$$

$$F_y = |\mathbf{F}|\cos\theta_y = 600\cos 40° \quad = 460 \text{ lb},$$

$$F_z = |\mathbf{F}|\cos\theta_z = 600\cos 74.9° = 156 \text{ lb}.$$

Example 2.9

Determining Scalar Components

The tether of the balloon in Fig. 2.28 exerts an 800-N force **F** on the hook at O. The vertical line AB intersects the x-z plane at point A. The angle between the z axis and the line OA is 60°, and the angle between the line OA and **F** is 45°. Express **F** in terms of components.

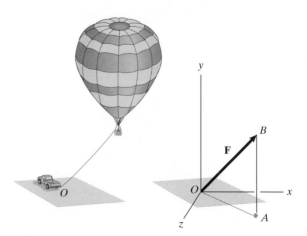

Figure 2.28

Strategy

We can determine the components of **F** from the given geometric information in two steps. First, we resolve **F** into two vector components parallel to the lines OA and AB. The component parallel to AB is the vector component $\mathbf{F}_y$. Then we can resolve the component parallel to OA to determine the vector components $\mathbf{F}_x$ and $\mathbf{F}_z$.

Solution

In Fig. a, we resolve **F** into its y component $\mathbf{F}_y$ and the component $\mathbf{F}_h$ parallel to OA. The magnitude of $\mathbf{F}_y$ is

$$|\mathbf{F}_y| = |\mathbf{F}| \sin 45° = 800 \sin 45° = 566 \text{ N},$$

and the magnitude of $\mathbf{F}_h$ is

$$|\mathbf{F}_h| = |\mathbf{F}| \cos 45° = 800 \cos 45° = 566 \text{ N},$$

In Fig. b, we resolve $\mathbf{F}_h$ into the vector components $\mathbf{F}_x$ and $\mathbf{F}_z$. The magnitude of $\mathbf{F}_x$ is

$$|\mathbf{F}_x| = |\mathbf{F}_h| \sin 60° = 566 \sin 60° = 490 \text{ N},$$

and the magnitude of $\mathbf{F}_z$ is

$$|\mathbf{F}_z| = |\mathbf{F}_h| \cos 60° = 566 \cos 60° = 283 \text{ N}.$$

The vector components $\mathbf{F}_x$, $\mathbf{F}_y$, and $\mathbf{F}_z$ all point in the positive axis directions, so the scalar components of **F** are positive:

$$\mathbf{F} = 490\mathbf{i} + 566\mathbf{j} + 283\mathbf{k} \text{ (N)}.$$

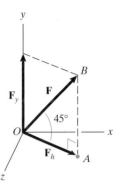

(a) Resolving **F** into vector components parallel to OA and OB.

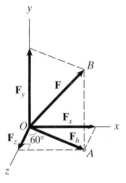

(b) Resolving $\mathbf{F}_h$ into vector components parallel to the x and z axes.

Example 2.10

Vector Whose Direction is Specified by Two Points

The bar AB in Fig. 2.29 exerts a 140-N force **F** on its support at A. The force is parallel to the bar and points toward B. Express **F** in terms of components.

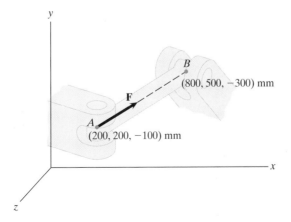

Figure 2.29

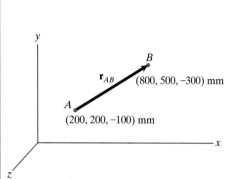

(a) The position vector $\mathbf{r}_{AB}$.

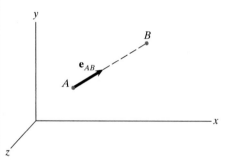

(b) The unit vector $\mathbf{e}_{AB}$ pointing from A toward B.

Strategy

Since we are given the coordinates of points A and B, we can determine the components of the position vector from A to B. By dividing the position vector by its magnitude, we can obtain a unit vector with the same direction as $\mathbf{F}$. Then by multiplying the unit vector by the magnitude of $\mathbf{F}$, we obtain $\mathbf{F}$ in terms of its components.

Solution

The position vector from A to B is (Fig. a)

$$
\begin{aligned}
\mathbf{r}_{AB} &= (x_B - x_A)\mathbf{i} + (y_B - y_A)\mathbf{j} + (z_B - z_A)\mathbf{k} \\
&= [(800) - (200)]\mathbf{i} + [(500) - (200)]\mathbf{j} + [(-300) - (-100)]\mathbf{k} \\
&= 600\mathbf{i} + 300\mathbf{j} - 200\mathbf{k} \text{ mm},
\end{aligned}
$$

and its magnitude is

$$
|\mathbf{r}_{AB}| = \sqrt{(600)^2 + (300)^2 + (-200)^2} = 700 \text{ mm}.
$$

By dividing $\mathbf{r}_{AB}$ by its magnitude, we obtain a unit vector with the same direction as $\mathbf{F}$ (Fig. b),

$$
\mathbf{e}_{AB} = \frac{\mathbf{r}_{AB}}{|\mathbf{r}_{AB}|} = \frac{6}{7}\mathbf{i} + \frac{3}{7}\mathbf{j} - \frac{2}{7}\mathbf{k}.
$$

Then, in terms of its scalar components, $\mathbf{F}$ is

$$
\mathbf{F} = |\mathbf{F}|\mathbf{e}_{AB} = (140)\left(\frac{6}{7}\mathbf{i} + \frac{3}{7}\mathbf{j} - \frac{2}{7}\mathbf{k}\right) = 120\mathbf{i} + 60\mathbf{j} - 40\mathbf{k} \text{ (N)}.
$$

Example 2.11

Determining Components in Three Dimensions

The rope in Fig. 2.30 extends from point B through a metal loop attached to the wall at A to point C. The rope exerts forces $\mathbf{F}_{AB}$ and $\mathbf{F}_{AC}$ on the loop at A

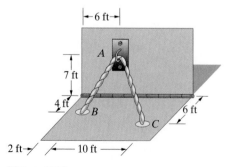

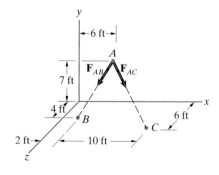

Figure 2.30

with magnitudes $|\mathbf{F}_{AB}| = |\mathbf{F}_{AC}| = 200$ lb. What is the magnitude of the total force $\mathbf{F} = \mathbf{F}_{AB} + \mathbf{F}_{AC}$ exerted on the loop by the rope?

Strategy

The force $\mathbf{F}_{AB}$ is parallel to the line from A to B, and the force $\mathbf{F}_{AC}$ is parallel to the line from A to C. Since we can determine the coordinates of points A, B, and C from the given dimensions, we can determine the components of unit vectors that have the same directions as the two forces and use them to express the forces in terms of scalar components.

Solution

Let $\mathbf{r}_{AB}$ be the position vector from point A to point B and let $\mathbf{r}_{AC}$ be the position vector from point A to point C (Fig. a). From the given dimensions, the coordinates of points A, B, and C are

$$A\colon (6, 7, 0) \text{ ft}, \qquad B\colon (2, 0, 4) \text{ ft}, \qquad C\colon (12, 0, 6) \text{ ft}.$$

Therefore the components of $\mathbf{r}_{AB}$ and $\mathbf{r}_{AC}$ are

$$\begin{aligned} \mathbf{r}_{AB} &= (x_B - x_A)\mathbf{i} + (y_B - y_A)\mathbf{j} + (z_B - z_A)\mathbf{k} \\ &= (2 - 6)\mathbf{i} + (0 - 7)\mathbf{j} + (4 - 0)\mathbf{k} \\ &= -4\mathbf{i} - 7\mathbf{j} + 4\mathbf{k} \text{ (ft)} \end{aligned}$$

and

$$\begin{aligned} \mathbf{r}_{AC} &= (x_C - x_A)\mathbf{i} + (y_C - y_A)\mathbf{j} + (z_C - z_A)\mathbf{k} \\ &= (12 - 6)\mathbf{i} + (0 - 7)\mathbf{j} + (6 - 0)\mathbf{k} \\ &= 6\mathbf{i} - 7\mathbf{j} + 6\mathbf{k} \text{ (ft)}. \end{aligned}$$

Their magnitudes are $|\mathbf{r}_{AB}| = 9$ ft and $|\mathbf{r}_{AC}| = 11$ ft. By dividing $\mathbf{r}_{AB}$ and $\mathbf{r}_{AC}$ by their magnitudes, we obtain unit vectors $\mathbf{e}_{AB}$ and $\mathbf{e}_{AC}$ that point in the directions of $\mathbf{F}_{AB}$ and $\mathbf{F}_{AC}$ (Fig. b):

$$\mathbf{e}_{AB} = \frac{\mathbf{r}_{AB}}{|\mathbf{r}_{AB}|} = -0.444\mathbf{i} - 0.778\mathbf{j} + 0.444\mathbf{k},$$

$$\mathbf{e}_{AC} = \frac{\mathbf{r}_{AC}}{|\mathbf{r}_{AC}|} = 0.545\mathbf{i} - 0.636\mathbf{j} + 0.545\mathbf{k}.$$

The forces $\mathbf{F}_{AB}$ and $\mathbf{F}_{AC}$ are

$$\mathbf{F}_{AB} = 200\mathbf{e}_{AB} = -88.9\mathbf{i} - 155.6\mathbf{j} + 88.9\mathbf{k} \text{ (lb)},$$

$$\mathbf{F}_{AC} = 200\mathbf{e}_{AC} = 109.1\mathbf{i} - 127.3\mathbf{j} + 109.1\mathbf{k} \text{ (lb)}.$$

The total force exerted on the loop by the rope is

$$\mathbf{F} = \mathbf{F}_{AB} + \mathbf{F}_{AC} = 20.2\mathbf{i} - 282.8\mathbf{j} + 198.0\mathbf{k} \text{ (lb)},$$

and its magnitude is

$$|\mathbf{F}| = \sqrt{(20.2)^2 + (-282.8)^2 + (198.0)^2} = 346 \text{ lb}.$$

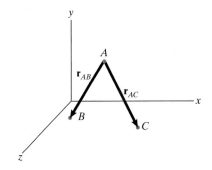

(a) The position vectors $\mathbf{r}_{AB}$ and $\mathbf{r}_{AC}$.

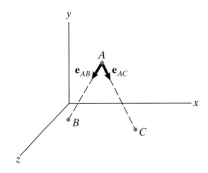

(b) The unit vector $\mathbf{e}_{AB}$ pointing and $\mathbf{e}_{AC}$.

Example 2.12

Determining Components of a Force

The cable AB in Fig. 2.31 exerts a 50-N force $\mathbf{T}$ on the collar at A. Express $\mathbf{T}$ in terms of components.

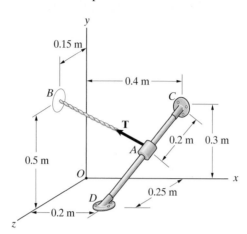

Figure 2.31

Strategy

Let $\mathbf{r}_{AB}$ be the position vector from A to B. We will divide $\mathbf{r}_{AB}$ by its magnitude to obtain a unit vector $\mathbf{e}_{AB}$ having the same direction as the force $\mathbf{T}$. Then we can obtain $\mathbf{T}$ in terms of scalar components by expressing it as the product of its magnitude and $\mathbf{e}_{AB}$. To begin this procedure, we must first determine the coordinates of the collar A. We will do so by obtaining a unit vector $\mathbf{e}_{CD}$ pointing from C toward D and multiplying it by 0.2 m to determine the position of the collar A relative to C.

Solution

Determining the Coordinates of Point A The position vector from C to D is

$$\mathbf{r}_{CD} = (0.2 - 0.4)\mathbf{i} + (0 - 0.3)\mathbf{j} + (0.25 - 0)\mathbf{k}$$
$$= -0.2\mathbf{i} - 0.3\mathbf{j} + 0.25\mathbf{k} \ (\text{m}).$$

Dividing this vector by its magnitude, we obtain the unit vector $\mathbf{e}_{CD}$ (Fig. a):

$$\mathbf{e}_{CD} = \frac{\mathbf{r}_{CD}}{|\mathbf{r}_{CD}|} = \frac{-0.2\mathbf{i} - 0.3\mathbf{j} + 0.25\mathbf{k}}{\sqrt{(-0.2)^2 + (-0.3)^2 + (0.25)^2}}$$
$$= -0.456\mathbf{i} - 0.684\mathbf{j} + 0.570\mathbf{k}.$$

Using this vector, we obtain the position vector from C to A:

$$\mathbf{r}_{CA} = (0.2 \text{ m})\mathbf{e}_{CD} = -0.091\mathbf{i} - 0.137\mathbf{j} + 0.114\mathbf{k} \ (\text{m}).$$

The position vector from the origin of the coordinate system to C is $\mathbf{r}_{OC} = 0.4\mathbf{i} + 0.3\mathbf{j}\,(\text{m})$, so the position vector from the origin to A is

$$\mathbf{r}_{OA} = \mathbf{r}_{OC} + \mathbf{r}_{CA} = (0.4\mathbf{i} + 0.3\mathbf{j}) + (-0.091\mathbf{i} - 0.137\mathbf{j} + 0.114\mathbf{k})$$
$$= 0.309\mathbf{i} + 0.163\mathbf{j} + 0.114\mathbf{k} \ (\text{m}).$$

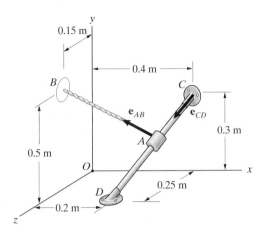

(a) The unit vectors $\mathbf{e}_{AB}$ and $\mathbf{e}_{CD}$.

The coordinates of A are $(0.309, 0.163, 0.114)$ m.

Determining the Components of T Using the coordinates of point A, the position vector from A to B is

$$\mathbf{r}_{AB} = (0 - 0.309)\mathbf{i} + (0.5 - 0.163)\mathbf{j} + (0.15 - 0.114)\mathbf{k}$$
$$= -0.309\mathbf{i} + 0.337\mathbf{j} + 0.036\mathbf{k} \text{ (m)}.$$

Dividing this vector by its magnitude, we obtain the unit vector $\mathbf{e}_{AB}$ [Fig. (a)]:

$$\mathbf{e}_{AB} = \frac{\mathbf{r}_{AB}}{|\mathbf{r}_{AB}|} = \frac{-0.309\mathbf{i} + 0.337\mathbf{j} + 0.036\mathbf{k}}{\sqrt{(-0.309)^2 + (0.337)^2 + (0.036)^2}}$$
$$= -0.674\mathbf{i} + 0.735\mathbf{j} + 0.079\mathbf{k}.$$

The force $\mathbf{T}$ is

$$\mathbf{T} = |\mathbf{T}|\mathbf{e}_{AB} = (50 \text{ N})(-0.674\mathbf{i} + 0.735\mathbf{j} + 0.079\mathbf{k})$$
$$= -33.7\mathbf{i} + 36.7\mathbf{j} + 3.9\mathbf{k} \text{ (N)}.$$

Problems

2.65 A vector $\mathbf{U} = 3\mathbf{i} - 4\mathbf{j} - 12\mathbf{k}$. What is its magnitude?

  Strategy: The magnitude of a vector is given in terms of its components by Eq. (2.14).

2.66 A force vector $\mathbf{F} = 20\mathbf{i} + 60\mathbf{j} - 90\mathbf{k}$ (N). Determine its magnitude.

2.67 An engineer determines that an attachment point will be subjected to a force $\mathbf{F} = 20\mathbf{i} + F_y\mathbf{j} - 45\mathbf{k}$ (kN). If the attachment point will safely support a force of 80-kN magnitude in any direction, what is the acceptable range of values of F_y?

2.68 A vector $\mathbf{U} = U_x\mathbf{i} + U_y\mathbf{j} + U_z\mathbf{k}$. Its magnitude $|\mathbf{U}| = 30$. Its components are related by the equations $U_y = -2U_x$ and $U_z = 4U_y$. Determine the components.

2.69 A vector $\mathbf{U} = 100\mathbf{i} + 200\mathbf{j} - 600\mathbf{k}$, and a vector $\mathbf{V} = -200\mathbf{i} + 450\mathbf{j} + 100\mathbf{k}$. Determine the magnitude of the vector $-2\mathbf{U} + 3\mathbf{V}$.

2.70 Two vectors $\mathbf{U} = 3\mathbf{i} - 2\mathbf{j} + 6\mathbf{k}$ and $\mathbf{V} = 4\mathbf{i} + 12\mathbf{j} - 3\mathbf{k}$.
(a) Determine the magnitudes of $\mathbf{U}$ and $\mathbf{V}$.
(b) Determine the magnitude of the vector $3\mathbf{U} + 2\mathbf{V}$.

2.71 A vector $\mathbf{U} = 40\mathbf{i} - 70\mathbf{j} - 40\mathbf{k}$.
(a) What is its magnitude?
(b) What are the angles θ_x, θ_y, and θ_z between $\mathbf{U}$ and the positive coordinate axes?

  Strategy: Since you know the components of $\mathbf{U}$, you can determine the angles θ_x, θ_y, and θ_z from Eqs. (2.15).

2.72 A force $\mathbf{F} = 600\mathbf{i} - 700\mathbf{j} + 600\mathbf{k}$ (lb). What are the angles θ_x, θ_y, and θ_z between the vector $\mathbf{F}$ and the positive coordinate axes?

2.73 The cable exerts a 50-lb force **F** on the metal hook at O. The angle between **F** and the x axis is $40°$, and the angle between **F** and the y axis is $70°$. The z component of **F** is positive.
(a) Express **F** in terms of components.
(b) What are the direction cosines of **F**?

   Strategy: Since you are given only two of the angles between **F** and the coordinate axes, you must first determine the third one. Then you can obtain the components of **F** from Eqs. (2.15).

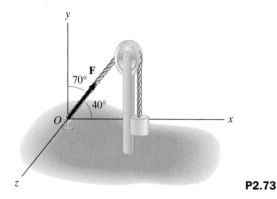

P2.73

2.74 A unit vector has direction cosines $\cos\theta_x = -0.5$ and $\cos\theta_y = 0.2$. Its z component is positive. Express it in terms of components.

2.75 The airplane's engines exert a total thrust force **T** of 200-kN magnitude. The angle between **T** and the x axis is $120°$, and the angle between **T** and the y axis is $130°$. The z component of **T** is positive.
(a) What is the angle between **T** and the z axis?
(b) Express **T** in terms of components.

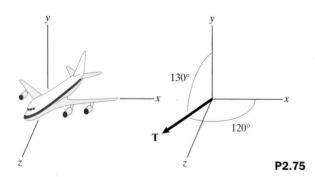

P2.75

2.76 The position vector from a point A to a point B is $3\mathbf{i} + 4\mathbf{j} - 4\mathbf{k}$ (ft). The position vector from point A to a point C is $-3\mathbf{i} + 13\mathbf{j} - 2\mathbf{k}$ (ft).
(a) What is the distance from point B to point C?
(b) What are the direction cosines of the position vector from point B to point C?

2.77 A vector $\mathbf{U} = 3\mathbf{i} - 2\mathbf{j} + 6\mathbf{k}$. Determine the components of the unit vector that has the same direction as **U**.

2.78 A force vector $\mathbf{F} = 3\mathbf{i} - 4\mathbf{j} - 2\mathbf{k}$ (N).
(a) What is the magnitude of **F**?
(b) Determine the components of the unit vector that has the same direction as **F**.

2.79 A force vector **F** points in the same direction as the unit vector $\mathbf{e} = \frac{2}{7}\mathbf{i} - \frac{6}{7}\mathbf{j} - \frac{3}{7}\mathbf{k}$. The magnitude of **F** is 700 lb. Express **F** in terms of components.

2.80 A force vector **F** points in the same direction as the position vector $\mathbf{r} = 4\mathbf{i} + 4\mathbf{j} - 7\mathbf{k}$ (m). The magnitude of **F** is 90 kN. Express **F** in terms of components.

2.81 Astronauts on the space shuttle use radar to determine the magnitudes and direction cosines of the position vectors of two satellites A and B. The vector $\mathbf{r}_A$ from the shuttle to satellite A has magnitude 2 km, and direction cosines $\cos\theta_x = 0.768$, $\cos\theta_y = 0.384$, $\cos\theta_z = 0.512$. The vector $\mathbf{r}_B$ from the shuttle to satellite B has magnitude 4 km and direction cosines $\cos\theta_x = 0.743$, $\cos\theta_y = 0.557$, $\cos\theta_z = -0.371$. What is the distance between the satellites?

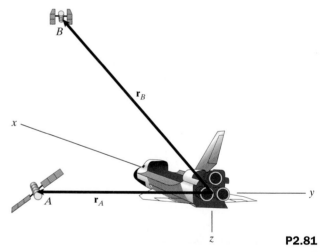

P2.81

2.82 Archaeologists measure a pre-Columbian ceremonial structure and obtain the dimensions shown. Determine (a) the magnitude and (b) the direction cosines of the position vector from point A to point B.

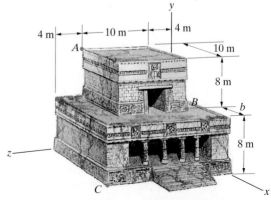

P2.82

2.83 Consider the structure described in Problem 2.82. After returning to the United States, an archaeologist discovers that he lost the notes containing the dimension b, but other notes indicate that the distance from point B to point C is 16.4 m. What are the direction cosines of the vector from B to C?

2.84 Observers at A and B use theodolites to measure the direction from their positions to a rocket in flight. If the coordinates of the rocket's position at a given instant are (4, 4, 2) km, determine the direction cosines of the vectors $\mathbf{r}_{AR}$ and $\mathbf{r}_{BR}$ that the observers would measure at that instant.

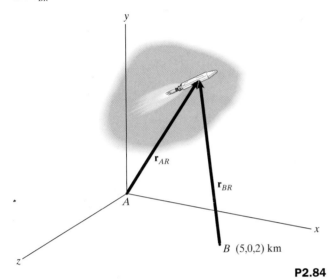

P2.84

2.85 In Problem 2.84, suppose that the coordinates of the rocket's position are unknown. At a given instant, the person at A determines that the direction cosines of $\mathbf{r}_{AR}$ are $\cos\theta_x = 0.535$, $\cos\theta_y = 0.802$, and $\cos\theta_z = 0.267$, and the person at B determines that the direction cosines of $\mathbf{r}_{BR}$ are $\cos\theta_x = -0.576$, $\cos\theta_y = 0.798$, and $\cos\theta_z = -0.177$. What are the coordinates of the rocket's position at that instant?

2.86 The height of Mount Everest was originally measured by a surveyor using the following procedure. He first measured the distance between two points A and B of equal altitude. Suppose that they are 10,000 ft above sea level and are 32,000 ft apart. He then used a theodolite to measure the direction cosines of the vectors from point A to the top of the mountain P and from point

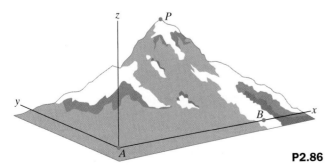

P2.86

B to P. Suppose that for $\mathbf{r}_{AP}$, the direction cosines are $\cos\theta_x = 0.509$, $\cos\theta_y = 0.509$, $\cos\theta_z = 0.694$, and for $\mathbf{r}_{BP}$ they are $\cos\theta_x = -0.605$, $\cos\theta_y = 0.471$, $\cos\theta_z = 0.642$. The z axis of the coordinate system is vertical. What is the height of Mount Everest above sea level?

2.87 The distance from point O to point A is 20 ft. The straight line AB is parallel to the y axis, and point B is in the x-z plane. Express the vector $\mathbf{r}_{OA}$ in terms of scalar components.

Strategy: You can resolve $\mathbf{r}_{OA}$ into a vector from O to B and a vector from B to A. You can then resolve the vector from O to B into vector components parallel to the x and z axes. See Example 2.9.

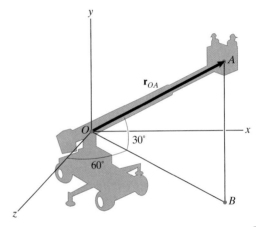

P2.87

2.88 The magnitude of $\mathbf{r}$ is 100 in. The straight line from the head of $\mathbf{r}$ to point A is parallel to the x axis, and point A is contained in the y-z plane. Express $\mathbf{r}$ in terms of scalar components.

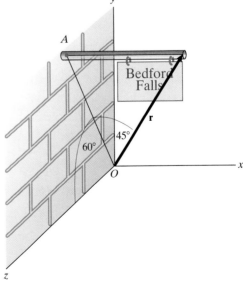

P2.88

2.89 The straight line from the head of **F** to point A is parallel to the y axis, and point A is contained in the x-z plane. The x component of **F** is $F_x = 100$ N.
(a) What is the magnitude of **F**?
(b) Determine the angles θ_x, θ_y, and θ_z between **F** and the positive coordinate axes.

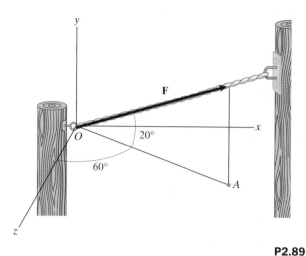

P2.89

2.90 The position of a point P on the surface of the earth is specified by the longitude λ, measured from the point G on the equator directly south of Greenwich, England, and the latitude L measured from the equator. Longitude is given as west (W) longitude or east (E) longitude, indicating whether the angle is measured west or east from point G. Latitude is given as north (N) latitude or south (S) latitude, indicating whether the angle is measured north or south from the equator. Suppose that P is at longitude 30°W and latitude 45°N. Let R_E be the radius of the earth. Using the coordinate system shown, determine the components of the position vector of P relative to the center of the earth. (Your answer will be in terms of R_E.)

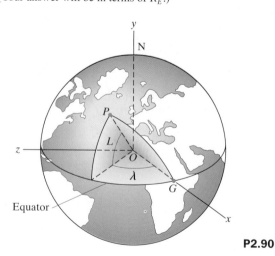

P2.90

2.91 An engineer calculates that the magnitude of the axial force in one of the beams of a geodesic dome is $|\mathbf{P}| = 7.65$ kN. The cartesian coordinates of the endpoints A and B of the straight beam are $(-12.4, 22.0, -18.4)$ m and $(-9.2, 24.4, -15.6)$ m, respectively. Express the force **P** in terms of scalar components.

P2.91

2.92 The cable BC exerts an 8-kN force **F** on the bar AB at B.
(a) Determine the components of a unit vector that points from point B toward point C.
(b) Express **F** in terms of components.

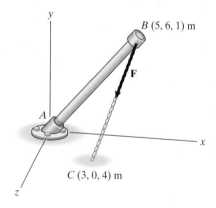

P2.92

2.93 A cable extends from point C to point E. It exerts a 50-lb force **T** on the plate at C that is directed along the line from C to E. Express **T** in terms of scalar components.

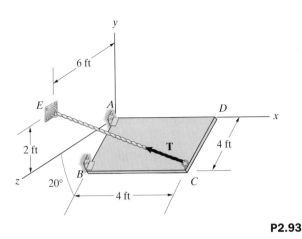

P2.93

2.94 What are the direction cosines of the force **T** in Problem 2.93?

2.95 The cable AB exerts a 200-lb force $\mathbf{F}_{AB}$ at point A that is directed along the line from A to B. Express $\mathbf{F}_{AB}$ in terms of scalar components.

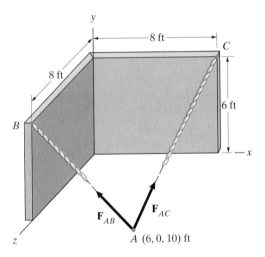

P2.95

2.96 Consider the cables and wall described in Problem 2.95. Cable AB exerts a 200-lb force $\mathbf{F}_{AB}$ at point A that is directed along the line from A to B. The cable AC exerts a 100-lb force $\mathbf{F}_{AC}$ at point A that is directed along the line from A to C. Determine the magnitude of the total force exerted at point A by the two cables.

2.97 The 70-m-tall tower is supported by three cables that exert forces $\mathbf{F}_{AB}$, $\mathbf{F}_{AC}$, and $\mathbf{F}_{AD}$ on it. The magnitude of each force is 2 kN. Express the total force exerted on the tower by the three cables in terms of scalar components.

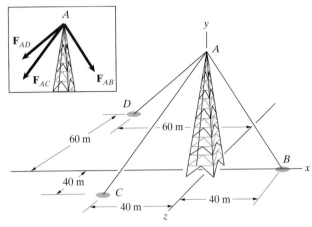

P2.97

2.98 Consider the tower described in Problem 2.97. The magnitude of the force $\mathbf{F}_{AB}$ is 2 kN. The x and z components of the vector sum of the forces exerted on the tower by the three cables are zero. What are the magnitudes of $\mathbf{F}_{AC}$ and $\mathbf{F}_{AD}$?

2.99 Express the position vector from point O to the collar at A in terms of scalar components.

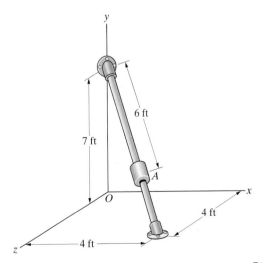

P2.99

2.100 The cable AB exerts a 32-lb force **T** on the collar at A. Express **T** in terms of scalar components.

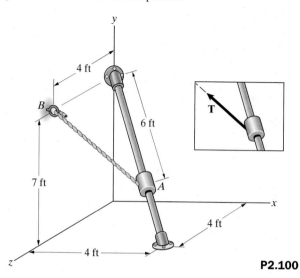

P2.100

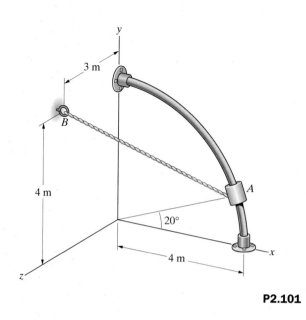

P2.101

2.101 The circular bar has a 4-m radius and lies in the x-y plane. Express the position vector from point B to the collar at A in terms of scalar components.

2.102 The cable AB in Problem 2.101 exerts a 60-N force **T** on the collar at A that is directed along the line from A toward B. Express **T** in terms of scalar components.

Products of Vectors

Two kinds of products of vectors, the dot and cross products, have been found to have applications in science and engineering, especially in mechanics and electromagnetic field theory. We use both of these products in Chapter 4 to evaluate moments of forces about points and lines. We discuss them here so that you can concentrate on mechanics when we introduce moments and not be distracted by the details of vector operations.

2.5 Dot Products

The dot product of two vectors has many uses, including resolving a vector into components parallel and perpendicular to a given line and determining the angle between two lines in space.

Definition

Consider two vectors **U** and **V** (Fig. 2.32a). The *dot product* of **U** and **V**, denoted by **U** · **V** (hence the name "dot product"), is defined to be the product of the magnitude of **U**, the magnitude of **V**, and the cosine of the angle θ between **U** and **V** when they are placed tail to tail (Fig. 2.32b):

$$\mathbf{U} \cdot \mathbf{V} = |\mathbf{U}||\mathbf{V}| \cos\theta. \tag{2.18}$$

Because the result of the dot product is a scalar, the dot product is sometimes called the scalar product. The units of the dot product are the product of the units of the two vectors. *Notice that the dot product of two nonzero vectors is equal to zero if and only if the vectors are perpendicular.*

The dot product has the properties

$$\mathbf{U} \cdot \mathbf{V} = \mathbf{V} \cdot \mathbf{U}, \quad \text{The dot product is commutative.} \qquad (2.19)$$

$$a(\mathbf{U} \cdot \mathbf{V}) = (a\mathbf{U}) \cdot \mathbf{V} = \mathbf{U} \cdot (a\mathbf{V}), \quad \begin{array}{l}\text{The dot product is} \\ \text{associative with} \\ \text{respect to scalar} \\ \text{multiplication.}\end{array} \qquad (2.20)$$

and

$$\mathbf{U} \cdot (\mathbf{V} + \mathbf{W}) = \mathbf{U} \cdot \mathbf{V} + \mathbf{U} \cdot \mathbf{W} \quad \begin{array}{l}\text{The dot product is} \\ \text{distributive with} \\ \text{respect to vector} \\ \text{addition.}\end{array} \qquad (2.21)$$

for any scalar a and vectors $\mathbf{U}$, $\mathbf{V}$, and $\mathbf{W}$.

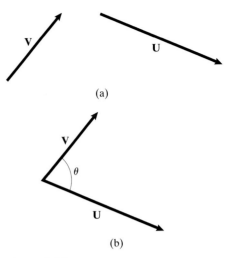

Figure 2.32
(a) The vectors $\mathbf{U}$ and $\mathbf{V}$.
(b) The angle θ between $\mathbf{U}$ and $\mathbf{V}$ when the two vectors are placed tail to tail.

Dot Products in Terms of Components

In this section we derive an equation that allows you to determine the dot product of two vectors if you know their scalar components. The derivation also results in an equation for the angle between the vectors. The first step is to determine the dot products formed from the unit vectors $\mathbf{i}$, $\mathbf{j}$, and $\mathbf{k}$. Let's evaluate the dot product $\mathbf{i} \cdot \mathbf{i}$. The magnitude $|\mathbf{i}| = 1$, and the angle between two identical vectors placed tail to tail is zero, so we obtain

$$\mathbf{i} \cdot \mathbf{i} = |\mathbf{i}||\mathbf{i}| \cos(0) = (1)(1)(1) = 1.$$

The dot product of $\mathbf{i}$ and $\mathbf{j}$ is

$$\mathbf{i} \cdot \mathbf{j} = |\mathbf{i}||\mathbf{j}| \cos(90°) = (1)(1)(0) = 0.$$

Continuing in this way, we obtain

$$\begin{array}{lll}
\mathbf{i} \cdot \mathbf{i} = 1, & \mathbf{i} \cdot \mathbf{j} = 0, & \mathbf{i} \cdot \mathbf{k} = 0, \\
\mathbf{j} \cdot \mathbf{i} = 0, & \mathbf{j} \cdot \mathbf{j} = 1, & \mathbf{j} \cdot \mathbf{k} = 0, \\
\mathbf{k} \cdot \mathbf{i} = 0, & \mathbf{k} \cdot \mathbf{j} = 0, & \mathbf{k} \cdot \mathbf{k} = 1.
\end{array} \qquad (2.22)$$

The dot product of two vectors $\mathbf{U}$ and $\mathbf{V}$ expressed in terms of their components is

$$\begin{aligned}
\mathbf{U} \cdot \mathbf{V} = {}& (U_x\mathbf{i} + U_y\mathbf{j} + U_z\mathbf{k}) \cdot (V_x\mathbf{i} + V_y\mathbf{j} + V_z\mathbf{k}) \\
= {}& U_xV_x(\mathbf{i} \cdot \mathbf{i}) + U_xV_y(\mathbf{i} \cdot \mathbf{j}) + U_xV_z(\mathbf{i} \cdot \mathbf{k}) \\
& + U_yV_x(\mathbf{j} \cdot \mathbf{i}) + U_yV_y(\mathbf{j} \cdot \mathbf{j}) + U_yV_z(\mathbf{j} \cdot \mathbf{k}) \\
& + U_zV_x(\mathbf{k} \cdot \mathbf{i}) + U_zV_y(\mathbf{k} \cdot \mathbf{j}) + U_zV_z(\mathbf{k} \cdot \mathbf{k}).
\end{aligned}$$

In obtaining this result, we used Eqs. (2.20) and (2.21). Substituting Eqs. (2.22) into this expression, we obtain an equation for the dot product in terms of the scalar components of the two vectors:

$$\mathbf{U} \cdot \mathbf{V} = U_xV_x + U_yV_y + U_zV_z. \qquad (2.23)$$

To obtain an equation for the angle θ in terms of the components of the vectors, we equate the expression for the dot product given by Eq. (2.23) to the definition of the dot product, Eq. (2.18), and solve for $\cos \theta$:

$$\cos \theta = \frac{\mathbf{U} \cdot \mathbf{V}}{|\mathbf{U}||\mathbf{V}|} = \frac{U_x V_x + U_y V_y + U_z V_z}{|\mathbf{U}||\mathbf{V}|}. \tag{2.24}$$

Vector Components Parallel and Normal to a Line

In some engineering applications you must resolve a vector into components that are parallel and normal (perpendicular) to a given line. The component of a vector parallel to a line is called the *projection* of the vector onto the line. For example, when the vector represents a force, the projection of the force onto a line is the component of the force in the direction of the line.

We can determine the components of a vector parallel and normal to a line by using the dot product. Consider a vector $\mathbf{U}$ and a straight line L (Fig. 2.33a). We can resolve $\mathbf{U}$ into components $\mathbf{U}_p$ and $\mathbf{U}_n$ that are parallel and normal to L (Fig. 2.33b).

Figure 2.33
(a) A vector $\mathbf{U}$ and line L.
(b) Resolving $\mathbf{U}$ into components parallel and normal to L.

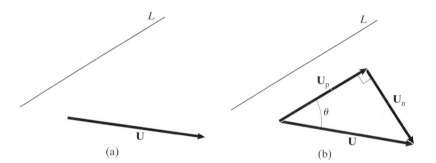

(a)   (b)

The Parallel Component   In terms of the angle θ between $\mathbf{U}$ and the component $\mathbf{U}_p$, the magnitude of $\mathbf{U}_p$ is

$$|\mathbf{U}_p| = |\mathbf{U}| \cos \theta. \tag{2.25}$$

Let $\mathbf{e}$ be a unit vector parallel to L (Fig. 2.34). The dot product of $\mathbf{e}$ and $\mathbf{U}$ is

$$\mathbf{e} \cdot \mathbf{U} = |\mathbf{e}||\mathbf{U}| \cos \theta = |\mathbf{U}| \cos \theta.$$

Comparing this result with Eq. (2.25), we see that the magnitude of $\mathbf{U}_p$ is

$$|\mathbf{U}_p| = \mathbf{e} \cdot \mathbf{U}.$$

Therefore the parallel component, or projection of $\mathbf{U}$ onto L, is

$$\mathbf{U}_p = (\mathbf{e} \cdot \mathbf{U})\mathbf{e}. \tag{2.26}$$

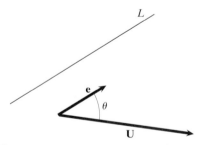

Figure 2.34
The unit vector $\mathbf{e}$ is parallel to L.

(This equation holds even if $\mathbf{e}$ doesn't point in the direction of $\mathbf{U}_p$. In that case, the angle $\theta > 90°$ and $\mathbf{e} \cdot \mathbf{U}$ is negative.) When the components of a vector and the components of a unit vector $\mathbf{e}$ parallel to a line L are known, we can use Eq. (2.26) to determine the component of the vector parallel to L.

The Normal Component Once the parallel component, has been determined, we can obtain the normal component from the relation $\mathbf{U} = \mathbf{U}_p + \mathbf{U}_n$:

$$\mathbf{U}_n = \mathbf{U} - \mathbf{U}_p. \tag{2.27}$$

Study Questions

1. What is the definition of the dot product?
2. The dot product is commutative. What does that mean?
3. If you know the components of two vectors $\mathbf{U}$ and $\mathbf{V}$, how can you determine their dot product?
4. How can you use the dot product to determine the components of a vector parallel and normal to a line?

Example 2.13

Calculating a Dot Product

The magnitude of the force $\mathbf{F}$ in Fig. 2.35 is 100 lb. The magnitude of the vector $\mathbf{r}$ from point O to point A is 8 ft.
(a) Use the definition of the dot product to determine $\mathbf{r} \cdot \mathbf{F}$.
(b) Use Eq. (2.23) to determine $\mathbf{r} \cdot \mathbf{F}$.

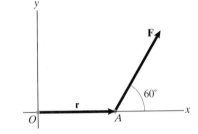

Figure 2.35

Strategy

(a) Since we know the magnitudes of $\mathbf{r}$ and $\mathbf{F}$ and the angle between them when they are placed tail to tail, we can determine $\mathbf{r} \cdot \mathbf{F}$ directly from the definition.
(b) We can determine the components of $\mathbf{r}$ and $\mathbf{F}$ and use Eq. (2.23) to determine their dot product.

Solution

(a) Using the definition of the dot product,

$$\mathbf{r} \cdot \mathbf{F} = |\mathbf{r}||\mathbf{F}| \cos\theta = (8)(100) \cos 60° = 400 \text{ ft-lb}.$$

(b) The vector $\mathbf{r} = 8\mathbf{i}$ (ft). The vector $\mathbf{F}$ in terms of scalar components is

$$\mathbf{F} = 100 \cos 60° \, \mathbf{i} + 100 \sin 60° \, \mathbf{j} \text{ (lb)}.$$

Therefore the dot product of $\mathbf{r}$ and $\mathbf{F}$ is

$$\mathbf{r} \cdot \mathbf{F} = r_x F_x + r_y F_y + r_z F_z$$
$$= (8)(100 \cos 60°) + (0)(100 \sin 60°) + (0)(0) = 400 \text{ ft-lb}.$$

Example 2.14

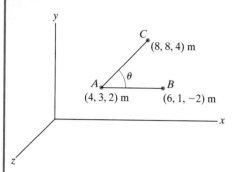

Figure 2.36

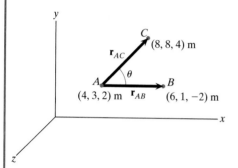

(a) The position vectors $\mathbf{r}_{AB}$ and $\mathbf{r}_{AC}$.

Using the Dot Product to Determine an Angle

What is the angle θ between the lines AB and AC in Fig. 2.36?

Strategy

We know the coordinates of the points A, B, and C, so we can determine the components of the vector $\mathbf{r}_{AB}$ from A to B and the vector $\mathbf{r}_{AC}$ from A to C (Fig. a). Then we can use Eq. (2.24) to determine θ.

Solution

The vectors $\mathbf{r}_{AB}$ and $\mathbf{r}_{AC}$ are

$$\mathbf{r}_{AB} = (6 - 4)\mathbf{i} + (1 - 3)\mathbf{j} + (-2 - 2)\mathbf{k} = 2\mathbf{i} - 2\mathbf{j} - 4\mathbf{k} \ (\mathrm{m}),$$

$$\mathbf{r}_{AC} = (8 - 4)\mathbf{i} + (8 - 3)\mathbf{j} + (4 - 2)\mathbf{k} = 4\mathbf{i} + 5\mathbf{j} + 2\mathbf{k} \ (\mathrm{m}).$$

Their magnitudes are

$$\left| \mathbf{r}_{AB} \right| = \sqrt{(2)^2 + (-2)^2 + (-4)^2} = 4.90 \ \mathrm{m},$$

$$\left| \mathbf{r}_{AC} \right| = \sqrt{(4)^2 + (5)^2 + (2)^2} = 6.71 \ \mathrm{m}.$$

The dot product of $\mathbf{r}_{AB}$ and $\mathbf{r}_{AC}$ is

$$\mathbf{r}_{AB} \cdot \mathbf{r}_{AC} = (2)(4) + (-2)(5) + (-4)(2) = -10 \ \mathrm{m}^2.$$

Therefore

$$\cos \theta = \frac{\mathbf{r}_{AB} \cdot \mathbf{r}_{AC}}{\left| \mathbf{r}_{AB} \right| \left| \mathbf{r}_{AC} \right|} = \frac{-10}{(4.90)(6.71)} = -0.304.$$

The angle $\theta = \arccos(-0.304) = 107.7°$.

Example 2.15

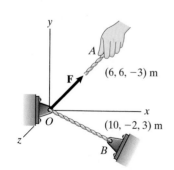

Figure 2.37

Components Parallel and Normal to a Line

Suppose that you pull on the cable OA in Fig. 2.37, exerting a 50-N force $\mathbf{F}$ at O. What are the components of $\mathbf{F}$ parallel and normal to the cable OB?

Strategy

Resolving $\mathbf{F}$ into components parallel and normal to OB (Fig. a), we can determine the components by using Eqs. (2.26) and (2.27). But to apply them, we must first express $\mathbf{F}$ in terms of scalar components and determine the components of a unit vector parallel to OB. We can obtain the components of

F by determining the components of the unit vector pointing from O toward A and multiplying them by $|\mathbf{F}|$.

Solution

The position vectors from O to A and from O to B are (Fig. b)

$$\mathbf{r}_{OA} = 6\mathbf{i} + 6\mathbf{j} - 3\mathbf{k} \text{ (m)},$$

$$\mathbf{r}_{OB} = 10\mathbf{i} - 2\mathbf{j} + 3\mathbf{k} \text{ (m)}.$$

Their magnitudes are $|\mathbf{r}_{OA}| = 9$ m and $|\mathbf{r}_{OB}| = 10.6$ m. Dividing these vectors by their magnitudes, we obtain unit vectors that point from the origin toward A and B (Fig. c):

$$\mathbf{e}_{OA} = \frac{\mathbf{r}_{OA}}{|\mathbf{r}_{OA}|} = \frac{6\mathbf{i} + 6\mathbf{j} - 3\mathbf{k}}{9} = 0.667\mathbf{i} + 0.667\mathbf{j} - 0.333\mathbf{k},$$

$$\mathbf{e}_{OB} = \frac{\mathbf{r}_{OB}}{|\mathbf{r}_{OB}|} = \frac{10\mathbf{i} - 2\mathbf{j} + 3\mathbf{k}}{10.6} = 0.941\mathbf{i} + 0.188\mathbf{j} - 0.282\mathbf{k}.$$

The force **F** in terms of scalar components is

$$\mathbf{F} = |\mathbf{F}|\mathbf{e}_{OA} = (50)(0.667\mathbf{i} + 0.667\mathbf{j} - 0.333\mathbf{k})$$

$$= 33.3\mathbf{i} + 33.3\mathbf{j} - 16.7\mathbf{k} \text{ (N)}.$$

Taking the dot product of $\mathbf{e}_{OB}$ and **F**, we obtain

$$\mathbf{e}_{OB} \cdot \mathbf{F} = (0.941)(33.3) + (-0.188)(33.3) + (0.282)(-16.7)$$

$$= 20.4 \text{ N}.$$

The parallel component of **F** is

$$\mathbf{F}_{p} = (\mathbf{e}_{OB} \cdot \mathbf{F})\mathbf{e}_{OB} = (20.4)(0.941\mathbf{i} - 0.188\mathbf{j} + 0.282\mathbf{k})$$

$$= 19.2\mathbf{i} - 3.8\mathbf{j} + 5.8\mathbf{k} \text{ (N)}.$$

and the normal component is

$$\mathbf{F}_{n} = \mathbf{F} - \mathbf{F}_{p} = 14.2\mathbf{i} + 37.2\mathbf{j} - 22.4\mathbf{k} \text{ (N)}.$$

Discussion

You can confirm that two vectors are perpendicular by making sure their dot product is zero. In this example,

$$\mathbf{F}_{p} \cdot \mathbf{F}_{n} = (19.2)(14.2) + (-3.8)(37.2) + (5.8)(-22.4) = 0.$$

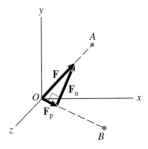

(a) The components of **F** parallel and normal to OB.

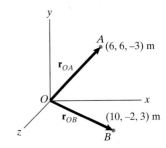

(b) The position vectors $\mathbf{r}_{OA}$ and $\mathbf{r}_{OB}$.

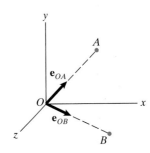

(c) The unit vectors $\mathbf{e}_{OA}$ and $\mathbf{e}_{OB}$.

Problems

2.103 Determine the dot product of the vectors $\mathbf{U} = 8\mathbf{i} - 6\mathbf{j} + 4\mathbf{k}$ and $\mathbf{V} = 3\mathbf{i} + 7\mathbf{j} + 9\mathbf{k}$.

   Strategy: Since the vectors are expressed in terms of their components, you can use Eq. (2.23) to determine their dot product.

2.104 Determine the dot product $\mathbf{U} \cdot \mathbf{V}$ of the vectors $\mathbf{U} = 40\mathbf{i} + 20\mathbf{j} + 60\mathbf{k}$ and $\mathbf{V} = -30\mathbf{i} + 15\mathbf{k}$.

2.105 What is the dot product of the position vector $\mathbf{r} = -10\mathbf{i} + 25\mathbf{j}$ (m) and the force $\mathbf{F} = 300\mathbf{i} + 250\mathbf{j} + 300\mathbf{k}$ (N)?

2.106 What is the dot product of the position vector $\mathbf{r} = 4\mathbf{i} - 12\mathbf{j} - 3\mathbf{k}$ (ft) and the force $\mathbf{F} = 20\mathbf{i} + 30\mathbf{j} - 10\mathbf{k}$ (lb)?

2.107 Two *perpendicular* vectors are given in terms of their components by $\mathbf{U} = U_x\mathbf{i} - 4\mathbf{j} + 6\mathbf{k}$ and $\mathbf{V} = 3\mathbf{i} + 2\mathbf{j} - 3\mathbf{k}$. Use the dot product to determine the component U_x.

2.108 The three vectors

$$\mathbf{U} = U_x\mathbf{i} + 3\mathbf{j} + 2\mathbf{k},$$
$$\mathbf{V} = -3\mathbf{i} + V_y\mathbf{j} + 3\mathbf{k},$$
$$\mathbf{W} = -2\mathbf{i} + 4\mathbf{j} + W_z\mathbf{k}$$

are mutually perpendicular. Use the dot product to determine the components U_x, V_y, and W_z.

2.109 The magnitudes $|\mathbf{U}| = 10$ and $|\mathbf{V}| = 20$.
(a) Use the definition of the dot product to determine $\mathbf{U} \cdot \mathbf{V}$.
(b) Use Eq. (2.23) to determine $\mathbf{U} \cdot \mathbf{V}$.

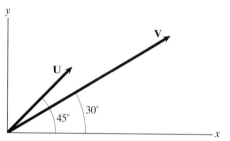

P2.109

2.110 By evaluating the dot product $\mathbf{U} \cdot \mathbf{V}$, prove the identity $\cos(\theta_1 - \theta_2) = \cos\theta_1 \cos\theta_2 + \sin\theta_1 \sin\theta_2$.
Strategy: Evaluate the dot product both by using the definition and by using Eq. (2.23).

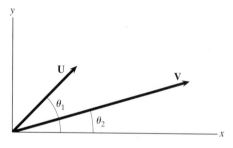

P2.110

2.111 Use the dot product to determine the angle between the forestay (cable AB) and the backstay (cable BC) of the sailboat in Problem 2.41.

2.112 What is the angle θ between the straight lines AB and AC?

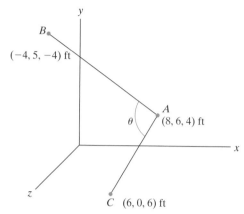

P2.112

2.113 The ship O measures the positions of the ship A and the airplane B and obtains the coordinates shown. What is the angle θ between the lines of sight OA and OB?

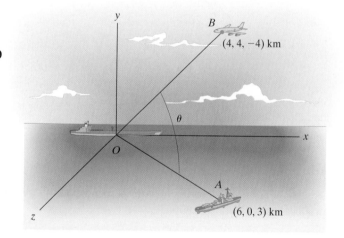

P2.113

2.114 Astronauts on the space shuttle use radar to determine the magnitudes and direction cosines of the position vectors of two satellites A and B. The vector $\mathbf{r}_A$ from the shuttle to satellite A has magnitude 2 km and direction cosines $\cos\theta_x = 0.768$, $\cos\theta_y = 0.384$, $\cos\theta_z = 0.512$. The vector $\mathbf{r}_B$ from the shuttle to satellite B has magnitude 4 km and direction cosines $\cos\theta_x = 0.743$, $\cos\theta_y = 0.557$, $\cos\theta_z = -0.371$. What is the angle θ between the vectors $\mathbf{r}_A$ and $\mathbf{r}_B$?

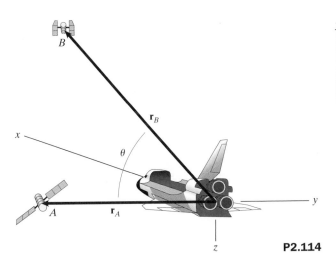

P2.114

2.115 The cable BC exerts an 800-N force $\mathbf{F}$ on the bar AB at B. Use Eq. (2.26) to determine the vector component of $\mathbf{F}$ parallel to the bar.

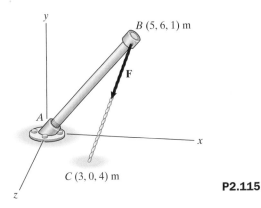

P2.115

2.116 The force $\mathbf{F} = 21\mathbf{i} + 14\mathbf{j}$ (kN). Resolve it into vector components parallel and normal to the line OA.

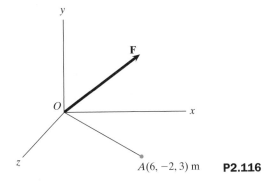

P2.116

2.117 At the instant shown, the Harrier's thrust vector is $\mathbf{T} = 3800\mathbf{i} + 15,300\mathbf{j} - 1800\mathbf{k}$ (lb), and its velocity vector is $\mathbf{v} = 24\mathbf{i} + 6\mathbf{j} - 2\mathbf{k}$ (ft/s). Resolve $\mathbf{T}$ into vector components parallel and normal to $\mathbf{v}$. (These are the components of the airplane's thrust parallel and normal to the direction of its motion.)

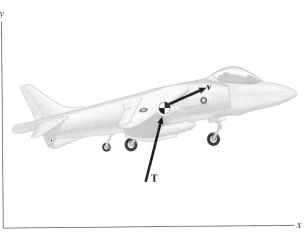

P2.117

2.118 Cables extend from A to B and from A to C. The cable AC exerts a 1000-lb force $\mathbf{F}$ at A.
(a) What is the angle between the cables AB and AC?
(b) Determine the vector component of $\mathbf{F}$ parallel to the cable AB.

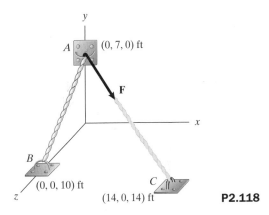

P2.118

2.119 Consider the cables AB and AC shown in Problem 2.118. Let $\mathbf{r}_{AB}$ be the position vector from point A to point B. Determine the vector component of $\mathbf{r}_{AB}$ parallel to the cable AC.

2.120 The force $\mathbf{F} = 10\mathbf{i} + 12\mathbf{j} - 6\mathbf{k}$ (N). Determine the vector components of $\mathbf{F}$ parallel and normal to the line OA.

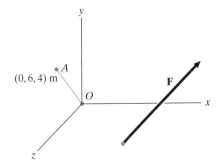

P2.120

2.121 The rope AB exerts a 50-N force $\mathbf{T}$ on collar A. Determine the vector component of $\mathbf{T}$ parallel to the bar CD.

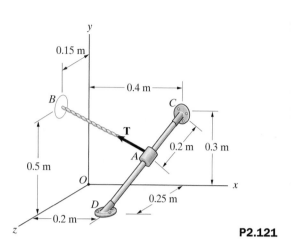

P2.121

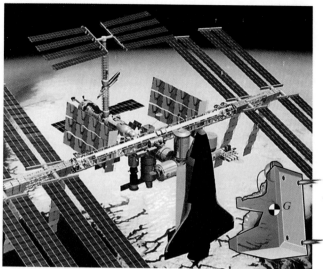

P2.125

2.122 In Problem 2.121, determine the vector component of $\mathbf{T}$ normal to the bar CD.

2.123 The disk A is at the midpoint of the sloped surface. The string from A to B exerts a 0.2-lb force $\mathbf{F}$ on the disk. If you resolve $\mathbf{F}$ into vector components parallel and normal to the sloped surface, what is the component normal to the surface?

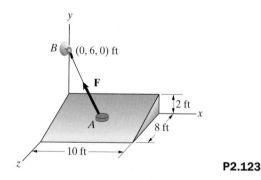

P2.123

2.124 In Problem 2.123, what is the vector component of $\mathbf{F}$ parallel to the surface?

2.125 An astronaut in a maneuvering unit approaches a space station. At the present instant, the station informs him that his position relative to the origin of the station's coordinate system is $\mathbf{r}_G = 50\mathbf{i} + 80\mathbf{j} + 180\mathbf{k}$ (m) and his velocity is $\mathbf{v} = -2.2\mathbf{j} - 3.6\mathbf{k}$ (m/s). The position of an airlock is $\mathbf{r}_A = -12\mathbf{i} + 20\mathbf{k}$ (m). Determine the angle between his velocity vector and the line from his position to the airlock's position.

2.126 In Problem 2.125, determine the vector component of the astronaut's velocity parallel to the line from his position to the airlock's position.

2.127 Point P is at longitude 30°W and latitude 45°N on the Atlantic Ocean between Nova Scotia and France. (See Problem 2.90.) Point Q is at longitude 60°E and latitude 20°N in the Arabian Sea. Use the dot product to determine the shortest distance along the surface of the earth from P to Q in terms of the radius of the earth R_E.

Strategy: Use the dot product to determine the angle between the lines OP and OQ; then use the definition of an angle in radians to determine the distance along the surface of the earth from P to Q.

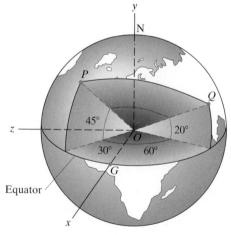

P2.127

2.6 Cross Products

Like the dot product, the cross product of two vectors has many applications, including determining the rate of rotation of a fluid particle and calculating the force exerted on a charged particle by a magnetic field. Because of its usefulness for determining moments of forces, the cross product is an indispensable tool in mechanics. In this section we show you how to evaluate cross products and give examples of simple applications.

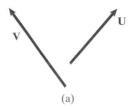

(a)

Definition

Consider two vectors **U** and **V** (Fig. 2.38a). The *cross product* of **U** and **V**, denoted **U** × **V**, is defined by

$$\mathbf{U} \times \mathbf{V} = |\mathbf{U}||\mathbf{V}| \sin\theta \, \mathbf{e}. \qquad (2.28)$$

The angle θ is the angle between **U** and **V** when they are placed tail to tail (Fig. 2.38b). The vector **e** is a unit vector defined to be perpendicular to both **U** and **V**. Since this leaves two possibilities for the direction of **e**, the vectors **U**, **V**, and **e** are defined to be a right-handed system. The *right-hand rule* for determining the direction of **e** is shown in Fig. 2.38c. When you point the four fingers of your right hand in the direction of the vector **U** (the first vector in the cross product) and close your fingers toward the vector **V** (the second vector in the cross product), your thumb points in the direction of **e**.

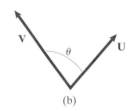

(b)

Because the result of the cross product is a vector, it is sometimes called the vector product. The units of the cross product are the product of the units of the two vectors. Notice that the cross product of two nonzero vectors is equal to zero if and only if the two vectors are parallel.

An interesting property of the cross product is that it is *not* commutative. Eq. (2.28) implies that the magnitude of the vector **U** × **V** is equal to the magnitude of the vector **V** × **U**, but the right-hand rule indicates that they are opposite in direction (Fig. 2.39). That is,

$$\mathbf{U} \times \mathbf{V} = -\mathbf{V} \times \mathbf{U}. \quad \text{The cross product is } \textit{not} \text{ commutative.} \qquad (2.29)$$

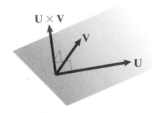

(c)

Figure 2.38
(a) The vectors **U** and **V**.
(b) The angle θ between the vectors when they are placed tail to tail.
(c) Determining the direction of **e** by the right-hand rule.

The cross product also satisfies the relations

$$a(\mathbf{U} \times \mathbf{V}) = (a\mathbf{U}) \times \mathbf{V} = \mathbf{U} \times (a\mathbf{V}) \quad \begin{array}{l}\text{The cross product is} \\ \text{associative with} \\ \text{respect to scalar} \\ \text{multiplication.}\end{array} \qquad (2.30)$$

and

$$\mathbf{U} \times (\mathbf{V} + \mathbf{W}) = (\mathbf{U} \times \mathbf{V}) + (\mathbf{U} \times \mathbf{W}) \quad \begin{array}{l}\text{The cross product} \\ \text{is distributive with} \\ \text{respect to vector} \\ \text{addition.}\end{array} \qquad (2.31)$$

for any scalar a and vectors **U**, **V**, and **W**.

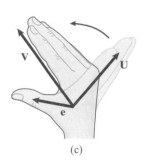

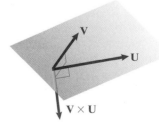

Figure 2.39
Directions of **U** × **V** and **V** × **U**.

Cross Products in Terms of Components

To obtain an equation for the cross product of two vectors in terms of their components, we must determine the cross products formed from the unit vectors **i**, **j**, and **k**. Since the angle between two identical vectors placed tail to tail is zero,

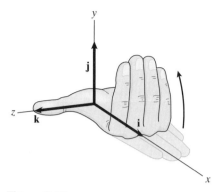

Figure 2.40
The right-hand rule indicates that
$\mathbf{i} \times \mathbf{j} = \mathbf{k}.$

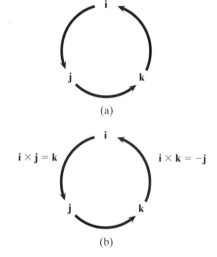

(a)

(b)

Figure 2.41
(a) Arrange the unit vectors in a circle with arrows to indicate their order.
(b) You can use the circle to determine their cross products.

$$\mathbf{i} \times \mathbf{i} = |\mathbf{i}||\mathbf{i}|\sin(0)\mathbf{e} = \mathbf{0}.$$

The cross product $\mathbf{i} \times \mathbf{j}$ is

$$\mathbf{i} \times \mathbf{j} = |\mathbf{i}||\mathbf{j}|\sin(90)°\mathbf{e} = \mathbf{e},$$

where $\mathbf{e}$ is a unit vector perpendicular to $\mathbf{i}$ and $\mathbf{j}$. Either $\mathbf{e} = \mathbf{k}$ or $\mathbf{e} = -\mathbf{k}$. Applying the right-hand rule, we find that $\mathbf{e} = \mathbf{k}$ (Fig. 2.40). Therefore

$$\mathbf{i} \times \mathbf{j} = \mathbf{k}.$$

Continuing in this way, we obtain

$$
\begin{array}{lll}
\mathbf{i} \times \mathbf{i} = \mathbf{0}, & \mathbf{i} \times \mathbf{j} = \mathbf{k}. & \mathbf{i} \times \mathbf{k} = -\mathbf{j}, \\
\mathbf{j} \times \mathbf{i} = -\mathbf{k}, & \mathbf{j} \times \mathbf{j} = \mathbf{0}, & \mathbf{j} \times \mathbf{k} = \mathbf{i}, \\
\mathbf{k} \times \mathbf{i} = \mathbf{j}, & \mathbf{k} \times \mathbf{j} = -\mathbf{i}, & \mathbf{k} \times \mathbf{k} = \mathbf{0}.
\end{array}
\tag{2.32}
$$

These results can be remembered easily by arranging the unit vectors in a circle, as shown in Fig. 2.41a. The cross product of adjacent vectors is equal to the third vector with a positive sign if the order of the vectors in the cross product is the order indicated by the arrows and a negative sign otherwise. For example, in Fig. 2.41b we see that $\mathbf{i} \times \mathbf{j} = \mathbf{k}$, but $\mathbf{i} \times \mathbf{k} = -\mathbf{j}$.

The cross product of two vectors $\mathbf{U}$ and $\mathbf{V}$ expressed in terms of their components is

$$
\begin{aligned}
\mathbf{U} \times \mathbf{V} &= \left(U_x\mathbf{i} + U_y\mathbf{j} + U_z\mathbf{k}\right) \times \left(V_x\mathbf{i} + V_y\mathbf{j} + V_z\mathbf{k}\right) \\
&= U_xV_x(\mathbf{i} \times \mathbf{i}) + U_xV_y(\mathbf{i} \times \mathbf{j}) + U_xV_z(\mathbf{i} \times \mathbf{k}) \\
&\quad + U_yV_x(\mathbf{j} \times \mathbf{i}) + U_yV_y(\mathbf{j} \times \mathbf{j}) + U_yV_z(\mathbf{j} \times \mathbf{k}) \\
&\quad + U_zV_x(\mathbf{k} \times \mathbf{i}) + U_zV_y(\mathbf{k} \times \mathbf{j}) + U_zV_z(\mathbf{k} \times \mathbf{k}).
\end{aligned}
$$

By substituting Eqs. (2.32) into this expression, we obtain the equation

$$
\begin{aligned}
\mathbf{U} \times \mathbf{V} &= \left(U_yV_z - U_zV_y\right)\mathbf{i} - \left(U_xV_z - U_zV_x\right)\mathbf{j} \\
&\quad + \left(U_xV_y - U_yV_x\right)\mathbf{k}.
\end{aligned}
\tag{2.33}
$$

This result can be compactly written as the determinant

$$
\mathbf{U} \times \mathbf{V} = \begin{vmatrix} \mathbf{i} & \mathbf{j} & \mathbf{k} \\ U_x & U_y & U_z \\ V_x & V_y & V_z \end{vmatrix}
\tag{2.34}
$$

This equation is based on Eqs. (2.32), which we obtained using a right-handed coordinate system. It gives the correct result for the cross product only if a right-handed coordinate system is used to determine the components of $\mathbf{U}$ and $\mathbf{V}$.

Evaluating a 3 × 3 Determinant

A 3 × 3 determinant can be evaluated by repeating its first two columns as shown and evaluating the products of the terms along the six diagonal lines.

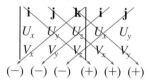

Adding the terms obtained from the diagonals that run downward to the right and subtracting the terms obtained from the diagonals that run downward to the left gives the value of the determinant:

$$\begin{vmatrix} \mathbf{i} & \mathbf{j} & \mathbf{k} \\ U_x & U_y & U_z \\ V_x & V_y & V_z \end{vmatrix} = U_y V_z \mathbf{i} + U_z V_x \mathbf{j} + U_x V_y \mathbf{k} \\ - U_y V_x \mathbf{k} - U_z V_y \mathbf{i} - U_x V_z \mathbf{j}.$$

A 3 × 3 determinant can also be evaluated by expressing it as

$$\begin{vmatrix} \mathbf{i} & \mathbf{j} & \mathbf{k} \\ U_x & U_y & U_z \\ V_x & V_y & V_z \end{vmatrix} = \mathbf{i} \begin{vmatrix} U_y & U_z \\ V_y & V_z \end{vmatrix} - \mathbf{j} \begin{vmatrix} U_x & U_z \\ V_x & V_z \end{vmatrix} + \mathbf{k} \begin{vmatrix} U_x & U_y \\ V_x & V_y \end{vmatrix}.$$

The terms on the right are obtained by multiplying each element of the first row of the 3 × 3 determinant by the 2 × 2 determinant obtained by crossing out that element's row and column. For example, the first element of the first row, **i**, is multiplied by the 2 × 2 determinant

$$\begin{vmatrix} \cancel{\mathbf{i}} & \cancel{\mathbf{j}} & \cancel{\mathbf{k}} \\ \cancel{U_x} & U_y & U_z \\ \cancel{V_x} & V_y & V_z \end{vmatrix}$$

Be sure to remember that the second term is subtracted. Expanding the 2 × 2 determinants, we obtain the value of the determinant:

$$\begin{vmatrix} \mathbf{i} & \mathbf{j} & \mathbf{k} \\ U_x & U_y & U_z \\ V_x & V_y & V_z \end{vmatrix} = (U_y V_z - U_z V_y)\mathbf{i} - (U_x V_z - U_z V_x)\mathbf{j} \\ + (U_x V_y - U_y V_x)\mathbf{k}.$$

2.7 Mixed Triple Products

In Chapter 4, when we discuss the moment of a force about a line, we will use an operation called the *mixed triple product*, defined by

$$\mathbf{U} \cdot (\mathbf{V} \times \mathbf{W}). \tag{2.35}$$

In terms of the scalar components of the vectors,

$$\mathbf{U} \cdot (\mathbf{V} \times \mathbf{W}) = (U_x \mathbf{i} + U_y \mathbf{j} + U_z \mathbf{k}) \cdot \begin{vmatrix} \mathbf{i} & \mathbf{j} & \mathbf{k} \\ V_x & V_y & V_z \\ W_x & W_y & W_z \end{vmatrix}$$

$$= (U_x \mathbf{i} + U_y \mathbf{j} + U_z \mathbf{k}) \cdot [(V_y W_z - V_z W_y)\mathbf{i} \\ - (V_x W_z - V_z W_x)\mathbf{j} + (V_x W_y - V_y W_x)\mathbf{k}]$$

$$= U_x(V_y W_z - V_z W_y) - U_y(V_x W_z - V_z W_x) \\ + U_z(V_x W_y - V_y W_x).$$

This result can be expressed as the determinant

$$\mathbf{U} \cdot (\mathbf{V} \times \mathbf{W}) = \begin{vmatrix} U_x & U_y & U_z \\ V_x & V_y & V_z \\ W_x & W_y & W_z \end{vmatrix}. \tag{2.36}$$

Interchanging any two of the vectors in the mixed triple product changes the sign but not the absolute value of the result. For example,

$$\mathbf{U} \cdot (\mathbf{V} \times \mathbf{W}) = -\mathbf{W} \cdot (\mathbf{V} \times \mathbf{U}).$$

If the vectors **U**, **V**, and **W** in Fig. 2.42 form a right-handed system, it can be shown that the volume of the parallelepiped equals $\mathbf{U} \cdot (\mathbf{V} \times \mathbf{W})$.

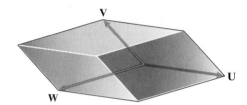

Figure 2.42
Parallelepiped defined by the vectors **U**, **V**, and **W**.

Study Questions

1. What is the definition of the cross product?
2. If you know the components of two vectors **U** and **V**, how can you determine their cross product?
3. If the cross product of two vectors is zero, what does that mean?

Example 2.16

Cross Product in Terms of Components

Determine the cross product $\mathbf{U} \times \mathbf{V}$ of the vectors $\mathbf{U} = -2\mathbf{i} + \mathbf{j}$ and $\mathbf{V} = 3\mathbf{i} - 4\mathbf{k}$.

Strategy

We can evaluate the cross product of the vectors in two ways: by evaluating the cross products of their components term by term and by using Eq. (2.34).

Solution

$$\begin{aligned} \mathbf{U} \times \mathbf{V} &= (-2\mathbf{i} + \mathbf{j}) \times (3\mathbf{i} - 4\mathbf{k}) \\ &= (-2)(3)(\mathbf{i} \times \mathbf{i}) + (-2)(-4)(\mathbf{i} \times \mathbf{k}) + (1)(3)(\mathbf{j} \times \mathbf{i}) \\ &\quad + (1)(-4)(\mathbf{j} \times \mathbf{k}) \\ &= (-6)(0) + (8)(-\mathbf{j}) + (3)(-\mathbf{k}) + (-4)(\mathbf{i}) \\ &= -4\mathbf{i} - 8\mathbf{j} - 3\mathbf{k}. \end{aligned}$$

Using Eq. (2.34), we obtain

$$\mathbf{U} \times \mathbf{V} = \begin{vmatrix} \mathbf{i} & \mathbf{j} & \mathbf{k} \\ U_x & U_y & U_z \\ V_x & V_y & V_z \end{vmatrix} = \begin{vmatrix} \mathbf{i} & \mathbf{j} & \mathbf{k} \\ -2 & 1 & 0 \\ 3 & 0 & -4 \end{vmatrix} = -4\mathbf{i} - 8\mathbf{j} - 3\mathbf{k}.$$

Example 2.17

Calculating the Cross Product

The magnitude of the force $\mathbf{F}$ in Fig. 2.43 is 100 lb. The magnitude of the vector $\mathbf{r}$ from point O to point A is 8 ft.
(a) Use the definition of the cross product to determine $\mathbf{r} \times \mathbf{F}$.
(b) Use Eq. (2.34) to determine $\mathbf{r} \times \mathbf{F}$.

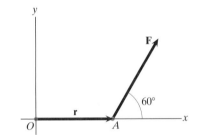

Figure 2.43

Strategy

(a) We know the magnitudes of $\mathbf{r}$ and $\mathbf{F}$ and the angle between them when they are placed tail to tail. Since both vectors lie in the x-y plane, the unit vector $\mathbf{k}$ is perpendicular to both $\mathbf{r}$ and $\mathbf{F}$. We therefore have all the information we need to determine $\mathbf{r} \times \mathbf{F}$ directly from the definition.
(b) We can determine the components of $\mathbf{r}$ and $\mathbf{F}$ and use Eq. (2.34) to determine $\mathbf{r} \times \mathbf{F}$.

Solution

(a) Using the definition of the cross product,

$$\mathbf{r} \times \mathbf{F} = |\mathbf{r}||\mathbf{F}| \sin\theta \, \mathbf{e} = (8)(100) \sin 60° \, \mathbf{e} = 693 \, \mathbf{e} \, (\text{ft-lb}).$$

Since $\mathbf{e}$ is defined to be perpendicular to $\mathbf{r}$ and $\mathbf{F}$, either $\mathbf{e} = \mathbf{k}$ or $\mathbf{e} = -\mathbf{k}$. Pointing the fingers of the right hand in the direction of $\mathbf{r}$ and closing them toward $\mathbf{F}$, the right-hand rule indicates that $\mathbf{e} = \mathbf{k}$. Therefore

$$\mathbf{r} \times \mathbf{F} = 693\mathbf{k} \, (\text{ft-lb}).$$

(b) The vector $\mathbf{r} = 8\mathbf{i}$ (ft). The vector $\mathbf{F}$ in terms of scalar components is

$$\mathbf{F} = 100 \cos 60° \, \mathbf{i} + 100 \sin 60° \, \mathbf{j} \, (\text{lb}).$$

From Eq. (2.34),

$$\mathbf{r} \times \mathbf{F} = \begin{vmatrix} \mathbf{i} & \mathbf{j} & \mathbf{k} \\ r_x & r_y & r_z \\ F_x & F_y & F_z \end{vmatrix} = \begin{vmatrix} \mathbf{i} & \mathbf{j} & \mathbf{k} \\ 8 & 0 & 0 \\ 100 \cos 60° & 100 \sin 60° & 0 \end{vmatrix}$$

$$= (8)(100 \cos 60°)\mathbf{k} = 693\mathbf{k} \, (\text{ft-lb}).$$

Example 2.18

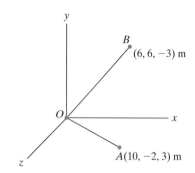

Figure 2.44

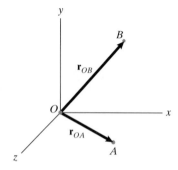

(a) The vectors $\mathbf{r}_{OA}$ and $\mathbf{r}_{OB}$.

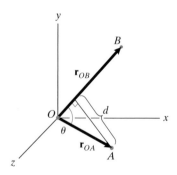

(b) The minimum distance d from A to the line OB.

Minimum Distance from a Point to a Line

Consider the straight lines OA and OB in Fig. 2.44.
(a) Determine the components of a unit vector that is perpendicular to both OA and OB.
(b) What is the minimum distance from point A to the line OB?

Strategy

(a) Let $\mathbf{r}_{OA}$ and $\mathbf{r}_{OB}$ be the position vectors from O to A and from O to B (Fig. a). Since the cross product $\mathbf{r}_{OA} \times \mathbf{r}_{OB}$ is perpendicular to $\mathbf{r}_{OA}$ and $\mathbf{r}_{OB}$ we will determine it and divide it by its magnitude to obtain a unit vector perpendicular to the lines OA and OB.
(b) The minimum distance from A to the line OB is the length d of the straight line from A to OB that is perpendicular to OB (Fig. b). We can see that $d = |\mathbf{r}_{OA}| \sin\theta$, where θ is the angle between $\mathbf{r}_{OA}$ and $\mathbf{r}_{OB}$. From the definition of the cross product, the magnitude of $\mathbf{r}_{OA} \times \mathbf{r}_{OB}$ is $|\mathbf{r}_{OA}||\mathbf{r}_{OB}| \sin\theta$, so we can determine d by dividing the magnitude of $\mathbf{r}_{OA} \times \mathbf{r}_{OB}$ by the magnitude of $\mathbf{r}_{OB}$.

Solution

(a) The components of $\mathbf{r}_{OA}$ and $\mathbf{r}_{OB}$ are

$$\mathbf{r}_{OA} = 10\mathbf{i} - 2\mathbf{j} + 3\mathbf{k} \text{ (m)},$$
$$\mathbf{r}_{OB} = 6\mathbf{i} + 6\mathbf{j} - 3\mathbf{k} \text{ (m)}.$$

By using Eq. (2.34), we obtain $\mathbf{r}_{OA} \times \mathbf{r}_{OB}$:

$$\mathbf{r}_{OA} \times \mathbf{r}_{OB} = \begin{vmatrix} \mathbf{i} & \mathbf{j} & \mathbf{k} \\ 10 & -2 & 3 \\ 6 & 6 & -3 \end{vmatrix} = -12\mathbf{i} + 48\mathbf{j} + 72\mathbf{k} \text{ (m}^2\text{)}.$$

This vector is perpendicular to $\mathbf{r}_{OA}$ and $\mathbf{r}_{OB}$. Dividing it by its magnitude, we obtain a unit vector $\mathbf{e}$ that is perpendicular to the lines OA and OB:

$$\mathbf{e} = \frac{\mathbf{r}_{OA} \times \mathbf{r}_{OB}}{|\mathbf{r}_{OA} \times \mathbf{r}_{OB}|} = \frac{-12\mathbf{i} + 48\mathbf{j} + 72\mathbf{k}}{\sqrt{(-12)^2 + (48)^2 + (72)^2}}$$
$$= -0.137\mathbf{i} + 0.549\mathbf{j} + 0.824\mathbf{k}.$$

(b) From Fig. b, the minimum distance d is

$$d = |\mathbf{r}_{OA}| \sin\theta.$$

The magnitude of $\mathbf{r}_{OA} \times \mathbf{r}_{OB}$ is

$$|\mathbf{r}_{OA} \times \mathbf{r}_{OB}| = |\mathbf{r}_{OA}||\mathbf{r}_{OB}| \sin\theta.$$

Solving this equation for $\sin\theta$, the distance d is

$$d = |\mathbf{r}_{OA}| \left(\frac{|\mathbf{r}_{OA} \times \mathbf{r}_{OB}|}{|\mathbf{r}_{OA}||\mathbf{r}_{OB}|} \right) = \frac{|\mathbf{r}_{OA} \times \mathbf{r}_{OB}|}{|\mathbf{r}_{OB}|}$$
$$= \frac{\sqrt{(-12)^2 + (48)^2 + (72)^2}}{\sqrt{(6)^2 + (6)^2 + (-3)^2}} = 9.71 \text{ m}.$$

Example 2.19°

Component of a Vector Perpendicular to a Plane

The rope CE in Fig. 2.45 exerts a 500-N force $\mathbf{T}$ on the door $ABCD$. What is the magnitude of the component of $\mathbf{T}$ perpendicular to the door?

Strategy

We are given the coordinates of the corners A, B, and C of the door. By taking the cross product of the position vector $\mathbf{r}_{CB}$ from C to B and the position vector $\mathbf{r}_{CA}$ from C to A, we will obtain a vector that is perpendicular to the door. We can divide the resulting vector by its magnitude to obtain a unit vector perpendicular to the door and then apply Eq. (2.26) to determine the component of $\mathbf{T}$ perpendicular to the door.

Solution

The components of $\mathbf{r}_{CB}$ and $\mathbf{r}_{CA}$ are

$$\mathbf{r}_{CB} = 0.35\mathbf{i} - 0.2\mathbf{j} + 0.2\mathbf{k} \ (\text{m}),$$

$$\mathbf{r}_{CA} = 0.5\mathbf{i} - 0.2\mathbf{j} \ (\text{m}).$$

Their cross product is

$$\mathbf{r}_{CB} \times \mathbf{r}_{CA} = \begin{vmatrix} \mathbf{i} & \mathbf{j} & \mathbf{k} \\ 0.35 & -0.2 & 0.2 \\ 0.5 & -0.2 & 0 \end{vmatrix} = 0.04\mathbf{i} + 0.1\mathbf{j} + 0.03\mathbf{k} \ (\text{m}^2).$$

Dividing this vector by its magnitude, we obtain a unit vector $\mathbf{e}$ that is perpendicular to the door (Fig. a):

$$\mathbf{e} = \frac{\mathbf{r}_{CB} \times \mathbf{r}_{CA}}{|\mathbf{r}_{CB} \times \mathbf{r}_{CA}|} = \frac{0.04\mathbf{i} + 0.1\mathbf{j} + 0.03\mathbf{k}}{\sqrt{(0.04)^2 + (0.1)^2 + (0.03)^2}}$$

$$= 0.358\mathbf{i} + 0.894\mathbf{j} + 0.268\mathbf{k}.$$

To use Eq. (2.26), we must express $\mathbf{T}$ in terms of its scalar components. The position vector from C to E is

$$\mathbf{r}_{CE} = 0.4\mathbf{i} + 0.05\mathbf{j} - 0.1\mathbf{k} \ (\text{m}),$$

so we can express the force $\mathbf{T}$ as

$$\mathbf{T} = |\mathbf{T}| \frac{\mathbf{r}_{CE}}{|\mathbf{r}_{CE}|} = (500) \frac{0.4\mathbf{i} + 0.05\mathbf{j} - 0.1\mathbf{k}}{\sqrt{(0.4)^2 + (0.05)^2 + (-0.1)^2}}$$

$$= 481.5\mathbf{i} + 60.2\mathbf{j} - 120.4\mathbf{k} \ (\text{N}).$$

The component of $\mathbf{T}$ parallel to the unit vector $\mathbf{e}$, which is the component perpendicular to the door, is

$$\mathbf{T}_p = (\mathbf{e} \cdot \mathbf{T})\mathbf{e} = \big[(0.358)(481.5) + (0.894)(60.2) + (0.268)(-120.4)\big]\mathbf{e}$$

$$= 194\mathbf{e} \ (\text{N}).$$

The magnitude of $\mathbf{T}_p$ is 194 N.

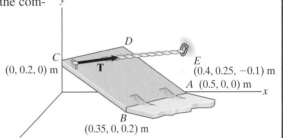

C $(0, 0.2, 0)$ m

D

E $(0.4, 0.25, -0.1)$ m

A $(0.5, 0, 0)$ m

B $(0.35, 0, 0.2)$ m

Figure 2.45

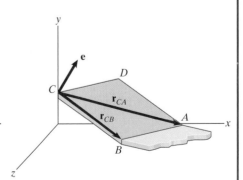

(a) Determining a unit vector perpendicular to the door.

Problems

2.128 Determine the cross product $\mathbf{U} \times \mathbf{V}$ of the vectors $\mathbf{U} = 8\mathbf{i} - 6\mathbf{j} + 4\mathbf{k}$ and $\mathbf{V} = 3\mathbf{i} + 7\mathbf{j} + 9\mathbf{k}$.

Strategy: Since the vectors are expressed in terms of their components, you can use Eq. (2.34) to determine their cross product.

2.129 Two vectors $\mathbf{U} = 3\mathbf{i} + 2\mathbf{j}$ and $\mathbf{V} = 2\mathbf{i} + 4\mathbf{j}$.
(a) What is the cross product $\mathbf{U} \times \mathbf{V}$?
(b) What is the cross product $\mathbf{V} \times \mathbf{U}$?

2.130 What is the cross product $\mathbf{r} \times \mathbf{F}$ of the position vector $\mathbf{r} = 2\mathbf{i} + 2\mathbf{j} + 2\mathbf{k}$ (m) and the force $\mathbf{F} = 20\mathbf{i} - 40\mathbf{k}$ (N)?

2.131 Determine the cross product $\mathbf{r} \times \mathbf{F}$ of the position vector $\mathbf{r} = 4\mathbf{i} - 12\mathbf{j} + 3\mathbf{k}$ (m) and the force $\mathbf{F} = 16\mathbf{i} - 22\mathbf{j} - 10\mathbf{k}$ (kN).

2.132 Consider the vectors $\mathbf{U} = 6\mathbf{i} - 2\mathbf{j} - 3\mathbf{k}$ and $\mathbf{V} = -12\mathbf{i} + 4\mathbf{j} + 6\mathbf{k}$.
(a) Determine the cross product $\mathbf{U} \times \mathbf{V}$.
(b) What can you conclude about $\mathbf{U}$ and $\mathbf{V}$ from the result of (a)?

2.133 The cross product of two vectors $\mathbf{U}$ and $\mathbf{V}$ is $\mathbf{U} \times \mathbf{V} = -30\mathbf{i} + 40\mathbf{k}$. The vector $\mathbf{V} = 4\mathbf{i} - 2\mathbf{j} + 3\mathbf{k}$. Determine the components of $\mathbf{U}$.

2.134 The magnitudes $|\mathbf{U}| = 10$ and $|\mathbf{V}| = 20$.
(a) Use the definition of the cross product to determine $\mathbf{U} \times \mathbf{V}$.
(b) Use the definition of the cross product to determine $\mathbf{V} \times \mathbf{U}$.
(c) Use Eq. (2.34) to determine $\mathbf{U} \times \mathbf{V}$.
(d) Use Eq. (2.34) to determine $\mathbf{V} \times \mathbf{U}$.

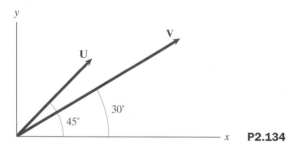

P2.134

2.135 The force $\mathbf{F} = 10\mathbf{i} - 4\mathbf{j}$ (N). Determine the cross product $\mathbf{r}_{AB} \times \mathbf{F}$.

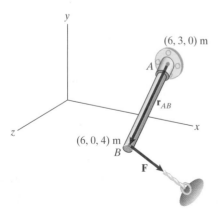

P2.135

2.136 By evaluating the cross product $\mathbf{U} \times \mathbf{V}$, prove the identity $\sin(\theta_1 - \theta_2) = \sin\theta_1 \cos\theta_2 - \cos\theta_1 \sin\theta_2$.

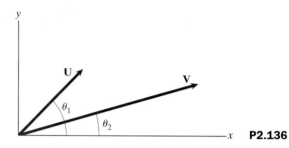

P2.136

2.137 Use the cross product to determine the components of a unit vector $\mathbf{e}$ that is normal to both of the vectors $\mathbf{U} = 8\mathbf{i} - 6\mathbf{j} + 4\mathbf{k}$ and $\mathbf{V} = 3\mathbf{i} + 7\mathbf{j} + 9\mathbf{k}$.

2.138 (a) What is the cross product $\mathbf{r}_{OA} \times \mathbf{r}_{OB}$?
(b) Determine a unit vector $\mathbf{e}$ that is perpendicular to $\mathbf{r}_{OA}$ and $\mathbf{r}_{OB}$.

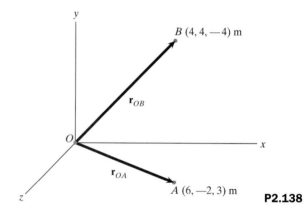

P2.138

2.139 For the points O, A, and B in Problem 2.138, use the cross product to determine the length of the shortest straight line from point B to the straight line that passes through points O and A.

2.140 The cable BC exerts a 1000-lb force $\mathbf{F}$ on the hook at B. Determine $\mathbf{r}_{AB} \times \mathbf{F}$.

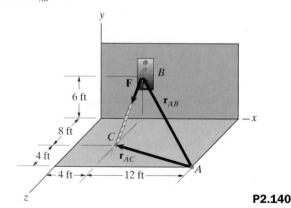

P2.140

2.141 The cable BC shown in Problem 2.140 exerts a 300-lb force $\mathbf{F}$ on the hook at B.
(a) Determine $\mathbf{r}_{AB} \times \mathbf{F}$ and $\mathbf{r}_{AC} \times \mathbf{F}$.
(b) Use the definition of the cross product to explain why the results of (a) are equal.

2.142 The rope AB exerts a 50-N force $\mathbf{T}$ on the collar at A. Let $\mathbf{r}_{CA}$ be the position vector from point C to point A. Determine the cross product $\mathbf{r}_{CA} \times \mathbf{T}$.

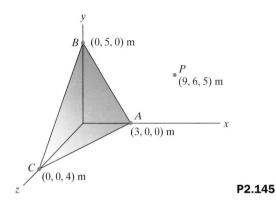

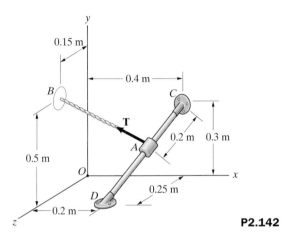

P2.142

2.143 In Problem 2.142, let $\mathbf{r}_{CB}$ be the position vector from point C to point B. Determine the cross product $\mathbf{r}_{CB} \times \mathbf{T}$ and compare your answer to the answer to Problem 2.142.

2.144 The bar AB is 6 m long and is perpendicular to the bars AC and AD. Use the cross product to determine the coordinates x_B, y_B, z_B of point B.

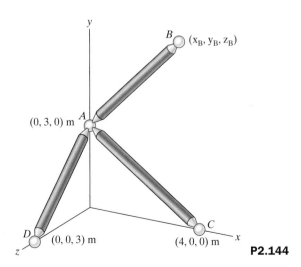

P2.144

2.145 Determine the minimum distance from point P to the plane defined by the three points A, B, and C.

P2.145

2.146 Consider vectors $\mathbf{U} = 3\mathbf{i} - 10\mathbf{j}$, $\mathbf{V} = -6\mathbf{j} + 2\mathbf{k}$, and $\mathbf{W} = 2\mathbf{i} + 6\mathbf{j} - 4\mathbf{k}$.
(a) Determine the value of the mixed triple product $\mathbf{U} \cdot (\mathbf{V} \times \mathbf{W})$ by first evaluating the cross product $\mathbf{V} \times \mathbf{W}$ and then taking the dot product of the result with the vector $\mathbf{U}$.
(b) Determine the value of the mixed triple product $\mathbf{U} \cdot (\mathbf{V} \times \mathbf{W})$ by using Eq. (2.36).

2.147 For the vectors $\mathbf{U} = 6\mathbf{i} + 2\mathbf{j} - 4\mathbf{k}$, $\mathbf{V} = 2\mathbf{i} + 7\mathbf{j}$, and $\mathbf{W} = 3\mathbf{i} + 2\mathbf{k}$, evaluate the following mixed triple products:
(a) $\mathbf{U} \cdot (\mathbf{V} \times \mathbf{W})$; (b) $\mathbf{W} \cdot (\mathbf{V} \times \mathbf{U})$; (c) $\mathbf{V} \cdot (\mathbf{W} \times \mathbf{U})$.

2.148 Use the mixed triple product to calculate the volume of the parallelepiped.

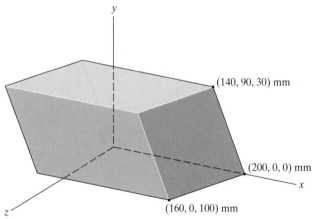

P2.148

2.149 By using Eqs. (2.23) and (2.34), show that

$$\mathbf{U} \cdot (\mathbf{V} \times \mathbf{W}) = \begin{vmatrix} U_x & U_y & U_z \\ V_x & V_y & V_z \\ W_x & W_y & W_z \end{vmatrix}.$$

2.150 The vectors $\mathbf{U} = \mathbf{i} + U_y\mathbf{j} + 4\mathbf{k}$, $\mathbf{V} = 2\mathbf{i} + \mathbf{j} - 2\mathbf{k}$, and $\mathbf{W} = -3\mathbf{i} + \mathbf{j} - 2\mathbf{k}$ are coplanar (they lie in the same plane). What is the component U_y?

Chapter Summary

In this chapter we have defined scalars, vectors, and vector operations. We showed how to express vectors in terms of cartesian components and carry out vector operations in terms of components. We introduced the definitions of the dot and cross products and the mixed triple product and demonstrated some applications of these operations, particularly the use of the dot product to resolve a vector into components parallel and perpendicular to a given direction. In Chapter 3 we will use vector operations to analyze forces acting on objects in equilibrium.

A physical quantity completely described by a real number is a *scalar*. A *vector* has both *magnitude* and *direction*. A vector is represented graphically by an arrow whose length is defined to be proportional to its magnitude.

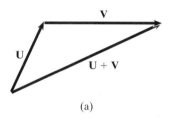

(a)

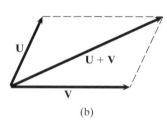

(b)

Rules for Manipulating Vectors

The sum of two vectors is defined by the *triangle rule* (Fig. a) or the equivalent *parallelogram rule* (Fig. b).

The product of a scalar a and a vector $\mathbf{U}$ is a vector $a\mathbf{U}$ with magnitude $|a||\mathbf{U}|$. Its direction is the same as $\mathbf{U}$ when a is positive and opposite to $\mathbf{U}$ when a is negative. The product $(-1)\mathbf{U}$ is written $-\mathbf{U}$ and is called the negative of $\mathbf{U}$. The division of $\mathbf{U}$ by a is the product $(1/a)\mathbf{U}$.

A *unit vector* is a vector whose magnitude is 1. A unit vector specifies a direction. Any vector $\mathbf{U}$ can be expressed as $|\mathbf{U}|\mathbf{e}$, where $\mathbf{e}$ is a unit vector with the same direction as $\mathbf{U}$. Dividing any vector by its magnitude yields a unit vector with the same direction as the vector.

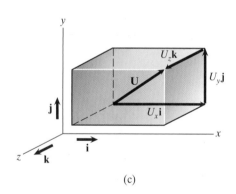

(c)

Cartesian Components

A vector $\mathbf{U}$ is expressed in terms of *scalar components* as

$$\mathbf{U} = U_x\mathbf{i} + U_y\mathbf{j} + U_z\mathbf{k} \quad \textbf{Eq. (2.12)}$$

(Fig. c). The coordinate system is *right-handed* (Fig. d): If the fingers of the right hand are pointed in the positive x direction and then closed toward the positive y direction, the thumb points in the z direction. The magnitude of $\mathbf{U}$ is

$$|\mathbf{U}| = \sqrt{U_x^2 + U_y^2 + U_z^2}. \quad \textbf{Eq. (2.14)}$$

Let θ_x, θ_y, and θ_z be the angles between $\mathbf{U}$ and the positive coordinate axes (Fig. e). Then the scalar components of $\mathbf{U}$ are

$$U_x = |\mathbf{U}|\cos\theta_x, \quad U_y = |\mathbf{U}|\cos\theta_y, \quad U_z = |\mathbf{U}|\cos\theta_z, \quad \textbf{Eq. (2.15)}$$

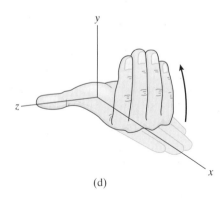

(d)

The quantities $\cos\theta_x$, $\cos\theta_y$, and $\cos\theta_z$ are the *direction cosines* of **U**. They satisfy the relation

$$\cos^2\theta_x + \cos^2\theta_y + \cos^2\theta_z = 1. \quad \text{Eq. (2.16)}$$

The *position vector* $\mathbf{r}_{AB}$ from a point A with coordinates (x_A, y_A, z_A) to a point B with coordinates (x_B, y_B, z_B) is given by

$$\mathbf{r}_{AB} = (x_B - x_A)\mathbf{i} + (y_B - y_A)\mathbf{j} + (z_B - z_A)\mathbf{k}. \quad \text{Eq. (2.17)}$$

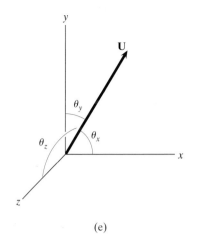

(e)

Dot Products

The dot product of two vectors **U** and **V** is

$$\mathbf{U} \cdot \mathbf{V} = |\mathbf{U}||\mathbf{V}|\cos\theta, \quad \text{Eq. (2.18)}$$

where θ is the angle between the vectors when they are placed tail to tail. The dot product of two nonzero vectors is equal to zero if and only if the two vectors are perpendicular.

In terms of scalar components,

$$\mathbf{U} \cdot \mathbf{V} = U_x V_x + U_y V_y + U_z V_z. \quad \text{Eq. (2.23)}$$

A vector **U** can be resolved into vector components $\mathbf{U}_p$ and $\mathbf{U}_n$ parallel and normal to a straight line L. In terms of a unit vector **e** that is parallel to L,

$$\mathbf{U}_p = (\mathbf{e} \cdot \mathbf{U})\mathbf{e}. \quad \text{Eq. (2.26)}$$

and

$$\mathbf{U}_n = \mathbf{U} - \mathbf{U}_p. \quad \text{Eq. (2.27)}$$

Cross Products

The cross product of two vectors **U** and **V** is

$$\mathbf{U} \times \mathbf{V} = |\mathbf{U}||\mathbf{V}|\sin\theta\,\mathbf{e}, \quad \text{Eq. (2.28)}$$

where θ is the angle between the vectors **U** and **V** when they are placed tail to tail and **e** is a unit vector perpendicular to **U** and **V**. The direction of **e** is specified by the *right-hand rule*: When the fingers of the right hand are pointed in the direction of **U** (the first vector in the cross product) and closed toward **V** (the second vector in the cross product), the thumb points in the direction of **e**. The cross product of two nonzero vectors is equal to zero if and only if the two vectors are parallel.

In terms of scalar components,

$$\mathbf{U} \times \mathbf{V} = \begin{vmatrix} \mathbf{i} & \mathbf{j} & \mathbf{k} \\ U_x & U_y & U_z \\ V_x & V_y & V_z \end{vmatrix} \quad \text{Eq. (2.34)}$$

Mixed Triple Products

The *mixed triple product* is the operation

$$\mathbf{U} \cdot (\mathbf{V} \times \mathbf{W}). \quad \text{Eq. (2.35)}$$

In terms of scalar components,

$$\mathbf{U} \cdot (\mathbf{V} \times \mathbf{W}) = \begin{vmatrix} U_x & U_y & U_z \\ V_x & V_y & V_z \\ W_x & W_y & W_z \end{vmatrix} \quad \text{Eq. (2.36)}$$

Review Problems

2.151 The magnitude of **F** is 8 kN. Express **F** in terms of scalar components.

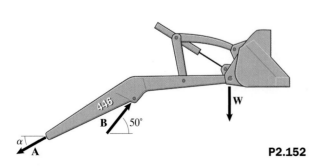

P2.151

2.152 The magnitude of the vertical force **W** is 600 lb, and the magnitude of the force **B** is 1500 lb. Given that **A** + **B** + **W** = 0, determine the magnitude of the force **A** and the angle α.

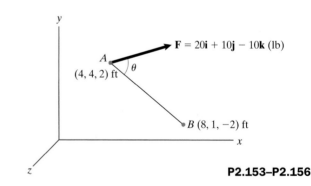

P2.152

Refer to the following diagram when solving Problems 2.153 through 2.156.

P2.153–P2.156

2.153 What are the direction cosines of **F**?

2.154 Determine the scalar components of a unit vector parallel to line *AB* that points from *A* toward *B*.

2.155 What is the angle θ between the line *AB* and the force **F**?

2.156 Determine the vector component of **F** that is parallel to the line *AB*.

2.157 The magnitude of $\mathbf{F}_B$ is 400 N and $|\mathbf{F}_A + \mathbf{F}_B| = 900$ N. Determine the components of $\mathbf{F}_A$.

2.159 The rope CE exerts a 500-N force $\mathbf{T}$ on the door $ABCD$. Determine the vector component of $\mathbf{T}$ in the direction parallel to the line from point A to point B.

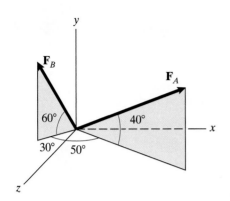

P2.157

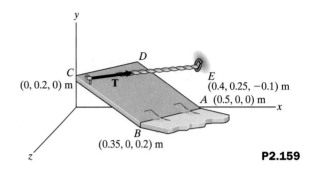

P2.159

2.158 Suppose that the forces $\mathbf{F}_A$ and $\mathbf{F}_B$ shown in Problem 2.157 have the same magnitude and $\mathbf{F}_A \cdot \mathbf{F}_B = 600$ N^2. What are $\mathbf{F}_A$ and $\mathbf{F}_B$?

2.160 In Problem 2.159, let $\mathbf{r}_{BC}$ be the position vector from point B to point C. Determine the cross product $\mathbf{r}_{BC} \times \mathbf{T}$.

The gravitational force on the climber is balanced by the forces exerted by the rope suspending him. In this chapter we use free-body diagrams to analyze forces on objects in equilibrium.

Forces

n Chapter 2 we represented forces by vectors and used vector addition to sum forces. In this chapter we discuss forces in more detail and introduce two of the most important concepts in mechanics, equilibrium and the free-body diagram. We will use free-body diagrams to identify the forces on objects and use equilibrium to determine unknown forces.

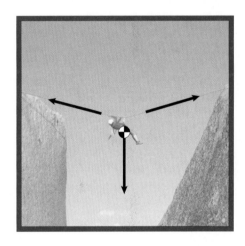

Types of Forces

Force is a familiar concept, as is evident from the words push, pull, and lift used in everyday conversation. In engineering we deal with different types of forces having a large range of magnitudes. In this section we introduce some terms used to describe forces and discuss particular forces that occur frequently in engineering applications.

Terminology

Line of Action When a force is represented by a vector, the straight line collinear with the vector is called the *line of action* of the force (Fig. 3.1).

Systems of Forces A *system of forces* is simply a particular set of forces. A system of forces is *coplanar*, or *two-dimensional,* if the lines of action of the forces lie in a plane. Otherwise it is *three-dimensional.* A system of forces is *concurrent* if the lines of action of the forces intersect at a point (Fig. 3.2a) and *parallel* if the lines of action are parallel (Fig. 3.2b).

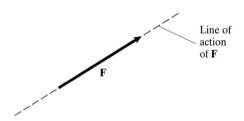

Figure 3.1
A force **F** and its line of action.

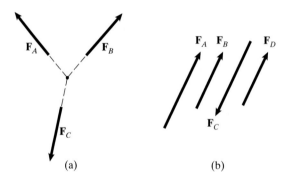

(a) (b)

Figure 3.2
(a) Concurrent forces.
(b) Parallel forces.

External and Internal Forces We say that a given object is subjected to an *external force* if the force is exerted by a different object. When one part of a given object is subjected to a force by another part of the same object, we say it is subjected to an *internal force.* These definitions require that you clearly define the object you are considering. For example, suppose that you are the object. When you are standing, the floor—a different object——exerts an external force on your feet. If you press your hands together, your left hand exerts an internal force on your right hand. However, if your right hand is the object you are considering, the force exerted by your left hand is an external force.

Body and Surface Forces A force acting on an object is called a *body force* if it acts on the volume of the object and a *surface force* if it acts on its surface. The gravitational force on an object is a body force. A surface force can be exerted on an object by contact with another object. Both body and surface forces can result from electromagnetic effects.

Gravitational Forces

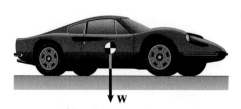

Figure 3.3
Representing an object's weight by a vector.

You are aware of the force exerted on an object by the earth's gravity whenever you pick up something heavy. We can represent the gravitational force, or weight, of an object by a vector (Fig. 3.3).

The magnitude of an object's weight is related to its mass m by

$$|\mathbf{W}| = mg,$$

where g is the acceleration due to gravity at sea level. We will use the values $g = 9.81 \text{ m/s}^2$ in SI units and $g = 32.2 \text{ ft/s}^2$ in U.S. Customary units.

Gravitational forces, and also electromagnetic forces, act at a distance. The objects they act on are not necessarily in contact with the objects exerting the forces. In the next section we discuss forces resulting from contacts between objects.

Contact Forces

Contact forces are the forces that result from contacts between objects. For example, you exert a contact force when you push on a wall (Fig. 3.4a). The surface of your hand exerts a force on the surface of the wall that can be represented by a vector $\mathbf{F}$ (Fig. 3.4b). The wall exerts an equal and opposite force $-\mathbf{F}$ on your hand (Fig. 3.4c). (Recall Newton's third law: The forces exerted on each other by any two particles are equal in magnitude and opposite in direction. If you have any doubt that the wall exerts a force on your hand, try pushing on the wall while standing on roller skates.)

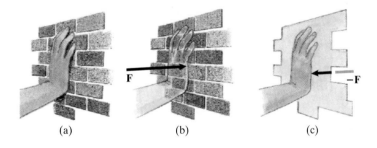

(a) (b) (c)

Figure 3.4
(a) Exerting a contact force on a wall by pushing on it.
(b) The vector $\mathbf{F}$ represents the force you exert on the wall.
(c) The wall exerts a force $-\mathbf{F}$ on your hand.

We will be concerned with contact forces exerted on objects by contact with the surfaces of other objects and by ropes, cables, and springs.

Surfaces   Consider two plane surfaces in contact (Fig. 3.5a). We represent the force exerted on the right surface by the left surface by the vector $\mathbf{F}$ in Fig. 3.5(b). We can resolve $\mathbf{F}$ into a component $\mathbf{N}$ that is normal to the surface and a component $\mathbf{f}$ that is parallel to the surface (Fig. 3.5c). The component $\mathbf{N}$ is called the *normal force*, and the component $\mathbf{f}$ is called the *friction force*. We sometimes assume that the friction force between two surfaces is negligible in comparison to the normal force, a condition we describe by saying that the surfaces are *smooth*. In this case we show only the normal force (Fig. 3.5d). When the friction force cannot be neglected, we say the surfaces are *rough*.

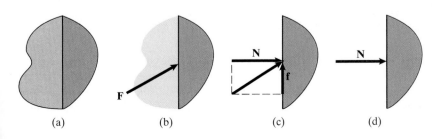

(a) (b) (c) (d)

Figure 3.5
(a) Two plane surfaces in contact.
(b) The force $\mathbf{F}$ exerted on the right surface.
(c) The force $\mathbf{F}$ resolved into components normal and parallel to the surface.
(d) Only the normal force is shown when friction is neglected.

If the contacting surfaces are curved (Fig. 3.6a), the normal force and the friction force are perpendicular and parallel to the plane tangent to the surfaces at their point of contact (Fig. 3.6b).

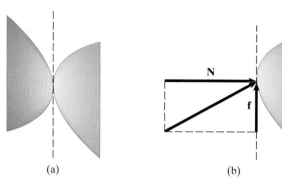

Figure 3.6
(a) Curved contacting surfaces. The dashed line indicates the plane tangent to the surfaces at their point of contact.
(b) The normal force and friction force on the right surface.

(a)   (b)

Ropes and Cables   You can exert a contact force on an object by attaching a rope or cable to the object and pulling on it. In Fig. 3.7a, the crane's cable is attached to a container of building materials. We can represent the force the cable exerts on the container by a vector **T** (Fig. 3.7b). The magnitude of **T** is called the *tension* in the cable, and the line of action of **T** is collinear with the cable. The cable exerts an equal and opposite force −**T** on the crane (Fig. 3.7c).

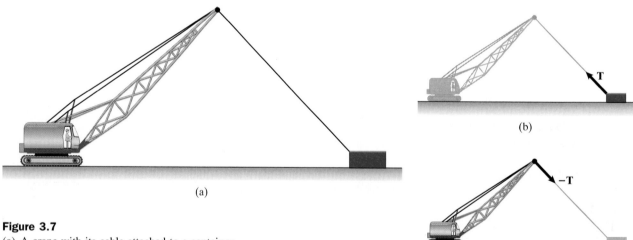

(a)

Figure 3.7
(a) A crane with its cable attached to a container.
(b) The force **T** exerted on the container by the cable.
(c) The force −**T** exerted on the crane by the cable.

(b)

(c)

Notice that we have assumed that the cable is straight and that the tension where the cable is connected to the container equals the tension near the crane. This is approximately true if the weight of the cable is small compared to the tension. Otherwise, the cable will sag significantly and the tension will vary along its length. You should assume that ropes and cables are straight and that their tensions are constant along their lengths.

A *pulley* is a wheel with a grooved rim that can be used to change the direction of a rope or cable (Fig. 3.8a). For now, we assume that the tension is the same on both sides of a pulley (Fig. 3.8b). This is true, or at least approximately true, when the pulley can turn freely and the rope or cable either is stationary or turns the pulley at a constant rate.

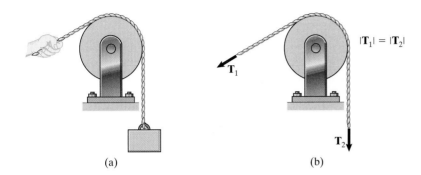

(a)   (b)

Figure 3.8
(a) A pulley changes the direction of a rope or cable.
(b) For now, you should assume that the tensions on each side of the pulley are equal.

Springs Springs are used to exert contact forces in mechanical devices, for example, in the suspensions of cars (Fig. 3.9). Let's consider a coil spring whose unstretched length, the length of the spring when its ends are free, is L_0 (Fig. 3.10a). When the spring is stretched to a length L greater than L_0 (Fig. 3.10b), it pulls on the object to which it is attached with a force $\mathbf{F}$ (Fig. 3.10c). The object exerts an equal and opposite force $-\mathbf{F}$ on the spring (Fig. 3.10d).

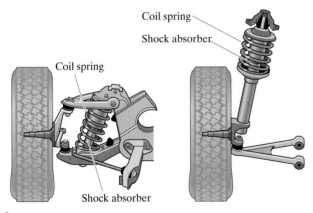

Figure 3.9
Coil springs in car suspensions. The arrangement on the right is called a MacPherson strut.

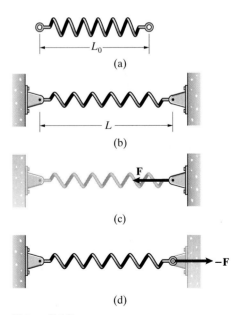

Figure 3.10
(a) A spring of unstretched length L_0.
(b) The spring stretched to a length $L > L_0$.
(c, d) The force $\mathbf{F}$ exerted by the spring and the force $-\mathbf{F}$ on the spring.

When the spring is compressed to a length L less than L_0 (Figs. 3.11a, b), the spring pushes on the object with a force $\mathbf{F}$ and the object exerts an equal and opposite force $-\mathbf{F}$ on the spring (Figs. 3.11c, d). If a spring is compressed too much, it may buckle (Fig. 3.11e). A spring designed to exert a force by being compressed is often provided with lateral support to prevent buckling, for example, by enclosing it in a cylindrical sleeve. In the car suspensions shown in Fig. 3.9, the shock absorbers within the coils prevent the springs from buckling.

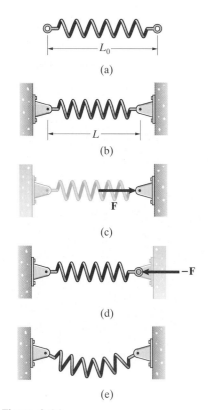

Figure 3.11
(a) A spring of length L_0.
(b) The spring compressed to a length $L < L_0$.
(c, d) The spring pushes on an object with a force $\mathbf{F}$, and the object exerts a force $-\mathbf{F}$ on the spring.
(e) A coil spring will buckle if it is compressed too much.

The magnitude of the force exerted by a spring depends on the material it is made of, its design, and how much it is stretched or compressed relative to its unstretched length. When the change in length is not too large compared to the unstretched length, the coil springs commonly used in mechanical devices exert a force approximately proportional to the change in length:

$$|\mathbf{F}| = k|L - L_0|. \tag{3.1}$$

Because the force is a linear function of the change in length (Fig. 3.12), a spring that satisfies this relation is called a *linear spring*. The value of the *spring constant* k depends on the material and design of the spring. Its dimensions are (force)/(length). Notice from Eq. (3.1) that k equals the magnitude of the force required to stretch or compress the spring a unit of length.

Suppose that the unstretched length of a spring is $L_0 = 1$ m and $k = 3000$ N/m. If the spring is stretched to a length $L = 1.2$ m, the magnitude of the pull it exerts is

$$k|L - L_0| = 3000(1.2 - 1) = 600 \text{ N}.$$

Although coil springs are commonly used in mechanical devices, we are also interested in them for a different reason. Springs can be used to *model* situations in which forces depend on displacements. For example, the force necessary to bend the steel beam in Fig. 3.13a is a linear function of the displacement δ,

$$|\mathbf{F}| = k\delta,$$

if δ is not too large. Therefore we can model the force-deflection behavior of the beam with a linear spring (Fig. 3.13b).

Study Questions

1. What is a two-dimensional system of forces?
2. What are internal and external forces?
3. If a surface is said to be smooth, what does that mean?
4. What is the relation between the magnitude of the force exerted by a linear spring and the change in its length?

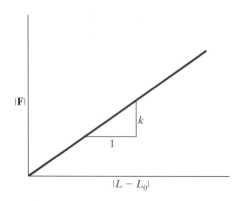

Figure 3.12
The graph of the force exerted by a linear spring as a function of its stretch or compression is a straight line with slope k.

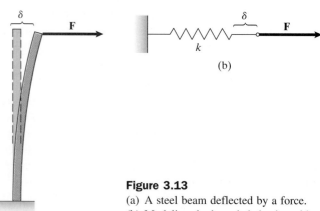

Figure 3.13
(a) A steel beam deflected by a force.
(b) Modeling the beam's behavior with a linear spring.

3.2 Equilibrium and Free-Body Diagrams

Statics is the study of objects in equilibrium. In everyday conversation, *equilibrium* means an unchanging state—a state of balance. Before we explain precisely what this term means in mechanics, let's consider some examples. Pieces of furniture sitting at rest in a room and a person standing stationary in the room are in equilibrium. If a train travels at constant speed on a straight track, objects that are at rest relative to the train, such as a person standing in the aisle, are in equilibrium (Fig. 3.14a). The person standing in the room and the person standing in the aisle of the train are not accelerating. If the train should start to increase or decrease its speed, however, the person standing in the aisle would no longer be in equilibrium and might lose his balance (Fig. 3.14b).

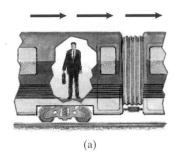

(a)

We say that an object is in *equilibrium* only if each point of the object has the same constant velocity, which is referred to as *steady translation*. The velocity must be measured relative to a frame of reference in which Newton's laws are valid, which is called an *inertial reference frame*. In most engineering applications, the velocity can be measured relative to the earth.

The vector sum of the external forces acting on an object in equilibrium is zero. We will use the symbol $\Sigma \mathbf{F}$ to denote the sum of the external forces. Thus when an object is in equilibrium,

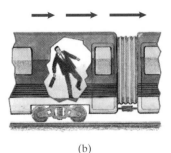

(b)

Figure 3.14
(a) While the train moves at a constant speed, a person standing in the aisle is in equilibrium.
(b) If the train starts to speed up, the person is no longer in equilibrium.

$$\Sigma \mathbf{F} = \mathbf{0}. \qquad (3.2)$$

In some situations we can use this *equilibrium equation* to determine unknown forces acting on an object in equilibrium. The first step will be to draw a *free-body diagram* of the object to identify the external forces acting on it. The free-body diagram is an essential tool in mechanics. It focuses attention on the object of interest and helps identify the external forces acting on it. Although in statics we will be concerned only with objects in equilibrium, free-body diagrams are also used in dynamics to analyze the motions of objects.

The free-body diagram is a simple concept. It is a drawing of an object and the external forces acting on it. Otherwise, nothing other than the object of interest is included. The drawing shows the object *isolated*, or *freed*, from its surroundings. Drawing a free-body diagram involves three steps:

1. *Identify the object you want to isolate.* As the following examples show, your choice is often dictated by particular forces you want to determine.

2. *Draw a sketch of the object isolated from its surroundings, and show relevant dimensions and angles.* Your drawing should be reasonably accurate, but it can omit irrelevant details.

3. *Draw vectors representing all of the external forces acting on the isolated object, and label them.* Don't forget to include the gravitational force if you are not intentionally neglecting it.

You will also need to choose a coordinate system so that you can express the forces on the isolated object in terms of components. Often you will find it convenient to choose the coordinate system before drawing the free-body diagram, but in some situations the best choice of coordinate system will not be apparent until after you have drawn it.

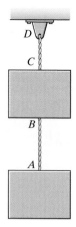

Figure 3.15
Stationary blocks suspended by cables.

A simple example demonstrates how you can choose free-body diagrams to determine particular forces and also that you must distinguish carefully between external and internal forces. Two stationary blocks of equal weight W are suspended by cables in Fig. 3.15. The system is in equilibrium. Suppose that we want to determine the tensions in the two cables.

To determine the tension in cable AB, we first isolate an "object" consisting of the lower block and part of cable AB (Fig. 3.16a). We then ask ourselves what forces can be exerted on our isolated object by objects not included in the diagram. The earth exerts a gravitational force of magnitude W on the block. Also, where we "cut" cable AB, the cable is subjected to a contact force equal to the tension in the cable (Fig. 3.16b). The arrows in this figure indicate the directions of the forces. The scalar W is the weight of the block and T_{AB} is the tension in cable AB. We assume that the weight of the part of cable AB included in the free-body diagram can be neglected in comparison to the weight of the block.

Since the free-body diagram is in equilibrium, the sum of the external forces equals zero. In terms of a coordinate system with the y axis upward (Fig. 3.16c), we obtain the equilibrium equation

$$\Sigma \mathbf{F} = T_{AB}\mathbf{j} - W\mathbf{j} = (T_{AB} - W)\mathbf{j} = \mathbf{0}.$$

Thus the tension in cable AB is $T_{AB} = W$.

Figure 3.16
(a) Isolating the lower block and part of cable AB.
(b) Indicating the external forces completes the free-body diagram.
(c) Introducing a coordinate system.

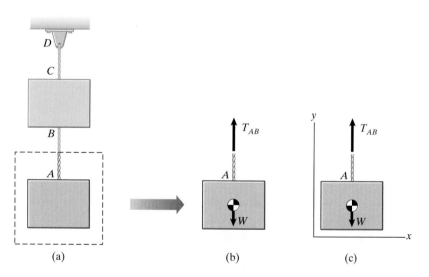

(a) (b) (c)

We can determine the tension in cable CD by isolating the upper block (Fig. 3.17a). The external forces are the weight of the upper block and the tensions in the two cables (Fig. 3.17b). In this case we obtain the equilibrium equation

$$\Sigma \mathbf{F} = T_{CD}\mathbf{j} - T_{AB}\mathbf{j} - W\mathbf{j} = (T_{CD} - T_{AB} - W)\mathbf{j} = \mathbf{0}.$$

Since $T_{AB} = W$, we find that $T_{CD} = 2W$.

We could also have determined the tension in cable CD by treating the two blocks and the cable AB as a single object (Figs. 3.18a, b). The equilibrium equation is

$$\Sigma \mathbf{F} = T_{CD}\mathbf{j} - W\mathbf{j} - W\mathbf{j} = (T_{CD} - 2W)\mathbf{j} = \mathbf{0},$$

and we again obtain $T_{CD} = 2W$.

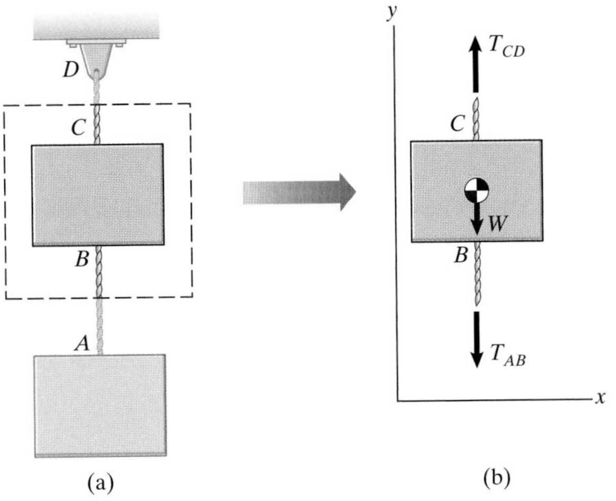

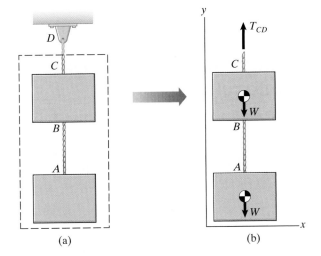

Figure 3.17
(a) Isolating the upper block to determine the tension in cable CD.
(b) Free-body diagram of the upper block.

Figure 3.18
(a) An alternative choice for determining the tension in cable CD.
(b) Free-body diagram including both blocks and cable AB.

Why doesn't the tension in cable AB appear on the free-body diagram in Fig. 3.18b? Remember that only external forces are shown on free-body diagrams. Since cable AB is part of the free-body diagram in this case, the forces it exerts on the upper and lower blocks are internal forces.

We have described the procedure for drawing free-body diagrams. In the next section we will draw free-body diagrams of objects subjected to two-dimensional systems of forces and use them to determine unknown forces acting on objects in equilibrium.

3.3 Two-Dimensional Force Systems

Suppose that the system of external forces acting on an object in equilibrium is two-dimensional (coplanar). By orienting a coordinate system so that the forces lie in the x-y plane, we can express the sum of the external forces as

$$\Sigma \mathbf{F} = (\Sigma F_x)\mathbf{i} + (\Sigma F_y)\mathbf{j} = \mathbf{0},$$

where ΣF_x and ΣF_y are the sums of the x and y components of the forces. Since a vector is zero only if each of its components is zero, we obtain two scalar equilibrium equations:

$$\Sigma F_x = 0, \qquad \Sigma F_y = 0. \tag{3.3}$$

The sums of the x and y components of the external forces acting on an object in equilibrium must each equal zero.

Study Questions

1. What do you know about the sum of the external forces acting on an object in equilibrium?
2. Is a free-body diagram only useful when an object is in equilibrium?
3. What are the steps in drawing a free-body diagram?

Example 3.1

Using Equilibrium to Determine Forces on an Object

For display at an automobile show, the 1440-kg car in Fig. 3.19 is held in place on the inclined surface by the horizontal cable from A to B. Determine the tension that the cable (and the fixture to which it is connected at B) must support. The car's brakes are not engaged, so the tires exert only normal forces on the inclined surface.

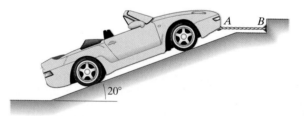

Figure 3.19

(a) Isolating the car.

Strategy

Since the car is in equilibrium, we can draw its free-body diagram and use Eqs. (3.3) to determine the forces exerted on the car by the cable and the inclined surface.

Solution

Draw the Free-Body Diagram We first draw a diagram of the car isolated from its surrounding (Fig. a) and then complete the free-body diagram by showing the force exerted by the car's weight, the force T exerted by the cable, and the normal force N exerted by the inclined surface (Fig. b).

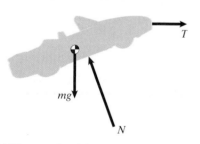

(b) The completed free-body diagram shows the known and unknown external forces.

Apply the Equilibrium Equations In Fig. c, we introduce a coordinate system and resolve the normal force into x and y components. The equilibrium equations are

$$\Sigma F_x = T - N \sin 20° = 0,$$

$$\Sigma F_y = N \cos 20° - mg = 0.$$

We can solve the second equilibrium equation for N,

$$N = \frac{mg}{\cos 20°} = \frac{(1440)(9.81)}{\cos 20°} = 15.0 \text{ kN},$$

and then solve the first equilibrium equation for the tension T:

$$T = N \sin 20° = 5.14 \text{ kN}.$$

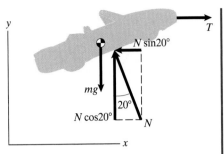

(c) Introducing a coordinate system and resolving N into its components.

Example 3.2

Choosing a Free-Body Diagram

The automobile engine block in Fig. 3.20 is suspended by a system of cables. The mass of the block is 200 kg. What are the tensions in cables AB and AC?

Strategy

We need a free-body diagram that is subjected to the forces we want to determine. By isolating part of the cable system near point A where the cables are joined, we can obtain a free-body diagram that is subjected to the weight of the block and the unknown tensions in cables AB and AC.

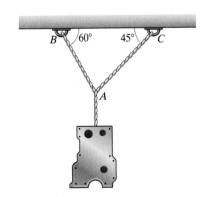

Figure 3.20

Solution

Draw the Free-Body Diagram Isolating part of the cable system near point A (Fig. a), we obtain a free-body diagram subjected to the weight of the block $W = mg = (200 \text{ kg})(9.81 \text{ m/s}^2) = 1962 \text{ N}$ and the tensions in cables AB and AC (Fig. b).

Apply the Equilibrium Equations We select the coordinate system shown in Fig. c and resolve the cable tensions into x and y components. The resulting equilibrium equations are

$$\Sigma F_x = T_{AC} \cos 45° - T_{AB} \cos 60° = 0,$$

$$\Sigma F_y = T_{AC} \sin 45° + T_{AB} \sin 60° - 1962 = 0.$$

Solving these equations, we find that the tensions in the cables are $T_{AB} = 1436 \text{ N}$ and $T_{AC} = 1016 \text{ N}$.

  Alternative Solution: We can determine the tensions in the cables in another way that will also help you visualize the conditions for equilibrium. Since the sum of the three forces acting on our free-body diagram is zero, the vectors

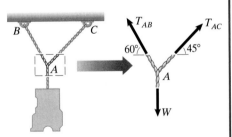

(a) Isolating part of the cable system.
(b) The completed free-body diagram.

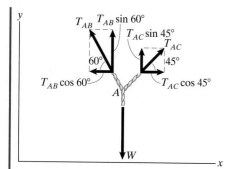

(c) Selecting a coordinate system and resolving the forces into components.

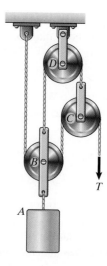

T_{AC} 45°

$W = 1962$ N

T_{AB} 30°

(d) The triangle formed by the sum of the three forces.

form a closed polygon when placed head to tail (Fig. d). You can see that the sum of the vertical components of the tensions supports the weight and that the horizontal components of the tensions must balance each other. The angle of the triangle opposite the weight W is $180° - 30° - 45° = 105°$. By applying the law of sines,

$$\frac{\sin 45°}{T_{AB}} = \frac{\sin 30°}{T_{AC}} = \frac{\sin 105°}{1962},$$

we obtain $T_{AB} = 1436$ N and $T_{AC} = 1016$ N.

Discussion

How were we able to choose a free-body diagram that permitted us to determine the unknown tensions in the cables? There are no definite rules for choosing free-body diagrams. You will learn what to do in many cases from the examples we present, but you will also encounter new situations. It may be necessary to try several free-body diagrams before finding one that provides the information you need. Remember that forces you want to determine should appear as external forces on your free-body diagram, and your objective is to obtain a number of equilibrium equations equal to the number of unknown forces.

Example 3.3

Applying Equilibrium to a System of Pulleys

The mass of each pulley of the system in Fig. 3.21 is m, and the mass of the suspended object A is m_A. Determine the force T necessary for the system to be in equilibrium.

Figure 3.21

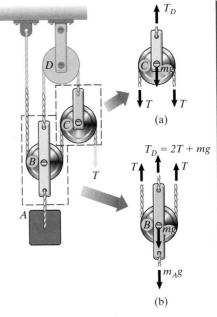

Strategy

By drawing free-body diagrams of the individual pulleys and applying equilibrium, we can relate the force T to the weights of the pulleys and the object A.

Solution

We first draw a free-body diagram of the pulley C to which the force T is applied (Fig. a). Notice that we assume the tension in the cable supported by the pulley to equal T on both sides (see Fig. 3.8). From the equilibrium equation

$$T_D - T - T - mg = 0,$$

we determine that the tension in the cable supported by pulley D is

$$T_D = 2T + mg.$$

We now know the tensions in the cables extending from pulleys C and D to pulley B in terms of T. Drawing the free-body diagram of pulley B (Fig. b), we obtain the equilibrium equation

$$T + T + 2T + mg - mg - m_A g = 0.$$

Solving, we obtain $T = m_A g/4$.

(a) Free-body diagram of pulley C.
(b) Free-body diagram of pulley B.

Problems

3.1 The figure shows the external forces acting on an object in equilibrium. The forces $F_1 = 32$ N and $F_3 = 50$ N. Determine F_2 and the angle α.

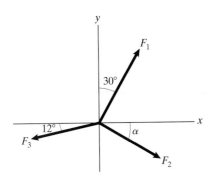

P3.1

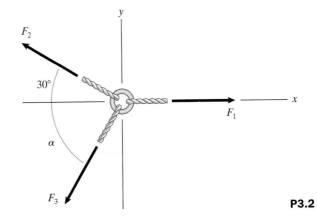

P3.2

3.2 The force $F_1 = 100$ N and the angle $\alpha = 60°$. The weight of the ring is negligible. Determine the forces F_2 and F_3.

3.3 Consider the forces shown in Problem 3.2. Suppose that $F_2 = 100$ N and you want to choose the angle α so that the magnitude of F_3 is a minimum. What is the resulting magnitude of F_3?
Strategy: Draw a vector diagram of the sum of the three forces.

3.4 The beam is in equilibrium. If $A_x = 77$ kN, $B = 400$ kN, and the beam's weight is negligible, what are the forces A_y and C?

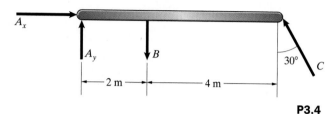

P3.4

3.5 Suppose that the mass of the beam shown in Problem 3.4 is 20 kg and it is in equilibrium. The force A_y points upward. If $A_y = 258$ kN and $B = 240$ kN, what are the forces A_x and C?

3.6 A zoologist estimates that the jaw of a predator, *Martes*, is subjected to a force P as large as 800 N. What forces T and M must be exerted by the temporalis and masseter muscles to support this value of P?

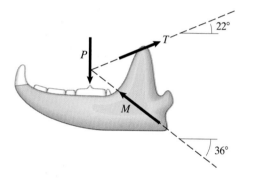

P3.6

3.7 The two springs are identical, with unstretched lengths 250 mm and spring constants $k = 1200$ N/m.

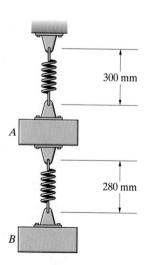

P3.7

(a) Draw the free-body diagram of block A.
(b) Draw the free-body diagram of block B.
(c) What are the masses of the two blocks?

3.8 The two springs in Problem 3.7 are identical, with unstretched lengths 250 mm and spring constants k. The sum of the masses of blocks A and B is 10 kg. Determine the value of k and the masses of the two blocks.

3.9 The 200-kg horizontal steel bar is suspended by the three springs. The stretch of each spring is 0.1 m. The constant of spring B is $k_B = 8000$ N/m. Determine the constants $k_A = k_C$ of springs A and C.

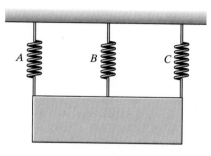

P3.9

3.10 The mass of the crane is 20 Mg (megagrams), and the tension in its cable is 1 kN. The crane's cable is attached to a caisson whose mass is 400 kg. Determine the magnitudes of the normal and friction forces exerted on the crane by the level ground.

Strategy: Draw the free-body diagram of the crane and the part of its cable within the dashed line.

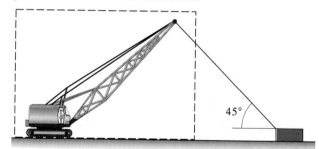

P3.10

3.11 What is the tension in the horizontal cable AB in Example 3.1 if the 20° angle is increased to 25°?

3.12 The 2400-lb car will remain in equilibrium on the sloping road only if the friction force exerted on the car by the road is not

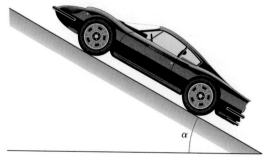

P3.12

greater than 0.6 times the normal force. What is the largest angle α for which the car will remain in equilibrium?

3.13 The crate is in equilibrium on the smooth surface. (Remember that "smooth" means that friction is negligible.) The spring constant is $k = 2500$ N/m and the stretch of the spring is 0.055 m. What is the mass of the crate?

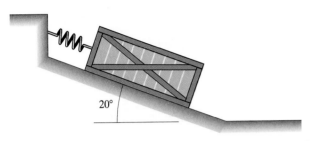

20°

P3.13

3.14 The 600-lb box is held in place on the smooth bed of the dump truck by the rope AB.
(a) If $\alpha = 25°$, what is the tension in the rope?
(b) If the rope will safely support a tension of 400 lb, what is the maximum allowable value of α?

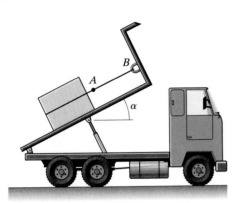

P3.14

3.15 Three forces act on the free-body diagram of a joint of a structure. If the structure is in equilibrium and $F_A = 4.20$ kN, what are F_B and F_C?

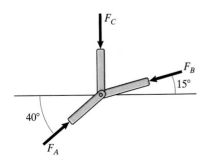

P3.15

3.16 The weights of the two blocks are $W_1 = 200$ lb and $W_2 = 50$ lb. Neglecting friction, determine the force the man must exert to hold the blocks in place.

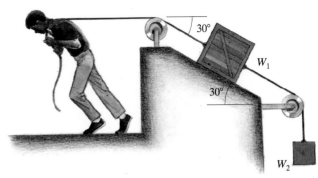

P3.16

3.17 The two springs have the same unstretched length, and the inclined surface is smooth. Show that the magnitudes of the forces exerted by the two springs are

$$F_1 = \frac{W \sin \alpha}{1 + k_2/k_1}, \quad F_2 = \frac{W \sin \alpha}{1 + k_1/k_2}.$$

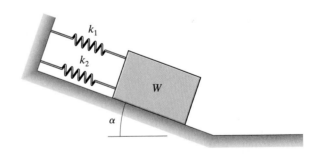

P3.17

3.18 A 10-kg painting is suspended by a wire. If $\alpha = 25°$, what is the tension in the wire?

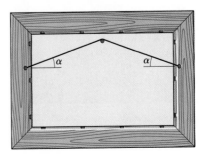

P3.18

3.19 If the wire supporting the suspended painting in Problem 3.18 breaks when the tension exceeds 150 N and you want a 100 percent safety factor (that is, you want the wire to be able to support twice the actual weight of the painting), what is the smallest value of α you can use?

3.20 Assume that the 150-lb climber is in equilibrium. What are the tensions in the rope on the left and right sides?

P3.20

3.21 If the mass of the climber shown in Problem 3.20 is 80 kg, what are the tensions in the rope on the left and right sides?

3.22 A construction worker holds a 180-kg crate in the position shown. What force must she exert on the cable?

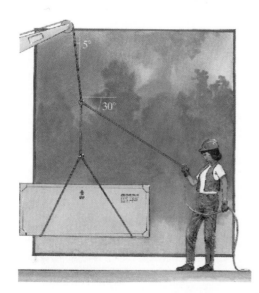

P3.22

3.23 A construction worker on the moon (acceleration due to gravity 1.62 m/s^2) holds the same crate described in Problem 3.22 in the position shown. What force must she exert on the cable?

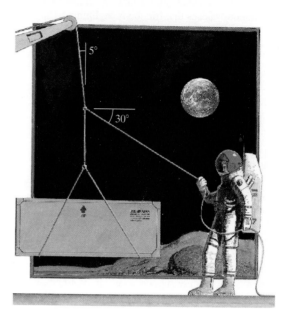

P3.23

3.24 A student on his summer job needs to pull a crate across the floor. Pulling as shown in Fig. a, he can exert a tension of 60 lb. He finds that the crate doesn't move, so he tries the arrangement in Fig. b, exerting a vertical force of 60 lb on the rope. What is the magnitude of the horizontal force he exerts on the crate in each case?

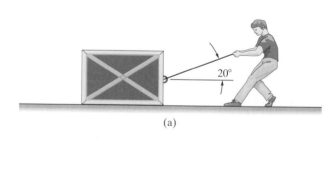

(a)

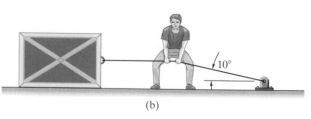

(b)

P3.24

3.25 The 140-kg traffic light is suspended above the street by two cables. What is the tension in the cables?

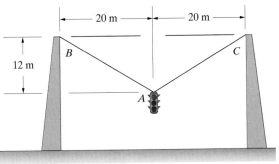

P3.25

3.26 Consider the suspended traffic light in Problem 3.25. To raise the light temporarily during a parade, an engineer wants to connect the 17-m length of cable DE to the midpoints of cables AB and AC as shown. However, for safety considerations, he doesn't want to subject any of the cables to a tension larger than 4 kN. Can he do it?

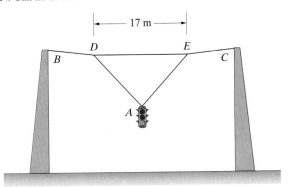

P3.26

3.27 The mass of the suspended crate is 5 kg. What are the tensions in cables AB and AC?

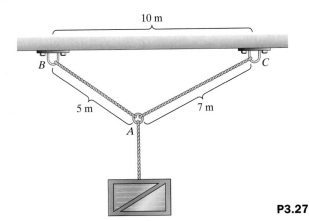

P3.27

3.28 What are the tensions in the upper and lower cables? (Your answers will be in terms of W. Neglect the weight of the pulley.)

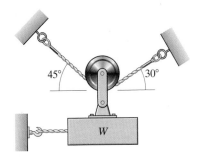

P3.28

3.29 Two tow trucks lift a motorcycle out of a ravine following an accident. If the 100-kg motorcycle is in equilibrium in the position shown, what are the tensions in the cables AB and AC?

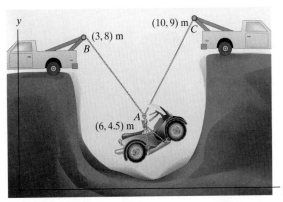

P3.29

3.30 An astronaut candidate conducts experiments on an airbearing platform. While he carries out calibrations, the platform is held in place by the horizontal tethers AB, AC, and AD.

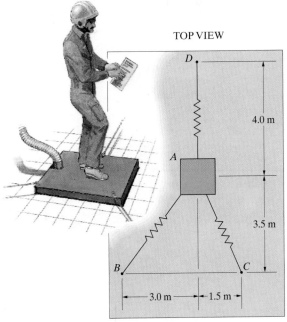

P3.30

The forces exerted by the tethers are the only horizontal forces acting on the platform. If the tension in tether AC is 2 N, what are the tensions in the other two tethers?

3.31 The forces exerted on the shoes and back of the 72-kg climber by the walls of the "chimney" are perpendicular to the walls exerting them. The tension in the rope is 640 N. What is the magnitude of the force exerted on his back?

P3.31

3.32 The slider A is in equilibrium and the bar is smooth. What is the mass of the slider?

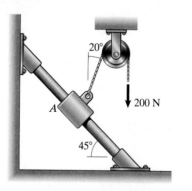

P3.32

3.33 The unstretched length of the spring AB is 660 mm, and the spring constant $k = 1000$ N/m. What is the mass of the suspended object?

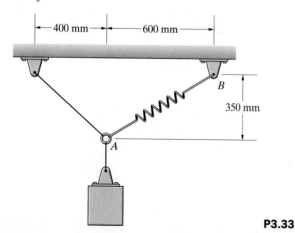

P3.33

3.34 The unstretched length of the spring in Problem 3.33 is 660 mm. If the mass of the suspended object is 10 kg and the system is in equilibrium in the position shown, what is the spring constant?

3.35 The collar A slides on the smooth vertical bar. The masses $m_A = 20$ kg and $m_B = 10$ kg. When $h = 0.1$ m, the spring is unstretched. When the system is in equilibrium, $h = 0.3$ m. Determine the spring constant k.

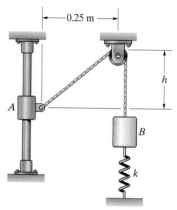

P3.35

3.36 You are designing a cable system to support a suspended object of weight W. The two wires must be identical, and the dimension b is fixed. The ratio of the tension T in each wire to its cross-sectional area A must equal a specified value $T/A = \sigma$. The "cost" of your design is the total volume of material in the two wires, $V = 2A\sqrt{b^2 + h^2}$. Determine the value of h that minimizes the cost.

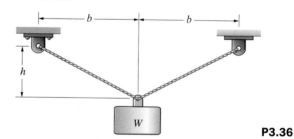

P3.36

3.37 The system of cables suspends a 1000-lb bank of lights above a movie set. Determine the tensions in cables AB, CD, and CE.

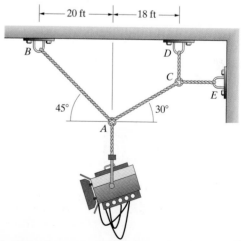

P3.37

3.38 Consider the 1000-lb bank of lights in Problem 3.37. A technician changes the position of the lights by removing the cable *CE*. What is the tension in cable *AB* after the change?

3.39 While working on another exhibit, a curator at the Smithsonian Institution pulls the suspended *Voyager* aircraft to one side by attaching three horizontal cables as shown. The mass of the aircraft is 1250 kg. Determine the tensions in the cable segments *AB*, *BC*, and *CD*.

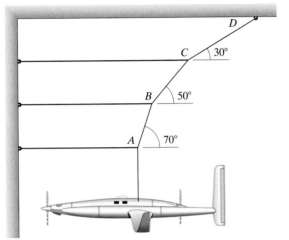

P3.39

3.40 A truck dealer wants to suspend a 4-Mg (megagram) truck as shown for advertising. The distance *b* = 15 m, and the sum of the lengths of the cables *AB* and *BC* is 42 m. What are the tensions in the cables?

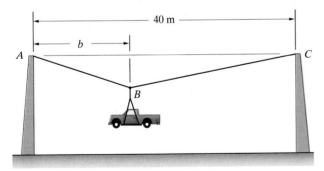

P3.40

3.41 The distance *h* = 12 in., and the tension in cable *AD* is 200 lb. What are the tensions in cables *AB* and *AC*?

3.42 You are designing a cable system to support a suspended object of weight *W*. Because your design requires points *A* and *B* to be placed as shown, you have no control over the angle α, but you can choose the angle β by placing point *C* wherever you wish.

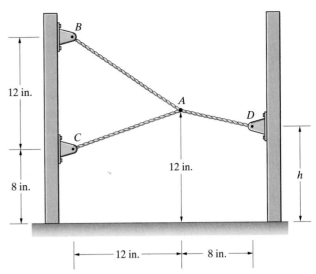

P3.41

Show that to minimize the tensions in cables *AB* and *BC*, you must choose β = α if the angle α ≥ 45°.

Strategy: Draw a diagram of the sum of the forces exerted by the three cables at *A*.

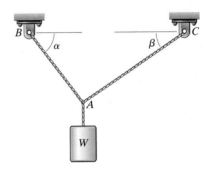

P3.42

3.43 In Problem 3.42, suppose that you have no control over the angle α and you want to design the cable system so that the tension in cable *AC* is a minimum. What is the required angle β?

3.44 The masses of the boxes on the left and right are 25 kg and 40 kg, respectively. The surfaces are smooth and the boxes are in equilibrium. Determine the tension in the cable and the angle α.

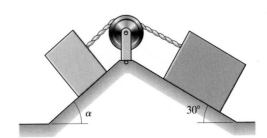

P3.44

3.45 Consider the system shown in Problem 3.44. The angle $\alpha = 45°$, the surfaces are smooth, and the boxes are in equilibrium. Determine the ratio of the mass of the right box to the mass of the left box.

3.46 The 3000-lb car and the 4600-lb tow truck are stationary. The muddy surface on which the car rests exerts a negligible friction force on the car. What is the tension in the tow cable?

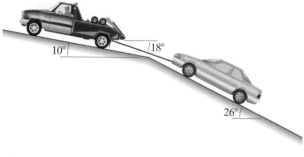

P3.46

3.47 The hydraulic cylinder is subjected to three forces. An 8-kN force is exerted on the cylinder at B that is parallel to the cylinder and points from B toward C. The link AC exerts a force at C that is parallel to the line from A to C. The link CD exerts a force at C that is parallel to the line from C to D.
(a) Draw the free-body diagram of the cylinder. (The cylinder's weight is negligible.)
(b) Determine the magnitudes of the forces exerted by the links AC and CD.

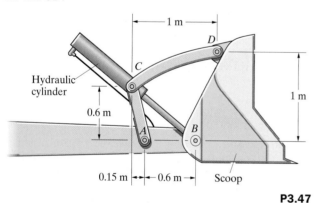

P3.47

3.48 The 50-lb cylinder rests on two smooth surfaces.
(a) Draw the free-body diagram of the cylinder.

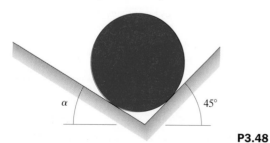

P3.48

(b) If $\alpha = 30°$, what are the magnitudes of the forces exerted on the cylinder by the left and right surfaces?

3.49 For the 50-lb cylinder in Problem 3.48, obtain an equation for the force exerted on the cylinder by the left surface in terms of the angle α in two ways: (a) using a coordinate system with the y axis vertical, (b) using a coordinate system with the y axis parallel to the right surface.

3.50 The 50-kg sphere is at rest on the smooth horizontal surface. The horizontal force $F = 500$ N. What is the normal force exerted on the sphere by the surface?

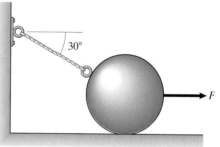

P3.50

3.51 Consider the stationary sphere in Problem 3.50.
(a) Draw a graph of the normal force exerted on the sphere by the surface as a function of the force F from $F = 0$ to $F = 1$ kN.
(b) In the result of (a), notice that the normal force decreases to zero and becomes negative as F increases. What does that mean?

3.52 The 1440-kg car is moving at constant speed on a road with the slope shown. The aerodynamic forces on the car are the drag $D = 530$ N, which is parallel to the road, and the lift $L = 360$ N, which is perpendicular to the road. Determine the magnitudes of the total normal and friction forces exerted on the car by the road.

P3.52

3.53 The device shown is towed beneath a ship to measure water temperature and salinity. The mass of the device is 130 kg. The angle $\alpha = 20°$. The motion of the water relative to the device causes a horizontal drag force D. The hydrostatic pressure distribution in the water exerts a vertical "buoyancy" force B. The magnitude of the buoyancy force is equal to the product of the volume of the device, $V = 0.075$ m³, and the weight density of the water, $\gamma = 9500$ N/m³. Determine the drag force D and the tension in the cable.

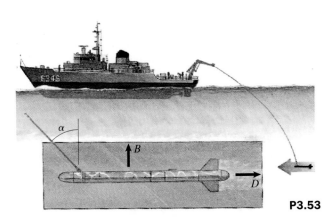

P3.53

3.54 The mass of each pulley of the system is m and the mass of the suspended object A is m_A. Determine the force T necessary for the system to be in equilibrium.

P3.54

3.55 The mass of each pulley of the system is m and the mass of the suspended object A is m_A. Determine the force T necessary for the system to be in equilibrium.

P3.55

3.56 The system is in equilibrium. What are the coordinates of point A?

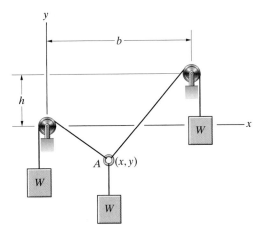

P3.56

3.57 The light fixture of weight W is suspended from a circular arch by a large number N of equally spaced cables. The tension T in each cable is the same. Show that

$$T = \frac{\pi W}{2N}.$$

Strategy: Consider an element of the arch defined by an angle $d\theta$ measured from the point where the cables join:

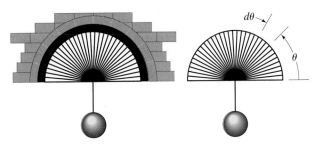

P3.57

Since the total angle described by the arch is π radians, the number of cables attached to the element is $(N/\pi)d\theta$. You can use this result to write the equilibrium equations for the part of the cable system where the cables join.

3.58 The solution to Problem 3.57 is an "asymptotic" result whose accuracy increases as N increases. Determine the exact tension T_{exact} for $N = 3, 5, 9$, and 17, and confirm the numbers in the following table. (For example, for $N = 3$, the cables are attached at $\theta = 0$, $\theta = 90°$, and $\theta = 180°$.)

N	3	5	9	17
$\dfrac{T_{\text{exact}}}{\pi W/2N}$	1.91	1.32	1.14	1.07

3.4 Three-Dimensional Force Systems

The equilibrium situations we have considered so far have involved only coplanar forces. When the system of external forces acting on an object in equilibrium is three-dimensional, we can express the sum of the external forces as

$$\Sigma \mathbf{F} = (\Sigma F_x)\mathbf{i} + (\Sigma F_y)\mathbf{j} + (\Sigma F_z)\mathbf{k} = \mathbf{0}.$$

Each component of this equation must equal zero, resulting in three scalar equilibrium equations:

$$\Sigma F_x = 0, \qquad \Sigma F_y = 0, \qquad \Sigma F_z = 0. \tag{3.4}$$

The sums of the x, y, and z components of the external forces acting on an object in equilibrium must each equal zero.

Example 3.4

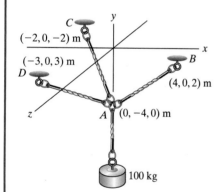

Figure 3.22

Applyng Equilibrium in Three Dimensions

The 100-kg cylinder in Fig. 3.22 is suspended from the ceiling by cables attached at points B, C, and D. What are the tensions in cables AB, AC, and AD?

Strategy

We can determine the tensions by the same approach we used for similar two-dimensional problems. By isolating part of the cable system near point A, we can obtain a free-body diagram subjected to forces due to the tensions in the cables. Since the sums of the x, y, and z components of the external forces must each equal zero, we obtain three equations for the three unknown tensions.

Solution

Draw the Free-Body Diagram We isolate part of the cable system near point A (Fig. a) and complete the free-body diagram by showing the forces exerted by the tensions in the cables (Fig. b). The magnitudes of the vectors $\mathbf{T}_{AB}$, $\mathbf{T}_{AC}$, and $\mathbf{T}_{AD}$ are the tensions in cables AB, AC, and AD, respectively.

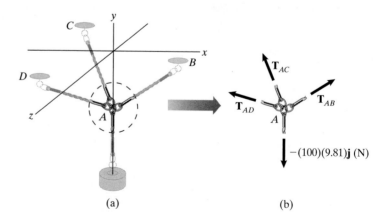

(a) Isolating part of the cable system.
(b) The completed free-body diagram showing the forces exerted by the tensions in the cables.

(a)   (b)

Apply the Equilibrium Equations The sum of the external forces acting on the free-body diagram is

$$\Sigma \mathbf{F} = \mathbf{T}_{AB} + \mathbf{T}_{AC} + \mathbf{T}_{AD} - 981\mathbf{j} = \mathbf{0}.$$

To solve this equation for the tensions in the cables, we need to express the vectors $\mathbf{T}_{AB}$, $\mathbf{T}_{AC}$, and $\mathbf{T}_{AD}$ in terms of their components.

We first determine the components of a unit vector that points in the direction of the vector $\mathbf{T}_{AB}$. Let $\mathbf{r}_{AB}$ be the position vector from point A to point B (Fig. c):

$$\mathbf{r}_{AB} = (x_B - x_A)\mathbf{i} + (y_B - y_A)\mathbf{j} + (z_B - z_A)\mathbf{k} = 4\mathbf{i} + 4\mathbf{j} + 2\mathbf{k} \text{ (m)}.$$

Dividing $\mathbf{r}_{AB}$ by its magnitude, we obtain a unit vector that has the same direction as $\mathbf{T}_{AB}$:

$$\mathbf{e}_{AB} = \frac{\mathbf{r}_{AB}}{|\mathbf{r}_{AB}|} = 0.667\mathbf{i} + 0.667\mathbf{j} + 0.333\mathbf{k}.$$

Now we can write the vector $\mathbf{T}_{AB}$ as the product of the tension T_{AB} in cable AB and $\mathbf{e}_{AB}$:

$$\mathbf{T}_{AB} = T_{AB}\mathbf{e}_{AB} = T_{AB}(0.667\mathbf{i} + 0.667\mathbf{j} + 0.333\mathbf{k}).$$

We now express the force vectors $\mathbf{T}_{AC}$ and $\mathbf{T}_{AD}$ in terms of the tensions T_{AC} and T_{AD} in cables AC and AD in the same way. The results are

$$\mathbf{T}_{AC} = T_{AC}(-0.408\mathbf{i} + 0.816\mathbf{j} - 0.408\mathbf{k}),$$

$$\mathbf{T}_{AD} = T_{AD}(-0.514\mathbf{i} + 0.686\mathbf{j} - 0.514\mathbf{k}).$$

We use these expressions to write the sum of the external forces in terms of the tensions T_{AB}, T_{AC}, and T_{AD}:

$$\begin{aligned}
\Sigma \mathbf{F} &= \mathbf{T}_{AB} + \mathbf{T}_{AC} + \mathbf{T}_{AD} - 981\mathbf{j} \\
&= (0.667T_{AB} - 0.408T_{AC} - 0.514T_{AD})\mathbf{i} \\
&\quad + (0.667T_{AB} + 0.816T_{AC} + 0.686T_{AD} - 981)\mathbf{j} \\
&\quad + (0.333T_{AB} - 0.408T_{AC} + 0.514T_{AD})\mathbf{k} \\
&= \mathbf{0}.
\end{aligned}$$

The sums of the forces in the x, y, and z directions must each equal zero:

$$\Sigma F_x = 0.667T_{AB} - 0.408T_{AC} - 0.514T_{AD} = 0,$$

$$\Sigma F_y = 0.667T_{AB} + 0.816T_{AC} + 0.686T_{AD} - 981 = 0,$$

$$\Sigma F_z = 0.333T_{AB} - 0.408T_{AC} + 0.514T_{AD} = 0.$$

Solving these equations, we find that the tensions are $T_{AB} = 519$ N, $T_{AC} = 636$ N, and $T_{AD} = 168$ N.

Discussion

Notice that this example required several of the techniques we covered in Chapter 2. In particular, we had to determine the components of a position vector, divide the position vector by its magnitude to obtain a unit vector with the same direction as a particular force, and express the force in terms of its components by writing it as the product of the unit vector and the magnitude of the force.

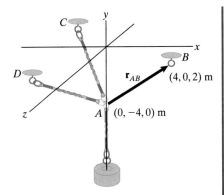

(c) The position vector $\mathbf{r}_{AB}$.

Example 3.5

Application of the Dot Product

The 100-lb "slider" C in Fig. 3.23 is held in place on the smooth bar by the cable AC. Determine the tension in the cable and the force exerted on the slider by the bar.

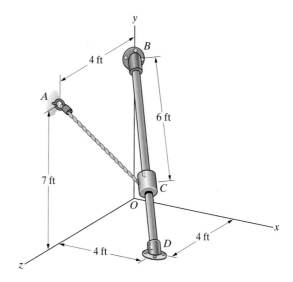

Figure 3.23

(a)

$-100\,\mathbf{j}$ (lb)

(b)

(a) Isolating the slider.
(b) Free-body diagram of the slider showing the forces exerted by its weight, the cable, and the bar.

Strategy

Since we want to determine forces that act on the slider, we need to draw its free-body diagram. The external forces acting on the slider are its weight and the forces exerted on it by the cable and the bar. If we approached this example as we did the previous one, our next step would be to express the forces in terms of their components. However, we don't know the direction of the force exerted on the slider by the bar. Since the smooth bar exerts negligible friction force, we do know that the force exerted by the bar is normal to its axis. Therefore we can eliminate this force from the equation $\Sigma\mathbf{F} = \mathbf{0}$ by taking the dot product of the equation with a unit vector that is parallel to the bar.

Solution

Draw the Free-Body Diagram We isolate the slider (Fig. a) and complete the free-body diagram by showing the weight of the slider, the force $\mathbf{T}$ exerted by the tension in the cable, and the normal force $\mathbf{N}$ exerted by the bar (Fig. b).

Apply the Equilibrium Equations The sum of the external forces acting on the free-body diagram is

$$\Sigma\mathbf{F} = \mathbf{T} + \mathbf{N} - 100\mathbf{j} = \mathbf{0}. \tag{3.5}$$

Let $\mathbf{e}_{BD}$ be the unit vector pointing from point B toward point D. Since $\mathbf{N}$ is perpendicular to the bar, $\mathbf{e}_{BD} \cdot \mathbf{N} = 0$. Therefore

$$\mathbf{e}_{BD} \cdot (\Sigma\mathbf{F}) = \mathbf{e}_{BD} \cdot (\mathbf{T} - 100\mathbf{j}) = 0. \tag{3.6}$$

This equation has a simple interpretation: The component of the slider's weight parallel to the bar is balanced by the component of **T** parallel to the bar.

Determining $\mathbf{e}_{BD}$: We determine the vector from point B to point D,

$$\mathbf{r}_{BD} = (4 - 0)\mathbf{i} + (0 - 7)\mathbf{j} + (4 - 0)\mathbf{k} = 4\mathbf{i} - 7\mathbf{j} + 4\mathbf{k} \text{ (ft)},$$

and divide it by its magnitude to obtain the unit vector $\mathbf{e}_{BD}$:

$$\mathbf{e}_{BD} = \frac{\mathbf{r}_{BD}}{|\mathbf{r}_{BD}|} = \frac{4}{9}\mathbf{i} - \frac{7}{9}\mathbf{j} + \frac{4}{9}\mathbf{k}.$$

Resolving **T** *into components*: To express **T** in terms of its components, we need to determine the coordinates of the slider C. We can write the vector from B to C in terms of the unit vector $\mathbf{e}_{BD}$,

$$\mathbf{r}_{BC} = 6\mathbf{e}_{BD} = 2.67\mathbf{i} - 4.67\mathbf{j} + 2.67\mathbf{k} \text{ (ft)},$$

and then add it to the vector from the origin O to B to obtain the vector from O to C:

$$\mathbf{r}_{OC} = \mathbf{r}_{OB} + \mathbf{r}_{BC} = 7\mathbf{j} + (2.67\mathbf{i} - 4.67\mathbf{j} + 2.67\mathbf{k})$$
$$= 2.67\mathbf{i} + 2.33\mathbf{j} + 2.67\mathbf{k} \text{ (ft)}.$$

The components of this vector are the coordinates of point C.

Now we can determine a unit vector with the same direction as **T**. The vector from C to A is

$$\mathbf{r}_{CA} = (0 - 2.67)\mathbf{i} + (7 - 2.33)\mathbf{j} + (4 - 2.67)\mathbf{k}$$
$$= -2.67\mathbf{i} + 4.67\mathbf{j} + 1.33\mathbf{k} \text{ (ft)},$$

and the unit vector that points from point C toward point A is

$$\mathbf{e}_{CA} = \frac{\mathbf{r}_{CA}}{|\mathbf{r}_{CA}|} = -0.482\mathbf{i} + 0.843\mathbf{j} + 0.241\mathbf{k}.$$

Let T be the tension in the cable AC. Then we can write the vector **T** as

$$\mathbf{T} = T\mathbf{e}_{CA} = T(-0.482\mathbf{i} + 0.843\mathbf{j} + 0.241\mathbf{k}).$$

Determining **T** *and* **N**: Substituting our expressions for $\mathbf{e}_{BD}$ and **T** in terms of their components into Eq. (3.6),

$$\mathbf{e}_{BD} \cdot (\mathbf{T} - 100\mathbf{j})$$
$$= \left[\frac{4}{9}\mathbf{i} - \frac{7}{9}\mathbf{j} + \frac{4}{9}\mathbf{k}\right] \cdot \left[-0.482T\mathbf{i} + (0.843T - 100)\mathbf{j} + 0.241T\mathbf{k}\right]$$
$$= -0.762T + 77.8 = 0,$$

we obtain the tension $T = 102$ lb.

Now we can determine the force exerted on the slider by the bar by using Eq. (3.5):

$$\mathbf{N} = -\mathbf{T} + 100\mathbf{j} = -102(-0.482\mathbf{i} + 0.843\mathbf{j} + 0.241\mathbf{k}) + 100\mathbf{j}$$
$$= 49.1\mathbf{i} + 14.0\mathbf{j} - 24.6\mathbf{k} \text{ (lb)}.$$

Problems

3.59 If the coordinates of point A in Example 3.4 are changed to $(0, -2, 0)$ m, what are the tensions in cables AB, AC, and AD?

3.60 The force $\mathbf{F} = 5\mathbf{i}$ (kN) acts on point A where the cables AB, AC, and AD are joined. What are the tensions in the three cables?

    Strategy: Isolate part of the cable system near point A. See Example 3.4.

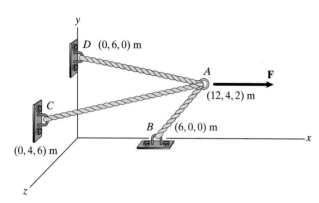

P3.60

3.61 The cables in Problem 3.60 will safely support a tension of 25 kN. Based on this criterion, what is the largest safe magnitude of the force $\mathbf{F} = F\mathbf{i}$?

3.62 To support the tent, the tension in the rope AB must be 40 lb. What are the tensions in the ropes AC, AD, and AE?

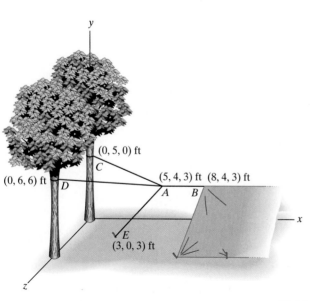

P3.62

3.63 The bulldozer exerts a force $\mathbf{F} = 2\mathbf{i}$ (kip) at A. What are the tensions in cables AB, AC, and AD?

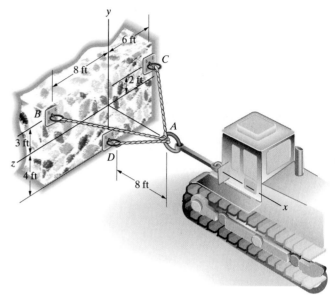

P3.63

3.64 Prior to its launch, a balloon carrying a set of experiments to high altitude is held in place by groups of student volunteers holding the tethers at B, C, and D. The mass of the balloon, experiments package, and the gas it contains is 90 kg, and the buoyancy force on the balloon is 1000 N. The supervising professor conservatively estimates that each student can exert at least a 40-N tension on the tether for the necessary length of time. Based on this estimate, what minimum numbers of students are needed at B, C, and D?

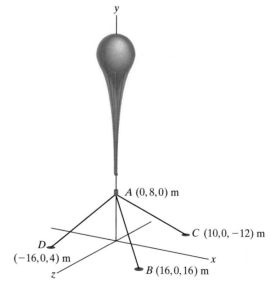

P3.64

3.65 The 20-kg mass is suspended by cables attached to three vertical 2-m posts. Point A is at $(0, 1.2, 0)$ m. Determine the tensions in cables AB, AC, and AD.

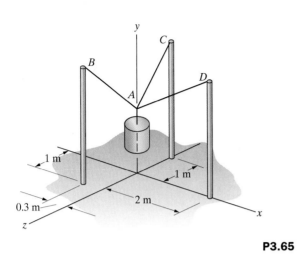

P3.65

3.66 The weight of the horizontal wall section is $W = 20,000$ lb. Determine the tensions in the cables AB, AC, and AD.

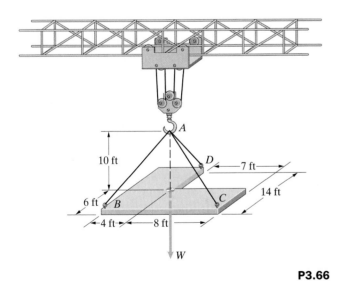

P3.66

3.67 In Problem 3.66, each cable will safely support a tension of 40,000 lb. Based on this criterion, what is the largest safe value of the weight W?

3.68 The 680-kg load suspended from the helicopter is in equilibrium. The aerodynamic drag force on the load is horizontal. The

y axis is vertical, and cable OA lies in the x-y plane. Determine the magnitude of the drag force and the tension in cable OA.

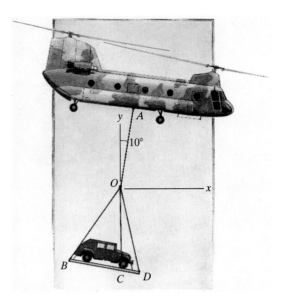

P3.68

3.69 In Problem 3.68, the coordinates of the three cable attachment points B, C, and D are $(-3.3, -4.5, 0)$ m, $(1.1, -5.3, 1)$ m, and $(1.6, -5.4, -1)$ m, respectively. What are the tensions in cables OB, OC, and OD?

3.70 The small sphere A weighs 20 lb, and its coordinates are $(4, 0, 6)$ ft. It is supported by two smooth flat plates labeled 1 and 2 and the cable AB. The unit vector $\mathbf{e}_1 = \frac{4}{9}\mathbf{i} + \frac{7}{9}\mathbf{j} + \frac{4}{9}\mathbf{k}$ is perpendicular to plate 1, and the unit vector $\mathbf{e}_2 = -\frac{9}{11}\mathbf{i} + \frac{2}{11}\mathbf{j} + \frac{6}{11}\mathbf{k}$ is perpendicular to plate 2. What is the tension in the cable?

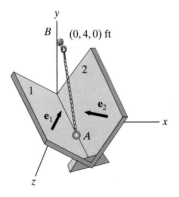

P3.70

3.71 The 1350-kg car is at rest on a plane surface. The unit vector $\mathbf{e}_n = 0.231\mathbf{i} + 0.923\mathbf{j} + 0.308\mathbf{k}$ is perpendicular to the surface.

The *y* axis points upward. Determine the magnitudes of the normal and friction forces the car's wheels exert on the surface.

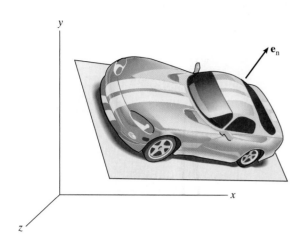

P3.71

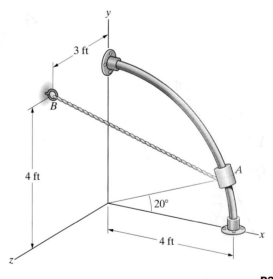

P3.74

3.72 The system shown anchors a stanchion of a cable-suspended roof. If the tension in cable *AB* is 900 kN, what are the tensions in cables *EF* and *EG*?

3.75 The 100-lb slider at *A* is held in place on the smooth circular bar by the cable *AB*. The circular bar is contained in the *x–y* plane.
(a) Determine the tension in the cable.
(b) Determine the normal force exerted on the slider by the bar.

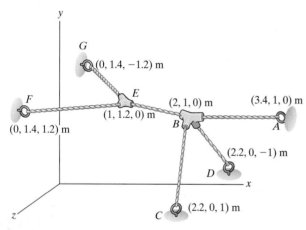

P3.72

3.73 The cables of the system in Problem 3.72 will each safely support a tension of 1500 kN. Based on this criterion, what is the largest safe value of the tension in cable *AB*?

3.74 The 200-kg slider at *A* is held in place on the smooth vertical bar by the cable *AB*.
(a) Determine the tension in the cable.
(b) Determine the force exerted on the slider by the bar.

P3.75

3.76 The cable *AB* keeps the 8-kg collar *A* in place on the smooth bar *CD*. The *y* axis points upward. What is the tension in the cable?

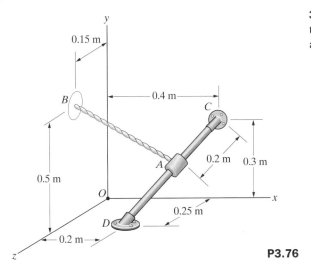

P3.76

3.77 In Problem 3.76, determine the magnitude of the normal force exerted on the collar *A* by the smooth bar.

3.78 The 10-kg collar *A* and 20-kg collar *B* are held in place on the smooth bars by the 3-m cable from *A* to *B* and the force *F* acting on *A*. The force *F* is parallel to the bar. Determine *F*.

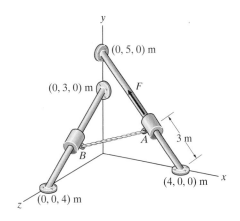

P3.78

Chapter Summary

In this chapter we discussed the forces that occur frequently in engineering applications and introduced two of the most important concepts in mechanics: the free-body diagram and equilibrium. By drawing free-body diagrams and applying the vector techniques developed in Chapter 2, we showed how unknown forces acting on objects in equilibrium can be determined from the condition that the sum of the external forces must equal zero. The sum of the moments of the external forces on an object in equilibrium must also equal zero, and this condition can be used to obtain additional information about unknown forces on objects. We will discuss moments of forces in Chapter 4. We will then apply equilibrium to individual objects in Chapter 5 and to structures in Chapter 6.

The straight line coincident with a force vector is called the *line of action* of the force. A system of forces is *coplanar*, or *two-dimensional*, if the lines of action of the forces lie in a plane. Otherwise, it is *three-dimensional*. A system of forces is *concurrent* if the lines of action of the forces intersect at a point and *parallel* if the lines of action are parallel.

An object is subjected to an *external force* if the force is exerted by a different object. When one part of an object is subjected to a force by another part of the same object, the force is *internal*.

A *body force* acts on the volume of an object, and a *surface* or *contact force* acts on its surface.

Gravitational Forces

The weight of an object is related to its mass by $W = mg$, where $g = 9.81$ m/s^2 in SI units and $g = 32.2$ ft/s^2 in U.S. Customary units.

Surfaces

Two surfaces in contact exert forces on each other that are equal in magnitude and opposite in direction. Each force can be resolved into the *normal force* and the *friction force*. If the friction force is negligible in comparison to the normal force, the surfaces are said to be *smooth*. Otherwise, they are *rough*.

Ropes and Cables

A rope or cable attached to an object exerts a force on the object whose magnitude is equal to the tension and whose line of action is parallel to the rope or cable at the point of attachment.

A *pulley* is a wheel with a grooved rim that can be used to change the direction of a rope or cable. When a pulley can turn freely and the rope or cable either is stationary or turns the pulley at a constant rate, the tension is approximately the same on both sides of the pulley.

Springs

The force exerted by a *linear spring* is

$$|\mathbf{F}| = k|L - L_0|, \qquad \text{Eq. (3.1)}$$

where k is the *spring constant*, L is the length of the spring, and L_0 is its unstretched length.

1. Choose an object to isolate.

Free-Body Diagrams

A free-body diagram is a drawing of an object in which the object is isolated from its surroundings and the external forces acting on the object are shown. Drawing a free-body diagram requires the steps shown in Figs. 1–3. A coordinate system must be chosen to express the forces on the isolated object in terms of components.

2. Draw the isolated object.

Equilibrium

If an object is in equilibrium, the sum of the external forces acting on it is zero:

$$\Sigma \mathbf{F} = \mathbf{0}. \qquad \text{Eq. (3.2)}$$

This implies that the sums of the external forces in the x, y, and z directions each equal zero:

3. Show the external forces.

$$\Sigma F_x = 0, \qquad \Sigma F_y = 0, \qquad \Sigma F_z = 0, \qquad \text{Eqs. (3.4)}$$

Review Problems

3.79 The 100-lb crate is held in place on the smooth surface by the rope AB. Determine the tension in the rope and the magnitude of the normal force exerted on the crate by the surface.

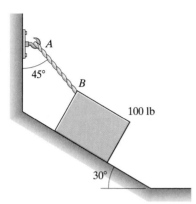

P3.79

3.80 The system shown is called Russell's traction. If the sum of the downward forces exerted at A and B by the patient's leg is 32.2 lb, what is the weight W?

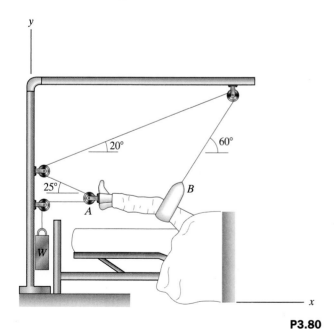

P3.80

3.81 A heavy rope used as a hawser for a cruise ship sags as shown. If it weighs 200 lb, what are the tensions in the rope at A and B?

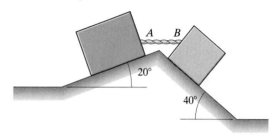

P3.81

3.82 The cable AB is horizontal, and the box on the right weighs 100 lb. The surfaces are smooth.

(a) What is the tension in the cable?
(b) What is the weight of the box on the left?

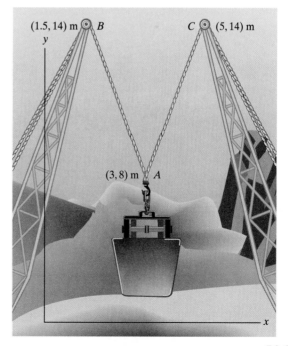

P3.82

3.83 A concrete bucket used at a construction site is supported by two cranes. The 100-kg bucket contains 500 kg of concrete. Determine the tensions in the cables AB and AC.

P3.83

3.84 The mass of the suspended object A is m_A and the masses of the pulleys are negligible. Determine the force T necessary for the system to be in equilibrium.

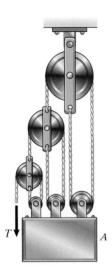

P3.84

3.85 The assembly A, including the pulley, weighs 60 lb. What force F is necessary for the system to be in equilibrium?

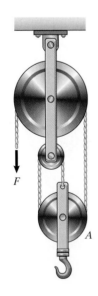

P3.85

3.86 The mass of block A is 42 kg, and the mass of block B is 50 kg. The surfaces are smooth. If the blocks are in equilibrium, what is the force F?

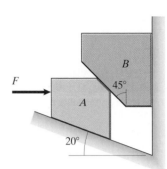

P3.86

3.87 Cable AB is attached to the top of the vertical 3-m post, and its tension is 50 kN. What are the tensions in cables AO, AC, and AD?

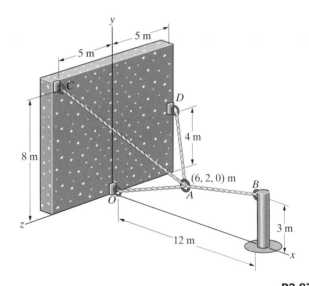

P3.87

3.88 The 1350-kg car is at rest on a plane surface with its brakes locked. The unit vector $\mathbf{e}_n = 0.231\mathbf{i} + 0.923\mathbf{j} + 0.308\mathbf{k}$ is perpendicular to the surface. The y axis points upward. The direction cosines of the cable from A to B are $\cos\theta_x = -0.816$, $\cos\theta_y = 0.408$, $\cos\theta_z = -0.408$, and the tension in the cable is 1.2 kN. Determine the magnitudes of the normal and friction forces the car's wheels exert on the surface.

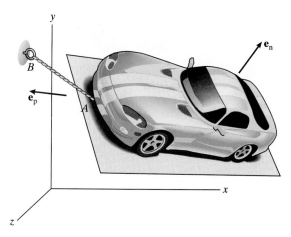

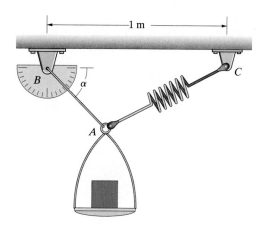

P3.88

𝒟esign Experience A possible design for a simple scale to weigh objects is shown. The length of the string AB is 0.5 m. When an object is placed in the pan, the spring stretches and the string AB rotates. The object's weight can be determined by observing the change in the angle α.

(a) Assume that objects with masses in the range 0.2–2 kg are to be weighed. Choose the unstretched length and spring constant

of the spring in order to obtain accurate readings for weights in the desired range. (Neglect the weights of the pan and spring. Notice that a significant change in the angle α is needed to determine the weight accurately.)

(b) Suppose that you can use the same components—the pan, protractor, a spring, string—and also one or more pulleys. Suggest another possible configuration for the scale. Use statics to analyze your proposed configuration and compare its accuracy with that of the configuration shown for objects with masses in the range 0.2–2 kg.

Loads lifted by a building crane can exert large moments that the crane's structure must support. In this chapter we calculate moments of forces and analyze systems of forces and moments.

Systems of Forces and Moments

The effects of forces can depend not only on their magnitudes and directions but also on the moments, or torques, they exert. If an object is in equilibrium, the moment about any point due to the forces acting on the object is zero. Before continuing our discussion of free-body diagrams and equilibrium, we must explain how to calculate moments and introduce the concept of equivalent systems of forces and moments.

4.1 Two-Dimensional Description of the Moment

Consider a force of magnitude F and a point P, and let's view them in the direction perpendicular to the plane containing the force vector and the point (Fig. 4.1a). The *magnitude of the moment* of the force about P is DF, where D is the perpendicular distance from P to the line of action of the force (Fig. 4.1b). In this example, the force would tend to cause counterclockwise rotation about point P. That is, if we imagine the force acts on an object that can rotate about point P, the force would tend to cause counterclockwise rotation (Fig. 4.1c). We say that the *sense of the moment* is counterclockwise. *We define counterclockwise moments to be positive and clockwise moments to be negative.* (This is the usual convention, although we occasionally encounter situations in which it is more convenient to define clockwise moments to be positive.) Thus the moment of the force about P is

$$M_P = DF. \tag{4.1}$$

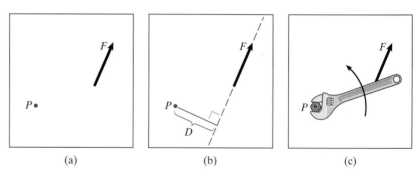

Figure 4.1
(a) The force and point P.
(b) The perpendicular distance D from point P to the line of action of F.
(c) The sense of the moment is counterclockwise.

Notice that if the line of action of F passes through P, the perpendicular distance $D = 0$ and the moment of F about P is zero.

The dimensions of the moment are (distance) × (force). For example, moments can be expressed in newton-meters in SI units and in foot-pounds in U.S. Customary units.

Suppose that you want to place a television set on a shelf, and you aren't certain the attachment of the shelf to the wall is strong enough to support it. Instinctively, you place it near the wall (Fig. 4.2a), knowing that the attachment is more likely to fail if you place it away from the wall (Fig. 4.2b). What is the difference in the two cases? The magnitude and direction of the force exerted on the shelf by the weight of the television are the same in each case, but the moments exerted on the attachment are different. The moment exerted about P by its weight when it is near the wall, $M_P = -D_1 W$, is smaller in magnitude than the moment about P when it is placed away from the wall, $M_P = -D_2 W$.

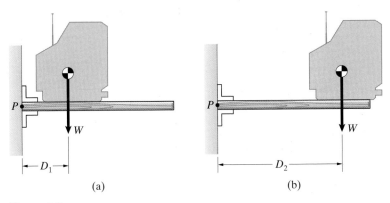

(a)   (b)

Figure 4.2
It is better to place the television near the wall (a) instead of away from it
(b) because the moment exerted on the support at P is smaller.

The method we describe in this section can be used to determine the sum
of the moments of a system of forces about a point if the forces are two-
dimensional (coplanar) and the point lies in the same plane. For example,
consider the construction crane shown in Fig. 4.3. The sum of the moments
exerted about point P by the load W_1 and the counterweight W_2 is

$$\Sigma M_P = D_1 W_1 - D_2 W_2.$$

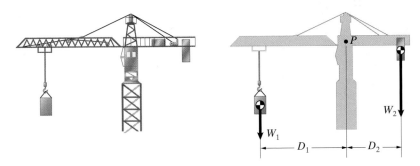

Figure 4.3
A tower crane used in the construction of high-rise buildings.

This moment tends to cause the top of the vertical tower to rotate and could
cause it to collapse. If the distance D_2 is adjusted so that $D_1 W_1 = D_2 W_2$, the
moment about point P due to the load and the counterweight is zero.
    If a force is resolved into components, the moment of the force about a
point P is equal to the sum of the moments of its components about P. We
prove this very useful result in the next section.

Study Questions

1. How do you determine the magnitude of the moment of a force about a point?
2. The moment of a force about a point is defined to be positive if its sense is
    counterclockwise. What does that mean?
3. If the line of action of a force passes through a point P, what do you know
    about the moment of the force about P?

Example 4.1

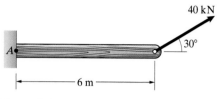

Figure 4.4

Determining the Moment of a Force

What is the moment of the 40-kN force in Fig. 4.4 about point A?

Strategy

We can calculate the moment in two ways: by determining the perpendicular distance from point A to the line of action of the force or by resolving the force into components and determining the sum of the moments of the components about A.

Solution

First Method From Fig. a, the perpendicular distance from A to the line of action of the force is

$$D = 6 \sin 30° = 3 \text{ m}.$$

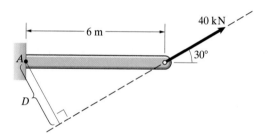

(a) Determining the perpendicular distance D.

The magnitude of the moment of the force about A is $(3 \text{ m})(40 \text{ kN}) = 120$ kN-m, and the sense of the moment about A is counterclockwise. Therefore the moment is

$$M_A = 120 \text{ kN-m}.$$

Second Method In Fig. b, we resolve the force into horizontal and vertical components. The perpendicular distance from A to the line of action of the horizontal component is zero, so the horizontal component exerts no moment about A. The magnitude of the moment of the vertical component about A is $(6 \text{ m})(40 \sin 30° \text{ kN}) = 120$ kN-m, and the sense of its moment about A is counterclockwise. The moment is

$$M_A = 120 \text{ kN-m}.$$

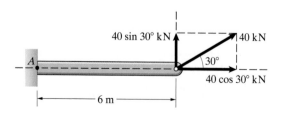

(b) Resolving the force into components.

Example 4.2

Moment of a System of Forces

Four forces act on the machine part in Fig. 4.5. What is the sum of the moments of the forces about the origin O?

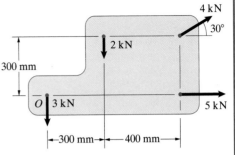

Figure 4.5

Strategy

We can determine the moments of the forces about point O directly from the given information except for the 4-kN force. We will determine its moment by resolving it into components and summing the moments of the components.

Solution

Moment of the 3-kN Force The line of action of the 3-kN force passes through O. It exerts no moment about O.

Moment of the 5-kN Force The line of action of the 5-kN force also passes through O. It too exerts no moment about O.

Moment of the 2-kN Force The perpendicular distance from O to the line of action of the 2-kN force is 0.3 m, and the sense of the moment about O is clockwise. The moment of the 2-kN force about O is

$$-(0.3 \text{ m})(2 \text{ kN}) = -0.600 \text{ kN-m}.$$

(Notice that we converted the perpendicular distance from millimeters into meters, obtaining the result in terms of kilonewton-meters.)

Moment of the 4-kN Force In Fig. a, we introduce a coordinate system and resolve the 4-kN force into x and y components. The perpendicular distance from O to the line of action of the x component is 0.3 m, and the sense of the moment about O is clockwise. The moment of the x component about O is

$$-(0.3 \text{ m})(4 \cos 30° \text{ kN}) = -1.039 \text{ kN-m}.$$

The perpendicular distance from point O to the line of action of the y component is 0.7 m, and the sense of the moment about O is counterclockwise. The moment of the y component about O is

$$(0.7 \text{ m})(4 \sin 30° \text{ kN}) = 1.400 \text{ kN-m}.$$

The sum of the moments of the four forces about point O is

$$\Sigma M_0 = -0.600 - 1.039 + 1.400 = -0.239 \text{ kN-m}.$$

The four forces exert a 0.239 kN-m clockwise moment about point O.

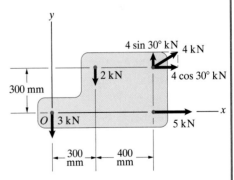

(a) Resolving the 4-kN force into components.

Example 4.3

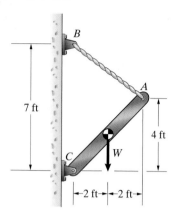

7 ft

4 ft

W

C

—2 ft—|—2 ft—

Figure 4.6

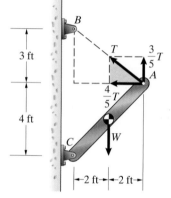

(a) Resolving the force exerted by the
cable into horizontal and vertical
components.

Summing Moments to Determine an Unknown Force

The weight $W = 300$ lb (Fig. 4.6). The sum of the moments about C due to the weight W and the force exerted on the bar CA by the cable AB is zero. What is the tension in the cable?

Strategy

Let T be the tension in cable AB. Using the given dimensions, we can express the horizontal and vertical components of the force exerted on the bar by the cable in terms of T. Then by setting the sum of the moments about C due to the weight of the bar and the force exerted by the cable equal to zero, we can obtain an equation for T.

Solution

Using similar triangles, we resolve the force exerted on the bar by the cable into horizontal and vertical components (Fig. a). The sum of the moments about C due to the weight of the bar and the force exerted by the cable AB is

$$\Sigma M_C = 4\left(\frac{4}{5}T\right) + 4\left(\frac{3}{5}T\right) - 2W = 0.$$

Solving for T, we obtain

$$T = 0.357W = 107.1 \text{ lb.}$$

Problems

4.1 Determine the moment of the 50-N force about (a) point A, (b) point B.

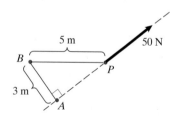

P4.1

4.2 The radius of the pulley is $r = 0.2$ m and it is not free to rotate. The magnitudes of the forces are $|\mathbf{F}_A| = 140$ N and $|\mathbf{F}_B| = 180$ N.

(a) What is the moment about the center of the pulley due to the force $\mathbf{F}_A$?
(b) What is the sum of the moments about the center of the pulley due to the forces $\mathbf{F}_A$ and $\mathbf{F}_B$?

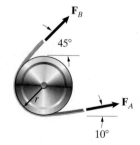

P4.2

4.3 The wheels of the overhead crane exert downward forces on the horizontal I-beam at B and C. If the force at B is 40 kip and the force at C is 44 kip, determine the sum of the moments of the forces on the beam about (a) point A, (b) point D.

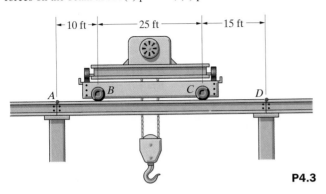

P4.3

4.4 If you exert a 90-N force on the wrench in the direction shown, what moment do you exert about the center of the nut? Compare your answer to the moment exerted if you exert the 90-N force perpendicular to the shaft of the wrench.

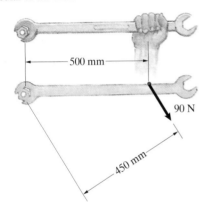

P4.4

4.5 If you exert a force F on the wrench in the direction shown and a 50 N-m moment is required to loosen the nut, what force F must you apply?

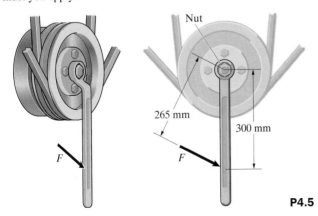

P4.5

4.6 The support at the left end of the beam will fail if the moment about P due to the 20-kN force exceeds 35 kN-m. Based

on this criterion, what is the maximum safe value of the angle α in the range $0 \le \alpha \le 90°$?

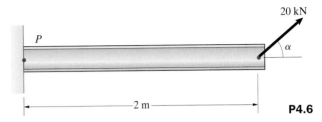

P4.6

4.7 The gears exert 200-N forces on each other at their point of contact.
(a) Determine the moment about A due to the force exerted on the left gear.
(b) Determine the moment about B due to the force exerted on the right gear.

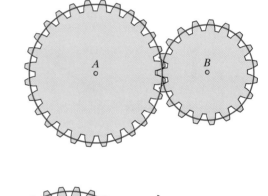

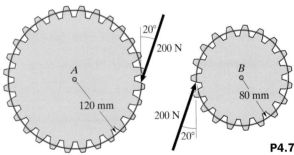

P4.7

4.8 The support at the left end of the beam will fail if the moment about A of the 15-kN force F exceeds 18 kN-m. Based on this criterion, what is the largest allowable length of the beam?

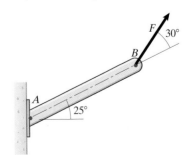

P4.8

4.9 Determine the moment of the 80-lb force about P.

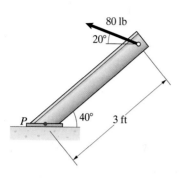

P4.9

4.10 The 20-N force F exerts a 20 N-m counterclockwise moment about P.
(a) What is the perpendicular distance from P to the line of action of F?
(b) What is the angle α?

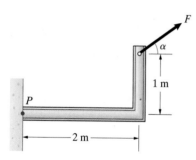

P4.10

4.11 The lengths of bars AB and AC are 350 mm and 450 mm respectively. The magnitude of the vertical force at A is $|\mathbf{F}| = 600$ N. Determine the moment of $\mathbf{F}$ about B and about C.

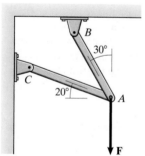

P4.11

4.12 Two students attempt to loosen a lug nut with a lug wrench. One of the students exerts the two 60-lb forces; the other, having to reach around his friend, can only exert the two 30-lb forces. What torque (moment) do they exert on the nut?

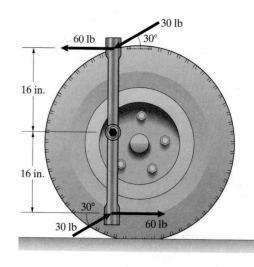

P4.12

4.13 The two students described in Problem 4.12, having failed to loosen the lug nut, try a different tactic. One of them stands on the lug wrench, exerting a 150-lb force on it. The other pulls on the wrench with the force F. If a torque of 245 ft-lb is required to loosen the lug nut, what force F must the student exert?

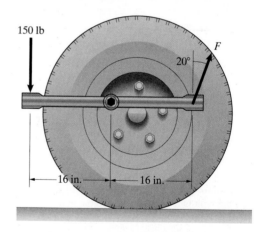

P4.13

4.14 The moment exerted about point E by the weight is 299 in-lb. What moment does the weight exert about point S?

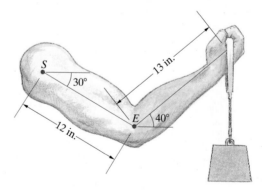

P4.14

4.15 Three forces act on the square plate. Determine the sum of the moments of the forces (a) about A, (b) about B, (c) about C.

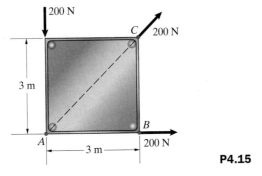

P4.15

4.16 Determine the sum of the moments of the three forces about (a) point A, (b) point B, (c) point C.

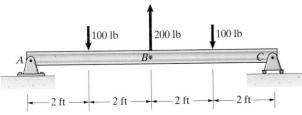

P4.16

4.17 Determine the sum of the moments of the five forces acting on the Howe truss about point A.

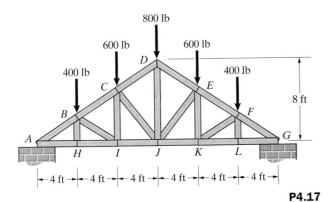

P4.17

4.18 The right support of the truss in Problem 4.17 exerts an upward force of magnitude G. (Assume that the force acts at the right end of the truss.) The sum of the moments about A due to the upward force G and the five downward forces exerted on the truss is zero. What is the force G?

4.19 The sum of the forces F_1 and F_2 is 250 N and the sum of the moments of F_1 and F_2 about B is 700 N-m. What are F_1 and F_2?

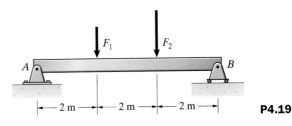

P4.19

4.20 Consider the beam shown in Problem 4.19. If the two forces exert a 140 kN-m clockwise moment about A and a 20 kN-m clockwise moment about B, what are F_1 and F_2?

4.21 The force $F = 140$ lb. The vector sum of the forces acting on the beam is zero, and the sum of the moments about the left end of the beam is zero.
(a) What are the forces A_x, A_y, and B?
(b) What is the sum of the moments about the right end of the beam?

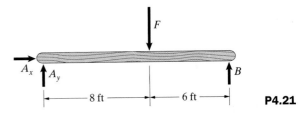

P4.21

4.22 The vector sum of the three forces is zero, and the sum of the moments of the three forces about A is zero.
(a) What are F_A and F_B?
(b) What is the sum of the moments of the three forces about B?

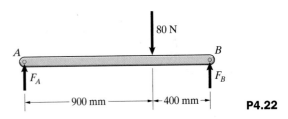

P4.22

4.23 The weights (in ounces) of fish A, B, and C are 2.7, 8.1, and 2.1, respectively. The sum of the moments due to the weights of the fish about the point where the mobile is attached to the ceiling is zero. What is the weight of fish D?

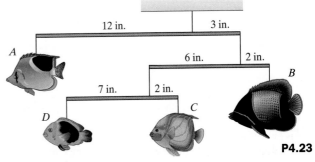

P4.23

4.24 The weight $W = 1.2$ kN. The sum of the moments about A due to W and the force exerted at the end of the bar by the rope is zero. What is the tension in the rope?

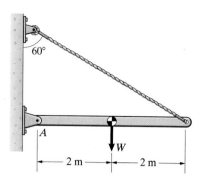

P4.24

4.25 The 160-N weights of the arms AB and BC of the robotic manipulator act at their midpoints. Determine the sum of the moments of the three weights about A.

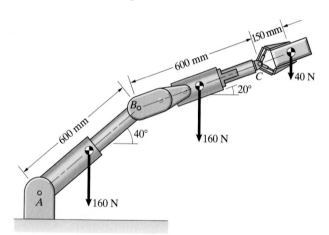

P4.25

4.26 The space shuttle's attitude thrusters exert two forces of magnitude $F = 7.70$ kN. What moment do the thrusters exert about the center of mass G?

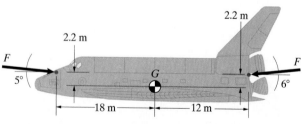

P4.26

4.27 The force F exerts a 200 ft-lb counterclockwise moment about A and a 100 ft-lb clockwise moment about B. What are F and θ?

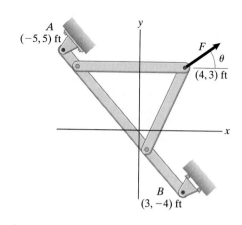

P4.27

4.28 Five forces act on a link in the gear-shifting mechanism of a lawn mower. The vector sum of the five forces on the bar is zero. The sum of their moments about the point where the forces A_x and A_y act is zero.
(a) Determine the forces A_x, A_y, and B.
(b) Determine the sum of the moments of the forces about the point where the force B acts.

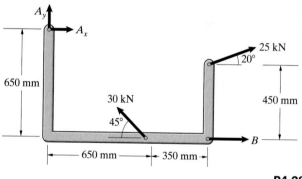

P4.28

4.29 Five forces act on a model truss built by a civil engineering student as part of a design project. The dimensions are $b = 300$ mm and $h = 400$ mm; $F = 100$ N. The sum of the moments of the forces about the point where A_x and A_y act is zero. If the weight of the truss is negligible, what is the force B?

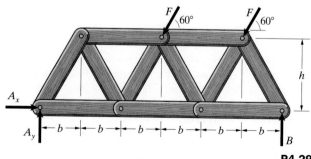

P4.29

4.30 Consider the truss shown in Problem 4.29. The dimensions are $b = 3$ ft and $h = 4$ ft; $F = 300$ lb. The vector sum of the forces acting on the truss is zero, and the sum of the moments of the forces about the point where A_x and A_y act is zero.
(a) Determine the forces A_x, A_y, and B.
(b) Determine the sum of the moments of the forces about the point where the force B acts.

4.31 The mass $m = 70$ kg. What is the moment about A due to the force exerted on the beam at B by the cable?

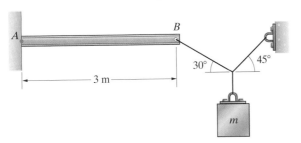

P4.31

4.32 Consider the system shown in Problem 4.31. The beam will collapse at A if the magnitude of the moment about A due to the force exerted on the beam at B by the cable exceeds 2 kN-m. What is the largest mass m that can be suspended?

4.33 The bar AB exerts a force at B that helps support the vertical retaining wall. The force is parallel to the bar. The civil engineer wants the bar to exert a 38 kN-m moment about O. What is the magnitude of the force the bar must exert?

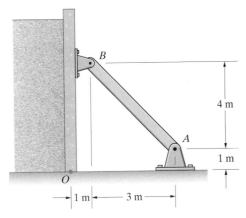

P4.33

4.34 A contestant in a fly-casting contest snags his line in some grass. If the tension in the line is 5 lb, what moment does the force exerted on the rod by the line exert about point H, where he holds the rod?

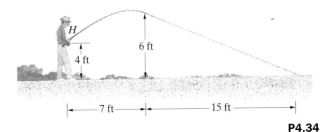

P4.34

4.35 The cables AB and AC help support the tower. The tension in cable AB is 5 kN. The points A, B, C, and O are contained in the same vertical plane.
(a) What is the moment about O due to the force exerted on the tower by cable AB?
(b) If the sum of the moments about O due to the forces exerted on the tower by the two cables is zero, what is the tension in cable AC?

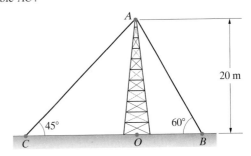

P4.35

4.36 The cable from B to A (the sailboat's forestay) exerts a 230-N force at B. The cable from B to C (the backstay) exerts a 660-N force at B. The bottom of the sailboat's mast is located at $x = 4$ m, $y = 0$. What is the sum of the moments about the bottom of the mast due to the forces exerted at B by the forestay and backstay?

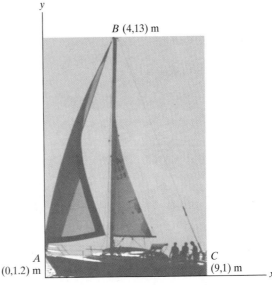

P4.36

4.37 The tension in each cable is the same. The forces exerted on the beam by the three cables exert a 1.2 kN-m counterclockwise moment about O. What is the tension in the cables?

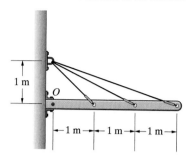

P4.37

4.38 The tension in cable AB is 300 lb. The sum of the moments about O due to the forces exerted on the beam by the two cables is zero. What is the magnitude of the sum of the forces exerted on the beam by the two cables?

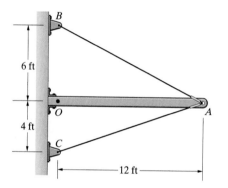

P4.38

4.39 The beam shown in Problem 4.38 will safely support the forces exerted by the two cables at A if the magnitude of the horizontal component of the total force exerted at A does not exceed 1000 lb and the sum of the moments about O due to the forces exerted by the cables equals zero. Based on these criteria, what are the maximum permissible tensions in the two cables?

4.40 The hydraulic cylinder BC exerts a 300-kN force on the boom of the crane at C. The force is parallel to the cylinder. What is the moment of the force about A?

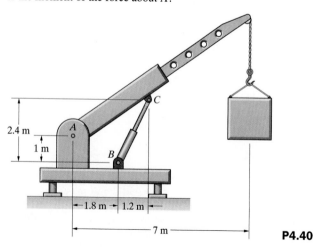

P4.40

4.41 The hydraulic cylinder BC exerts a 2200-lb force on the boom of the crane at C. The force is parallel to the cylinder. The angle $\alpha = 40°$. What is the moment of the force about A?

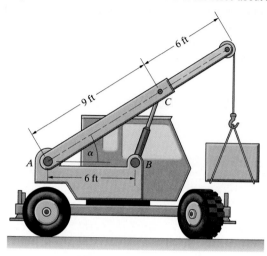

P4.41

4.42 The hydraulic cylinder BC in Problem 4.41 exerts a 2200-lb force on the boom of the crane at C. The force is parallel to the cylinder. The cable supporting the suspended crate exerts a downward force at the end of the boom equal to the weight of the crate. The angle $\alpha = 35°$. If the sum of the moments about A due to the two forces exerted on the boom is zero, what is the weight of the crate?

4.43 The unstretched length of the spring is 1 m, and the spring constant is $k = 20$ N/m. If $\alpha = 30°$, what is the moment about A due to the force exerted by the spring on the circular bar at B?

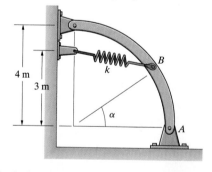

P4.43

4.44 The hydraulic cylinder exerts an 8-kN force at B that is parallel to the cylinder and points from C toward B. Determine the moments of the force about points A and D.

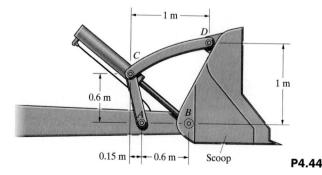

P4.44

4.2 The Moment Vector

The moment of a force about a point is a vector. In this section we define this vector and explain how it is evaluated. We then show that when we use the two-dimensional description of the moment described in Section 4.1, we are specifying the magnitude and direction of the moment vector.

Consider a force vector **F** and point P (Fig. 4.7a). The *moment* of **F** about P is the vector

$$\mathbf{M}_P = \mathbf{r} \times \mathbf{F}. \qquad (4.2)$$

where **r** is a position vector from P to *any* point on the line of action of **F** (Fig. 4.7b).

(a)

Magnitude of the Moment

From the definition of the cross product, the magnitude of $\mathbf{M}_P$ is

$$|\mathbf{M}_P| = |\mathbf{r}||\mathbf{F}| \sin\theta,$$

where θ is the angle between the vectors **r** and **F** when they are placed tail to tail. The perpendicular distance from P to the line of action of **F** is $D = |\mathbf{r}| \sin\theta$ (Fig. 4.7c). Therefore the magnitude of the moment $\mathbf{M}_P$ equals the product of the perpendicular distance from P to the line of action of **F** and the magnitude of **F**:

$$|\mathbf{M}_P| = D|\mathbf{F}|. \qquad (4.3)$$

Notice that if you know the vectors $\mathbf{M}_P$ and **F**, you can solve this equation for the perpendicular distance D.

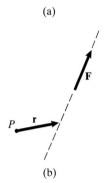

(b)

Sense of the Moment

We know from the definition of the cross product that $\mathbf{M}_P$ is perpendicular to both **r** and **F**. That means that $\mathbf{M}_P$ is perpendicular to the plane containing P and **F** (Fig. 4.8a). Notice in this figure that we denote a moment by a circular arrow around the vector.

The direction of $\mathbf{M}_P$ also indicates the sense of the moment: If you point the thumb of your right hand in the direction of $\mathbf{M}_P$, the "arc" of your fingers indicates the sense of the rotation that **F** tends to cause about P (Fig. 4.8b).

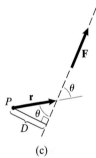

(c)

Figure 4.7
(a) The force **F** and point P.
(b) A vector **r** from P to a point on the line of action of **F**.
(c) The angle θ and the perpendicular distance D.

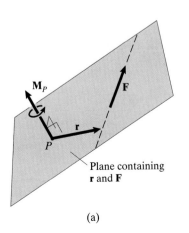

Plane containing
r and **F**

(a)

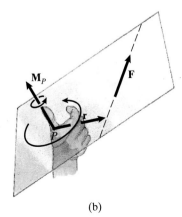

(b)

Figure 4.8
(a) $\mathbf{M}_P$ is perpendicular to the plane containing P and **F**.
(b) The direction of $\mathbf{M}_P$ indicates the sense of the moment.

The result obtained from Eq. (4.2) doesn't depend on where the vector **r** intersects the line of action of **F**. Instead of using the vector **r** in Fig. 4.9a, we could use the vector **r'** in Fig. 4.9b. The vector **r** = **r'** + **u**, where **u** is parallel to **F** (Fig. 4.9c). Therefore

$$\mathbf{r} \times \mathbf{F} = (\mathbf{r'} + \mathbf{u}) \times \mathbf{F} = \mathbf{r'} \times \mathbf{F}$$

because the cross product of the parallel vectors **u** and **F** is zero.

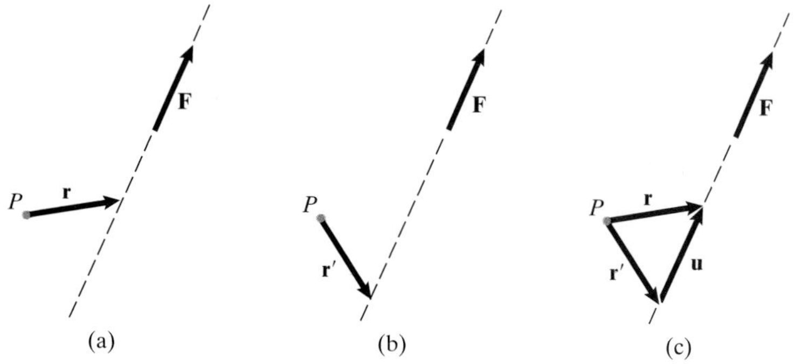

Figure 4.9
(a) A vector **r** from P to the line of action of **F**.
(b) A different vector **r'**.
(c) **r** = **r'** + **u**.

(a)   (b)   (c)

In summary, the moment of a force **F** about a point P has three properties:

1. The magnitude of $\mathbf{M}_P$ is equal to the product of the magnitude of **F** and the perpendicular distance from P to the line of action of **F**. If the line of action of **F** passes through P, $\mathbf{M}_P = \mathbf{0}$.

2. $\mathbf{M}_P$ is perpendicular to the plane containing P and **F**.

3. The direction of $\mathbf{M}_P$ indicates the sense of the moment through a right-hand rule (Fig. 4.8b). Since the cross product is not commutative, you must be careful to maintain the correct sequence of the vectors in the equation $\mathbf{M}_P = \mathbf{r} \times \mathbf{F}$.

Let us determine the moment of the force **F** in Fig. 4.10a about the point P. Since the vector **r** in Eq. (4.2) can be a position vector to any point on the line of action of **F**, we can use the vector from P to the point of application of **F** (Fig. 4.10b):

$$\mathbf{r} = (12 - 3)\mathbf{i} + (6 - 4)\mathbf{j} + (-5 - 1)\mathbf{k} = 9\mathbf{i} + 2\mathbf{j} - 6\mathbf{k} \text{ (ft).}$$

The moment is

$$\mathbf{M}_P = \mathbf{r} \times \mathbf{F} = \begin{vmatrix} \mathbf{i} & \mathbf{j} & \mathbf{k} \\ 9 & 2 & -6 \\ 4 & 4 & 7 \end{vmatrix} = 38\mathbf{i} - 87\mathbf{j} + 28\mathbf{k} \text{ (ft-lb).}$$

The magnitude of $\mathbf{M}_P$,

$$|\mathbf{M}_P| = \sqrt{(38)^2 + (-87)^2 + (28)^2} = 99.0 \text{ ft-lb,}$$

equals the product of the magnitude of **F** and the perpendicular distance D from point P to the line of action of **F**. Therefore

$$D = \frac{|\mathbf{M}_P|}{|\mathbf{F}|} = \frac{99.0 \text{ ft-lb}}{9 \text{ lb}} = 11.0 \text{ ft.}$$

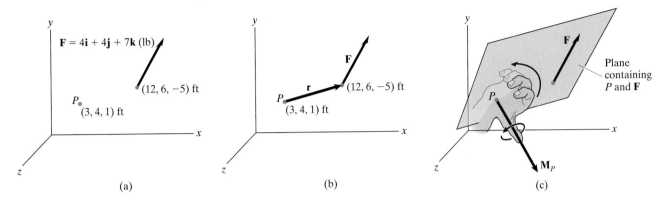

Figure 4.10
(a) A force **F** and point P.
(b) The vector **r** from P to the point of application of **F**.
(c) $\mathbf{M}_P$ is perpendicular to the plane containing P and **F**. The right-hand rule
  indicates the sense of the moment.

The direction of $\mathbf{M}_P$ tells us both the orientation of the plane containing P
and **F** and the sense of the moment (Fig. 4.10c).

Relation to the Two-Dimensional Description

If our view is perpendicular to the plane containing the point P and the force **F**,
the two-dimensional description of the moment we used in Section 4.1 speci-
fies both the magnitude and direction of $\mathbf{M}_P$. In this situation, $\mathbf{M}_P$ is perpendicular to the page, and the right-hand rule indicates whether it points out of
or into the page.

For example, in Fig. 4.11a, the view is perpendicular to the x–y plane
and the 10-N force is contained in the x–y plane. Suppose that we want to

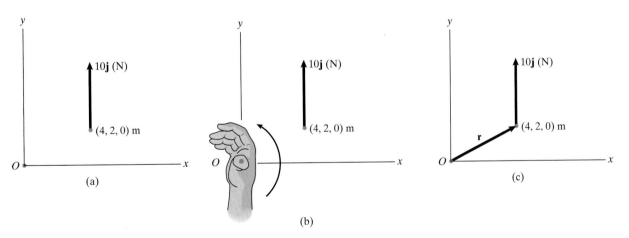

Figure 4.11
(a) The force is contained in the x–y plane.
(b) The sense of the moment indicates that $\mathbf{M}_O$ points out of the page.
(c) The vector **r** from O to the point of application of **F**.

determine the moment of the force about the origin O. The perpendicular distance from O to the line of action of the force is 4 m. The two-dimensional description of the moment of the force about O is that its magnitude is $(4 \text{ m})(10 \text{ N}) = 40$ N-m and its sense is counterclockwise, or

$$M_O = 40 \text{ N-m}.$$

That tells us that the magnitude of the vector $\mathbf{M}_O$ is 40 N-m, and the right-hand rule (Fig. 4.11b) indicates that it points out of the page. Therefore

$$\mathbf{M}_O = 40\mathbf{k} \text{ (N-m)}.$$

We can confirm this result by using Eq. (4.2). If we let $\mathbf{r}$ be the vector from O to the point of application of the force (Fig. 4.11c),

$$\mathbf{M}_O = \mathbf{r} \times \mathbf{F} = (4\mathbf{i} + 2\mathbf{j}) \times 10\mathbf{j} = 40\mathbf{k} \text{ (N-m)}.$$

As this example illustrates, the two-dimensional description of the moment determines the moment vector. The converse is also true. The magnitude of $\mathbf{M}_O$ equals the product of the magnitude of the force and the perpendicular distance from O to the line of action of the force, 40 N-m, and the direction of $\mathbf{M}_O$ indicates that the sense of the moment is counterclockwise (Fig. 4.11b).

Varignon's Theorem

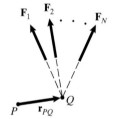

Figure 4.12
A system of concurrent forces and a point P.

Let $\mathbf{F}_1, \mathbf{F}_2, \ldots, \mathbf{F}_N$ be a concurrent system of forces whose lines of action intersect at a point Q. The moment of the system about a point P is

$$\left(\mathbf{r}_{PQ} \times \mathbf{F}_1\right) + \left(\mathbf{r}_{PQ} \times \mathbf{F}_2\right) + \cdots + \left(\mathbf{r}_{PQ} \times \mathbf{F}_N\right)$$
$$= \mathbf{r}_{PQ} \times \left(\mathbf{F}_1 + \mathbf{F}_2 + \cdots + \mathbf{F}_N\right),$$

where $\mathbf{r}_{PQ}$ is the vector from P to Q (Fig. 4.12). This result, known as *Varignon's theorem*, follows from the distributive property of the cross product, Eq. (2.31). It confirms that the moment of a force about a point P is equal to the sum of the moments of its components about P.

Study Questions

1. When you use the equation $\mathbf{M}_P = \mathbf{r} \times \mathbf{F}$ to determine the moment of a force $\mathbf{F}$ about a point P, how do you choose the vector $\mathbf{r}$?

2. If you know the components of the vector $\mathbf{M}_P = \mathbf{r} \times \mathbf{F}$, how can you determine the product of the magnitude of $\mathbf{F}$ and the perpendicular distance from P to the line of action of $\mathbf{F}$?

3. How does the direction of the vector $\mathbf{M}_P = \mathbf{r} \times \mathbf{F}$ indicate the sense of the moment of $\mathbf{F}$ about P?

Example 4.4

Two-Dimensional Description and the Moment Vector

Determine the moment of the 400-N force in Fig. 4.13 about O.
(a) What is the two-dimensional description of the moment?
(b) Express the moment as a vector without using Eq. (4.2).
(c) Use Eq. (4.2) to determine the moment.

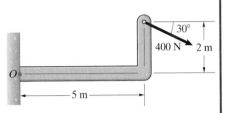

Figure 4.13

Solution

(a) Resolving the force into horizontal and vertical components (Fig. a), the two-dimensional description of the moment is

$$M_O = -(2 \text{ m})(400 \cos 30° \text{ N}) - (5 \text{ m})(400 \sin 30° \text{ N})$$

$$= -1.69 \text{ kN-m}.$$

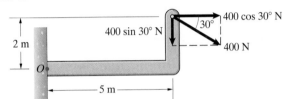

(a) Resolving the force into components.

(b) To express the moment as a vector, we introduce the coordinate system shown in Fig. b. The magnitude of the moment is 1.69 kN-m, and its sense is clockwise. Pointing the arc of the fingers of the right- hand clockwise, the thumb points into the page. Therefore

$$\mathbf{M}_O = -1.69\mathbf{k} \text{ (kN-m)}.$$

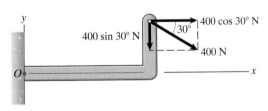

(b) Introducing a coordinate system.

(c) We apply Eq. (4.2):

Choose the Vector r   We can let $\mathbf{r}$ be the vector from O to the point of application of the force (Fig. c):

$$\mathbf{r} = 5\mathbf{i} + 2\mathbf{j} \text{ (m)}.$$

Evaluate r × F   The moment is

$$\mathbf{M}_O = \mathbf{r} \times \mathbf{F} = (5\mathbf{i} + 2\mathbf{j}) \times (400 \cos 30°\mathbf{i} - 400 \sin 30°\mathbf{j})$$

$$= -1.69\mathbf{k} \text{ (kN-m)}.$$

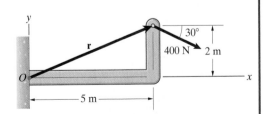

(c) The vector $\mathbf{r}$ from O to the point of application of the force.

Example 4.5

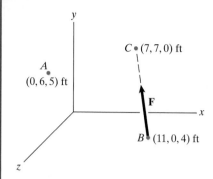

Figure 4.14

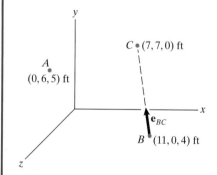

(a) The unit vector $\mathbf{e}_{BC}$.

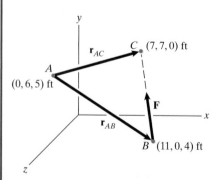

(b) The moment can be determined using either $\mathbf{r}_{AB}$ or $\mathbf{r}_{AC}$.

Determining the Moment and the Perpendicular Distance to the Line of Action

The line of action of the 90-lb force $\mathbf{F}$ in Fig. 4.14 passes through points B and C.
(a) What is the moment of $\mathbf{F}$ about point A?
(b) What is the perpendicular distance from point A to the line of action of $\mathbf{F}$?

Strategy

(a) We must use Eq. (4.2) to determine the moment. Since $\mathbf{r}$ is a vector from A to any point on the line of action of $\mathbf{F}$, we can use either the vector from A to B or the vector from A to C. To demonstrate that we obtain the same result, we will determine the moment using both.
(b) Since the magnitude of the moment is equal to the product of the magnitude of $\mathbf{F}$ and the perpendicular distance from A to the line of action of $\mathbf{F}$, we can use the result of (a) to determine the perpendicular distance.

Solution

(a) To evaluate the cross product in Eq. (4.2), we need the components of $\mathbf{F}$. The vector from B to C is

$$(7 - 11)\mathbf{i} + (7 - 0)\mathbf{j} + (0 - 4)\mathbf{k} = -4\mathbf{i} + 7\mathbf{j} - 4\mathbf{k} \text{ (ft)}.$$

Dividing this vector by its magnitude, we obtain a unit vector $\mathbf{e}_{BC}$ that has the same direction as $\mathbf{F}$ (Fig. a):

$$\mathbf{e}_{BC} = -\frac{4}{9}\mathbf{i} + \frac{7}{9}\mathbf{j} - \frac{4}{9}\mathbf{k}.$$

Now we express $\mathbf{F}$ as the product of its magnitude and $\mathbf{e}_{BC}$:

$$\mathbf{F} = 90\mathbf{e}_{BC} = -40\mathbf{i} + 70\mathbf{j} - 40\mathbf{k} \text{ (lb)}.$$

Choose the Vector r   The position vector from A to B (Fig. b) is

$$\mathbf{r}_{AB} = (11 - 0)\mathbf{i} + (0 - 6)\mathbf{j} + (4 - 5)\mathbf{k} = 11\mathbf{i} - 6\mathbf{j} - \mathbf{k} \text{ (ft)}.$$

Evaluate r × F   The moment of $\mathbf{F}$ about A is

$$\mathbf{M}_A = \mathbf{r}_{AB} \times \mathbf{F} = \begin{vmatrix} \mathbf{i} & \mathbf{j} & \mathbf{k} \\ 11 & -6 & -1 \\ -40 & 70 & -40 \end{vmatrix}$$

$$= 310\mathbf{i} + 480\mathbf{j} + 530\mathbf{k} \text{ (ft-lb)}.$$

Alternative Choice of Position Vector   If we use the vector from A to C instead,

$$\mathbf{r}_{AC} = (7 - 0)\mathbf{i} + (7 - 6)\mathbf{j} + (0 - 5)\mathbf{k} = 7\mathbf{i} + \mathbf{j} - 5\mathbf{k} \text{ (ft)},$$

we obtain the same result:

$$M_A = r_{AC} \times F = \begin{vmatrix} i & j & k \\ 7 & 1 & -5 \\ -40 & 70 & -40 \end{vmatrix}$$

$$= 310i + 480j + 530k \text{ (ft-lb)}.$$

(b) The perpendicular distance is

$$\frac{|M_A|}{|F|} = \frac{\sqrt{(310)^2 + (480)^2 + (530)^2}}{\sqrt{(-40)^2 + (70)^2 + (-40)^2}} = 8.66 \text{ ft}.$$

Example 4.6

Applying the Moment Vector

The cables *AB* and *AC* in Fig. 4.15 extend from an attachment point *A* on the floor to attachment points *B* and *C* in the walls. The tension in cable *AB* is 10 kN, and the tension in cable *AC* is 20 kN. What is the sum of the moments about *O* due to the forces exerted on *A* by the two cables?

Solution

Let F_{AB} and F_{AC} be the forces exerted on the attachment point *A* by the two cables (Fig. a). To express F_{AB} in terms of its components, we determine the position vector from *A* to *B*,

$$(0 - 4)i + (4 - 0)j + (8 - 6)k = -4i + 4j + 2k \text{ (m)},$$

and divide it by its magnitude to obtain a unit vector e_{AB} with the same direction as F_{AB} (Fig. b):

$$e_{AB} = \frac{-4i + 4j + 2k}{6} = -\frac{2}{3}i + \frac{2}{3}j + \frac{1}{3}k.$$

Now we write F_{AB} as

$$F_{AB} = 10e_{AB} = -6.67i + 6.67j + 3.33k \text{ (kN)}.$$

We express the force F_{AC} in terms of its components in the same way:

$$F_{AC} = 5.71i + 8.57j - 17.14k \text{ (kN)}.$$

Choose the Vector r Since the lines of action of both forces pass through point *A*, we can use the vector from *O* to *A* to determine the moments of both forces about point *O* (Fig. a):

$$r = 4i + 6k \text{ (m)}.$$

Evaluate r × F The sum of the moments is

$$\Sigma M_O = (r \times F_{AB}) + (r \times F_{AC})$$

$$= \begin{vmatrix} i & j & k \\ 4 & 0 & 6 \\ -6.67 & 6.67 & 3.33 \end{vmatrix} + \begin{vmatrix} i & j & k \\ 4 & 0 & 6 \\ 5.71 & 8.57 & -17.14 \end{vmatrix}$$

$$= -91.4i + 49.5j + 61.0k \text{ (kN-m)}.$$

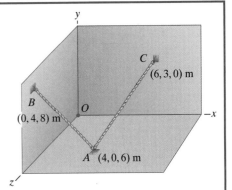

Figure 4.15

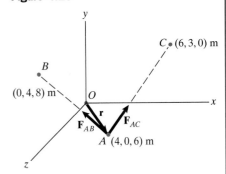

(a) The forces F_{AB} and F_{AC} exerted at *A* by the cables.

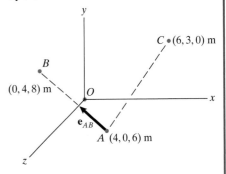

(b) The unit vector e_{AB} has the same direction as F_{AB}.

Problems

4.45 Use Eq. (4.2) to determine the moment of the 50-lb force about the origin O. Compare your answer with the two-dimensional description of the moment.

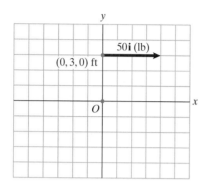

P4.45

4.46 Use Eq. (4.2) to determine the moment of the 80-N force about the origin O letting $\mathbf{r}$ be the vector (a) from O to A; (b) from O to B.

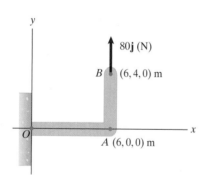

P4.46

4.47 A bioengineer studying an injury sustained in throwing the javelin estimates that the magnitude of the maximum force exerted was $|\mathbf{F}| = 360$ N and the perpendicular distance from O to the line of action of $\mathbf{F}$ was 550 mm. The vector $\mathbf{F}$ and point O are contained in the x–y plane. Express the moment of $\mathbf{F}$ about the shoulder joint at O as a vector.

P4.47

4.48 Use Eq.(4.2) to determine the moment of the 100-kN force (a) about A, (b) about B.

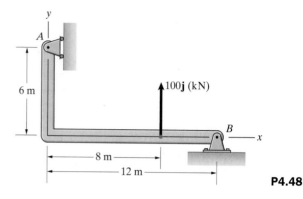

P4.48

4.49 The line of action of the 100-lb force is contained in the x–y plane.
(a) Use Eq.(4.2) to determine the moment of the force about the origin O.
(b) Use the result of (a) to determine the perpendicular distance from O to the line of action of the force.

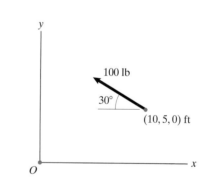

P4.49

4.50 The line of action of $\mathbf{F}$ is contained in the x–y plane. The moment of $\mathbf{F}$ about O is $140\mathbf{k}$ (N-m), and the moment of $\mathbf{F}$ about A is $280\mathbf{k}$ (N-m). What are the components of $\mathbf{F}$?

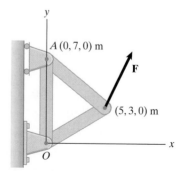

P4.50

4.51 To test the bending stiffness of a light composite beam, engineering students subject it to the vertical forces shown. Use Eq. (4.2) to determine the moment of the 6-kN force about A.

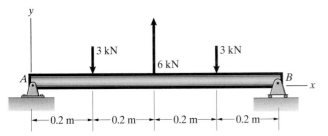

P4.51

4.52 Consider the beam and forces shown in Problem 4.51. Use Eq. (4.2) to determine the sum of the moments of the three forces (a) about A, (b) about B.

4.53 Three forces are applied to the plate. Use Eq.(4.2) to determine the sum of the moments of the three forces about the origin O.

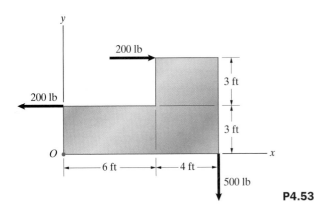

P4.53

4.54 (a) Determine the magnitude of the moment of the 150-N force about A by calculating the perpendicular distance from A to the line of action of the force.
(b) Use Eq. (4.2) to determine the magnitude of the moment of the 150-N force about A.

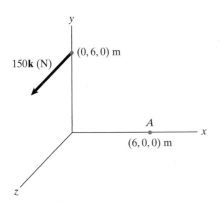

P4.54

4.55 A force $\mathbf{F} = -4\mathbf{i} + 6\mathbf{j} - 2\mathbf{k}$ (kN) is applied at the point (8, 4, −4) m. What is the magnitude of the moment of **F** about the point P with coordinates (2, 2, 2) m? What is the perpendicular distance D from P to the line of action of **F**?

4.56 A force $\mathbf{F} = 20\mathbf{i} - 30\mathbf{j} + 60\mathbf{k}$ (N) is applied at the point (2, 3, 6) m. What is the magnitude of the moment of **F** about the point P with coordinates (−2, −1, −1) m? What is the perpendicular distance D from P to the line of action of **F**?

4.57 A force $\mathbf{F} = 20\mathbf{i} - 30\mathbf{j} + 60\mathbf{k}$ (lb). The moment of **F** about a point P is $\mathbf{M}_P = 450\mathbf{i} - 100\mathbf{j} - 200\mathbf{k}$ (ft-lb). What is the perpendicular distance from point P to the line of action of **F**?

4.58 A force **F** is applied at the point (8, 6, 13) m. Its magnitude is $|\mathbf{F}| = 90$ N, and the moment of **F** about the point (4, 2, 6) is zero. What are the components of **F**?

4.59 The force $\mathbf{F} = 30\mathbf{i} + 20\mathbf{j} - 10\mathbf{k}$ (N).
(a) Determine the magnitude of the moment of **F** about A.
(b) Suppose that you can change the direction of **F** while keeping its magnitude constant, and you want to choose a direction that maximizes the moment of **F** about A. What is the magnitude of the resulting maximum moment?

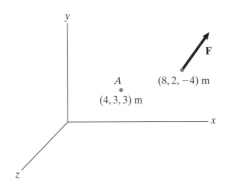

P4.59

4.60 The direction cosines of the force **F** are $\cos\theta_x = 0.818$, $\cos\theta_y = 0.182$, and $\cos\theta_z = -0.545$. The support of the beam at O will fail if the magnitude of the moment of **F** about O exceeds 100 kN-m. Determine the magnitude of the largest force **F** that can safely be applied to the beam.

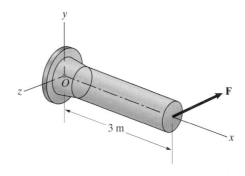

P4.60

4.61 The force **F** exerted on the grip of the exercise machine points in the direction of the unit vector $\mathbf{e} = \frac{2}{3}\mathbf{i} - \frac{2}{3}\mathbf{j} + \frac{1}{3}\mathbf{k}$ and its magnitude is 120 N. Determine the magnitude of the moment of **F** about the origin O.

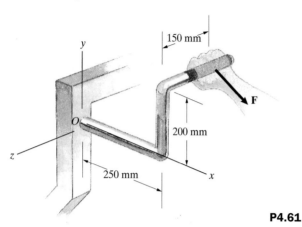

P4.61

4.62 The force **F** in Problem 4.61 points in the direction of the unit vector $\mathbf{e} = \frac{2}{3}\mathbf{i} - \frac{2}{3}\mathbf{j} + \frac{1}{3}\mathbf{k}$. The support at O will safely support a moment of 560 N-m magnitude.
(a) Based on this criterion, what is the largest safe magnitude of **F**?
(b) If the force **F** may be exerted in any direction, what is its largest safe magnitude?

4.63 An engineer estimates that under the most adverse expected weather conditions, the total force on the highway sign will be $\mathbf{F} = \pm1.4\mathbf{i} - 2.0\mathbf{j}$ (kN). What moment does this force exert about the base O?

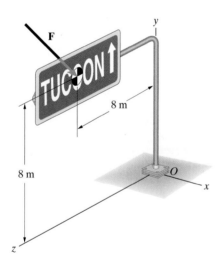

P4.63

4.64 The weights of the arms OA and AB of the robotic manipulator act at their midpoints. The direction cosines of the centerline of arm OA are $\cos\theta_x = 0.500$, $\cos\theta_y = 0.866$, and $\cos\theta_z = 0$, and the direction cosines of the centerline of arm AB are $\cos\theta_x = 0.707$, $\cos\theta_y = 0.619$, and $\cos\theta_z = -0.342$. What is the sum of the moments about O due to the two forces?

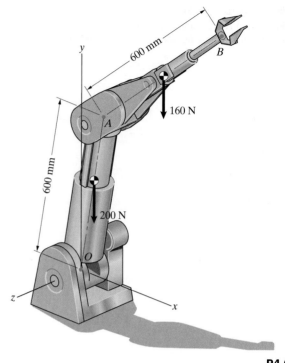

P4.64

4.65 The tension in cable AC is 100 lb. Determine the moment about the origin O due to the force exerted at A by cable AC. Use the cross product, letting **r** be the vector (a) from O to A, (b) from O to C.

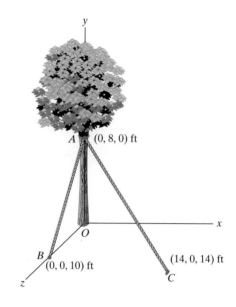

P4.65

4.66 Consider the tree in Problem 4.65. The tension in cable AB is 100 lb, and the tension in cable AC is 140 lb. Determine the magnitude of the sum of the moments about O due to the forces exerted at A by the two cables.

4.67 The force $\mathbf{F} = 5\mathbf{i}$ (kN) acts on the ring A where the cables AB, AC, and AD are joined. What is the sum of the moments about point D due to the force $\mathbf{F}$ and the three forces exerted on the ring by the cables?

Strategy: The ring is in equilibrium. Use what you know about the four forces acting on it.

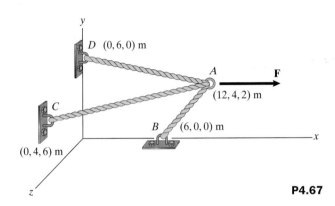

P4.67

4.68 In Problem 4.67, determine the moment about point D due to the force exerted on the ring A by the cable AB.

4.69 The tower is 70 m tall. The tensions in cables AB, AC, and AD are 4 kN, 2 kN, and 2 kN, respectively. Determine the sum of the moments about the origin O due to the forces exerted by the cables at point A.

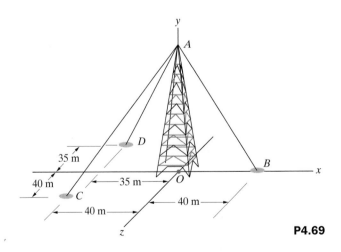

P4.69

4.70 Consider the 70-m tower in Problem 4.69. Suppose that the tension in cable AB is 4 kN, and you want to adjust the tensions in cables AC and AD so that the sum of the moments about the origin O due to the forces exerted by the cables at point A is zero. Determine the tensions.

4.71 The tension in cable AB is 150 N. The tension in cable AC is 100 N. Determine the sum of the moments about D due to the forces exerted on the wall by the cables.

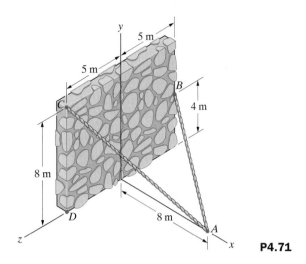

P4.71

4.72 Consider the wall shown in Problem 4.71. The total force exerted by the two cables in the direction perpendicular to the wall is 2 kN. The magnitude of the sum of the moments about D due to the forces exerted on the wall by the cables is 18 kN-m. What are the tensions in the cables?

4.73 The force $F = 800$ lb. The sum of the moments about O due to the force F and the forces exerted at A by the cables AB and AC is zero. What are the tensions in the cables?

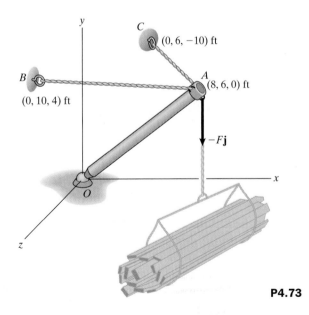

P4.73

4.74 In Problem 4.73, the sum of the moments about O due to the force F and the forces exerted at A by the cables AB and AC is zero. Each cable will safely support a tension of 2000 lb. Based on this criterion, what is the largest safe value of the force F?

4.75 The 200-kg slider at A is held in place on the smooth vertical bar by the cable AB. Determine the moment about the bottom of the bar (point C with coordinates $x = 2$ m, $y = z = 0$) due to the force exerted on the slider by the cable.

4.76 To evaluate the adequacy of the design of the vertical steel post, you must determine the moment about the bottom of the post due to the force exerted on the post at B by the cable AB. A calibrated strain gauge mounted on cable AC indicates that the tension in cable AC is 22 kN. What is the moment?

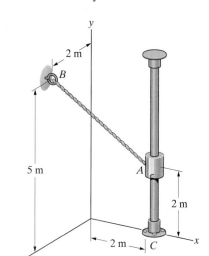

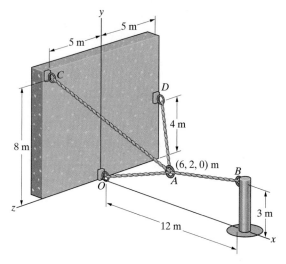

P4.75   P4.76

4.3 Moment of a Force About a Line

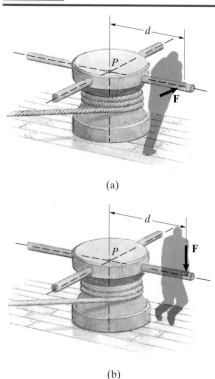

(a)

(b)

Figure 4.16
(a) Turning a capstan.
(b) A vertical force does not turn the capstan.

The device in Fig. 4.16, called a *capstan*, was used in the days of square-rigged sailing ships. Crewmen turned it by pushing on the handles as shown in Fig. 4.16a, providing power for such tasks as raising anchors and hoisting yards. A vertical force $\mathbf{F}$ applied to one of the handles as shown in Fig. 4.16b does not cause the capstan to turn, even though the magnitude of the moment about point P is $d|\mathbf{F}|$ in both cases.

The measure of the tendency of a force to cause rotation about a line, or axis, is called the moment of the force about the line. Suppose that a force $\mathbf{F}$ acts on an object such as a turbine that rotates about an axis L, and we resolve $\mathbf{F}$ into components in terms of the coordinate system shown in Fig. 4.17. The components F_x and F_z do not tend to rotate the turbine, just as the force parallel to the axis of the capstan did not cause it to turn. It is the component F_y that tends to cause rotation, by exerting a moment of magnitude aF_y about the turbine's axis. In this example we can determine the moment of $\mathbf{F}$ about L easily because the coordinate system is conveniently placed. We now introduce an expression that determines the moment of a force about any line.

Definition

Consider a line L and force $\mathbf{F}$ (Fig. 4.18a). Let $\mathbf{M}_P$ be the moment of $\mathbf{F}$ about an arbitrary point P on L (Fig. 4.18b). The moment of $\mathbf{F}$ about L is the component of $\mathbf{M}_P$ parallel to L, which we denote by $\mathbf{M}_L$ (Fig. 4.18c). The magnitude of the moment of $\mathbf{F}$ about L is $|\mathbf{M}_L|$, and when the thumb of the right hand is pointed in the direction of $\mathbf{M}_L$, the arc of the fingers indicates the sense of the moment about L.

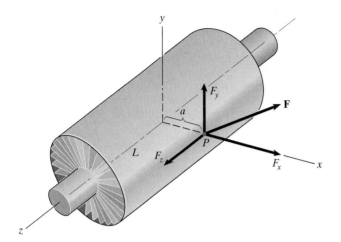

Figure 4.17
Applying a force to a turbine with axis of rotation L.

In terms of a unit vector $\mathbf{e}$ along L (Fig. 4.18d), $\mathbf{M}_L$ is given by

$$\mathbf{M}_L = \left(\mathbf{e} \cdot \mathbf{M}_P\right)\mathbf{e}. \tag{4.4}$$

(The unit vector $\mathbf{e}$ can point in either direction. See our discussion of vector components parallel and normal to a line in Section 2.5.) The moment $\mathbf{M}_P = \mathbf{r} \times \mathbf{F}$, so we can also express $\mathbf{M}_L$ as

$$\mathbf{M}_L = \left[\mathbf{e} \cdot (\mathbf{r} \times \mathbf{F})\right]\mathbf{e}. \tag{4.5}$$

The mixed triple product in this expression is given in terms of the components of the three vectors by

$$\mathbf{e} \cdot \left(\mathbf{r} \times \mathbf{F}\right) = \begin{vmatrix} e_x & e_y & e_z \\ r_x & r_y & r_z \\ F_x & F_y & F_z \end{vmatrix}. \tag{4.6}$$

Notice that the value of the scalar $\mathbf{e} \cdot \mathbf{M}_P = \mathbf{e} \cdot (\mathbf{r} \times \mathbf{F})$ determines both the magnitude and direction of $\mathbf{M}_L$. The absolute value of $\mathbf{e} \cdot \mathbf{M}_P$ is the magnitude of $\mathbf{M}_L$. If $\mathbf{e} \cdot \mathbf{M}_P$ is positive, $\mathbf{M}_L$ points in the direction of $\mathbf{e}$, and if $\mathbf{e} \cdot \mathbf{M}_P$ is negative, $\mathbf{M}_L$ points in the direction opposite to $\mathbf{e}$.

The result obtained with Eq. (4.4) or (4.5) doesn't depend on which point on L is chosen to determine $\mathbf{M}_P = \mathbf{r} \times \mathbf{F}$. If we use point P in Fig. 4.19 to determine the moment of $\mathbf{F}$ about L, we get the result given by Eq. (4.5). If we use P' instead, we obtain the same result,

$$\left[\mathbf{e} \cdot (\mathbf{r}' \times \mathbf{F})\right]\mathbf{e} = \left\{\mathbf{e} \cdot \left[(\mathbf{r} + \mathbf{u}) \times \mathbf{F}\right]\right\}\mathbf{e}$$

$$= \left[\mathbf{e} \cdot (\mathbf{r} \times \mathbf{F}) + \mathbf{e} \cdot (\mathbf{u} \times \mathbf{F})\right]\mathbf{e}$$

$$= \left[\mathbf{e} \cdot (\mathbf{r} \times \mathbf{F})\right]\mathbf{e},$$

because $\mathbf{u} \times \mathbf{F}$ is perpendicular to $\mathbf{e}$.

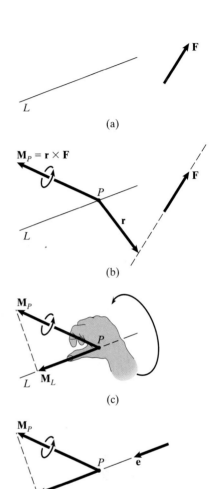

Figure 4.18
(a) The line L and force $\mathbf{F}$.
(b) $\mathbf{M}_P$ is the moment of $\mathbf{F}$ about any point P on L.
(c) The component $\mathbf{M}_L$ is the moment of $\mathbf{F}$ about L.
(d) A unit vector $\mathbf{e}$ along L.

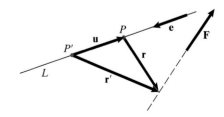

Figure 4.19
Using different points P and P' to determine the moment of $\mathbf{F}$ about L.

Applying the Definition

To demonstrate that M_L is the measure of the tendency of $\mathbf{F}$ to cause rotation about L, we return to the turbine in Fig. 4.17. Let Q be a point on L at an arbitrary distance b from the origin (Fig. 4.20a). The vector $\mathbf{r}$ from Q to P is $\mathbf{r} = a\mathbf{i} - b\mathbf{k}$, so the moment of $\mathbf{F}$ about Q is

$$\mathbf{M}_Q = \mathbf{r} \times \mathbf{F} = \begin{vmatrix} \mathbf{i} & \mathbf{j} & \mathbf{k} \\ a & 0 & -b \\ F_x & F_y & F_z \end{vmatrix} = bF_y\mathbf{i} - (aF_z + bF_x)\mathbf{j} + aF_y\mathbf{k}.$$

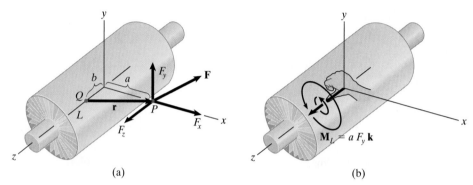

Figure 4.20
(a) An arbitrary point Q on L and the vector $\mathbf{r}$ from Q to P.
(b) $\mathbf{M}_L$ and the sense of the moment about L.

Since the z axis is coincident with L, the unit vector $\mathbf{k}$ is along L. Therefore the moment of $\mathbf{F}$ about L is

$$\mathbf{M}_L = (\mathbf{k} \cdot \mathbf{M}_Q)\mathbf{k} = aF_y\mathbf{k}.$$

The components F_x and F_z exert no moment about L. If we assume that F_y is positive, it exerts a moment of magnitude aF_y about the turbine's axis in the direction shown in Fig. 4.20b.

Now let's determine the moment of a force about an arbitrary line L (Fig. 4.21a). The first step is to choose a point on the line. If we choose point A (Fig. 4.21b), the vector $\mathbf{r}$ from A to the point of application of $\mathbf{F}$ is

$$\mathbf{r} = (8 - 2)\mathbf{i} + (6 - 0)\mathbf{j} + (4 - 4)\mathbf{k} = 6\mathbf{i} + 6\mathbf{j} \text{ (m)}.$$

The moment of $\mathbf{F}$ about A is

$$\mathbf{M}_A = \mathbf{r} \times \mathbf{F} = \begin{vmatrix} \mathbf{i} & \mathbf{j} & \mathbf{k} \\ 6 & 6 & 0 \\ 10 & 60 & -20 \end{vmatrix}$$

$$= -120\mathbf{i} + 120\mathbf{j} + 300\mathbf{k} \text{ (N-m)}.$$

The next step is to determine a unit vector along L. The vector from A to B is

$$(-7 - 2)\mathbf{i} + (6 - 0)\mathbf{j} + (2 - 4)\mathbf{k} = -9\mathbf{i} + 6\mathbf{j} - 2\mathbf{k} \text{ (m)}.$$

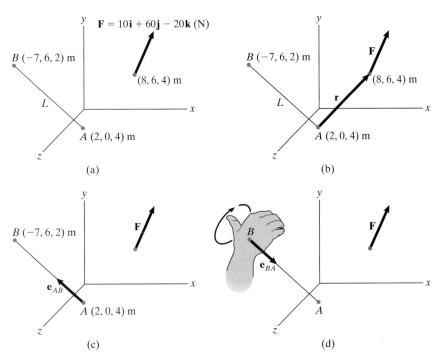

(a)

(b)

(c)

(d)

Figure 4.21
(a) A force **F** and line L.
(b) The vector **r** from A to the point of application of **F**.
(c) $\mathbf{e}_{AB}$ points from A toward B.
(d) The right-hand rule indicates the sense of the moment.

Dividing this vector by its magnitude, we obtain a unit vector $\mathbf{e}_{AB}$ that points from A toward B (Fig. 4.21c):

$$\mathbf{e}_{AB} = -\frac{9}{11}\mathbf{i} + \frac{6}{11}\mathbf{j} - \frac{2}{11}\mathbf{k}.$$

The moment of **F** about L is

$$\mathbf{M}_L = (\mathbf{e}_{AB} \cdot \mathbf{M}_A)\mathbf{e}_{AB}$$
$$= \left[\left(-\frac{9}{11}\right)(-120) + \left(\frac{6}{11}\right)(120) + \left(-\frac{2}{11}\right)(300)\right]\mathbf{e}_{AB}$$
$$= 109\mathbf{e}_{AB} \ (\text{N-m}).$$

The magnitude of $\mathbf{M}_L$ is 109 N-m; pointing the thumb of the right hand in the direction of $\mathbf{e}_{AB}$ indicates the direction.

If we calculate $\mathbf{M}_L$ using the unit vector $\mathbf{e}_{BA}$ that points from B toward A instead, we obtain

$$\mathbf{M}_L = -109\mathbf{e}_{BA} \ (\text{N-m}).$$

We obtain the same magnitude, and the minus sign indicates that $\mathbf{M}_L$ points in the direction opposite to $\mathbf{e}_{BA}$, so the direction of $\mathbf{M}_L$ is the same. Therefore the right-hand rule indicates the same sense (Fig. 4.21d).

The preceding examples demonstrate three useful results that we can state in more general terms:

- When the line of action of **F** is perpendicular to a plane containing L (Fig. 4.22a), the magnitude of the moment of **F** about L is equal to the product of the magnitude of **F** and the perpendicular distance D from L to the point where the line of action intersects the plane: $|\mathbf{M}_L| = |\mathbf{F}|D$.
- When the line of action of **F** is parallel to L (Fig. 4.22b), the moment of **F** about L is zero: $\mathbf{M}_L = 0$. Since $\mathbf{M}_P = \mathbf{r} \times \mathbf{F}$ is perpendicular to **F**, $\mathbf{M}_P$ is perpendicular to L and the vector component of $\mathbf{M}_P$ parallel to L is zero.

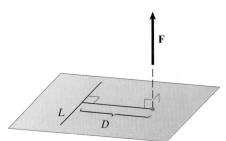

(a)

(b)

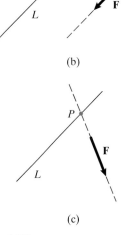

(c)

Figure 4.22
(a) **F** is perpendicular to a plane containing L.
(b) **F** is parallel to L.
(c) The line of action of **F** intersects L at P.

• When the line of action of **F** intersects L (Fig. 4.22c), the moment of **F** about L is zero. Since we can choose any point on L to evaluate $\mathbf{M}_P$, we can use the point where the line of action of **F** intersects L. The moment $\mathbf{M}_P$ about that point is zero, so its vector component parallel to L is zero.

In summary, determining the moment of a force **F** about a point P using Eqs. (4.4)–(4.6) requires three steps:

1. Determine a vector **r**—Choose any point P on L, and determine the components of a vector **r** from P to any point on the line of action of **F**.
2. Determine a vector **e**—Determine the components of a unit vector along L. It doesn't matter in which direction along L it points.
3. Evaluate $\mathbf{M}_L$—You can calculate $\mathbf{M}_P = \mathbf{r} \times \mathbf{F}$ and determine $\mathbf{M}_L$ by using Eq. (4.4), or you can use Eq. (4.6) to evaluate the mixed triple product and substitute the result into Eq. (4.5).

Study Questions

1. When you use Eq. (4.5) to determine the moment of a force **F** about a line L, how do you choose the vector **r**? What is the definition of the vector **e**?
2. Explain how the direction of the vector $\mathbf{M}_L$ in Eq. (4.5) indicates the sense of the moment of **F** about L.
3. What is the moment of a force **F** about a line L if the line of action of **F** passes through L? What is the moment if the line of action of **F** is parallel to L?

Example 4.7

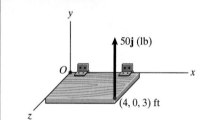

Figure 4.23

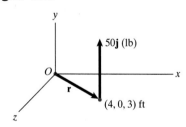

(a) The vector **r** from O to the point of application of the force.

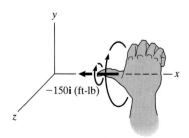

(b) The sense of the moment.

Moment of a Force About the x Axis

What is the moment of the 50-lb force in Fig. 4.23 about the x axis?

Strategy

We can determine the moment in two ways.

First Method We can use Eqs. (4.5) and (4.6). Since **r** can extend from any point on the x axis to the line of action of the force, we can use the vector from O to the point of application of the force. The vector **e** must be a unit vector along the x axis, so we can use either **i** or $-\mathbf{i}$.

Second Method This example is the first of the special cases we just discussed, because the 50-lb force is perpendicular to the x–z plane. We can determine the magnitude and direction of the moment directly from the given information.

Solution

First Method *Determine a vector* **r**. The vector from O to the point of application of the force is (Fig. a)
$$\mathbf{r} = 4\mathbf{i} + 3\mathbf{k} \text{ (ft)}.$$

Determine a vector **e**. We can use the unit vector **i**.
Evaluate $\mathbf{M}_L$. Using Eq. (4.6), the mixed triple product is

$$\mathbf{i} \cdot (\mathbf{r} \times \mathbf{F}) = \begin{vmatrix} 1 & 0 & 0 \\ 4 & 0 & 3 \\ 0 & 50 & 0 \end{vmatrix} = -150 \text{ ft-lb.}$$

Then from Eq. (4.5), the moment of the force about the x axis is
$$\mathbf{M}_{(x\,\text{axis})} = \left[\mathbf{i} \cdot (\mathbf{r} \times \mathbf{F})\right]\mathbf{i} = -150\mathbf{i} \text{ (ft-lb).}$$

The magnitude of the moment is 150 ft-lb, and its sense is as shown in Fig. b.

Second Method Since the 50-lb force is perpendicular to a plane (the x–z plane) containing the x axis, the magnitude of the moment about the x axis is equal to the perpendicular distance from the x axis to the point where the line of action of the force intersects the x–z plane (Fig. c):

$$\left|\mathbf{M}_{(x\,\text{axis})}\right| = (3\,\text{ft})(50\,\text{lb}) = 150\,\text{ft-lb}.$$

Pointing the arc of the fingers in the direction of the sense of the moment about the x axis (Fig. c), we find that the right-hand rule indicates that $\mathbf{M}_{(x\,\text{axis})}$ points in the negative x-axis direction. Therefore

$$\mathbf{M}_{(x\,\text{axis})} = -150\mathbf{i}\,(\text{ft-lb}).$$

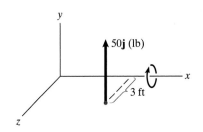

(c) The distance from the x axis to the point where the line of action of the force intersects the x–z plane is 3 ft. The arrow indicates the sense of the moment about the x axis.

Example 4.8

Moment of a Force About a Line

What is the moment of the force $\mathbf{F}$ in Fig. 4.24 about the bar BC?

Strategy

We can use Eqs. (4.5) and (4.6) to determine the moment. Since we know the coordinates of points B and C, we can determine the components of a vector $\mathbf{r}$ that extends either from B to the point of application of the force or from C to the point of application. We can also use the coordinates of points B and C to determine a unit vector along the line BC.

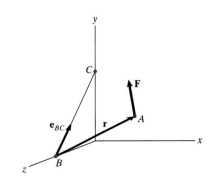

Figure 4.24

Solution

Determine a Vector r We need a vector from any point on the line BC to any point on the line of action of the force. We can let $\mathbf{r}$ be the vector from B to the point of application of $\mathbf{F}$ (Fig. a):

$$\mathbf{r} = (4 - 0)\mathbf{i} + (2 - 0)\mathbf{j} + (2 - 3)\mathbf{k} = 4\mathbf{i} + 2\mathbf{j} - \mathbf{k}\,(\text{m}).$$

Determine a Vector e To obtain a unit vector along the bar BC, we determine the vector from B to C,

$$(0 - 0)\mathbf{i} + (4 - 0)\mathbf{j} + (0 - 3)\mathbf{k} = 4\mathbf{j} - 3\mathbf{k}\,(\text{m}),$$

and divide it by its magnitude (Fig. a):

$$\mathbf{e}_{BC} = \frac{4\mathbf{j} - 3\mathbf{k}}{5} = 0.8\mathbf{j} - 0.6\mathbf{k}.$$

Evaluate M_L Using Eq. (4.6), the mixed triple product is

$$\mathbf{e}_{BC} \cdot (\mathbf{r} \times \mathbf{F}) = \begin{vmatrix} 0 & 0.8 & -0.6 \\ 4 & 2 & -1 \\ -2 & 6 & 3 \end{vmatrix} = -24.8\,\text{kN-m}.$$

Substituting this result into Eq. (4.5), the moment of $\mathbf{F}$ about the bar BC is

$$\mathbf{M}_{BC} = \left[[\mathbf{e}_{BC} \cdot (\mathbf{r} \times \mathbf{F})]\mathbf{e}_{BC} = -24.8\mathbf{e}_{BC}\,(\text{kN-m}).\right.$$

The magnitude of $\mathbf{M}_{BC}$ is 24.8 kN-m, and its direction is opposite to that of $\mathbf{e}_{BC}$. The sense of the moment is shown in Fig. b.

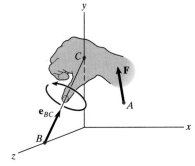

(a) The vectors $\mathbf{r}$ and $\mathbf{e}_{BC}$.

(b) The right-hand rule indicates the sense of the moment about BC.

Problems

4.77 Use Eqs. (4.5) and (4.6) to determine the moment of the 40-N force about the z axis. (First see if you can write down the result without using the equations.)

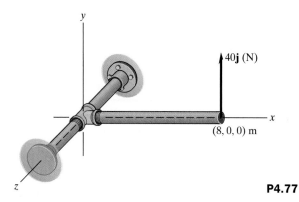

P4.77

4.78 Use Eqs. (4.5) and (4.6) to determine the moment of the 20-N force about (a) the x axis, (b) the y axis, (c) the z axis. (First see if you can write down the results without using the equations.)

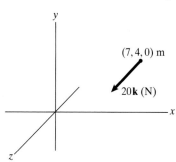

P4.78

4.79 Three forces parallel to the y axis act on the rectangular plate. Use Eqs. (4.5) and (4.6) to determine the sum of the moments of the forces about the x axis. (First see if you can write down the result without using the equations.)

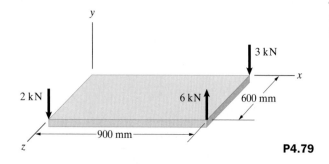

P4.79

4.80 Consider the rectangular plate shown in Problem 4.79. The three forces are parallel to the y axis. Determine the sum of the moments of the forces (a) about the y axis, (b) about the z axis.

4.81 The person exerts a force $\mathbf{F} = 0.2\mathbf{i} - 0.4\mathbf{j} + 1.2\mathbf{k}$ (lb) on the gate at C. Point C lies in the x–y plane. What moment does the person exert about the gate's hinge axis, which is coincident with the y axis?

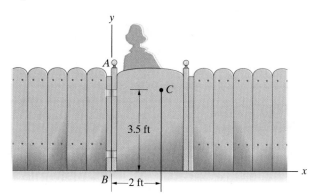

P4.81

4.82 Four forces parallel to the y axis act on the rectangular plate. The sum of the forces in the positive y direction is 200 lb. The sum of the moments of the forces about the x axis is $-300\mathbf{i}$ (ft-lb) and the sum of the moments about the z axis is $400\mathbf{k}$ (ft-lb). What are the magnitudes of the forces?

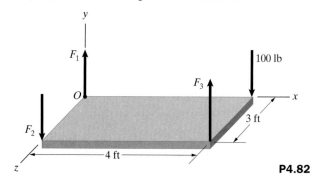

P4.82

4.83 The force $\mathbf{F} = 100\mathbf{i} + 60\mathbf{j} - 40\mathbf{k}$ (lb). What is the moment of $\mathbf{F}$ about the y axis? Draw a sketch to indicate the sense of the moment.

P4.83

4.84 Suppose that the moment of the force **F** shown in Problem 4.83 about the x axis is $-80\mathbf{i}$ (ft-lb), the moment about the y axis is zero, and the moment about the z axis is $160\mathbf{k}$ (ft-lb). If $F_y = 80$ lb, what are F_x and F_z?

4.85 The robotic manipulator is stationary. The weights of the arms AB and BC act at their midpoints. The direction cosines of the centerline of arm AB are $\cos\theta_x = 0.500$, $\cos\theta_y = 0.866$, $\cos\theta_z = 0$, and the direction cosines of the centerline of arm BC are $\cos\theta_x = 0.707$, $\cos\theta_y = 0.619$, $\cos\theta_z = -0.342$. What total moment is exerted about the z axis by the weights of the arms?

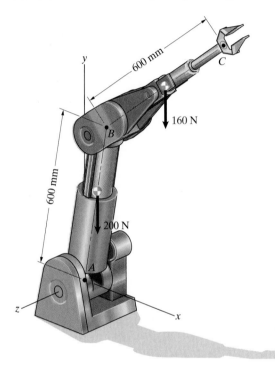

P4.85

4.86 In Problem 4.85, what total moment is exerted about the x axis by the weights of the arms?

4.87 Two forces are exerted on the crankshaft by the connecting rods. The direction cosines of F_A are $\cos\theta_x = -0.182$, $\cos\theta_y = 0.818$, and $\cos\theta_z = 0.545$, and its magnitude is 4 kN.

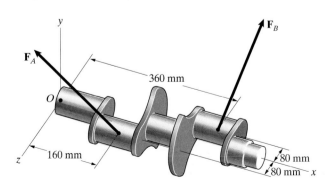

P4.87

The direction cosines of F_B are $\cos\theta_x = 0.182$, $\cos\theta_y = 0.818$, and $\cos\theta_z = -0.545$, and its magnitude is 2 kN. What is the sum of the moments of the two forces about the x axis? (This is the moment that causes the crankshaft to rotate.)

4.88 Determine the moment of the 20-N force about the line AB. Use Eqs. (4.5) and (4.6), letting the unit vector **e** point (a) from A toward B, (b) from B toward A.

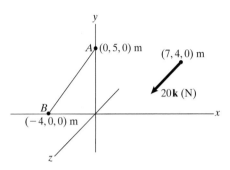

P4.88

4.89 The force $\mathbf{F} = -10\mathbf{i} + 5\mathbf{j} - 5\mathbf{k}$ (kip). Determine the moment of **F** about the line AB. Draw a sketch to indicate the sense of the moment.

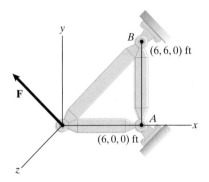

P4.89

4.90 The force $\mathbf{F} = 10\mathbf{i} + 12\mathbf{j} - 6\mathbf{k}$ (N). What is the moment of **F** about the line AO? Draw a sketch to indicate the sense of the moment.

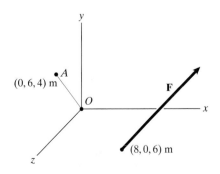

P4.90

4.91 The tension in the cable AB is 1 kN. Determine the moment about the x axis due to the force exerted on the hatch by the cable at point B. Draw a sketch to indicate the sense of the moment.

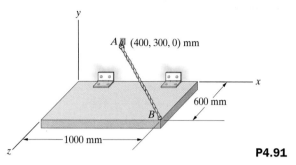

P4.91

4.92 Determine the moment of the force applied at D about the straight line through the hinges A and B. (The line through A and B lies in the y–z plane.)

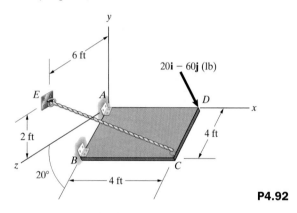

P4.92

4.93 In Problem 4.92, the tension in the cable CE is 160 lb. Determine the moment of the force exerted by the cable on the hatch at C about the straight line through the hinges A and B.

4.94 The coordinates of A are $(-2.4, 0, -0.6)$ m, and the coordinates of B are $(-2.2, 0.7, -1.2)$ m. The force exerted at B by the sailboat's main sheet AB is 130 N. Determine the moment

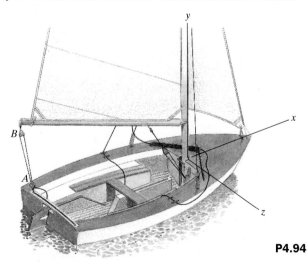

P4.94

of the force about the centerline of the mast (the y axis). Draw a sketch to indicate the sense of the moment.

4.95 The tension in cable AB is 200 lb. Determine the moments about each of the coordinate axes due to the force exerted on point B by the cable. Draw sketches to indicate the senses of the moments.

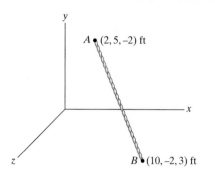

P4.95

4.96 The total force exerted on the blades of the turbine by the steam nozzle is $\mathbf{F} = 20\mathbf{i} - 120\mathbf{j} + 100\mathbf{k}$ (N), and it effectively acts at the point $(100, 80, 300)$ mm. What moment is exerted about the axis of the turbine (the x axis)?

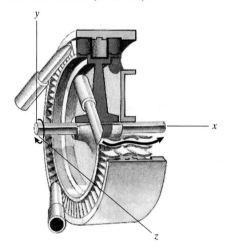

P4.96

4.97 The tension in cable AB is 50 N. Determine the moment about the line OC due to the force exerted by the cable at B. Draw a sketch to indicate the sense of the moment.

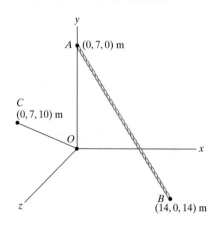

P4.97

4.98 The tension in cable AB is 80 lb. What is the moment about the line CD due to the force exerted by the cable on the wall at B?

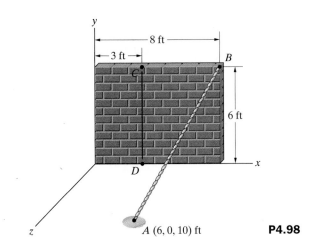

P4.98

4.99 The universal joint is connected to the drive shaft at A and A'. The coordinates of A are $(0, 40, 0)$ mm, and the coordinates of A' are $(0, -40, 0)$ mm. The forces exerted on the drive shaft by the universal joint are $-30\mathbf{j} + 400\mathbf{k}$ (N) at A and $30\mathbf{j} - 400\mathbf{k}$ (N) at A'. What is the magnitude of the torque (moment) exerted by the universal joint on the drive shaft about the shaft axis O-O'?

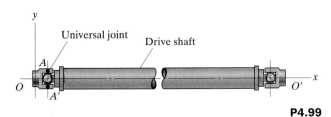

P4.99

4.100 A motorist applies the two forces shown to loosen a lug nut. The direction cosines of $\mathbf{F}$ are $\cos\theta_x = \frac{4}{13}$, $\cos\theta_y = \frac{12}{13}$, and $\cos\theta_z = \frac{3}{13}$. If the magnitude of the moment about the x axis must be 32 ft-lb to loosen the nut, what is the magnitude of the forces the motorist must apply?

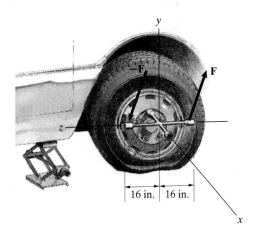

P4.100

4.101 The tension in cable AB is 2 kN. What is the magnitude of the moment about the shaft CD due to the force exerted by the cable at A? Draw a sketch to indicate the sense of the moment about the shaft.

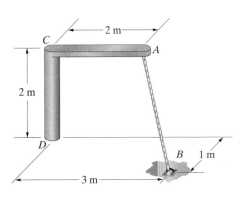

P4.101

4.102 The axis of the car's wheel passes through the origin of the coordinate system and its direction cosines are $\cos\theta_x = 0.940$, $\cos\theta_y = 0$, $\cos\theta_z = 0.342$. The force exerted on the tire by the road effectively acts at the point $x = 0$, $y = -0.36$ m, $z = 0$ and has components $\mathbf{F} = -720\mathbf{i} + 3660\mathbf{j} + 1240\mathbf{k}$ (N). What is the moment of $\mathbf{F}$ about the wheel's axis?

P4.102

4.103 The direction cosines of the centerline OA are $\cos\theta_x = 0.500$, $\cos\theta_y = 0.866$, and $\cos\theta_z = 0$, and the direction cosines of the line AG are $\cos\theta_x = 0.707$, $\cos\theta_y = 0.619$, and

$\cos\theta_z = -0.342$. What is the moment about OA due to the 250-N weight? Draw a sketch to indicate the sense of the moment about the shaft.

4.105 Consider the steering wheel in Problem 4.104. Determine the moment of **F** about the shaft OC of the steering wheel if $\alpha=30°$. Draw a sketch to indicate the sense of the moment about the shaft.

4.106 The weight W causes a tension of 100 lb in cable CD. If $d = 2$ ft, what is the moment about the z axis due to the force exerted by the cable CD at point C?

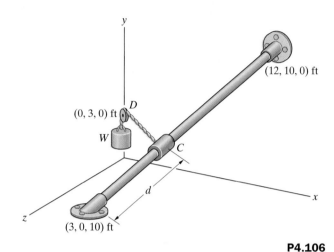

P4.106

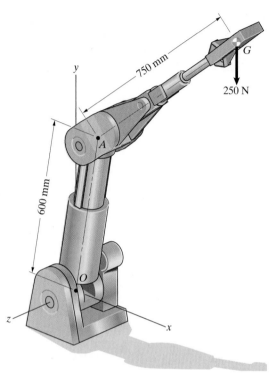

P4.103

4.107 The rod AB supports the open hood of the car. The force exerted by the rod on the hood at B is parallel to the rod. If the rod must exert a moment of 100 ft-lb magnitude about the x axis to support the hood and the distance $d = 2$ ft, what is the magnitude of the force the rod must exert on the hood?

4.104 The radius of the steering wheel is 200 mm. The distance from O to C is 1 m. The center C of the steering wheel lies in the x–y plane. The driver exerts a force $\mathbf{F} = 10\mathbf{i} + 10\mathbf{j} - 5\mathbf{k}$ (N) on the wheel at A. If the angle $\alpha = 0$, what is the magnitude of the moment about the shaft OC? Draw a sketch to indicate the sense of the moment about the shaft.

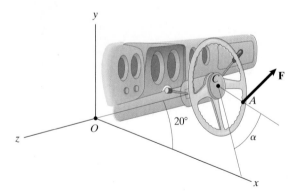

P4.104

P4.107

4.4 Couples

Now that we have described how to calculate the moment due to a force, consider this question: Is it possible to exert a moment on an object without subjecting it to a net force? The answer is yes, and it occurs when a compact disk begins rotating or a screw is turned by a screwdriver. Forces are exerted on these objects, but in such a way that the net force is zero while the net moment is not zero.

Two forces that have equal magnitudes, opposite directions, and different lines of action are called a *couple* (Fig. 4.25a). A couple tends to cause rotation of an object even though the vector sum of the forces is zero, and it has the remarkable property that *the moment it exerts is the same about any point.*

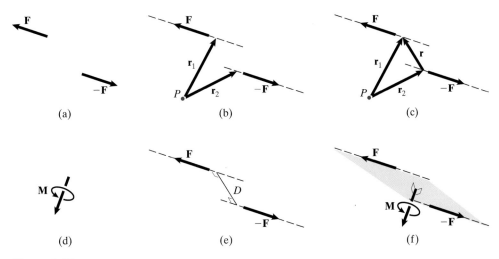

(a)   (b)   (c)

(d)   (e)   (f)

Figure 4.25
(a) A couple. (b) Determining the moment about P. (c) The vector $\mathbf{r} = \mathbf{r}_1 - \mathbf{r}_2$.
(d) Representing the moment of the couple. (e) The distance D between the lines of action.
(f) $\mathbf{M}$ is perpendicular to the plane containing $\mathbf{F}$ and $-\mathbf{F}$.

The moment of a couple is simply the sum of the moments of the forces about a point P (Fig. 4.25b):

$$\mathbf{M} = \left[\mathbf{r}_1 \times \mathbf{F}\right] + \left[\mathbf{r}_2 \times (-\mathbf{F})\right] = (\mathbf{r}_1 - \mathbf{r}_2) \times \mathbf{F}.$$

The vector $\mathbf{r}_1 - \mathbf{r}_2$ is equal to the vector $\mathbf{r}$ shown in Fig. 4.25c, so we can express the moment as

$$\mathbf{M} = \mathbf{r} \times \mathbf{F}.$$

Since $\mathbf{r}$ doesn't depend on the position of P, the moment $\mathbf{M}$ is the same for *any* point P.

Because a couple exerts a moment but the sum of the forces is zero, it is often represented in diagrams simply by showing the moment (Fig. 4.25d). Like the Cheshire cat in *Alice's Adventures in Wonderland*, which vanished except for its grin, the forces don't appear; you see only the moment they exert. But we recognize the origin of the moment by referring to it as a *moment of a couple*, or simply a *couple*.

Notice in Fig. 4.25c that $\mathbf{M} = \mathbf{r} \times \mathbf{F}$ is the moment of $\mathbf{F}$ about a point on the line of action of the force $-\mathbf{F}$. The magnitude of the moment of a force about a point equals the product of the magnitude of the force and the perpendicular distance from the point to the line of action of the force, so $|\mathbf{M}| = D|\mathbf{F}|$, where D is the perpendicular distance between the lines of action of the two forces (Fig. 4.25e). The cross product $\mathbf{r} \times \mathbf{F}$ is perpendicular to $\mathbf{r}$ and $\mathbf{F}$, which means that $\mathbf{M}$ is perpendicular to the plane containing $\mathbf{F}$ and $-\mathbf{F}$ (Fig. 4.25f). Pointing the thumb of the right hand in the direction of $\mathbf{M}$, the arc of the fingers indicates the sense of the moment.

In Fig. 4.26a, our view is perpendicular to the plane containing the two forces. The distance between the lines of action of the forces is 4 m, so the magnitude of the moment of the couple is $|\mathbf{M}| = (4 \text{ m})(2 \text{ kN}) = 8$ kN-m. The moment $\mathbf{M}$ is perpendicular to the plane containing the two forces. Pointing the arc of the fingers of the right hand counterclockwise, we find that the right-hand rule indicates that $\mathbf{M}$ points out of the page. Therefore the moment of the couple is

$$\mathbf{M} = 8\mathbf{k} \text{ (kN-m)}.$$

We can also determine the moment of the couple by calculating the sum of the moments of the two forces about *any* point. The sum of the moments of the forces about the origin O is (Fig. 4.26b)

$$
\begin{aligned}
\mathbf{M} &= \left[\mathbf{r}_1 \times (2\mathbf{j})\right] + \left[\mathbf{r}_2 \times (-2\mathbf{j})\right] \\
&= \left[(7\mathbf{i} + 2\mathbf{j}) \times (2\mathbf{j})\right] + \left[(3\mathbf{i} + 7\mathbf{j}) \times (-2\mathbf{j})\right] \\
&= 8\mathbf{k} \text{ (kN-m)}.
\end{aligned}
$$

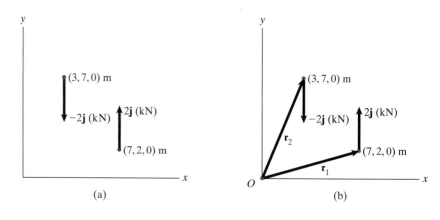

(a)

(b)

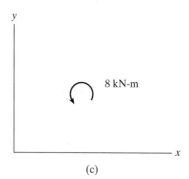

(c)

Figure 4.26
(a) A couple consisting of 2-kN forces.
(b) Determining the sum of the moments of the forces about O.
(c) Representing a couple in two dimensions.

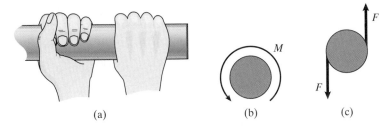

Figure 4.27
(a) Twisting a bar.
(b) The moment about the axis of the bar.
(c) The same effect is obtained by applying
    two equal and opposite forces.

In a two-dimensional situation like this example, it isn't convenient to represent a couple by showing the moment vector, because the vector is perpendicular to the page. Instead, we represent the couple by showing its magnitude and a circular arrow that indicates its sense (Fig. 4.26c).

By grasping a bar and twisting it (Fig. 4.27a), a moment can be exerted about its axis (Fig. 4.27b). Although the system of forces exerted is distributed over the surface of the bar in a complicated way, the effect is the same as if two equal and opposite forces are exerted (Fig. 4.27c). When we represent a couple as in Fig. 4.27b, or by showing the moment vector **M**, we imply that some system of forces exerts that moment. The system of forces (such as the forces exerted in twisting the bar, or the forces on the crankshaft that exert a moment on the drive shaft of a car) is nearly always more complicated than two equal and opposite forces, but the effect is the same. For this reason, we can *model* the actual system as a simple system of two forces.

Study Questions

1. How do you determine the moment exerted about a point P by a couple consisting of forces **F** and $-$**F**?

2. If you know the moment of a couple about a point P, what do you know about the moment of the couple about a different point P'?

3. A couple consists of forces **F** and $-$**F**. The perpendicular distance between the lines of action of the forces is D. What is the magnitude of the moment of the couple?

Example 4.9

Determining the Moment of a Couple

The force **F** in Fig. 4.28 is $10\mathbf{i} - 4\mathbf{j}$ (N). Determine the moment of the couple and represent it as shown in Fig. 4.27b.

Strategy

We can determine the moment in two ways: We can calculate the sum of the moments of the forces about a point, or we can sum the moments of the two couples formed by the x and y components of the forces.

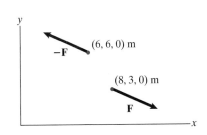

Figure 4.28

Solution

First Method   If we calculate the sum of the moments of the forces about a point on the line of action of one of the forces, the moment of that force is

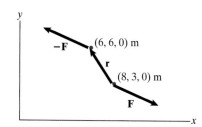

(a) Determining the moment about the point of application of **F**.

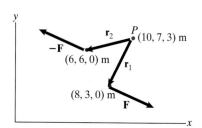

(b) Determining the moment about P.

(c) The x and y components form two couples.

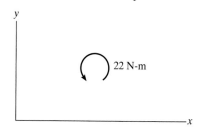

(d) Representing the moment.

zero and we only need to calculate the moment of the other force. Choosing the point of application of **F** (Fig. a), we calculate the moment as

$$\mathbf{M} = \mathbf{r} \times (-\mathbf{F}) = (-2\mathbf{i} + 3\mathbf{j}) \times (-10\mathbf{i} + 4\mathbf{j}) = 22\mathbf{k} \text{ (N-m)}.$$

We would obtain the same result by calculating the sum of the moments about any point. For example, the sum of the moments about the point P in Fig. b is

$$\mathbf{M} = \left[\mathbf{r}_1 \times \mathbf{F}\right] + \left[\mathbf{r}_2 \times (-\mathbf{F})\right]$$

$$= \begin{vmatrix} \mathbf{i} & \mathbf{j} & \mathbf{k} \\ -2 & -4 & -3 \\ 10 & -4 & 0 \end{vmatrix} + \begin{vmatrix} \mathbf{i} & \mathbf{j} & \mathbf{k} \\ -4 & -1 & -3 \\ -10 & 4 & 0 \end{vmatrix}$$

$$= 22\mathbf{k} \text{ (N-m)}.$$

Second Method The x and y components of the forces form two couples (Fig. c). We determine the moment of the original couple by summing the moments of the couples formed by the components.

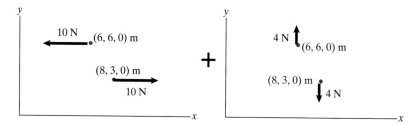

Consider the 10-N couple. The magnitude of its moment is (3 m)(10 N) = 30 N-m, and its sense is counterclockwise, indicating that the moment vector points out of the page. Therefore the moment is $30\mathbf{k}$ N-m.

The 4-N couple causes a moment of magnitude (2 m)(4 N) = 8 N-m and its sense is clockwise, so the moment is $-8\mathbf{k}$ N-m. The moment of the original couple is

$$\mathbf{M} = 30\mathbf{k} - 8\mathbf{k} = 22\mathbf{k} \text{ (N-m)}.$$

Its magnitude is 22 N-m, and its sense is counterclockwise (Fig. d).

Example 4.10

Determining Unknown Forces

Two forces A and B and a 200 ft-lb couple act on the beam in Fig. 4.29. The sum of the forces is zero, and the sum of the moments about the left end of the beam is zero. What are the forces A and B?

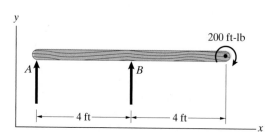

Figure 4.29

Solution

The sum of the forces is

$$\Sigma F_y = A + B = 0.$$

The moment of the couple (200 ft-lb clockwise) is the same about any point, so the sum of the moments about the left end of the beam is

$$\Sigma M_{(\text{left end})} = 4B - 200 = 0.$$

The forces are $B = 50$ lb and $A = -50$ lb.

Discussion

Notice that A and B form a couple (Fig. a). It causes a moment of magnitude (4 ft)(50 lb) = 200 ft-lb, and its sense is counterclockwise, so the sum of the moments of the couple formed by A and B and the 200 ft-lb clockwise couple is zero.

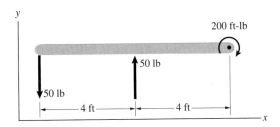

(a) The forces on the beam form a couple.

Example 4.11

Sum of the Moments Due to Two Couples

Determine the sum of the moments exerted on the pipe in Fig. 4.30 by the two couples.

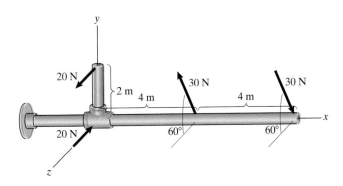

Figure 4.30

Solution

Consider the 20-N couple. The magnitude of the moment of the couple is (2 m)(20 N) = 40 N-m. The direction of the moment vector is perpendicular to the y–z plane, and the right-hand rule indicates that it points in the positive x-axis direction. The moment of the 20-N couple is $40\mathbf{i}$ (N-m).

$30 \sin 60°$ N $30 \sin 60°$ N

4 m

x

$30 \cos 60°$ N $30 \cos 60°$ N

z

(a) Resolving the 30-N forces into y and z components.

By resolving the 30-N forces into y and z components, we obtain the two couples in Fig. a. The moment of the couple formed by the y components is $-(30 \sin 60°)(4)\mathbf{k}$ (N-m), and the moment of the couple formed by the z components is $(30 \cos 60°)(4)\mathbf{j}$ (N-m).

The sum of the moments is

$$\Sigma\mathbf{M} = 40\mathbf{i} + (30\cos 60°)(4)\mathbf{j} - (30\sin 60°)(4)\mathbf{k}$$
$$= 40\mathbf{i} + 60\mathbf{j} - 103.9\mathbf{k} \text{ (N-m)}.$$

Discussion

Although the method we used in this example helps you recognize the contributions of the individual couples to the sum of the moments, it is convenient only when the orientations of the forces and their points of application relative to the coordinate system are fairly simple. When that is not the case, you can determine the sum of the moments by choosing any point and calculating the sum of the moments of the forces about that point.

Example 4.12

Distance Between the Lines of Action

The force $\mathbf{F}$ in Fig. 4.31 is $-20\mathbf{i} + 20\mathbf{j} + 10\mathbf{k}$ (lb).
(a) What moment does the couple exert on the bracket?
(b) What is the perpendicular distance D between the lines of action of the two forces?

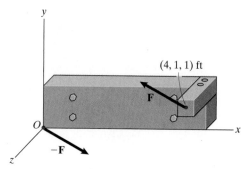

y

$(4, 1, 1)$ ft

$\mathbf{F}$

O

x

$-\mathbf{F}$

z

Figure 4.31

Strategy

(a) We can choose a point and determine the sum of the moments of the forces about that point.

(b) The magnitude of the moment of the couple equals $D\,|\mathbf{F}|$, so we can use the result of (a) to determine D.

Solution

(a) If we determine the sum of the moments of the forces about the origin O, the moment of the force $-\mathbf{F}$ is zero. The moment of the couple is (Fig. a)

$$\mathbf{M} = \mathbf{r} \times \mathbf{F} = \begin{vmatrix} \mathbf{i} & \mathbf{j} & \mathbf{k} \\ 4 & 1 & 1 \\ -20 & 20 & 10 \end{vmatrix} = -10\mathbf{i} - 60\mathbf{j} + 100\mathbf{k}\ \text{(ft-lb)}.$$

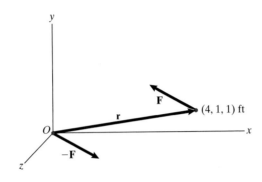

(a) Determining the sum of the moments about O.

(b) The perpendicular distance is

$$D = \frac{|\mathbf{M}|}{|\mathbf{F}|} = \frac{\sqrt{(-10)^2 + (-60)^2 + (100)^2}}{\sqrt{(-20)^2 + (20)^2 + (10)^2}} = 3.90\ \text{ft}.$$

Problems

4.108 Determine the moment of the couple and represent it as shown in Fig. 4.26c.

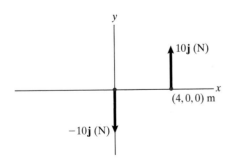

P4.108

4.109 The forces are contained in the x–y plane.
(a) Determine the moment of the couple and represent it as shown in Fig. 4.26c.
(b) What is the sum of the moments of the two forces about the point $(10, -40, 20)$ ft?

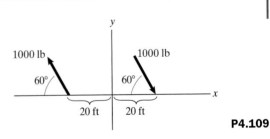

P4.109

4.110 The forces are contained in the x–y plane and the moment of the couple is $-110\mathbf{k}$ (N-m).
(a) What is the distance b?
(b) What is the sum of the moments of the two forces about the point $(3, -3, 2)$ m?

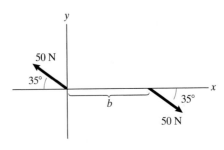

P4.110

4.111 Point P is contained in the x–y plane, $|\mathbf{F}| = 100$ N, and the moment of the couple is $-500\mathbf{k}$ (N-m). What are the coordinates of P?

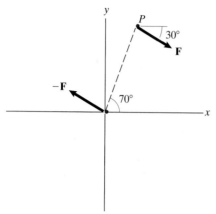

P4.111

4.112 The forces are contained in the x–y plane.
(a) Determine the sum of the moments of the two couples.
(b) What is the sum of the moments of the four forces about the point $(-6, -6, 2)$ m?
(c) Represent the result of (a) as shown in Fig. 4.26c.

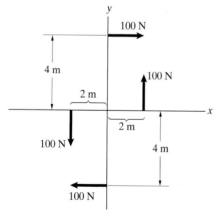

P4.112

4.113 The moment of the couple is 40 kN-m counterclockwise.
(a) Express the moment of the couple as a vector.
(b) Draw a sketch showing two equal and opposite forces that exert the given moment.

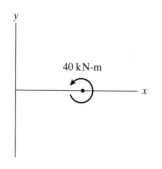

P4.113

4.114 The moments of two couples are shown. What is the sum of the moments about point P?

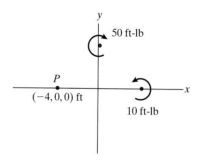

P4.114

4.115 Determine the sum of the moments exerted on the plate by the two couples.

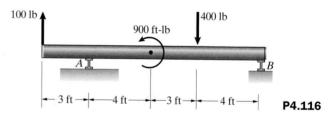

P4.115

4.116 Determine the sum of the moments exerted about A by the couple and the two forces.

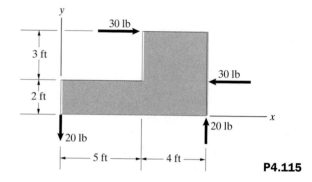

P4.116

4.117 Determine the sum of the moments exerted about A by the couple and the two forces.

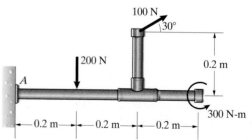

P4.117

4.118 What is the sum of the moments exerted on the object?

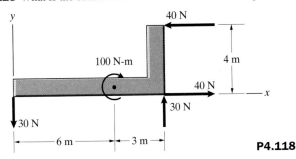

P4.118

4.119 Four forces and a couple act on the beam. The vector sum of the forces is zero, and the sum of the moments about the left end of the beam is zero. What are the forces A_x, A_y, and B?

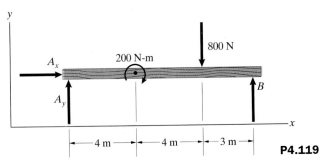

P4.119

4.120 The force $\mathbf{F} = 40\mathbf{i} + 24\mathbf{j} + 12\mathbf{k}$ (N).
(a) What is the moment of the couple?
(b) Determine the perpendicular distance between the lines of action of the two forces.

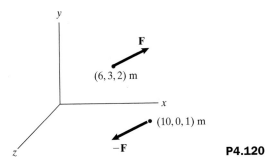

P4.120

4.121 Determine the sum of the moments exerted on the plate by the three couples. (The 80-lb forces are contained in the x–z plane.)

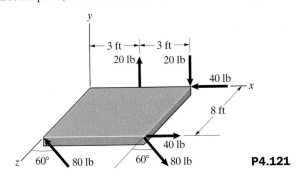

P4.121

4.122 What is the magnitude of the sum of the moments exerted on the T-shaped structure by the two couples?

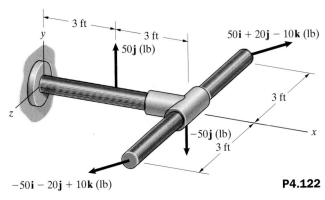

P4.122

4.123 The tension in cables AB and CD is 500 N.
(a) Show that the two forces exerted by the cables on the rectangular hatch at B and C form a couple.
(b) What is the moment exerted on the plate by the cables?

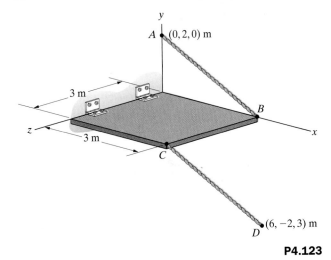

P4.123

4.124 Determine the sum of the moments exerted about P by the couple and two forces acting on the cube.

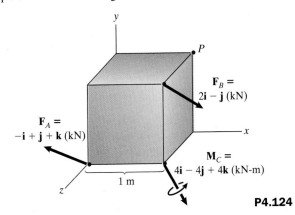

P4.124

4.125 The bar is loaded by the forces

$$\mathbf{F}_B = 2\mathbf{i} + 6\mathbf{j} + 3\mathbf{k} \ (\text{kN}),$$
$$\mathbf{F}_C = \mathbf{i} - 2\mathbf{j} + 2\mathbf{k} \ (\text{kN}),$$

and the couple

$$\mathbf{M}_C = 2\mathbf{i} + \mathbf{j} - 2\mathbf{k} \ (\text{kN-m}).$$

Determine the sum of the moments of the two forces and the couple about A.

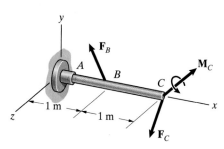

P4.125

4.126 In Problem 4.125, the forces

$$\mathbf{F}_B = 2\mathbf{i} + 6\mathbf{j} + 3\mathbf{k} \ (\text{kN}),$$
$$\mathbf{F}_C = \mathbf{i} - 2\mathbf{j} + 2\mathbf{k} \ (\text{kN}),$$

and the couple

$$\mathbf{M}_C = M_{Cy}\mathbf{j} + M_{Cz}\mathbf{k} \ (\text{kN-m}).$$

Determine the values of M_{Cy} and M_{Cz} so that the sum of the moments of the two forces and the couple about A is zero.

4.127 Two wrenches are used to tighten an elbow fitting. The force $\mathbf{F} = 10\mathbf{k}$ (lb) on the right wrench is applied at $(6, -5, -3)$ in., and the force $-\mathbf{F}$ on the left wrench is applied at $(4, -5, 3)$ in.
(a) Determine the moment about the x axis due to the force exerted on the right wrench.
(b) Determine the moment of the couple formed by the forces exerted on the two wrenches.
(c) Based on the results of (a) and (b), explain why two wrenches are used.

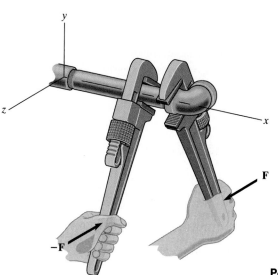

P4.127

4.5 **Equivalent Systems**

A *system of forces and moments* is simply a particular set of forces and moments of couples. The systems of forces and moments dealt with in engineering can be complicated. This is especially true in the case of distributed forces, such as the pressure forces exerted by water on a dam. Fortunately, if we are concerned only with the total force and moment exerted, we can represent complicated systems of forces and moments by much simpler systems.

Conditions for Equivalence

We define two systems of forces and moments, designated as system 1 and system 2, to be *equivalent* if the sums of the forces are equal,

$$(\Sigma \mathbf{F})_1 = (\Sigma \mathbf{F})_2, \tag{4.7}$$

and the sums of the moments about a point P are equal,

$$(\Sigma \mathbf{M}_P)_1 = (\Sigma \mathbf{M}_P)_2. \tag{4.8}$$

Demonstration of Equivalence

To see what the conditions for equivalence mean, consider the systems of forces and moments in Fig. 4.32a. In system 1, an object is subjected to two forces $\mathbf{F}_A$ and $\mathbf{F}_B$ and a couple $\mathbf{M}_C$. In system 2, the object is subjected to a force $\mathbf{F}_D$ and two couples $\mathbf{M}_E$ and $\mathbf{M}_F$. The first condition for equivalence is

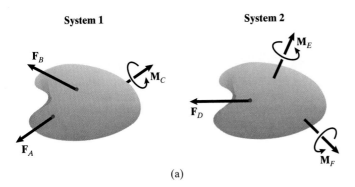

System 1   System 2

(a)

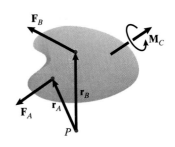

System 1

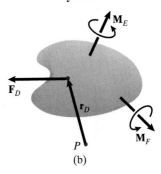

System 2

(b)

Figure 4.32

(a) Different systems of forces and moments applied to an object.

(b) Determining the sum of the moments about a point P for each system.

$$(\Sigma \mathbf{F})_1 = (\Sigma \mathbf{F})_2:$$

$$\mathbf{F}_A + \mathbf{F}_B = \mathbf{F}_D. \qquad (4.9)$$

If we determine the sums of the moments about the point P in Fig. 4.32b, the second condition for equivalence is

$$(\Sigma \mathbf{M}_P)_1 = (\Sigma \mathbf{M}_P)_2:$$

$$(\mathbf{r}_A \times \mathbf{F}_A) + (\mathbf{r}_B \times \mathbf{F}_B) + \mathbf{M}_C = (\mathbf{r}_D \times \mathbf{F}_D) + \mathbf{M}_E + \mathbf{M}_F. \qquad (4.10)$$

If these conditions are satisfied, systems 1 and 2 are equivalent.

We will use this example to demonstrate that *if the sums of the forces are equal for two systems of forces and moments and the sums of the moments about one point P are equal, then the sums of the moments about any point are equal.* Suppose that Eq. (4.9) is satisfied, and Eq. (4.10) is satisfied for the point P in Fig. 4.32b. For a different point P' (Fig. 4.33), we will show that

$$(\Sigma \mathbf{M}_{P'})_1 = (\Sigma \mathbf{M}_{P'})_2:$$

$$(\mathbf{r}'_A \times \mathbf{F}_A) + (\mathbf{r}'_B \times \mathbf{F}_B) + \mathbf{M}_C = (\mathbf{r}'_D \times \mathbf{F}_D) + \mathbf{M}_E + \mathbf{M}_F. \qquad (4.11)$$

In terms of the vector $\mathbf{r}$ from P' to P, the relations between the vectors $\mathbf{r}'_A$, $\mathbf{r}'_B$, and $\mathbf{r}'_D$ in Fig. 4.33 and the vectors $\mathbf{r}_A$, $\mathbf{r}_B$, and $\mathbf{r}_D$ in Fig. 4.32b are

$$\mathbf{r}'_A = \mathbf{r} + \mathbf{r}_A, \qquad \mathbf{r}'_B = \mathbf{r} + \mathbf{r}_B, \qquad \mathbf{r}'_D = \mathbf{r} + \mathbf{r}_D.$$

Substituting these expressions into Eq. (4.11), we obtain

$$[(\mathbf{r} + \mathbf{r}_A) \times \mathbf{F}_A] + [(\mathbf{r} + \mathbf{r}_B) \times \mathbf{F}_B] + \mathbf{M}_C$$

$$= [(\mathbf{r} + \mathbf{r}_D) \times \mathbf{F}_D] + \mathbf{M}_E + \mathbf{M}_F.$$

Rearranging terms, we can write this equation as

$$[\mathbf{r} \times (\Sigma \mathbf{F})_1] + (\Sigma \mathbf{M}_P)_1 = [\mathbf{r} \times (\Sigma \mathbf{F})_2] + (\Sigma \mathbf{M}_P)_2,$$

which holds in view of Eqs. (4.9) and (4.10). The sums of the moments of the two systems about any point are equal.

Study Questions

1. What conditions must be satisfied for two systems of forces and moments to be equivalent?

2. If the sums of the forces in two systems of forces and moments are the same, and the sums of the moments about a point P are the same, what do you know about the sums of the moments about a different point P'?

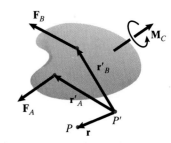

System 1

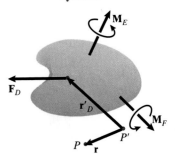

System 2

Figure 4.33

Determining the sum of the moments about a different point P' for each system.

Example 4.13

Determining Whether Systems are Equivalent

Three systems of forces and moments act on the beam in Fig. 4.34. Are they equivalent?

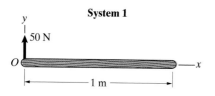

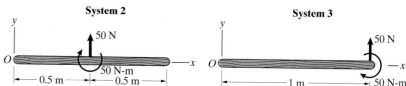

Figure 4.34

Solution

Are the Sums of the Forces Equal?   The sums of the forces are

$$(\Sigma \mathbf{F})_1 = 50\mathbf{j} \ (\text{N}),$$

$$(\Sigma \mathbf{F})_2 = 50\mathbf{j} \ (\text{N}),$$

$$(\Sigma \mathbf{F})_3 = 50\mathbf{j} \ (\text{N}).$$

Are the Sums of the Moments About an Arbitrary Point Equal?   The sums of the moments about the origin O are

$$(\Sigma M_O)_1 = 0,$$

$$(\Sigma M_O)_2 = (50 \ \text{N})(0.5 \ \text{m}) - (50 \ \text{N-m}) = -25 \ \text{N-m},$$

$$(\Sigma M_O)_3 = (50 \ \text{N})(1 \ \text{m}) - (50 \ \text{N-m}) = 0.$$

Systems 1 and 3 are equivalent.

Discussion

Remember that you can choose any convenient point to determine whether the sums of the moments are equal. For example, the sums of the moments about the right end of the beam are

$$(\Sigma M_{\text{right end}})_1 = -(50 \ \text{N})(1 \ \text{m}) = 50 \ \text{N-m},$$

$$(\Sigma M_{\text{right end}})_2 = -(50 \ \text{N})(0.5 \ \text{m}) - (50 \ \text{N-m}) = -75 \ \text{N-m},$$

$$(\Sigma M_{\text{right end}})_3 = -50 \ \text{N-m}.$$

Example 4.14

Determining Whether Systems are Equivalent

Two systems of forces and moments act on the rectangular plate in Fig. 4.35. Are they equivalent?

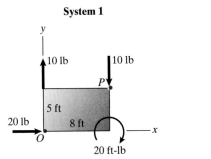

System 1

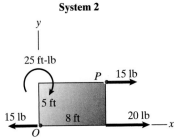

System 2

Figure 4.35

Solution

Are the Sums of the Forces Equal? The sums of the forces are

$$\left(\Sigma \mathbf{F}\right)_1 = 20\mathbf{i} + 10\mathbf{j} - 10\mathbf{j} = 20\mathbf{i} \text{ (lb)},$$

$$\left(\Sigma \mathbf{F}\right)_2 = 20\mathbf{i} + 15\mathbf{i} - 15\mathbf{i} = 20\mathbf{i} \text{ (lb)}.$$

Are the Sums of the Moments About an Arbitrary Point Equal? The sums of the moments about the origin O are

$$\left(\Sigma M_O\right)_1 = -(8 \text{ ft})(10 \text{ lb}) - (20 \text{ ft-lb}) = -100 \text{ ft-lb},$$

$$\left(\Sigma M_O\right)_2 = -(5 \text{ ft})(15 \text{ lb}) - (25 \text{ ft-lb}) = -100 \text{ ft-lb}.$$

The systems are equivalent.

Discussion

Let's confirm that the sums of the moments of the two systems about a different point are equal. The sums of the moments about P are

$$\left(\Sigma M_P\right)_1 = -(8 \text{ ft})(10 \text{ lb}) + (5 \text{ ft})(20 \text{ lb}) - (20 \text{ ft-lb}) = 0,$$

$$\left(\Sigma M_P\right)_2 = -(5 \text{ ft})(15 \text{ lb}) + (5 \text{ ft})(20 \text{ lb}) - (25 \text{ ft-lb}) = 0.$$

Example 4.15

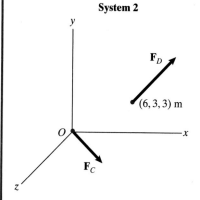

System 1

$(6, 0, 0)$ m

System 2

$(6, 3, 3)$ m

Figure 4.36

Determining Whether Systems are Equivalent

Two systems of forces and moments are shown in Fig. 4.36, where

$$\mathbf{F}_A = -10\mathbf{i} + 10\mathbf{j} - 15\mathbf{k} \text{ (kN)},$$

$$\mathbf{F}_B = 30\mathbf{i} + 5\mathbf{j} + 10\mathbf{k} \text{ (kN)},$$

$$\mathbf{M} = -90\mathbf{i} + 150\mathbf{j} + 60\mathbf{k} \text{ (kN-m)},$$

$$\mathbf{F}_C = 10\mathbf{i} - 5\mathbf{j} + 5\mathbf{k} \text{ (kN)},$$

$$\mathbf{F}_D = 10\mathbf{i} + 20\mathbf{j} - 10\mathbf{k} \text{ (kN)}.$$

Are they equivalent?

Solution

Are the Sums of the Forces Equal? The sums of the forces are

$$(\Sigma \mathbf{F})_1 = \mathbf{F}_A + \mathbf{F}_B = 20\mathbf{i} + 15\mathbf{j} - 5\mathbf{k} \text{ (kN)}.$$

$$(\Sigma \mathbf{F})_2 = \mathbf{F}_C + \mathbf{F}_D = 20\mathbf{i} + 15\mathbf{j} - 5\mathbf{k} \text{ (kN)}.$$

Are the Sums of the Moments About an Arbitrary Point Equal? The sum of the moments about the origin O in system 1 is

$$(\Sigma \mathbf{M}_O)_1 = (6\mathbf{i} \times \mathbf{F}_B) + \mathbf{M}$$

$$= \begin{vmatrix} \mathbf{i} & \mathbf{j} & \mathbf{k} \\ 6 & 0 & 0 \\ 30 & 5 & 10 \end{vmatrix} + (-90\mathbf{i} + 150\mathbf{j} + 60\mathbf{k})$$

$$= -90\mathbf{i} + 90\mathbf{j} + 90\mathbf{k} \text{ (kN-m)}.$$

The sum of the moments about O in system 2 is

$$(\Sigma \mathbf{M}_O)_2 = (6\mathbf{i} + 3\mathbf{j} + 3\mathbf{k}) \times \mathbf{F}_D = \begin{vmatrix} \mathbf{i} & \mathbf{j} & \mathbf{k} \\ 6 & 3 & 3 \\ 10 & 20 & -10 \end{vmatrix}$$

$$= -90\mathbf{i} + 90\mathbf{j} + 90\mathbf{k} \text{ (kN-m)}.$$

The systems are equivalent.

4.6 Representing Systems by Equivalent Systems

If we are concerned only with the total force and total moment exerted on an object by a given system of forces and moments, we can *represent* the system by an equivalent one. By this we mean that instead of showing the actual forces and couples acting on an object, we would show a different system that exerts the same total force and moment. In this way, we can replace a given system by a less complicated one to simplify the analysis of the forces and moments acting on an object and to gain a better intuitive understanding of their effects on the object.

Representing a System by a Force and a Couple

Let's consider an arbitrary system of forces and moments and a point P (system 1 in Fig. 4.37). We can represent this system by one consisting of a single force acting at P and a single couple (system 2). The conditions for equivalence are

$$(\Sigma \mathbf{F})_2 = (\Sigma \mathbf{F})_1:$$

$$\mathbf{F} = (\Sigma \mathbf{F})_1$$

and

$$(\Sigma \mathbf{M}_P)_2 = (\Sigma \mathbf{M}_P)_1:$$

$$\mathbf{M} = (\Sigma \mathbf{M}_P)_1.$$

These conditions are satisfied if $\mathbf{F}$ equals the sum of the forces in system 1 and $\mathbf{M}$ equals the sum of the moments about P in system 1.

Thus *no matter how complicated a system of forces and moments may be, we can represent it by a single force acting at a given point and a single couple*. Three particular cases occur frequently in practice:

Representing a Force by a Force and a Couple We can represent a force $\mathbf{F}_P$ acting at a point P (system 1 in Fig. 4.38a) by a force $\mathbf{F}$ acting at a different point Q and a couple $\mathbf{M}$ (system 2). The moment of system 1 about point Q is $\mathbf{r} \times \mathbf{F}_P$, where $\mathbf{r}$ is the vector from Q to P (Fig. 4.38b). The conditions for equivalence are

$$(\Sigma \mathbf{F})_2 = (\Sigma \mathbf{F})_1:$$

$$\mathbf{F} = \mathbf{F}_P$$

and

$$(\Sigma \mathbf{M}_Q)_2 = (\Sigma \mathbf{M}_Q)_1:$$

$$\mathbf{M} = \mathbf{r} \times \mathbf{F}_P.$$

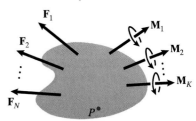

System 1

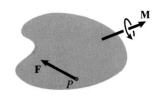

System 2

Figure 4.37
(a) An arbitrary system of forces and moments.
(b) A force acting at P and a couple.

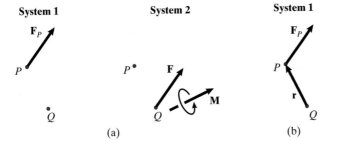

(a)   (b)

Figure 4.38
(a) System 1 is a force $\mathbf{F}_P$ acting at point P. System 2 consists of a force $\mathbf{F}$ acting at point Q and a couple $\mathbf{M}$.
(b) Determining the moment of system 1 about point Q.

The systems are equivalent if the force $\mathbf{F}$ equals the force $\mathbf{F}_P$ and the couple $\mathbf{M}$ equals the moment of $\mathbf{F}_P$ about Q.

Concurrent Forces Represented by a Force We can represent a system of concurrent forces whose lines of action intersect at a point P (system 1 in Fig. 4.39) by a single force whose line of action intersects P (system 2). The sums of the forces in the two systems are equal if

$$\mathbf{F} = \mathbf{F}_1 + \mathbf{F}_2 + \cdots + \mathbf{F}_N.$$

The sum of the moments about P equals zero for each system, so the systems are equivalent if the force $\mathbf{F}$ equals the sum of the forces in system 1.

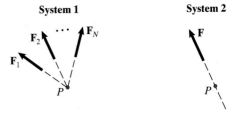

Figure 4.39
A system of concurrent forces and a system consisting of a single force $\mathbf{F}$.

Parallel Forces Represented by a Force   We can represent a system of parallel forces whose sum is not zero by a single force **F** (Fig. 4.40). We demonstrate this result in Example 4.19.

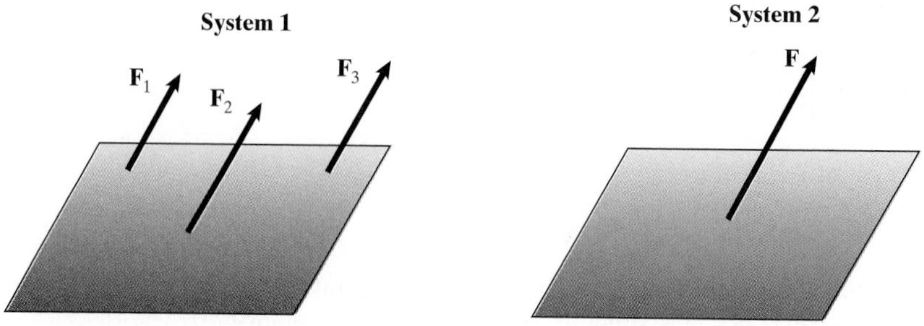

System 1

$\mathbf{F}_1$ $\mathbf{F}_2$ $\mathbf{F}_3$

System 2

F

Figure 4.40
A system of parallel forces and a system consisting of a single force **F**.

Study Questions

1. If you represent a system of forces and moments by a force **F** acting at a point P and a couple **M**, how do you determine **F** and **M**?
2. If you represent a system of concurrent forces by a single force **F**, what condition must be satisfied by the line of action of **F**?

Example 4.16

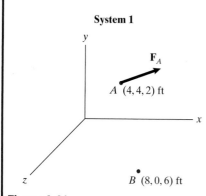

System 1

$\mathbf{F}_A$

A (4, 4, 2) ft

x

z

B (8, 0, 6) ft

Figure 4.41

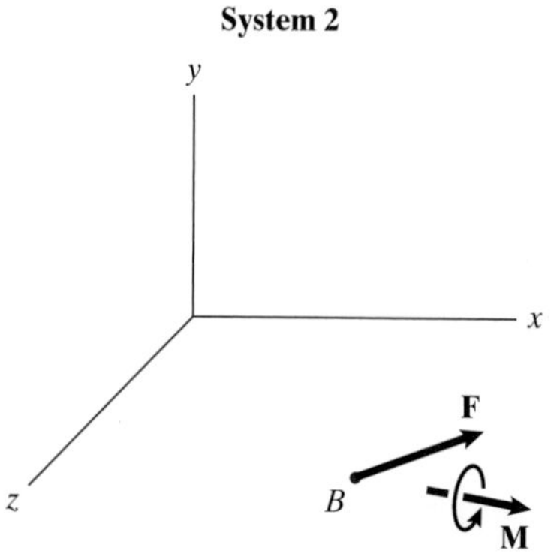

System 2

y

x

z

F

B

M

(a) A force acting at B and a couple.

Representing a Force by a Force and Couple

System 1 in Fig. 4.41 consists of a force $\mathbf{F}_A = 10\mathbf{i} + 4\mathbf{j} - 3\mathbf{k}$ (lb) acting at **A**. Represent it by a force acting at B and a couple.

Strategy

We want to represent the force $\mathbf{F}_A$ by a force **F** acting at B and a couple **M** (system 2 in Fig. a). We can determine **F** and **M** by using the two conditions for equivalence.

Solution

The sums of the forces must be equal:

$$(\Sigma \mathbf{F})_2 = (\Sigma \mathbf{F})_1:$$

$$\mathbf{F} = \mathbf{F}_A = 10\mathbf{i} + 4\mathbf{j} - 3\mathbf{k} \text{ (lb)}.$$

The sums of the moments about an arbitrary point must be equal: The vector from B to A is

$$\mathbf{r}_{BA} = (4 - 8)\mathbf{i} + (4 - 0)\mathbf{j} + (2 - 6)\mathbf{k} = -4\mathbf{i} + 4\mathbf{j} - 4\mathbf{k} \text{ (ft)},$$

so the moment about B in system 1 is

$$\mathbf{r}_{BA} \times \mathbf{F}_A = \begin{vmatrix} \mathbf{i} & \mathbf{j} & \mathbf{k} \\ -4 & 4 & -4 \\ 10 & 4 & -3 \end{vmatrix} = 4\mathbf{i} - 52\mathbf{j} - 56\mathbf{k} \text{ (ft-lb)}.$$

The sums of the moments about B must be equal:

$$(\mathbf{M}_B)_2 = (\mathbf{M}_B)_1:$$

$$\mathbf{M} = 4\mathbf{i} - 52\mathbf{j} - 56\mathbf{k} \text{ (ft-lb)}.$$

Example 4.17

Representing a System by a Simpler Equivalent System

System 1 in Fig. 4.42 consists of two forces and a couple acting on a pipe. Represent system 1 by (a) a single force acting at the origin O of the coordinate system and a single couple and (b) a single force.

Strategy

(a) We can represent system 1 by a force $\mathbf{F}$ acting at the origin and a couple M (system 2 in Fig. a) and use the conditions for equivalence to determine $\mathbf{F}$ and $\mathbf{M}$.

(b) Suppose that we place the force $\mathbf{F}$ with its point of application a distance D along the x axis (system 3 in Fig. b). The sums of the forces in systems 2 and 3 are equal. If we can choose the distance D so that the moment about O in system 3 equals $\mathbf{M}$, system 3 will be equivalent to system 2 and therefore equivalent to system 1.

Solution

(a) The conditions for equivalence are

$$(\Sigma \mathbf{F})_2 = (\Sigma \mathbf{F})_1:$$

$$\mathbf{F} = 30\mathbf{j} + (20\mathbf{i} + 20\mathbf{j}) = 20\mathbf{i} + 50\mathbf{j} \text{ (kN)},$$

and

$$(\Sigma M_O)_2 = (\Sigma M_O)_1:$$

$$M = (30 \text{ kN})(3 \text{ m}) + (20 \text{ kN})(5 \text{ m}) + 210 \text{ kN-m}$$

$$= 400 \text{ kN-m}.$$

(b) The sums of the forces in systems 2 and 3 are equal. Equating the sums of the moments about O,

$$(\Sigma M_O)_3 = (\Sigma M_O)_2:$$

$$(50 \text{ kN})D = 400 \text{ kN-m}.$$

we find that system 3 is equivalent to system 2 if $D = 8$ m.

Discussion

To represent the system by a single force in (b), we needed to place the line of action of the force so that the force exerted a 400 kN-m counterclockwise moment about O. Placing the point of application of the force a distance D along the x axis was simply a convenient way to accomplish that.

System 1

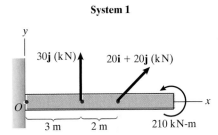

Figure 4.42

System 2

(a) A force $\mathbf{F}$ acting at O and a couple M.

System 3

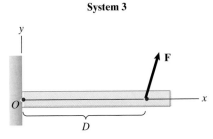

(b) A system consisting of the force $\mathbf{F}$ acting at a point on the x axis.

Example 4.18

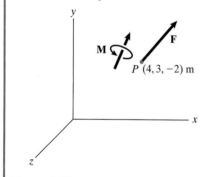

Figure 4.43

Representing a System by a Force and Couple

System 1 in Fig. 4.43 consists of the following forces and couple:

$$\mathbf{F}_A = -10\mathbf{i} + 10\mathbf{j} - 15\mathbf{k} \text{ (kN)},$$

$$\mathbf{F}_B = 30\mathbf{i} + 5\mathbf{j} + 10\mathbf{k} \text{ (kN)},$$

$$\mathbf{M}_C = -90\mathbf{i} + 150\mathbf{j} + 60\mathbf{k} \text{ (kN-m)}.$$

Suppose you want to represent it by a force $\mathbf{F}$ acting at P and a couple $\mathbf{M}$ (system 2). Determine $\mathbf{F}$ and $\mathbf{M}$.

Solution

The sums of the forces must be equal:

$$(\Sigma \mathbf{F})_2 = (\Sigma \mathbf{F})_1:$$

$$\mathbf{F} = \mathbf{F}_A + \mathbf{F}_B = 20\mathbf{i} + 15\mathbf{j} - 5\mathbf{k} \text{ (kN)}.$$

The sums of the moments about an arbitrary point must be equal: The sums of the moments about point P must be equal:

$$(\Sigma \mathbf{M}_P)_2 = (\Sigma \mathbf{M}_P)_1:$$

$$\mathbf{M} = \begin{vmatrix} \mathbf{i} & \mathbf{j} & \mathbf{k} \\ -4 & -3 & 2 \\ -10 & 10 & -15 \end{vmatrix} + \begin{vmatrix} \mathbf{i} & \mathbf{j} & \mathbf{k} \\ 2 & -3 & 2 \\ 30 & 5 & 10 \end{vmatrix}$$

$$+ (-90\mathbf{i} + 150\mathbf{j} + 60\mathbf{k})$$

$$= -105\mathbf{i} + 110\mathbf{j} + 90\mathbf{k} \text{ (kN-m)}.$$

Example 4.19

Representing Parallel Forces by a Single Force

System 1 in Fig. 4.44 consists of parallel forces. Suppose you want to represent it by a force $\mathbf{F}$ (system 2). What is $\mathbf{F}$, and where does its line of action intersect the x–z plane?

Strategy

We can determine $\mathbf{F}$ from the condition that the sums of the forces in the two systems must be equal. For the two systems to be equivalent, we must choose the point of application P so that the sums of the moments about a point are equal. This condition will tell us where the line of action intersects the x–z plane.

Solution

The sums of the forces must be equal:

$$(\Sigma \mathbf{F})_2 = (\Sigma \mathbf{F})_1:$$

$$\mathbf{F} = 30\mathbf{j} + 20\mathbf{j} - 10\mathbf{j} = 40\mathbf{j} \ (\text{lb}).$$

The sums of the moments about an arbitrary point must be equal: Let the co-ordinates of point P be (x, y, z). The sums of the moments about the origin O must be equal.

$$(\Sigma M_O)_2 = (\Sigma M_O)_1:$$

$$\begin{vmatrix} \mathbf{i} & \mathbf{j} & \mathbf{k} \\ x & y & z \\ 0 & 40 & 0 \end{vmatrix} = \begin{vmatrix} \mathbf{i} & \mathbf{j} & \mathbf{k} \\ 6 & 0 & 2 \\ 0 & 30 & 0 \end{vmatrix} + \begin{vmatrix} \mathbf{i} & \mathbf{j} & \mathbf{k} \\ 2 & 0 & 4 \\ 0 & -10 & 0 \end{vmatrix}$$

$$+ \begin{vmatrix} \mathbf{i} & \mathbf{j} & \mathbf{k} \\ -3 & 0 & -2 \\ 0 & 20 & 0 \end{vmatrix}.$$

Expanding the determinants, we obtain

$$(20 + 40z)\mathbf{i} + (100 - 40x)\mathbf{k} = \mathbf{0}.$$

The sums of the moments about the origin are equal if

$$x = 2.5 \text{ ft},$$

$$z = -0.5 \text{ ft}.$$

The systems are equivalent if $\mathbf{F} = 40\mathbf{j}$ (lb) and its line of action intersects the x–z plane at $x = 2.5$ ft and $z = -0.5$ ft. Notice that we did not obtain an equation for the y coordinate of P. The systems are equivalent if $\mathbf{F}$ is applied at any point along the line of action.

Discussion

We could have determined the x and z coordinates of point P in a simpler way. Since the sums of the moments about any point must be equal for the systems to be equivalent, the sums of the moments about any *line* must also be equal. Equating the sums of the moments about the x axis,

$$\left(\Sigma M_{x\,\text{axis}}\right)_2 = \left(\Sigma M_{x\,\text{axis}}\right)_1:$$

$$-40z = -(30)(2) + (10)(4) + (20)(2),$$

we obtain $z = -0.5$ ft, and equating the sums of the moments about the z axis,

$$\left(\Sigma M_{z\,\text{axis}}\right)_2 = \left(\Sigma M_{z\,\text{axis}}\right)_1:$$

$$40x = (30)(6) - (10)(2) - (20)(3),$$

we obtain $x = 2.5$ ft.

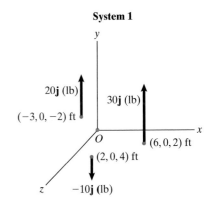

System 1

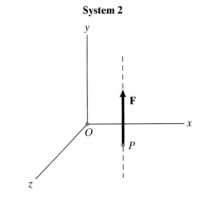

System 2

Figure 4.44

Representing a System by a Wrench

We have shown that any system of forces and moments can be represented by a single force acting at a given point and a single couple. This raises an interesting question: What is the simplest system that can be equivalent to any system of forces and moments?

To consider this question, let's begin with an arbitrary force **F** acting at a point P and an arbitrary couple **M** (system 1 in Fig. 4.45a) and see whether we can represent this system by a simpler one. For example, can we represent it by the force **F** acting at a different point Q and no couple (Fig 4.45b)? The sum of the forces is the same as in system 1. If we can choose the point Q so that $\mathbf{r} \times \mathbf{F} = \mathbf{M}$, where **r** is the vector from P to Q (Fig. 4.45c), the sum of the moments about P is the same as in system 1 and the systems are equivalent. But the vector $\mathbf{r} \times \mathbf{F}$ is perpendicular to **F**, so it can equal **M** only if **M** is perpendicular to **F**. That means that, in general, we can't represent system 1 by the force **F** alone.

However, we can represent system 1 by the force **F** acting at a point Q and the component of **M** that is parallel to **F**. Figure 4.45d shows system 1 with a coordinate system placed so that **F** is along the y axis and **M** is contained in the $x–y$ plane. In terms of this coordinate system, we can express

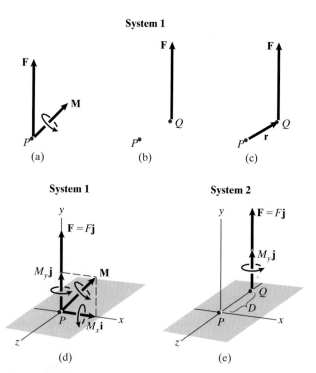

Figure 4.45
(a) System 1 is a single force and a single couple.
(b) Can system 1 be represented by a single force and no couple?
(c) The moment of **F** about P is $\mathbf{r} \times \mathbf{F}$.
(d) **F** is along the y axis, and **M** is contained in the $x–y$ plane.
(e) System 2 is the force **F** and the component of **M** parallel to **F**.

the force and couple as $\mathbf{F} = F\mathbf{j}$ and $\mathbf{M} = M_x\mathbf{i} + M_y\mathbf{j}$. System 2 in Fig. 4.45e consists of the force $\mathbf{F}$ acting at a point on the z axis and the component of $\mathbf{M}$ parallel to $\mathbf{F}$. If we choose the distance D so that $D = M_x/F$, system 2 is equivalent to system 1. The sum of the forces in each system is $\mathbf{F}$. The sum of the moments about P in system 1 is $\mathbf{M}$, and the sum of the moments about P in system 2 is

$$\left(\Sigma\,\mathbf{M}_P\right)_2 = \left[(-D\mathbf{k}) \times (F\mathbf{j})\right] + M_y\mathbf{j} = M_x\mathbf{i} + M_y\mathbf{j} = \mathbf{M}.$$

A force $\mathbf{F}$ and a couple $\mathbf{M}_p$ that is parallel to $\mathbf{F}$ is called a *wrench*; it is the simplest system that can be equivalent to an arbitrary system of forces and moments.

How can you represent a given system of forces and moments by a wrench? If the system is a single force or a single couple or if it consists of a force $\mathbf{F}$ and a couple that is parallel to $\mathbf{F}$, it is a wrench, and you can't simplify it further. If the system is more complicated than a single force and a single couple, begin by choosing a convenient point P and representing the system by a force $\mathbf{F}$ acting at P and a couple $\mathbf{M}$ (Fig. 4.46a). Then representing this system by a wrench requires two steps:

1. Determine the components of $\mathbf{M}$ parallel and normal to $\mathbf{F}$ (Fig. 4.46b).
2. The wrench consists of the force $\mathbf{F}$ acting at a point Q and the parallel component $\mathbf{M}_P$ (Fig. 4.46c). To achieve equivalence, you must choose the point Q so that the moment of $\mathbf{F}$ about P equals the normal component $\mathbf{M}_n$ (Fig. 4.46d)—that is, so that $\mathbf{r}_{PQ} \times \mathbf{F} = \mathbf{M}_n$.

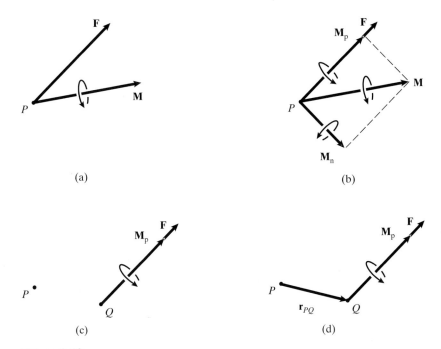

(a)

(b)

(c)

(d)

Figure 4.46
(a) If necessary, first represent the system by a single force and a single couple.
(b) The components of $\mathbf{M}$ parallel and normal to $\mathbf{F}$.
(c) The wrench.
(d) Choose Q so that the moment of $\mathbf{F}$ about P equals the normal component of $\mathbf{M}$.

Example 4.20

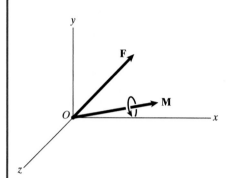

Figure 4.47

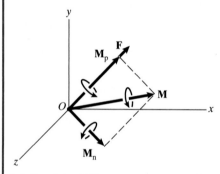

(a) Resolving **M** into components parallel and normal to **F**.

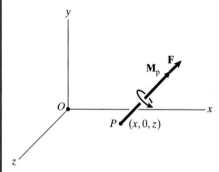

(b) The wrench acting at a point in the x–z plane.

Representing a Force and Couple by a Wrench

The system in Fig. 4.47 consists of the force and couple

$$\mathbf{F} = 3\mathbf{i} + 6\mathbf{j} + 2\mathbf{k} \ (\text{N}),$$

$$\mathbf{M} = 12\mathbf{i} + 4\mathbf{j} + 6\mathbf{k} \ (\text{N-m}).$$

Represent it by a wrench, and determine where the line of action of the wrench's force intersects the x–z plane.

Strategy

The wrench is the force **F** and the component of **M** parallel to **F** (Figs. a, b). We must choose the point of application P so that the moment of **F** about O equals the normal component $\mathbf{M}_n$. By letting P be an arbitrary point of the x–z plane, we can determine where the line of action of **F** intersects that plane.

Solution

Dividing **F** by its magnitude, we obtain a unit vector **e** with the same direction as **F**:

$$\mathbf{e} = \frac{\mathbf{F}}{|\mathbf{F}|} = \frac{3\mathbf{i} + 6\mathbf{j} + 2\mathbf{k}}{\sqrt{(3)^2 + (6)^2 + (2)^2}} = 0.429\mathbf{i} + 0.857\mathbf{j} + 0.286\mathbf{k}.$$

We can use **e** to calculate the component of **M** parallel to **F**:

$$\mathbf{M}_p = (\mathbf{e} \cdot \mathbf{M})\mathbf{e} = \big[(0.429)(12) + (0.857)(4) + (0.286)(6)\big]\mathbf{e}$$

$$= 4.408\mathbf{i} + 8.816\mathbf{j} + 2.939\mathbf{k} \ (\text{N-m}).$$

The component of **M** normal to **F** is

$$\mathbf{M}_n = \mathbf{M} - \mathbf{M}_p = 7.592\mathbf{i} - 4.816\mathbf{j} + 3.061\mathbf{k} \ (\text{N-m}).$$

The wrench is shown in Fig. b. Let the coordinates of P be $(x, 0, z)$. The moment of **F** about O is

$$\mathbf{r}_{OP} \times \mathbf{F} = \begin{vmatrix} \mathbf{i} & \mathbf{j} & \mathbf{k} \\ x & 0 & z \\ 3 & 6 & 2 \end{vmatrix} = -6z\mathbf{i} - (2x - 3z)\mathbf{j} + 6x\mathbf{k}.$$

By equating this moment to $\mathbf{M}_n$,

$$-6z\mathbf{i} - (2x - 3z)\mathbf{j} + 6x\mathbf{k} = 7.592\mathbf{i} - 4.816\mathbf{j} + 3.061\mathbf{k},$$

we obtain the equations

$$-6z = 7.592,$$

$$-2x + 3z = -4.816,$$

$$6x = 3.061.$$

Solving these equations, we find the coordinates of point P are $x = 0.510$ m, $z = -1.265$ m.

Problems

4.128 Two systems of forces act on the beam. Are they equivalent?

Strategy: Check the two conditions for equivalence. The sums of the forces must be equal, and the sums of the moments about an arbitrary point must be equal.

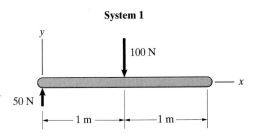

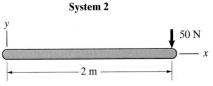

P4.128

4.129 Two systems of forces and moments act on the beam. Are they equivalent?

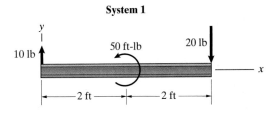

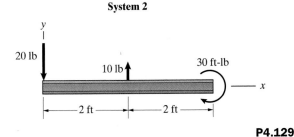

P4.129

4.130 Four systems of forces and moments act on an 8-m beam. Which systems are equivalent?

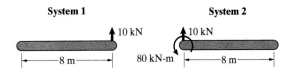

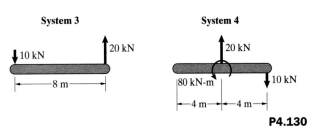

P4.130

4.131 The four systems shown in Problem 4.130 can be made equivalent by adding a couple to one of the systems. Which system is it, and what couple must be added?

4.132 System 1 is a force **F** acting at a point O. System 2 is the force **F** acting at a different point O' along the same line of action. Explain why these systems are equivalent. (This simple result is called the *principle of transmissibility.*)

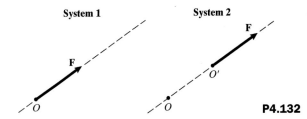

P4.132

4.133 The vector sum of the forces exerted on the log by the cables is the same in the two cases. Show that the systems of forces exerted on the log are equivalent.

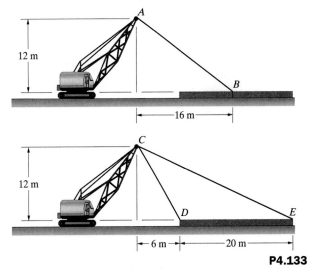

P4.133

4.134 Systems 1 and 2 each consist of a couple. If they are equivalent, what is F?

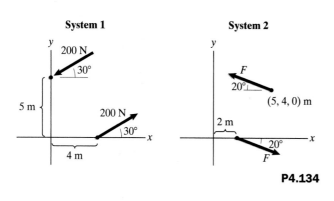

P4.134

4.135 Two equivalent systems of forces and moments act on the L-shaped bar. Determine the forces F_A and F_B and the couple M.

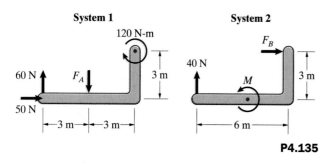

P4.135

4.136 Two equivalent systems of forces and moments act on the plate. Determine the force F and the couple M.

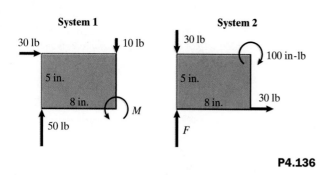

P4.136

4.137 In system 1, four forces act on the rectangular flat plate. The forces are perpendicular to the plate and the 400-kN force acts at its midpoint. In system 2, no forces or couples act on the plate. Systems 1 and 2 are equivalent. What are the forces F_1, F_2, and F_3?

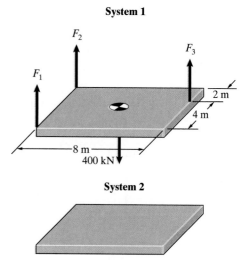

P4.137

4.138 Three forces and a couple are applied to a beam (system 1). (a) If you represent system 1 by a force applied at A and a couple (system 2), what are $\mathbf{F}$ and M? (b) If you represent system 1 by the force $\mathbf{F}$ (system 3), what is the distance D?

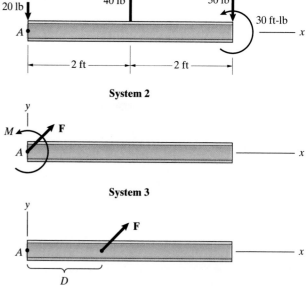

P4.138

4.139 Represent the two forces and couple acting on the beam by a force $\mathbf{F}$. Determine $\mathbf{F}$ and determine where its line of action intersects the x axis.

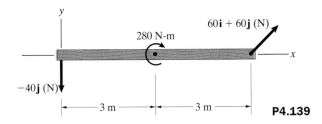

P4.139

4.140 The vector sum of the forces acting on the beam is zero, and the sum of the moments about the left end of the beam is zero.
(a) Determine the forces A_x, A_y, and B.
(b) If you represent the forces A_x, A_y, and B by a force **F** acting at the right end of the beam and a couple M, what are **F** and M?

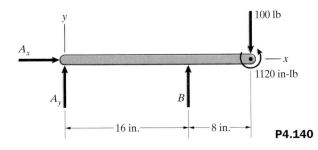

P4.140

4.141 The vector sum of the forces acting on the beam is zero, and the sum of the moments about the left end of the beam is zero.
(a) Determine the forces A_x and A_y, and the couple M_A.
(b) Determine the sum of the moments about the right end of the beam.
(c) If you represent the 600-N force, the 200-N force, and the 30 N-m couple by a force **F** acting at the left end of the beam and a couple M, what are **F** and M?

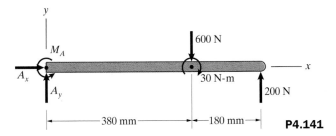

P4.141

4.142 The vector sum of the forces acting on the truss is zero, and the sum of the moments about the origin O is zero.
(a) Determine the forces A_x, A_y, and B.
(b) If you represent the 2-kip, 4-kip, and 6-kip forces by a force **F**, what is **F**, and where does its line of action intersect the y axis?
(c) If you replace the 2-kip, 4-kip, and 6-kip forces by the force you determined in (b), what are the vector sum of the forces acting on the truss and the sum of the moments about O?

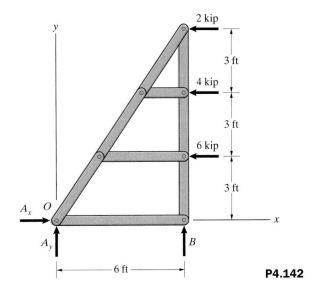

P4.142

4.143 The distributed force exerted on part of a building foundation by the soil is represented by five forces. If you represent them by a force **F**, what is **F**, and where does its line of action intersect the x axis?

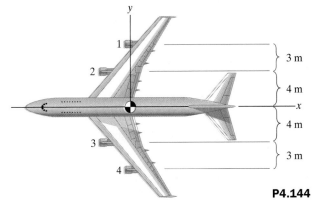

P4.143

4.144 After landing, the pilot engages the airplane's thrust reversers and engines 1, 2, 3, and 4 exert forces toward the right of magnitudes 39 kN, 40 kN, 42 kN, and 40 kN, respectively. If you represent the four forces by an equivalent force **F**, what is **F**, and what is the y coordinate of its line of action?

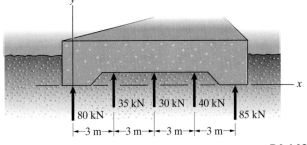

P4.144

4.145 The pilot of the airplane in Problem 4.144 wants to adjust engine 2 so that the forces exerted by the engines can be represented by an equivalent force whose line of action intersects the z axis. When this is done, what force is exerted by engine 2?

4.146 The system is in equilibrium. If you represent the forces $\mathbf{F}_{AB}$ and $\mathbf{F}_{AC}$ by a force $\mathbf{F}$ acting at A and a couple $\mathbf{M}$, what are $\mathbf{F}$ and $\mathbf{M}$?

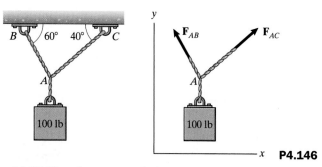

P4.146

4.147 Three forces act on the beam.
(a) Represent the system by a force $\mathbf{F}$ acting at the origin O and a couple M.
(b) Represent the system by a single force. Where does the line of action of the force intersect the x axis?

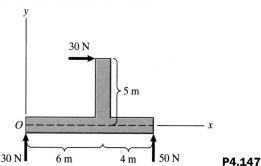

P4.147

4.148 The tension in cable AB is 400 N, and the tension in cable CD is 600 N.
(a) If you represent the forces exerted on the left post by the cables by a force $\mathbf{F}$ acting at the origin O and a couple M, what are $\mathbf{F}$ and M?

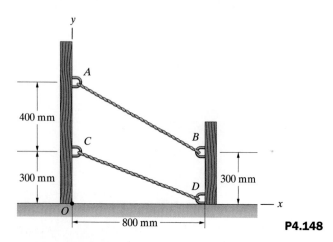

P4.148

(b) If you represent the forces exerted on the left post by the cables by the force $\mathbf{F}$ alone, where does its line of action intersect the y axis?

4.149 Consider the system shown in Problem 4.148. The tension in each of the cables AB and CD is 400 N. If you represent the forces exerted on the right post by the cables by a force $\mathbf{F}$, what is $\mathbf{F}$, and where does its line of action intersect the y axis?

4.150 If you represent the three forces acting on the beam cross section by a force $\mathbf{F}$, what is $\mathbf{F}$, and where does its line of action intersect the x axis?

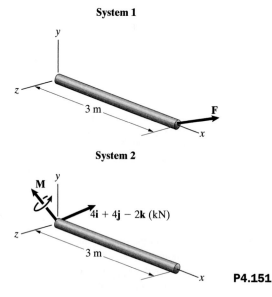

P4.150

4.151 The two systems of forces and moments acting on the beam are equivalent. Determine the force $\mathbf{F}$ and the couple $\mathbf{M}$.

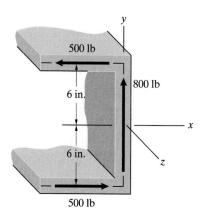

P4.151

4.152 The wall bracket is subjected to the force shown.
(a) Determine the moment exerted by the force about the z axis.
(b) Determine the moment exerted by the force about the y axis.
(c) If you represent the force by a force $\mathbf{F}$ acting at O and a couple $\mathbf{M}$, what are $\mathbf{F}$ and $\mathbf{M}$?

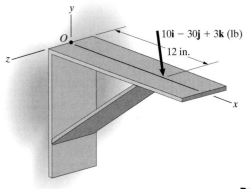

$10\mathbf{i} - 30\mathbf{j} + 3\mathbf{k}$ (lb)

12 in.

P4.152

4.153 A basketball player executes a "slam dunk" shot, then hangs momentarily on the rim, exerting the two 100-lb forces shown. The dimensions are $h = 14\frac{1}{2}$ in., and $r = 9\frac{1}{2}$ in., and the angle $\alpha = 120°$.
(a) If you represent the forces he exerts by a force $\mathbf{F}$ acting at O and a couple $\mathbf{M}$, what are $\mathbf{F}$ and $\mathbf{M}$?
(b) The glass backboard will shatter if $|\mathbf{M}| > 4000$ in-lb. Does it break?

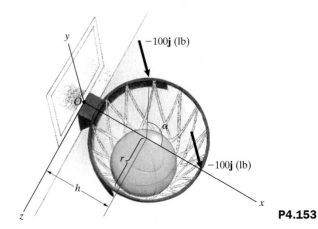

$-100\mathbf{j}$ (lb)

$-100\mathbf{j}$ (lb)

P4.153

4.154 The three forces are parallel to the x axis.
(a) If you represent the three forces by a force $\mathbf{F}$ acting at the origin O and a couple $\mathbf{M}$, what are $\mathbf{F}$ and $\mathbf{M}$?
(b) If you represent the forces by a single force, what is the force, and where does its line of action intersect the y–z plane?

Strategy: In (b), assume that the force acts at a point $(0, y, z)$ of the y–z plane, and use the conditions for equivalence to determine the force and the coordinates y and z. (See Example 4.19.)

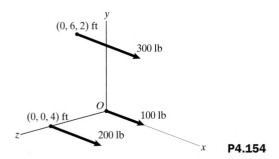

(0, 6, 2) ft

300 lb

(0, 0, 4) ft

O

100 lb

200 lb

P4.154

4.155 The positions and weights of three particles are shown. If you represent the weights by a single force $\mathbf{F}$, determine $\mathbf{F}$ and show that its line of action intersects the x–z plane at

$$x = \frac{\sum\limits_{i=1}^{3} x_i W_i}{\sum\limits_{i=1}^{3} W_i}, \qquad z = \frac{\sum\limits_{i=1}^{3} z_i W_i}{\sum\limits_{i=1}^{3} W_i}.$$

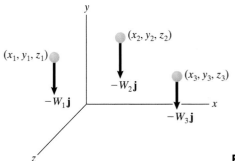

(x_1, y_1, z_1)

(x_2, y_2, z_2)

(x_3, y_3, z_3)

$-W_1\mathbf{j}$

$-W_2\mathbf{j}$

$-W_3\mathbf{j}$

P4.155

4.156 Two forces act on the beam. If you represent them by a force $\mathbf{F}$ acting at C and a couple $\mathbf{M}$, what are $\mathbf{F}$ and $\mathbf{M}$?

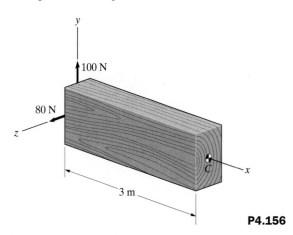

100 N

80 N

3 m

C

P4.156

4.157 An axial force of magnitude P acts on the beam. If you represent it by a force $\mathbf{F}$ acting at the origin O and a couple $\mathbf{M}$, what are $\mathbf{F}$ and $\mathbf{M}$?

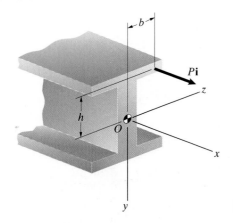

b

$P\mathbf{i}$

h

O

P4.157

4.158 The brace is being used to remove a screw.
(a) If you represent the forces acting on the brace by a force **F** acting at the origin O and a couple **M**, what are **F** and **M**?
(b) If you represent the forces acting on the brace by a force **F'** acting at a point P with coordinates (x_P, y_P, z_P) and a couple **M'**, what are **F'** and **M'**?

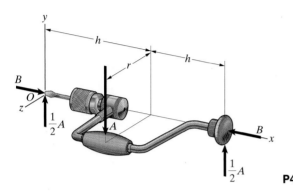

P4.158

4.159 Two forces and a couple act on the cube. If you represent them by a force **F** acting at point P and a couple **M**, what are **F** and **M**?

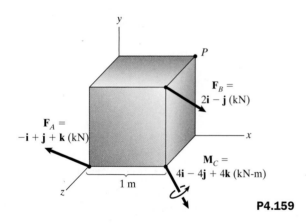

P4.159

4.160 The two shafts are subjected to the torques (couples) shown.
(a) If you represent the two couples by a force **F** acting at the origin O and a couple **M**, what are **F** and **M**?

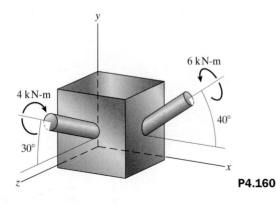

P4.160

(b) What is the magnitude of the total moment exerted by the two couples?

4.161 The persons A and B support a bar to which three dogs are tethered. The forces and couples they exert are

$$\mathbf{F}_A = -5\mathbf{i} + 15\mathbf{j} - 10\mathbf{k} \ (\text{lb}),$$

$$\mathbf{M}_A = 15\mathbf{j} + 10\mathbf{k} \ (\text{ft-lb}),$$

$$\mathbf{F}_B = 5\mathbf{i} + 10\mathbf{j} - 10\mathbf{k} \ (\text{lb}),$$

$$\mathbf{M}_B = -10\mathbf{j} - 15\mathbf{k} \ (\text{ft-lb}).$$

If person B let go, person A would have to exert a force **F** and couple **M** equivalent to the system both of them were exerting together. What are **F** and **M**?

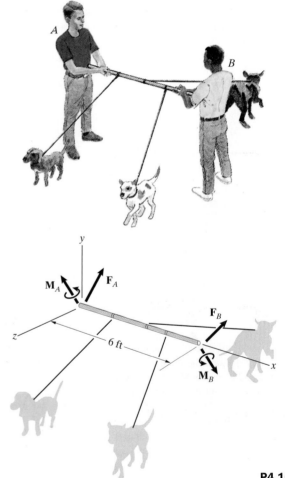

P4.161

4.162 Point G is at the center of the block. The forces are

$$\mathbf{F}_A = -20\mathbf{i} + 10\mathbf{j} + 20\mathbf{k} \ (\text{lb}),$$

$$\mathbf{F}_B = 10\mathbf{j} - 10\mathbf{k} \ (\text{lb}).$$

If you represent the two forces by a force **F** acting at G and a couple **M**, what are **F** and **M**?

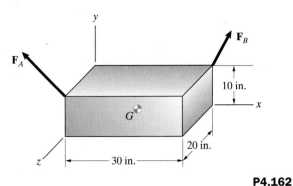

P4.162

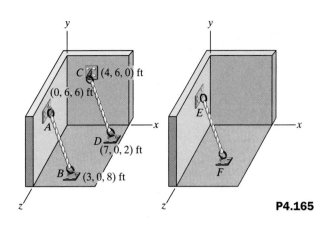

P4.165

4.163 The engine above the airplane's fuselage exerts a thrust $T_0 = 16$ kip, and each of the engines under the wings exerts a thrust $T_U = 12$ kip. The dimensions are $h = 8$ ft, $c = 12$ ft, and $b = 16$ ft. If you represent the three thrust forces by a force **F** acting at the origin O and a couple **M**, what are **F** and **M**?

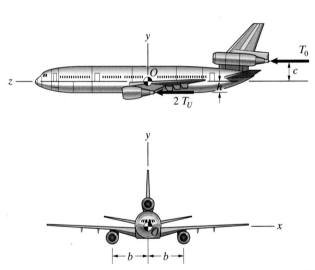

P4.163

4.164 Consider the airplane described in Problem 4.163 and suppose that the engine under the wing to the pilot's right loses thrust.
(a) If you represent the two remaining thrust forces by a force **F** acting at the origin O and a couple **M**, what are **F** and **M**?
(b) If you represent the two remaining thrust forces by the force **F** alone, where does its line of action intersect the x–y plane?

4.165 The tension in cable AB is 100 lb, and the tension in cable CD is 60 lb. Suppose that you want to replace these two cables by a single cable EF so that the force exerted on the wall at E is equivalent to the two forces exerted by cables AB and CD on the walls at A and C. What is the tension in cable EF, and what are the coordinates of points E and F?

4.166 The distance $s = 4$m. If you represent the force and the 200-N-m couple by a force **F** acting at the origin O and a couple **M**, what are **F** and **M**?

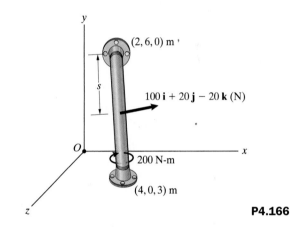

P4.166

4.167 The force **F** and couple **M** in system 1 are

$$\mathbf{F} = 12\mathbf{i} + 4\mathbf{j} - 3\mathbf{k} \text{ (lb)},$$

$$\mathbf{M} = 4\mathbf{i} + 7\mathbf{j} + 4\mathbf{k} \text{ (ft-lb)}.$$

Suppose you want to represent system 1 by a wrench (system 2). Determine the couple $\mathbf{M}_p$ and the coordinates x and z where the line of action of the force intersects the x–z plane.

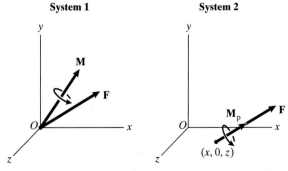

P4.167

4.168 A system consists of a force **F** acting at the origin O and a couple **M**, where

$$\mathbf{F} = 10\mathbf{i} \text{ (lb)}, \qquad \mathbf{M} = 20\mathbf{j} \text{ (ft-lb)}.$$

If you represent the system by a wrench consisting of the force **F** and a parallel couple $\mathbf{M}_\mathrm{p}$, what is $\mathbf{M}_\mathrm{p}$, and where does the line of action of **F** intersect the y–z plane?

4.169 A system consists of a force **F** acting at the origin O and a couple **M**, where

$$\mathbf{F} = \mathbf{i} + 2\mathbf{j} + 5\mathbf{k} \text{ (N)}, \qquad \mathbf{M} = 10\mathbf{i} + 8\mathbf{j} - 4\mathbf{k} \text{ (N-m)}.$$

If you represent it by a wrench consisting of the force **F** and a parallel couple $\mathbf{M}_\mathrm{p}$, (a) determine $\mathbf{M}_\mathrm{p}$, and determine where the line of action of **F** intersects (b) the x–z plane, (c) the y–z plane.

4.170 Consider the force **F** acting at the origin O and the couple **M** given in Example 4.20. If you represent this system by a wrench, where does the line of action of the force intersect the x–y plane?

4.171 Consider the force **F** acting at the origin O and the couple **M** given in Example 4.20. If you represent this system by a wrench, where does the line of action of the force intersect the plane $y = 3$ m?

4.172 A wrench consists of a force of magnitude 100 N acting at the origin O and a couple of magnitude 60 N-m. The force and couple point in the direction from O to the point $(1, 1, 2)$ m. If you represent the wrench by a force **F** acting at the point $(5, 3, 1)$ m and a couple **M**, what are **F** and **M**?

4.173 System 1 consists of two forces and a couple. Suppose that you want to represent it by a wrench (system 2). Determine the force **F**, the couple $\mathbf{M}_\mathrm{p}$, and the coordinates x and z where the line of action of **F** intersects the x–z plane.

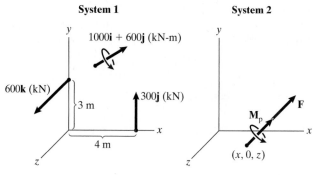

P4.173

4.174 A plumber exerts the two forces shown to loosen a pipe. (a) What total moment does he exert about the axis of the pipe? (b) If you represent the two forces by a force **F** acting at O and a couple **M**, what are **F** and **M**? (c) If you represent the two forces by a wrench consisting of the force **F** and a parallel couple $\mathbf{M}_\mathrm{p}$, what is $\mathbf{M}_\mathrm{p}$, and where does the line of action of **F** intersect the x–y plane?

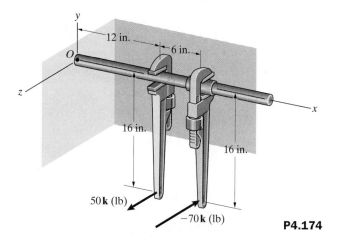

P4.174

Chapter Summary

In this chapter we have defined the moment of a force about a point and about a line and explained how to evaluate them. We introduced the concept of a couple and defined equivalent systems of forces and moments. We can now apply two consequences of equilibrium: The sum of the forces equals zero, and the sum of the moments about any point equals zero. We will consider individual objects in Chapter 5 and structures in Chapter 6.

Moment of a Force About a Point

The moment of a force about a point is the measure of the tendency of the force to cause rotation about the point. The *moment* of a force **F** about a point P is the vector

$$\mathbf{M}_P = \mathbf{r} \times \mathbf{F}, \qquad \text{Eq. (4.2)}$$

where $\mathbf{r}$ is a position vector from P to *any* point on the line of action of $\mathbf{F}$. The magnitude of $\mathbf{M}_P$ is equal to the product of the perpendicular distance D from P to the line of action of $\mathbf{F}$ and the magnitude of $\mathbf{F}$:

$$|\mathbf{M}_P| = D|\mathbf{F}|. \qquad \text{Eq. (4.3)}$$

The vector $\mathbf{M}_P$ is perpendicular to the plane containing P and $\mathbf{F}$. When the thumb of the right hand points in the direction of $\mathbf{M}_P$, the arc of the fingers indicates the sense of the rotation that $\mathbf{F}$ tends to cause about P. The dimensions of the moment are (distance) $\times$ (force).

If a force is resolved into components, the moment of the force about a point P is equal to the sum of the moments of its components about P. If the line of action of a force passes through a point P, the moment of the force about P is zero.

When the view is perpendicular to the plane containing the force and the point (Fig. a), the two-dimensional description of the moment is

$$M_p = DF. \qquad \text{Eq. (4.1)}$$

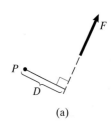

(a)

Figure (a)

Moment of a Force About a Line

The moment of a force about a line is the measure of the tendency of the force to cause rotation about the line. Let P be any point on a line L and let $\mathbf{M}_P$ be the moment about P of a force $\mathbf{F}$ (Fig. b). The moment $\mathbf{M}_L$ of $\mathbf{F}$ about L is the vector component of $\mathbf{M}_P$ parallel to L. If $\mathbf{e}$ is a unit vector along L,

$$\mathbf{M}_L = (\mathbf{e} \cdot \mathbf{M}_P)\mathbf{e} = \big[\mathbf{e} \cdot (\mathbf{r} \times \mathbf{F})\big]\mathbf{e}. \qquad \text{Eq. (4.4), (4.5)}$$

When the line of action of $\mathbf{F}$ is perpendicular to a plane containing L, $|\mathbf{M}_L|$ is equal to the product of the magnitude of $\mathbf{F}$ and the perpendicular distance D from L to the point where the line of action intersects the plane. When the line of action of $\mathbf{F}$ is parallel to L or intersects L, $\mathbf{M}_L = 0$.

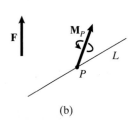

(b)

Figure (b)

Couples

Two forces that have equal magnitudes, opposite directions, and do not have the same line of action are called a *couple*. The moment $\mathbf{M}$ of a couple is the same about any point. The magnitude of $\mathbf{M}$ is equal to the product of the magnitude of one of the forces and the perpendicular distance between the lines of action, and its direction is perpendicular to the plane containing the lines of action.

Because a couple exerts a moment but no net force, it can be represented by showing the moment vector (Fig. c), or it can be represented in two dimensions by showing the magnitude of the moment and a circular arrow to indicate the sense (Fig. d). The moment represented in this way is called the *moment of a couple*, or simply a *couple*.

(c)

Figure (c)

(d)

Figure (d)

Equivalent Systems

Two systems of forces and moments are defined to be *equivalent* if the sums of the forces are equal,

$$(\Sigma \mathbf{F})_1 = (\Sigma \mathbf{F})_2, \qquad \text{Eq. (4.7)}$$

and the sums of the moments about a point P are equal,

$$(\Sigma \mathbf{M}_P)_1 = (\Sigma \mathbf{M}_P)_2. \qquad \text{Eq. (4.8)}$$

If the sums of the forces are equal and the sums of the moments about one point are equal, the sums of the moments about any point are equal.

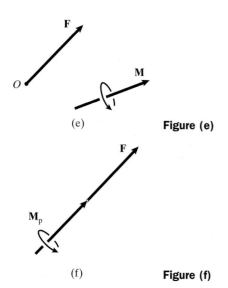

(e)

Figure (e)

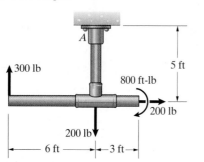

(f)

Figure (f)

Representing Systems by Equivalent Systems

If the system of forces and moments acting on an object is represented by an equivalent system, the equivalent system exerts the same total force and total moment on the object.

Any system can be represented by an equivalent system consisting of a force **F** acting at a given point *P* and a couple **M** (Fig. e). The simplest system that can be equivalent to any system of forces and moments is the *wrench*, which is a force **F** and a couple M_p that is parallel to **F** (Fig. f).

A system of concurrent forces can be represented by a single force. A system of parallel forces whose sum is not zero can be represented by a single force.

Review Problems

4.175 Determine the sum of the moments exerted about *A* by the three forces and the couple.

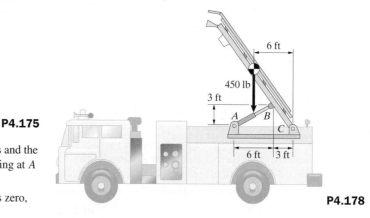

P4.175

4.176 In Problem 4.175, if you represent the three forces and the couple by an equivalent system consisting of a force **F** acting at *A* and a couple **M**, what are the magnitudes of **F** and **M**?

4.177 The vector sum of the forces acting on the beam is zero, and the sum of the moments about *A* is zero.
(a) What are the forces A_x, A_y, and *B*?
(b) What is the sum of the moments about *B*?

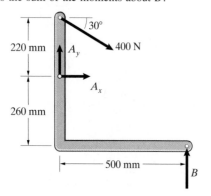

P4.177

4.178 To support the ladder, the force exerted at *B* by the hydraulic piston *AB* must exert a moment about *C* equal in magnitude to the moment about *C* due to the ladder's 450-lb weight. What is the magnitude of the force exerted at *B*?

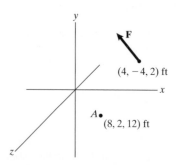

P4.178

4.179 The force $\mathbf{F} = -60\mathbf{i} + 60\mathbf{j}$ (lb).
(a) Determine the moment of **F** about point *A*.
(b) What is the perpendicular distance from point *A* to the line of action of **F**?

P4.179

4.180 The 20-kg mass is suspended by cables attached to three vertical 2-m posts. Point A is at $(0, 1.2, 0)$ m. Determine the moment about the base E due to the force exerted on the post BE by the cable AB.

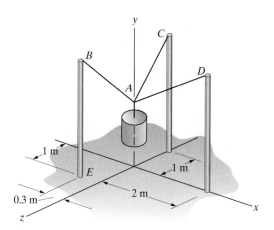

P4.180

4.181 Determine the moment of the vertical 800-lb force about point C.

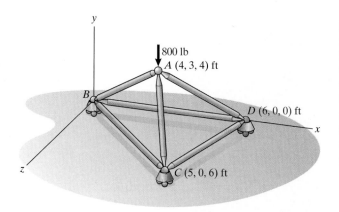

P4.181

4.182 In Problem 4.181, determine the moment of the vertical 800-lb force about the straight line through points C and D.

4.183 The tugboats A and B exert forces $F_A = 1$ kN and $F_B = 1.2$ kN on the ship. The angle $\theta = 30°$. If you represent the two forces by a force $\mathbf{F}$ acting at the origin O and a couple M, what are $\mathbf{F}$ and M?

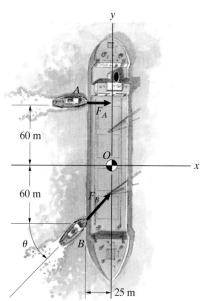

P4.183

4.184 The tugboats A and B in Problem 4.183 exert forces $F_A = 600$ N and $F_B = 800$ N on the ship. The angle $\theta = 45°$. If you represent the two forces by a force $\mathbf{F}$, what is $\mathbf{F}$, and where does its line of action intersect the y axis?

*𝒟***esign Experience** A relatively primitive device for exercising the biceps muscle is shown. Suggest an improved configuration for the device. You can use elastic cords (which behave like linear springs), weights, and pulleys. Seek a design such that the variation of the moment about the elbow joint as the device is used is small in comparison to the design shown. Give consideration to the safety of your device, its reliability, and the requirement to accommodate users having a range of dimensions and strengths. Choosing specific dimensions, determine the range of the magnitude of the moment exerted about the elbow joint as your device is used.

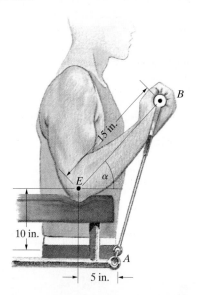

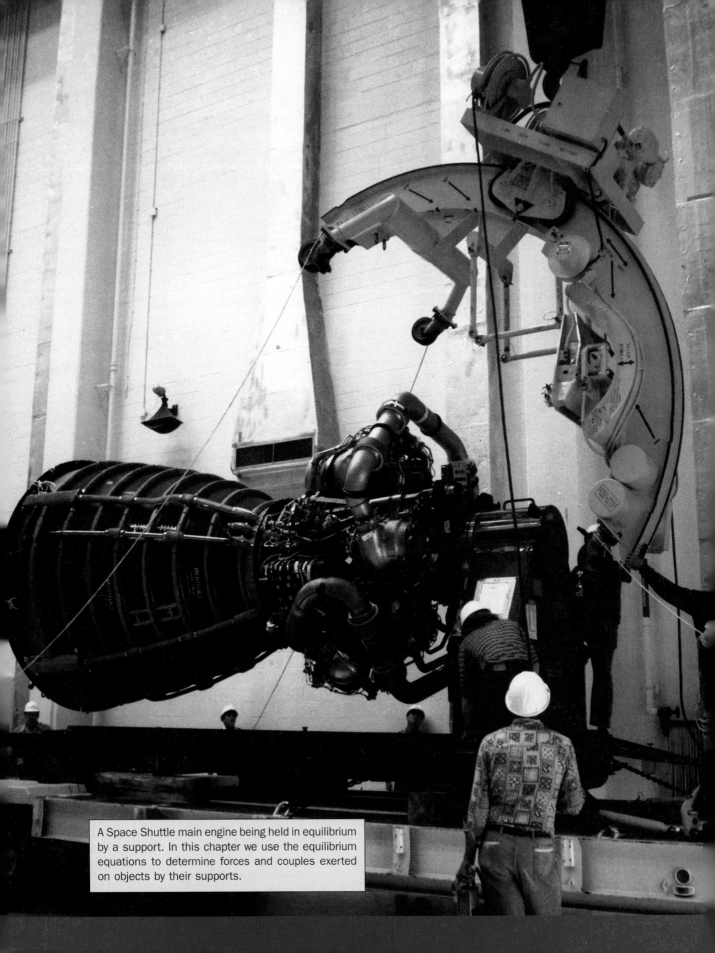

A Space Shuttle main engine being held in equilibrium by a support. In this chapter we use the equilibrium equations to determine forces and couples exerted on objects by their supports.

CHAPTER 5

Objects in Equilibrium

B y applying the techniques developed in Chapters 3 and 4, we can now analyze many of the equilibrium problems that arise in engineering applications. After stating the equilibrium equations, we describe the various types of supports that are used. We then show how free-body diagrams and equilibrium are used to determine unknown forces and couples acting on objects.

5.1 The Equilibrium Equations

In Chapter 3 we defined an object to be in equilibrium when it is stationary or in steady translation relative to an inertial reference frame. When an object acted upon by a system of forces and moments is in equilibrium, the following conditions are satisfied.

1. The sum of the forces is zero:

$$\Sigma \mathbf{F} = \mathbf{0}. \tag{5.1}$$

2. The sum of the moments about any point is zero:

$$\Sigma \mathbf{M}_{(\text{any point})} = \mathbf{0}. \tag{5.2}$$

Before we consider specific applications, some general observations about these equations are in order.

From our discussion of equivalent systems of forces and moments in Chapter 4, Eqs. (5.1) and (5.2) imply that the system of forces and moments acting on an object in equilibrium is equivalent to a system consisting of no forces and no couples. This provides insight into the nature of equilibrium. From the standpoint of the total force and total moment exerted on an object in equilibrium, the effects are the same as if no forces or couples acted on the object. This observation also makes it clear that if the sum of the forces on an object is zero and the sum of the moments about one point is zero, then the sum of the moments about every point is zero.

Figure 5.1 shows an object subjected to concurrent forces $\mathbf{F}_1, \mathbf{F}_2, \ldots, \mathbf{F}_N$ and no couples. If the sum of these forces is zero,

$$\mathbf{F}_1 + \mathbf{F}_2 + \cdots + \mathbf{F}_N = \mathbf{0}, \tag{5.3}$$

the conditions for equilibrium are satisfied, because the moment about point P is zero. The only condition imposed by equilibrium on a set of concurrent forces is that their sum is zero.

To determine the sum of the moments about a line L due to a system of forces and moments acting on an object, we choose any point P on the line and determine the sum of the moments $\Sigma \mathbf{M}_P$ about P (Fig. 5.2). Then the sum of the moments about the line is the component of $\Sigma \mathbf{M}_P$ parallel to the line. If the object is in equilibrium, $\Sigma \mathbf{M}_P = \mathbf{0}$. We see that the sum of the moments about any line due to the forces and couples acting on an object in equilibrium is zero. This result is useful in certain types of problems.

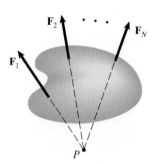

Figure 5.1
An object subjected to concurrent forces.

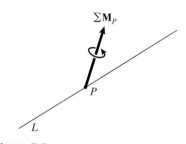

Figure 5.2
The sum of the moments $\Sigma \mathbf{M}_P$ about a point P on the line L.

5.2 Two-Dimensional Applications

Many engineering applications involve two-dimensional systems of forces and moments. These include the forces and moments exerted on many beams and planar structures, pliers, some cranes and other machines, and some types of bridges and dams. In this section we discuss supports, free-body diagrams, and the equilibrium equations for two-dimensional applications.

Supports

When you are standing, the floor supports you. When you sit in a chair, the chair supports you. In this section we are concerned with the ways objects are held in place or are attached to other objects. Forces and couples exerted on an object by its supports are called *reactions*, expressing the fact that the supports "react" to the other forces and couples, or *loads*, acting on the object. For example, a bridge is held up by the reactions exerted by its supports, and the loads are the forces exerted by the weight of the bridge itself, the traffic crossing it, and the wind.

Some very common kinds of supports are represented by stylized models called support conventions. Actual supports often closely resemble the support conventions, but even when they don't, we represent them by these conventions if the actual supports exert the same (or approximately the same) reactions as the models.

The Pin Support Figure 5.3a shows a *pin support*. The diagram represents a bracket to which an object (such as a beam) is attached by a smooth pin that passes through the bracket and the object. The side view is shown in Fig. 5.3b.

To understand the reactions that a pin support can exert, it's helpful to imagine holding a bar attached to a pin support (Fig. 5.3c). If you try to move the bar without rotating it (that is, translate the bar), the support exerts a reactive force that prevents this movement. However, you can rotate the bar about the axis of the pin. The support cannot exert a couple about the pin axis to prevent rotation. Thus a pin support can't exert a couple about the pin axis, but it can exert a force on an object in any direction, which is usually expressed by representing the force in terms of components (Fig. 5.3d). The arrows indicate the directions of the reactions if A_x and A_y are positive. If you determine A_x or A_y to be negative, the reaction is in the direction opposite to that of the arrow.

The pin support is used to represent any real support capable of exerting a force in any direction but not exerting a couple. Pin supports are used in many common devices, particularly those designed to allow connected parts to rotate relative to each other (Fig. 5.4).

The Roller Support The convention called a *roller support* (Fig. 5.5a) represents a pin support mounted on wheels. Like the pin support, it cannot exert

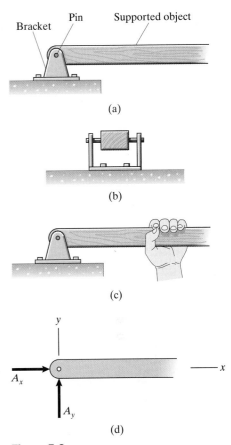

Figure 5.3
(a) A pin support.
(b) Side view showing the pin passing through the beam.
(c) Holding a supported bar.
(d) The pin support is capable of exerting two components of force.

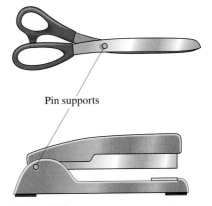

Figure 5.4
Pin supports in a pair of scissors and a stapler.

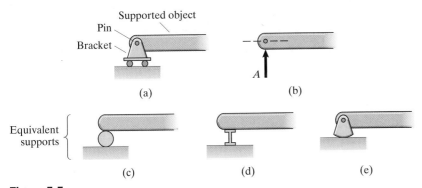

Figure 5.5
(a) A roller support.
(b) The reaction consists of a force normal to the surface.
(c)–(e) Supports equivalent to the roller support.

Figure 5.6
Supporting an object with a plane smooth surface.

a couple about the axis of the pin. Since it can move freely in the direction parallel to the surface on which it rolls, it can't exert a force parallel to the surface but can only exert a force normal (perpendicular) to this surface (Fig. 5.5b). Figures 5.5c–e are other commonly used conventions equivalent to the roller support. The wheels of vehicles and wheels supporting parts of machines are roller supports if the friction forces exerted on them are negligible in comparison to the normal forces. A plane smooth surface can also be modeled by a roller support (Fig. 5.6). Beams and bridges are sometimes supported in this way so that they will be free to undergo thermal expansion and contraction.

The supports shown in Fig. 5.7 are similar to the roller support in that they cannot exert a couple and can only exert a force normal to a particular direction. (Friction is neglected.) In these supports, the supported object is attached to a pin or slider that can move freely in one direction but is constrained in the perpendicular direction. Unlike the roller support, these supports can exert a normal force in either direction.

Figure 5.7
Supports similar to the roller support except that the normal force can be exerted in either direction.
(a) Pin in a slot.
(b) Slider in a slot.
(c) Slider on a shaft.

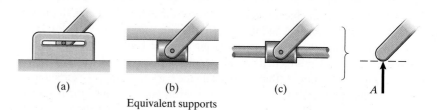

(a) (b) (c) A

Equivalent supports

The Built-In Support   The *built-in support* shows the supported object literally built into a wall (Fig. 5.8a). This convention is also called a *fixed support*. To understand the reactions, imagine holding a bar attached to a built-in support (Fig. 5.8b). If you try to translate the bar, the support exerts a reactive force that prevents translation, and if you try to rotate the bar, the support exerts a reactive couple that prevents rotation. A built-in support can exert two components of force and a couple (Fig. 5.8c). The term M_A is the couple exerted by the support, and the curved arrow indicates its direction. Fence posts and lampposts have built-in supports. The attachments of parts connected so that they cannot move or rotate relative to each other, such as the head of a hammer and its handle, can be modeled as built-in supports.

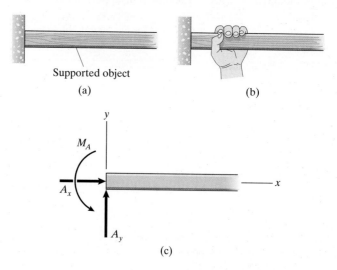

Supported object
(a)

(b)

y

M_A

A_x

x

A_y

(c)

Figure 5.8
(a) Built-in support.
(b) Holding a supported bar.
(c) The reactions a built-in support is capable of exerting.

Table 5.1 summarizes the support conventions commonly used in two-dimensional applications, including those we discussed in Chapter 3. Although the number of conventions may appear daunting, the examples and problems

Table 5.1 Supports used in two-dimensional applications.

Supports	Reactions
Rope or Cable Spring	One Collinear Force
Contact with a Smooth Surface	One Force Normal to the Supporting Surface
Contact with a Rough Surface	Two Force Components
Pin Support	Two Force Components
Roller Support Equivalents	One Force Normal to the Supporting Surface
Constrained Pin or Slider	One Normal Force
Built-in (Fixed) Support	Two Force Components and One Couple

will help you become familiar with them. You should also observe how various objects you see in your everyday experience are supported and think about whether each support could be represented by one of the conventions.

Free-Body Diagrams

We introduced free-body diagrams in Chapter 3 and used them to determine forces acting on simple objects in equilibrium. By using the support conventions, we can model more elaborate objects and construct their free-body diagrams in a systematic way.

For example, the beam in Fig. 5.9a has a pin support at the left end and a roller support at the right end and is loaded by a force F. The roller support rests on a surface inclined at 30° to the horizontal. To obtain the free-body diagram of the beam, we first isolate it from its supports (Fig. 5.9b), since the free-body diagram must contain no object other than the beam. We complete the free-body diagram by showing the reactions that may be exerted on the beam by the supports (Fig. 5.9c). Notice that the reaction B exerted by the roller support is normal to the surface on which the support rests.

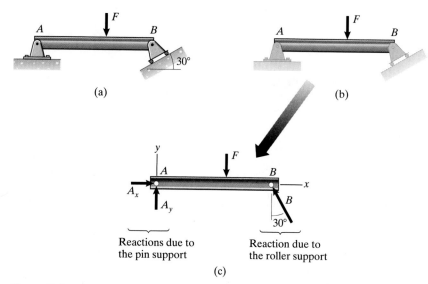

Figure 5.9
(a) A beam with pin and roller supports.
(b) Isolating the beam from its supports.
(c) The completed free-body diagram.

The object in Fig. 5.10a has a fixed support at the left end. A cable passing over a pulley is attached to the object at two points. We isolate it from its supports (Fig. 5.10b) and complete the free-body diagram by showing the reactions at the built-in support and the forces exerted by the cable (Fig. 5.10c). *Don't forget the couple at a built-in support.* Since we assume the tension in the cable is the same on both sides of the pulley, the two forces exerted by the cable have the same magnitude T.

Once you have obtained the free-body diagram of an object in equilibrium to identify the loads and reactions acting on it, you can apply the equilibrium equations.

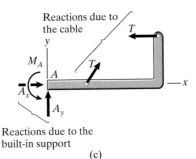

Reactions due to the cable

Reactions due to the built-in support

(a)   (b)   (c)

Figure 5.10
(a) An object with a built-in support.
(b) Isolating the object.
(c) The completed free-body diagram.

The Scalar Equilibrium Equations

When the loads and reactions on an object in equilibrium form a two-dimensional system of forces and moments, they are related by three scalar equilibrium equations:

$$\Sigma F_x = 0, \tag{5.4}$$
$$\Sigma F_y = 0, \tag{5.5}$$
$$\Sigma M_{(\text{any point})} = 0. \tag{5.6}$$

A natural question is whether more than one equation can be obtained from Eq. (5.6) by evaluating the sum of the moments about more than one point. The answer is yes, and in some cases it is convenient to do so. But there is a catch—the additional equations will not be independent of Eqs. (5.4)–(5.6). In other words, *more than three independent equilibrium equations cannot be obtained from a two-dimensional free-body diagram, which means we can solve for at most three unknown forces or couples.* We discuss this point further in Section 5.3.

The seesaw found on playgrounds, consisting of a board with a pin support at the center that allows it to rotate, is a simple and familiar example that illustrates the role of Eq. (5.6). If two people of unequal weight sit at the seesaw's ends, the heavier person sinks to the ground (Fig. 5.11a). To obtain equilibrium, that person must move closer to the center (Fig. 5.11b).

We draw the free-body diagram of the seesaw in Fig. 5.11c, showing the weights of the people W_1 and W_2 and the reactions at the pin support. Evaluating the sum of the moments about A, the equilibrium equations are

$$\Sigma F_x = A_x = 0, \tag{5.7}$$
$$\Sigma F_y = A_y - W_1 - W_2 = 0, \tag{5.8}$$
$$\Sigma M_{(\text{point } A)} = D_1 W_1 - D_2 W_2 = 0. \tag{5.9}$$

Thus $A_x = 0$, $A_y = W_1 + W_2$, and $D_1 W_1 = D_2 W_2$. The last condition indicates the relation between the positions of the two persons necessary for equilibrium.

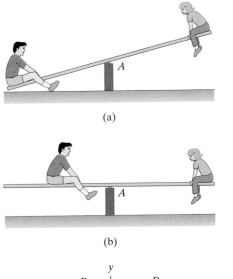

(a)

(b)

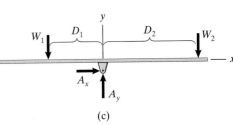

(c)

Figure 5.11
(a) If both people sit at the ends of the seesaw, the heavier one sinks.
(b) The seesaw and people in equilibrium.
(c) The free-body diagram of the seesaw, showing the weights of the people and the reactions at the pin support.

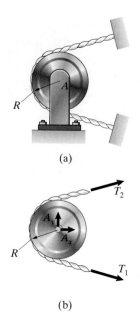

(a)

(b)

Figure 5.12
(a) A pulley of radius R.
(b) Free-body diagram of the pulley and part of the cable.

To demonstrate that an additional independent equation is not obtained by evaluating the sum of the moments about a different point, we can sum the moments about the right end of the seesaw:

$$\Sigma M_{(\text{right end})} = (D_1 + D_2)W_1 - D_2 A_y = 0.$$

This equation is a linear combination of Eqs. (5.8) and (5.9):

$$(D_1 + D_2)W_1 - D_2 A_y = \underbrace{-D_2(A_y - W_1 - W_2)}_{\textbf{Eq. (5.8)}}$$

$$+ \underbrace{(D_1 W_1 - D_2 W_2)}_{\textbf{Eq. (5.9)}} = 0.$$

Until now we have assumed in examples and problems that the tension in a rope or cable is the same on both sides of a pulley. Consider the pulley in Fig. 5.12a. In its free-body diagram in Fig. 5.12b, we do not assume that the tensions are equal. Summing the moments about the center of the pulley, we obtain the equilibrium equation

$$\Sigma M_{(\text{point } A)} = RT_1 - RT_2 = 0.$$

The tensions must be equal if the pulley is in equilibrium. However, notice that we have assumed that the pulley's support behaves like a pin support and cannot exert a couple on the pulley. When that is not true—for example, due to friction between the pulley and the support—the tensions are not necessarily equal.

Study Questions

1. What is a pin support? What reactions can it exert on an object subjected to a two-dimensional system of forces and moments?
2. What is a roller support? What reactions can it exert on an object subjected to a two-dimensional system of forces and moments?
3. How many independent equilibrium equations can you obtain from a two-dimensional free-body diagram?

Example 5.1

Reactions at Pin and Roller Supports

The beam in Fig. 5.13 has pin and roller supports and is subjected to a 2-kN force. What are the reactions at the supports?

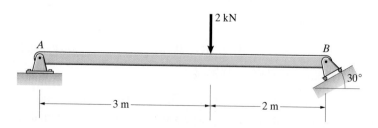

Figure 5.13

Solution

Draw the Free-Body Diagram We isolate the beam from its supports and show the loads and the reactions that may be exerted by the pin and roller supports (Fig. a). There are three unknown reactions: two components of force A_x and A_y at the pin support and a force B at the roller support.

Apply the Equilibrium Equations Summing the moments about point A, the equilibrium equations are

$$\Sigma F_x = A_x - B \sin 30° = 0,$$
$$\Sigma F_y = A_y + B \cos 30° - 2 = 0,$$
$$\Sigma M_{(\text{point } A)} = (5)(B \cos 30°) - (3)(2) = 0.$$

Solving these equations, the reactions are $A_x = 0.69$ kN, $A_y = 0.80$ kN, and $B = 1.39$ kN. The load and reactions are shown in Fig. b. It is good practice to show your answers in this way and confirm that the equilibrium equations are satisfied:

$$\Sigma F_x = 0.69 - 1.39 \sin 30° = 0,$$
$$\Sigma F_y = 0.80 + 1.39 \cos 30° - 2 = 0,$$
$$\Sigma M_{(\text{point } A)} = (5)(1.39 \cos 30°) - (3)(2) = 0.$$

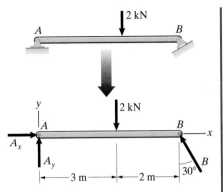

(a) Drawing the free-body diagram of the beam.

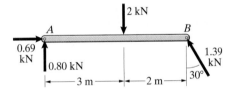

(b) The load and reaction.

Discussion

We drew the arrows indicating the directions of the reactions A_x and A_y in the positive x and y axis directions, but we could have drawn them in either direction. In Fig. c we draw the free-body diagram of the beam with the component A_y pointed downward. From this free-body diagram we obtain the equilibrium equations

$$\Sigma F_x = A_x - B \sin 30° = 0,$$
$$\Sigma F_y = -A_y + B \cos 30° - 2 = 0,$$
$$\Sigma M_{(\text{point } A)} = (5)(B \cos 30°) - (2)(3) = 0.$$

The solutions are $A_x = 0.69$ kN, $A_y = -0.80$ kN, and $B = 1.39$ kN. The negative value of A_y indicates that the vertical force exerted on the beam by the pin support is in the direction opposite to that of the arrow in Fig. c; that is, the force is 0.80 kN upward. Thus we again obtain the reactions shown in Fig. b.

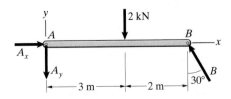

(c) An alternative free-body diagram.

Example 5.2

Reactions at a Built-In Support

The object in Fig. 5.14 has a built-in support and is subjected to two forces and a couple. What are the reactions at the support?

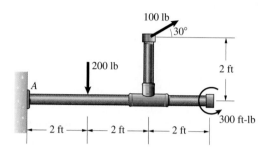

Figure 5.14

Solution

Draw the Free-Body Diagram We isolate the object from its support and show the reactions at the built-in support (Fig. a). There are three unknown reactions: two force components A_x and A_y and a couple M_A. (Remember that we can choose the directions of these arrows arbitrarily.) We also resolve the 100-lb force into its components.

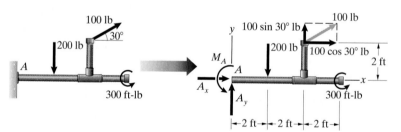

(a) Drawing the free-body diagram.

Apply the Equilibrium Equations Summing the moments about point A, the equilibrium equations are

$$\Sigma F_x = A_x + 100 \cos 30° = 0,$$
$$\Sigma F_y = A_y - 200 + 100 \sin 30° = 0,$$
$$\Sigma M_{(\text{point } A)} = M_A + 300 - (200)(2) - (100 \cos 30°)(2)$$
$$+ (100 \sin 30°)(4) = 0.$$

Solving these equations, we obtain the reactions $A_x = -86.6$ lb, $A_y = 150.0$ lb, and $M_A = 73.2$ ft-lb.

Discussion

Notice that the 300-ft-lb couple and the couple M_A exerted by the built-in support don't appear in the first two equilibrium equations because a couple exerts no net force. Also, since the moment due to a couple is the same about any point, the moment about point A due to the 300-ft-lb counterclockwise couple is 300 ft-lb counterclockwise.

Example 5.3

Reactions on a Car's Tires

The 2800-lb car in Fig. 5.15 is stationary. Determine the normal forces exerted on the front and rear tires by the road.

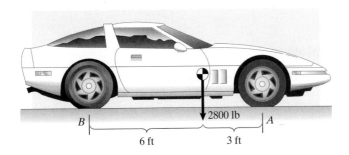

2800 lb

B A

6 ft   3 ft

Figure 5.15

Solution

Draw the Free-Body Diagram In Fig. a we isolate the car and show its weight and the reactions exerted by the road. There are two unknown reactions: the forces A and B exerted on the front and rear tires.

Apply the Equilibrium Equations The forces have no x components. Summing the moments about point B, the equilibrium equations are

$$\Sigma F_y = A + B - 2800 = 0,$$

$$\Sigma M_{(\text{point } B)} = (6)(2800) - 9A = 0.$$

Solving these equations, the reactions are $A = 1867$ lb and $B = 933$ lb.

Discussion

This example doesn't fall within our definition of a two-dimensional system of forces and moments because the forces acting on the car are not coplanar. Let's examine why you can analyze problems of this kind as if they were two-dimensional.

In Fig. b we show an oblique view of the free-body diagram of the car. In this view you can see the forces acting on the individual tires. The total normal force on the front tires is $A_L + A_R = A$, and the total normal force on the rear tires is $B_L + B_R = B$. The sum of the forces in the y direction is

$$\Sigma F_y = A_L + A_R + B_L + B_R - 2800 = A + B - 2800 = 0.$$

Since the sum of the moments about any line due to the forces and couples acting on an object in equilibrium is zero, the sum of the moments about the z axis due to the forces acting on the car is zero:

$$\Sigma M_{(z \text{ axis})} = (9)(A_L + A_R) - (6)(2800) = 9A - (6)(2800) = 0.$$

Thus we obtain the same equilibrium equations we did when we solved the problem using a two-dimensional analysis.

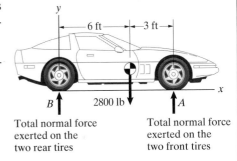

y

6 ft   3 ft

B 2800 lb A

Total normal force exerted on the two rear tires

Total normal force exerted on the two front tires

(a) The free-body diagram.

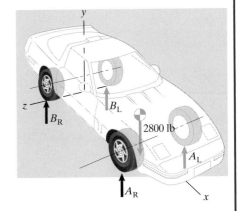

(b) An oblique view showing the forces on the individual tires.

Example 5.4

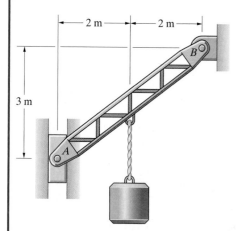

3 m

Figure 5.16

Choosing the Point About Which to Evaluate Moments

The structure AB in Fig. 5.16 supports a suspended 2-Mg (megagram) mass. The structure is attached to a slider in a vertical slot at A and has a pin support at B. What are the reactions at A and B?

Solution

Draw the Free-Body Diagram We isolate the structure and mass from the supports and show the reactions at the supports and the force exerted by the weight of the 2000-kg mass (Fig. a). The slot at A can exert only a horizontal force on the slider.

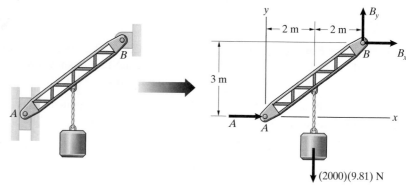

(a) Drawing the free-body diagram.

Apply the Equilibrium Equations Notice that if we sum the moments about point B, we obtain an equation containing only one unknown reaction, the force A. The equilibrium equations are

$$\Sigma F_x = A + B_x = 0,$$

$$\Sigma F_y = B_y - (2000)(9.81) = 0,$$

$$\Sigma M_{(\text{point } B)} = A(3) + (2000)(9.81)(2) = 0.$$

The reactions are $A = -13.1$ kN, $B_x = 13.1$ kN, and $B_y = 19.6$ kN.

Discussion

You can often simplify equilibrium equations by a careful choice of the point about which you sum moments. For example, when you can choose a point where the lines of action of unknown forces intersect, those forces will not appear in your moment equation.

Problems

Assume that objects are in equilibrium. In the statements of the answers, x components are positive to the right and y components are positive upward.

5.1 The beam has pin and roller supports and is subjected to a 4-kN load.
(a) Draw the free-body diagram of the beam.
(b) Determine the reactions at the supports.
   Strategy: (a) Draw a diagram of the beam isolated from its supports. Complete the free-body diagram of the beam by adding the 4-kN load and the reactions due to the pin and roller supports (see Table 5.1). (b) Use the scalar equilibrium equations (5.4)–(5.6) to determine the reactions.

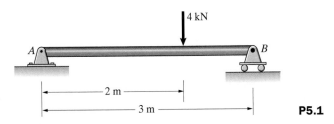

P5.1

5.2 The beam has a built-in support and is loaded by a 2-kN force and a 6 kN-m couple.
(a) Draw the free-body diagram of the beam.
(b) Determine the reactions at the supports.

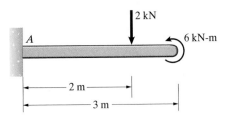

P5.2

5.3 The beam is subjected to a load $F = 400$ N and is supported by the rope and the smooth surfaces at A and B.
(a) Draw the free-body diagram of the beam.
(b) What are the magnitudes of the reactions at A and B?

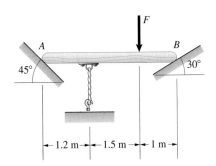

P5.3

5.4 (a) Draw the free-body diagram of the beam.
(b) Determine the reactions at the supports.

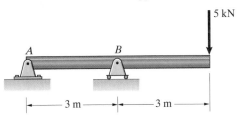

P5.4

5.5 (a) Draw the free-body diagram of the 60-lb drill press, assuming that the surfaces at A and B are smooth.
(b) Determine the reactions at A and B.

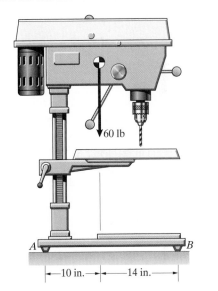

P5.5

5.6 The masses of the person and the diving board are 54 kg and 36 kg, respectively. Assume that they are in equilibrium.
(a) Draw the free-body diagram of the diving board.
(b) Determine the reactions at the supports A and B.

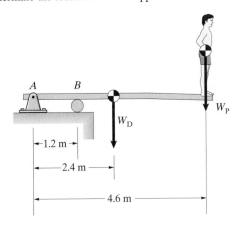

P5.6

5.7 The ironing board has supports at A and B that can be modeled as roller supports.
(a) Draw the free-body diagram of the ironing board.
(b) Determine the reactions at A and B.

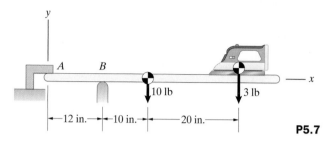

P5.7

5.8 The distance $x = 2$ m.
(a) Draw the free-body diagram of the beam.
(b) Determine the reactions at the supports.

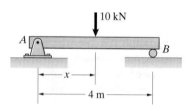

P5.8

5.9 Consider the beam in Problem 5.8. An engineer determines that each support will safely support a force of 7.5 kN. What is the range of values of the distance x at which the 10-kN force can safely be applied?

5.10 (a) Draw the free-body diagram of the beam.
(b) Determine the reactions at the supports.

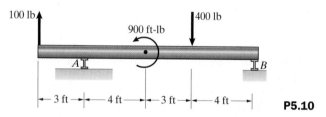

P5.10

5.11 Consider the beam in Problem 5.10. First represent the loads (the 100-lb force, the 400-lb force, and the 900 ft-lb couple) by a single equivalent force; then determine the reactions at the supports.

5.12 (a) Draw the free-body diagram of the beam.
(b) Determine the reactions at the supports.

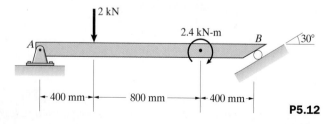

P5.12

5.13 Consider the beam in Problem 5.12. First represent the loads (the 2-kN force and 2.4-kN-m couple) by a single equivalent force; then determine the reactions at the supports.

5.14 If the force $F = 40$ kN, what are the reactions at A and B?

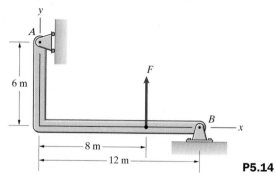

P5.14

5.15 In Problem 5.14, the structural designer determines that the magnitude of the force exerted on the support A by the beam must not exceed 80 kN, and the magnitude of the force exerted on the support B must not exceed 140 kN. Based on these criteria, what is the largest allowable value of the upward load F?

5.16 The person doing push-ups pauses in the position shown. His mass is 80 kg. Assume that his weight W acts at the point shown. The dimensions shown are $a = 250$ mm, $b = 740$ mm, and $c = 300$ mm. Determine the normal force exerted by the floor (a) on each hand, (b) on each foot.

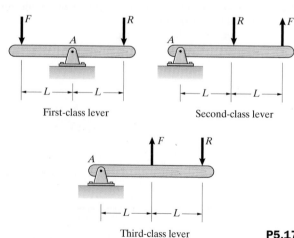

P5.16

5.17 With each of the devices shown you can support a load R by applying a force F. They are called levers of the first, second, and third class.

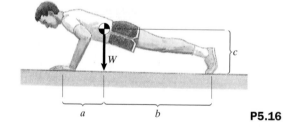

First-class lever

Second-class lever

Third-class lever

P5.17

(a) The ratio R/F is called the *mechanical advantage*. Determine the mechanical advantage of each lever.
(b) Determine the magnitude of the reaction at A for each lever. (Express your answers in terms of F.)

5.18 (a) Draw the free-body diagram of the beam. (b) Determine the reactions at the support.

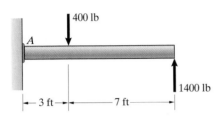

P5.18

5.19 The force $F = 12$ kN. Determine the reactions at A.

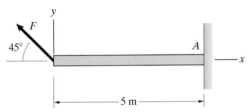

P5.19

5.20 The built-in support of the beam shown in Problem 5.19 will fail if the magnitude of the total force exerted on the beam by the support exceeds 20 kN or if the magnitude of the couple exerted by the support exceeds 65 kN-m. Based on these criteria, what is the maximum force F that can be applied?

5.21 The mobile is in equilibrium. The fish B weighs 27 oz. Determine the weights of the fish A, C, and D. (The weights of the crossbars are negligible.)

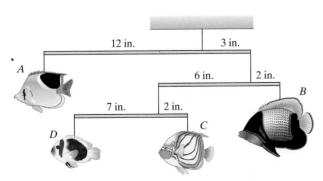

P5.21

5.22 The car's wheelbase (the distance between the wheels) is 2.82 m. The mass of the car is 1760 kg and its weight acts at the point $x = 2.00$ m, $y = 0.68$ m. If the angle $\alpha = 15°$, what is the total normal force exerted on the two rear tires by the sloped ramp?

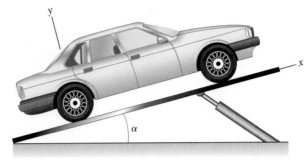

P5.22

5.23 The car in Problem 5.22 can remain in equilibrium on the sloped ramp only if the total friction force exerted on its tires does not exceed 0.8 times the total normal force exerted on the two rear tires. What is the largest angle α for which it can remain in equilibrium?

5.24 The 14.5-lb chain saw is subjected to the loads at A by the log it cuts. Determine the reactions R, B_x, and B_y that must be applied by the person using the saw to hold it in equilibrium.

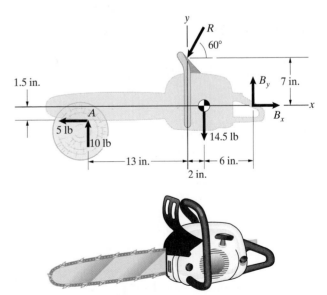

P5.24

5.25 The mass of the trailer is 2.2 Mg (megagrams). The distances $a = 2.5$ m and $b = 5.5$ m. The truck is stationary, and the wheels of

the trailer can turn freely, which means the road exerts no horizontal force on them. The hitch at B can be modeled as a pin support.
(a) Draw the free-body diagram of the trailer.
(b) Determine the total normal force exerted on the rear tires at A and the reactions exerted on the trailer at the pin support B.

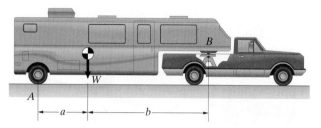

P5.25

5.26 The total weight of the wheelbarrow and its load is $W = 100$ lb.
(a) If $F = 0$, what are the vertical reactions at A and B?
(b) What force F is necessary to lift the support at A off the ground?

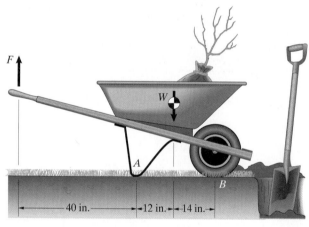

P5.26

5.27 The airplane's weight is $W = 2400$ lb. Its brakes keep the rear wheels locked. The front (nose) wheel can turn freely, and so

the ground exerts no horizontal force on it. The force T exerted by the airplane's propeller is horizontal.
(a) Draw the free-body diagram of the air-plane. Determine the reaction exerted on the nose wheel and the total normal reaction exerted on the rear wheels
(b) when $T = 0$;
(c) when $T = 250$ lb.

5.28 The forklift is stationary. The front wheels are free to turn, and the rear wheels are locked. The distances are $a = 1.25$ m, $b = 0.50$ m, and $c = 1.40$ m. The weight of the load is $W_L = 2$ kN, and the weight of the truck and operator is $W_F = 8$ kN. What are the reactions at A and B?

P5.28

5.29 Consider the stationary forklift shown in Problem 5.28. The front wheels are free to turn, and the rear wheels are locked. The distances are $a = 45$ in., $b = 20$ in., and $c = 50$ in. The weight of the truck and operator is $W_F = 3000$ lb. For safety reasons, a rule is established that the reaction at the rear wheels must be at least 400 lb. If the weight W_L of the load acts at the position shown, what is the maximum safe load?

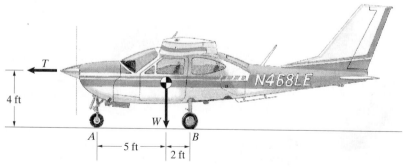

P5.27

5.30 The weight of the fan is $W = 20$ lb. Its base has four equally spaced legs of length $b = 12$ in., and $h = 36$ in. What is the largest thrust T exerted by the fan's propeller for which the fan will remain in equilibrium?

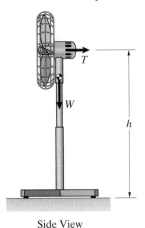

Side View

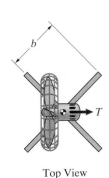

Top View

P5.30

5.31 Consider the fan described in Problem 5.30. As a safety criterion, an engineer decides that the vertical reaction on any of the fan's legs should not be less than 20% of the fan's weight. If the thrust T is 1 lb when the fan is set on its highest speed, what is the maximum safe value of h?

5.32 To decrease costs, an engineer considers supporting a fan with three equally spaced legs instead of the four-leg configuration shown in Problem 5.30. For the same values of b, h, and W, show that the largest thrust T for which the fan will remain in equilibrium with three legs is related to the value with four legs by

$$T_{\text{(three legs)}} = \left(1/\sqrt{2}\right)T_{\text{(four legs)}}.$$

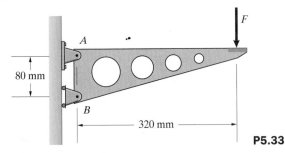

P5.32

5.33 A force $F = 400$ N acts on the bracket. What are the reactions at A and B?

80 mm

A

B

320 mm

F

P5.33

5.34 The hanging sign exerts vertical 25-lb forces at A and B. Determine the tension in the cable and the reactions at the support at C.

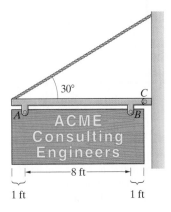

30°

C

A ACME
Consulting
Engineers B

|← 8 ft →|

1 ft   1 ft

P5.34

5.35 This device, called a *swape* or *shadoof*, is used to help a person lift a heavy load. (It was used in Egypt at least as early as 1550 B.C. and is still in use in various parts of the world today.) The distances are $a = 12$ ft and $b = 4$ ft. If the load being lifted weighs 100 lb and $W = 200$ lb, determine the vertical force the person must exert to support the stationary load (a) when the load is just above the ground (the position shown); (b) when the load is 3 ft above the ground. (Assume that the rope remains vertical.)

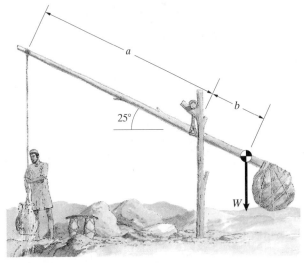

a

b

25°

W

P5.35

5.36 This structure, called a *truss*, has a pin support at A and a roller support at B and is loaded by two forces. Determine the reactions at the supports.

Strategy: Draw a free-body diagram, treating the entire truss as a single object.

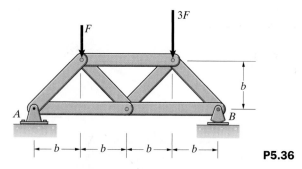

P5.36

5.37 An Olympic gymnast is stationary in the "iron cross" position. The weight of his left arm and the weight of his body *not including his arms* are shown. The distances are $a = b = 9$ in. and $c = 13$ in. Treat his shoulder S as a built-in support, and determine the magnitudes of the reactions at his shoulder. That is, determine the force and couple his shoulder must support.

P5.37

5.38 Determine the reactions at A.

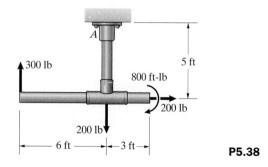

P5.38

5.39 The car's brakes keep the rear wheels locked, and the front wheels are free to turn. Determine the forces exerted on the front and rear wheels by the road when the car is parked (a) on an upslope with $\alpha = 15°$; (b) on a downslope with $\alpha = -15°$.

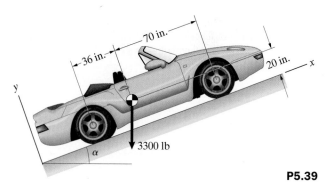

P5.39

5.40 The weight W of the bar acts at its center. The surfaces are smooth. What is the tension in the horizontal string?

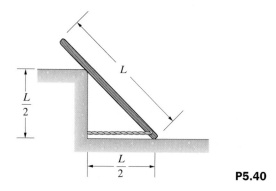

P5.40

5.41 The mass of the bar is 36 kg and its weight acts at its midpoint. The spring is unstretched when $\alpha = 0$. The bar is in equilibrium when $\alpha = 30°$. Determine the spring constant k.

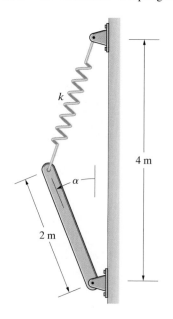

P5.41

5.42 The plate is supported by a pin in a smooth slot at B. What are the reactions at the supports?

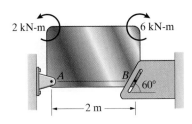

P5.42

5.43 The force $F = 800$ N, and the couple $M = 200$ N-m. The distance $L = 2$ m. What are the reactions at A and B?

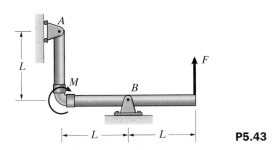

P5.43

5.44 The mass of the bar is 40 kg and its weight acts at its midpoint. Determine the tension in the cable and the reactions at A.

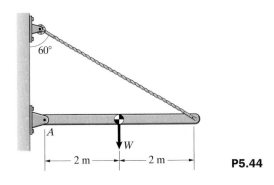

P5.44

5.45 If the length of the cable in Problem 5.44 is increased by 1 m, what are the tension in the cable and the reactions at A?

5.46 The mass of each of the suspended boxes is 80 kg. Determine the reactions at the supports at A and E.

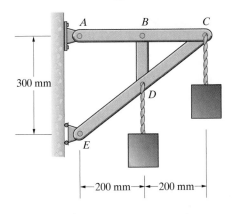

P5.46

5.47 The suspended boxes in Problem 5.46 are each of mass m. The supports at A and E will each safely support a force of 6 kN magnitude. Based on this criterion, what is the largest safe value of m?

5.48 The tension in cable BC is 100 lb. Determine the reactions at the built-in support.

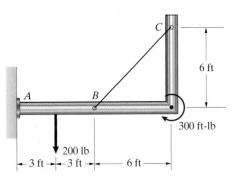

P5.48

5.49 The tension in cable AB is 2 kN. What are the reactions at C in the two cases?

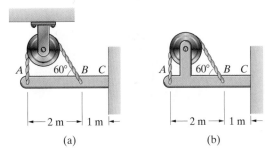

P5.49

5.50 Determine the reactions at the supports.

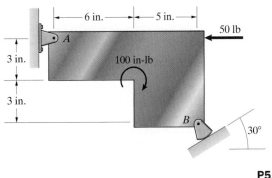

P5.50

5.51 The weight $W = 2$ kN. Determine the tension in the cable and the reactions at A.

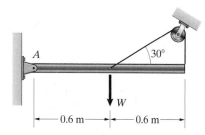

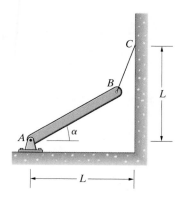

P5.51

5.52 The cable shown in Problem 5.51 will safely support a tension of 6 kN. Based on this criterion, what is the largest safe value of the weight W?

5.53 The spring constant is $k = 9600$ N/m and the unstretched length of the spring is 30 mm. Treat the bolt at A as a pin support and assume that the surface at C is smooth. Determine the reactions at A and the normal force at C.

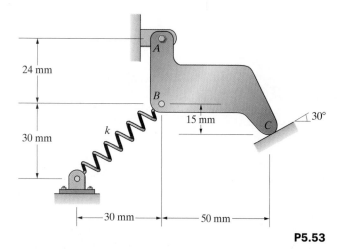

P5.53

5.54 The engineer designing the release mechanism shown in Problem 5.53 wants the normal force exerted at C to be 120 N. If the unstretched length of the spring is 30 mm, what is the necessary value of the spring constant k?

5.55 Suppose that you want to design the safety valve to open when the difference between the pressure p in the circular pipe (diameter $= 150$ mm) and atmospheric pressure is 10 MPa (megapascals; a pascal is 1 N/m²). The spring is compressed 20 mm when the valve is closed. What should the value of the spring constant be?

P5.55

5.56 The bar AB is of length L and weight W, and the weight acts at its midpoint. The angle $\alpha = 30°$. What is the tension in the string?

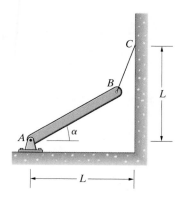

P5.56

5.57 The crane's arm has a pin support at A. The hydraulic cylinder BC exerts a force on the arm at C in the direction parallel to BC. The crane's arm has a mass of 200 kg, and its weight can be assumed to act at a point 2 m to the right of A. If the mass of the suspended box is 800 kg and the system is in equilibrium, what is the magnitude of the force exerted by the hydraulic cylinder?

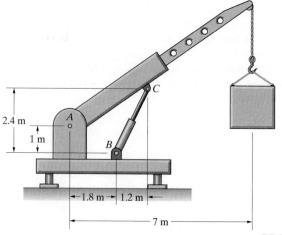

P5.57

5.58 In Problem 5.57, what is the magnitude of the force exerted on the crane's arm by the pin support at A?

5.59 A speaker system is suspended by the cables attached at D and E. The mass of the speaker system is 130 kg, and its weight acts at G. Determine the tensions in the cables and the reactions at A and C.

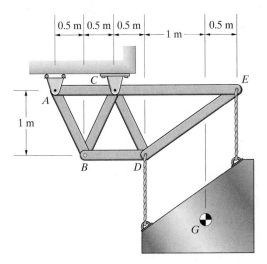

P5.59

5.60 The weight $W_1 = 1000$ lb. Neglect the weight of the bar AB. The cable goes over a pulley at C. Determine the weight W_2 and the reactions at the pin support A.

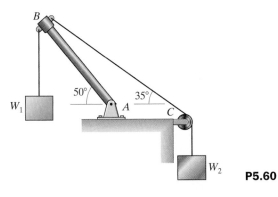

P5.60

5.61 The dimensions $a = 2$ m and $b = 1$ m. The couple $M = 2400$ N-m. The spring constant is $k = 6000$ N/m, and the spring would be unstretched if $h = 0$. The system is in equilibrium when $h = 2$ m and the beam is horizontal. Determine the force F and the reactions at A.

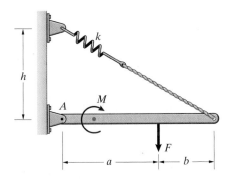

P5.61

5.62 The bar is 1 m long, and its weight W acts at its midpoint. The distance $b = 0.75$ m, and the angle $\alpha = 30°$. The spring constant is $k = 100$ N/m, and the spring is unstretched when the bar is vertical. Determine W and the reactions at A.

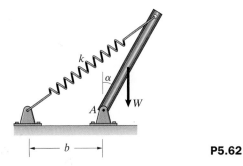

P5.62

5.63 The boom derrick supports a suspended 15-kip load. The booms BC and DE are each 20 ft long. The distances are $a = 15$ ft and $b = 2$ ft, and the angle $\theta = 30°$. Determine the tension in cable AB and the reactions at the pin supports C and D.

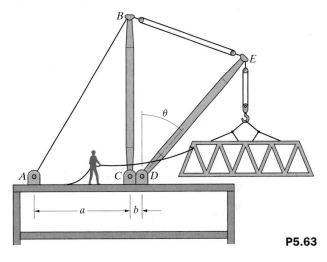

P5.63

5.64 The arrangement shown controls the elevators of an airplane. (The elevators are the horizontal control surfaces in the airplane's tail.) The elevators are attached to member EDG. Aerodynamic pressures on the elevators exert a clockwise couple of 120 in.-lb. Cable BG is slack, and its tension can be neglected. Determine the force F and the reactions at the pin support A.

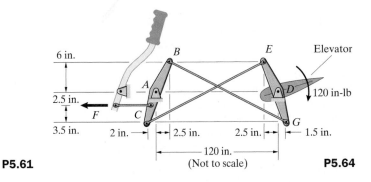

P5.64

5.3 Statically Indeterminate Objects

In Section 5.2 we discussed examples in which we were able to use the equilibrium equations to determine unknown forces and couples acting on objects in equilibrium. You need to be aware of two common situations in which this procedure doesn't lead to a solution.

First, the free-body diagram of an object can have more unknown forces or couples than the number of independent equilibrium equations you can obtain. Since you can write no more than three such equations for a given free-body diagram in a two-dimensional problem, when there are more than three unknowns you can't determine them from the equilibrium equations alone. This occurs, for example, when an object has more supports than the minimum number necessary to maintain it in equilibrium. Such an object is said to have *redundant supports*. The second situation is when the supports of an object are improperly designed such that they cannot maintain equilibrium under the loads acting on it. The object is said to have *improper supports*. In either situation, the object is said to be *statically indeterminate*.

Engineers use redundant supports whenever possible for strength and safety. Some designs, however, require that the object be incompletely supported so that it is free to undergo certain motions. These two situations— more supports than necessary for equilibrium or not enough—are so common that we consider them in detail.

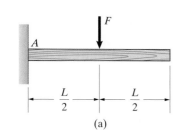

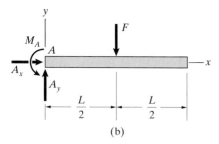

Figure 5.17
(a) A beam with a built-in support.
(b) The free-body diagram has three
   unknown reactions.

Redundant Supports

Let's consider a beam with a built-in support (Fig. 5.17a). From its free-body diagram (Fig. 5.17b), we obtain the equilibrium equations

$$\Sigma F_x = A_x = 0,$$

$$\Sigma F_y = A_y - F = 0,$$

$$\Sigma M_{(\text{point } A)} = M_A - \left(\frac{L}{2}\right)F = 0.$$

Assuming we know the load F, we have three equations and three unknown reactions, for which we obtain the solutions $A_x = 0$, $A_y = F$, and $M_A = FL/2$.

Now suppose we add a roller support at the right end of the beam (Fig. 5.18a). From the new free-body diagram (Fig. 5.18b), we obtain the equilibrium equations

$$\Sigma F_x = A_x = 0, \tag{5.10}$$

$$\Sigma F_y = A_y - F + B = 0, \tag{5.11}$$

$$\Sigma M_{(\text{point } A)} = M_A - \left(\frac{L}{2}\right)F + LB = 0. \tag{5.12}$$

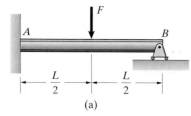

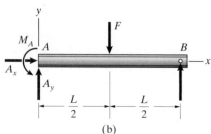

Figure 5.18
(a) A beam with built-in and roller
   supports.
(b) The free-body diagram has four
   unknown reactions.

Now we have three equations and four unknown reactions. Although the first equation tells us that $A_x = 0$, we can't solve the two equations (5.11) and (5.12) for the three reactions A_y, B, and M_A.

When faced with this situation, students often attempt to sum the moments about another point, such as point B, to obtain an additional equation:

$$\Sigma M_{(\text{point } B)} = M_A + \left(\frac{L}{2}\right)F - LA_y = 0.$$

Unfortunately, this doesn't help. This is not an independent equation but is a linear combination of Eqs. (5.11) and (5.12):

$$\Sigma M_{(\text{point } B)} = M_A + \left(\frac{L}{2}\right)F - LA_y$$

$$= \underbrace{M_A - \left(\frac{L}{2}\right)F + LB}_{\text{Eq. (5.12)}} - \underbrace{L(A_y - F + B)}_{\text{Eq. (5.11)}}.$$

As this example demonstrates, each support added to an object results in additional reactions. The difference between the number of reactions and the number of independent equilibrium equations is called the *degree of redundancy*.

Even if an object is statically indeterminate due to redundant supports, it may be possible to determine some of the reactions from the equilibrium equations. Notice that in our previous example we were able to determine the reaction A_x even though we could not determine the other reactions.

Since redundant supports are so ubiquitous, you may wonder why we devote so much effort to teaching you how to analyze objects whose reactions can be determined with the equilibrium equations. We want to develop your understanding of equilibrium and give you practice writing equilibrium equations. The reactions on an object with redundant supports *can* be determined by supplementing the equilibrium equations with additional equations that relate the forces and couples acting on the object to its deformation, or change in shape. Thus obtaining the equilibrium equations is the first step of the solution.

Example 5.5

Recognizing a Statically Indeterminate Object

The beam in Fig. 5.19 has two pin supports and is loaded by a 2-kN force.
(a) Show that the beam is statically indeterminate.
(b) Determine as many reactions as possible.

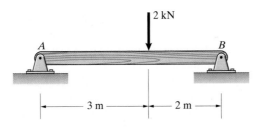

Figure 5.19

Strategy

The beam is statically indeterminate if its free-body diagram has more unknown reactions than the number of independent equilibrium equations we can obtain. But even if this is the case, we may be able to solve the equilibrium equations for some of the reactions.

Solution

Draw the Free-Body Diagram We draw the free-body diagram of the beam in Fig. a. There are four unknown reactions—A_x, A_y, B_x, and B_y—and we can write only three independent equilibrium equations. Therefore the beam is statically indeterminate.

Apply the Equilibrium Equations Summing the moments about point A, the equilibrium equations are

$$\Sigma F_x = A_x + B_x = 0,$$

$$\Sigma F_y = A_y + B_y - 2 = 0,$$

$$\Sigma M_{(\text{point } A)} = 5B_y - (2)(3) = 0.$$

We can solve the third equation for B_y and then solve the second equation for A_y. The results are $A_y = 0.8$ kN and $B_y = 1.2$ kN. The first equation tells us that $B_x = -A_x$, but we can't solve for their values.

Discussion

This example can give you insight into why the reactions on objects with redundant constraints can't be determined from the equilibrium equations alone. The two pin supports can exert horizontal reactions on the beam even in the absence of loads (Fig. b), and these reactions satisfy the equilibrium equations for any value of T $\left(\Sigma F_x = -T + T = 0\right)$.

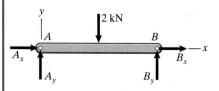

(a) The free-body diagram of the beam.

(b) The supports can exert reactions on the beam.

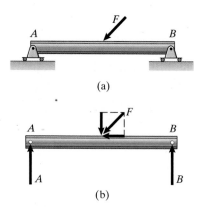

(a)

(b)

Figure 5.20
(a) A beam with two roller supports is not in equilibrium when subjected to the load shown.
(b) The sum of the forces in the horizontal direction is not zero.

Improper Supports

We say that an object has improper supports if it will not remain in equilibrium under the action of the loads exerted on it. Thus an object with improper supports will move when the loads are applied. In two-dimensional problems, this can occur in two ways:

1. *The supports can exert only parallel forces.* This leaves the object free to move in the direction perpendicular to the support forces. If the loads exert a component of force in that direction, the object is not in equilibrium. Figure 5.20a shows an example of this situation. The two roller supports can exert only vertical forces, while the force F has a horizontal component. The beam will move horizontally when F is applied. This is particularly apparent from the free-body diagram (Fig. 5.20b). The sum of the forces in the horizontal direction cannot be zero because the roller supports can exert only vertical reactions.

2. *The supports can exert only concurrent forces.* If the loads exert a moment about the point where the lines of action of the support forces intersect, the object is not in equilibrium. For example, consider the beam in Fig. 5.21a. From its free-body diagram (Fig. 5.21b) we see that the reactions A and B exert no moment about the point P, where their lines of action intersect, but the load F does. The sum of the moments about point P is not zero, and the beam will rotate when the load is applied.

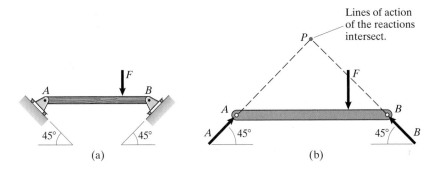

(a)   (b)

Figure 5.21
(a) A beam with roller supports on sloped surfaces.
(b) The sum of the moments about point P is not zero.

Except for problems that deal explicitly with improper supports, objects in our examples and problems have proper supports. You should develop the habit of examining objects in equilibrium and thinking about why they are properly supported for the loads acting on them.

Study Questions

1. What does it mean when an object is said to have redundant supports?
2. How can you recognize if an object is statically indeterminate due to redundant supports?
3. What is the "degree of redundancy" of an object?

Example 5.6

Proper and Improper Supports

State whether each L-shaped bar in Fig. 5.22 is properly or improperly supported. If a bar is properly supported, determine the reactions at its supports.

Solution

We draw the free-body diagrams of the bars in Fig. 5.23.

Bar (a)   The lines of action of the reactions due to the two roller supports intersect at P, and the load F exerts a moment about P. This bar is improperly supported.

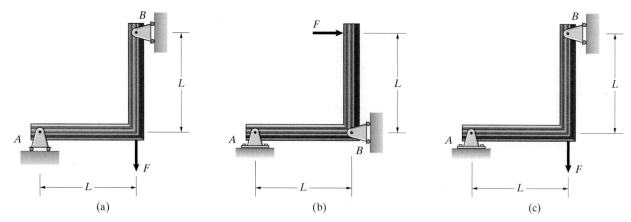

Figure 5.22

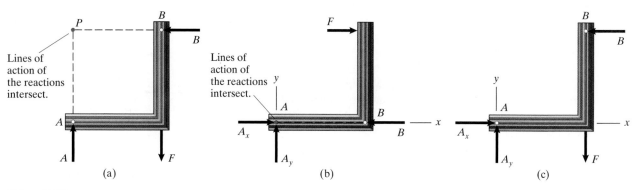

Figure 5.23
Free-body diagrams of the three bars.

Bar (b) The lines of action of the reactions intersect at A, and the load F exerts a moment about A. This bar is also improperly supported.

Bar (c) The three support forces are neither parallel nor concurrent. This bar is properly supported. The equilibrium equations are

$$\Sigma F_x = A_x - B = 0,$$

$$\Sigma F_y = A_y - F = 0,$$

$$\Sigma M_{(\text{point } A)} = BL - FL = 0.$$

Solving these equations, the reactions are $A_x = F$, $A_y = F$, and $B = F$.

Problems

5.65 (a) Draw the free-body diagram of the beam and show that it is statically indeterminate.
(b) Determine as many of the reactions as possible.

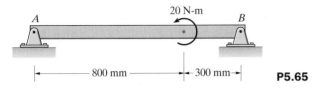

P5.65

5.66 Consider the beam in Problem 5.65. Choose supports at A and B so that it is not statically indeterminate. Determine the reactions at the supports.

5.67 (a) Draw the free-body diagram of the beam and show that it is statically indeterminate. (The external couple M_0 is known.)
(b) By an analysis of the beam's deflection, it is determined that the vertical reaction B exerted by the roller support is related to the couple M_0 by $B = 2M_0/L$. What are the reactions at A?

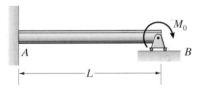

P5.67

5.68 Consider the beam in Problem 5.67. Choose supports at A and B so that it is not statically indeterminate. Determine the reactions at the supports.

5.69 Draw the free-body diagram of the L-shaped pipe assembly and show that it is statically indeterminate. Determine as many of the reactions as possible.
   Strategy: Place the coordinate system so that the x axis passes through points A and B.

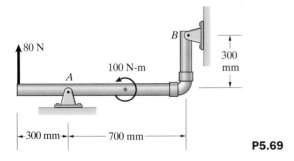

P5.69

5.70 Consider the pipe assembly in Problem 5.69. Choose supports at A and B so that it is not statically indeterminate. Determine the reactions at the supports.

5.71 State whether each of the L-shaped bars shown is properly or improperly supported. If a bar is properly supported, determine the reactions at its supports.

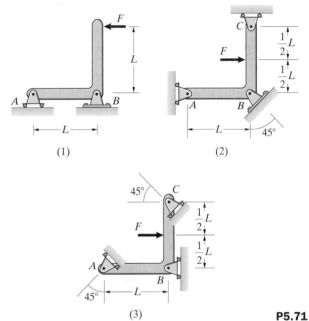

(1)   (2)

(3)   **P5.71**

5.72 State whether each of the L-shaped bars shown is properly or improperly supported. If a bar is properly supported, determine the reactions at its supports.

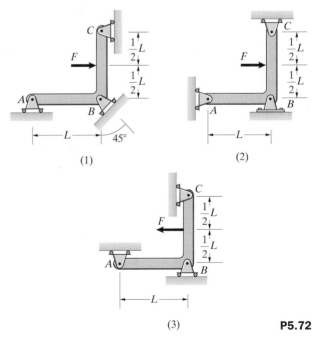

(1)   (2)

(3)   **P5.72**

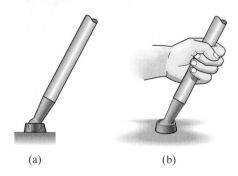

5.4 Three-Dimensional Applications

You have seen that when an object in equilibrium is subjected to a two-dimensional system of forces and moments, you can obtain no more than three independent equilibrium equations. In the case of a three-dimensional system of forces and moments, you can obtain up to six independent equilibrium equations: The three components of the sum of the forces must equal zero, and the three components of the sum of the moments about any point must equal zero. Your procedure for determining the reactions on objects subjected to three-dimensional systems of forces and moments—drawing the free-body diagram and applying the equilibrium equations—is the same as in two-dimensions. You just need to become familiar with the support conventions used in three-dimensional applications.

Supports

We present five conventions frequently used in three-dimensional problems. Again, even when actual supports do not physically resemble these models, we represent them by the models if they exert the same (or approximately the same) reactions.

The Ball and Socket Support In the *ball and socket support*, the supported object is attached to a ball enclosed within a spherical socket (Fig. 5.24a). The socket permits the ball to rotate freely (friction is neglected) but prevents it from translating in any direction.

Imagine holding a bar attached to a ball and socket support (Fig. 5.24b). If you try to translate the bar (move it without rotating it) in any direction, the support exerts a reactive force to prevent the motion. However, you can rotate the bar about the support. The support cannot exert a couple to prevent rotation. Thus a ball and socket support can't exert a couple but can exert three components of force (Fig. 5.24c). It is the three-dimensional analog of the two-dimensional pin support.

The human hip joint is an example of a ball and socket support (Fig. 5.25). The support of the gear shift lever of a car can be modeled as a ball and socket support within the lever's range of motion.

The Roller Support The *roller support* (Fig. 5.26a) is a ball and socket support that can roll freely on a supporting surface. A roller support can exert only a force normal to the supporting surface (Fig. 5.26b). The rolling "casters" sometimes used to support furniture legs are supports of this type.

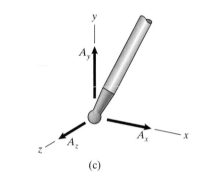

Figure 5.24
(a) A ball and socket support.
(b) Holding a supported bar.
(c) The ball and socket support can exert three components of force.

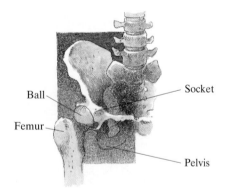

Figure 5.25
The human femur is attached to the pelvis by a ball and socket support.

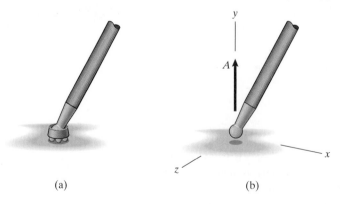

Figure 5.26
(a) A roller support.
(b) The reaction is normal to the supporting surface.

The Hinge   The hinge support is the familiar device used to support doors. It permits the supported object to rotate freely about a line, the *hinge axis*. An object is attached to a hinge in Fig. 5.27a. The z axis of the coordinate system is aligned with the hinge axis.

If you imagine holding a bar attached to a hinge (Fig. 5.27b), notice that you can rotate the bar about the hinge axis. The hinge cannot exert a couple about the hinge axis (the z axis) to prevent rotation. However, you can't rotate the bar about the x or y axis because the hinge can exert couples about those axes to resist the motion. In addition, you can't translate the bar in any direction. The reactions a hinge can exert on an object are shown in Fig. 5.27c. There are three components of force, A_x, A_y, and A_z, and couples about the x and y axes, M_{Ax} and M_{Ay}.

In some situations, either a hinge exerts no couples on the object it supports, or they are sufficiently small to neglect. An example of the latter case is when the axes of the hinges supporting a door are properly aligned (the axes of the individual hinges coincide). In these situations the hinge exerts only forces on an object (Fig. 5.27d). Situations also arise in which a hinge exerts no couples on an object and exerts no force in the direction of the hinge axis. (The hinge may actually be designed so that it cannot support a force parallel to the hinge axis.) Then the hinge exerts forces only in the directions perpendicular to the hinge axis (Fig. 5.27e). In examples and problems, we indicate when a hinge does not exert all five of the reactions in Fig. 5.27c.

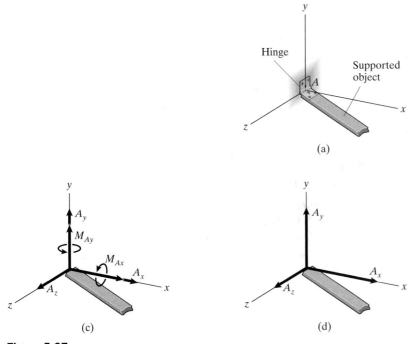

(a)

(b)

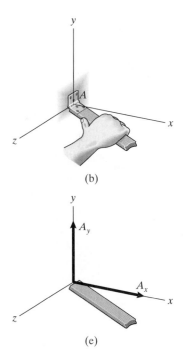

(c)

(d)

(e)

Figure 5.27

(a) A hinge. The z axis is aligned with the hinge axis.

(b) Holding a supported bar.

(c) In general, a hinge can exert five reactions: three force components and two couple components.

(d) The reactions when the hinge exerts no couples.

(e) The reactions when the hinge exerts neither couples nor a force parallel to the hinge axis.

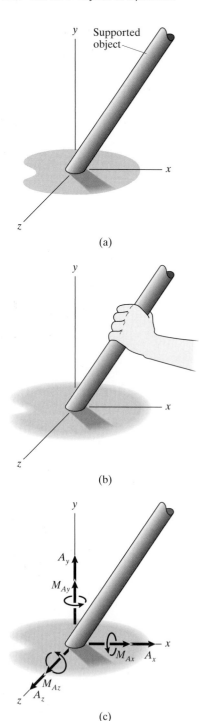

(a)

(b)

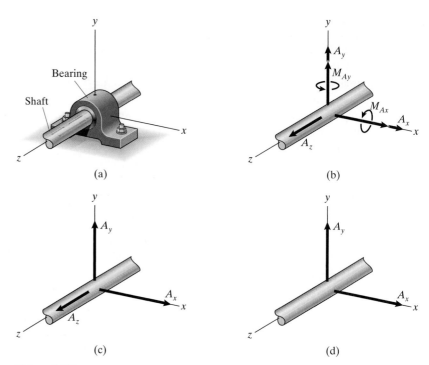

(a)

(b)

(c)

(d)

Figure 5.28
(a) A bearing. The z axis is aligned with the axis of the shaft.
(b) In general, a bearing can exert five reactions: three force components and two couple components.
(c) The reactions when the bearing exerts no couples.
(d) The reactions when the bearing exerts neither couples nor a force parallel to the axis of the shaft.

(c)

Figure 5.29
(a) A built-in support.
(b) Holding a supported bar.
(c) A built-in support can exert six reactions: three force components and three couple components.

The Bearing The type of bearing shown in Fig. 5.28a supports a circular shaft while permitting it to rotate about its axis. The reactions are identical to those exerted by a hinge. In the most general case (Fig. 5.28b), the bearing can exert a force on the supported shaft in each coordinate direction and can exert couples about axes perpendicular to the shaft but cannot exert a couple about the axis of the shaft.

As in the case of the hinge, situations can occur in which the bearing exerts no couples (Fig. 5.28c) or exerts no couples and no force parallel to the shaft axis (Fig. 5.28d). Some bearings are designed in this way for specific applications. In examples and problems, we indicate when a bearing does not exert all of the reactions in Fig. 5.28b.

The Built-In Support You are already familiar with the built-in, or fixed, support (Fig. 5.29a). Imagine holding a bar with a built-in support (Fig. 5.29b). You cannot translate it in any direction, and you cannot rotate it about any axis. The support is capable of exerting forces A_x, A_y, and A_z in each coordinate direction and couples M_{Ax}, M_{Ay}, and M_{Az} about each coordinate axis (Fig. 5.29c).

Table 5.2 summarizes the support conventions commonly used in three-dimensional applications.

Table 5.2 Supports used in three-dimensional applications.

Supports	Reactions
 Rope or Cable	 One Collinear Force
 Contact with a Smooth Surface	 One Normal Force
 Contact with a Rough Surface	 Three Force Components
 Ball and Socket Support	 Three Force Components
 Roller Support	 One Normal Force

(continues on next page)

Table 5.2 *(cont.)*

Supports	Reactions
Hinge (The z axis is parallel to the hinge axis.)	Three Force Components, Two Couple Components
Bearing (The z axis is parallel to the axis of the supported shaft.)	(When no couples are exerted) (When no couples and no axial force are exerted)
Built-in (Fixed) Support	Three Force Components, Three Couple Components

The Scalar Equilibrium Equations

The loads and reactions on an object in equilibrium satisfy the six scalar equilibrium equations

$$\Sigma F_x = 0, \tag{5.13}$$
$$\Sigma F_y = 0, \tag{5.14}$$
$$\Sigma F_z = 0, \tag{5.15}$$
$$\Sigma M_x = 0, \tag{5.16}$$
$$\Sigma M_y = 0, \tag{5.17}$$
$$\Sigma M_z = 0. \tag{5.18}$$

You can evaluate the sums of the moments about any point. Although you can obtain other equations by summing the moments about additional points, they will not be independent of these equations. *More than six independent equilibrium equations cannot be obtained from a given free-body diagram, so we can solve for at most six unknown forces or couples.*

The steps required to determine reactions in three dimensions are familiar from your experience with two-dimensional applications. You must first obtain a free-body diagram by isolating an object and showing the loads and reactions acting on it, then use Eqs. (5.13)–(5.18) to determine the reactions.

Study Questions

1. What is a ball and socket support? What reactions can it exert on an object?
2. In general, a hinge support can exert five reactions on an object. What are they?
3. If an object has a built-in support and any additional supports, it is statically indeterminate. Why is this true?

Example 5.7

Determining Reactions in Three Dimensions

The bar AB in Fig. 5.30 is supported by the cables BC and BD and a ball and socket support at A. Cable BC is parallel to the z axis, and cable BD is parallel to the x axis. The 200-N weight of the bar acts at its midpoint. What are the tensions in the cables and the reactions at A?

Strategy

We must obtain the free-body diagram of the bar AB by isolating it from the support at A and the two cables. Then we can use the equilibrium equations to determine the reactions at A and the tensions in the cables.

Solution

Draw the Free-Body Diagram In Fig. a we isolate the bar and show the reactions that may be exerted on it. The ball and socket support can exert three components of force, A_x, A_y, and A_z. The terms T_{BC} and T_{BD} represent the tensions in the cables.

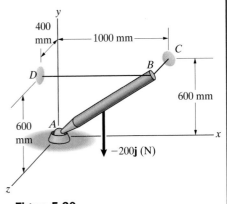

Figure 5.30

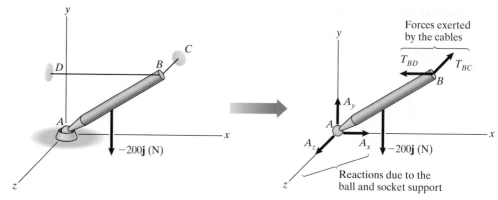

(a) Obtaining the free-body diagram of the bar.

Apply the Equilibrium Equations The sums of the forces in each coordinate direction equal zero:

$$\Sigma F_x = A_x - T_{BD} = 0,$$
$$\Sigma F_y = A_y - 200 = 0,$$
$$\Sigma F_z = A_z - T_{BC} = 0. \qquad (5.19)$$

Let $\mathbf{r}_{AB}$ be the position vector from A to B. The sum of the moments about A is

$$\Sigma \mathbf{M}_{(\text{point } A)} = \left[\mathbf{r}_{AB} \times (-T_{BC}\mathbf{k})\right] + \left[\mathbf{r}_{AB} \times (-T_{BD}\mathbf{i})\right]$$
$$+ \left[\tfrac{1}{2}\mathbf{r}_{AB} \times (-200\mathbf{j})\right]$$

$$= \begin{vmatrix} \mathbf{i} & \mathbf{j} & \mathbf{k} \\ 1 & 0.6 & 0.4 \\ 0 & 0 & -T_{BC} \end{vmatrix} + \begin{vmatrix} \mathbf{i} & \mathbf{j} & \mathbf{k} \\ 1 & 0.6 & 0.4 \\ -T_{BD} & 0 & 0 \end{vmatrix}$$

$$+ \begin{vmatrix} \mathbf{i} & \mathbf{j} & \mathbf{k} \\ 0.5 & 0.3 & 0.2 \\ 0 & -200 & 0 \end{vmatrix}$$

$$= (-0.6T_{BC} + 40)\mathbf{i} + (T_{BC} - 0.4T_{BD})\mathbf{j} + (0.6T_{BD} - 100)\mathbf{k}.$$

The components of this vector (the sums of the moments about the three coordinate axes) each equal zero:

$$\Sigma M_x = -0.6T_{BC} + 40 = 0,$$
$$\Sigma M_y = T_{BC} - 0.4T_{BD} = 0,$$
$$\Sigma M_z = 0.6T_{BD} - 100 = 0.$$

Solving these equations, we obtain the tensions in the cables:

$$T_{BC} = 66.7 \text{ N}, \qquad T_{BD} = 166.7 \text{ N}.$$

(Notice that we needed only two of the three equations to obtain the two tensions. The third equation is redundant.)

Then from Eqs. (5.19) we obtain the reactions at the ball and socket support:

$$A_x = 166.7 \text{ N}, \qquad A_y = 200 \text{ N}, \qquad A_z = 66.7 \text{ N}.$$

Discussion

Notice that by summing moments about A we obtained equations in which the unknown reactions A_x, A_y, and A_z did not appear. You can often simplify your solutions in this way.

Example 5.8

Reactions at a Hinge Support

The bar AC in Fig. 5.31 is 4 ft long and is supported by a hinge at A and the cable BD. The hinge axis is along the z axis. The centerline of the bar lies in the x–y plane, and the cable attachment point B is the midpoint of the bar. Determine the tension in the cable and the reactions exerted on the bar by the hinge.

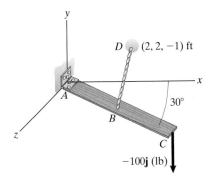

Figure 5.31

Solution

Draw the Free-Body Diagram We isolate the bar from the hinge support and the cable and show the reactions they exert (Fig. a). The terms A_x, A_y, and A_z are the components of force exerted by the hinge, and the terms M_{Ax} and M_{Ay} are the couples exerted by the hinge about the x and y axes. (Remember that the hinge cannot exert a couple on the bar about the hinge axis.) The term T is the tension in the cable.

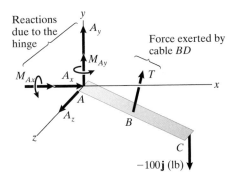

(a) The free-body diagram of the bar.

Apply the Equilibrium Equations To write the equilibrium equations, we must first express the cable force in terms of its components. The coordinates of point B are $(2\cos 30°, -2\sin 30°, 0)$ ft, so the position vector from B to D is

$$\mathbf{r}_{BD} = (2 - 2\cos 30°)\mathbf{i} + [2 - (-2\sin 30°)]\mathbf{j} + (-1 - 0)\mathbf{k}$$

$$= 0.268\mathbf{i} + 3\mathbf{j} - \mathbf{k}.$$

We divide this vector by its magnitude to obtain a unit vector $\mathbf{e}_{BD}$ that points from point B toward point D:

$$\mathbf{e}_{BD} = \frac{\mathbf{r}_{BD}}{|\mathbf{r}_{BD}|} = 0.084\mathbf{i} + 0.945\mathbf{j} - 0.315\mathbf{k}.$$

Now we can write the cable force as the product of its magnitude and $\mathbf{e}_{BD}$:

$$T\mathbf{e}_{BD} = T(0.084\mathbf{i} + 0.945\mathbf{j} - 0.315\mathbf{k}).$$

The sums of the forces in each coordinate direction must equal zero:

$$\Sigma F_x = A_x + 0.084T = 0,$$

$$\Sigma F_y = A_y + 0.945T - 100 = 0,$$

$$\Sigma F_z = A_z - 0.315T = 0. \tag{5.20}$$

If we sum moments about A, the resulting equations do not contain the unknown reactions A_x, A_y, and A_z. The position vectors from A to B and from A to C are

$$\mathbf{r}_{AB} = 2\cos 30°\mathbf{i} - 2\sin 30°\mathbf{j}.$$

$$\mathbf{r}_{AC} = 4\cos 30°\mathbf{i} - 4\sin 30°\mathbf{j}.$$

The sum of the moments about A is

$$\Sigma\mathbf{M}_{(\text{point }A)} = M_{Ax}\mathbf{i} + M_{Ay}\mathbf{j} + \left[\mathbf{r}_{AB} \times (T\mathbf{e}_{BD})\right] + \left[\mathbf{r}_{AC} \times (-100\mathbf{j})\right]$$

$$= M_{Ax}\mathbf{i} + M_{Ay}\mathbf{j} + \begin{vmatrix} \mathbf{i} & \mathbf{j} & \mathbf{k} \\ 1.732 & -1 & 0 \\ 0.084T & 0.945T & -0.315T \end{vmatrix}$$

$$+ \begin{vmatrix} \mathbf{i} & \mathbf{j} & \mathbf{k} \\ 3.464 & -2 & 0 \\ 0 & -100 & 0 \end{vmatrix}$$

$$= (M_{Ax} + 0.315T)\mathbf{i} + (M_{Ay} + 0.546T)\mathbf{j}$$

$$+ (1.72T - 346)\mathbf{k} = 0.$$

From this vector equation we obtain the scalar equations

$$\Sigma M_x = M_{Ax} + 0.315T = 0,$$

$$\Sigma M_y = M_{Ay} + 0.546T = 0,$$

$$\Sigma M_z = 1.72T - 346 = 0.$$

Solving these equations, we obtain the reactions

$$T = 201 \text{ lb}, \qquad M_{Ax} = -63.4 \text{ ft-lb}, \qquad M_{Ay} = -109.8 \text{ ft-lb}.$$

Then from Eqs. (5.20) we obtain the forces exerted on the bar by the hinge:

$$A_x = -17.0 \text{ lb}, \qquad A_y = -90.2 \text{ lb}, \qquad A_z = 63.4 \text{ lb}.$$

Example 5.9

Reactions at Properly Aligned Hinges

The plate in Fig. 5.32 is supported by hinges at A and B and the cable CE. The properly aligned hinges do not exert couples on the plate, and the hinge at A does not exert a force on the plate in the direction of the hinge axis. Determine the reactions at the hinges and the tension in the cable.

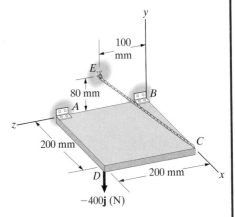

Figure 5.32

Solution

Draw the Free-Body Diagram We isolate the plate and show the reactions at the hinges and the force exerted by the cable (Fig. a). The term T is the force exerted on the plate by cable CE.

Apply the Equilibrium Equations Since we know the coordinates of points C and E, we can express the cable force as the product of its magnitude T and a unit vector directed from C toward E. The result is

$$T(-0.842\mathbf{i} + 0.337\mathbf{j} + 0.421\mathbf{k}).$$

The sums of the forces in each coordinate direction equal zero:

$$\Sigma F_x = A_x + B_x - 0.842T = 0,$$

$$\Sigma F_y = A_y + B_y + 0.337T - 400 = 0,$$

$$\Sigma F_z = B_z + 0.421T = 0.$$

If we sum the moments about B, the resulting equations will not contain the three unknown reactions at B. The sum of the moments about B is

$$\Sigma\mathbf{M}_{(\text{point } B)} = \begin{vmatrix} \mathbf{i} & \mathbf{j} & \mathbf{k} \\ 0.2 & 0 & 0 \\ -0.842T & 0.337T & 0.421T \end{vmatrix} + \begin{vmatrix} \mathbf{i} & \mathbf{j} & \mathbf{k} \\ 0 & 0 & 0.2 \\ A_x & A_y & 0 \end{vmatrix}$$

$$+ \begin{vmatrix} \mathbf{i} & \mathbf{j} & \mathbf{k} \\ 0.2 & 0 & 0.2 \\ 0 & -400 & 0 \end{vmatrix}$$

$$= (-0.2A_y + 80)\mathbf{i} + (-0.0842T + 0.2A_x)\mathbf{j}$$

$$+ (0.0674T - 80)\mathbf{k} = 0.$$

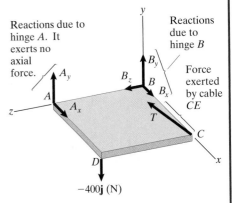

(a) The free-body diagram of the plate.

The scalar equations are

$$\Sigma M_x = -0.2A_y + 80 = 0,$$

$$\Sigma M_y = -0.0842T + 0.2A_x = 0,$$

$$\Sigma M_z = 0.0674T - 80 = 0.$$

Solving these equations, we obtain the reactions

$$T = 1187 \text{ N}, \qquad A_x = 500 \text{ N}, \qquad A_y = 400 \text{ N}.$$

Then from Eqs. (5.24), the reactions at B are

$$B_x = 500 \text{ N}, \qquad B_y = -400 \text{ N}, \qquad B_z = -500 \text{ N}.$$

Discussion

If our only objective had been to determine the tension T, we could have done so easily by setting the sum of the moments about the line AB (the z axis) equal to zero. Since the reactions at the hinges exert no moment about the z axis, we obtain the equation

$$(0.2)(0.337T) - (0.2)(400) = 0,$$

which yields the result $T = 1187$ N.

Problems

5.73 The bar AB has a built-in support at A and is loaded by the forces

$$\mathbf{F}_B = 2\mathbf{i} + 6\mathbf{j} + 3\mathbf{k} \text{ (kN)},$$
$$\mathbf{F}_C = \mathbf{i} - 2\mathbf{j} + 2\mathbf{k} \text{ (kN)}.$$

(a) Draw the free-body diagram of the bar.
(b) Determine the reactions at A.

Strategy: (a) Draw a diagram of the bar isolated from its supports. Complete the free-body diagram of the bar by adding the two external forces and the reactions due to the built-in support (see Table 5.2). (b) Use the scalar equilibrium equations (5.13)–(5.18) to determine the reactions.

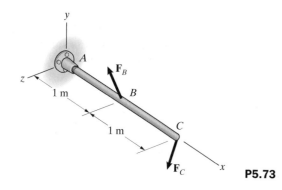

P5.73

5.74 The bar AB has a built-in support at A. The tension in cable BC is 8 kN. Determine the reactions at A.

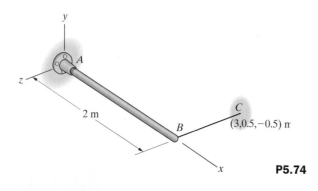

P5.74

5.75 The bar AB has a built-in support at A. The collar at B is fixed to the bar. The tension in the cable BC is 10 kN.
(a) Draw the free-body diagram of the bar.
(b) Determine the reactions at A.

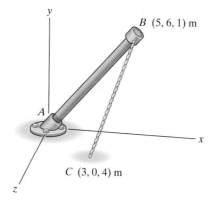

P5.75

5.76 Consider the bar in Problem 5.75. The magnitude of the couple exerted on the bar by the built-in support is 100 kN-m. What is the tension in the cable?

5.77 The force exerted on the highway sign by wind and the sign's weight is $\mathbf{F} = 800\mathbf{i} - 600\mathbf{j}$ (N). Determine the reactions at the built-in support at O.

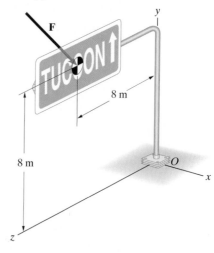

P5.77

5.78 In Problem 5.77, the force exerted on the sign by wind and the sign's weight is $\mathbf{F} = \pm 4.4v^2\mathbf{i} - 600\mathbf{j}$ (N), where v is the component of the wind's velocity perpendicular to the sign in meters per second (m/s). If you want to design the sign to remain standing in hurricane winds with velocities v as high as 70 m/s, what reactions must the built-in support at O be designed to withstand?

5.79 The tension in cable AB is 24 kN. Determine the reactions at the built-in support D.

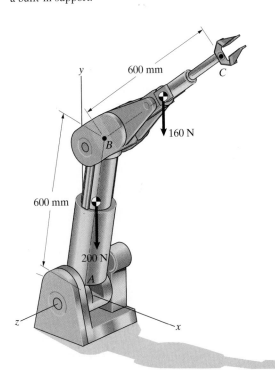

P5.79

5.80 The robotic manipulator is stationary and the y axis is vertical. The weights of the arms AB and BC act at their mid-points. The direction cosines of the centerline of arm AB are $\cos\theta_x = 0.174$, $\cos\theta_y = 0.985$, $\cos\theta_z = 0$, and the direction cosines of the centerline of arm BC are $\cos\theta_x = 0.743$, $\cos\theta_y = 0.557$, $\cos\theta_z = -0.371$. The support at A behaves like a built-in support.

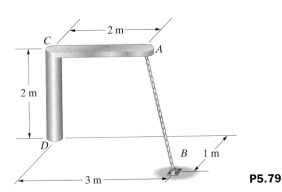

(a) What is the sum of the moments about A due to the weights of the two arms?
(b) What are the reactions at A?

5.81 The force exerted on the grip of the exercise machine is $\mathbf{F} = 260\mathbf{i} - 130\mathbf{j}$ (N). What are the reactions at the built-in support at O?

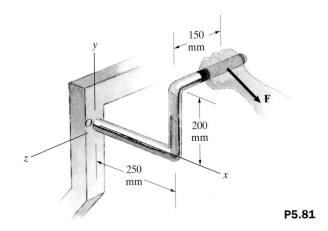

P5.81

5.82 The designer of the exercise machine in Problem 5.81 assumes that the force $\mathbf{F}$ exerted on the grip will be parallel to the x–y plane and that its magnitude will not exceed 900 N. Based on these criteria, what reactions must the built-in support at O be designed to withstand?

5.83 The boom ABC is subjected to a force $\mathbf{F} = -8\mathbf{j}$ (kN) at C and is supported by a ball and socket at A and the cables BD and BE.
(a) Draw the free-body diagram of the boom.
(b) Determine the tensions in the cables and the reactions at A.

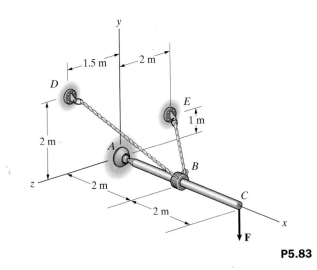

P5.83

5.84 The cables supporting the boom ABC in Problem 5.83 will each safely support a tension of 25 kN. Based on this criterion, what is the largest safe magnitude of the downward force $\mathbf{F}$?

P5.80

5.85 The suspended load exerts a force $F = 600$ lb at A, and the weight of the bar OA is negligible. Determine the tensions in the cables and the reactions at the ball and socket support O.

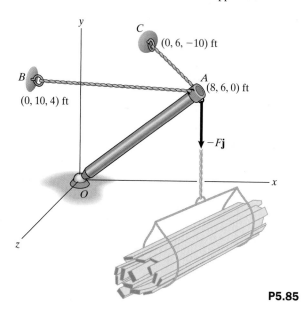

P5.85

5.86 In Problem 5.85, suppose that the suspended load exerts a force $F = 600$ lb at A and bar OA weighs 200 lb. Assume that the bar's weight acts at its midpoint. Determine the tensions in the cables and the reactions at the ball and socket support O.

5.87 The 158,000-kg airplane is at rest on the ground ($z = 0$ is ground level). The landing gear carriages are at A, B, and C. The coordinates of the point G at which the weight of the plane acts are $(3, 0.5, 5)$ m. What are the magnitudes of the normal reactions exerted on the landing gear by the ground?

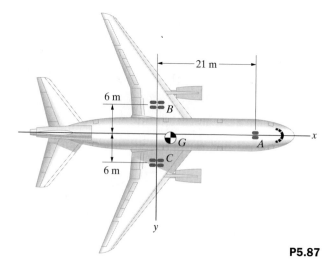

P5.87

5.88 The 800-kg horizontal wall section is supported by the three vertical cables A, B, and C. What are the tensions in the cables?

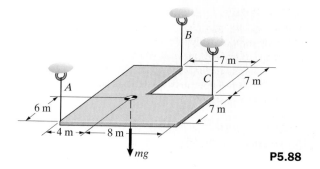

P5.88

5.89 The cables in Problem 5.88 will each safely support a tension of 10 kN. Based on this criterion, what is the largest safe mass of the horizontal wall section?

5.90 An engineer designs a system of pulleys to pull his model trains up and out of the way when they aren't in use. What are the tensions in the three ropes when the system is in equilibrium?

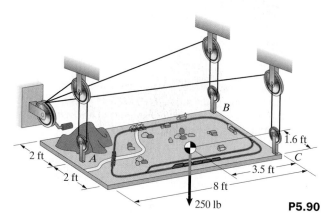

P5.90

5.91 The L-shaped bar is supported by a bearing at A and rests on a smooth horizontal surface at B. The vertical force $F = 4$ kN and the distance $b = 0.15$ m. Determine the reactions at A and B.

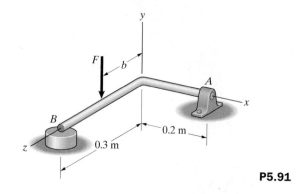

P5.91

5.92 In Problem 5.91, the vertical force $F = 4$ kN and the distance $b = 0.15$ m. If you represent the reactions at A and B by an equivalent system consisting of a single force, what is the force and where does its line of action intersect the x–z plane?

5.93 In Problem 5.91, the vertical force $F = 4$ kN. The bearing at A will safely support a force of 2.5-kN magnitude and a couple of 0.5 kN-m magnitude. Based on these criteria, what is the allowable range of the distance b?

5.94 The 1.1-m bar is supported by a ball and socket support at A and the two smooth walls. The tension in the vertical cable CD is 1 kN.
(a) Draw the free-body diagram of the bar.
(b) Determine the reactions at A and B.

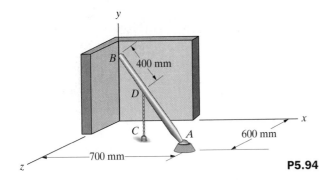

P5.94

5.95 The 8-ft bar is supported by a ball and socket support at A, the cable BD, and a roller support at C. The collar at B is fixed to the bar at its midpoint. The force $\mathbf{F} = -50\mathbf{k}$ (lb). Determine the tension in cable BD and the reactions at A and C.

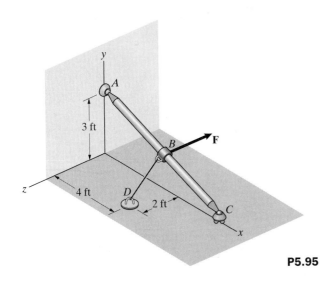

P5.95

5.96 Consider the 8-ft bar in Problem 5.95. The force $\mathbf{F} = F_y\mathbf{j} - 50\mathbf{k}$ (lb). What is the largest value of F_y for which the roller support at C will remain on the floor?

5.97 The tower is 70 m tall. The tension in each cable is 2 kN. Treat the base of the tower A as a built-in support. What are the reactions at A?

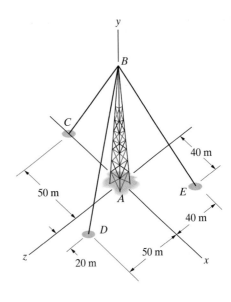

P5.97

5.98 Consider the tower in Problem 5.97. If the tension in cable BC is 2 kN, what must the tensions in cables BD and BE be if you want the couple exerted on the tower by the built-in support at A to be zero? What are the resulting reactions at A?

5.99 The space truss has roller supports at B, C, and D and is subjected to a vertical force $F = 20$ kN at A. What are the reactions at the roller supports?

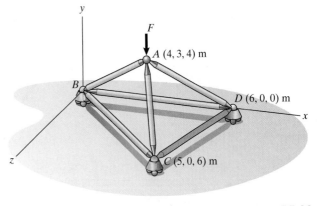

P5.99

5.100 In Problem 5.99, suppose that you don't want the reaction at any of the roller supports to exceed 15 kN. What is the largest force F the truss can support?

5.101 The 40-lb door is supported by hinges at A and B. The y axis is vertical. The hinges do not exert couples on the door, and the hinge at B does not exert a force parallel to the hinge axis. The weight of the door acts at its midpoint. What are the reactions at A and B?

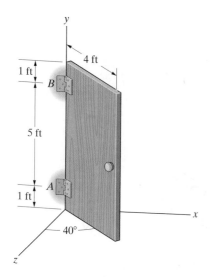

P5.101

5.102 The vertical cable is attached at A. Determine the tension in the cable and the reactions at the bearing B due to the force $\mathbf{F} = 10\mathbf{i} - 30\mathbf{j} - 10\mathbf{k}$ (N).

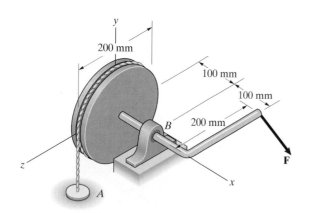

P5.102

5.103 In Problem 5.102, suppose that the z component of the force $\mathbf{F}$ is zero, but otherwise $\mathbf{F}$ is unknown. If the couple exerted on the shaft by the bearing at B is $\mathbf{M}_B = 6\mathbf{j} - 6\mathbf{k}$ N-m, what are the force $\mathbf{F}$ and the tension in the cable?

5.104 The device in Problem 5.102 is badly designed because of the couples that must be supported by the bearing at B, which would cause the bearing to "bind." (Imagine trying to open a door supported by only one hinge.) In this improved design, the bearings at B and C support no couples, and the bearing at C does

not exert a force in the x direction. If the force $\mathbf{F} = 10\mathbf{i} - 30\mathbf{j} - 10\mathbf{k}$ (N), what are the tension in the vertical cable and the reactions at the bearings B and C?

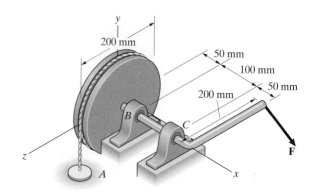

P5.104

5.105 The rocket launcher is supported by the hydraulic jack DE and the bearings A and B. The bearings lie on the x axis and support shafts parallel to the x axis. The hydraulic cylinder DE exerts a force on the launcher that points along the line from D to E. The coordinates of D are $(7, 0, 7)$ ft, and the coordinates of E are $(9, 6, 4)$ ft. The weight $W = 30$ kip acts at $(4.5, 5, 2)$ ft. What is the magnitude of the reaction on the launcher at E?

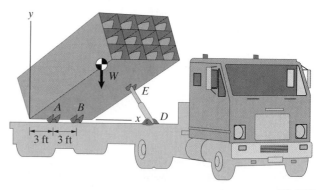

P5.105

5.106 Consider the rocket launcher described in Problem 5.105. The bearings at A and B do not exert couples, and the bearing B does not exert a force in the x direction. Determine the reactions at A and B.

5.107 The crane's cable CD is attached to a stationary object at D. The crane is supported by the bearings E and F and the horizontal cable AB. The tension in cable AB is 8 kN. Determine the tension in the cable CD.

Strategy: Since the reactions exerted on the crane by the bearings do not exert moments about the z axis, the sum of the moments about the z axis due to the forces exerted on the crane by the cables AB and CD equals zero. (See the discussion at the end of Example 5.9.)

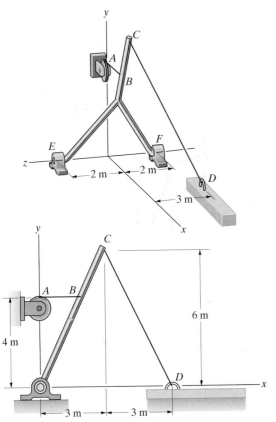

P5.107

5.108 The crane in Problem 5.107 is supported by the horizontal cable *AB* and the bearings at *E* and *F*. The bearings do not exert couples, and the bearing at *F* does not exert a force in the *z* direction. The tension in cable *AB* is 8 kN. Determine the reactions at *E* and *F*.

5.109 The plate is supported by hinges at *A* and *B* and the cable *CE*, and it is loaded by the force at *D*. The edge of the plate to which the hinges are attached lies in the *y–z* plane, and the axes of the hinges are parallel to the line through points *A* and *B*. The hinges do not exert couples on the plate. What is the tension in cable *CE*?

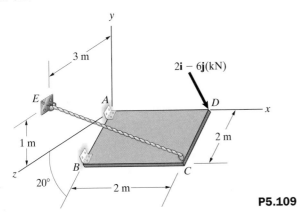

P5.109

5.110 In Problem 5.109, the hinge at *B* does not exert a force on the plate in the direction of the hinge axis. What are the magnitudes of the forces exerted on the plate by the hinges at *A* and *B*?

5.111 The bar *ABC* is supported by ball and socket supports at *A* and *C* and the cable *BD*, and is loaded by the 200-lb suspended weight. What is the tension in cable *BD*?

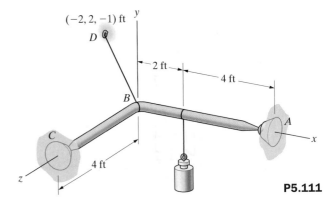

P5.111

5.112 In Problem 5.111, determine the *y* components of the reactions exerted on the bar *ABC* by the ball and socket supports at *A* and *C*.

5.113 The bearings at *A*, *B*, and *C* do not exert couples on the bar and do not exert forces in the direction of the axis of the bar. Determine the reactions at the bearings due to the two forces on the bar.

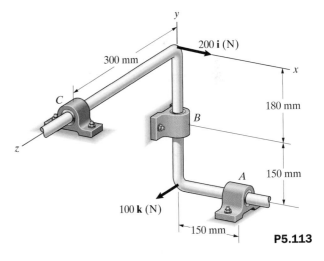

P5.113

5.114 The support that attaches the sailboat's mast to the deck behaves like a ball and socket support. The line that attaches the spinnaker (the sail) to the top of the mast exerts a 200-lb force on the mast. The force is in the horizontal plane at 15° from the centerline of the boat. (See the top view.) The spinnaker pole exerts a 50-lb force on the mast at *P*. The force is in the horizontal plane at 45° from the centerline. (See the top view.) The mast is supported by two cables, the back stay *AB* and the port shroud *ACD*. (The fore stay *AE* and the starboard shroud *AFG* are slack, and their tensions can be neglected.) Determine the tensions in the cables AB and CD and the reactions at the bottom of the mast.

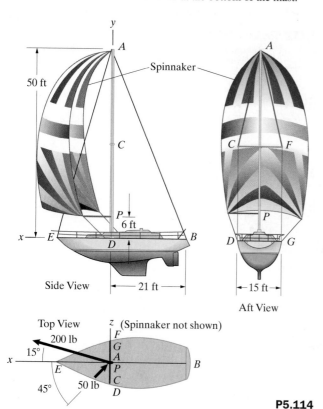

Side View

Top View z (Spinnaker not shown)

P5.114

5.115 The door is supported by the cable DE and hinges at A and B, and is subjected to a 2-kN force at C. The door's weight is negligible. The hinges do not exert couples on the door, and their axes are aligned with the line from A to B. Determine the tension in the cable.

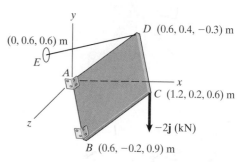

P5.115

5.116 Determine the reactions at the hinges supporting the door in Problem 5.115. Assume that the hinge at B exerts no force parallel to the hinge axis.

Strategy: Express the reactions at the hinges as $\mathbf{A} = A_x\mathbf{i} + A_y\mathbf{j} + A_z\mathbf{k}$ and $\mathbf{B} = B_x\mathbf{i} + B_y\mathbf{j} + B_z\mathbf{k}$. Let $\mathbf{e}_{AB}$ be a unit vector parallel to the hinge axes. Since the hinge at B exerts no force parallel to the hinge axis, you know that $\mathbf{e}_{AB} \cdot \mathbf{B} = 0$.

5.5   Two-Force and Three-Force Members

You have seen how the equilibrium equations are used to analyze objects supported and loaded in different ways. Here we discuss two particular cases that occur so frequently you need to be familiar with them. The first one is especially important and plays a central role in our analysis of structures in the next chapter.

Two-Force Members

If the system of forces and moments acting on an object is equivalent to two forces acting at different points, we refer to the object as a *two-force member*. For example, the object in Fig. 5.33a is subjected to two sets of concurrent forces whose lines of action intersect at A and B. Since we can represent them by single forces acting at A and B (Fig. 5.33b), where $\mathbf{F} = \mathbf{F}_1 + \mathbf{F}_2 + \cdots + \mathbf{F}_N$ and $\mathbf{F}' = \mathbf{F}'_1 + \mathbf{F}'_2 + \cdots + \mathbf{F}'_M$, this object is a two-force member.

If the object is in equilibrium, what can we infer about the forces $\mathbf{F}$ and $\mathbf{F}'$? The sum of the forces equals zero only if $\mathbf{F}' = -\mathbf{F}$ (Fig. 5.33c). Furthermore, the forces $\mathbf{F}$ and $-\mathbf{F}$ form a couple, so the sum of the moments is not zero unless the lines of action of the forces lie along the line through the points A and

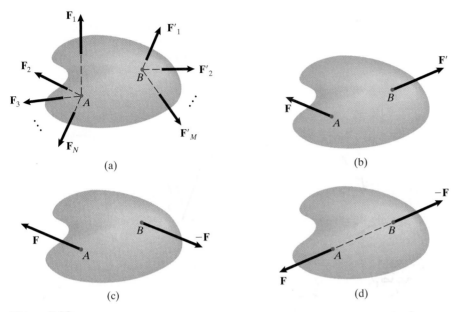

(a)   (b)

(c)   (d)

Figure 5.33
(a) An object subjected to two sets of concurrent forces.
(b) Representing the concurrent forces by two forces **F** and **F′**.
(c) If the object is in equilibrium, the forces must be equal and opposite.
(d) The forces form a couple unless they have the same line of action.

B (Fig. 5.33d). Thus equilibrium tells us that *the two forces are equal in magnitude, are opposite in direction, and have the same line of action.* However, without additional information, we cannot determine their magnitude.

A cable attached at two points (Fig. 5.34a) is a familiar example of a two-force member (Fig. 5.34b). The cable exerts forces on the attachment points that are directed along the line between them (Fig. 5.34c).

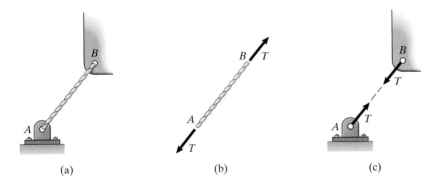

(a)   (b)   (c)

Figure 5.34
(a) A cable attached at A and B.
(b) The cable is a two-force member.
(c) The forces exerted by the cable.

A bar that has two supports that exert only forces on it (no couples) and is not subjected to any loads is a two-force member (Fig. 5.35a). Such bars are often used as supports for other objects. Because the bar is a two-force member, the lines of action of the forces exerted on the bar must lie along the line between the supports (Fig. 5.35b). Notice that, unlike the cable, the bar can exert forces at A and B either in the directions shown in Fig. 5.35c or in

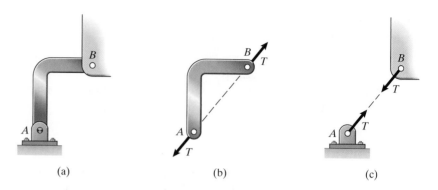

Figure 5.35
(a) The bar AB attaches the object to the pin support.
(b) The bar AB is a two-force member.
(c) The force exerted on the supported object by the bar AB.

(a) (b) (c)

the opposite directions. (In other words, the cable can only pull on its supports, while the bar can either pull or push.)

In these examples we assumed that the weights of the cable and the bar could be neglected in comparison with the forces exerted on them by their supports. When that is not the case, they are clearly not two-force members.

Three-Force Members

If the system of forces and moments acting on an object is equivalent to three forces acting at different points, we call it a *three-force member*. We can show that if a three-force member is in equilibrium, the three forces are coplanar and are either parallel or concurrent.

We first prove that the forces are coplanar. Let them be called $\mathbf{F}_1$, $\mathbf{F}_2$, and $\mathbf{F}_3$, and let P be the plane containing the three points of application (Fig. 5.36a). Let L be the line through the points of application of $\mathbf{F}_1$ and $\mathbf{F}_2$. Since the moments due to $\mathbf{F}_1$ and $\mathbf{F}_2$ about L are zero, the moment due to $\mathbf{F}_3$ about L must equal zero (Fig. 5.36b):

$$\big[\mathbf{e} \cdot (\mathbf{r} \times \mathbf{F}_3)\big]\mathbf{e} = \big[\mathbf{F}_3 \cdot (\mathbf{e} \times \mathbf{r})\big]\mathbf{e} = \mathbf{0}.$$

This equation requires that $\mathbf{F}_3$ be perpendicular to $\mathbf{e} \times \mathbf{r}$, which means that $\mathbf{F}_3$ is contained in P. The same procedure can be used to show that $\mathbf{F}_1$ and $\mathbf{F}_2$ are contained in P, so the forces are coplanar. (A different proof is required if the points of application lie on a straight line, but the result is the same.)

If the three coplanar forces are not parallel, there will be points where their lines of action intersect. Suppose that the lines of action of two of the forces intersect at a point Q. Then the moments of those two forces about Q are zero, and the sum of the moments about Q is zero only if the line of action of the third force also passes through Q. Therefore either the forces are parallel or they are concurrent (Fig. 5.36c).

You can often simplify the analysis of an object in equilibrium by recognizing that it is a two-force or three-force member. However, you are not getting something for nothing. Once you have drawn the free-body diagram of a two-force member as shown in Figs. 5.34b and 5.35b, you cannot obtain any further information about the forces from the equilibrium equations. When you require that the lines of action of nonparallel forces acting on a three-force member be coincident, you have used the fact that the sum of the moments about a point must be zero and cannot obtain any further information from that condition.

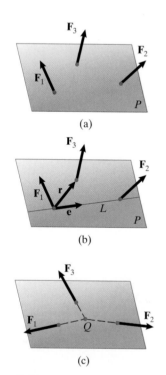

(a)

(b)

(c)

Figure 5.36
(a) The three forces and the plane P.
(b) Determining the moment due to force $\mathbf{F}_3$ about L.
(c) If the forces are not parallel, they must be concurrent.

Example 5.10

A Two-Force Member

The L-shaped bar in Fig. 5.37 has a pin support at A and is loaded by a 6-kN force at B. Neglect the weight of the bar. Determine the angle α and the reactions at A.

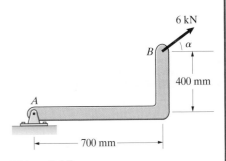

Figure 5.37

Strategy

The bar is a two-force member because it is subjected only to the 6-kN force at B and the force exerted by the pin support. (If we could not neglect the weight of the bar, it would not be a two-force member.) We will determine the angle α and the reactions at A in two ways, first by applying the equilibrium equations in the usual way and then by using the fact that the bar is a two-force member.

Solution

Applying the Equilibrium Equations We draw the free-body diagram of the bar in Fig. a, showing the reactions at the pin support. Summing moments about point A, the equilibrium equations are

$$\Sigma F_x = A_x + 6\cos\alpha = 0,$$
$$\Sigma F_y = A_y + 6\sin\alpha = 0,$$
$$\Sigma M_{(\text{point } A)} = (6\sin\alpha)(0.7) - (6\cos\alpha)(0.4) = 0.$$

From the third equation we see that $\alpha = \arctan(0.4/0.7)$. In the range $0 \leq \alpha \leq 360°$, this equation has the two solutions $\alpha = 29.7°$ and $\alpha = 209.7°$. Knowing α, we can determine A_x and A_y from the first two equilibrium equations. The solutions for the two values of α are

$$\alpha = 29.7°, \qquad A_x = -5.21 \text{ kN}, \qquad A_y = -2.98 \text{ kN},$$

and

$$\alpha = 209.7°, \qquad A_x = 5.21 \text{ kN}, \qquad A_y = 2.98 \text{ kN}.$$

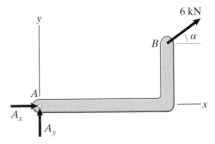

(a) The free-body diagram of the bar.

Treating the Bar as a Two-Force Member We know that the 6-kN force at B and the force exerted by the pin support must be equal in magnitude, opposite in direction, and directed along the line between points A and B. The two possibilities are shown in Figs. b and c. Thus by recognizing that the bar is a two-force member, we immediately know the possible directions of the forces and the magnitude of the reaction at A.

In Fig. b we can see that $\tan\alpha = 0.4/0.7$, so $\alpha = 29.7°$ and the components of the reaction at A are

$$A_x = -6\cos 29.7° = -5.21 \text{ kN}$$
$$A_y = -6\sin 29.7° = -2.98 \text{ kN}.$$

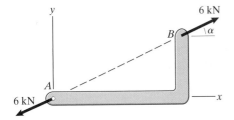
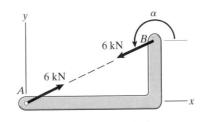

(b), (c) The possible directions of the forces.

In Fig. c, $\alpha = 180° + 29.7° = 209.7°$, and the components of the reaction at A are

$$A_x = 6\cos 29.7° = 5.21 \text{ kN},$$
$$A_y = 6\sin 29.7° = 2.98 \text{ kN}.$$

Example 5.11

Two and Three Force Members

The 100-lb weight of the rectangular plate in Fig. 5.38 acts at its midpoint. Determine the reactions exerted on the plate at B and C.

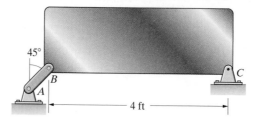

Figure 5.38

Strategy

The plate is subjected to its weight and the reactions exerted by the pin supports at B and C, so it is a three-force member. Furthermore, the bar AB is a two-force member, so we know that the line of action of the reaction it exerts on the plate at B is directed along the line between A and B. We can use this information to simplify the free-body diagram of the plate.

Solution

The reaction exerted on the plate by the two-force member AB must be directed along the line between A and B, and the line of action of the weight is vertical. Since the three forces on the plate must be either parallel or concurrent, their lines of action must intersect at the point P shown in Fig. a. From the equilibrium equations

$$\Sigma F_x = B\sin 45° - C\sin 45° = 0,$$
$$\Sigma F_y = B\cos 45° + C\cos 45° - 100 = 0,$$

we obtain the reactions $B = C = 70.7$ lb.

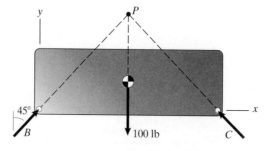

(a) The free-body diagram of the plate. The three forces must be concurrent.

Problems

5.117 The horizontal bar has a mass of 10 kg. Its weight acts at the midpoint of the bar, and it is supported by a roller support at A and the cable BC. Use the fact that the bar is a three-force member to determine the angle α, the tension in the cable BC, and the magnitude of the reaction at A.

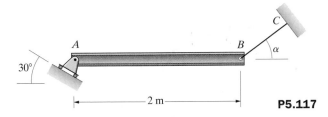

P5.117

5.118 The horizontal bar is of negligible weight. Use the fact that the bar is a three-force member to determine the angle α necessary for equilibrium.

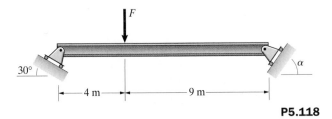

P5.118

5.119 The suspended load weighs 1000 lb. If you neglect its weight, the structure is a three-force member. Use this fact to determine the magnitudes of the reactions at A and B.

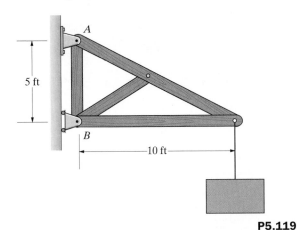

P5.119

5.120 The weight $W = 50$ lb acts at the center of the disk. Use the fact that the disk is a three-force member to determine the tension in the cable and the magnitude of the reaction at the pin support.

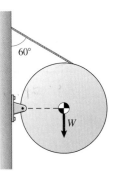

P5.120

5.121 The weight $W = 40$ N acts at the center of the disk. The surfaces are rough. What force F is necessary to lift the disk off the floor?

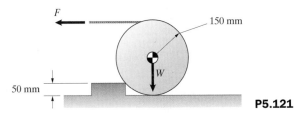

P5.121

5.122 Use the fact that the horizontal bar is a three-force member to determine the angle α and the magnitudes of the reactions at A and B.

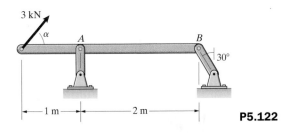

P5.122

5.123 The suspended load weighs 600 lb. Use the fact that ABC is a three-force member to determine the magnitudes of the reactions at A and B.

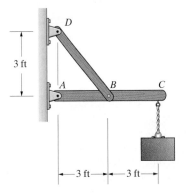

P5.123

5.124 (a) Is the L-shaped bar a three-force member?
(b) Determine the magnitudes of the reactions at *A* and *B*.
(c) Are the three forces acting on the L-shaped bar concurrent?

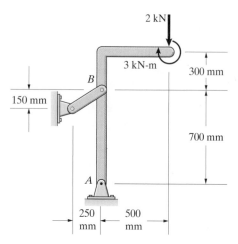

P5.124

5.125 The bucket of the excavator is supported by the two-force member *AB* and the pin support at *C*. Its weight is $W = 1500$ lb. What are the reactions at *C*?

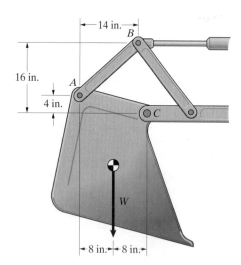

P5.125

5.126 The member *ACG* of the front-end loader is subjected to a load $W = 2$ kN and is supported by a pin support at *A* and the hydraulic cylinder *BC*. Treat the hydraulic cylinder as a two-force member.
(a) Draw the free-body diagrams of the hydraulic cylinder and the member *ACG*.
(b) Determine the reactions on the member *ACG*.

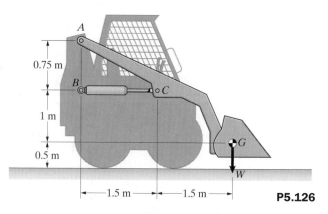

P5.126

5.127 In Problem 5.126, determine the reactions on the member *ACG* by using the fact that it is a three-force member.

5.128 A rectangular plate is subjected to two forces *A* and *B* (Fig. a). In Fig. b, the two forces are resolved into components. By writing equilibrium equations in terms of the components A_x, A_y, B_x, and B_y, show that the two forces *A* and *B* are equal in magnitude, opposite in direction, and directed along the line between their points of application.

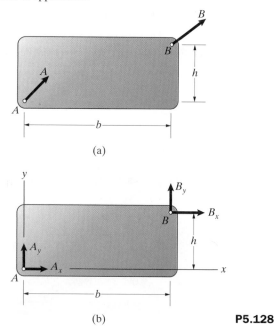

P5.128

5.129 An object in equilibrium is subjected to three forces whose points of application lie on a straight line. Prove that the forces are coplanar.

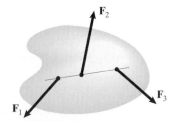

P5.129

Chapter Summary

Building on our discussions of forces in Chapter 3 and moments in Chapter 4, in this chapter we have used the equilibrium equations to analyze the forces and couples acting on many types of objects. We defined the support conventions commonly used in engineering and presented examples of their use. We discussed situations that can result in an object's being statically indeterminate. Finally, we defined two-force and three-force members. In Chapter 6 we will use the concepts and methods developed in this chapter to analyze the individual members of structures, beginning with structures consisting entirely of two-force members.

When an object is in equilibrium, the following conditions are satisfied:

1. The sum of the forces is zero,

$$\Sigma \mathbf{F} = \mathbf{0}. \qquad \text{Eq. (5.1)}$$

2. The sum of the moments about any point is zero,

$$\Sigma \mathbf{M}_{(\text{any point})} = \mathbf{0}. \qquad \text{Eq. (5.2)}$$

Forces and couples exerted on an object by its supports are called *reactions*. The other forces and couples on the object are the *loads*. Common supports are represented by models called *support conventions*.

Two-Dimensional Applications

When the loads and reactions on an object in equilibrium form a two-dimensional system of forces and moments, they are related by three scalar equilibrium equations:

$$\Sigma F_x = 0,$$
$$\Sigma F_y = 0, \qquad \text{Eqs. (5.4)–(5.6)}$$
$$\Sigma M_{(\text{any point})} = 0.$$

No more than three independent equilibrium equations can be obtained from a given two-dimensional free-body diagram.

Support conventions commonly used in two-dimensional applications are summarized in Table 5.1.

Three-Dimensional Applications

The loads and reactions on an object in equilibrium satisfy the six scalar equilibrium equations

$$\Sigma F_x = 0, \qquad \Sigma F_y = 0, \qquad \Sigma F_z = 0,$$
$$\Sigma M_x = 0, \qquad \Sigma M_y = 0, \qquad \Sigma M_z = 0. \qquad \text{Eqs. (5.13)–(5.18)}$$

No more than six independent equilibrium equations can be obtained from a given free-body diagram.

Support conventions commonly used in three-dimensional applications are summarized in Table 5.2.

Statically Indeterminate Objects

An object has *redundant supports* when it has more supports than the minimum number necessary to maintain it in equilibrium and *improper supports* when its supports are improperly designed to maintain equilibrium under the applied loads. In either situation, the object is *statically indeterminate*. The difference between the number of reactions and the number of independent equilibrium equations is called the *degree of redundancy*. Even if an object is statically indeterminate due to redundant supports, it may be possible to determine some of the reactions from the equilibrium equations.

Two-Force and Three-Force Members

If the system of forces and moments acting on an object is equivalent to two forces acting at different points, the object is a *two-force member*. If the object is in equilibrium, the two forces are equal in magnitude, opposite in direction, and directed along the line through their points of application. If the system of forces and moments acting on an object is equivalent to three forces acting at different points, it is a *three-force member*. If the object is in equilibrium, the three forces are coplanar and either parallel or concurrent.

Review Problems

5.130 (a) Draw the free-body diagram of the 50-lb plate, and explain why it is statically indeterminate.
(b) Determine as many of the reactions at *A* and *B* as possible.

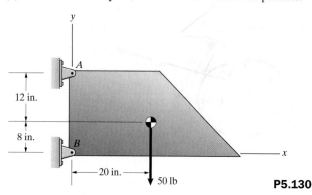

P5.130

5.131 The mass of the truck is 4 Mg. Its wheels are locked, and the tension in its cable is $T = 10$ kN.

(a) Draw the free-body diagram of the truck.
(b) Determine the normal forces exerted on the truck's wheels by the road.

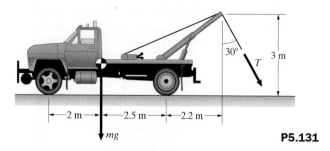

P5.131

5.132 Assume that the force exerted on the head of the nail by the hammer is vertical, and neglect the hammer's weight.

(a) Draw the free-body diagram of the hammer.
(b) If $F = 10$ lb, what are the magnitudes of the force exerted on the nail by the hammer and the normal and friction forces exerted on the floor by the hammer?

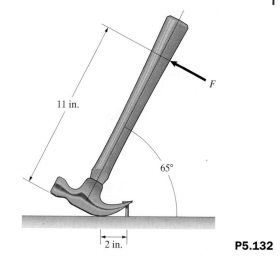

P5.132

5.133 (a) Draw the free-body diagram of the beam.
(b) Determine the reactions at the supports.

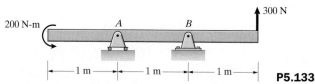

P5.133

5.134 Consider the beam shown in Problem 5.133. First represent the loads (the 300-N force and the 200-N-m couple) by a single equivalent force; then determine the reactions at the supports.

5.135 The truss supports a 90-kg suspended object. What are the reactions at the supports *A* and *B*?

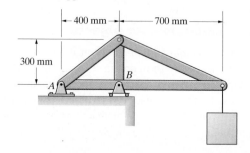

P5.135

5.136 The trailer is parked on a 15° slope. Its wheels are free to turn. The hitch *H* behaves like a pin support. Determine the reactions at *A* and *H*.

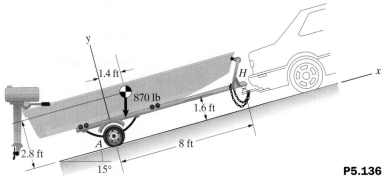

P5.136

5.137 To determine the location of the point where the weight of a car acts (the *center of mass*), an engineer places the car on scales and measures the normal reactions at the wheels for two values of α, obtaining the following results.

α	A_y (kN)	B (kN)
10°	10.134	4.357
20°	10.150	3.677

What are the distances *b* and *h*?

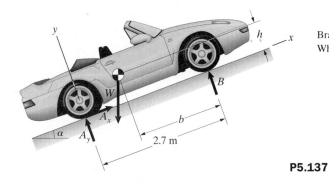

P5.137

5.138 The horizontal bar of weight *W* is supported by a roller support at *A* and the cable *BC*. Use the fact that the bar is a three-force member to determine the angle α, the tension in the cable, and the magnitude of the reaction at *A*.

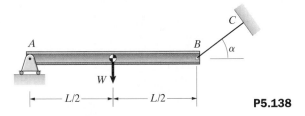

P5.138

5.139 The bicycle brake on the right is pinned to the bicycle's frame at *A*. Determine the force exerted by the brake pad on the wheel rim at *B* in terms of the cable tension *T*.

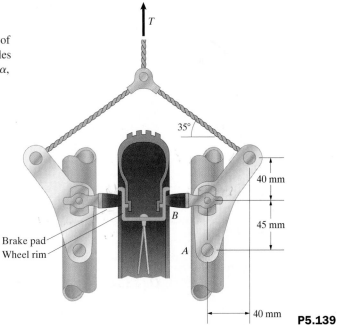

P5.139

𝒟esign Experience The traditional wheelbarrow shown is designed to transport a load *W* while being supported by an upward force *F* applied to the handles by the user. (a) Use statics to analyze the effects of a range of choices of the dimensions *a* and *b* on the size of load that could be carried. Also consider the implications of these dimensions on the wheelbarrow's ease and practicality of use. (b) Suggest a different design for this classic device that achieves the same function. Use statics to compare your design to the wheelbarrow with respect to load-carrying ability and ease of use.

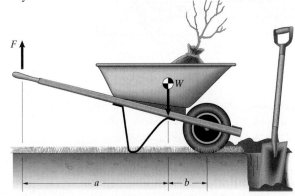

The highway bridge is supported by a truss structure. In this chapter we describe techniques for determining the forces and couples acting on the individual members of structures.

Structures in Equilibrium

6

I n engineering, the term *structure* can refer to any object that has the capacity to support and exert loads. In this chapter we consider structures composed of interconnected parts, or *members*. To design such a structure, or to determine whether an existing one is adequate, you must determine the forces and couples acting on the structure as a whole as well as on its individual members. We first demonstrate how this is done for the structures called trusses, which are composed entirely of two-force members. The familiar frameworks of steel members that support some highway bridges are trusses. We then consider other structures, called *frames* if they are designed to remain stationary and support loads and *machines* if they are designed to move and exert loads.

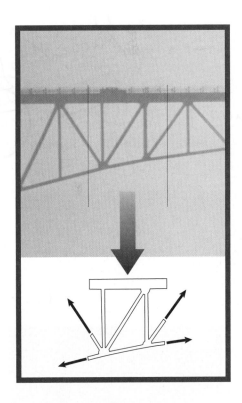

6.1 Trusses

We can explain the nature of truss structures such as the beams supporting a house (Fig. 6.1) by starting with very simple examples. Suppose we pin three bars together at their ends to form a triangle. If we add supports as shown in Fig. 6.2a, we obtain a structure that will support a load F. We can construct more elaborate structures by adding more triangles (Figs. 6.2b and c). The bars are the members of these structures, and the places where the bars are pinned together are called the *joints*. Even though these examples are quite simple, you can see that Fig. 6.2c, which is called a Warren truss, begins to resemble the structures used to support bridges and the roofs of houses (Fig. 6.3). If these structures are supported and loaded at their joints and we neglect the weights of the bars, each bar is a two-force member. We call such a structure a *truss*.

We draw the free-body diagram of a member of a truss in Fig. 6.4a. Because it is a two-force member, the forces at the ends, which are the sums of the forces exerted on the member at its joints, must be equal in magnitude, opposite in direction, and directed along the line between the joints. We call the force T the *axial force* in the member. When T is positive in the direction shown (that is, when the forces are directed away from each other), the member is in *tension*. When the forces are directed toward each other, the member is in *compression*.

Figure 6.1
A typical house is supported by trusses made of wood beams.

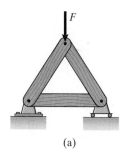

(a)

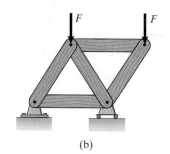

(b)

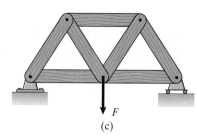

(c)

Figure 6.2
Making structures by pinning bars together to form triangles.

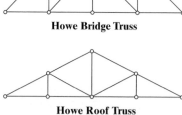

Howe Bridge Truss

Pratt Bridge Truss

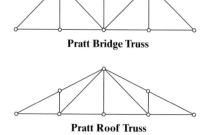

Howe Roof Truss

Pratt Roof Truss

Figure 6.3
Simple examples of bridge and roof structures. (The lines represent members, and the circles represent joints.)

In Fig. 6.4b, we "cut" the member by a plane and draw the free-body diagram of the part of the member on one side of the plane. We represent the system of internal forces and moments exerted by the part not included in the free-body diagram by a force **F** acting at the point P where the plane intersects the axis of the member and a couple **M**. The sum of the moments about P must equal zero, so $\mathbf{M} = \mathbf{0}$. Therefore we have a two-force member, which means that **F** must be equal in magnitude and opposite in direction to the force T acting at the joint (Fig. 6.4c). The internal force is a tension or compression equal to the tension or compression exerted at the joint. Notice the similarity to a rope or cable, in which the internal force is a tension equal to the tension applied at the ends.

Although many actual structures, including "roof trusses" and "bridge trusses," consist of bars connected at the ends, very few have pinned joints. For example, if you examine a joint of a bridge truss, you will see that the members are bolted or riveted together so that they are not free to rotate at the joint (Fig. 6.5). It is obvious that such a joint can exert couples on the members. Why are these structures called trusses?

The reason is that they are designed to function as trusses, meaning that they support loads primarily by subjecting their members to axial forces. They can usually be *modeled* as trusses, treating the joints as pinned connections under the assumption that couples they exert on the members are small in comparison to axial forces. When we refer to structures with riveted joints as trusses in problems, we mean that you can model them as trusses.

In the following sections we describe two methods for determining the axial forces in the members of trusses. The method of joints is usually the preferred approach when you need to determine the axial forces in all members of a truss. When you only need to determine the axial forces in a few members, the method of sections often results in a faster solution than the method of joints.

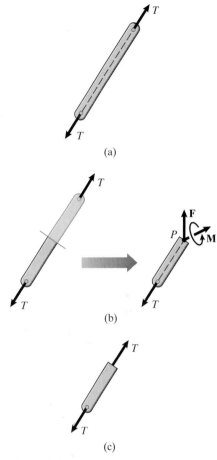

(a)

(b)

(c)

Figure 6.4
(a) Each member of a truss is a two-force member.
(b) Obtaining the free-body diagram of part of the member.
(c) The internal force is equal and opposite to the force acting at the joint, and the internal couple is zero.

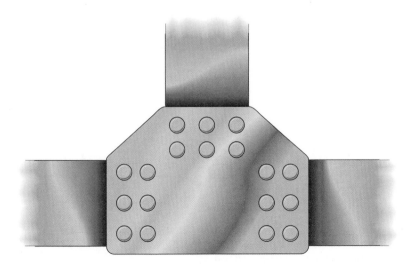

Figure 6.5
A joint of a bridge truss.

6.2 The Method of Joints

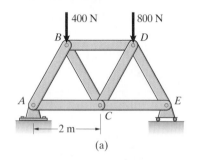

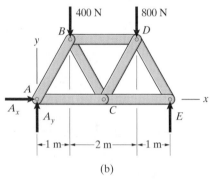

(a)

(b)

Figure 6.6
(a) A Warren truss supporting two loads.
(b) Free-body diagram of the truss.

The method of joints involves drawing free-body diagrams of the joints of a truss one by one and using the equilibrium equations to determine the axial forces in the members. Before beginning, it is usually necessary to draw a free-body diagram of the entire truss (that is, treat the truss as a single object) and determine the reactions at its supports. For example, let's consider the Warren truss in Fig. 6.6a, which has members 2 m in length and supports loads at B and D. We draw its free-body diagram in Fig. 6.6b. From the equilibrium equations,

$$\Sigma F_x = A_x = 0,$$

$$\Sigma F_y = A_y + E - 400 - 800 = 0,$$

$$\Sigma M_{(\text{point } A)} = -(1)(400) - (3)(800) + 4E = 0,$$

we obtain the reactions $A_x = 0$, $A_y = 500$ N, and $E = 700$ N.

Our next step is to choose a joint and draw its free-body diagram. In Fig. 6.7a, we isolate joint A by cutting members AB and AC. The terms T_{AB} and T_{AC} are the axial forces in members AB and AC, respectively. Although the directions of the arrows representing the unknown axial forces can be chosen arbitrarily, notice that we have chosen them so that a member is in tension if we obtain a positive value for the axial force. We feel that consistently choosing the directions in this way helps avoid errors.

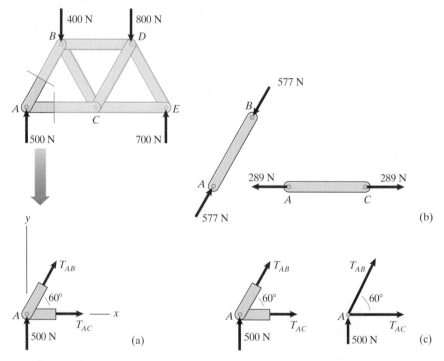

Figure 6.7
(a) Obtaining the free-body diagram of joint A.
(b) The axial forces on members AB and AC.
(c) Realistic and simple free-body diagrams of joint A.

The equilibrium equations for joint A are

$$\Sigma F_x = T_{AC} + T_{AB} \cos 60° = 0,$$

$$\Sigma F_y = T_{AB} \sin 60° + 500 = 0.$$

Solving these equations, we obtain the axial forces $T_{AB} = -577$ N and $T_{AC} = 289$ N. Member AB is in compression, and member AC is in tension (Fig. 6.7b).

Although we use a realistic figure for the joint in Fig. 6.7a to help you understand the free-body diagram, in your own work you can use a simple figure showing only the forces acting on the joint (Fig. 6.7c).

We next obtain a free-body diagram of joint B by cutting members AB, BC, and BD (Fig. 6.8a). From the equilibrium equations for joint B,

$$\Sigma F_x = T_{BD} + T_{BC} \cos 60° + 577 \cos 60° = 0,$$

$$\Sigma F_y = -400 + 577 \sin 60° - T_{BC} \sin 60° = 0,$$

we obtain $T_{BC} = 115$ N and $T_{BD} = -346$ N. Member BC is in tension, and member BD is in compression (Fig. 6.8b). By continuing to draw free-body diagrams of the joints, we can determine the axial forces in all of the members.

In two dimensions, you can obtain only two independent equilibrium equations from the free-body diagram of a joint. Summing the moments about a point does not result in an additional independent equation because the forces are concurrent. Therefore when applying the method of joints, you should choose joints to analyze that are subjected to no more than two unknown forces. In our example, we analyzed joint A first because it was subjected to the known reaction exerted by the pin support and two unknown forces, the axial forces T_{AB} and T_{AC} (Fig. 6.7a). We could then analyze joint B because it was subjected to two known forces and two unknown forces, T_{BC} and T_{BD} (Fig. 6.8a). If we had attempted to analyze joint B first, there would have been three unknown forces.

When you determine the axial forces in the members of a truss, your task will often be simpler if you are familiar with three particular types of joints.

- **Truss joints with two collinear members and no load** (Fig. 6.9). The sum of the forces must equal zero, $T_1 = T_2$. The axial forces are equal.

- **Truss joints with two noncollinear members and no load** (Fig. 6.10). Because the sum of the forces in the x direction must equal zero, $T_2 = 0$. Therefore T_1 must also equal zero. The axial forces are zero.

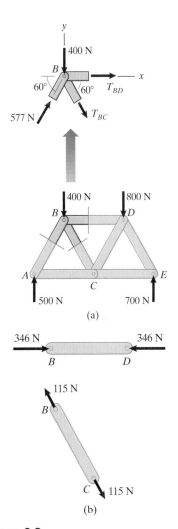

Figure 6.8
(a) Obtaining the free-body diagram of joint B.
(b) Axial forces in members BD and BC.

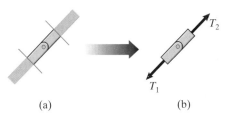

Figure 6.9
(a) A joint with two collinear members and no load.
(b) Free-body diagram of the joint.

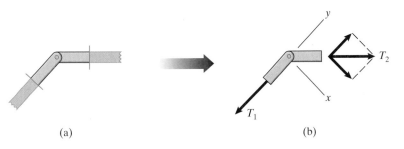

Figure 6.10
(a) A joint with two noncollinear members and no load.
(b) Free-body diagram of the joint.

• **Truss joints with three members, two of which are collinear, and no load** (Fig. 6.11). Because the sum of the forces in the x direction must equal zero, $T_3 = 0$. The sum of the forces in the y direction must equal zero, so $T_1 = T_2$. The axial forces in the collinear members are equal, and the axial force in the third member is zero.

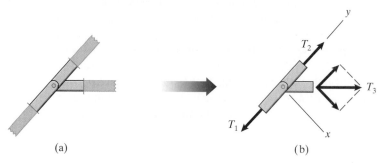

Figure 6.11
(a) A joint with three members, two of which are collinear, and no load.
(b) Free-body diagram of the joint.

(a)   (b)

Study Questions

1. What is a truss?
2. What is the method of joints?
3. How many independent equilibrium equations can you obtain from the free-body diagram of a joint?

Example 6.1

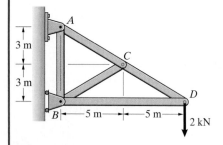

Figure 6.12

Applying the Method of Joints

Determine the axial forces in the members of the truss in Fig. 6.12.

Solution

Determine the Reactions at the Supports We first draw the free-body diagram of the entire truss (Fig. a). From the equilibrium equations,

$$\Sigma F_x = A_x + B = 0,$$
$$\Sigma F_y = A_y - 2 = 0,$$
$$\Sigma M_{(\text{point } B)} = -6A_x - (10)(2) = 0,$$

we obtain the reactions $A_x = -3.33$ kN, $A_y = 2$ kN, and $B = 3.33$ kN.

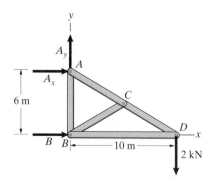

(a) Free-body diagram of the entire truss.

Identify Special Joints   Because joint C has three members, two of which are collinear, and no load, the axial force in member BC is zero, $T_{BC} = 0$, and the axial forces in the collinear members AC and CD are equal, $T_{AC} = T_{CD}$.

Draw Free-Body Diagrams of the Joints   We know the reaction exerted on joint A by the support, and joint A is subjected to only two unknown forces, the axial forces in members AB and AC. We draw its free-body diagram in Fig. b. The angle $\alpha = \arctan(5/3) = 59.0°$. The equilibrium equations for joint A are

$$\Sigma F_x = T_{AC} \sin \alpha - 3.33 = 0,$$
$$\Sigma F_y = 2 - T_{AB} - T_{AC} \cos \alpha = 0.$$

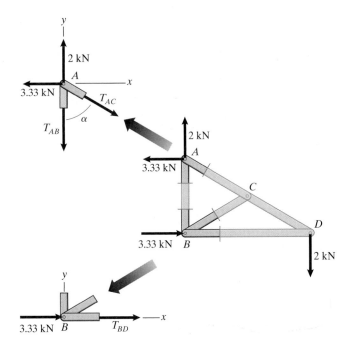

(b) Free-body diagram of joint A.

(c) Free-body diagram of joint B.

Solving these equations, we obtain $T_{AB} = 0$ and $T_{AC} = 3.89$ kN. Because the axial forces in members AC and CD are equal, $T_{CD} = 3.89$ kN.

Now we draw the free-body diagram of joint B in Fig. c. (We already know that the axial forces in members AB and BC are zero.) From the equilibrium equation

$$\Sigma F_x = T_{BD} + 3.33 = 0,$$

we obtain $T_{BD} = -3.33$ kN. The negative sign indicates that member BD is in compression.

The axial forces in the members are

AB: 0,

AC: 3.89 kN in tension (T),

BC: 0,

BD: 3.33 kN in compression (C),

CD: 3.89 kN in tension (T).

Example 6.2

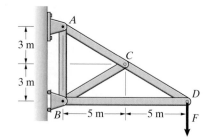

Figure 6.13

Determining the Largest Force a Truss Will Support

Each member of the truss in Fig. 6.13 will safely support a tensile force of 10 kN and a compressive force of 2 kN. What is the largest downward load F that the truss will safely support?

Strategy

This truss is identical to the one we analyzed in Example 6.1. By applying the method of joints in the same way, the axial forces in the members can be determined in terms of the load F. The smallest value of F that will cause a tensile force of 10 kN or a compressive force of 2 kN in any of the members is the largest value of F that the truss will support.

Solution

By using the method of joints in the same way as in Example 6.1, we obtain the axial forces

$$AB: 0,$$
$$AC: 1.94F \ (T),$$
$$BC: 0,$$
$$BD: 1.67F \ (C),$$
$$CD: 1.94F \ (T).$$

For a given load F, the largest tensile force is $1.94F$ (in members AC and CD) and the largest compressive force is $1.67F$ (in member BD). The largest safe tensile force would occur when $1.94F = 10$ kN or when $F = 5.14$ kN. The largest safe compressive force would occur when $1.67F = 2$ kN or when $F = 1.20$ kN. Therefore the largest load F that the truss will safely support is 1.20 kN.

Problems

6.1 Determine the axial forces in the members of the truss and indicate whether they are in tension (T) or compression (C).

Strategy: Draw a free-body diagram of joint A. By writing the equilibrium equations for the joint, you can determine the axial forces in the two members.

6.2 The truss supports a 10-kN load at C.
(a) Draw the free-body diagram of the entire truss, and determine the reactions at its supports.
(b) Determine the axial forces in the members. Indicate whether they are in tension (T) or compression (C).

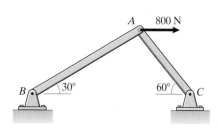

P6.1

P6.2

6.3 In Example 6.1, suppose that the 2-kN load is applied at D in the horizontal direction, pointing from D toward B. What are the axial forces in the members?

6.4 Determine the axial forces in the members of the truss.

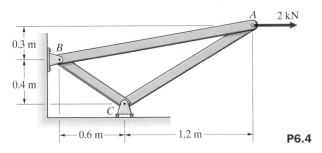

P6.4

6.5 (a) Let the dimension $h = 0.1$ m. Determine the axial forces in the members, and show that in this case this truss is equivalent to the one in Problem 6.4.
(b) Let the dimension $h = 0.5$ m. Determine the axial forces in the members. Compare the results to (a), and observe the dramatic effect of this simple change in design on the maximum tensile and compressive forces to which the members are subjected.

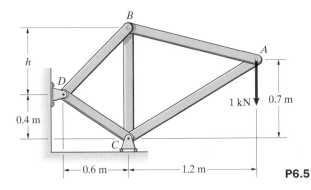

P6.5

6.6 The load $F = 10$ kN. Determine the axial forces in the members.

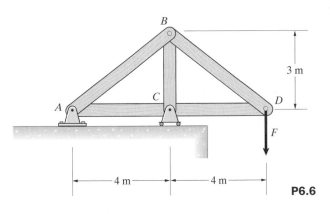

P6.6

6.7 Consider the truss in Problem 6.6. Each member will safely support a tensile force of 150 kN and a compressive force of 30 kN. What is the largest downward load F that the truss will safely support at D?

6.8 The Howe and Pratt bridge trusses are subjected to identical loads.
(a) In which truss does the largest tensile force occur? In what member(s) does it occur, and what is its value?
(b) In which truss does the largest compressive force occur? In what member(s) does it occur, and what is its value?

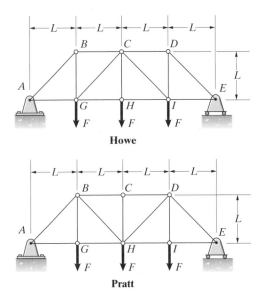

P6.8

6.9 The truss shown is part of an airplane's internal structure. Determine the axial forces in members BC, BD, and BE.

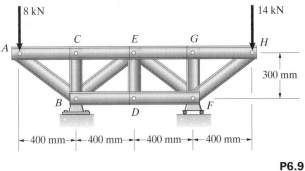

P6.9

6.10 For the truss in Problem 6.9, determine the axial forces in members DF, EF, and FG.

6.11 The loads $F_1 = F_2 = 8$ kN. Determine the axial forces in members BD, BE, and BG.

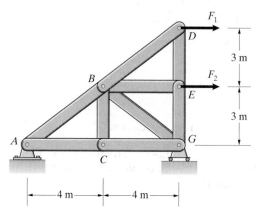

P6.11

6.12 If the loads on the truss shown in Problem 6.11 are $F_1 = 6$ kN and $F_2 = 10$ kN, what are the axial forces in members AB, BC, and BD?

6.13 The truss supports loads at C and E. If $F = 3$ kN, what are the axial forces in members BC and BE?

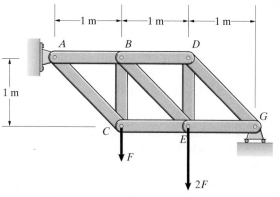

P6.13

6.14 Consider the truss in Problem 6.13. Each member will safely support a tensile force of 28 kN and a compressive force of 12 kN. Taking this criterion into account, what is the largest safe (positive) value of F?

6.15 The truss is a preliminary design for a structure to attach one end of a stretcher to a rescue helicopter. Based on dynamic simulations, the design engineer estimates that the downward forces the stretcher will exert will be no greater than 360 lb at A and at B. What are the resulting axial forces in members CF, DF, and FG?

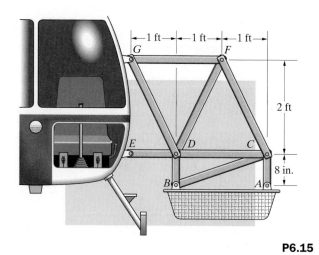

P6.15

6.16 Upon learning of an upgrade in the helicopter's engine, the engineer designing the truss shown in Problem 6.15 does new simulations and concludes that the downward forces the stretcher will exert at A and at B may be as large as 400 lb. What are the resulting axial forces in members DE, DF, and DG?

6.17 Determine the axial forces in the members in terms of the weight W.

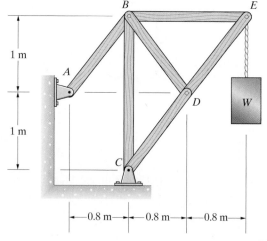

P6.17

6.18 Consider the truss in Problem 6.17. Each member will safely support a tensile force of 6 kN and a compressive force of 2 kN. Use this criterion to determine the largest weight W the truss will safely support.

6.19 The loads $F_1 = 600$ lb and $F_2 = 300$ lb. Determine the axial forces in members AE, BD, and CD.

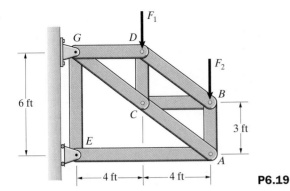

P6.19

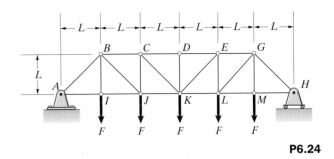

P6.24

6.20 Consider the truss in Problem 6.19. The loads $F_1 = 450$ lb and $F_2 = 150$ lb. Determine the axial forces in members AB, AC, and BC.

6.21 Each member of the truss will safely support a tensile force of 4 kN and a compressive force of 1 kN. Determine the largest mass m that can safely be suspended.

6.25 For the Pratt bridge truss in Problem 6.24, determine the axial forces in members CD, CJ, and CK.

6.26 The Howe truss helps support a roof. Model the supports at A and G as roller supports. Determine the axial forces in members AB, BC, and CD.

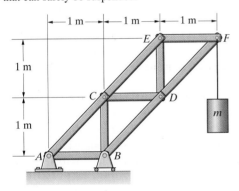

P6.21

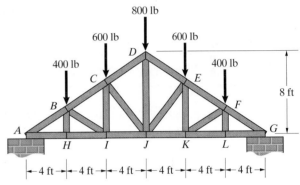

P6.26

6.22 The Warren truss supporting the walkway is designed to support vertical 50-kN loads at B, D, F, and H. If the truss is subjected to these loads, what are the resulting axial forces in members BC, CD, and CE?

6.27 The plane truss forms part of the supports of a crane on an offshore oil platform. The crane exerts vertical 75-kN forces on the truss at B, C, and D. You can model the support at A as a pin support and model the support at E as a roller support that can exert a force normal to the dashed line but cannot exert a force parallel to it. The angle $\alpha = 45°$. Determine the axial forces in the members of the truss.

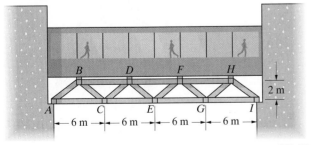

P6.22

6.23 For the Warren truss in Problem 6.22, determine the axial forces in members DF, EF, and FG.

6.24 The Pratt bridge truss supports five forces ($F = 300$ kN). The dimension $L = 8$ m. Determine the axial forces in members BC, BI, and BJ.

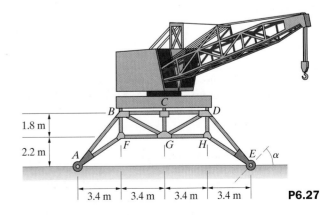

P6.27

6.28 (a) Design a truss attached to the supports A and B that supports the loads applied at points C and D.
(b) Determine the axial forces in the members of the truss you designed in (a).

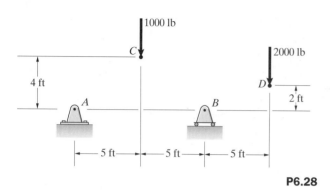

P6.28

6.29 (a) Design a truss attached to the supports A and B that supports the loads applied at points C and D.
(b) Determine the axial forces in the members of the truss you designed in (a).

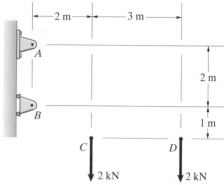

P6.29

6.3 The Method of Sections

When we need to know the axial forces only in certain members of a truss, we often can determine them more quickly using the method of sections than using the method of joints. For example, let's reconsider the Warren truss we used to introduce the method of joints (Fig. 6.14a). It supports loads at B and D, and each member is 2 m in length. Suppose that we need to determine only the axial force in member BC.

Just as in the method of joints, we begin by drawing a free-body diagram of the entire truss and determining the reactions at the supports. The results of this step are shown in Fig. 6.14b. Our next step is to cut the members AC, BC, and BD to obtain a free-body diagram of a part, or **section**, of the truss

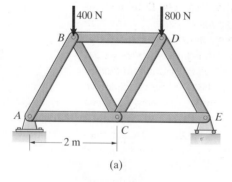

(a)

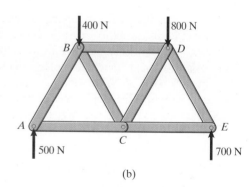

(b)

Figure 6.14
(a) A Warren truss supporting two loads.
(b) Free-body diagram of the truss, showing the reactions at the supports.

(Fig. 6.15). Summing moments about point B, the equilibrium equations for the section are

$$\Sigma F_x = T_{AC} + T_{BD} + T_{BC}\cos 60° = 0,$$

$$\Sigma F_y = 500 - 400 - T_{BC}\sin 60° = 0,$$

$$\Sigma M_{(\text{point } B)} = T_{AC}(2\sin 60°) - (500)(2\cos 60°) = 0.$$

Solving them, we obtain $T_{AC} = 289$ N, $T_{BC} = 115$ N, and $T_{BD} = -346$ N.

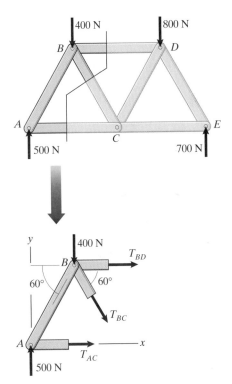

Figure 6.15

Obtaining a free-body diagram of a section of the truss.

Notice how similar this method is to the method of joints. Both methods involve cutting members to obtain free-body diagrams of parts of a truss. In the method of joints, we move from joint to joint, drawing free-body diagrams of the joints and determining the axial forces in the members as we go. In the method of sections, we try to obtain a single free-body diagram that allows us to determine the axial forces in specific members. In our example, we obtained a free-body diagram by cutting three members, including the one (member BC) whose axial force we wanted to determine.

In contrast to the free-body diagrams of joints, the forces on the free-body diagrams used in the method of sections are not usually concurrent, and as in our example, we can obtain three independent equilibrium equations. Although there are exceptions, it is usually necessary to choose a section that requires cutting no more than three members, or there will be more unknown axial forces than equilibrium equations.

Example 6.3

Applying the Method of Sections

The truss in Fig. 6.16 supports a 100-kN load. The horizontal members are each 1 m in length. Determine the axial force in member CJ, and state whether it is in tension or compression.

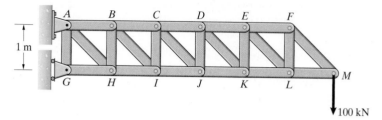

Figure 6.16

Strategy

We need to obtain a section by cutting members that include member CJ. By cutting members CD, CJ, and IJ, we will obtain a free-body diagram with three unknown axial forces.

Solution

To obtain a section (Fig. a), we cut members CD, CJ, and IJ and draw the free-body diagram of the part of the truss on the right side of the cuts. From the equilibrium equation

$$\Sigma F_y = T_{CJ} \sin 45° - 100 = 0,$$

we obtain $T_{CJ} = 141.4$ kN. The axial force in member CJ is 141.4 kN (T).

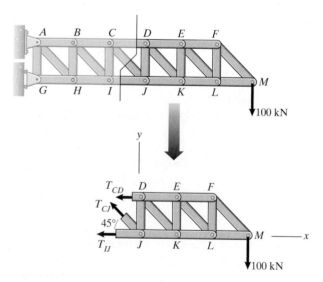

(a) Obtaining the section.

Discussion

Notice that by using the section on the right side of the cuts, we did not need to determine the reactions at the supports A and G.

Example 6.4

Choosing an Appropriate Section

Determine the axial forces in members DG and BE of the truss in Fig. 6.17.

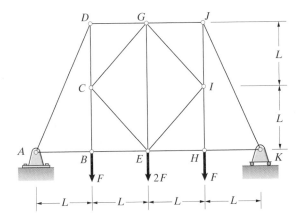

Figure 6.17

Strategy

An appropriate choice of section is not obvious, and it isn't clear beforehand that we can determine the requested information by the method of sections. We can't obtain a section that involves cutting members DG and BE without cutting more than three members. However, cutting members DG, BE, CD, and BC results in a section with which we can determine the axial forces in members DG and BE even though the resulting free-body diagram is statically indeterminate.

Solution

Determine the Reactions at the Supports We draw the free-body diagram of the entire truss in Fig. a. From the equilibrium equations,

$$\Sigma F_x = A_x = 0,$$
$$\Sigma F_y = A_y + K - F - 2F - F = 0,$$
$$\Sigma M_{(\text{point } A)} = -FL - 2F(2L) - F(3L) + K(4L) = 0,$$

we obtain the reactions $A_x = 0$, $A_y = 2F$, and $K = 2F$.

Choose a Section In Fig. b, we obtain a section by cutting members DG, CD, BC, and BE. Because the lines of action of T_{BE}, T_{BC}, and T_{CD} pass through point B, we can determine T_{DG} by summing moments about B:

$$\Sigma M_{(\text{point } B)} = -2FL - T_{DG}(2L) = 0.$$

The axial force $T_{DG} = -F$. Then from the equilibrium equation

$$\Sigma F_x = T_{DG} + T_{BE} = 0,$$

we see that $T_{BE} = -T_{DG} = F$. Member DG is in compression, and member BE is in tension.

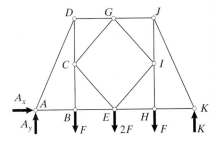

(a) Free-body diagram of the entire truss.

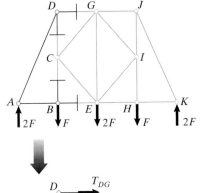

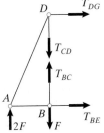

(b) A section of the truss obtained by passing planes through members DG, CD, BC, and BE.

Problems

6.30 The truss supports a 100-kN load at J. The horizontal members are each 1 m in length.
(a) Use the method of joints to determine the axial force in member DG.
(b) Use the method of sections to determine the axial force in member DG.

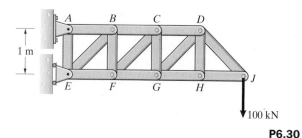

100 kN

P6.30

6.31 For the truss in Problem 6.30, use the method of sections to determine the axial forces in members BC, CF, and FG.

6.32 Use the method of sections to determine the axial forces in members AB, BC, and CE.

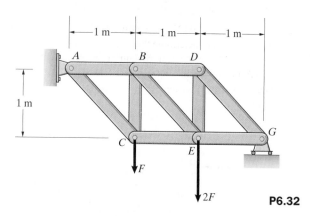

P6.32

6.33 The truss supports loads at A and H. Use the method of sections to determine the axial forces in members CE, BE, and BD.

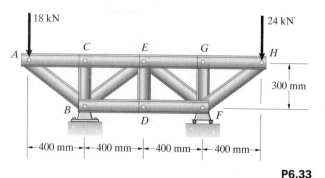

P6.33

6.34 For the truss in Problem 6.33, use the method of sections to determine the axial forces in members EG, EF, and DF.

6.35 For the Howe and Pratt trusses, use the method of sections to determine the axial force in member BC.

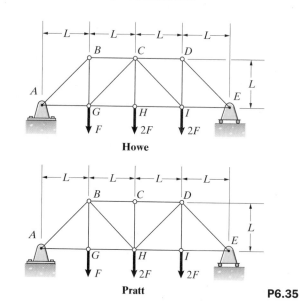

Howe

Pratt

P6.35

6.36 For the Howe and Pratt trusses in Problem 6.35, determine the axial force in member HI.

6.37 The Pratt bridge truss supports five forces $F = 340$ kN. The dimension $L = 8$ m. Use the method of sections to determine the axial force in member JK.

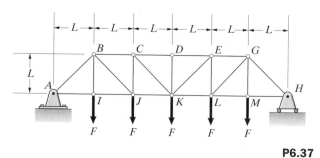

P6.37

6.38 For the Pratt bridge truss in Problem 6.37, use the method of sections to determine the axial force in member EK.

6.39 The walkway exerts vertical 50-kN loads on the Warren truss at B, D, F, and H. Use the method of sections to determine the axial force in member CE.

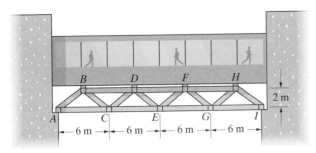

P6.39

6.40 The walkway in Problem 6.39 exerts equal vertical loads on the Warren truss at B, D, F, and H. Use the method of sections to determine the maximum allowable value of each vertical load if the magnitude of the axial force in member FG is not to exceed 100 kN.

6.41 The mass $m = 120$ kg. Use the method of sections to determine the axial forces in members BD, CD, and CE.

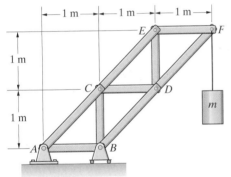

P6.41

6.42 For the truss in Problem 6.41, use the method of sections to determine the axial forces in members AC, BC, and BD.

6.43 The Howe truss helps support a roof. Model the supports at A and G as roller supports.
(a) Use the method of joints to determine the axial force in member BI.
(b) Use the method of sections to determine the axial force in member BI.

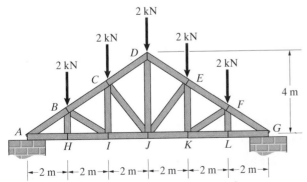

P6.43

6.44 Consider the truss in Problem 6.43. Use the method of sections to determine the axial force in member EJ.

6.45 Use the method of sections to determine the axial force in member EF.

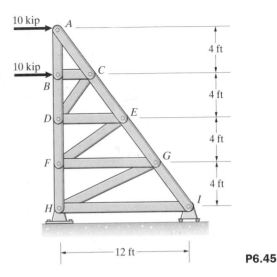

P6.45

6.46 Consider the truss in Problem 6.45. Use the method of sections to determine the axial force in member FG.

6.47 The load $F = 20$ kN and the dimension $L = 2$ m. Use the method of sections to determine the axial force in member HK.
  Strategy: Obtain a section by cutting members HK, HI, IJ, and JM. You can determine the axial forces in members HK and JM even though the resulting free-body diagram is statically indeterminate.

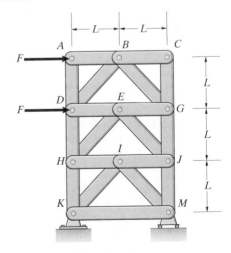

P6.47

6.48 The weight of the bucket is $W = 1000$ lb. The cable passes over pulleys at A and D.
(a) Determine the axial forces in members FG and HI.
(b) By drawing free-body diagrams of sections, explain why the axial forces in members FG and HI are equal.

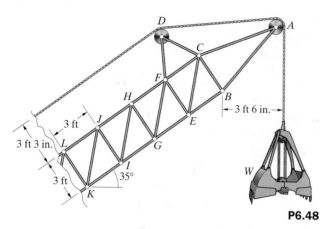

P6.48

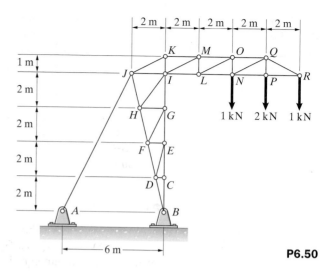

P6.50

6.49 Consider the truss in Problem 6.48. The weight of the bucket is $W = 1000$ lb. The cable passes over pulleys at A and D. Determine the axial forces in members IK and JL.

6.50 The truss supports loads at N, P, and R. Determine the axial forces in members IL and KM.

6.51 Consider the truss in Problem 6.50. Determine the axial forces in members HJ and GI.

6.4   **Frames and Machines**

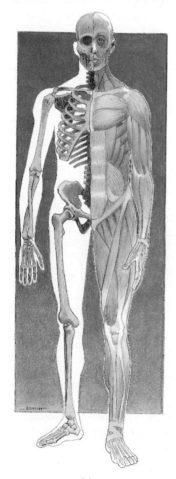

Many structures, such as the frame of a car and the human structure of bones, tendons, and muscles (Fig. 6.18), are not composed entirely of two-force members and thus cannot be modeled as trusses. In this section we consider structures of interconnected members that do not satisfy the definition of a truss. Such structures are called *frames* if they are designed to remain stationary and support loads and *machines* if they are designed to move and apply loads.

When trusses are analyzed by cutting members to obtain free-body diagrams of joints or sections, the internal forces acting at the "cuts" are simple axial forces (see Fig. 6.4). This is not generally true for frames or machines, and a different method of analysis is necessary. Instead of cutting members, you isolate entire members, or in some cases combinations of members, from the structure.

Figure 6.18
The internal structure of a person (a) and a car's frame (b) are not trusses.

(b)

(a)

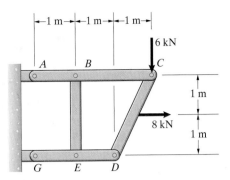

To begin analyzing a frame or machine, we draw a free-body diagram of the entire structure (that is, treat the structure as a single object) and determine the reactions at its supports. In some cases the entire structure will be statically indeterminate, but it is helpful to determine as many of the reactions as possible. We then draw free-body diagrams of individual members, or selected combinations of members, and apply the equilibrium equations to determine the forces and couples acting on them. For example, let's consider the stationary structure in Fig. 6.19. Member *BE* is a two-force member, but the other three members—*ABC*, *CD*, and *DEG*—are not. This structure is a frame. Our objective is to determine the forces on its members.

Figure 6.19
A frame supporting two loads.

Analyzing the Entire Structure

We draw the free-body diagram of the entire frame in Fig. 6.20. It is statically indeterminate: There are four unknown reactions, A_x, A_y, G_x, and G_y, whereas we can write only three independent equilibrium equations. However, notice that the lines of action of three of the unknown reactions intersect at *A*. By summing moments about *A*,

$$\Sigma M_{\text{(point } A)} = 2G_x + (1)(8) - (3)(6) = 0,$$

we obtain the reaction $G_x = 5$ kN. Then from the equilibrium equation

$$\Sigma F_x = A_x + G_x + 8 = 0,$$

we obtain the reaction $A_x = -13$ kN. Although we cannot determine A_y or G_y from the free-body diagram of the entire structure, we can do so by analyzing the individual members.

Analyzing the Members

Our next step is to draw free-body diagrams of the members. To do so, we treat the attachment of a member to another member just as if it were a support. Looked at in this way, we can think of each member as a supported object of the kind analyzed in Chapter 5. Furthermore, the forces and couples the members exert on one another are *equal in magnitude and opposite in direction*. A simple demonstration is instructive. If you clasp your hands as shown in Fig. 6.21a and exert a force on your left hand with your right hand, your left hand exerts an equal and opposite force on your right hand (Fig. 6.21b). Similarly, if you exert a couple on your left hand, your left hand exerts an equal and opposite couple on your right hand.

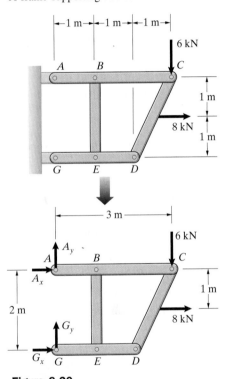

Figure 6.20
Obtaining the free-body diagram of the entire frame.

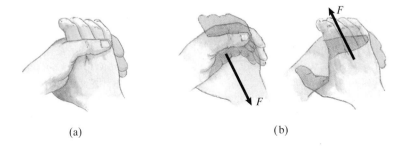

(a)   (b)

Figure 6.21
Demonstrating Newton's third law:
(a) Clasp your hands and pull on your left hand.
(b) Your hands exert equal and opposite forces.

In Fig. 6.22 we "disassemble" the frame and draw free-body diagrams of its members. Observe that the forces exerted on one another by the members are equal and opposite. For example, at point C on the free-body diagram of member ABC, the force exerted by member CD is denoted by the components C_x, and C_y. We can choose the directions of these unknown forces arbitrarily, but once we have done so, the forces exerted by member ABC on member CD at point C must be equal and opposite, as shown.

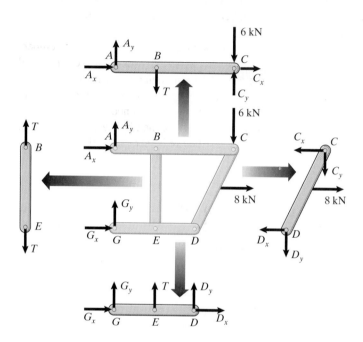

Figure 6.22
Obtaining the free-body diagrams of the members.

We need to discuss two important aspects of these free-body diagrams before completing the analysis.

Two-Force Members Member BE is a two-force member, and we have taken this into account in drawing its free-body diagram in Fig. 6.22. The force T is the axial force in member BE, and an equal and opposite force is subjected on member ABC at B and on member GED at E.

Recognizing two-force members in frames and machines and drawing their free-body diagrams as we have done will reduce the number of unknowns and will greatly simplify the analysis. In our example, if we did not treat member BE as a two-force member, its free-body diagram would have four unknown forces (Fig. 6.23a). By treating it as a two-force member (Fig. 6.23b), we reduce the number of unknown forces by three.

Loads Applied at Joints A question arises when a load is applied at a joint: Where does the load appear on the free-body diagrams of the individual members? The answer is that you can place the load on *any one* of the members attached at the joint. For example, in Fig. 6.19, the 6-kN load acts at the joint where members ABC and CD are connected. In drawing the free-body diagrams of the individual members (Fig. 6.22), we assumed that the 6-kN load acted on member ABC. The force components C_x and C_y on the free-body diagram of member ABC are the forces exerted by the member CD.

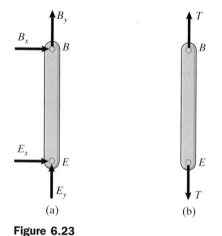

(a)   (b)

Figure 6.23
Free-body diagram of member BE:
(a) Not treating it as a two-force member.
(b) Treating it as a two-force member.

To explain why we can draw the free-body diagrams in this way, let us assume that the 6-kN force acts on the pin connecting members *ABC* and *CD*, and draw separate free-body diagrams of the pin and the two members (Fig. 6.24a). The force components C'_x and C'_y are the forces exerted by the pin on member *ABC*, and C_x and C_y are the forces exerted by the pin on member *CD*. If we superimpose the free-body diagrams of the pin and member *ABC*, we obtain the two free-body diagrams in Fig. 6.24b, which is the way we drew them in Fig. 6.22. Alternatively, by superimposing the free-body diagrams of the pin and member *CD*, we obtain the two free-body diagrams in Fig. 6.24c.

Thus if a load acts at a joint, it can be placed on any one of the members attached at the joint when drawing the free-body diagrams of the individual members. Just make sure not to place it on more than one member.

To detect errors in the free-body diagrams of the members, it is helpful to "reassemble" them (Fig. 6.25a). The forces at the connections between the members cancel (they are internal forces once the members are reassembled), and the free-body diagram of the entire structure is recovered (Fig. 6.25b).

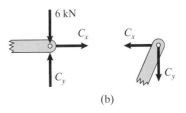

(a)

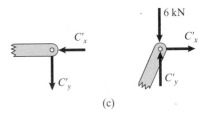

(b)

(c)

Figure 6.24
(a) Drawing free-body diagrams of the pin and the two members.
(b) Superimposing the pin on member *ABC*.
(c) Superimposing the pin on member *CD*.

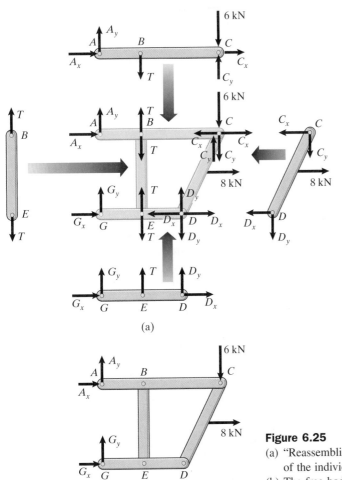

(a)

(b)

Figure 6.25
(a) "Reassembling" the free-body diagrams of the individual members.
(b) The free-body diagram of the entire frame is recovered.

Our final step is to apply the equilibrium equations to the free-body diagrams of the members (Fig. 6.26). In two dimensions, we can obtain three independent equilibrium equations from the free-body diagram of each member of a structure that we do not treat as a two-force member. (By assuming that the forces on a two-force member are equal and opposite axial forces, we have already used the three equilibrium equations for that member.) In this example, there are three members in addition to the two-force member, so we can write $(3)(3) = 9$ independent equilibrium equations, and there are 9 unknown forces: A_x, A_y, C_x, C_y, D_x, D_y, G_x, G_y, and T.

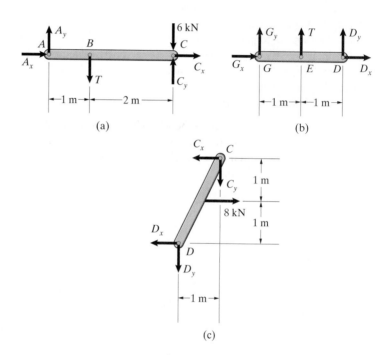

Figure 6.26
Free-body diagrams of the members.

Recall that we determined that $A_x = -13$ kN and $G_x = 5$ kN from our analysis of the entire structure. The equilibrium equations we obtained from the free-body diagram of the entire structure are not independent of the equilibrium equations obtained from the free-body diagrams of the members, but by using them to determine A_x and G_x, we get a head start on solving the equations for the members. Consider the free-body diagram of member ABC (Fig. 6.26a). Because we know A_x, we can determine C_x from the equation

$$\Sigma F_x = A_x + C_x = 0,$$

obtaining $C_x = -A_x = 13$ kN. Now consider the free-body diagram of GED (Fig. 6.26b). We can determine D_x from the equation

$$\Sigma F_x = G_x + D_x = 0,$$

obtaining $D_x = -G_x = -5$ kN. Now consider the free-body diagram of member CD (Fig. 6.26c). Because we know C_x, we can determine C_y by summing moments about D:

$$\Sigma M_{(\text{point } D)} = (2)C_x - (1)C_y - (1)(8) = 0.$$

We obtain $C_y = 18$ kN. Then from the equation

$$\Sigma F_y = -C_y - D_y = 0,$$

we find that $D_y = -C_y = -18$ kN. Now we can return to the free-body diagrams of members ABC and GED to determine A_y and G_y. Summing moments about point B of member ABC,

$$\Sigma M_{(\text{point } B)} = -(1)A_y + (2)C_y - (2)(6) = 0,$$

we obtain $A_y = 2C_y - 12 = 24$ kN. Then by summing moments about point E of member GED,

$$\Sigma M_{(\text{point } E)} = (1)D_y - (1)G_y = 0,$$

we obtain $G_y = D_y = -18$ kN. Finally, from the free-body diagram of member GED, we use the equilibrium equation

$$\Sigma F_y = D_y + G_y + T = 0,$$

which gives us the result $T = -D_y - G_y = 36$ kN. The forces on the members are shown in Fig. 6.27. As this example demonstrates, determination of the forces on the members can often be simplified by carefully choosing the order in which the equations are solved.

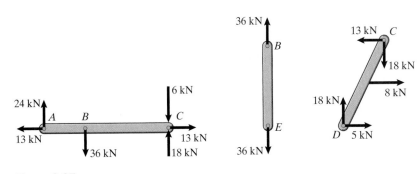

Figure 6.27
Forces on the members of the frame.

We see that determining the forces and couples on the members of frames and machines requires two steps:

1. **Determine the reactions at the supports**—Draw the free-body diagram of the entire structure, and determine the reactions at its supports. This step can greatly simplify your analysis of the members. If the free-body diagram is statically indeterminant, determine as many of the reactions as possible.

2. **Analyze the members**—Draw free-body diagrams of the members, and apply the equilibrium equations to determine the forces acting on them. You can simplify this step by identifying two-force members. If a load acts at a joint of the structure, you can place the load on the free-body diagram of any one of the members attached at that joint.

Example 6.5

Analyzing a Frame

The frame in Fig. 6.28 is subjected to a 200-N-m couple. Determine the forces and couples on its members.

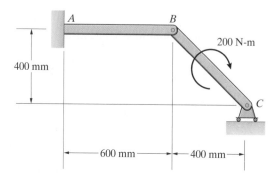

Figure 6.28

Solution

Determine the Reactions at the Supports We draw the free-body diagram of the entire frame in Fig. a. The term M_A is the couple exerted by the built-in support. From the equilibrium equations

$$\Sigma F_x = A_x = 0,$$

$$\Sigma F_y = A_y + C = 0,$$

$$\Sigma M_{(\text{point } A)} = M_A - 200 + (1)C = 0,$$

we obtain the reaction $A_x = 0$. We can't determine A_y, M_A, or C from this free-body diagram.

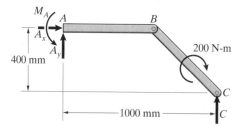

(a) Free-body diagram of the entire frame.

Analyze the Members We "disassemble" the frame to obtain the free-body diagrams of the members in Fig. b. The equilibrium equations for member BC are

$$\Sigma F_x = -B_x = 0,$$

$$\Sigma F_y = -B_y + C = 0,$$

$$\Sigma M_{(\text{point } B)} = -200 + (0.4)C = 0.$$

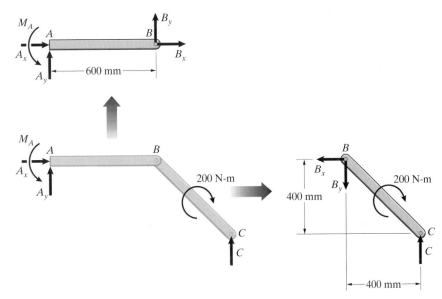

(b) Obtaining the free-body diagrams of the members.

Solving these equations, we obtain $B_x = 0$, $B_y = 500$ N, and $C = 500$ N. The equilibrium equations for member AB are

$$\Sigma F_x = A_x + B_x = 0,$$

$$\Sigma F_y = A_y + B_y = 0,$$

$$\Sigma M_{(\text{point } A)} = M_A + (0.6)B_y = 0.$$

Because we already know A_x, B_x, and B_y, we can solve these equations for A_y and M_A. The results are $A_y = -500$ N and $M_A = -300$ N-m. This completes the solution (Fig. c).

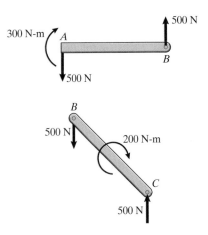

(c) Forces and couples on the members.

Discussion

We were able to solve the equilibrium equations for member BC without having to consider the free-body diagram of member AB. We were then able to solve the equilibrium equations for member AB. By choosing the members with the fewest unknowns to analyze first, you will often be able to solve them sequentially, but in some cases you will have to solve the equilibrium equations for the members simultaneously.

Even though we were unable to determine the four reactions A_x, A_y, M_A, and C with the three equilibrium equations obtained from the free-body diagram of the entire frame, we were able to determine them from the free-body diagrams of the individual members. By drawing free-body diagrams of the members, we gained three equations because we obtained three equilibrium equations from each member but only two new unknowns, B_x and B_y.

Example 6.6

Determining Forces on Members of a Frame

The frame in Fig. 6.29 supports a suspended weight $W = 40$ lb. Determine the forces on members $ABCD$ and CEG.

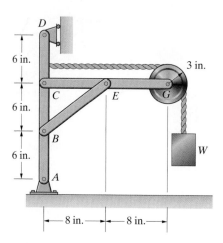

Figure 6.29

Solution

Determine the Reactions at the Supports We draw the free-body diagram of the entire frame in Fig. a. From the equilibrium equations

$$\Sigma F_x = A_x - D = 0,$$

$$\Sigma F_y = A_y - 40 = 0,$$

$$\Sigma M_{(\text{point } A)} = (18)D - (19)(40) = 0,$$

we obtain the reactions $A_x = 42.2$ lb, $A_y = 40$ lb, and $D = 42.2$ lb.

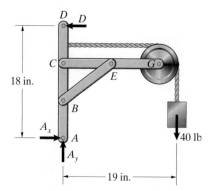

(a) Free-body diagram of the entire frame.

Analyze the Members We obtain the free-body diagrams of the members in Fig. b. Notice that BE is a two-force member. The angle $\alpha = \arctan(6/8) = 36.9°$.

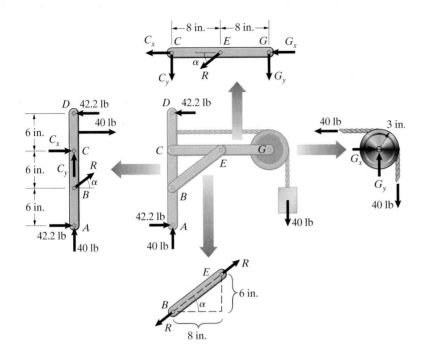

(b) Obtaining the free-body diagrams of the members.

The free-body diagram of the pulley has only two unknown forces. From the equilibrium equations

$$\Sigma F_x = G_x - 40 = 0,$$

$$\Sigma F_y = G_y - 40 = 0,$$

we obtain $G_x = 40$ lb and $G_y = 40$ lb. There are now only three unknown forces on the free-body diagram of member CEG. From the equilibrium equations

$$\Sigma F_x = -C_x - R \cos \alpha - 40 = 0,$$

$$\Sigma F_y = -C_y - R \sin \alpha - 40 = 0,$$

$$\Sigma M_{(\text{point } C)} = -(8)R \sin \alpha - (16)(40) = 0,$$

we obtain $C_x = 66.7$ lb, $C_y = 40$ lb, and $R = -133.3$ lb, completing the solution (Fig. c).

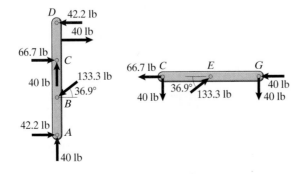

(c) Forces on members $ABCD$ and CEG.

Example 6.7

Free-Body Diagrams for Three Joined Members

Determine the forces on the members of the frame in Fig. 6.30.

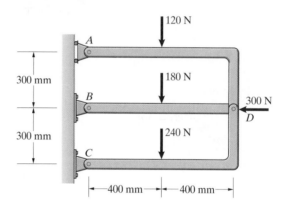

Figure 6.30

Strategy

You can confirm that no information can be obtained from the free-body diagram of the entire frame. To analyze the members, we must deal with an interesting challenge at joint D, where a load acts and three members are connected. We will obtain the free-body diagrams of the members by first isolating member AD, then separating members BD and CD.

Solution

We first isolate member AD from the rest of the structure, introducing the reactions D_x and D_y (Fig. a). We then separate members BD and CD, introducing equal and opposite forces E_x and E_y (Fig. b). In this step we could have placed the 300-N load and the forces D_x and D_y on either free-body diagram.

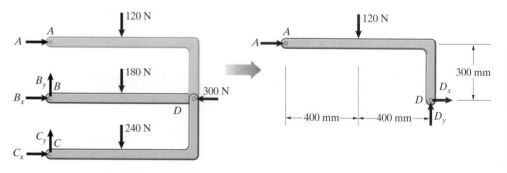

(a) Isolating member AD.

Only three unknown forces act on member AD. From the equilibrium equations

$$\Sigma F_x = A + D_x = 0,$$

$$\Sigma F_y = D_y - 120 = 0,$$

$$\Sigma M_{(\text{point } D)} = -(0.3)A + (0.4)(120) = 0,$$

we obtain $A = 160$ N, $D_x = -160$ N, and $D_y = 120$ N. Now we consider the free-body diagram of member BD. From the equation

$$\Sigma M_{(\text{point } D)} = -(0.8)B_y + (0.4)(180) = 0,$$

we obtain $B_y = 90$ N. Now we use the equation

$$\Sigma F_y = B_y - D_y + E_y - 180 = 90 - 120 + E_y - 180 = 0,$$

obtaining $E_y = 210$ N. Now that we know E_y, there are only three unknown forces on the free-body diagram of member CD. From the equilibrium equations

$$\Sigma F_x = C_x - E_x = 0,$$

$$\Sigma F_y = C_y - E_y - 240 = C_y - 210 - 240 = 0,$$

$$\Sigma M_{(\text{point } C)} = (0.3)E_x - (0.8)E_y - (0.4)(240)$$

$$= (0.3)E_x - (0.8)(210) - (0.4)(240) = 0,$$

we obtain $C_x = 880$ N, $C_y = 450$ N, and $E_x = 880$ N. Finally, we return to the free-body diagram of member BD and use the equation

$$\Sigma F_x = B_x + E_x - D_x - 300 = B_x + 880 + 160 - 300 = 0$$

to obtain $B_x = -740$ N, completing the solution (Fig. c).

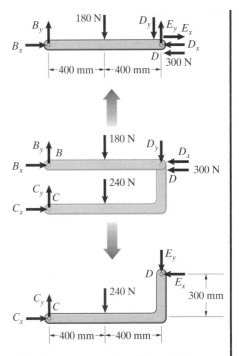

(b) Separating members BD and CD.

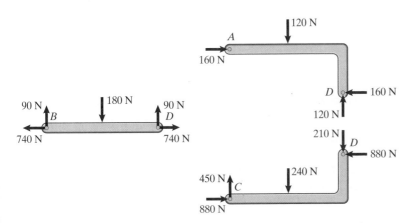

(c) Solutions for the forces on the members.

Example 6.8

Analyzing a Truck and Trailer as a Frame

The truck in Fig. 6.31 is parked on a 10° slope. Its brakes prevent the wheels at B from turning, but the wheels at C and the wheels of the trailer at A can turn freely. The trailer hitch at D behaves like a pin support. Determine the forces exerted on the truck at B, C, and D.

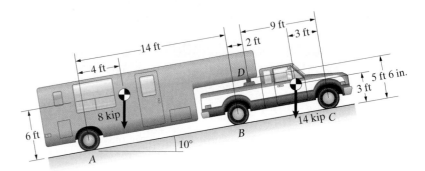

Figure 6.31

Strategy

We can treat this example as a structure whose "members" are the truck and trailer. We must isolate the truck and trailer and draw their individual free-body diagrams to determine the forces acting on the truck.

Solution

Determine the Reactions at the Supports The reactions in this example are the forces exerted on the truck and trailer by the road. We draw the free-body diagram of the connected truck and trailer in Fig. a. Because the tires at B are locked, the road can exert both a normal force and a friction force, but only normal forces are exerted at A and C. The equilibrium equations are

$$\Sigma F_x = B_x - 8 \sin 10° - 14 \sin 10° = 0,$$
$$\Sigma F_y = A + B_y + C - 8 \cos 10° - 14 \cos 10° = 0,$$
$$\Sigma M_{(\text{point } A)} = 14B_y + 25C + (6)(8 \sin 10°)$$
$$- (4)(8 \cos 10°) + (3)(14 \sin 10°)$$
$$- (22)(14 \cos 10°) = 0.$$

From the first equation we obtain the reaction $B_x = 3.82$ kip, but we can't solve the other two equations for the three reactions A, B_y, and C.

Analyze the Members We draw the free-body diagrams of the trailer and truck in Figs. b and c, showing the forces D_x and D_y exerted at the hitch. Only three unknown forces appear on the free-body diagram of the trailer. From the equilibrium equations for the trailer,

$$\Sigma F_x = D_x - 8 \sin 10° = 0,$$
$$\Sigma F_y = A + D_y - 8 \cos 10° = 0,$$
$$\Sigma M_{(\text{point } D)} = (0.5)(8 \sin 10°) + (12)(8 \cos 10°) - 16A = 0.$$

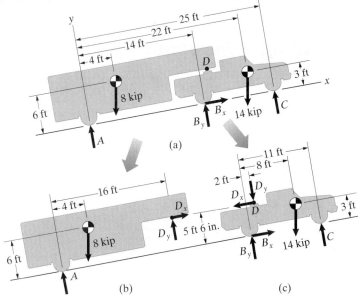

(a)

(b)

(c)

(a) Free-body diagram of the combined truck and trailer.

(b), (c) The individual free-body diagrams.

we obtain $A = 5.95$ kip, $D_x = 1.39$ kip, and $D_y = 1.93$ kip. (Notice that by summing moments about D, we obtained an equation containing only one unknown force.)

The equilibrium equations for the truck are

$$\Sigma F_x = B_x - D_x - 14 \sin 10° = 0,$$
$$\Sigma F_y = B_y + C - D_y - 14 \cos 10° = 0,$$
$$\Sigma M_{(\text{point } B)} = 11C + 5.5D_x - 2D_y + (3)(14 \sin 10°)$$
$$- (8)(14 \cos 10°) = 0.$$

Using the known values of D_x and D_y, we can solve these equations, obtaining $B_x = 3.82$ kip, $B_y = 6.69$ kip, and $C = 9.02$ kip.

Discussion

We were unable to solve two of the equilibrium equations for the connected truck and trailer. When that happens, you can use the equilibrium equations for the entire structure to check your results:

$$\Sigma F_x = B_x - 8 \sin 10° - 14 \sin 10°$$
$$= 3.82 - 8 \sin 10° - 14 \sin 10° = 0,$$
$$\Sigma F_y = A + B_y + C - 8 \cos 10° - 14 \cos 10°$$
$$= 5.95 + 6.69 + 9.02 - 8 \cos 10° - 14 \cos 10° = 0,$$
$$\Sigma M_{(\text{point } A)} = 14B_y + 25C + (6)(8 \sin 10°)$$
$$- (4)(8 \cos 10°) + (3)(14 \sin 10°) - (22)(14 \cos 10°)$$
$$= (14)(6.69) + (25)(9.02) + (6)(8 \sin 10°)$$
$$- (4)(8 \cos 10°) + (3)(14 \sin 10°)$$
$$- (22)(14 \cos 10°) = 0.$$

Example 6.9

Analyzing a Machine

What forces are exerted on the bolt at E in Fig. 6.32 as a result of the 150-N forces on the pliers?

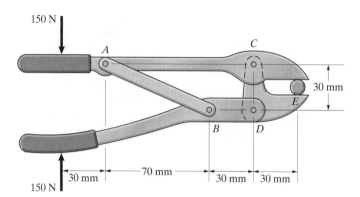

Figure 6.32

Strategy

A pair of pliers is a simple example of a machine, a structure designed to move and exert forces. The interconnections of the members are designed to create a mechanical advantage, subjecting an object to forces greater than the forces exerted by the user.

In this case there is no information to be gained from the free-body diagram of the entire structure. We must determine the forces exerted on the bolt by drawing free-body diagrams of the members.

Solution

We "disassemble" the pliers in Fig. a to obtain the free-body diagrams of the members, labeled (1), (2), and (3). The force R on free-body diagrams (1) and (3) is exerted by the two-force member AB. The angle $\alpha = \arctan(30/70) = 23.2°$. Our objective is to determine the force E exerted by the bolt.

The free-body diagram of member (3) has only three unknown forces and the 150-N load, so we can determine R, D_x, and D_y from this free-body diagram alone. The equilibrium equations are

$$\Sigma F_x = D_x + R\cos\alpha = 0,$$

$$\Sigma F_y = D_y - R\sin\alpha + 150 = 0,$$

$$\Sigma M_{(\text{point } B)} = 30D_y - (100)(150) = 0.$$

Solving these equations, we obtain $D_x = -1517$ N, $D_y = 500$ N, and $R = 1650$ N. Knowing D_x, we can determine E from the free-body diagram of member (2) by summing moments about C,

$$\Sigma M_{(\text{point } C)} = -30E - 30D_x = 0.$$

The force exerted on the bolt by the pliers is $E = -D_x = 1517$ N. The mechanical advantage of the pliers is $(1517\text{ N})/(150\text{ N}) = 10.1$.

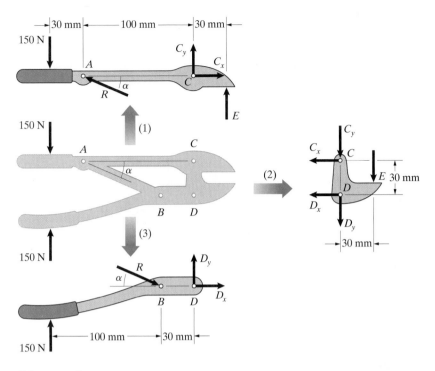

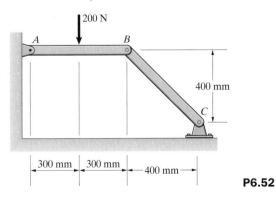

(a) Obtaining the free-body diagrams of the members.

Discussion

Notice that we did not need to use the free-body diagram of member (1) to determine E. When this happens, you can use the "leftover" free-body diagram to check your work. Using our results for R and E, we can confirm that the sum of the moments about point C of member (1) is zero:

$$\Sigma M_{(\text{point } C)} = (130)(150) - 100R \sin \alpha + 30E$$
$$= (130)(150) - (100)(1650) \sin 23.2° + (30)(1517) = 0.$$

Problems

6.52 Determine the reactions on member AB at A. (Notice that BC is a two-force member.)

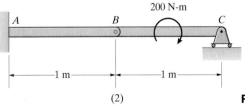

P6.52

6.53 (a) Determine the forces and couples on member AB for cases (1) and (2).

(b) You know that the moment of a couple is the same about any point. Explain why the answers are not the same in cases (1) and (2).

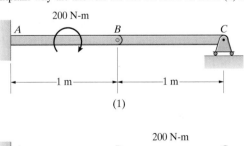

P6.53

6.54 For the frame shown, determine the reactions at the built-in support *A* and the forces exerted on member *AB* at *B*.

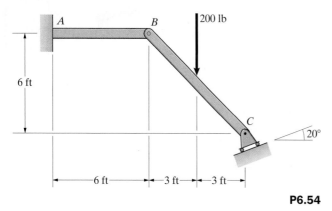

P6.54

6.55 The force $F = 10$ kN. Determine the forces on member *ABC*, presenting your answers as shown in Fig. 6.27.

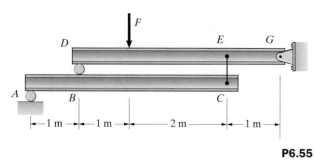

P6.55

6.56 Consider the frame in Problem 6.55. The cable *CE* will safely support a tension of 10 kN. Based on this criterion, what is the largest downward force *F* that can be applied to the frame?

6.57 The hydraulic actuator *BD* exerts a 6-kN force on member *ABC*. The force is parallel to *BD*, and the actuator is in compression. Determine the forces on member *ABC*, presenting your answers as shown in Fig. 6.27.

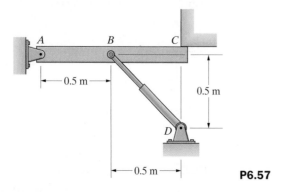

P6.57

6.58 The simple hydraulic jack shown in Problem 6.57 is designed to exert a vertical force at point *C*. The hydraulic actuator *BD* exerts a force on the beam *ABC* that is parallel to *BD*.

The largest lifting force the jack can exert is limited by the pin support *A*, which will safely support a force of magnitude 20 kN. What is the largest lifting force the jack can exert at *C*, and what is the resulting axial force in the hydraulic actuator?

6.59 Determine the forces on member *BC* and the axial force in member *AC*.

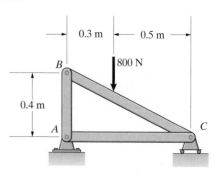

P6.59

6.60 An athlete works out with a squat thrust machine. To rotate the bar *ABD*, he must exert a vertical force at *A* that causes the magnitude of the axial force in the two-force member *BC* to be 1800 N. When the bar *ABD* is on the verge of rotating, what are the reactions on the vertical bar *CDE* at *D* and *E*?

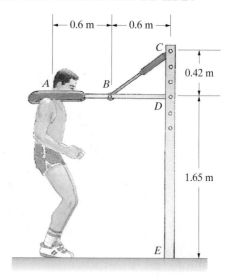

P6.60

6.61 The frame supports a 6-kN load at *C*. Determine the reactions on the frame at *A* and *D*.

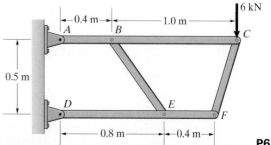

P6.61

6.62 The mass $m = 120$ kg. Determine the forces on member *ABC*, presenting your answers as shown in Fig. 6.27.

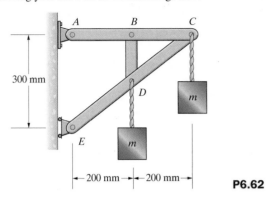

P6.62

6.63 The tension in cable *BD* is 500 lb. Determine the reactions at *A* for cases (1) and (2).

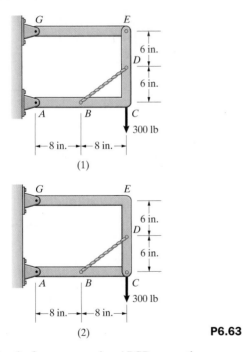

(1)

(2)

P6.63

6.64 Determine the forces on member *ABCD*, presenting your answers as shown in Fig. 6.27.

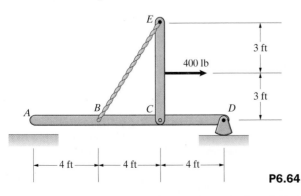

P6.64

6.65 The mass $m = 50$ kg. Determine the forces on member *ABCD*, presenting your answers as shown in Fig. 6.27.

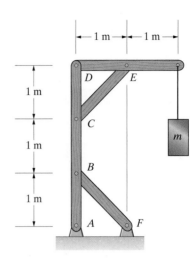

P6.65

6.66 Determine the forces on member *BCD*.

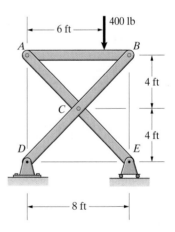

P6.66

6.67 Determine the forces on member *ABC*.

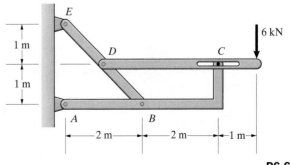

P6.67

6.68 Determine the forces on member *ABD*.

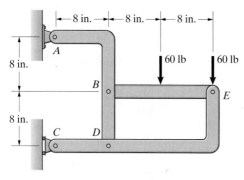

P6.68

6.69 The mass *m* = 12 kg. Determine the forces on member *CDE*.

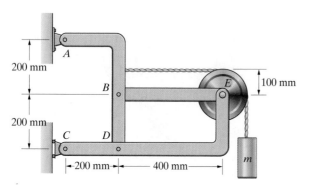

P6.69

6.70 The weight *W* = 80 lb. Determine the forces on member *ABCD*.

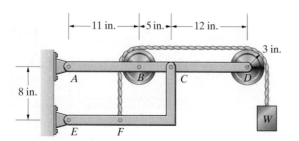

P6.70

6.71 The man using the exercise machine is holding the 80-lb weight stationary in the position shown. What are the reactions at the built-in support *E* and the pin support *F*? (*A* and *C* are pinned connections.)

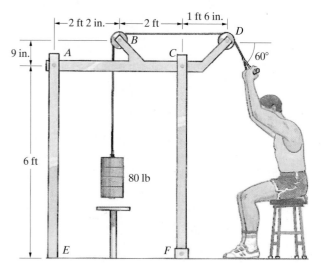

P6.71

6.72 The frame supports a horizontal load *F* at *C*. The resulting compressive axial force in the two-force member *CD* is 2400 N. Determine the magnitude of the reaction exerted on member *ABC* at *B*.

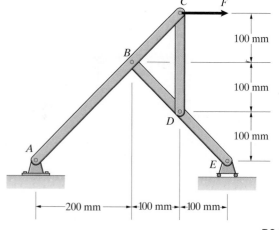

P6.72

6.73 The two-force member *CD* of the frame shown in Problem 6.72 will safely support a compressive axial load of 3 kN. Based on this criterion, what is the largest safe magnitude of the horizontal load *F*?

6.74 The unstretched length of the spring is L_0. Show that when the system is in equilibrium the angle α satisfies the relation $\sin\alpha = 2(L_0 - 2F/k)/L$.

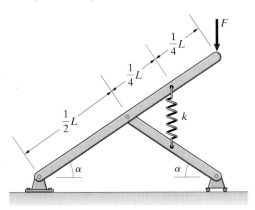

P6.74

6.75 The pin support B will safely support a force of 24-kN magnitude. Based on this criterion, what is the largest mass m that the frame will safely support?

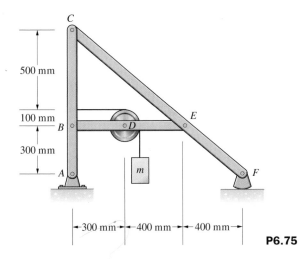

P6.75

6.76 Determine the reactions at A and C.

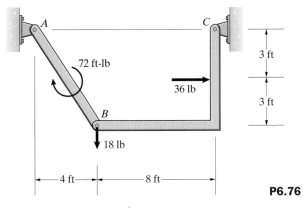

P6.76

6.77 Determine the forces on member AD.

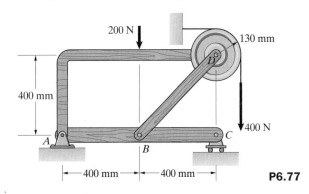

P6.77

6.78 The frame shown is used to support high-tension wires. If $b = 3$ ft, $\alpha = 30°$, and $W = 200$ lb, what is the axial force in member HJ?

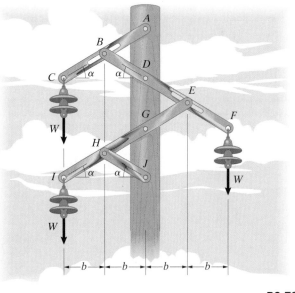

P6.78

6.79 What are the magnitudes of the forces exerted by the pliers on the bolt at A when 30-lb forces are applied as shown? (B is a pinned connection.)

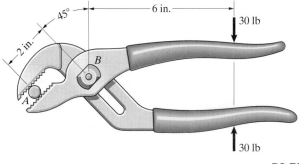

P6.79

6.80 The weight $W = 60$ kip. What is the magnitude of the force the members exert on each other at D?

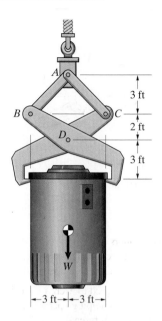

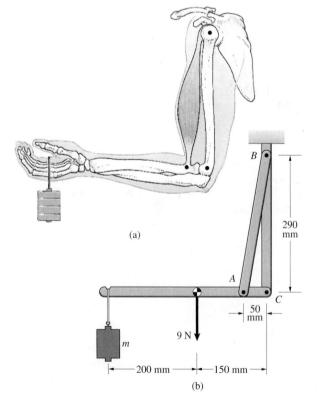

P6.80

6.81 Figure a is a diagram of the bones and biceps muscle of a person's arm supporting a mass. Tension in the biceps muscle

holds the forearm in the horizontal position, as illustrated in the simple mechanical model in Fig. b. The weight of the forearm is 9 N, and the mass $m = 2$ kg.
(a) Determine the tension in the biceps muscle AB.
(b) Determine the magnitude of the force exerted on the upper arm by the forearm at the elbow joint C.

6.82 The clamp presses two blocks of wood together. Determine the magnitude of the force the members exert on each other at C if the blocks are pressed together with a force of 200 N.

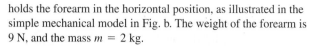

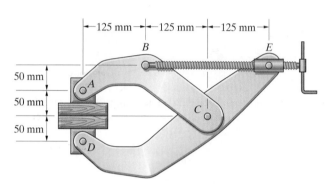

P6.82

6.83 The pressure force exerted on the piston is 2 kN toward the left. Determine the couple M necessary to keep the system in equilibrium.

P6.83

6.84 In Problem 6.83, determine the forces on member AB at A and B.

6.85 This mechanism is used to weigh mail. A package placed at A causes the weighted pointer to rotate through an angle α. Neglect the weights of the members except for the counterweight at B, which has a mass of 4 kg. If $\alpha = 20°$, what is the mass of the package at A?

P6.81

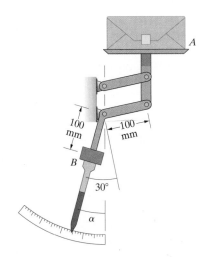

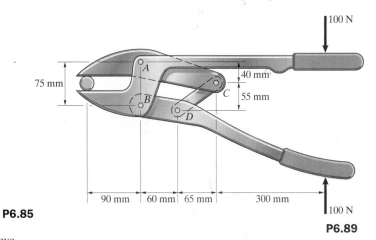

6.89 Determine the force exerted on the bolt by the bolt cutters.

P6.89

P6.85

6.86 The scoop C of the front-end loader is supported by two identical arms, one on each side of the loader. One of the two arms (ABC) is visible in the figure. It is supported by a pin support at A and the hydraulic actuator BD. The sum of the other loads exerted on the arm, including its own weight, is $F = 1.6$ kN. Determine the axial force in the actuator BD and the magnitude of the reaction at A.

6.90 For the bolt cutters in Problem 6.89, determine the magnitude of the force the members exert on each other at the pin connection B and the axial force in the two-force member CD.

6.91 This device is designed to exert a large force on the horizontal bar at A for a stamping operation. If the hydraulic cylinder DE exerts an axial force of 800 N and $\alpha = 80°$, what horizontal force is exerted on the horizontal bar at A?

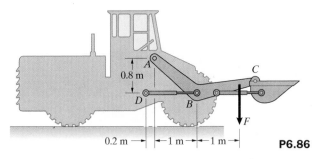

P6.86

6.87 The mass of the scoop is 220 kg, and its weight acts at G. Both the scoop and the hydraulic actuator BC are pinned to the horizontal member at B. The hydraulic actuator can be treated as a two-force member. Determine the forces exerted on the scoop at B and D.

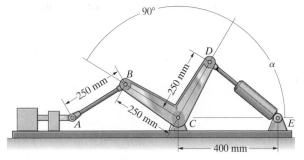

P6.91

6.92 This device raises a load W by extending the hydraulic actuator DE. The bars AD and BC are 4 ft long, and the distances $b = 2.5$ ft and $h = 1.5$ ft. If $W = 300$ lb, what force must the actuator exert to hold the load in equilibrium?

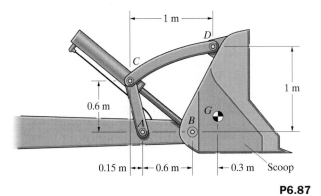

P6.87

6.88 In Problem 6.87, determine the axial force in the hydraulic actuator BC.

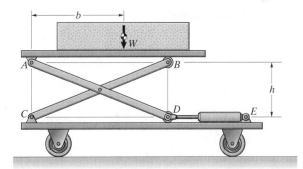

P6.92

6.93 The linkage is in equilibrium under the action of the couples M_A and M_B. If $\alpha_A = 60°$ and $\alpha_B = 70°$, what is the ratio M_A/M_B?

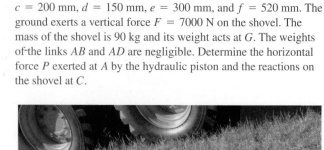
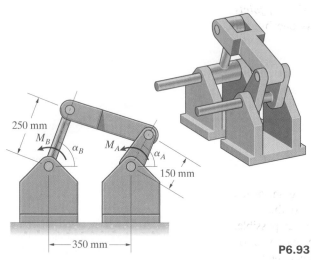

P6.93

6.94 A load $W = 2$ kN is supported by the member ACG and the hydraulic actuator BC. Determine the reactions at A and the compressive axial force in the actuator BC.

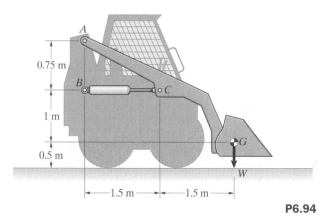

P6.94

6.95 The dimensions are $a = 260$ mm, $b = 300$ mm, $c = 200$ mm, $d = 150$ mm, $e = 300$ mm, and $f = 520$ mm. The ground exerts a vertical force $F = 7000$ N on the shovel. The mass of the shovel is 90 kg and its weight acts at G. The weights of the links AB and AD are negligible. Determine the horizontal force P exerted at A by the hydraulic piston and the reactions on the shovel at C.

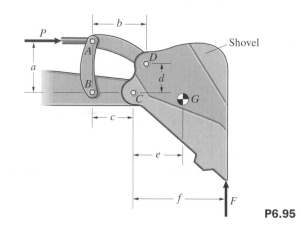

P6.95

Chapter Summary

A structure of *members* interconnected at *joints* is a *truss* if it is composed entirely of two-force members. Otherwise, it is a *frame* if it is designed to remain stationary and support loads and a *machine* if it is designed to move and exert loads.

Trusses

A member of a truss is in *tension* if the *axial forces* at the ends are directed away from each other and is in *compression* if the axial forces are directed toward each other. Before beginning to determine the axial forces in the members of a truss, it is usually necessary to draw a free-body diagram of the entire truss and determine the reactions at its supports. The axial forces in the members can be determined by two methods. The *method of joints* involves drawing free-body diagrams of the joints of a truss one by one and using the equilibrium equations to determine the axial forces in the members. In two dimensions, choose joints to analyze that are subjected to known forces and no more than two unknown forces. The *method of sections* involves drawing free-body diagrams of parts, or *sections*, of a truss and using the equilibrium equations to determine the axial forces in selected members.

Frames and Machines

Begin analyzing a frame or machine by drawing a free-body diagram of the entire structure and determining the reactions at its supports. If the entire structure is statically indeterminate, determine as many reactions as possible. Then draw free-body diagrams of individual members, or selected combinations of members, and apply the equilibrium equations to determine the forces and couples acting on them. Recognizing two-force members will reduce the number of unknown forces that must be determined. If a load is applied at a joint, it can be placed on the free-body diagram of *any one* of the members attached at the joint.

Review Problems

6.96 The truss supports a load $F = 10$ kN. Determine the axial forces in members AB, AC, and BC.

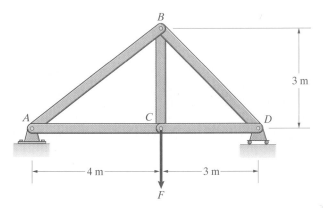

P6.96

6.97 Each member of the truss shown in Problem 6.96 will safely support a tensile force of 40 kN and a compressive force of 32 kN. Based on this criterion, what is the largest downward load F that can safely be applied at C?

6.98 The Pratt bridge truss supports loads at F, G, and H. Determine the axial forces in members BC, BG, and FG.

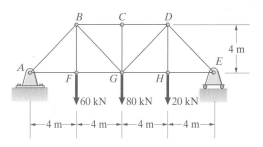

P6.98

6.99 Consider the truss in Problem 6.98. Determine the axial forces in members CD, GD, and GH.

6.100 The truss supports a 400-N load at G. Determine the axial forces in members AC, CD, and CF.

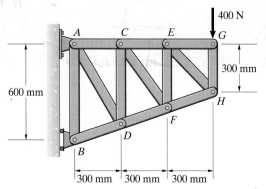

P6.100

6.101 Consider the truss in Problem 6.100. Determine the axial forces in members CE, EF, and EH.

6.102 The mass $m = 120$ kg. Determine the forces on member ABC.

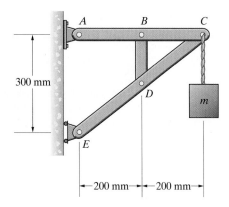

P6.102

6.103 Determine the forces on member ABC, presenting your answers as shown in Fig. 6.27.

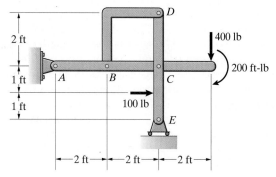

P6.103

6.104 Determine the force exerted on the bolt by the bolt cutters and the magnitude of the force the members exert on each other at the pin connection A.

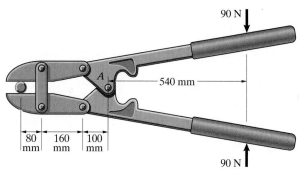

P6.104

6.105 The 600-lb weight of the scoop acts at a point 1 ft 6 in. to the right of the vertical line CE. The line ADE is horizontal. The hydraulic actuator AB can be treated as a two-force member. Determine the axial force in the hydraulic actuator AB and the forces exerted on the scoop at C and E.

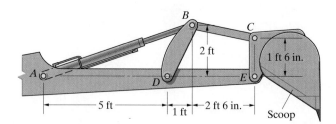

P6.105

$\mathscr{D}$**esign Experience** Design a truss structure to support a foot bridge with an unsupported span (width) of 8 m. Make conservative estimates of the loads the structure will need to support if the pathway supported by the truss is made of wood. Consider two options: (1) Your client wants the bridge to be supported by a truss below the bridge so that the upper surface will be unencumbered by structure. (2) The client wants the truss to be above the bridge and designed so that it can serve as handrails. For each option, use statics to estimate the maximum axial forces to which the members of the structure will be subjected. Investigate alternative designs and compare the resulting axial loads.

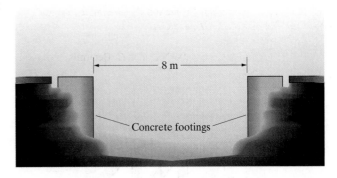

The loads that the legs of the piano must be designed to support depend not only on the piano's weight but also on the position of its center of mass—the point at which the weight effectively acts.

Centroids and Moments of Inertia

CHAPTER

7

In this chapter, we introduce definitions that can be interpreted as the average positions of areas, volumes, lines, and masses. These average positions are called centroids. The weight of an object can be represented by a single equivalent force acting at the centroid of its mass, which is called the center of mass of the object. We also define quantities called moments and products of inertia of areas. These quantities, which arise repeatedly in engineering applications, are used in analyzing the internal forces and deflections of beams.

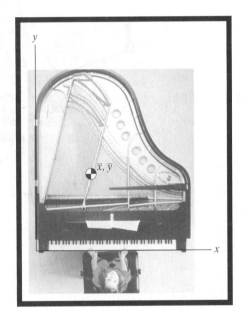

7.1 Centroids

Because centroids have such varied applications, we first define them using the general concept of a *weighted average*. Let's begin with the familiar idea of an average position. Suppose we want to determine the average position of a group of students sitting in a room. First, we introduce a coordinate system so that we can specify the position of each student. For example, we can align the axes with the walls of the room (Fig. 7.1 a). We number the students from 1 to N and denote the position of student 1 by x_1, y_1, the position of student 2 by x_2, y_2, and so on. The average x coordinate $\bar{x}$ is the sum of their x coordinates divided by N,

$$\bar{x} = \frac{x_1 + x_2 + \cdots + x_N}{N} = \frac{\sum_i x_i}{N}, \tag{7.1}$$

where the symbol $\sum\limits_i$ stands for "sum over the range of i". The average y coordinate is

$$\bar{y} = \frac{\sum_i y_i}{N}. \tag{7.2}$$

We indicate the average position by the symbol shown in Fig. 7.1b.

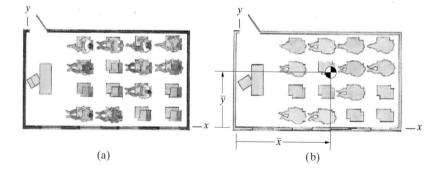

Figure 7.1
(a) A group of students in a classroom.
(b) Their average position.

(a)   (b)

Now suppose that we pass out some pennies to the students. Let the number of coins given to student 1 be c_1, the number given to student 2 be c_2, and so on. What is the average position of the coins in the room? Clearly, the average position of the coins may not be the same as the average position of the students. For example, if the students in the front of the room have more coins, the average position of the coins will be closer to the front of the room than the average position of the students.

To determine the x coordinate of the average position of the coins, we need to sum the x coordinates of the coins and divide by the number of coins. We can obtain the sum of the x coordinates of the coins by multiplying the number of coins each student has by his or her x coordinate and summing.

We can obtain the number of coins by summing the numbers $c_1, c_2, \ldots$. Thus the average x coordinate of the coins is

$$\bar{x} = \frac{\sum_i x_i c_i}{\sum_i c_i}. \tag{7.3}$$

We can determine the average y coordinate of the coins in the same way:

$$\bar{y} = \frac{\sum_i y_i c_i}{\sum_i c_i}. \tag{7.4}$$

By assigning other meanings to $c_1, c_2, \ldots$, we can determine the average positions of other measures associated with the students. For example, we could determine the average position of their age or the average position of their height.

More generally, we can use Eqs. (7.3) and (7.4) to determine the average position of any set of quantities with which we can associate positions. An average position obtained from these equations is called a *weighted average position*, or *centroid*. The "weight" associated with position x_1, y_1, is c_1, the weight associated with position x_2, y_2 is c_2, and so on. In Eqs. (7.1) and (7.2), the weight associated with the position of each student is 1. When the census is taken, the centroid of the population of the United States—the average position of the population—is determined in this way. In the next section we use Eqs. (7.3) and (7.4) to determine centroids of areas.

7.2 Centroids of Areas

Consider an arbitrary area A in the x–y plane (Fig. 7.2a). Let us divide the area into parts $A_1, A_2, \ldots, A_N$ (Fig. 7.2b) and denote the positions of the parts by $(x_1, y_1), (x_2, y_2), \ldots, (x_N, y_N)$. We can obtain the centroid, or average position of the area, by using Eqs. (7.3) and (7.4) with the areas of the parts as the weights:

$$\bar{x} = \frac{\sum_i x_i A_i}{\sum_i A_i}, \qquad \bar{y} = \frac{\sum_i y_i A_i}{\sum_i A_i}. \tag{7.5}$$

A question arises if we try to carry out this procedure: What are the exact positions of the areas $A_1, A_2, \ldots, A_N$? We could reduce the uncertainty in their positions by dividing A into smaller parts, but we would still obtain only

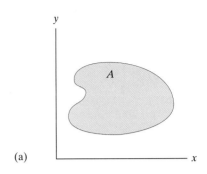
(a)

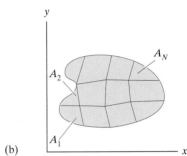
(b)

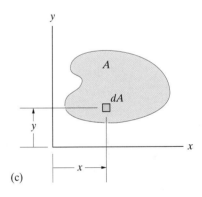
(c)

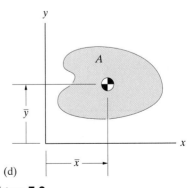
(d)

Figure 7.2
(a) The area A.
(b) Dividing A into N parts.
(c) A differential element of area dA with coordinates x, y.
(d) The centroid of the area.

approximate values for $\bar{x}$ and $\bar{y}$. To determine the exact location of the centroid, we must take the limit as the sizes of the parts approach zero. We obtain this limit by replacing Eqs. (7.5) by the integrals

$$\bar{x} = \frac{\displaystyle\int_A x\, dA}{\displaystyle\int_A dA}, \qquad (7.6)$$

$$\bar{y} = \frac{\displaystyle\int_A y\, dA}{\displaystyle\int_A dA}, \qquad (7.7)$$

where x and y are the coordinates of the differential element of area dA (Fig. 7.2c). The subscript A on the integral signs means the integration is carried out over the entire area. The centroid of the area is shown in Fig. 7.2d.

Keeping in mind that the centroid of an area is its average position will often help you locate it. For example, the centroid of a circular area or a rectangular area obviously lies at the center of the area. If an area has "mirror image" symmetry about an axis, the centroid lies on the axis (Fig. 7.3a), and if an area is symmetric about two axes, the centroid lies at their intersection (Fig. 7.3b).

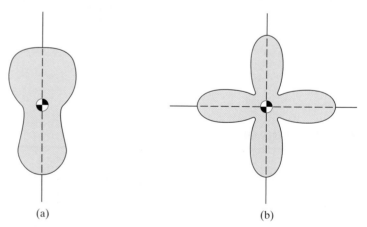

(a)   (b)

Figure 7.3
(a) An area that is symmetric about an axis.
(b) An area with two axes of symmetry.

Study Questions

1. How is a weighted average position defined?
2. How is the concept of a weighted average used to define the centroid of a plane area?
3. Why is integration generally needed to determine the exact position of the centroid of an area?

Example 7.1

Centroid of an Area by Integration

Determine the centroid of the triangular area in Fig. 7.4.

Strategy

We will determine the coordinates of the centroid by using an element of area dA in the form of a "strip" of width dx.

Solution

Let dA be the vertical strip in Fig. a. The height of the strip is $(h/b)x$, so $dA = (h/b)x\,dx$. To integrate over the entire area, we must integrate with respect to x from $x = 0$ to $x = b$. The x coordinate of the centroid is

$$\bar{x} = \frac{\int_A x\,dA}{\int_A dA} = \frac{\int_0^b x\left(\dfrac{h}{b}x\,dx\right)}{\int_0^b \dfrac{h}{b}x\,dx} = \frac{\dfrac{h}{b}\left[\dfrac{x^3}{3}\right]_0^b}{\dfrac{h}{b}\left[\dfrac{x^2}{2}\right]_0^b} = \frac{2}{3}b.$$

To determine $\bar{y}$, we let y in Eq. (7.7) be the y coordinate of the midpoint of the strip (Fig. b):

$$\bar{y} = \frac{\int_A y\,dA}{\int_A dA} = \frac{\int_0^b \dfrac{1}{2}\left(\dfrac{h}{b}x\right)\left(\dfrac{h}{b}x\,dx\right)}{\int_0^b \dfrac{h}{b}x\,dx} = \frac{\dfrac{1}{2}\left(\dfrac{h}{b}\right)^2\left[\dfrac{x^3}{3}\right]_0^b}{\dfrac{h}{b}\left[\dfrac{x^2}{2}\right]_0^b} = \frac{1}{3}h.$$

The centroid is shown in Fig. c.

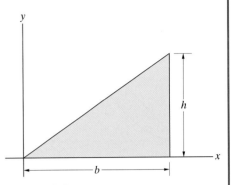

Figure 7.4

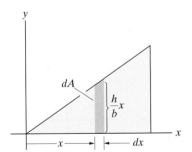

(a) An element dA in the form of a strip.

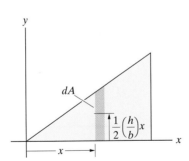

(b) The y coordinate of the midpoint of the strip is $\frac{1}{2}(h/b)x$.

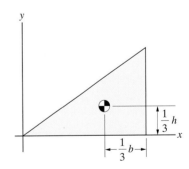

(c) Centroid of the area.

Discussion

You should always be alert for opportunities to check your results. In this example we should make sure that our integration procedure gives the correct result for the area of the triangle:

$$\int_A dA = \int_0^b \frac{h}{b}x\,dx = \frac{h}{b}\left[\frac{x^2}{2}\right]_0^b = \frac{1}{2}bh.$$

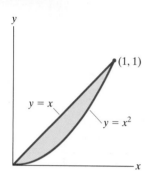

Example 7.2

Figure 7.5

Area Defined by Two Equations

Determine the centroid of the area in Fig. 7.5.

Solution

Let dA be the vertical strip in Fig. a. The height of the strip is $x - x^2$, so $dA = (x - x^2)\,dx$. The x coordinate of the centroid is

$$\bar{x} = \frac{\int_A x\,dA}{\int_A dA} = \frac{\int_0^1 x(x - x^2)\,dx}{\int_0^1 (x - x^2)\,dx} = \frac{\left[\dfrac{x^3}{3} - \dfrac{x^4}{4}\right]_0^1}{\left[\dfrac{x^2}{2} - \dfrac{x^3}{3}\right]_0^1} = \frac{1}{2}.$$

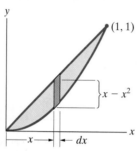

(a) A vertical strip of width dx. The height of the strip is equal to the difference in the two functions.

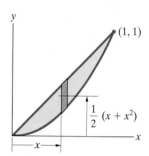

(b) The y coordinate of the midpoint of the strip.

The y coordinate of the midpoint of the strip is $x^2 + \frac{1}{2}(x - x^2) = \frac{1}{2}(x + x^2)$ (Fig. b). Substituting this expression for y in Eq. (7.7), we obtain the y coordinate of the centroid:

$$\bar{y} = \frac{\int_A y\,dA}{\int_A dA} = \frac{\int_0^1 \left[\frac{1}{2}(x + x^2)\right](x - x^2)\,dx}{\int_0^1 (x - x^2)\,dx} = \frac{\frac{1}{2}\left[\dfrac{x^3}{3} - \dfrac{x^5}{5}\right]_0^1}{\left[\dfrac{x^2}{2} - \dfrac{x^3}{3}\right]_0^1} = \frac{2}{5}.$$

Problems

7.1 If $a = 2$, what is the x coordinate of the centroid of the area?

Strategy: The x coordinate of the centroid is given by Eq. (7.6). For the element of area dA, use a vertical strip of width dx. (See Example 7.1.)

7.2 Determine the y coordinate of the centroid of the area shown in Problem 7.1 if $a = 3$.

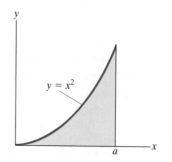

P7.1

7.3 If the x coordinate of the centroid of the area is $\bar{x} = 2$, what is the value of a?

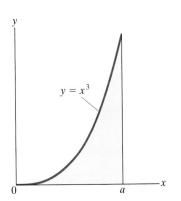

P7.3

7.4 The x coordinate of the centroid of the area shown in Problem 7.3 is $\bar{x} = 2$. What is the y coordinate of the centroid?

7.5 Consider the area in Problem 7.3. The "center of area" is defined to be the point for which there is as much area to the right of the point as to the left of it and as much area above the point as below it. If $a = 4$, what are the x coordinate of the center of area and the x coordinate of the centroid?

7.6 Determine the x coordinate of the centroid of the area and compare your answer to the value given in Appendix B.

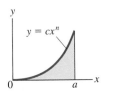

P7.6

7.7 Determine the y coordinate of the centroid of the area and compare your answer to the value given in Appendix B.

7.8 Suppose that an art student wants to paint a panel of wood as shown, with the horizontal and vertical lines passing through the centroid of the painted area, and asks you to determine the coordinates of the centroid. What are they?

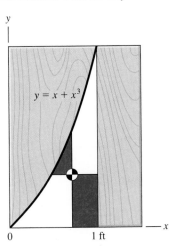

P7.8

7.9 The y coordinate of the centroid of the area is $\bar{y} = 1.063$. Determine the value of the constant c and the x coordinate of the centroid.

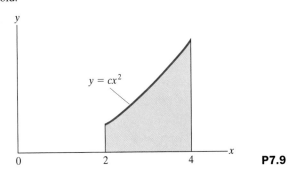

P7.9

7.10 Determine the coordinates of the centroid of the metal plate's cross-sectional area.

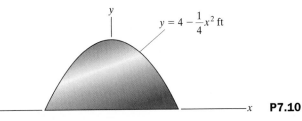

P7.10

7.11 An architect wants to build a wall with the profile shown. To estimate the effects of wind loads, he must determine the wall's area and the coordinates of its centroid. What are they?

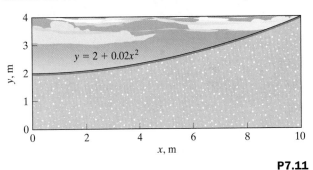

P7.11

7.12 Determine the x coordinate of the centroid of the area.

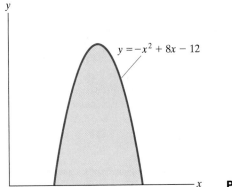

P7.12

7.13 Determine the y coordinate of the centroid of the area shown in Problem 7.12.

7.14 Determine the x coordinate of the centroid of the area.

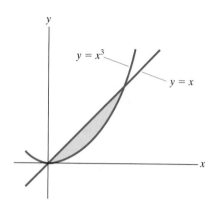

P7.14

7.15 Determine the y coordinate of the centroid of the area shown in Problem 7.14.

7.16 Determine the coordinates of the centroid of the area.

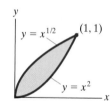

P7.16

7.17 Determine the x coordinate of the centroid of the area.

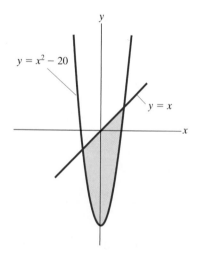

P7.17

7.18 Determine the y coordinate of the centroid of the area in Problem 7.17.

7.19 Determine the y coordinate of the centroid of the area.

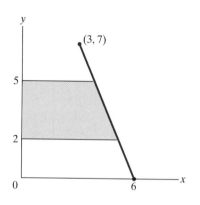

P7.19

7.20 Determine the x coordinate of the centroid of the area in Problem 7.19.

7.21 An agronomist wants to measure the rainfall at the centroid of a plowed field between two roads. What are the coordinates of the point where the rain gauge should be placed?

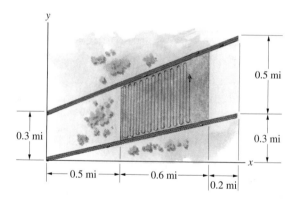

P7.21

7.22 The cross section of an earth-fill dam is shown. Determine the coefficients a and b so that the y coordinate of the centroid of the cross section is 10 m.

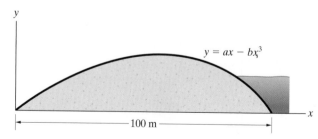

P7.22

7.23 The Supermarine Spitfire used by Great Britain in World War II had a wing with an elliptical profile. Determine the coordinates of its centroid.

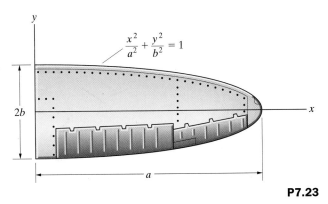

$$\frac{x^2}{a^2} + \frac{y^2}{b^2} = 1$$

P7.23

7.24 Determine the coordinates of the centroid of the area.
Strategy: Write the equation for the circular boundary in the form $y = (R^2 - x^2)^{1/2}$ and use a vertical "strip" of width dx as the element of area dA.

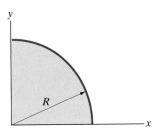

P7.24

7.25 Determine the x coordinate of the centroid of the area. By setting $h = 0$, confirm the answer to Problem 7.24.

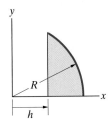

P7.25

7.26 Determine the y coordinate of the centroid of the area in Problem 7.25.

7.3 Centroids of Composite Areas

Although centroids of areas can be determined by integration, the process becomes difficult and tedious for complicated areas. In this section we describe a much easier approach that can be used if an area consists of a combination of simple areas, which we call a *composite area*. We can determine the centroid of a composite area without integration if the centroids of its parts are known.

The area in Fig. 7.6a consists of a triangle, a rectangle, and a semicircle, which we call parts 1, 2, and 3. The x coordinate of the centroid of the composite area is

$$\bar{x} = \frac{\displaystyle\int_A x\,dA}{\displaystyle\int_A dA} = \frac{\displaystyle\int_{A_1} x\,dA + \int_{A_2} x\,dA + \int_{A_3} x\,dA}{\displaystyle\int_{A_1} dA + \int_{A_2} dA + \int_{A_3} dA}. \quad (7.8)$$

The x coordinates of the centroids of the parts are shown in Fig. 7.6b. From the equation for the x coordinate of the centroid of part 1,

$$\bar{x}_1 = \frac{\displaystyle\int_{A_1} x\,dA}{\displaystyle\int_{A_1} dA},$$

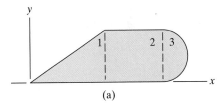

(a)

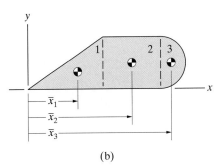

(b)

Figure 7.6
(a) A composite area composed of three simple areas.
(b) The centroids of the parts.

we obtain

$$\int_{A_1} x \, dA = \bar{x}_1 A_1.$$

Using this equation and equivalent equations for parts 2 and 3, we can write Eq. (7.8) as

$$\bar{x} = \frac{\bar{x}_1 A_1 + \bar{x}_2 A_2 + \bar{x}_3 A_3}{A_1 + A_2 + A_3}.$$

We have obtained an equation for the x coordinate of the composite area in terms of those of its parts. The coordinates of the centroid of a composite area with an arbitrary number of parts are

$$\bar{x} = \frac{\sum_i \bar{x}_i A_i}{\sum_i A_i}, \qquad \bar{y} = \frac{\sum_i \bar{y}_i A_i}{\sum_i A_i}. \tag{7.9}$$

When you can divide an area into parts whose centroids are known, you can use these expressions to determine its centroid. The centroids of some simple areas are tabulated in Appendix B.

We began our discussion of the centroid of an area by dividing an area into finite parts and writing equations for its weighted average position. The results, Eqs. (7.5), are approximate because of the uncertainty in the positions of the parts of the area. The exact Eqs. (7.9) are identical except that the positions of the parts are their centroids.

The area in Fig. 7.7a consists of a triangular area with a circular hole, or cutout. Designating the triangular area (without the cutout) as part 1 of the composite area (Fig. 7.7b) and the area of the cutout as part 2 (Fig. 7.7c), we obtain the x coordinate of the centroid of the composite area:

$$\bar{x} = \frac{\displaystyle\int_{A_1} x \, dA - \int_{A_2} x \, dA}{\displaystyle\int_{A_1} dA - \int_{A_2} dA} = \frac{\bar{x}_1 A_1 - \bar{x}_2 A_2}{A_1 - A_2}.$$

This equation is identical in form to the first of Eqs. (7.9) except that the terms corresponding to the cutout are negative. As this example demonstrates, you can use Eqs. (7.9) to determine the centroids of composite areas containing cutouts by treating the cutouts as negative areas.

We see that determining the centroid of a composite area requires three steps:

1. Choose the parts—Try to divide the composite area into parts whose centroids you know or can easily determine.

2. Determine the values for the parts—Determine the centroid and the area of each part. Watch for instances of symmetry that can simplify your task.

3. Calculate the centroid—Use Eqs. (7.9) to determine the centroid of the composite area.

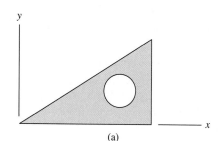

(a)

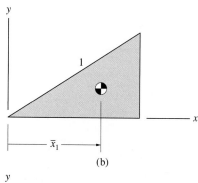

(b)

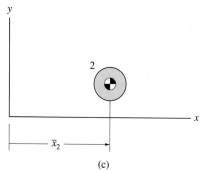

(c)

Figure 7.7
(a) An area with a cutout.
(b) The triangular area.
(c) The area of the cutout.

Example 7.3

Centroid of a Composite Area

Determine the centroid of the area in Fig. 7.8.

Solution

Choose the Parts We can divide the area into a triangle, a rectangle, and a semicircle, which we call parts 1, 2, and 3, respectively.

Determine the Values for the Parts The x coordinates of the centroids of the parts are shown in Fig. a. The x coordinates, the areas of the parts, and their products are summarized in Table 1.

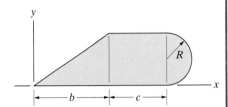

Figure 7.8

Table 1 Information for determining the x coordinate of the centroid

	$\bar{x}_i$	A_i	$\bar{x}_i A_i$
Part 1 (triangle)	$\frac{2}{3}b$	$\frac{1}{2}b(2R)$	$\left(\frac{2}{3}b\right)\left[\frac{1}{2}b(2R)\right]$
Part 2 (rectangle)	$b + \frac{1}{2}c$	$c(2R)$	$\left(b + \frac{1}{2}c\right)\left[c(2R)\right]$
Part 3 (semicircle)	$b + c + \dfrac{4R}{3\pi}$	$\frac{1}{2}\pi R^2$	$\left(b + c + \dfrac{4R}{3\pi}\right)\left(\frac{1}{2}\pi R^2\right)$

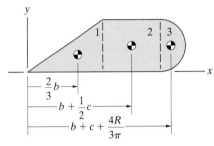

(a) The x coordinates of the centroids of the parts.

Calculate the Centroid The x coordinate of the centroid of the composite area is

$$\bar{x} = \frac{\bar{x}_1 A_1 + \bar{x}_2 A_2 + \bar{x}_3 A_3}{A_1 + A_2 + A_3}$$

$$= \frac{\left(\frac{2}{3}b\right)\left[\frac{1}{2}b(2R)\right] + \left(b + \frac{1}{2}c\right)\left[c(2R)\right] + \left(b + c + \dfrac{4R}{3\pi}\right)\left(\frac{1}{2}\pi R^2\right)}{\frac{1}{2}b(2R) + c(2R) + \frac{1}{2}\pi R^2}.$$

We repeat the last two steps to determine the y coordinate of the centroid. The y coordinates of the centroids of the parts are shown in Fig. b. Using the information summarized in Table 2, we obtain

$$\bar{y} = \frac{\bar{y}_1 A_1 + \bar{y}_2 A_2 + \bar{y}_3 A_3}{A_1 + A_2 + A_3}$$

$$= \frac{\left[\frac{1}{3}(2R)\right]\left[\frac{1}{2}b(2R)\right] + R\left[c(2R)\right] + R\left(\frac{1}{2}\pi R^2\right)}{\frac{1}{2}b(2R) + c(2R) + \frac{1}{2}\pi R^2}.$$

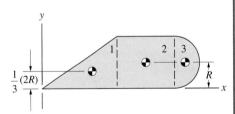

(b) The y coordinates of the centroids of the parts.

Table 2 Information for determining the y coordinate of the centroid

	$\bar{y}_i$	A_i	$\bar{y}_i A_i$
Part 1 (triangle)	$\frac{1}{3}(2R)$	$\frac{1}{2}b(2R)$	$\left[\frac{1}{3}(2R)\right]\left[\frac{1}{2}b(2R)\right]$
Part 2 (rectangle)	R	$c(2R)$	$R\left[c(2R)\right]$
Part 3 (semicircle)	R	$\frac{1}{2}\pi R^2$	$R\left(\frac{1}{2}\pi R^2\right)$

Example 7.4

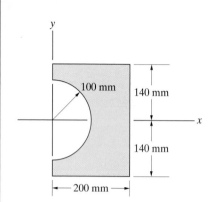

Figure 7.9

Centroid of an Area with a Cutout

Determine the centroid of the area in Fig. 7.9.

Solution

Choose the Parts We will treat the area as a composite area consisting of the rectangle without the semicircular cutout and the area of the cutout, which we call parts 1 and 2, respectively (Fig. a).

Determine the Values for the Parts From Appendix B, the x coordinate of the centroid of the cutout is

$$\bar{x}_2 = \frac{4R}{3\pi} = \frac{4(100)}{3\pi} \text{ mm.}$$

The information for determining the x coordinate of the centroid is summarized in the Table. Notice that we treat the cutout as a negative area.

Table Information for determining $\bar{x}$

	$\bar{x}_i(\text{mm})$	$A_i(\text{mm}^2)$	$\bar{x}_i A_i(\text{mm}^3)$
Part 1 (rectangle)	100	$(200)(280)$	$(100)\big[(200)(280)\big]$
Part 2 (cutout)	$\dfrac{4(100)}{3\pi}$	$-\tfrac{1}{2}\pi(100)^2$	$-\dfrac{4(100)}{3\pi}\big[\tfrac{1}{2}\pi(100)^2\big]$

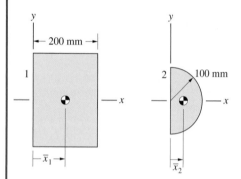

(a) The rectangle and the semicircular cutout.

Calculate the Centroid The x coordinate of the centroid is

$$\bar{x} = \frac{\bar{x}_1 A_1 + \bar{x}_2 A_2}{A_1 + A_2} = \frac{(100)\big[(200)(280)\big] - \dfrac{4(100)}{3\pi}\big[\tfrac{1}{2}\pi(100)^2\big]}{(200)(280) - \tfrac{1}{2}\pi(100)^2} = 122 \text{ mm}$$

Because of the symmetry of the area, $\bar{y} = 0$.

Problems

For Problems 7.27–7.36, determine the coordinates of the centroids.

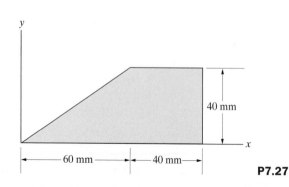

P7.27

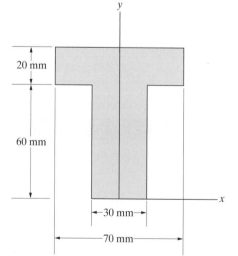

P7.28

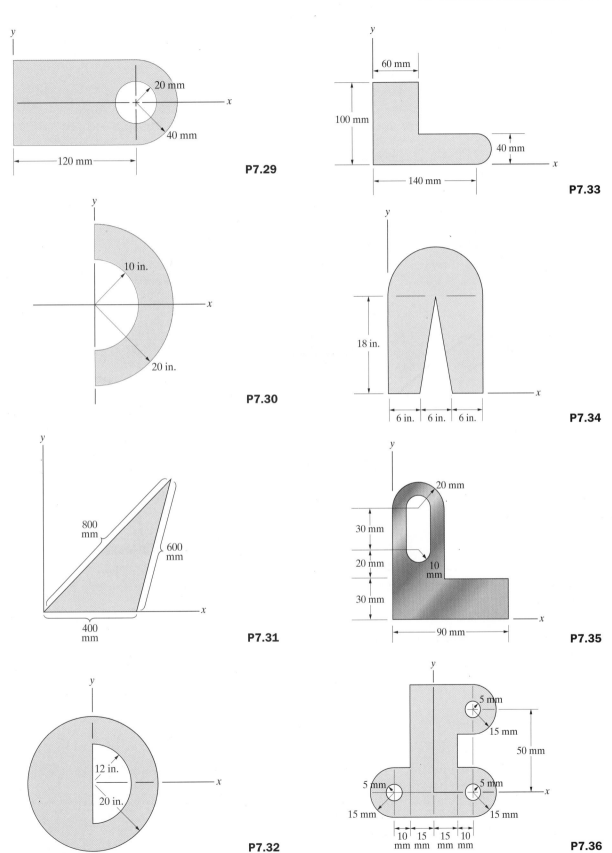

P7.29

P7.30

P7.31

P7.32

P7.33

P7.34

P7.35

P7.36

7.37 The dimensions $b = 42$ mm and $h = 22$ mm. Determine the y coordinate of the centroid of the beam's cross section.

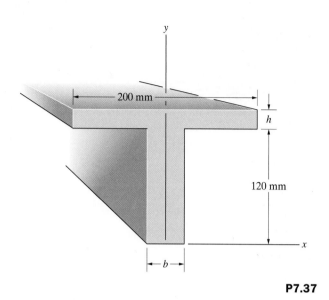

P7.37

7.38 If the cross-sectional area of the beam shown in Problem 7.37 is 8400 mm^2 and the y coordinate of the centroid of the area is $\bar{y} = 90$ mm, what are the dimensions b and h?

7.39 Determine the x coordinate of the centroid of the Boeing 747's vertical stabilizer.

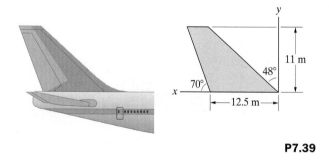

P7.39

7.40 Determine the y coordinate of the centroid of the vertical stabilizer in Problem 7.39.

7.41 The area has elliptical boundaries. If $a = 30$ mm, $b = 15$ mm, and $\varepsilon = 6$ mm, what is the x coordinate of the centroid of the area?

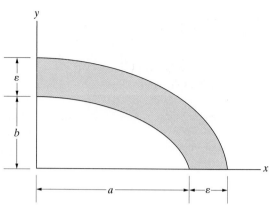

P7.41

7.42 By determining the x coordinate of the centroid of the area shown in Problem 7.41 in terms of a, b, and ε, and evaluating its limit as $\varepsilon \to 0$, show that the x coordinate of the centroid of a quarter-elliptical line is

$$\bar{x} = \frac{4a(a + 2b)}{3\pi(a + b)}.$$

7.43 Three sails of a New York pilot schooner are shown. The coordinates of the points are in feet. Determine the centroid of sail 1.

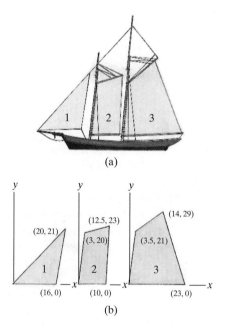

P7.43

7.44 Determine the centroid of sail 2 in Problem 7.43.

7.45 Determine the centroid of sail 3 in Problem 7.43.

7.4 Distributed Loads

The load exerted on a beam (stringer) supporting a floor of a building is distributed over the beam's length (Fig. 7.10a). The load exerted by wind on a television transmission tower is distributed along the tower's height (Fig. 7.10b). In many engineering applications, loads are continuously distributed along lines. We will show that the concept of the centroid of an area can be useful in the analysis of objects subjected to such loads.

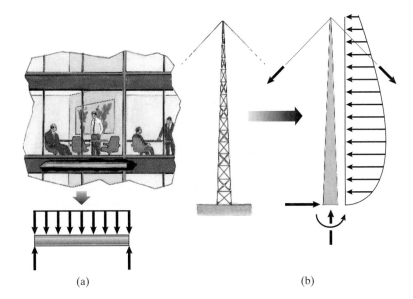

(a)   (b)

Figure 7.10
Examples of distributed forces:
(a) Uniformly distributed load exerted on a beam of a building's frame by the floor.
(b) Wind load distributed along the height of a tower.

Describing a Distributed Load

We can use a simple example to demonstrate how such loads are expressed analytically. Suppose that we pile bags of sand on a beam, as shown in Fig. 7.11a. You can see that the load exerted by the bags is distributed over the length of the beam and that its magnitude at a given position x depends on how high the bags are piled at that position. To describe the load, we define a function w such that the *downward* force exerted on an infinitesimal element dx of the beam is $w\,dx$. With this function we can model the varying magnitude of the load exerted by the sand bags (Fig. 7.11b). The arrows in the figure indicate that the load acts in the downward direction. Loads distributed along lines, from simple examples such as a beam's own weight to complicated ones such as the lift distributed along the length of an airplane's wing, are modeled by the function w. Since the product of w, and dx is a force, the dimensions of w are (force)/(length). For example, w, can be expressed in newtons per meter in SI units or in pounds per foot in U.S. Customary units.

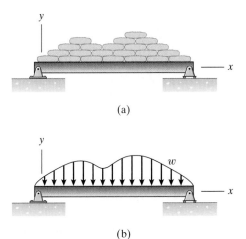

Figure 7.11
(a) Loading a beam with bags of sand.
(b) The distributed load w models the load exerted by the bags.

Determining Force and Moment

Let's assume that the function w describing a particular distributed load is known (Fig. 7.12a). The graph of w, is called the *loading curve*. Since the force

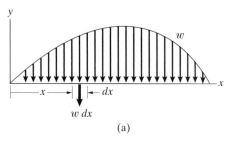

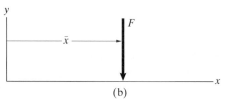

Figure 7.12
(a) A distributed load and the force exerted on a differential element dx.
(b) The equivalent force.

acting on an element dx of the line is $w \, dx$, we can determine the total force F exerted by the distributed load by integrating the loading curve with respect to x:

$$F = \int_L w \, dx. \tag{7.10}$$

We can also integrate to determine the moment about a point exerted by the distributed load. For example, the moment about the origin due to the force exerted on the element dx is $xw \, dx$, so the total moment about the origin due to the distributed load is

$$M = \int_L xw \, dx. \tag{7.11}$$

When you are concerned only with the total force and moment exerted by a distributed load, you can represent it by a single equivalent force F (Fig. 7.12b). For equivalence, the force must act at a position $\bar{x}$ on the x axis such that the moment of F about the origin is equal to the moment of the distributed load about the origin:

$$\bar{x} F = \int_L xw \, dx.$$

Therefore the force F is equivalent to the distributed load if we place it at the position

$$\bar{x} = \frac{\displaystyle\int_L xw \, dx}{\displaystyle\int_L w \, dx}. \tag{7.12}$$

The Area Analogy

Notice that the term $w \, dx$ is equal to an element of "area" dA between the loading curve and the x axis (Fig. 7.13a). (We use quotation marks because $w \, dx$ is actually a force and not an area.) Interpreted in this way, Eq. (7.10) states that the total force exerted by the distributed load is equal to the "area" A between the loading curve and the x axis:

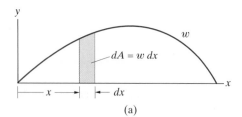

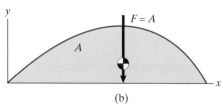

Figure 7.13
(a) Determining the "area" between the function w and the x axis.
(b) The equivalent force is equal to the "area," and the line of action passes through its centroid.

$$F = \int_L w \, dx = \int_A dA = A. \tag{7.13}$$

Substituting $w \, dx = dA$ into Eq. (7.12), we obtain

$$\bar{x} = \frac{\displaystyle\int_L xw \, dx}{\displaystyle\int_L w \, dx} = \frac{\displaystyle\int_A x \, dA}{\displaystyle\int_A dA}. \tag{7.14}$$

The force F is equivalent to the distributed load if it acts at the centroid of the "area" between the loading curve and the x axis (Fig. 7.13b). Using this analogy to represent a distributed load by an equivalent force can be very useful when the loading curve is relatively simple (see Example 7.5).

Study Questions

1. What is the definition of the function w?
2. How is the force exerted by a distributed load determined from the loading curve?
3. How is the moment exerted by a distributed load determined from the loading curve?

Example 7.5

Beam with a Triangular Distributed Load

The beam in Fig. 7.14 is subjected to a "triangular" distributed load whose value at B is 100 N/m.
(a) Represent the distributed load by a single equivalent force.
(b) Determine the reactions at A and B.

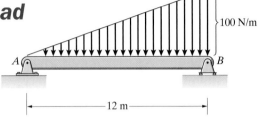

Figure 7.14

Strategy

(a) The magnitude of the force is equal to the "area" under the triangular loading curve, and the equivalent force acts at the centroid of the triangular "area."
(b) Once the distributed load is represented by a single equivalent force, we can apply the equilibrium equations to determine the reactions.

Solution

(a) The "area" of the triangular distributed load is one-half its base times its height, or $\frac{1}{2}(12 \text{ m}) \times (100 \text{ N/m}) = 600$ N. The centroid of the triangular "area" is located at $\bar{x} = \frac{2}{3}(12 \text{ m}) = 8$ m. We can therefore represent the distributed load by an equivalent downward force of 600-N magnitude acting at $x = 8$ m (Fig. a).

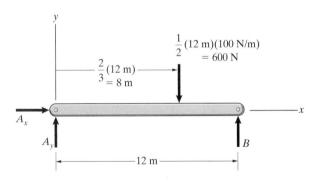

(a) Representing the distributed load by an equivalent force.

(b) From the equilibrium equations

$$\Sigma F_x = A_x = 0,$$
$$\Sigma F_y = A_y + B - 600 = 0,$$
$$\Sigma M_{(\text{point } A)} = 12B - (8)(600) = 0,$$

we obtain $A_x = 0$, $A_y = 200$ N, and $B = 400$ N.

Discussion

The loading curve in this example was sufficiently simple that we did not need to integrate to determine its area and centroid. In the following example we must integrate to determine the area and centroid.

Example 7.6

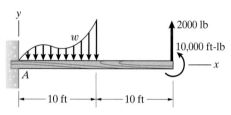

Figure 7.15

Beam with a Distributed Load

The beam in Fig. 7.15 is subjected to a distributed load, a force, and a couple. The distributed load is $w = 300x - 50x^2 + 0.3x^4$ lb/ft.
(a) Represent the distributed load by a single equivalent force.
(b) Determine the reactions at the built-in support A.

Strategy

(a) Since we know the function w, we can use Eq. (7.13) to determine the "area" under the loading curve, which is equal to the total force exerted by the distributed load. The x coordinate of the centroid is given by Eq. (7.14).
(b) Once the distributed load is represented by a single equivalent force, we can apply the equilibrium equations to determine the reactions at the built-in support.

Solution

(a) The downward force exerted by the distributed load is

$$F = \int_L w \, dx = \int_0^{10} (300x - 50x^2 + 0.3x^4) \, dx = 4330 \text{ lb}.$$

The x coordinate of the centroid of the distributed load is

$$\bar{x} = \frac{\int_L xw \, dx}{\int_L w \, dx} = \frac{\int_0^{10} x(300x - 50x^2 + 0.3x^4) \, dx}{\int_0^{10} (300x - 50x^2 + 0.3x^4) \, dx}$$

$$= \frac{25{,}000}{4330} = 5.77 \text{ ft}.$$

The distributed load is equivalent to a downward force of 4330-lb magnitude acting at $x = 5.77$ ft.
(b) In Fig. a, we draw the free-body diagram of the beam with the distributed force represented by the single equivalent force. From the equilibrium equations

$$\Sigma F_x = A_x = 0,$$
$$\Sigma F_y = A_y + 2000 - 4330 = 0,$$
$$\Sigma M_{(\text{point } A)} = (20)(2000) + 10{,}000 - (5.77)(4330) + M_A = 0,$$

we obtain $A_x = 0$, $A_y = 2330$ lb, and $M_A = -25{,}000$ ft-lb.

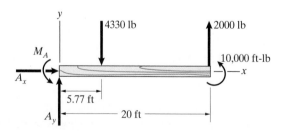

(a) Free-body diagram of the beam.

Example 7.7

Beam Subjected to Distributed Loads

The beam in Fig. 7.16 is subjected to two distributed loads. Determine the reactions at A and B.

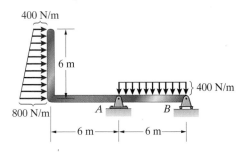

Figure 7.16

Strategy

We can easily represent the uniform distributed load on the right by an equivalent force. We can treat the distributed load on the left as the sum of uniform and triangular distributed loads and represent each load by an equivalent force.

Solution

We draw the free-body diagram of the beam in Fig. a, expressing the left distributed load as the sum of uniform and triangular loads. In Fig. b, we represent the three distributed loads by equivalent forces. The "area" of the uniform distributed load on the right is $(6 \text{ m}) \times (400 \text{ N/m}) = 2400 \text{ N}$, and its centroid is 3 m from B. The area of the uniform distributed load on the vertical part of the beam is $(6 \text{ m}) \times (400 \text{ N/m}) = 2400 \text{ N}$, and its centroid is located at $y = 3$ m. The area of the triangular distributed load is $\frac{1}{2}(6 \text{ m}) \times (400 \text{ N/m}) = 1200 \text{ N}$, and its centroid is located at $y = \frac{1}{3}(6 \text{ m}) = 2$ m.

From the equilibrium equations

$$\Sigma F_x = A_x + 1200 + 2400 = 0,$$
$$\Sigma F_y = A_y + B - 2400 = 0,$$
$$\Sigma M_{(\text{point } A)} = 6B - (3)(2400) - (2)(1200) - (3)(2400) = 0,$$

we obtain $A_x = -3600$ N, $A_y = -400$ N, and $B = 2800$ N.

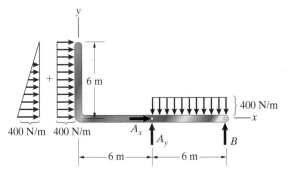

(a) Free-body diagram of the beam.

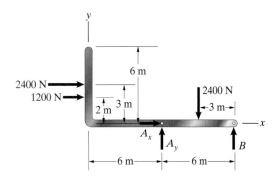

(b) Representing the distributed loads by equivalent forces.

Problems

7.46 The value of the distributed load w at $x = 6$ m is 240 N/m.
(a) The equation for the loading curve is $w = 40x$ N/m. Use Eq. (7.10) to determine the magnitude of the total force exerted on the beam by the distributed load.
(b) If you use the area analogy to represent the distributed load by an equivalent force, what is the magnitude of the force and where does it act?
(c) Determine the reactions at A and B.

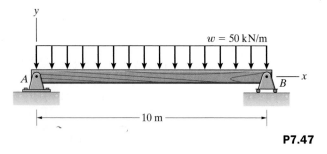

P7.46

7.47 In a preliminary design study for a pedestrian bridge, an engineer models the combined weight of the bridge and maximum expected load due to traffic by the distributed load shown.
(a) Use Eq. (7.10) to determine the magnitude of the total force exerted on the bridge by the distributed load.
(b) If you use the area analogy to represent the distributed load by an equivalent force, what is the magnitude of the force and where does it act?
(c) Determine the reactions at A and B.

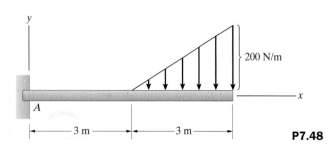

P7.47

7.48 Determine the reactions at the built-in support A.

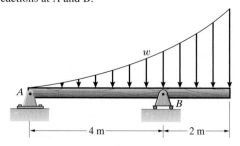

P7.48

7.49 Determine the reactions at A and B.

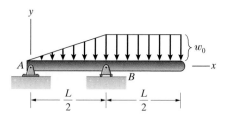

P7.49

7.50 Determine the reactions at the built-in support A.

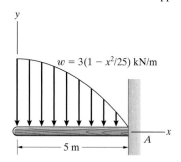

P7.50

7.51 An engineer measures the forces exerted by the soil on a 10-m section of a building foundation and finds that they are described by the distributed load $w = -10x - x^2 + 0.2x^3$ kN/m.
(a) Determine the magnitude of the total force exerted on the foundation by the distributed load.
(b) Determine the magnitude of the moment about A due to the distributed load.

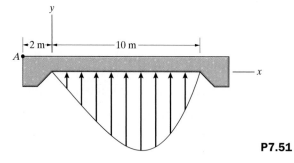

P7.51

7.52 The distributed load is $w = 6x + 0.4x^3$ N/m. Determine the reactions at A and B.

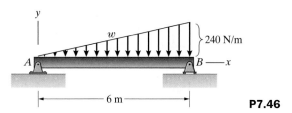

P7.52

7.53 The aerodynamic lift of the wing is described by the distributed load $w = -300\sqrt{1 - 0.04x^2}$ N/m. The mass of the wing is 27 kg, and its center of mass is located 2 m from the wing root R.
(a) Determine the magnitudes of the force and the moment about R exerted by the lift of the wing.
(b) Determine the reactions on the wing at R.

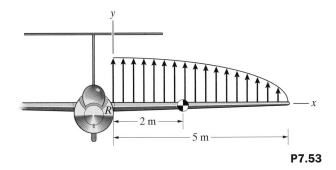

P7.53

7.54 The force $F = 2000$ lb. Determine the reactions at A and B.

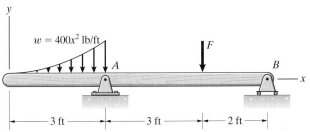

P7.54

7.55 Determine the reactions at A and B.

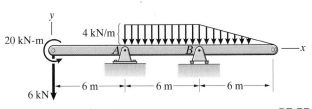

P7.55

7.56 Determine the reactions on member AB at A and B.

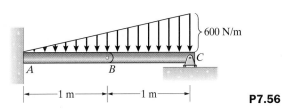

P7.56

7.57 Determine the reactions on member $ABCD$ at A and D.

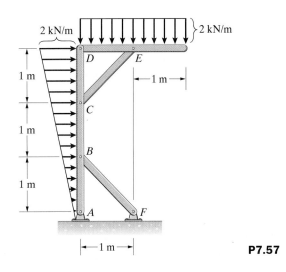

P7.57

7.58 Determine the forces on member ABC of the frame.

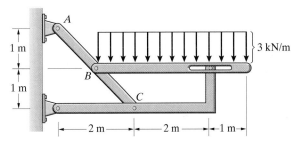

P7.58

| 7.5 | **Centroids of Volumes and Lines** |

Here we define the centroids, or average positions, of volumes and lines, and show how to determine the centroids of composite volumes and lines. We will show in Section 7.6 that knowing the centroids of volumes and lines allows you to determine the centers of mass of certain types of objects, which tells you where their weights effectively act.

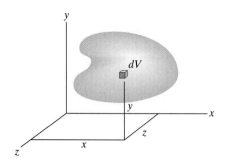

Figure 7.17
A volume V and differential element dV.

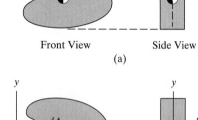

Front View   Side View

(a)

(b)

Figure 7.18
(a) A volume of uniform thickness.
(b) Obtaining dV by projecting dA through the volume.

Definitions

Volumes   Consider a volume V, and let dV be a differential element of V with coordinates x, y, and z (Fig. 7.17). By analogy with Eqs. (7.6) and (7.7), the coordinates of the centroid of V are

$$\bar{x} = \frac{\displaystyle\int_V x\, dV}{\displaystyle\int_V dV}, \qquad \bar{y} = \frac{\displaystyle\int_V y\, dV}{\displaystyle\int_V dV}, \qquad \bar{z} = \frac{\displaystyle\int_V z\, dV}{\displaystyle\int_V dV}. \tag{7.15}$$

The subscript V on the integral signs means that the integration is carried out over the entire volume.

If a volume has the form of a plate with uniform thickness and cross-sectional area A (Fig. 7.18a), its centroid coincides with the centroid of A and lies at the midpoint between the two faces. To show that this is true, we obtain a volume element dV by projecting an element dA of the cross-sectional area through the thickness T of the volume, so that $dV = T\, dA$ (Fig. 7.18b). Then the x and y coordinates of the centroid of the volume are

$$\bar{x} = \frac{\displaystyle\int_V x\, dV}{\displaystyle\int_V dV} = \frac{\displaystyle\int_A xT\, dA}{\displaystyle\int_A T\, dA} = \frac{\displaystyle\int_A x\, dA}{\displaystyle\int_A dA},$$

$$\bar{y} = \frac{\displaystyle\int_V y\, dV}{\displaystyle\int_V dV} = \frac{\displaystyle\int_A yT\, dA}{\displaystyle\int_A T\, dA} = \frac{\displaystyle\int_A y\, dA}{\displaystyle\int_A dA}.$$

The coordinate $\bar{z} = 0$ by symmetry. Thus you know the centroid of this type of volume if you know (or can determine) the centroid of its cross-sectional area.

Lines   The coordinates of the centroid of a line L are

$$\bar{x} = \frac{\displaystyle\int_L x\, dL}{\displaystyle\int_L dL}, \qquad \bar{y} = \frac{\displaystyle\int_L y\, dL}{\displaystyle\int_L dL}, \qquad \bar{z} = \frac{\displaystyle\int_L z\, dL}{\displaystyle\int_L dL}, \tag{7.16}$$

where dL is a differential length of the line with coordinates x, y, and z. (Fig. 7.19).

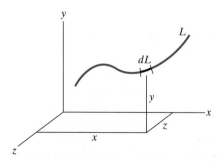

Figure 7.19
A line L and differential element dL.

Example 7.8

Centroid of a Cone by Integration

Determine the centroid of the cone in Fig. 7.20.

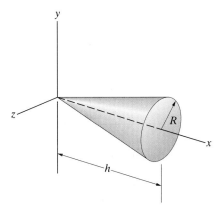

Figure 7.20

Strategy

The centroid must lie on the x axis because of symmetry. We will determine its x coordinate by using an element of volume dV in the form of a "disk" of width dx.

Solution

Let dV be the disk in Fig. a. The radius of the disk is $(R/h)x$ (Fig. b), and its volume equals the product of the area of the disk and its thickness, $dV = \pi\left[(R/h)x\right]^2 dx$. To integrate over the entire volume, we must integrate with respect to x from $x = 0$ to $x = h$. The x coordinate of the centroid is

$$\bar{x} = \frac{\displaystyle\int_V x\, dV}{\displaystyle\int_V dV} = \frac{\displaystyle\int_0^h x\pi\, \frac{R^2}{h^2}\, x^2\, dx}{\displaystyle\int_0^h \pi\, \frac{R^2}{h^2}\, x^2\, dx} = \frac{3}{4}\, h.$$

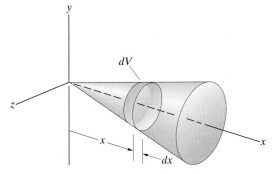

(a) An element dV in the form of a disk.

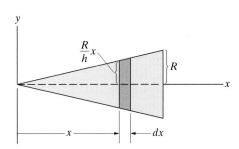

(b) The radius of the element is $(R/h)x$.

Centroid of a Line by Integration

The line L in Fig. 7.21 is defined by the function $y = x^2$. Determine the x coordinate of its centroid.

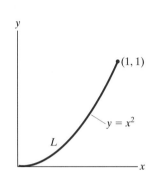

Figure 7.21

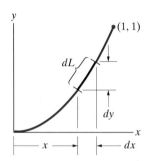

(a) A differential line element dL.

Solution

We can express a differential element dL of the line (Fig. a) in terms of dx and dy:

$$dL = \sqrt{dx^2 + dy^2} = \sqrt{1 + \left(\frac{dy}{dx}\right)^2}\, dx.$$

From the equation describing the line, the derivative $dy/dx = 2x$, so we obtain an expression for dL in terms of x:

$$dL = \sqrt{1 + 4x^2}\, dx.$$

To integrate over the entire line, we must integrate from $x = 0$ to $x = 1$. The x coordinate of the centroid is

$$\bar{x} = \frac{\displaystyle\int_L x\, dL}{\displaystyle\int_L dL} = \frac{\displaystyle\int_0^1 x\sqrt{1 + 4x^2}\, dx}{\displaystyle\int_0^1 \sqrt{1 + 4x^2}\, dx} = 0.574.$$

Centroid of a Semicircular Line by Integration

Determine the centroid of the semicircular line in Fig. 7.22.

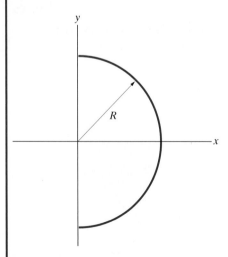

Figure 7.22

Strategy

Because of the symmetry of the line, the centroid lies on the x axis. To determine $\bar{x}$, we will integrate in terms of polar coordinates.

Solution

By letting θ change by an amount $d\theta$, we obtain a differential line element of length $dL = R\, d\theta$ (Fig. a). The x coordinate of dL is $x = R\cos\theta$. To integrate over the entire line, we must integrate with respect to θ from $\theta = -\pi/2$ to $\theta = +\pi/2$:

$$\bar{x} = \frac{\displaystyle\int_L x\,dL}{\displaystyle\int_L dL} = \frac{\displaystyle\int_{-\pi/2}^{\pi/2} (R\cos\theta)R\,d\theta}{\displaystyle\int_{-\pi/2}^{\pi/2} R\,d\theta} = \frac{R^2[\,\sin\theta\,]_{-\pi/2}^{\pi/2}}{R[\theta]_{-\pi/2}^{\pi/2}} = \frac{2R}{\pi}.$$

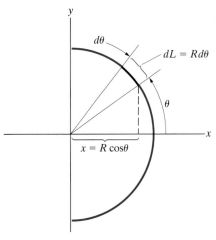

Discussion

Notice that our integration procedure gives the correct length of the line:

$$\int_L dL = \int_{-\pi/2}^{\pi/2} R\,d\theta = R[\theta]_{-\pi/2}^{\pi/2} = \pi R.$$

(a) A differential line element $dL = R\,d\theta$.

Problems

7.59 Determine the coordinates of the centroid of the truncated conical volume.

    Strategy: Use the method described in Example 7.8.

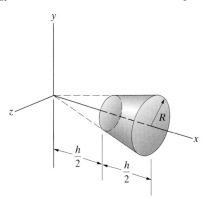

P7.59

7.60 A grain storage tank has the form of a surface of revolution with the profile shown. The height of the tank is 7 m and its diameter at ground level is 10 m. Determine the volume of the tank and the height *above ground level* of the centroid of its volume.

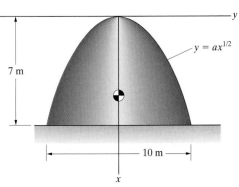

P7.60

7.61 The object shown, designed to serve as a pedestal for a speaker, has a profile obtained by revolving the curve $y = 0.167x^2$ about the x axis. What is the x coordinate of the centroid of the object?

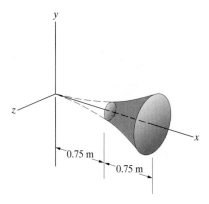

P7.61

7.62 Determine the volume and centroid of the pyramid.

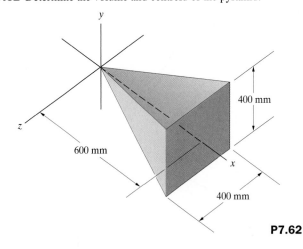

P7.62

7.63 Determine the centroid of the hemispherical volume.

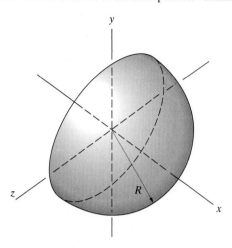

P7.63

7.64 The volume consists of a segment of a sphere of radius R. Determine its centroid.

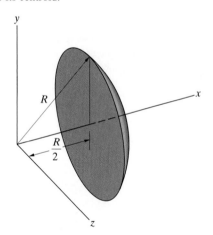

P7.64

7.65 A volume of revolution is obtained by revolving the curve $x^2/a^2 + y^2/b^2 = 1$ about the x axis. Determine its centroid.

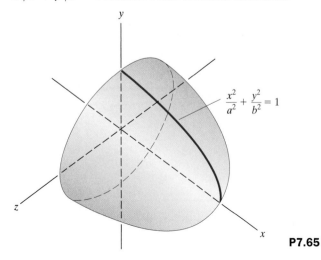

P7.65

7.66 The volume of revolution has a cylindrical hole of radius R. Determine its centroid.

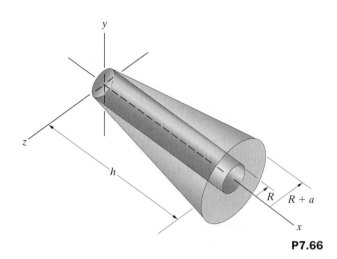

P7.66

7.67 Determine the y coordinate of the centroid of the line (see Example 7.9).

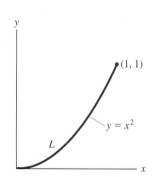

P7.67

7.68 Determine the x coordinate of the centroid of the line.

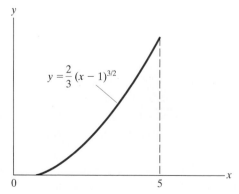

P7.68

7.69 Determine the x coordinate of the centroid of the line.

7.70 Determine the centroid of the circular arc.

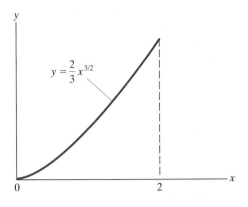

P7.69

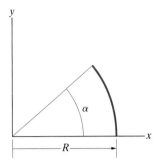

P7.70

Centroids of Composite Volumes and Lines

The centroids of composite volumes and lines can be derived using the same approach we applied to areas. The coordinates of the centroid of a composite volume are

$$\bar{x} = \frac{\sum_i \bar{x}_i V_i}{\sum_i V_i}, \qquad \bar{y} = \frac{\sum_i \bar{y}_i V_i}{\sum_i V_i}, \qquad \bar{z} = \frac{\sum_i \bar{z}_i V_i}{\sum_i V_i}, \qquad (7.17)$$

and the coordinates of the centroid of a composite line are

$$\bar{x} = \frac{\sum_i \bar{x}_i L_i}{\sum_i L_i}, \qquad \bar{y} = \frac{\sum_i \bar{y}_i L_i}{\sum_i L_i}, \qquad \bar{z} = \frac{\sum_i \bar{z}_i L_i}{\sum_i L_i}. \qquad (7.18)$$

The centroids of some simple volumes and lines are tabulated in Appendices B and C.

Determining the centroid of a composite volume or line requires three steps:

1. **Choose the parts**—Try to divide the composite into parts whose centroids you know or can easily determine.
2. **Determine the values for the parts**—Determine the centroid and the volume or length of each part. Watch for instances of symmetry that can simplify your task.
3. **Calculate the centroid**—Use Eqs. (7.17) or (7.18) to determine the centroid of the composite volume or line.

Example 7.11

Centroid of a Composite Volume

Determine the centroid of the volume in Fig. 7.23.

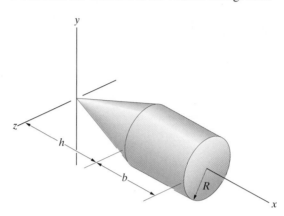

Figure 7.23

Solution

Choose the Parts   The volume consists of a cone and a cylinder, which we call parts 1 and 2, respectively.

Determine the Values for the Parts   The centroid and volume of the cone are given in Appendix C. The x coordinates of the centroids of the parts are shown in Fig. a, and the information for determining the x coordinate of the centroid is summarized in the Table.

Table Information for determining $\bar{x}$

	$\bar{x}_i$	V_i	$\bar{x}_i V_i$
Part 1 (cone)	$\frac{3}{4}h$	$\frac{1}{3}\pi R^2 h$	$\left(\frac{4}{3}h\right)\left(\frac{1}{3}\pi R^2 h\right)$
Part 2 (cylinder)	$h + \frac{1}{2}b$	$\pi R^2 b$	$\left(h + \frac{1}{2}b\right)\left(\pi R^2 b\right)$

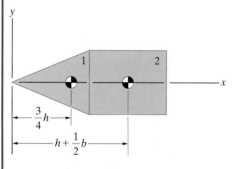

(a) The x coordinates of the centroids of the cone and cylinder.

Calculate the Centroid   The x coordinate of the centroid of the composite volume is

$$\bar{x} = \frac{\bar{x}_1 V_1 + \bar{x}_2 V_2}{V_1 + V_2} = \frac{\left(\frac{3}{4}h\right)\left(\frac{1}{3}\pi R^2 h\right) + \left(h + \frac{1}{2}b\right)\left(\pi R^2 b\right)}{\frac{1}{3}\pi R^2 h + \pi R^2 b}.$$

Because of symmetry, $\bar{y} = 0$ and $\bar{z} = 0$.

Example 7.12

Centroid of a Volume Containing a Cutout

Determine the centroid of the volume in Fig. 7.24.

Solution

Choose the Parts   We can divide the volume into the five simple parts shown in Fig. a. Part 5 is the volume of the 20-mm-diameter hole.

Determine the Values for the Parts The centroids of parts 1 and 3 are located at the centroids of their semicircular cross sections (Fig. b). The information for determining the x coordinate of the centroid is summarized in the Table. Part 5 is a negative volume.

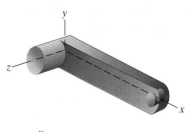

Table Information for determining $\bar{x}$.

	$\bar{x}_i\,(\text{mm})$	$V_i\,(\text{mm}^3)$	$\bar{x}_i V_i\,(\text{mm}^4)$
Part 1	$-\dfrac{4(25)}{3\pi}$	$\dfrac{\pi(25)^2}{2}(20)$	$\left[-\dfrac{4(25)}{3\pi}\right]\left[\dfrac{\pi(25)^2}{2}(20)\right]$
Part 2	100	$(200)(50)(20)$	$(100)\big[(200)(50)(20)\big]$
Part 3	$200 + \dfrac{4(25)}{3\pi}$	$\dfrac{\pi(25)^2}{2}(20)$	$\left[200 + \dfrac{4(25)}{3\pi}\right]\left[\dfrac{\pi(25)^2}{2}(20)\right]$
Part 4	0	$\pi(25)^2(40)$	0
Part 5	200	$-\pi(10)^2(20)$	$-(200)\big[\pi(10)^2(20)\big]$

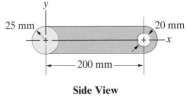

Side View

Calculate the Centroid The x coordinate of the centroid of the composite volume is

$$\bar{x} = \frac{\bar{x}_1 V_1 + \bar{x}_2 V_2 + \bar{x}_3 V_3 + \bar{x}_4 V_4 + \bar{x}_5 V_5}{V_1 + V_2 + V_3 + V_4 + V_5}$$

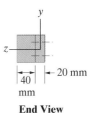

End View

Figure 7.24

$$= \frac{\left[-\dfrac{4(25)}{3\pi}\right]\left[\dfrac{\pi(25)^2}{2}(20)\right] + (100)\big[(200)(50)(20)\big] + \left[200 + \dfrac{4(25)}{3\pi}\right]\left[\dfrac{\pi(25)^2}{2}(20)\right] + 0 - (200)\big[\pi(10)^2(20)\big]}{\dfrac{\pi(25)^2}{2}(20) + (200)(50)(20) + \dfrac{\pi(25)^2}{2}(20) + \pi(25)^2(40) - \pi(10)^2(20)}$$

$$= 72.77 \text{ mm.}$$

The z coordinates of the centroids of the parts are zero except $\bar{z}_4 = 30$ mm. Therefore the z coordinate of the centroid of the composite volume is

$$\bar{z} = \frac{\bar{z}_4 V_4}{V_1 + V_2 + V_3 + V_4 + V_5}$$

$$= \frac{30\big[\pi(25)^2(40)\big]}{\dfrac{\pi(25)^2}{2}(20) + (200)(50)(20) + \dfrac{\pi(25)^2}{2}(20) + \pi(25)^2(40) - \pi(10)^2(20)}$$

$$= 7.56 \text{ mm.}$$

Because of symmetry, $\bar{y} = 0$.

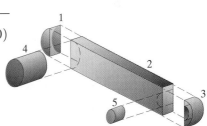

(a) Dividing the volume into five parts.

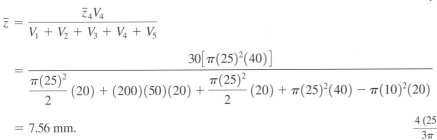

(b) Positions of the centroids of parts 1 and 3.

Example 7.13

Centroid of a Composite Line

Determine the centroid of the line in Fig. 7.25. The quarter-circular arc lies in the y–z plane.

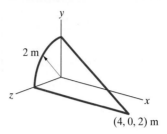

Figure 7.25

Solution

Choose the Parts The line consists of a quarter-circular arc and two straight segments, which we call parts 1, 2, and 3 (Fig. a).

Determine the Values for the Parts From Appendix B, the coordinates of the centroid of the quarter-circular arc are $\bar{x}_1 = 0$, $\bar{y}_1 = \bar{z}_1 = 2(2)/\pi$ m. The centroids of the straight segments lie at their midpoints. For segment 2, $\bar{x}_2 = 2$ m, $\bar{y}_2 = 0$, and $\bar{z}_2 = 2$ m, and for segment 3, $\bar{x}_3 = 2$ m, $\bar{y}_3 = 1$ m, and $\bar{z}_3 = 1$ m. The length of segment 3 is $L_3 = \sqrt{(4)^2 + (2)^2 + (2)^2} = 4.90$ m. This information is summarized in the Table.

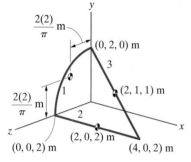

(a) Dividing the line into three parts.

Table Information for determining the centroid.

	$\bar{x}_i$	$\bar{y}_i$	$\bar{z}_i$	L_i
Part 1	0	$2(2)/\pi$	$2(2)/\pi$	$\pi(2)/2$
Part 2	2	0	2	4
Part 3	2	1	1	4.90

Calculate the Centroid The coordinates of the centroid of the composite line are

$$\bar{x} = \frac{\bar{x}_1 L_1 + \bar{x}_2 L_2 + \bar{x}_3 L_3}{L_1 + L_2 + L_3} = \frac{0 + (2)(4) + (2)(4.90)}{\pi + 4 + 4.90} = 1.478 \text{ m},$$

$$\bar{y} = \frac{\bar{y}_1 L_1 + \bar{y}_2 L_2 + \bar{y}_3 L_3}{L_1 + L_2 + L_3} = \frac{[2(2)/\pi][\pi(2)/2] + 0 + (1)(4.90)}{\pi + 4 + 4.90} = 0.739 \text{ m},$$

$$\bar{z} = \frac{\bar{z}_1 L_1 + \bar{z}_2 L_2 + \bar{z}_3 L_3}{L_1 + L_2 + L_3} = \frac{[2(2)/\pi][\pi(2)/2] + (2)(4) + (1)(4.90)}{\pi + 4 + 4.90} = 1.404 \text{ m}.$$

Problems

For Problems 7.71–7.78, determine the centroids of the volumes.

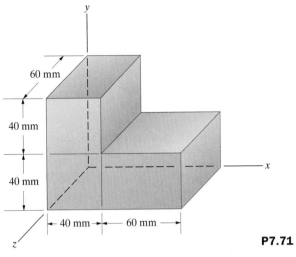

P7.71

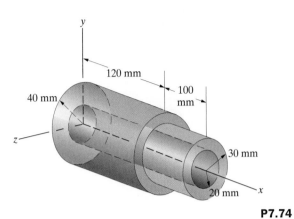

P7.74

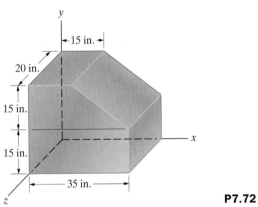

P7.72

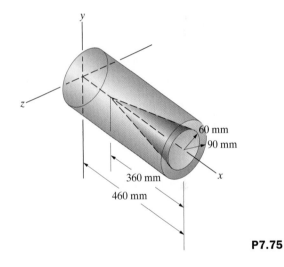

P7.75

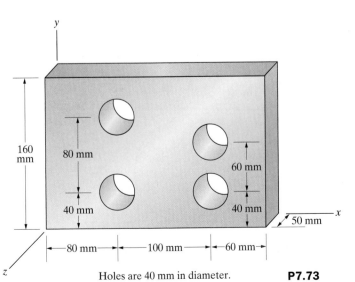

Holes are 40 mm in diameter.   P7.73

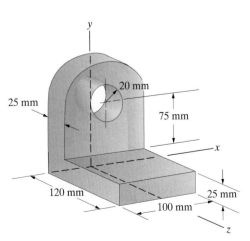

P7.76

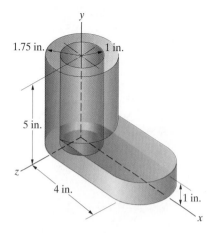

P7.77

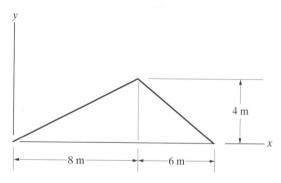

P7.78

7.79 The dimensions of the *Gemini* spacecraft (in meters) are $a = 0.70$, $b = 0.88$, $c = 0.74$, $d = 0.98$, $e = 1.82$, $f = 2.20$, $g = 2.24$, and $h = 2.98$. Determine the centroid of its volume.

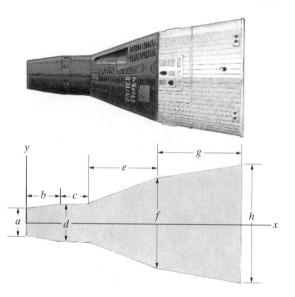

P7.79

7.80 Two views of a machine element are shown. Determine the centroid of its volume.

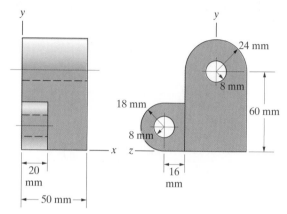

P7.80

For Problems 7.81–7.83, determine the centroids of the lines.

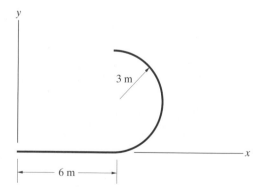

P7.81

P7.82

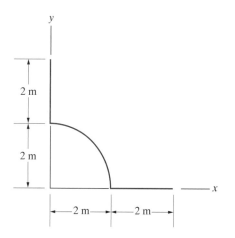

P7.83

7.84 The semicircular part of the line lies in the x–z plane. Determine the centroid of the line.

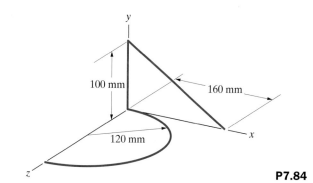

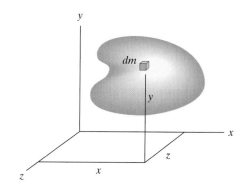

P7.84

7.6 **Centers of Mass**

7.6 Centers of Mass

The *center of mass* of an object is the centroid, or average position, of its mass. We give the analytical definition of the center of mass and demonstrate one of its most important properties: An object's weight can be represented by a single equivalent force acting at its center of mass. We then discuss how to locate centers of mass and show that for particular classes of objects, the center of mass coincides with the centroid of a volume, area, or line.

The center of mass of an object is defined by

$$\bar{x} = \frac{\int_m x \, dm}{\int_m dm}, \qquad \bar{y} = \frac{\int_m y \, dm}{\int_m dm}, \qquad \bar{z} = \frac{\int_m z \, dm}{\int_m dm}, \qquad (7.19)$$

where x, y, and z are the coordinates of the differential element of mass dm (Fig. 7.26). The subscripts m indicate that the integration must be carried out over the entire mass of the object.

Before considering how to determine the center of mass of an object, we will demonstrate that the weight of an object can be represented by a single equivalent force acting at its center of mass. Consider an element of mass dm of an object (Fig. 7.27a). If the y axis of the coordinate system points upward, the weight of dm is $-dm \, g\mathbf{j}$. Integrating this expression over the mass m, we obtain the total weight of the object,

$$\int_m - g\mathbf{j} \, dm = -mg\mathbf{j} = -W\mathbf{j}.$$

The moment of the weight of the element dm about the origin is

$$(x\mathbf{i} + y\mathbf{j} + z\mathbf{k}) \times (-dm \, g\mathbf{j}) = gz\mathbf{i} \, dm - gx\mathbf{k} \, dm.$$

Figure 7.26
An object and differential element of mass dm.

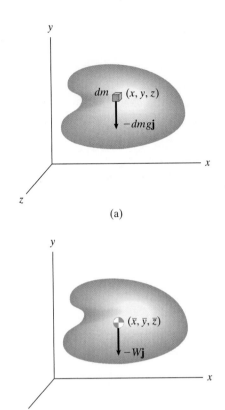

Figure 7.27
(a) Weight of the element dm.
(b) Representing the weight by a single
    force at the center of mass.

Integrating this expression over m, we obtain the total moment about the origin due to the weight of the object:

$$\int_m (gz\mathbf{i}\, dm - gx\mathbf{k}\, dm) = mg\bar{z}\mathbf{i} - mg\bar{x}\mathbf{k} = W\bar{z}\mathbf{i} - W\bar{x}\mathbf{k}.$$

If we represent the weight of the object by the force $-W\mathbf{j}$ acting at the center of mass (Fig. 7.27b), the moment of this force about the origin is equal to the total moment due to the weight:

$$(\bar{x}\mathbf{i} + \bar{y}\mathbf{j} + \bar{z}\mathbf{k}) \times (-W\mathbf{j}) = W\bar{z}\mathbf{i} - W\bar{x}\mathbf{k}.$$

This result shows that when you are concerned only with the total force and total moment exerted by the weight of an object, you can assume that its weight acts at the center of mass.

To apply Eqs. (7.19) to specific objects, we will change the variable of integration from mass to volume by introducing the mass density.

The *mass density* ρ of an object is defined such that the mass of a differential element of its volume is $dm = \rho\, dV$. The dimensions of ρ are therefore (mass)/(volume). For example, it can be expressed in kg/m^3 in SI units or in $slug/ft^3$ in U.S. Customary units. The total mass of an object is

$$m = \int_m dm = \int_V \rho\, dV. \tag{7.20}$$

An object whose mass density is uniform throughout its volume is said to be *homogeneous*. In this case, the total mass equals the product of the mass density and the volume:

$$m = \rho \int_V dV = \rho V. \qquad \textbf{Homogeneous object} \tag{7.21}$$

The *weight density* $\gamma = g\rho$. It can be expressed in N/m^3 in SI units or in lb/ft^3 in U.S. Customary units. The weight of an element of volume dV of an object is $dW = \gamma\, dV$, and the total weight of a homogeneous object equals γV.

By substituting $dm = \rho\, dV$ into Eqs. (7.19), we can express the coordinates of the center of mass in terms of volume integrals:

$$\bar{x} = \frac{\int_V \rho x\, dV}{\int_V \rho\, dV}, \qquad \bar{y} = \frac{\int_V \rho y\, dV}{\int_V \rho\, dV}, \qquad \bar{z} = \frac{\int_V \rho z\, dV}{\int_V \rho\, dV}. \tag{7.22}$$

If ρ is known as a function of position in an object, these integrals determine its center of mass. Furthermore, we can use them to show that the centers of mass of particular classes of objects coincide with centroids of volumes, areas, and lines:

- **The center of mass of a homogeneous object coincides with the centroid of its volume.** If an object is homogeneous, $\rho = $ constant and Eqs. (7.22) become the equations for the centroid of the volume,

$$\bar{x} = \frac{\displaystyle\int_V x\, dV}{\displaystyle\int_V dV}, \qquad \bar{y} = \frac{\displaystyle\int_V y\, dV}{\displaystyle\int_V dV}, \qquad \bar{z} = \frac{\displaystyle\int_V z\, dV}{\displaystyle\int_V dV}.$$

- **The center of mass of a homogeneous plate of uniform thickness coincides with the centroid of its cross-sectional area** (Fig. 7.28). The center of mass of the plate coincides with the centroid of its volume, and we showed in Section 7.5 that the centroid of the volume of a plate of uniform thickness coincides with the centroid of its cross-sectional area.

- **The center of mass of a homogeneous slender bar of uniform cross-sectional area coincides approximately with the centroid of the axis of the bar** (Fig. 7.29a). The axis of the bar is defined to be the line through the centroid of its cross section. Let $dm = \rho A\, dL$, where A is the cross-sectional area of the bar and dL is a differential element of length of its axis (Fig. 7.29b). If we substitute this expression into Eqs. (7.22), they become the equations for the centroid of the axis:

$$\bar{x} = \frac{\displaystyle\int_L x\, dL}{\displaystyle\int_L dL}, \qquad \bar{y} = \frac{\displaystyle\int_L y\, dL}{\displaystyle\int_L dL}, \qquad \bar{z} = \frac{\displaystyle\int_L z\, dL}{\displaystyle\int_L dL}.$$

This result is approximate because the center of mass of the element dm does not coincide with the centroid of the cross section in regions where the bar is curved.

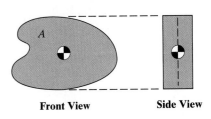

Front View   **Side View**

Figure 7.28
A plate of uniform thickness.

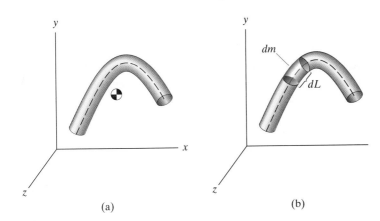

(a)   (b)

Figure 7.29
(a) A slender bar and the centroid of its axis.
(b) The element dm.

Study Questions

1. If you want to represent the weight of an object as a single equivalent force, at what point must the force act?
2. How is the mass density of an object defined?
3. What is the relationship between the mass density ρ and the weight density γ?
4. If an object is homogeneous, what do you know about the position of its center of mass?

Example 7.14

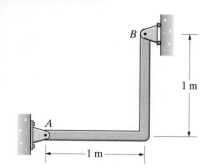

B

$1\ \text{m}$

A

$1\ \text{m}$

Figure 7.30

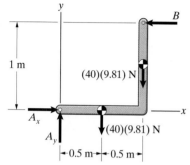

y

B

$1\ \text{m}$

$(40)(9.81)\ \text{N}$

A_x

A_y

$(40)(9.81)\ \text{N}$

x

$0.5\ \text{m}$ $0.5\ \text{m}$

(a) Placing the weights of the straight segments at their centers of mass.

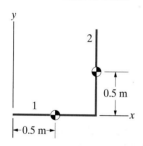

y

2

$0.5\ \text{m}$

1

x

$0.5\ \text{m}$

(b) Centroids of the straight segments of the axis.

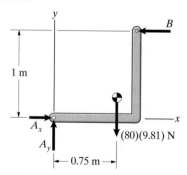

y

B

$1\ \text{m}$

A_x

A_y

$(80)(9.81)\ \text{N}$

$0.75\ \text{m}$

x

(c) Placing the weight of the bar at its center of mass.

Representing the Weight of an L-Shaped Bar

The mass of the homogeneous slender bar in Fig. 7.30 is 80 kg. What are the reactions at A and B?

Strategy

We determine the reactions in two ways.

First Method We represent the weight of each straight segment of the bar by a force acting at the center of mass of the segment.

Second Method We determine the center of mass of the bar by determining the centroid of its axis and represent the weight of the bar by a single force acting at the center of mass.

Solution

First Method In the free-body diagram in Fig. a, we place half of the weight of the bar at the center of mass of each straight segment. From the equilibrium equations

$$\Sigma F_x = A_x - B = 0,$$

$$\Sigma F_y = A_y - (40)(9.81) - (40)(9.81) = 0,$$

$$\Sigma M_{(\text{point } A)} = (1)B - (1)(40)(9.81) - (0.5)(40)(9.81) = 0,$$

we obtain $A_x = 589\ \text{N}$, $A_y = 785\ \text{N}$, and $B = 589\ \text{N}$.

Second Method We can treat the centerline of the bar as a composite line composed of two straight segments (Fig. b). The coordinates of the centroid of the composite line are

$$\bar{x} = \frac{\bar{x}_1 L_1 + \bar{x}_2 L_2}{L_1 + L_2} = \frac{(0.5)(1) + (1)(1)}{1 + 1} = 0.75\ \text{m},$$

$$\bar{y} = \frac{\bar{y}_1 L_1 + \bar{y}_2 L_2}{L_1 + L_2} = \frac{(0)(1) + (0.5)(1)}{1 + 1} = 0.25\ \text{m}.$$

In the free-body diagram in Fig. c, we place the weight of the bar at its center of mass. From the equilibrium equations

$$\Sigma F_x = A_x - B = 0,$$

$$\Sigma F_y = A_y - (80)(9.81) = 0$$

$$\Sigma M_{(\text{point } A)} = (1)B - (0.75)(80)(9.81) = 0,$$

we again obtain $A_x = 589\ \text{N}$, $A_y = 785\ \text{N}$, and $B = 589\ \text{N}$.

Example 7.15

Cylinder with Nonuniform Density

Determine the mass of the cylinder in Fig. 7.31 and the position of its center of mass if (a) it is homogeneous with mass density ρ_0; (b) its density is given by the equation $\rho = \rho_0(1 + x/L)$.

Strategy

In (a), the mass of the cylinder is simply the product of its mass density and its volume and the center of mass is located at the centroid of its volume. In (b), the cylinder is nonhomogeneous and we must use Eqs. (7.20) and (7.22) to determine its mass and center of mass.

Solution

(a) The volume of the cylinder is LA, so its mass is $\rho_0 LA$. Since the center of mass is coincident with the centroid of the volume of the cylinder, the coordinates of the center of mass are $\bar{x} = \frac{1}{2}L, \bar{y} = 0, \bar{z} = 0$.

(b) We can determine the mass of the cylinder by using an element of volume dV in the form of a disk of thickness dx (Fig. a). The volume $dV = A\, dx$. The mass of the cylinder is

$$m = \int_V \rho\, dV = \int_0^L \rho_0\left(1 + \frac{x}{L}\right)A\, dx = \frac{3}{2}\rho_0 AL.$$

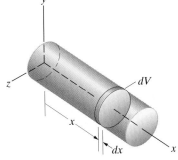

Figure 7.31

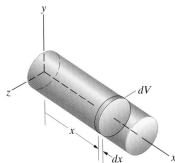

(a) An element of volume dV in the form of a disk.

The x coordinate of the center of mass is

$$\bar{x} = \frac{\displaystyle\int_V x\rho\, dV}{\displaystyle\int_V \rho\, dV} = \frac{\displaystyle\int_0^L \rho_0\left(x + \frac{x^2}{L}\right)A\, dx}{\dfrac{3}{2}\rho_0 AL} = \frac{5}{9}L.$$

Because the density does not depend on y or z, we know from symmetry that $\bar{y} = 0$ and $\bar{z} = 0$.

Discussion

Notice that the center of mass of the nonhomogeneous cylinder is *not* located at the centroid of its volume.

Problems

7.85 The mass of the homogeneous flat plate is 450 kg. What are the reactions at *A* and *B*?

Strategy: The center of mass of the plate is coincident with the centroid of its area. Determine the horizontal coordinate of the centroid and assume that the plate's weight acts there.

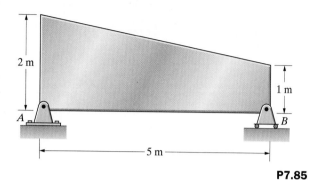

P7.85

7.86 The mass of the homogeneous flat plate is 50 kg. Determine the reactions at the supports *A* and *B*.

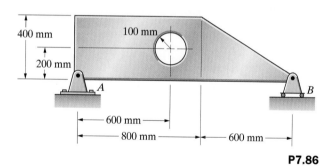

P7.86

7.87 The suspended sign is a homogeneous flat plate that has a mass of 130 kg. Determine the axial forces in members *AD* and *CE*. (Notice that the *y* axis is positive downward.)

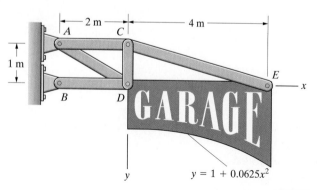

P7.87

7.88 The bar has a mass of 80 kg. What are the reactions at *A* and *B*?

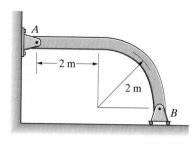

P7.88

7.89 The semicircular part of the homogeneous slender bar lies in the *x*–*z* plane. Determine the center of mass of the bar.

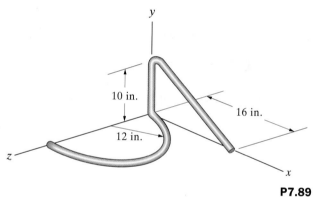

P7.89

7.90 When the truck is unloaded, the total reactions at the front and rear wheels are $A = 54$ kN and $B = 36$ kN. The density of the load of gravel is $\rho = 1600$ kg/m^3. The dimension of the load in the *z* direction is 3 m, and its surface profile, given by the function shown, does not depend on *z*. What are the total reactions at the front and rear wheels of the loaded truck?

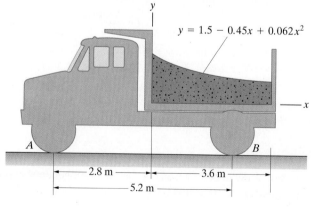

P7.90

7.91 The 10-ft horizontal cylinder with 1-ft radius is supported at A and B. Its weight density is $\gamma = 100(1 - 0.002x^2)$ lb/ft^3. What are the reactions at A and B?

7.92 A horizontal cone with 800-mm length and 200-mm radius has a built-in support at A. Its mass density is $\rho = 6000(1 + 0.4x^2))$ kg/m^3, where x is in meters. What are the reactions at A?

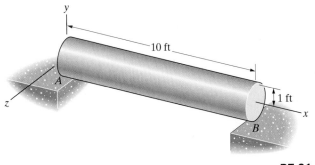

P7.91

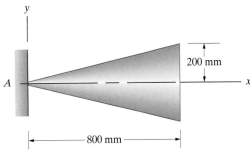

P7.92

7.7 Centers of Mass of Composite Objects

You can easily determine the center of mass of an object consisting of a combination of parts if you know the centers of mass of its parts. The coordinates of the center of mass of a composite object composed of parts with masses $m_1, m_2, \ldots$, are

$$\bar{x} = \frac{\sum_i \bar{x}_i m_i}{\sum_i m_i}, \qquad \bar{y} = \frac{\sum_i \bar{y}_i m_i}{\sum_i m_i}, \qquad \bar{z} = \frac{\sum_i \bar{z}_i m_i}{\sum_i m_i}, \qquad (7.23)$$

where $\bar{x}_i, \bar{y}_i, \bar{z}_i$ are the coordinates of the centers of mass of the parts. Because the weights of the parts are related to their masses by $W_i = gm_i$, Eqs. (7.23) can also be expressed as

$$\bar{x} = \frac{\sum_i \bar{x}_i W_i}{\sum_i W_i}, \qquad \bar{y} = \frac{\sum_i \bar{y}_i W_i}{\sum_i W_i}, \qquad \bar{z} = \frac{\sum_i \bar{z}_i W_i}{\sum_i W_i}. \qquad (7.24)$$

When you know the masses or weights and the centers of mass of the parts of a composite object, you can use these equations to determine its center of mass.

Determining the center of mass of a composite object requires three steps:

1. Choose the parts—Try to divide the object into parts whose centers of mass you know or can easily determine.

2. Determine the values for the parts—Determine the center of mass and the mass or weight of each part. Watch for instances of symmetry that can simplify your task.

3. Calculate the center of mass—Use Eqs. (7.23) or (7.24) to determine the center of mass of the composite object.

Example 7.16

Center of Mass of a Composite Object

The L-shaped machine part in Fig. 7.32 is composed of two homogeneous bars. Bar 1 is tungsten alloy with mass density 14,000 kg/m³, and bar 2 is steel with mass density 7800 kg/m³. Determine the center of mass of the machine part.

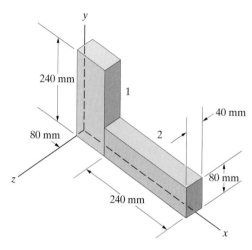

Figure 7.32

Solution

The volume of bar 1 is

$$(80)(240)(40) = 7.68 \times 10^5 \text{ mm}^3 = 7.68 \times 10^{-4} \text{ m}^3,$$

so its mass is $(7.68 \times 10^{-4})(1.4 \times 10^4) = 10.75$ kg. The center of mass of bar 1 coincides with the centroid of its volume: $\bar{x}_1 = 40$ mm, $\bar{y}_1 = 120$ mm, $\bar{z}_1 = 0$.

Bar 2 has the same volume as bar 1, so its mass is (7.68×10^{-4}) $(7.8 \times 10^3) = 5.99$ kg. The coordinates of its center of mass are $\bar{x}_2 = 200$ mm, $\bar{y}_2 = 40$ mm, $\bar{z}_2 = 0$. Using the information summarized in the Table, we obtain the x coordinate of the center of mass,

$$\bar{x} = \frac{\bar{x}_1 m_1 + \bar{x}_2 m_2}{m_1 + m_2} = \frac{(40)(10.75) + (200)(5.99)}{10.75 + 5.99} = 97.2 \text{ mm},$$

and the y coordinate,

$$\bar{y} = \frac{\bar{y}_1 m_1 + \bar{y}_2 m_2}{m_1 + m_2} = \frac{(120)(10.75) + (40)(5.99)}{10.75 + 5.99} = 91.4 \text{ mm}.$$

Because of the symmetry of the object, $\bar{z} = 0$.

Table Information for determining the center of mass

	m_1 (kg)	$\bar{x}_i$ (mm)	$\bar{x}_i m_i$ (mm-kg)	$\bar{y}_i$ (mm)	$\bar{y}_i m_i$ (mm-kg)
Bar 1	10.75	40	(40)(10.75)	120	(120)(10.75)
Bar 2	5.99	200	(200)(5.99)	40	(40)(5.99)

Example 7.17

Center of Mass of a Composite Object

The composite object in Fig. 7.33 consists of a bar welded to a cylinder. The homogeneous bar is aluminum (weight density 168 lb/ft³), and the homogeneous cylinder is bronze (weight density 530 lb/ft³). Determine the center of mass of the object.

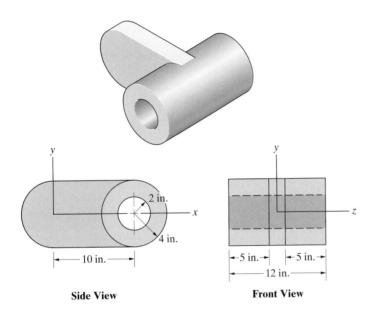

Side View   Front View   **Figure 7.33**

Strategy

We can determine the weight of each homogeneous part by multiplying its volume by its weight density. We also know that the center of mass of each part coincides with the centroid of its volume. The centroid of the cylinder is located at its center, but we must determine the location of the centroid of the bar by treating it as a composite volume.

Solution

The volume of the cylinder is $12[\pi(4)^2 - \pi(2)^2] = 452$ in.³ $= 0.262$ ft³, so its weight is

$$W_{(cylinder)} = (0.262)(530) = 138.8 \text{ lb.}$$

The x coordinate of its center of mass is $\bar{x}_{(cylinder)} = 10$ in.

The volume of the bar is $(10)(8)(2) + \frac{1}{2}\pi(4)^2(2) - \frac{1}{2}\pi(4)^2(2) = 160$ in.³ $= 0.0926$ ft³, and its weight is

$$W_{(bar)} = (0.0926)(168) = 15.6 \text{ lb.}$$

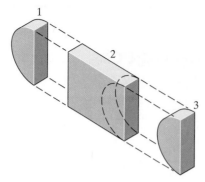

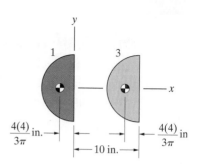

(a) Dividing the bar into three parts.

(b) The centroids of the two semicircular parts.

We can determine the centroid of the volume of the bar by treating it as a composite volume consisting of three parts (Fig. a). Part 3 is a semicircular "cutout." The centroids of part 1 and the semicircular cutout 3 are located at the centroids of their semicircular cross sections (Fig b). Using the information summarized in the Table, we have

$$\bar{x}_{(\text{bar})} = \frac{\bar{x}_1 V_1 + \bar{x}_2 V_2 + \bar{x}_3 V_3}{V_1 + V_2 + V_3}$$

$$= \frac{-\dfrac{4(4)}{3\pi}\left[\frac{1}{2}\pi(4)^2(2)\right] + 5\left[(10)(8)(2)\right] - \left[10 - \dfrac{4(4)}{3\pi}\right]\left[\frac{1}{2}\pi(4)^2(2)\right]}{\frac{1}{2}\pi(4)^2(2) + (10)(8)(2) - \frac{1}{2}\pi(4)^2(2)}$$

$$= 1.86 \text{ in.}$$

Table Information for determining the x coordinate of the centroid of the bar

	$\bar{x}_i$ (in.)	V_i (in^3)	$\bar{x}_i V_i$ (in^4)
Part 1	$-\dfrac{4(4)}{3\pi}$	$\frac{1}{2}\pi(4)^2(2)$	$-\dfrac{4(4)}{3\pi}\left[\frac{1}{2}\pi(4)^2(2)\right]$
Part 2	5	$(10)(8)(2)$	$5\left[(10)(8)(2)\right]$
Part 3	$10 - \dfrac{4(4)}{3\pi}$	$-\frac{1}{2}\pi(4)^2(2)$	$-\left[10 - \dfrac{4(4)}{3\pi}\right]\left[\frac{1}{2}\pi(4)^2(2)\right]$

Therefore the x coordinate of the center of mass of the composite object is

$$\bar{x} = \frac{\bar{x}_{(\text{bar})} W_{(\text{bar})} + \bar{x}_{(\text{cylinder})} W_{(\text{cylinder})}}{W_{(\text{bar})} + W_{(\text{cylinder})}}$$

$$= \frac{(1.86)(15.6) + (10)(138.8)}{15.6 + 138.8} = 9.18 \text{ in.}$$

Because of the symmetry of the bar, the y and z coordinates of its center of mass are $\bar{y} = 0$ and $\bar{z} = 0$.

Problems

7.93 The circular cylinder is made of aluminum (Al) with mass density 2700 kg/m³ and iron (Fe) with mass density 7860 kg/m³.
(a) Determine the centroid of the volume of the cylinder.
(b) Determine the center of mass of the cylinder.

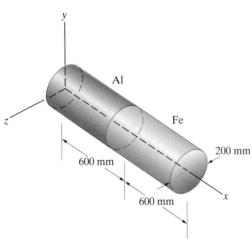

P7.93

7.94 The cylindrical tube is made of aluminum with mass density 2700 kg/m³. The cylindrical plug is made of steel with mass density 7800 kg/m³. Determine the coordinates of the center of mass of the composite object.

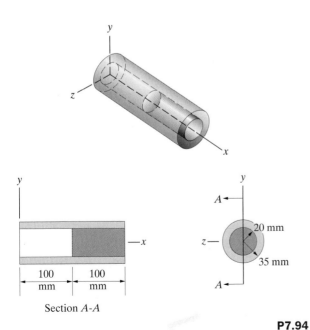

Section *A-A*

P7.94

7.95 A machine consists of three parts. The masses and the locations of the centers of mass of the parts are

Part	Mass (kg)	$\bar{x}$ (mm)	$\bar{y}$ (mm)	$\bar{z}$ (mm)
1	2.0	100	50	−20
2	4.5	150	70	0
3	2.5	180	30	0

Determine the coordinates of the center of mass of the machine.

7.96 A machine consists of three parts. The masses and the locations of the centers of mass of two of the parts are

Part	Mass (kg)	$\bar{x}$ (mm)	$\bar{y}$ (mm)	$\bar{z}$ (mm)
1	2.0	100	50	−20
2	4.5	150	70	0

The mass of part 3 is 2.5 kg. The design engineer wants to position part 3 so that the center of mass location of the machine is $\bar{x} = 120$ mm, $\bar{y} = 80$ mm, $\bar{z} = 0$. Determine the necessary position of the center of mass of part 3.

7.97 Two views of a machine element are shown. Part 1 is aluminum alloy with mass density 2800 kg/m³, and part 2 is steel with mass density 7800 kg/m³. Determine the *x* coordinate of its center of mass.

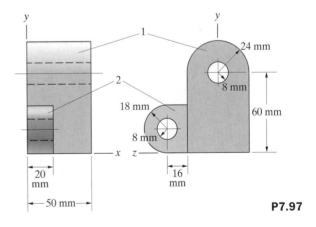

P7.97

7.98 Determine the *y* and *z* coordinates of the center of mass of the machine element in Problem 7.97.

7.99 With its engine removed, the mass of the car is 1100 kg and its center of mass is at *C*. The mass of the engine is 220 kg.
(a) Suppose that you want to place the center of mass *E* of the engine so that the center of mass of the car is midway between the front wheels *A* and the rear wheels *B*. What is the distance *b*?

(b) If the car is parked on a 15° slope facing up the slope, what total normal force is exerted by the road on the rear wheels B?

147 kN, respectively. Determine the mass and the x and y coordinates of the center of mass of the crate.

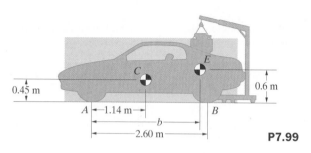

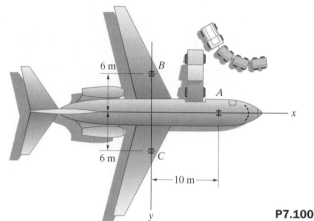

P7.99

7.100 The airplane is parked with its landing gear resting on scales. The weights measured at A, B, and C are 30 kN, 140 kN, and 146 kN, respectively. After a crate is loaded onto the plane, the weights measured at A, B, and C are 31 kN, 142 kN, and

P7.100

7.8 Moments of Inertia of Areas

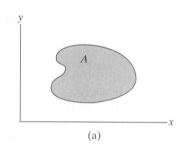

(a)

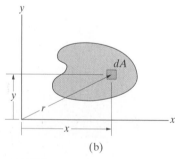

(b)

Figure 7.34
(a) An area A in the x–y plane.
(b) A differential element of A.

The moments of inertia of an area are integrals similar in form to those used to determine the centroid of an area. Consider an area A in the x–y plane (Fig. 7.34a). Four moments of inertia of A are defined:

1. **Moment of inertia about the x axis:**

$$I_x = \int_A y^2 \, dA, \tag{7.25}$$

where y is the y coordinate of the differential element of area dA (Fig. 7.34b). This moment of inertia is sometimes expressed in terms of the *radius of gyration* about the x axis, k_x, defined by

$$I_x = k_x^2 A. \tag{7.26}$$

2. **Moment of inertia about the y axis:**

$$I_y = \int_A x^2 \, dA, \tag{7.27}$$

where x is the x coordinate of the element dA (Fig. 7.34b). The radius of gyration about the y axis, k_y, is defined by

$$I_y = k_y^2 A. \tag{7.28}$$

3. **Product of inertia:**

$$I_{xy} = \int_A xy \, dA. \tag{7.29}$$

4. **Polar moment of inertia:**

$$J_O = \int_A r^2 \, dA, \tag{7.30}$$

where r is the radial distance from the origin of the coordinate system to dA (Fig. 7.34b). The radius of gyration about the origin, k_O, is defined by

$$J_O = k_O^2 A. \tag{7.31}$$

The polar moment of inertia is equal to the sum of the moments of inertia about the x and y axes:

$$J_O = \int_A r^2 \, dA = \int_A (y^2 + x^2) \, dA = I_x + I_y.$$

Substituting the expressions for the moments of inertia in terms of the radii of gyration into this equation, we obtain

$$k_O^2 = k_x^2 + k_y^2.$$

Example 7.18

Moments of Inertia of a Triangular Area

Determine the moments of inertia and radii of gyration of the triangular area in Fig. 7.35.

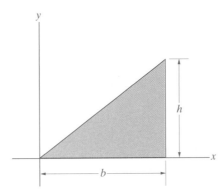

Figure 7.35

Strategy

Equation (7.27) for the moment of inertia about the y axis is very similar in form to the equation for the x coordinate of the centroid of an area, and we can evaluate it for this triangular area in exactly the same way: by using a differential element of area dA in the form of a vertical strip of width dx. We can then show that I_x and I_{xy} can be evaluated using the same element of area. The polar moment of inertia J_O is the sum of I_x and I_y.

Solution

Let dA be the vertical strip in Fig. a. The height of the strip is $(h/b)x$, so $dA = (h/b)x \, dx$. To integrate over the entire area, we must integrate with respect to x from $x = 0$ to $x = b$.

Moment of Inertia About the y Axis

$$I_y = \int_A x^2 \, dA = \int_0^b x^2 \left(\frac{h}{b} x \right) dx = \frac{h}{b} \left[\frac{x^4}{4} \right]_0^b = \frac{1}{4} hb^3.$$

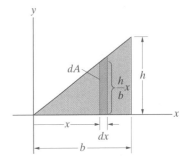

(a) An element dA in the form of a strip.

The radius of gyration k_y is

$$k_y = \sqrt{\frac{I_y}{A}} = \sqrt{\frac{(1/4)hb^3}{(1/2)bh}} = \frac{1}{\sqrt{2}}b.$$

Moment of Inertia About the x Axis   We will first determine the moment of inertia of the strip dA about the x axis while holding x and dx fixed. In terms of the element of area $dA_s = dx\,dy$ shown in Fig. b,

$$\left(I_x\right)_{\text{strip}} = \int_{\text{strip}} y^2\,dA_s = \int_0^{(h/b)x} \left(y^2\,dx\right)dy$$

$$= \left[\frac{y^3}{3}\right]_0^{(h/b)x} dx = \frac{h^3}{3b^3}\,x^3\,dx.$$

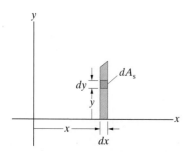

(b) An element of the strip element dA.

Integrating this expression with respect to x from $x = 0$ to $x = b$, we obtain the value of I_x for the entire area:

$$I_x = \int_0^b \frac{h^3}{3b^3}\,x^3\,dx = \frac{1}{12}\,bh^3.$$

The radius of gyration k_x is

$$k_x = \sqrt{\frac{I_x}{A}} = \sqrt{\frac{(1/12)bh^3}{(1/2)bh}} = \frac{1}{\sqrt{6}}h.$$

Product of Inertia   We can determine I_{xy} the same way we determined I_x. We first evaluate the product of inertia of the strip dA, holding x and dx fixed (Fig. b):

$$\left(I_{xy}\right)_{\text{strip}} = \int_{\text{strip}} xy\,dA_s = \int_0^{(h/b)x} \left(xy\,dx\right)dy$$

$$= \left[\frac{y^2}{2}\right]_0^{(h/b)x} x\,dx = \frac{h^2}{2b^2}\,x^3\,dx.$$

Integrating this expression with respect to x from $x = 0$ to $x = b$, we obtain the value of I_{xy} for the entire area:

$$I_{xy} = \int_0^b \frac{h^2}{2b^2}\,x^3\,dx = \frac{1}{8}\,b^2h^2.$$

Polar Moment of Inertia

$$J_O = I_x + I_y = \frac{1}{12}\,bh^3 + \frac{1}{4}\,hb^3.$$

The radius of gyration k_O is

$$k_O = \sqrt{k_x^2 + k_y^2} = \sqrt{\frac{1}{6}\,h^2 + \frac{1}{2}\,b^2}.$$

Discussion

As this example demonstrates, the integrals defining the moments and products of inertia are so similar in form to the integrals used to determine centroids of areas (Section 7.2) that you can use the same methods to evaluate them.

Example 7.19

Moments of Inertia of a Circular Area

Determine the moments of inertia and radii of gyration of the circular area in Fig. 7.36.

Strategy

We will first determine the polar moment of inertia J_O by integrating in terms of polar coordinates. We know from the symmetry of the area that $I_x = I_y$, and since $I_x + I_y = J_O$, the moments of inertia I_x and I_y are each equal to $\frac{1}{2}J_O$. We also know from the symmetry of the area that $I_{xy} = 0$.

Solution

By letting r change by an amount dr, we obtain an annular element of area $dA = 2\pi r\, dr$ (Fig. a). The polar moment of inertia is

$$J_O = \int_A r^2\, dA = \int_0^R 2\pi r^3\, dr = 2\pi\left[\frac{r^4}{4}\right]_0^R = \frac{1}{2}\pi R^4,$$

and the radius of gyration about O is

$$k_O = \sqrt{\frac{J_O}{A}} = \sqrt{\frac{(1/2)\pi R^4}{\pi R^2}} = \frac{1}{\sqrt{2}}R.$$

The moments of inertia about the x and y axes are

$$I_x = I_y = \frac{1}{2}J_O = \frac{1}{4}\pi R^4,$$

and the radii of gyration about the x and y axes are

$$k_x = k_y = \sqrt{\frac{I_x}{A}} = \sqrt{\frac{(1/4)\pi R^4}{\pi R^2}} = \frac{1}{2}R.$$

The product of inertia is zero:

$$I_{xy} = 0.$$

Discussion

The symmetry of this example saved us from having to integrate to determine I_x, I_y, and I_{xy}. Be alert for symmetry that can shorten your work. In particular, remember that $I_{xy} = 0$ if the area is symmetric about either the x or the y axis.

Figure 7.36

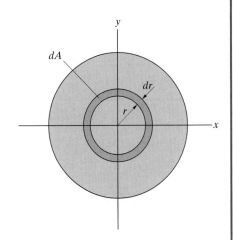

(a) An annular element dA.

Problems

7.101 Determine I_y and k_y.

7.102 Determine I_x and k_x by letting dA be (a) a horizontal strip of height dy; (b) a vertical strip of width dx.

7.103 Determine I_{xy}.

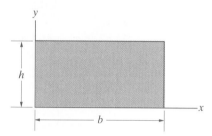

P7.101–P7.103

7.104 Determine I_x, k_x, I_y, and k_y for the beam's rectangular cross section.

7.105 Determine I_{xy} and J_O for the beam's rectangular cross section.

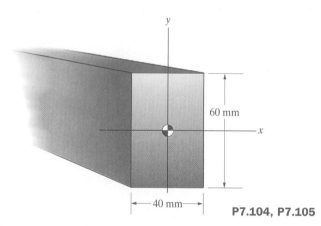

P7.104, P7.105

7.106 Determine I_y and k_y.

7.107 Determine J_O and k_O.

7.108 Determine I_{xy}.

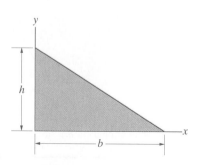

P7.106–P7.108

7.109 Determine I_y.

7.110 Determine I_x.

7.111 Determine J_O.

7.112 Determine I_{xy}.

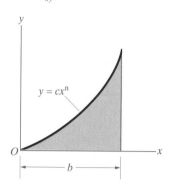

$y = cx^n$

P7.109–P7.112

7.113 Determine I_y and k_y.

7.114 Determine I_x and k_x.

7.115 Determine J_O and k_O.

7.116 Determine I_{xy}.

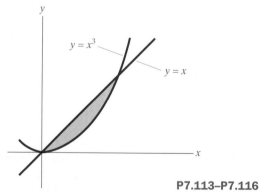

$y = x^3$

$y = x$

P7.113–P7.116

7.117 Determine the moment of inertia I_y of the metal plate's cross-sectional area.

7.118 Determine the moment of inertia I_x and the radius of gyration k_x of the cross-sectional area of the metal plate.

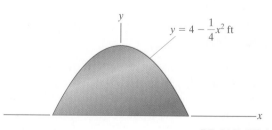

$y = 4 - \dfrac{1}{4}x^2$ ft

P7.117, P7.118

7.119 (a) Determine I_y and k_y by letting dA be a vertical strip of width dx.

(b) The polar moment of inertia of a circular area with its center at the origin is $J_O = \frac{1}{2}\pi R^4$. Explain how you can use this information to confirm your answer to (a).

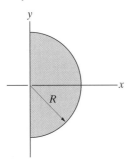

P7.119

7.120 (a) Determine I_x and k_x for the area in Problem 7.119 by letting dA be a horizontal strip of height dy.

(b) The polar moment of inertia of a circular area with its center at the origin is $J_O = \frac{1}{2}\pi R^4$. Explain how you can use this information to confirm your answer to (a).

7.121 Determine the moments of inertia I_x and I_y.

Strategy: Use the procedure described in Example 7.19 to determine J_O, then use the symmetry of the area to determine I_x and I_y.

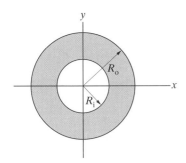

P7.121

7.122 If $a = 5$ m and $b = 1$ m, what are the values of I_y and k_y for the elliptical area of the airplane's wing?

7.123 What are the values of I_x and k_x for the elliptical area of the airplane's wing in Problem 7.122?

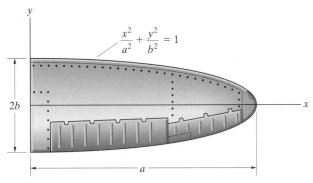

P7.122, P7.123

7.124 Determine I_y and k_y.

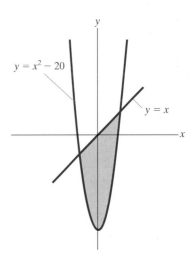

P7.124

7.125 Determine I_x and k_x for the area in Problem 7.124.

7.126 A vertical plate of area A is beneath the surface of a stationary body of water. The pressure of the water subjects each element dA of the surface of the plate to a force $(p_0 + \gamma y)\,dA$, where p_0 is the pressure at the surface of the water and γ is the weight density of the water. Show that the magnitude of the moment about the x axis due to the pressure on the front face of the plate is

$$M_{(x\,\text{axis})} = p_0 \bar{y} A + \gamma I_x,$$

where $\bar{y}$ is the y coordinate of the centroid of A and I_x is the moment of inertia of A about the x axis.

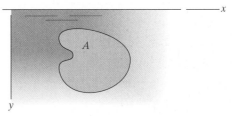

P7.126

7.9 Parallel-Axis Theorems

In some situations the moments of inertia of an area are known in terms of a particular coordinate system but we need their values in terms of a different coordinate system. When the coordinate systems are parallel, the desired moments of inertia can be obtained by using the theorems we describe in this section. Furthermore, these theorems make it possible for us to determine the moments of inertia of a composite area when the moments of inertia of its parts are known.

Suppose that we know the moments of inertia of an area A in terms of a coordinate system $x'y'$ with its origin at the centroid of the area, and we wish to determine the moments of inertia in terms of a parallel coordinate system xy (Fig. 7.37a). We denote the coordinates of the centroid of A in the xy coordinate system by (d_x, d_y), and $d = \sqrt{d_x^2 + d_y^2}$ is the distance from the origin of the xy coordinate system to the centroid (Fig. 7.37b).

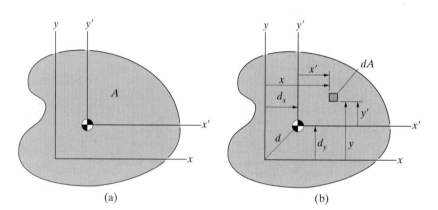

Figure 7.37
(a) The area A and the coordinate systems $x'y'$ and xy.
(b) The differential element dA.

(a)   (b)

We need to obtain two preliminary results before deriving the parallel-axis theorems. In terms of the $x'y'$ coordinate system, the coordinates of the centroid of A are

$$\bar{x}' = \frac{\int_A x'\, dA}{\int_A dA}, \qquad \bar{y}' = \frac{\int_A y'\, dA}{\int_A dA}.$$

But the origin of the $x'y'$ coordinate system is located at the centroid of A, so $\bar{x}' = 0$ and $\bar{y}' = 0$. Therefore

$$\int_A x'\, dA = 0, \qquad \int_A y'\, dA = 0. \tag{7.32}$$

Moment of Inertia About the x Axis   In terms of the xy coordinate system, the moment of inertia of A about the x axis is

$$I_x = \int_A y^2\, dA, \tag{7.33}$$

where y is the coordinate of the element of area dA relative to the xy coordinate system. From Fig. 7.37b, $y = y' + d_y$, where y' is the coordinate of dA

relative to the $x'y'$ coordinate system. Substituting this expression into Eq. (7.33), we obtain

$$I_x = \int_A (y' + d_y)^2\, dA = \int_A (y')^2\, dA + 2d_y \int_A y'\, dA + d_y^2 \int_A dA.$$

The first integral on the right is the moment of inertia of A about the x' axis. From Eq. (7.32), the second integral on the right equals zero. Therefore we obtain

$$I_x = I_{x'} + d_y^2 A. \tag{7.34}$$

This is a *parallel-axis theorem*. It relates the moment of inertia of A about the x' axis through the centroid to the moment of inertia about the parallel axis x (Fig. 7.38).

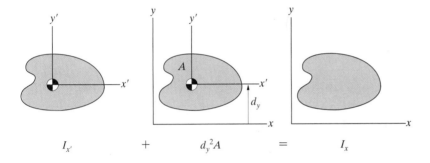

$$I_{x'} \qquad + \qquad d_y^2 A \qquad = \qquad I_x$$

Figure 7.38
The parallel-axis theorem for the moment of inertia about the x axis.

Moment of Inertia About the y Axis   In terms of the xy coordinate system, the moment of inertia of A about the y axis is

$$I_y = \int_A x^2\, dA = \int_A (x' + d_x)^2\, dA$$

$$= \int_A (x')^2\, dA + 2d_x \int_A x'\, dA + d_x^2 \int_A dA.$$

From Eq. (7.32), the second integral on the right equals zero. Therefore the parallel-axis theorem that relates the moment of inertia of A about the y' axis through the centroid to the moment of inertia about the parallel axis y is

$$I_y = I_{y'} + d_x^2 A. \tag{7.35}$$

Product of Inertia   The parallel-axis theorem for the product of inertia is

$$I_{xy} = I_{x'y'} + d_x d_y A. \tag{7.36}$$

Polar Moment of Inertia   The parallel-axis theorem for the polar moment of inertia is

$$J_O = J'_O + (d_x^2 + d_y^2)A = J'_O + d^2 A, \tag{7.37}$$

where d is the distance from the origin of the $x'y'$ coordinate system to the origin of the xy coordinate system.

How can the parallel-axis theorems be used to determine the moments of inertia of a composite area? Suppose that we want to determine the moment of inertia about the y axis of the area in Fig. 7.39a. We can divide it into a triangle, a semicircle, and a circular cutout, denoted parts 1, 2, and 3 (Fig. 7.39b). By using the parallel-axis theorem for I_y, can determine the moment of inertia of each part about the y axis. For example, the moment of inertia of part 2 (the semicircle) about the y axis is (Fig. 7.39c)

$$(I_y)_2 = (I_{y'})_2 + (d_x)_2^2 A_2.$$

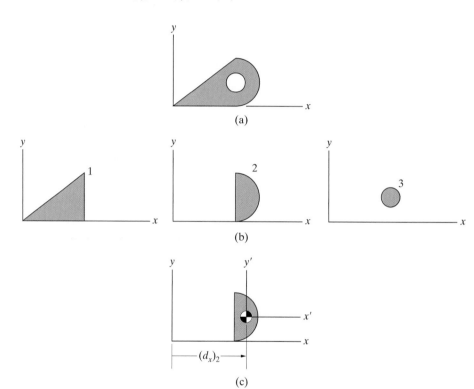

Figure 7.39
(a) A composite area.
(b) The three parts of the area.
(c) Determining $(I_y)_2$.

We must determine the values of $(I_{y'})_2$ and $(d_x)_2$. Moments of inertia and centroid locations for some simple areas are tabulated in Appendix B. Once this procedure is carried out for each part, the moment of inertia of the composite area is

$$I_y = (I_y)_1 + (I_y)_2 - (I_y)_3.$$

Notice that the moment of inertia of the circular cutout is subtracted.

We see that determining a moment of inertia of a composite area in terms of a given coordinate system involves three steps:

1. Choose the parts—Try to divide the composite area into parts whose moments of inertia you know or can easily determine.
2. Determine the moments of inertia of the parts—Determine the moment of inertia of each part in terms of a parallel coordinate system with its origin at the centroid of the part, and then use the parallel-axis theorem to determine the moment of inertia in terms of the given coordinate system.
3. Sum the results—Sum the moments of inertia of the parts (or subtract in the case of a cutout) to obtain the moment of inertia of the composite area.

Example 7.20

Demonstration of the Parallel Axis Theorems

The moments of inertia of the rectangular area in Fig. 7.40 in terms of the $x'y'$ coordinate system are $I_{x'} = \frac{1}{12}bh^3$, $I_{y'} = \frac{1}{12}hb^3$, $I_{x'y'} = 0$, and $J'_O = \frac{1}{12}(bh^3 + hb^3)$ (see Appendix B). Determine its moments of inertia in terms of the xy coordinate system.

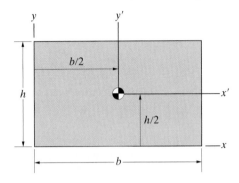

Figure 7.40

Strategy

The $x'y'$ coordinate system has its origin at the centroid of the area and is parallel to the xy coordinate system. We can use the parallel-axis theorems to determine the moments of inertia of A in terms of the xy coordinate system.

Solution

The coordinates of the centroid in terms of the xy coordinate system are $d_x = b/2$, $d_y = h/2$. The moment of inertia about the x axis is

$$I_x = I_{x'} + d_y^2 A = \frac{1}{12}bh^3 + \left(\frac{1}{2}h\right)^2 bh = \frac{1}{3}bh^3.$$

The moment of inertia about the y axis is

$$I_y = I_{y'} + d_x^2 A = \frac{1}{12}hb^3 + \left(\frac{1}{2}b\right)^2 bh = \frac{1}{3}hb^3.$$

The product of inertia is

$$I_{xy} = I_{x'y'} + d_x d_y A = 0 + \left(\frac{1}{2}b\right)\left(\frac{1}{2}h\right)bh = \frac{1}{4}b^2h^2.$$

The polar moment of inertia is

$$J_O = J'_O + d^2 A = \frac{1}{12}(bh^3 + hb^3) + \left[\left(\frac{1}{2}b\right)^2 + \left(\frac{1}{2}h\right)^2\right]bh$$

$$= \frac{1}{3}(bh^3 + hb^3).$$

Discussion

Notice that we could also have determined J_O by using the relation

$$J_O = I_x + I_y = \frac{1}{3}bh^3 + \frac{1}{3}hb^3.$$

Example 7.21

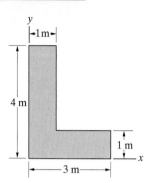

Figure 7.41

Moments of Inertia of a Composite Area

Determine I_x, k_x, and I_{xy} for the composite area in Fig. 7.41.

Solution

Choose the Parts We can determine the moments of inertia by dividing the area into the rectangular parts 1 and 2 shown in Fig. a.

Determine the Moments of Inertia of the Parts For each part, we introduce a coordinate system $x'y'$ with its origin at the centroid of the part (Fig. b). The moments of inertia of the rectangular parts in terms of these coordinate systems are given in Appendix B. We then use the parallel-axis theorem to determine the moment of inertia of each part about the x axis (Table 1).

Table 1 Determining the moments of inertia of the parts about the x axis.

	d_y (m)	A (m²)	$I_{x'}$ (m⁴)	$I_x = I_{x'} + d_y^2 A$ (m⁴)
Part 1	2	$(1)(4)$	$\frac{1}{12}(1)(4)^3$	21.33
Part 2	0.5	$(2)(1)$	$\frac{1}{12}(2)(1)^3$	0.67

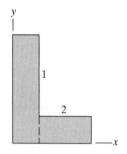

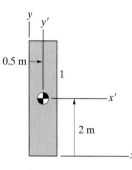

(a) Dividing the area into rectangles 1 and 2.

Sum the Results The moment of inertia of the composite area about the x axis is

$$I_x = (I_x)_1 + (I_x)_2 = 21.33 + 0.67 = 22.00 \text{ m}^4.$$

The sum of the areas is $A = A_1 + A_2 = 6 \text{ m}^2$, so the radius of gyration about the x axis is

$$k_x = \sqrt{\frac{I_x}{A}} = \sqrt{\frac{22}{6}} = 1.91 \text{ m}.$$

Repeating this procedure, we determine I_{xy} for each part in Table 2. The product of inertia of the composite area is

$$I_{xy} = (I_{xy})_1 + (I_{xy})_2 = 4 + 2 = 6 \text{ m}^4.$$

Table 2 Determining the products of inertia of the parts in terms of the xy coordinate system

	d_x (m)	d_y (m)	A (m²)	$I_{x'y'}$	$I_{xy} = I_{x'y'} + d_x d_y A$ (m⁴)
Part 1	0.5	2	$(1)(4)$	0	4
Part 2	2	0.5	$(2)(1)$	0	2

Discussion

The moments of inertia you obtain do not depend on how you divide a composite area into parts, and you will often have a choice of convenient ways to divide a given area. See Problem 7.127 and 7.128, in which we divide the composite area in this example in a different way.

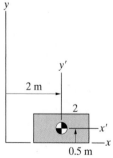

(b) Parallel coordinate systems $x'y'$ with origins at the centroids of the parts.

Example 7.22

Moments of Inertia of a Composite Area

Determine I_y and k_y for the composite area in Fig. 7.42.

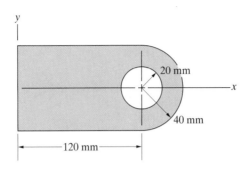

Figure 7.42

Solution

Choose the Parts We divide the area into a rectangle, a semicircle, and the circular cutout, calling them parts 1, 2, and 3, respectively (Fig. a).

Determine the Moments of Inertia of the Parts The moments of inertia of the parts in terms of the $x'y'$ coordinate systems and the location of the centroid of the semicircular part are given in Appendix B. In the Table we use the parallel-axis theorem to determine the moment of inertia of each part about the y axis.

Table Determining the moments of inertia of the parts.

	$d_x \, (\text{mm})$	$A \, (\text{mm}^2)$	$I_{y'} \, (\text{mm}^4)$	$I_y = I_{y'} + d_x^2 A \, (\text{mm}^4)$
Part 1	60	$(120)(80)$	$\frac{1}{12}(80)(120)^3$	4.608×10^7
Part 2	$120 + \dfrac{4(40)}{3\pi}$	$\frac{1}{2}\pi(40)^2$	$\left(\dfrac{\pi}{8} - \dfrac{8}{9\pi}\right)(40)^4$	4.744×10^7
Part 3	120	$\pi(20)^2$	$\frac{1}{4}\pi(20)^4$	1.822×10^7

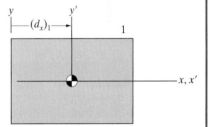

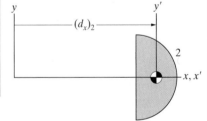

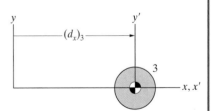

(a) Parts 1, 2, and 3.

Sum the Results The moment of inertia of the composite area about the y axis is

$$I_y = (I_y)_1 + (I_y)_2 - (I_y)_3 = (4.608 + 4.744 - 1.822) \times 10^7$$

$$= 7.530 \times 10^7 \text{ mm}^4.$$

The total area is

$$A = A_1 + A_2 - A_3 = (120)(80) + \frac{1}{2}\pi(40)^2 - \pi(20)^2$$

$$= 1.086 \times 10^4 \text{ mm}^2,$$

so the radius of gyration about the y axis is

$$k_y = \sqrt{\frac{I_y}{A}} = \sqrt{\frac{7.530 \times 10^7}{1.086 \times 10^4}} = 83.3 \text{ mm}.$$

Problems

7.127 Determine I_x and k_x for the composite area by dividing it into rectangles 1 and 2 as shown, and compare your results to those of Example 7.21.

7.128 Determine I_y and k_y for the composite area.

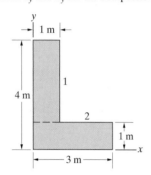

P7.127, P7.128

7.129 Determine I_x and k_x.

7.130 Determine I_y and k_y.

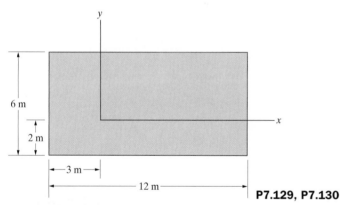

P7.129, P7.130

7.131 Determine I_x and k_x.

7.132 Determine I_y and k_y.

7.133 Determine J_O and k_O.

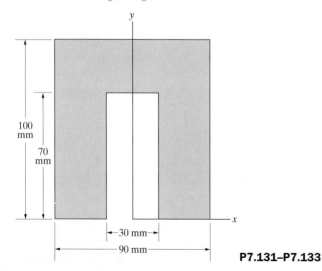

P7.131–P7.133

7.134 If you design the beam cross section so that $I_x = 6.4 \times 10^5$ mm⁴, what are the resulting values of I_y and J_O?

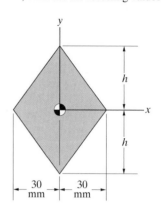

P7.134

7.135 Determine I_y and k_y.

7.136 Determine I_x and k_x.

7.137 Determine I_{xy}.

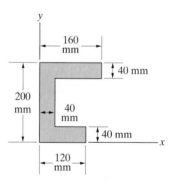

P7.135–P7.137

7.138 Determine I_x and k_x.

7.139 Determine I_y and k_y.

7.140 Determine I_{xy}.

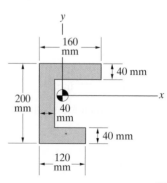

P7.138–P7.140

7.141 Determine I_x and k_x.

7.142 Determine J_O and k_O.

7.143 Determine I_{xy}.

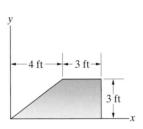

P7.141–P7.143

7.144 Determine I_x and k_x.

7.145 Determine J_O and k_O.

7.146 Determine I_{xy}.

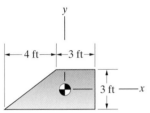

P7.144–P7.146

7.147 Determine I_x and k_x.

7.148 Determine J_O and k_O.

7.149 Determine I_{xy}.

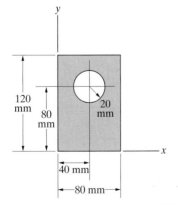

P7.147–P7.149

7.150 Determine I_x and k_x.

7.151 Determine I_y and k_y.

7.152 Determine J_O and k_O.

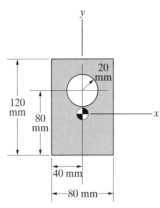

P7.150–P7.152

7.153 Determine I_y and k_y.

7.154 Determine J_O and k_O.

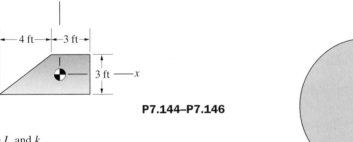

P7.153, P7.154

7.155 Determine I_y and k_y if $h = 3$ m.

7.156 Determine I_x and k_x if $h = 3$ m.

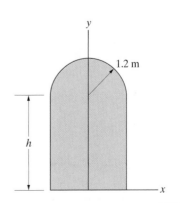

P7.155–P7.156

Chapter Summary

Centroids

A *centroid* is a weighted average position. The coordinates of the centroid of an area A in the x–y plane are

$$\bar{x} = \frac{\displaystyle\int_A x \, dA}{\displaystyle\int_A dA}, \qquad \bar{y} = \frac{\displaystyle\int_A y \, dA}{\displaystyle\int_A dA}. \qquad\qquad \text{Eqs. (7.6), (7.7)}$$

The coordinates of the centroid of a *composite area* composed of parts $A_1, A_2, \ldots$, are

$$\bar{x} = \frac{\displaystyle\sum_i \bar{x}_i A_i}{\displaystyle\sum_i A_i}, \qquad \bar{y} = \frac{\displaystyle\sum_i \bar{y}_i A_i}{\displaystyle\sum_i A_i}. \qquad\qquad \text{Eq. (7.9)}$$

Similar equations define the centroids of volumes [Eqs. (7.15) and (7.17)] and lines [(Eqs. (7.16) and (7.18)].

Distributed Forces

A force distributed along a line is described by a function w, defined such that the force on a differential element dx of the line is $w \, dx$. The force exerted by a distributed load is

$$F = \int_L w \, dx, \qquad\qquad \text{Eq. (7.10)}$$

and the moment about the origin is

$$M = \int_L xw \, dx. \qquad\qquad \text{Eq. (7.11)}$$

The force F is equal to the "area" between the function w, and the x axis and is equivalent to the distributed load if it is placed at the centroid of the "area."

Centers of Mass

The *center of mass* of an object is the centroid of its mass. The weight of an object can be represented by a single equivalent force acting at its center of mass.

The *mass density* ρ is defined such that the mass of a differential element of volume is $dm = \rho \, dV$. An object whose mass density is uniform throughout its volume is said to be *homogeneous*. The *weight density* $\gamma = g\rho$.

The coordinates of the center of mass of an object are

$$\bar{x} = \frac{\displaystyle\int_V \rho x \, dV}{\displaystyle\int_V \rho \, dV}, \qquad \bar{y} = \frac{\displaystyle\int_V \rho y \, dV}{\displaystyle\int_V \rho \, dV}, \qquad \bar{z} = \frac{\displaystyle\int_V \rho z \, dV}{\displaystyle\int_V \rho \, dV}. \qquad \text{Eq. (7.22)}$$

The center of mass of a homogeneous object coincides with the centroid of its volume. The center of mass of a homogeneous plate of uniform thickness coincides with the centroid of its cross-sectional area. The center of mass of a homogeneous slender bar of uniform cross-sectional area coincides approximately with the centroid of the axis of the bar.

Moments of Inertia of Areas

Four *area moments of inertia* are defined (Fig. a):

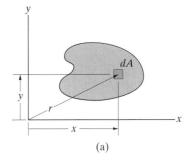

(a)

1. The moment of inertia about the x axis:

$$I_x = \int_A y^2 \, dA. \qquad \text{Eq. (7.25)}$$

2. The moment of inertia about the y axis:

$$I_y = \int_A x^2 \, dA. \qquad \text{Eq. (7.27)}$$

3. The product of inertia:

$$I_{xy} = \int_A xy \, dA. \qquad \text{Eq. (7.29)}$$

4. The polar moment of inertia:

$$J_O = \int_A r^2 \, dA. \qquad \text{Eq. (7.30)}$$

The *radii of gyration* about the x and y axes are defined by $k_x = \sqrt{I_x/A}$ and $k_y = \sqrt{I_y/A}$, respectively, and the radius of gyration about the origin O is defined by $k_O = \sqrt{J_O/A}$.

The polar moment of inertia is equal to the sum of the moments of inertia about the x and y axes: $J_O = I_x + I_y$. If an area is symmetric about either the x axis or the y axis, its product of inertia is zero.

Let $x'y'$ be a coordinate system with its origin at the centroid of an area A, and let xy be a parallel coordinate system. The moments of inertia of A in terms of the two systems are related by the *parallel-axis theorems* [Eqs. (7.34)–(7.37)]:

$$I_x = I_{x'} + d_y^2 A,$$

$$I_y = I_{y'} + d_x^2 A,$$

$$I_{xy} = I_{x'y'} + d_x d_y A,$$

$$J_O = J_O' + (d_x^2 + d_y^2)A = J_O' + d^2 A,$$

where d_x and d_y are the coordinates of the centroid of A in the xy coordinate system.

Review Problems

7.157 Determine the centroid of the area.

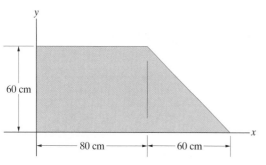

P7.157

7.158 Determine the centroid of the area.

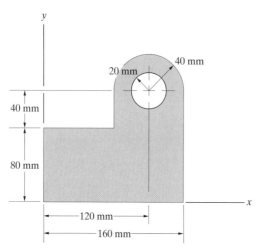

P7.158

7.159 The cantilever beam is subjected to a triangular distributed load. What are the reactions at A?

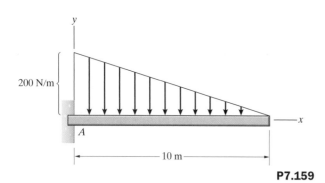

P7.159

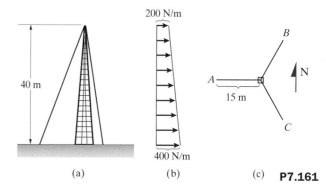

P7.160

7.161 An engineer estimates that the maximum wind load on the 40-m tower in Fig. a is described by the distributed load in Fig. b. The tower is supported by three cables, A, B, and C, from the top of the tower to equally spaced points 15 m from the bottom of the tower (Fig. c). If the wind blows from the west and cables B and C are slack, what is the tension in cable A? (Model the base of the tower as a ball and socket support.)

(a)   (b)   (c)   **P7.161**

7.162 If the wind in Problem 7.161 blows from the east and cable A is slack, what are the tensions in cables B and C?

7.163 Determine the y coordinate of the center of mass of the homogeneous steel plate.

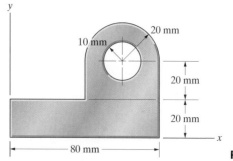

P7.163

7.164 Determine I_y and k_y.

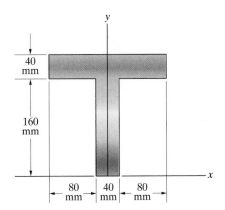

P7.164

7.166 Determine I_x and k_x.

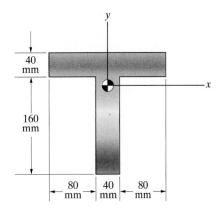

P7.166

7.165 Determine I_x and k_x for the area in Problem 7.164.

7.167 Determine J_O and k_O for the area in Problem 7.166.

Project Construct a homogeneous thin flat plate with the shape shown. (Use the cardboard back of a pad of paper to construct the plate. Choose your dimensions so that the plate is as large as possible.) Calculate the location of the center of mass of the plate. Measuring as carefully as possible, mark the center of mass clearly on both sides of the plate. Then carry out the following experiments.

(a) Balance the plate on your finger (Fig. a) and observe that it balances at its center of mass. Explain the result of this experiment by drawing a free-body diagram of the plate.

(b) This experiment requires a needle or slender nail, a length of string, and a small weight. Tie the weight to one end of the string and make a small loop at the other end. Stick the needle through the plate at any point other than its center of mass. Hold the needle horizontal so that the plate hangs freely from it (Fig. b). Use the loop to hang the weight from the needle, and let the weight hang freely so that the string lies along the face of the plate. Observe that the string passes through the center of mass of the plate. Repeat this experiment several times, sticking the needle through various points on the plate. Explain the results of this experiment by drawing a free-body diagram of the plate.

(c) Hold the plate so that the plane of the plate is vertical, and throw the plate upward, spinning it like a Frisbee. Observe that the plate spins about its center of mass.

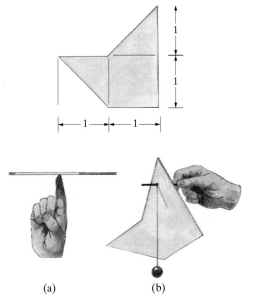

(a)   (b)

Shoe soles are designed to support the friction forces necessary to prevent slipping. In this chapter we analyze friction forces between surfaces in contact.

Friction

Friction forces have many important effects, both desirable and undesirable, in engineering applications. The Coulomb theory of friction allows us to estimate the maximum friction forces that can be exerted by contacting surfaces and the friction forces exerted by sliding surfaces. This opens the path to the analysis of important new classes of supports and machines, including wedges (shims), threaded connections, bearings, and belts.

8.1 Theory of Dry Friction

When you climb a ladder, it remains in place because of the friction force exerted on it by the floor (Fig. 8.1a). If you remain stationary on the ladder, the equilibrium equations determine the friction force. But an important question cannot be answered by the equilibrium equations alone: Will the ladder remain in place, or will it slip on the floor? If a truck is parked on an incline, the friction force exerted on it by the road prevents it from sliding down the incline (Fig. 8.1b). Here too there is another question: What is the steepest incline on which the truck can be parked?

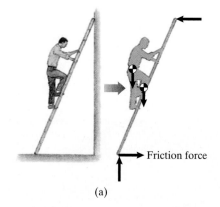

Friction force

(a)

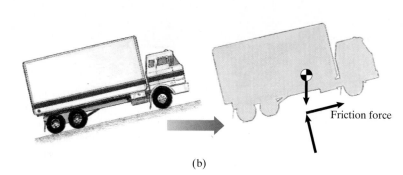

Friction force

(b)

Figure 8.1
Objects supported by friction forces.

To answer these questions, we must examine the nature of friction forces in more detail. Place a book on a table and push it with a small horizontal force, as shown in Fig. 8.2a. If the force you exert is sufficiently small, the book does not move. The free-body diagram of the book is shown in Fig. 8.2b. The force W is the book's weight, and N is the normal force exerted by the table. The force F is the horizontal force you apply, and f is the friction force exerted by the table. Because the book is in equilibrium, $f = F$.

Figure 8.2
(a) Exerting a horizontal force on a book.
(b) The free-body diagram of the book.

(a)

(b)

Now slowly increase the force you apply to the book. As long as the book remains in equilibrium, the friction force must increase correspondingly, since it equals the force you apply. When the force you apply becomes too large, the book moves. It slips on the table. After reaching some maximum value, the friction force can no longer maintain the book in equilibrium. Also, notice that the force you must apply to keep the book moving on the table is smaller than the force required to cause it to slip. (You are familiar with this phenomenon if you've ever pushed a piece of furniture across a floor.)

How does the table exert a friction force on the book? Why does the book slip? Why is less force required to slide the book across the table than is required to start it moving? If the surfaces of the table and the book are magnified sufficiently, they will appear rough (Fig. 8.3). Friction forces arise in part from the interactions of the roughnesses, or *asperities*, of the contacting surfaces. On a still smaller scale, contacting surfaces tend to form atomic bonds that "glue" them together (Fig. 8.4). The fact that more force is required to start an object sliding on a surface than to keep it sliding is explained in part by the necessity to break these bonds before sliding can begin.

In the following sections we present a theory that predicts the basic phenomena we have described and has been found useful for approximating friction forces between dry surfaces in engineering applications. (Friction between lubricated surfaces is a hydrodynamic phenomenon and must be analyzed in the context of fluid mechanics.)

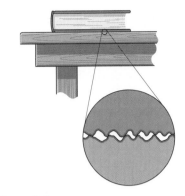

Figure 8.3
The roughnesses of the surfaces can be seen in a magnified view.

Coefficients of Friction

The theory of dry friction, or *Coulomb friction*, predicts the maximum friction forces that can be exerted by dry, contacting surfaces that are stationary relative to each other. It also predicts the friction forces exerted by the surfaces when they are in relative motion, or sliding. We first consider surfaces that are not in relative motion.

The Static Coefficient The magnitude of the *maximum* friction force that can be exerted between two plane dry surfaces in contact is

$$f = \mu_s N, \tag{8.1}$$

where N is the normal component of the contact force between the surfaces and μ_s is a constant called the *coefficient of static friction*.

The value of μ_s is assumed to depend only on the materials of the contacting surfaces and the conditions (smoothness and degree of contamination by other materials) of the surfaces. Typical values of μ_s for various materials are shown in Table 8.1. The relatively large range of values for each pair of materials reflects the sensitivity of μ_s to the conditions of the surfaces. In engineering applications it is usually necessary to measure the value of μ_s for the actual surfaces used.

Figure 8.4
Computer simulation of a bond or "neck" of atoms formed between a nickel tip and a gold surface.

Table 8.1 Typical values of the coefficient of static friction.

Materials	Coefficient of Static Friction μ_s
Metal on metal	0.15–0.20
Masonry on masonry	0.60–0.70
Wood on wood	0.25–0.50
Metal on masonry	0.30–0.70
Metal on wood	0.20–0.60
Rubber on concrete	0.50–0.90

Let's return to the example of the book on the table (Fig. 8.2). If the force F exerted on the book is small enough that the book does not move, the condition for equilibrium requires that the friction force $f = F$. Why do we need the theory of dry friction? If we begin to increase F, the friction force f will increase until the book slips. Equation (8.1) gives the *maximum* friction force that the two surfaces can exert and thus tells us the largest force F that can be applied to the book without causing it to slip. Suppose that we know the coefficient of static friction μ_s between the book and the table and the weight W of the book. Since the normal force $N = W$, the largest value of F that can be applied to the book without causing it to slip is $F = f = \mu_s W$.

Equation (8.1) determines the magnitude of the maximum friction force but not its direction. The friction force is a maximum, and Eq. (8.1) is applicable, when two surfaces are on the verge of slipping relative to each other. We say that slip is *impending*, and the friction forces resist the impending motion. In Fig. 8.5a, suppose that the lower surface is fixed and slip of the upper surface toward the right is impending. The friction force on the upper surface resists its impending motion (Fig. 8.5b). The friction force on the lower surface is in the opposite direction.

The Kinetic Coefficient   According to the theory of dry friction, the magnitude of the friction force between two plane dry contacting surfaces that are in motion (sliding) relative to each other is

$$f = \mu_k N, \tag{8.2}$$

where N is the normal force between the surfaces and μ_k is the *coefficient of kinetic friction*. The value of μ_k is assumed to depend only on the compositions of the surfaces and their conditions. For a given pair of surfaces, its value is generally smaller than that of μ_s.

Once you have caused the book in Fig. 8.2 to begin sliding on the table, the friction force $f = \mu_k N = \mu_k W$. Therefore the force you must exert to keep the book in uniform motion is $F = f = \mu_k W$.

When two surfaces are sliding relative to each other, the friction forces resist the relative motion. In Fig. 8.6a, suppose that the lower surface is fixed and the upper surface is moving to the right. The friction force on the upper surface acts in the direction opposite to its motion (Fig. 8.6b). The friction force on the lower surface is in the opposite direction.

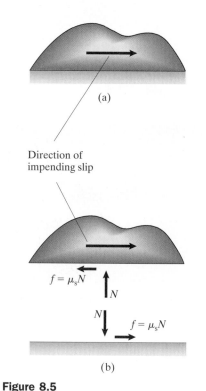

Direction of impending slip

Figure 8.5
(a) The upper surface is on the verge of slipping to the right.
(b) Directions of the friction forces.

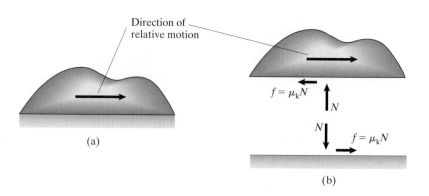

Direction of relative motion

Figure 8.6
(a) The upper surface is moving to the right relative to the lower surface.
(b) Directions of the friction forces.

Angles of Friction

Instead of resolving the reaction exerted on a surface due to its contact with another surface into the normal force N and friction force f (Fig. 8.7a), we can express it in terms of its magnitude R and the *angle of friction* θ between the force and the normal to the surface (Fig. 8.7b). The normal force and friction force are related to R and θ by

$$f = R \sin \theta, \tag{8.3}$$

$$N = R \cos \theta. \tag{8.4}$$

The value of θ when slip is impending is called the *angle of static friction* θ_s, and its value when the surfaces are sliding relative to each other is called the *angle of kinetic friction* θ_k. By using Eqs. (8.1)–(8.4), we can express the angles of static and kinetic friction in terms of the coefficients of friction:

$$\tan \theta_s = \mu_s, \tag{8.5}$$

$$\tan \theta_k = \mu_k. \tag{8.6}$$

In summary, if slip is impending, the magnitude of the friction force is given by Eq. (8.1) and the angle of friction by Eq. (8.5). If surfaces are sliding relative to each other, the magnitude of the friction force is given by Eq. (8.2) and the angle of friction by Eq. (8.6). Otherwise, the friction force must be determined from the equilibrium equations. The sequence of decisions in evaluating the friction force and angle of friction is summarized in Fig. 8.8.

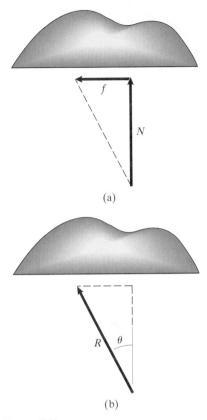

(a)

(b)

Figure 8.7
(a) The normal force N and the friction force f.
(b) The magnitude R and the angle of friction θ.

Study Questions

1. How is the coefficient of static friction defined?
2. How is the coefficient of kinetic friction defined?
3. If relative slip of two dry surfaces in contact is impending, what do you know about the friction forces they exert on each other?
4. If two dry surfaces in contact are sliding relative to each other, what do you know about the resulting friction forces?

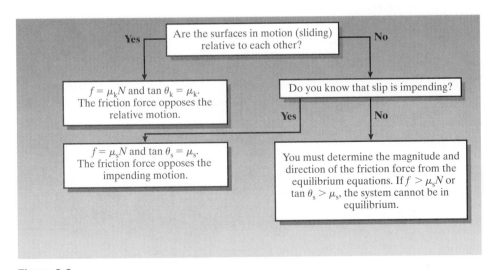

Figure 8.8
Evaluating the friction force.

Example 8.1

Determining the Friction Force

The arrangement in Fig. 8.9 exerts a horizontal force on the stationary 80-kg crate. The coefficient of static friction between the crate and the ramp is $\mu_s = 0.4$.

(a) If the rope exerts a 400-N force on the crate, what is the friction force exerted on the crate by the ramp?

(b) What is the largest force the rope can exert on the crate without causing it to slide up the ramp?

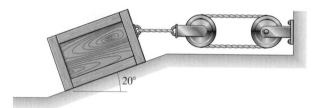

Figure 8.9

Strategy

(a) We can follow the logic in Fig. 8.8 to decide how to evaluate the friction force. The crate is not sliding on the ramp, and we don't know whether slip is impending, so we must determine the friction force by using the equilibrium equations.

(b) We want to determine the value of the force exerted by the rope that causes the crate to be on the verge of slipping up the ramp. When slip is impending, the magnitude of the friction force is $f = \mu_s N$ and the friction force opposes the impending slip. We can use the equilibrium equations to determine the force exerted by the rope.

Solution

(a) We draw the free-body diagram of the crate in Fig. a, showing the force T exerted by the rope, the weight mg of the crate, and the normal force N and friction force f exerted by the ramp. We can choose the direction of f arbitrarily, and our solution will indicate the actual direction of the friction force. By aligning the coordinate system with the ramp as shown, we obtain the equilibrium equation

$$\Sigma F_x = f + T \cos 20° - mg \sin 20° = 0.$$

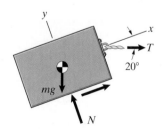

(a) Free-body diagram of the crate.

Solving for the friction force, we obtain

$$f = -T \cos 20° + mg \sin 20° = -(400) \cos 20° + (80)(9.81) \sin 20°$$

$$= -107 \text{ N}.$$

The minus sign indicates that the direction of the friction force on the crate is down the ramp.

(b) The friction force is $f = \mu_s N$, and it opposes the impending slip. To simplify our solution for T, we align the coordinate system as shown in Fig. b, obtaining the equilibrium equations

$$\Sigma F_x = T - N \sin 20° - \mu_s N \cos 20° = 0,$$

$$\Sigma F_y = N \cos 20° - \mu_s N \sin 20° - mg = 0.$$

Solving the second equilibrium equation for N, we obtain

$$N = \frac{mg}{\cos 20° - \mu_s \sin 20°} = \frac{(80)(9.81)}{\cos 20° - (0.4) \sin 20°} = 977 \text{ N}.$$

Then from the first equilibrium equation, T is

$$T = N(\sin 20° + \mu_s \cos 20°) = (977)\left[\sin 20° + (0.4) \cos 20°\right]$$

$$= 702 \text{ N}.$$

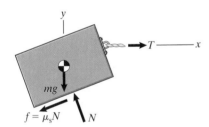

(b) The free-body diagram when slip up the ramp is impending.

Alternative Solution   We can also determine T by representing the reaction exerted on the crate by the ramp as a single force (Fig. c). Because slip of the crate up the ramp is impending, R opposes the impending motion and the friction angle is $\theta_s = \arctan \mu_s = \arctan (0.4) = 21.8°$. From the triangle formed by the sum of the forces acting on the crate (Fig. d), we obtain

$$T = mg \tan(20° + \theta_s) = (80)(9.81) \tan(20° + 21.8°) = 702 \text{ N}.$$

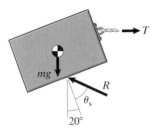

(c) Representing the reaction exerted by the ramp as a single force.

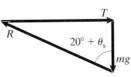

(d) The forces on the crate.

Example 8.2

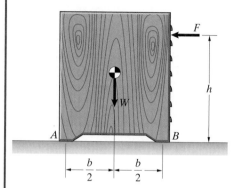

Figure 8.10

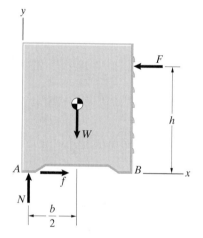

(a) The free-body diagram when the chest is on the verge of tipping over.

Determining Whether an Object Will Tip Over

Suppose that we want to push the tool chest in Fig. 8.10 across the floor by applying the horizontal force F. If we apply the force at too great a height h, the chest will tip over before it slips. If the coefficient of static friction between the floor and the chest is μ_s, what is the largest value of h for which the chest will slip before it tips over?

Strategy

When the chest is on the verge of tipping over, it is in equilibrium with no reaction at B. We can use this condition to determine F in terms of h. Then, by determining the value of F that will cause the chest to slip, we will obtain the value of h that causes the chest to be on the verge of tipping over *and* on the verge of slipping.

Solution

We draw the free-body diagram of the chest when it is on the verge of tipping over in Fig. a. Summing moments about A, we obtain

$$\Sigma M_{(\text{point }A)} = Fh - W\left(\frac{1}{2}b\right) = 0.$$

Equilibrium also requires that $f = F$ and $N = W$.
When the chest is on the verge of slipping,

$$f = \mu_s N,$$

so

$$F = f = \mu_s N = \mu_s W.$$

Substituting this expression into the moment equation, we obtain

$$\mu_s Wh - W\left(\frac{1}{2}b\right) = 0.$$

Solving this equation for h, we find that the chest is on the verge of tipping over *and* on the verge of slipping when

$$h = \frac{b}{2\mu_s}.$$

If h is smaller than this value, the chest will begin sliding before it tips over.

Discussion

Notice that the largest value of h for which the chest will slip before it tips over is independent of F. Whether the chest will tip over depends only on where the force is applied, not how large it is.

Example 8.3

Analyzing a Friction Brake

The motion of the disk in Fig. 8.11 is controlled by the friction force exerted at C by the brake ABC. The hydraulic actuator BE exerts a horizontal force of magnitude F on the brake at B. The coefficients of friction between the disk and the brake are μ_s and μ_k. What couple M is necessary to rotate the disk at a constant rate in the counterclockwise direction?

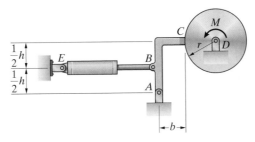

Figure 8.11

Strategy

We can use the free-body diagram of the disk to obtain a relation between M and the reaction exerted on the disk by the brake, then use the free-body diagram of the brake to determine the reaction in terms of F.

Solution

We draw the free-body diagram of the disk in Fig. a, representing the force exerted by the brake by a single force R. The force R opposes the counterclockwise rotation of the disk, and the friction angle is the angle of kinetic friction $\theta_k = \arctan \mu_k$. Summing moments about D, we obtain

$$\Sigma M_{(\text{point } D)} = M - \left(R \sin \theta_k\right)r = 0.$$

Then, from the free-body diagram of the brake (Fig. b), we obtain

$$\Sigma M_{(\text{point } A)} = -F\left(\frac{1}{2}h\right) + \left(R \cos \theta_k\right)h - \left(R \sin \theta_k\right)b = 0.$$

We can solve these two equations for M and R. The solution for the couple M is

$$M = \frac{(1/2)hr\,F\sin\theta_k}{h\cos\theta_k - b\sin\theta_k} = \frac{(1/2)hr\,F\mu_k}{h - b\mu_k}.$$

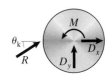

(a) The free-body diagram of the disk.

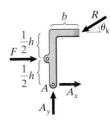

(b) The free-body diagram of the brake.

Discussion

If μ_k is sufficiently small, then the denominator of the solution for the couple, $\left(h\cos\theta_k - b\sin\theta_k\right)$, is positive. As μ_k becomes larger, the denominator becomes smaller, because $\cos\theta_k$ decreases and $\sin\theta_k$ increases. As the denominator approaches zero, the couple required to rotate the disk approaches infinity. To understand this result, notice that the denominator equals zero when $\tan\theta_k = h/b$, which means that the line of action of R passes through point A (Fig. c). As μ_k becomes larger and the line of action of R approaches point A, the magnitude of R necessary to balance the moment of F about A approaches infinity and, as a result, M approaches infinity.

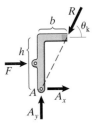

(c) The line of action of R passing through point A.

Example 8.4

A Friction Problem in Three Dimensions

The 80-kg climber at A in Fig. 8.12 is being helped up an icy slope by friends. The tensions in ropes AB and AC are 130 N and 220 N, respectively. The y axis is vertical, and the unit vector $\mathbf{e} = -0.182\mathbf{i} + 0.818\mathbf{j} + 0.545\mathbf{k}$ is perpendicular to the ground where the climber stands. What minimum coefficient of static friction between the climber's shoes and the ground is necessary to prevent him from slipping?

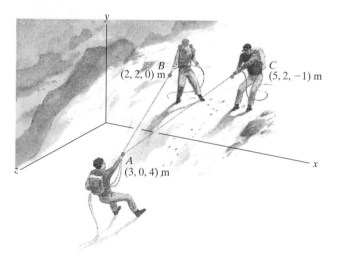

Figure 8.12

Strategy

We know the forces exerted on the climber by the two ropes and by his weight, so we can use equilibrium to determine the force $\mathbf{R}$ exerted on him by the ground. When slip is impending, the angle between $\mathbf{R}$ and the unit vector $\mathbf{e}$ is equal to the angle of static friction θ_s. We can use this condition to calculate the coefficient of static friction for impending slip.

Solution

We draw the free-body diagram of the climber in Fig. a, showing the forces $\mathbf{T}_{AB}$ and $\mathbf{T}_{AC}$ exerted by the ropes, the force $\mathbf{R}$ exerted by the ground, and his weight. The sum of the forces equals zero:

$$\mathbf{R} + \mathbf{T}_{AB} + \mathbf{T}_{AC} - mg\mathbf{j} = \mathbf{0}.$$

By expressing $\mathbf{T}_{AB}$ and $\mathbf{T}_{AC}$ in terms of their components, we can solve this equation for the components of $\mathbf{R}$. The force $\mathbf{T}_{AB}$ is

$$\mathbf{T}_{AB} = |\mathbf{T}_{AB}| \left[\frac{(2-3)\mathbf{i} + (2-0)\mathbf{j} + (0-4)\mathbf{k}}{\sqrt{(2-3)^2 + (2-0)^2 + (0-4)^2}} \right]$$

$$= (130)(-0.218\mathbf{i} + 0.436\mathbf{j} - 0.873\mathbf{k})$$

$$= -28.4\mathbf{i} + 56.7\mathbf{j} - 113.5\mathbf{k} \ (\text{N}),$$

and the force $\mathbf{T}_{AC}$ is

$$\mathbf{T}_{AC} = |\mathbf{T}_{AC}| \left[\frac{(5-3)\mathbf{i} + (2-0)\mathbf{j} + (-1-4)\mathbf{k}}{\sqrt{(5-3)^2 + (2-0)^2 + (-1-4)^2}} \right]$$

$$= (220)(0.348\mathbf{i} + 0.348\mathbf{j} - 0.870\mathbf{k})$$

$$= 76.6\mathbf{i} + 76.6\mathbf{j} - 191.5\mathbf{k} \ (\text{N}).$$

Substituting these expressions into the equilibrium equation and solving for $\mathbf{R}$, we obtain

$$\mathbf{R} = -48.2\mathbf{i} + 651.5\mathbf{j} + 305.0\mathbf{k} \ (\text{N}).$$

To determine the angle θ between $\mathbf{R}$ and the unit vector $\mathbf{e}$ that is normal to the surface on which the climber stands (Fig. b), we use the dot product. From the definition $\mathbf{R} \cdot \mathbf{e} = |\mathbf{R}| |\mathbf{e}| \cos\theta$, we obtain

$$\cos\theta = \frac{\mathbf{R} \cdot \mathbf{e}}{|\mathbf{R}| |\mathbf{e}|} = \frac{(-48.2)(-0.182) + (651.5)(0.818) + (305.0)(0.545)}{\sqrt{(-48.2)^2 + (651.5)^2 + (305.0)^2}}$$

$$= 0.982.$$

The angle $\theta = 10.9°$. Setting this angle equal to the angle of static friction, we obtain the coefficient of static friction for impending slip:

$$\mu_s = \tan\theta_s = \tan 10.9° = 0.193.$$

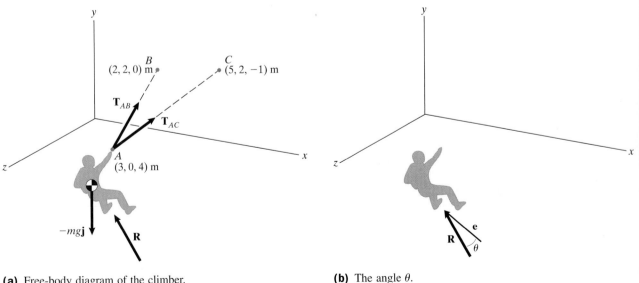

(a) Free-body diagram of the climber.

(b) The angle θ.

Problems

8.1 The coefficients of static and kinetic friction between the 0.4-kg book and the table are $\mu_s = 0.30$ and $\mu_k = 0.28$. A person exerts a horizontal force on the book as shown.
(a) If the magnitude of the force is 1 N and the book remains stationary, what is the magnitude of the friction force exerted on the book by the table?
(b) What is the largest force the person can exert without causing the book to slip?
(c) If the person pushes the book across the table at a constant speed, what is the magnitude of the friction force?

P8.1

8.2 The 10.5-kg Sojourner rover, placed on the surface of Mars by the Pathfinder Lander on July 4, 1997, was designed to negotiate a 45° slope without tipping over.
(a) What minimum static coefficient of friction between the wheels of the rover and the surface is necessary for it to rest on a 45° slope? The acceleration due to gravity at the surface of Mars is 3.69 m/s².
(b) Engineers testing the Sojourner on Earth want to confirm that it will negotiate a 45° slope without tipping over. What minimum static coefficient of friction between the wheels of the rover and the surface is necessary for it to rest on a 45° slope on Earth?

P8.2

8.3 The coefficient of static friction between the tires of the 8000-kg truck and the road is $\mu_s = 0.6$.
(a) If the truck is stationary on the incline and $\alpha = 15°$, what is the magnitude of the total friction force exerted on the tires by the road?
(b) What is the largest value of α for which the truck will not slip?

P8.3

8.4 The coefficient of static friction between the 5-kg box and the inclined surface is $\mu_s = 0.3$. The force F is horizontal and the box is stationary.
(a) If $F = 40$ N, what friction force is exerted on the box by the inclined surface?
(b) What is the largest value of F for which the box will not slip?

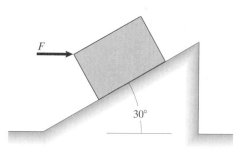

P8.4

8.5 In Problem 8.4, what is the smallest value of the force F for which the box will not slip?

8.6 The device shown is designed to position pieces of luggage on a ramp. It exerts a force parallel to the ramp. The mass of the suitcase S is 9 kg. The coefficients of friction between the suitcase and ramp are $\mu_s = 0.20$ and $\mu_k = 0.18$.
(a) Will the suitcase remain stationary on the ramp when the device exerts no force on it?
(b) What force must the device exert to start the suitcase moving up the ramp?
(c) What force must the device exert to move the suitcase up the ramp at a constant speed?

P8.6

8.7 The mass of the stationary crate is 40 kg. The length of the spring is 180 mm, its unstretched length is 200 mm, and the spring constant is $k = 2500$ N/mm. The coefficient of static friction between the crate and the inclined surface is $\mu_s = 0.6$. Determine the magnitude of the friction force exerted on the crate.

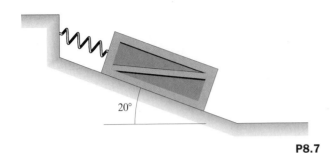

P8.7

8.8 The coefficient of kinetic friction between the 40-kg crate and the floor is $\mu_k = 0.3$. If the angle $\alpha = 20°$, what tension must the person exert on the rope to move the crate at constant speed?

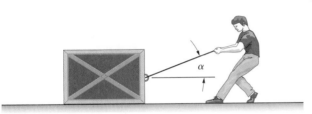

P8.8

8.9 In Problem 8.8, for what angle α is the tension necessary to move the crate at constant speed a minimum? What is the necessary tension?

8.10 Box A weighs 100 lb, and box B weighs 30 lb. The coefficients of friction between box A and the ramp are $\mu_s = 0.30$ and $\mu_k = 0.28$. What is the magnitude of the friction force exerted on box A by the ramp?

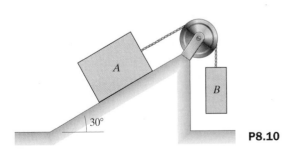

P8.10

8.11 In Problem 8.10, box A weighs 100 lb, and the coefficients of friction between box A and the ramp are $\mu_s = 0.30$ and $\mu_k = 0.28$. For what range of weights of the box B will the system remain stationary?

8.12 The mass of the box on the left is 30 kg, and the mass of the box on the right is 40 kg. The coefficient of static friction between each box and the inclined surface is $\mu_s = 0.2$. Determine the minimum angle α for which the boxes will remain stationary.

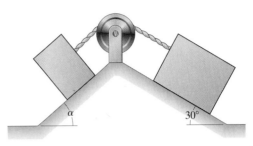

P8.12

8.13 In Problem 8.12, determine the maximum angle α for which the boxes will remain stationary.

8.14 The box is stationary on the inclined surface. The coefficient of static friction between the box and the surface is μ_s.
(a) If the mass of the box is 10 kg, $\alpha = 20°$, $\beta = 30°$, and $\mu_s = 0.24$, what force T is necessary to start the box sliding up the surface?
(b) Show that the force T necessary to start the box sliding up the surface is a minimum when $\tan \beta = \mu_s$.

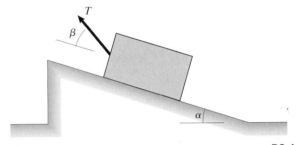

P8.14

8.15 To explain observations of ship launchings at the port of Rochefort in 1779, Coulomb analyzed the system shown in Problem 8.14 to determine the minimum force T necessary to hold the box stationary on the inclined surface. Show that the result is

$$T = \frac{(\sin \alpha - \mu_s \cos \alpha)mg}{\cos \beta - \mu_s \sin \beta}.$$

8.16 Two sheets of plywood A and B lie on the bed of the truck. They have the same weight W, and the coefficient of static friction between the two sheets of wood and between sheet B and the truck bed is μ_s.
(a) If you apply a horizontal force to sheet A and apply no force to sheet B, can you slide sheet A off the truck without causing sheet B to move? What force is necessary to cause sheet A to start moving?

(b) If you prevent sheet *A* from moving by exerting a horizontal force on it, what horizontal force on sheet *B* is necessary to start it moving?

P8.16

8.17 Suppose that the truck in Problem 8.16 is loaded with *N* sheets of plywood of the same weight *W*, labeled (from the top) sheets 1, 2, ..., *N*. The coefficient of static friction between the sheets of wood and between the bottom sheet and the truck bed is μ_s. If you apply a horizontal force to the sheets above it to prevent them from moving, can you pull out the *i*th sheet, $1 \le i \le N$, without causing any of the sheets below it to move? What force must you apply to cause it to start moving?

8.18 The masses of the two boxes are $m_1 = 45$ kg and $m_2 = 20$ kg. The coefficients of friction between the left box and the inclined surface are $\mu_s = 0.12$ and $\mu_k = 0.10$. Determine the tension the man must exert on the rope to pull the boxes upward at a constant rate.

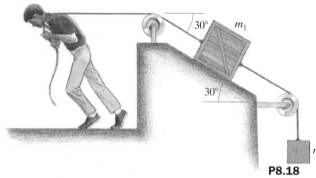

P8.18

8.19 In Problem 8.18, for what range of tensions exerted on the rope by the man will the boxes remain stationary?

8.20 The coefficient of static friction between the two boxes is $\mu_s = 0.2$, and between the lower box and the inclined surface it is

$\mu_s = 0.32$. What is the largest angle α for which the lower box will not slip?

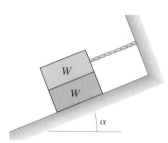

P8.20

8.21 The coefficient of static friction between the two boxes and between the lower box and the inclined surface is μ_s. What is the largest force *F* that will not cause the boxes to slip?

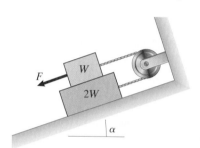

P8.21

8.22 Consider the system shown in Problem 8.21. The coefficient of static friction between the two boxes and between the lower box and the inclined surface is μ_s. If $F = 0$, the lower box will slip down the inclined surface. What is the smallest force *F* for which the boxes will not slip?

8.23 A sander consists of a rotating cylinder with sandpaper bonded to the outer surface. The normal force exerted on the workpiece *A* by the sander is 30 lb. The workpiece *A* weighs 50 lb. The coefficients of friction between the sander and the workpiece *A* are $\mu_s = 0.65$ and $\mu_k = 0.60$. The coefficients of friction between the workpiece *A* and the table are $\mu_s = 0.35$ and $\mu_k = 0.30$. Will the workpiece remain stationary while it is being sanded?

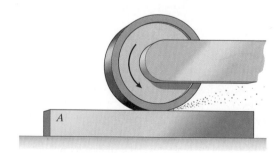

P8.23

8.24 Suppose that you want the bar of length L to act as a simple brake that will allow the workpiece A to slide to the left but will not allow it to slide to the right no matter how large a horizontal force is applied to it. The weight of the bar is W, and the coefficient of static friction between it and the workpiece A is μ_s. You can neglect friction between the workpiece and the surface it rests on.
(a) What is the largest angle α for which the bar will prevent the workpiece from moving to the right?
(b) If α has the value determined in (a), what horizontal force is necessary to slide the workpiece A toward the left at a constant rate?

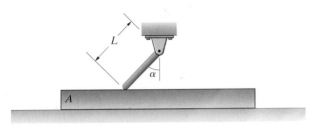

P8.24

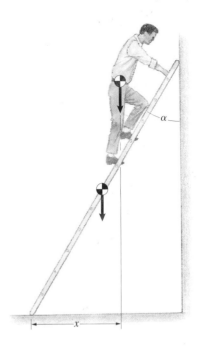

P8.26

8.25 The coefficient of static friction between the 20-lb bar and the floor is $\mu_s = 0.3$. Neglect friction between the bar and the wall.
(a) If $\alpha = 20°$, what is the magnitude of the friction force exerted on the bar by the floor?
(b) What is the maximum value of α for which the bar will not slip?

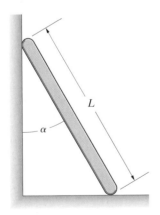

P8.25

8.26 The masses of the ladder and the person are 18 kg and 90 kg, respectively. The center of mass of the 4-m ladder is at its midpoint. If $\alpha = 30°$, what is the minimum coefficient of static friction between the ladder and the floor necessary for the person to climb to the top of the ladder? Neglect friction between the ladder and the wall.

8.27 In Problem 8.26, the coefficient of static friction between the ladder and the floor is $\mu_s = 0.6$. The masses of the ladder and the person are 18 kg and 100 kg, respectively. The center of mass of the 4-m ladder is at its midpoint. What is the maximum value of α for which the person can climb to the top of the ladder? Neglect friction between the ladder and the wall.

8.28 In Problem 8.26, the coefficient of static friction between the ladder and the floor is $\mu_s = 0.6$, and $\alpha = 35°$. The center of mass of the 4-m ladder is at its midpoint, and its mass is 18 kg.
(a) If a football player with a mass of 140 kg attempts to climb the ladder, what maximum value of x will he reach? Neglect friction between the ladder and the wall.
(b) What minimum friction coefficient would be required for him to reach the top of the ladder?

8.29 The disk weighs 50 lb. Neglect the weight of the bar. The coefficients of friction between the disk and the floor are $\mu_s = 0.6$ and $\mu_k = 0.4$.

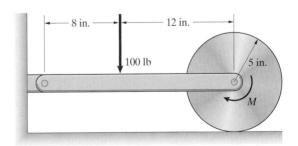

P8.29

(a) What is the largest couple M that can be applied to the stationary disk without causing it to start rotating?

(b) What couple M is necessary to rotate the disk at a constant rate?

8.30 The cylinder has weight W. The coefficient of static friction between the cylinder and the floor and between the cylinder and the wall is μ_s. What is the largest couple M that can be applied to the stationary cylinder without causing it to rotate?

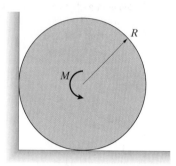

P8.30

8.31 The cylinder has weight W. The coefficient of static friction between the cylinder and the floor and between the cylinder and the wall is μ_s. What is the largest couple M that can be applied to the stationary cylinder without causing it to rotate?

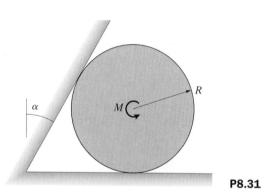

P8.31

8.32 Suppose that $\alpha = 30°$ in Problem 8.31 and that a couple $M = 0.5RW$ is required to turn the cylinder at a constant rate. What is the coefficient of kinetic friction?

8.33 The disk of weight W and radius R is held in equilibrium on the circular surface by a couple M. The coefficient of static friction between the disk and the surface is μ_s. Show that the largest value M can have without causing the disk to slip is

$$M = \frac{\mu_s RW}{\sqrt{1 + \mu_s^2}}.$$

P8.33

8.34 The coefficient of static friction between the jaws of the pliers and the gripped object is μ_s. What is the largest value of the angle α for which the gripped object will not slip out? (Neglect the object's weight.)

Strategy: Draw the free-body diagram of the gripped object, and assume that slip is impending.

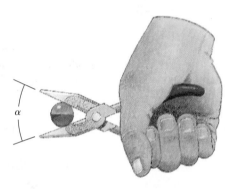

P8.34

8.35 The stationary disk, of 300-mm radius, is attached to a pin support at D. The disk is held in place by the brake ABC in contact with the disk at C. The hydraulic actuator BE exerts a horizontal 400-N force on the brake at B. The coefficients of friction between the disk and the brake are $\mu_s = 0.6$ and $\mu_k = 0.5$. What couple must be applied to the stationary disk to cause it to slip in the counterclockwise direction?

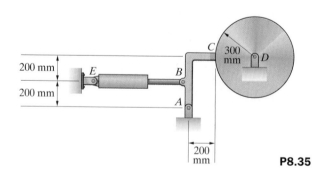

P8.35

8.36 What couple must be applied to the stationary disk in Problem 8.35 to cause it to slip in the clockwise direction?

8.37 The mass of block B is 8 kg. The coefficient of static friction between the surfaces of the clamp and the block is $\mu_s = 0.2$. When the clamp is aligned as shown, what minimum force must the spring exert to prevent the block from slipping out?

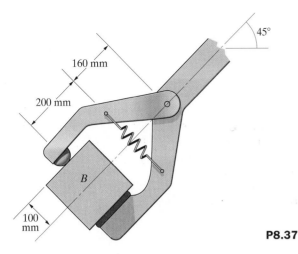

P8.37

8.38 By altering its dimensions, redesign the clamp in Problem 8.37 so that the minimum force the spring must exert to prevent the block from slipping out is 180 N. Draw a sketch of your new design.

8.39 The horizontal bar is attached to a collar that slides on the smooth vertical bar. The collar at P slides on the smooth

horizontal bar. The total mass of the horizontal bar and the two collars is 12 kg. The system is held in place by the pin in the circular slot. The pin contacts only the lower surface of the slot, and the coefficient of static friction between the pin and the slot is 0.8. If the system is in equilibrium and y = 260 mm, what is the magnitude of the friction force exerted on the pin by the slot?

8.40 In Problem 8.39, what is the minimum height y at which the system can be in equilibrium?

8.41 The rectangular 100-lb plate is supported by the pins A and B. If friction can be neglected at A and the coefficient of static friction between the pin at B and the slot is $\mu_s = 0.4$, what is the largest angle α for which the plate will not slip?

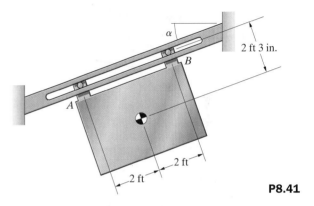

P8.41

8.42 If you can neglect friction at B in Problem 8.41 and the coefficient of static friction between the pin at A and the slot is $\mu_s = 0.4$, what is the largest angle α for which the plate will not slip?

8.43 The airplane's weight is W = 2400 lb. Its brakes keep the rear wheels locked, and the coefficient of static friction between the wheels and the runway is $\mu_s = 0.6$. The front (nose) wheel can turn freely and so exerts only a normal force on the runway. Determine the largest horizontal thrust force T the plane's propeller can generate without causing the rear wheels to slip.

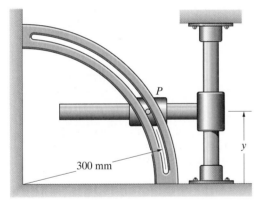

P8.39

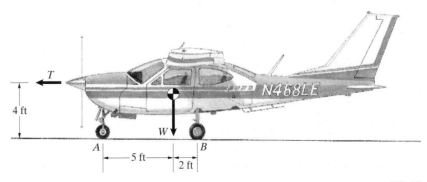

P8.43

8.44 The refrigerator weighs 350 lb. The distances $h = 60$ in. and $b = 14$ in. The coefficient of static friction at A and B is $\mu_s = 0.24$.
(a) What force F is necessary for impending slip?
(b) Will the refrigerator tip over before it slips?

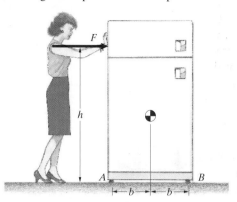

P8.44

8.45 If you want the refrigerator in Problem 8.44 to slip before it tips over, what is the maximum height h at which you can push it?

8.46 To obtain a preliminary evaluation of the stability of a turning car, imagine subjecting the stationary car to an increasing lateral force F at the height of its center of mass, and determine whether the car will slip (skid) laterally before it tips over. Show that this will be the case if $b/h > 2\mu_s$. (Notice the importance of the height of the center of mass relative to the width of the car. This reflects on recent discussions of the stability of sport utility vehicles and vans that have relatively high centers of mass.)

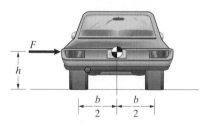

P8.46

8.47 The man exerts a force P on the car at an angle $\alpha = 20°$. The 1760-kg car has front wheel drive. The driver spins the front wheels, and the coefficient of kinetic friction is $\mu_k = 0.02$. Snow behind the rear tires exerts a horizontal resisting force S. Getting the car to move requires overcoming a resisting force $S = 420$ N. What force P must the man exert?

8.48 In Problem 8.47, what value of the angle α minimizes the magnitude of the force P the man must exert to overcome the resisting force $S = 420$ N exerted on the rear tires by the snow? What force must he exert?

8.49 The coefficient of static friction between the 3000-lb car's tires and the road is $\mu_s = 0.5$. Determine the steepest grade (the largest value of the angle α) the car can drive up at constant speed if the car has (a) rear-wheel drive; (b) front-wheel drive; (c) fourwheel drive.

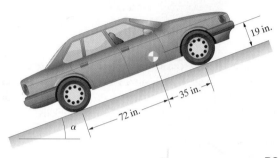

P8.49

8.50 The stationary cabinet has weight W. Determine the force F that must be exerted to cause it to move if (a) the coefficient of static friction at A and at B is μ_s; (b) the coefficient of static friction at A is μ_{sA} and the coefficient of static friction at B is μ_{sB}.

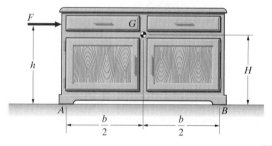

P8.50

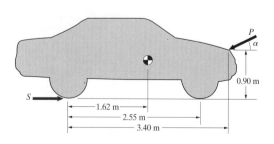

P8.47

Effects of friction forces, such as wear, loss of energy, and generation of heat, are often undesirable. But many devices cannot function properly without friction forces and may actually be designed to create them. A car's brakes work by exerting friction forces on the rotating wheels, and its tires are designed to maximize the friction forces they exert on the road under various weather conditions. In this section we analyze several types of devices in which friction forces play important roles.

Wedges

A *wedge* is a bifacial tool with the faces set at a small acute angle (Figs. 8.13a and b). When a wedge is pushed forward, the faces exert large lateral forces as a result of the small angle between them (Fig. 8.13c). In various forms, wedges are used in many engineering applications.

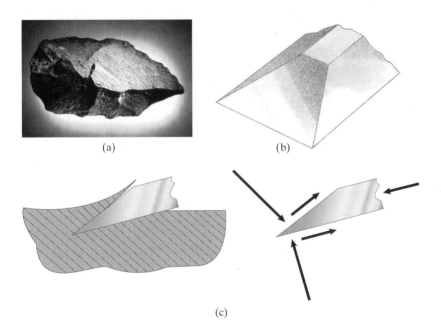

(a)

(b)

(c)

Figure 8.13
(a) An early wedge tool—a bifacial "hand axe" from Olduvai Gorge, Tanzania.
(b) A modern chisel blade.
(c) The faces of a wedge can exert large lateral forces.

The large lateral force generated by a wedge can be used to lift a load (Fig. 8.14a). Let W_L be the weight of the load and W_W the weight of the wedge. To determine the force F necessary to start raising the load, we assume that slip of the load and wedge are impending (Fig. 8.14b). From the free-body diagram of the load, we obtain the equilibrium equations

$$\Sigma F_x = Q - N \sin\alpha - \mu_s N \cos\alpha = 0,$$

$$\Sigma F_y = N \cos\alpha - \mu_s N \sin\alpha - \mu_s Q - W_L = 0.$$

From the free-body diagram of the wedge, we obtain the equations

$$\Sigma F_x = N \sin\alpha + \mu_s N \cos\alpha + \mu_s P - F = 0,$$

$$\Sigma F_y = P - N \cos\alpha + \mu_s N \sin\alpha - W_W = 0.$$

These four equations determine the three normal forces Q, N, and P and the force F. The solution for F is

$$F = \mu_s W_W + \left[\frac{(1 - \mu_s^2) \tan\alpha + 2\mu_s}{(1 - \mu_s^2) - 2\mu_s \tan\alpha} \right] W_L.$$

Suppose that $W_W = 0.2W_L$ and $\alpha = 10°$. If $\mu_s = 0$, the force necessary to lift the load is only $0.176W_L$. But if $\mu_s = 0.2$, the force becomes $0.680W_L$, and if $\mu_s = 0.4$, it becomes $1.44W_L$. From this standpoint, friction is undesirable. But if there were no friction, the wedge would not remain in place when the force F is removed.

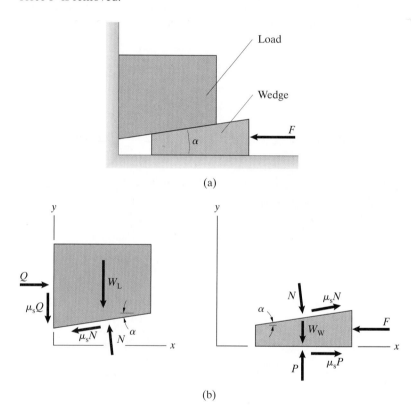

(a)

(b)

Figure 8.14
(a) Raising a load with a wedge.
(b) Free-body diagrams of the load and the wedge when slip is impending.

Example 8.5

Forces on a Wedge

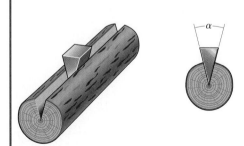

Figure 8.15

Splitting a log must have been among the first applications of the wedge (Fig. 8.15). Although it is a dynamic process—the wedge is hammered into the wood—you can get an idea of the forces involved from a static analysis. Suppose that $\alpha = 10°$ and the coefficients of friction between the surfaces of the wedge and the log are $\mu_s = 0.22$ and $\mu_k = 0.20$. Neglect the weight of the wedge.
(a) If the wedge is driven into the log at a constant rate by a vertical force F, what are the magnitudes of the normal forces exerted on the log by the wedge?
(b) Will the wedge remain in place in the log when the force is removed?

Strategy

(a) The friction forces resist the motion of the wedge into the log and are equal to $\mu_k N$, where N is the normal force the log exerts on the faces. We can use equilibrium to determine N in terms of F.

(b) By assuming that the wedge is on the verge of slipping out of the log, we can determine the minimum value of μ_s necessary for the wedge to stay in place.

Solution

(a) In Fig. a we draw the free-body diagram of the wedge as it is pushed into the log by a force F. The faces of the wedge are subjected to normal forces and friction forces by the log. The friction forces resist the motion of the wedge. From the equilibrium equation

$$2N \sin\left(\frac{\alpha}{2}\right) + 2\mu_k N \cos\left(\frac{\alpha}{2}\right) - F = 0,$$

we obtain the normal force N:

$$N = \frac{F}{2\left[\sin(\alpha/2) + \mu_k \cos(\alpha/2)\right]} = \frac{F}{2\left[\sin(10°/2) + (0.20)\cos(10°/2)\right]}$$

$$= 1.75F.$$

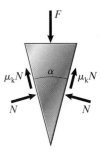

(a) Free-body diagram of the wedge with a vertical force F applied to it.

(b) In Fig. b we draw the free-body diagram when $F = 0$ and the wedge is on the verge of slipping out. From the equilibrium equation

$$2N \sin\left(\frac{\alpha}{2}\right) - 2\mu_s N \cos\left(\frac{\alpha}{2}\right) = 0,$$

we obtain the minimum coefficient of friction necessary for the wedge to remain in place:

$$\mu_s = \tan\left(\frac{\alpha}{2}\right) = \tan\left(\frac{10°}{2}\right) = 0.087.$$

(b) Free-body diagram of the wedge when it is on the verge of slipping out.

We can also obtain this result by representing the reaction exerted on the wedge by the log as a single force (Fig. c). When the wedge is on the verge of slipping out, the friction angle is the angle of static friction θ_s. The sum of the forces in the vertical direction is zero only if

$$\theta_s = \arctan(\mu_s) = \frac{\alpha}{2} = 5°,$$

so $\mu_s = \tan 5° = 0.087$. Thus we conclude that the wedge will remain in place.

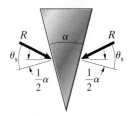

(c) Representing the reactions by a single force.

Threads

Threads are familiar from their use on wood screws, machine screws, and other machine elements. We show a shaft with square threads in Fig. 8.16a. The axial distance p from one thread to the next is called the *pitch* of the thread, and the angle α is its *slope*. We will consider only the case in which the shaft has a single continuous thread, so the relation between the pitch and slope is

$$\tan \alpha = \frac{p}{2\pi r}, \tag{8.7}$$

where r is the mean radius of the thread.

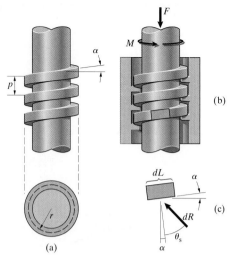

(b)

(c)

Figure 8.16
(a) A shaft with a square thread.
(b) The shaft within a sleeve with a mating groove and the direction of M that can cause the shaft to start moving in the axial direction opposite to F.
(c) A differential element of the thread when slip is impending.

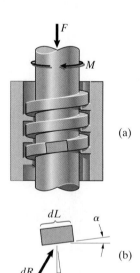

(a)

(b)

Figure 8.17
(a) The direction of M that can cause the shaft to move in the axial direction of F.
(b) A differential element of the thread when slip is impending.

Suppose that the threaded shaft is enclosed in a fixed sleeve with a mating groove and is subjected to an axial load F (Fig. 8.16b). Applying a couple M in the direction shown will tend to cause the shaft to start rotating and moving in the axial direction opposite to F. Our objective is to determine the couple M necessary to cause the shaft to start rotating.

We draw the free-body diagram of a differential element of the thread of length dL in Fig. 8.16c, representing the reaction exerted by the mating groove by the force dR. If the shaft is on the verge of rotating, dR resists the impending motion and the friction angle is the angle of static friction θ_s. The vertical component of the reaction on the element is $dR\cos(\theta_s + \alpha)$. To determine the total vertical force on the thread, we must integrate this expression over the length L of the thread. For equilibrium, the result must equal the axial force F acting on the shaft:

$$\cos(\theta_s + \alpha)\int_L dR = F. \tag{8.8}$$

The moment about the center of the shaft due to the reaction on the element is $r\,dR\sin(\theta_s + \alpha)$. The total moment must equal the couple M exerted on the shaft:

$$r\sin(\theta_s + \alpha)\int_L dR = M.$$

Dividing this equation by Eq. (8.8), we obtain the couple M necessary for the shaft to be on the verge of rotating and moving in the axial direction opposite to F:

$$M = rF\tan(\theta_s + \alpha). \tag{8.9}$$

Replacing the angle of static friction θ_s in this expression with the angle of kinetic friction θ_k gives the couple required to cause the shaft to rotate at a constant rate.

If the couple M is applied to the shaft in the opposite direction (Fig. 8.17a), the shaft tends to start rotating and moving in the axial direction of the load F. Figure 8.17b shows the reaction on a differential element of the thread of length dL when slip is impending. The direction of the reaction opposes the rotation of the shaft. In this case, the vertical component of the reaction on the element is $dR\cos(\theta_s - \alpha)$. Equilibrium requires that

$$\cos(\theta_s - \alpha)\int_L dR = F. \tag{8.10}$$

The moment about the center of the shaft due to the reaction is $r\,dR\sin(\theta_s - \alpha)$, so

$$r\sin(\theta_s - \alpha)\int_L dR = M.$$

Dividing this equation by Eq. (8.10), we obtain the couple M necessary for the shaft to be on the verge of rotating and moving in the direction of the force F:

$$M = rF\tan(\theta_s - \alpha). \tag{8.11}$$

Replacing θ_s with θ_k in this expression gives the couple necessary to rotate the shaft at a constant rate.

Notice in Eq. (8.11) that the couple required for impending motion is zero when $\theta_s = \alpha$. When the angle of static friction is less than this value, the shaft will rotate and move in the direction of the force F with no couple applied.

Study Questions

1. How is the slope α of a thread defined?
2. If you know the pitch and mean radius of a thread, how do you determine its slope?
3. If a threaded shaft is subjected to an axial load, how do you determine the couple necessary to rotate the shaft at a constant rate and cause it to move in the direction opposite to the direction of the axial load?

Example 8.6

Rotating a Threaded Collar

The right end of bar AB in Fig. 8.18 is pinned to an unthreaded collar B that rests on the threaded collar C. The mean radius of the thread is $r = 40$ mm and its pitch is $p = 5$ mm. The coefficients of static and kinetic friction between the threads of the collar C and those of the threaded shaft are $\mu_s = 0.25$ and $\mu_k = 0.22$. The 180-kg suspended object can be raised or lowered by turning the collar C.

(a) When the system is in the position shown, what couple must be applied to the collar C to rotate it at a constant rate and cause the suspended object to move upward?

(b) Will the system remain in equilibrium in the position shown if no couple is applied to the collar C?

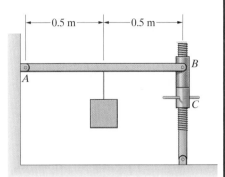

Figure 8.18

Strategy

(a) By drawing the free-body diagram of the bar and collar B, we can determine the axial force exerted on collar C. Then we can use Eq. (8.9), with θ_s replaced by θ_k, to determine the required couple.

(b) From Eq. (8.11), the collar C is on the verge of rotating and moving in the direction of the axial load when no couple is exerted on it if $\theta_s = \alpha$. If the angle of static friction θ_s is greater than or equal to the slope α, the system will remain in equilibrium with no couple applied.

Solution

(a) We draw the free-body diagram of the bar and collar B in Fig. a, where F is the force exerted on the collar B by the collar C. From the equilibrium equation

$$\Sigma M_{(\text{point } A)} = (1.0)F - (0.5)mg = 0,$$

we obtain $F = \frac{1}{2}mg = \frac{1}{2}(180)(9.81) = 883$ N. This is the axial force exerted on collar C (Fig. b). Replacing θ_s by θ_k in Eq. (8.9), the couple necessary to rotate the collar at a constant rate is

$$M = rF \tan(\theta_k + \alpha).$$

The slope α is related to the pitch and mean radius of the thread by Eq. (8.7):

$$\tan \alpha = \frac{p}{2\pi r} = \frac{0.005}{2\pi(0.04)} = 0.0199.$$

We obtain $\alpha = \arctan(0.0199) = 1.14°$. The angle of kinetic friction is

$$\theta_k = \arctan(\mu_k) = \arctan(0.22) = 12.41°.$$

Using these values, the required couple is

$$
\begin{aligned}
M &= rF \tan(\theta_k + \alpha) \\
&= (0.04)(883) \tan(12.41° + 1.14°) \\
&= 8.51 \text{ N-m.}
\end{aligned}
$$

(b) The angle of static friction is

$$\theta_s = \arctan(\mu_s) = \arctan(0.25) = 14.04°.$$

Therefore θ_s is greater than the slope α, and we conclude from Eq. (8.11) that the system will remain in equilibrium with no couple applied to collar C.

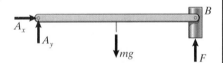

(a) Free-body diagram of bar AB and the collar B.

(b) The threaded shaft and the collar C.

Problems

8.51 A force $F = 200$ N is necessary to raise the block A at a constant rate. The mass of the wedge B is negligible. Between all of the contacting surfaces, $\mu_s = 0.28$ and $\mu_k = 0.26$. What is the mass of block A?

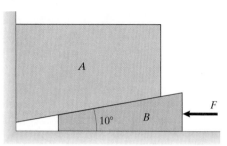

P8.51

8.52 In Problem 8.51, suppose that the mass of block A is 30 kg and the mass of the wedge B is 5 kg. What force F is necessary to start the wedge B moving to the left?

8.53 The wedge shown is being used to split the log. The wedge weighs 20 lb and the angle α equals 30°. The coefficient of kinetic friction between the faces of the wedge and the log is 0.28. If the normal force exerted by each face of the wedge must equal 150 lb to split the log, what vertical force F is necessary to drive the wedge into the log at a constant rate?

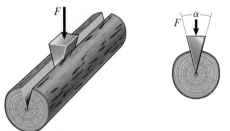

P8.53

8.54 The coefficient of static friction between the faces of the wedge and the log in Problem 8.53 is 0.30. Will the wedge remain in place in the log when the vertical force F is removed?

8.55 The masses of A and B are 42 kg and 50 kg, respectively. Between all contacting surfaces, $\mu_s = 0.05$. What force F is required to start A moving to the right?

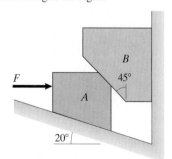

P8.55

8.56 The stationary blocks A, B, and C each have a mass of 200 kg. Between all contacting surfaces, $\mu_s = 0.6$. What force F is necessary to start B moving downward?

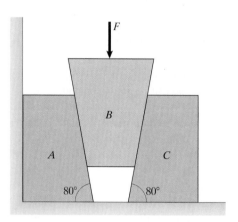

P8.56

8.57 Small wedges called *shims* can be used to hold an object in place. The coefficient of kinetic friction between the contacting surfaces is 0.4. What force F is needed to push the shim downward until the horizontal force exerted on the object A is 200 N?

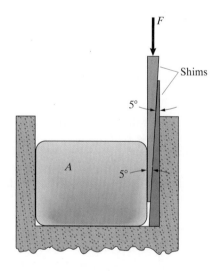

P8.57

8.58 The coefficient of static friction between the contacting surfaces in Problem 8.57 is 0.44. If the shims are in place and exert a 200-N horizontal force on the object A, what upward force must be exerted on the left shim to loosen it?

8.59 The crate A weighs 600 lb. Between all contacting surfaces, $\mu_s = 0.32$ and $\mu_k = 0.30$. Neglect the weights of the wedges. What force F is required to move A to the right at a constant rate?

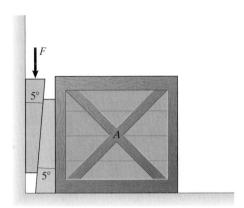

P8.59

8.60 Suppose that between all contacting surfaces in Problem 8.59, $\mu_s = 0.32$ and $\mu_k = 0.30$. Neglect the weights of the $5°$ wedges. If a force $F = 800$ N is required to move A to the right at a constant rate, what is the mass of A?

8.61 The box A has a mass of 80 kg, and the wedge B has a mass of 40 kg. Between all contacting surfaces, $\mu_s = 0.15$ and $\mu_k = 0.12$. What force F is required to raise A at a constant rate?

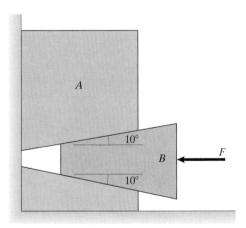

P8.61

8.62 Suppose that in Problem 8.61, A weighs 800 lb and B weighs 400 lb. The coefficients of friction between all of the contacting surfaces are $\mu_s = 0.15$ and $\mu_k = 0.12$. Will B remain in place if the force F is removed?

8.63 Between A and B, $\mu_s = 0.20$, and between B and C, $\mu_s = 0.18$. Between C and the wall, $\mu_s = 0.30$. The weights $W_B = 20$ lb and $W_C = 80$ lb. What force F is required to start C moving upward?

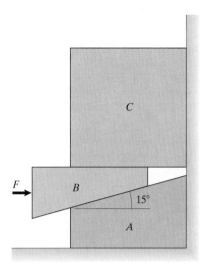

P8.63

8.64 The masses of A, B, and C are 8 kg, 12 kg, and 80 kg, respectively. Between all contacting surfaces, $\mu_s = 0.4$. What force F is required to start C moving upward?

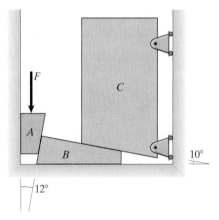

P8.64

8.65 The vertical threaded shaft fits into a mating groove in the tube C. The pitch of the threaded shaft is $p = 0.1$ in., and the mean radius of the thread is $r = 0.5$ in. The coefficients of friction between the thread and the mating groove are $\mu_s = 0.15$ and

P8.65

$\mu_k = 0.10$. The weight $W = 200$ lb. Neglect the weight of the threaded shaft.

(a) Will the stationary threaded shaft support the weight if no couple is applied to the shaft?

(b) What couple must be applied to the threaded shaft to raise the weight at a constant rate?

8.66 Suppose that in Problem 8.65, the pitch of the threaded shaft is $p = 2$ mm and the mean radius of the thread is $r = 20$ mm. The coefficients of friction between the thread and the mating groove are $\mu_s = 0.22$ and $\mu_k = 0.20$. The weight $W = 500$ N. Neglect the weight of the threaded shaft. What couple must be applied to the threaded shaft to lower the weight at a constant rate?

8.67 The position of the horizontal beam can be adjusted by turning the machine screw A. Neglect the weight of the beam. The pitch of the screw is $p = 1$ mm, and the mean radius of the thread is $r = 4$ mm. The coefficients of friction between the thread and the mating groove are $\mu_s = 0.20$ and $\mu_k = 0.18$. If the system is initially stationary, determine the couple that must be applied to the screw to cause the beam to start moving (a) upward; (b) downward.

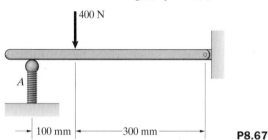

P8.67

8.68 Suppose that in Problem 8.67, the pitch of the machine screw is $p = 1$ mm and the mean radius of the thread is $r = 4$ mm. What minimum value of the coefficient of static friction between the thread and the mating groove is necessary for the beam to remain in the position shown with no couple applied to the screw?

8.69 The mass of block A is 60 kg. Neglect the weight of the 5° wedge. The coefficient of kinetic friction between the contacting surfaces of the block A, the wedge, the table, and the wall is $\mu_k = 0.4$. The pitch of the threaded shaft is 5 mm, the mean radius of the thread is 15 mm, and the coefficient of kinetic friction between the thread and the mating groove is 0.2. What couple must be exerted on the threaded shaft to raise the block A at a constant rate?

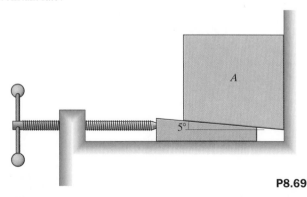

P8.69

8.70 The vise exerts 80-lb forces on A. The threaded shafts are subjected only to axial loads by the jaws of the vise. The pitch of their threads is $p = 1/8$ in., the mean radius of the threads is $r = 1$ in., and the coefficient of static friction between the threads and the mating grooves is 0.2. Suppose that you want to loosen the vise by turning one of the shafts. Determine the couple you must apply (a) to shaft B; (b) to shaft C.

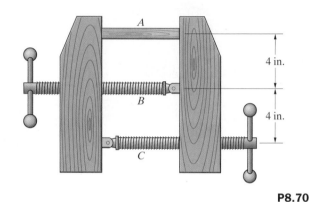

P8.70

8.71 Suppose that you want to tighten the vise in Problem 8.70 by turning one of the shafts. Determine the couple you must apply (a) to shaft B; (b) to shaft C.

8.72 The threaded shaft has a ball and socket support at B. The 400-lb load A can be raised or lowered by rotating the threaded shaft, causing the threaded collar at C to move relative to the shaft. Neglect the weights of the members. The pitch of the shaft is $p = \frac{1}{4}$ in., the mean radius of the thread is $r = 1$ in., and the coefficient of static friction between the thread and the mating groove is 0.24. If the system is stationary in the position shown, what couple is necessary to start the shaft rotating to raise the load?

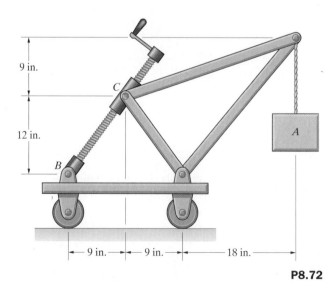

P8.72

8.73 In Problem 8.72, if the system is stationary in the position shown, what couple is necessary to start the shaft rotating to lower the load?

8.74 The car jack is operated by turning the threaded shaft at A. The threaded shaft fits into a mating groove in the collar at B, causing the collar to move relative to the shaft as the shaft turns. As a result, points B and D move closer together or farther apart, causing point C (where the jack is in contact with the car) to move up or down. The pitch of the threaded shaft is $p = 5$ mm, the mean radius of the thread is $r = 10$ mm, and the coefficient of kinetic friction between the thread and the mating groove is 0.15. What couple is necessary to turn the shaft at a constant rate and raise the jack when it is in the position shown if $F = 6.5$ kN?

8.75 In Problem 8.74, what couple is necessary to turn the threaded shaft at a constant rate and lower the jack when it is in the position shown if the force $F = 6.5$ kN?

8.76 A *turnbuckle*, used to adjust the length or tension of a bar or cable, is threaded at both ends. Rotating it draws threaded segments of a bar or cable together or moves them apart. Suppose that the pitch of the threads is $p = 3$ mm their mean radius is $r = 25$ mm, and the coefficient of static friction between the threads and the mating grooves is 0.24. If $T = 800$ N, what couple must be exerted on the turnbuckle to start tightening it?

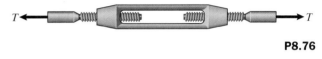

P8.76

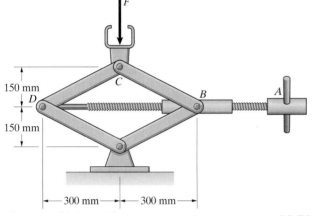

P8.74

Journal Bearings

A *bearing* is a support. This term usually refers to supports designed to allow the supported object to move. For example, in Fig. 8.19a, a horizontal shaft is supported by two *journal bearings*, which allow the shaft to rotate. The shaft can then be used to support a load perpendicular to its axis, such as that subjected by a pulley (Fig. 8.19b).

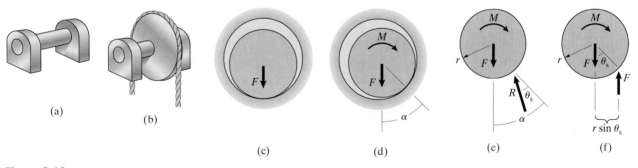

Figure 8.19
(a) A shaft supported by journal bearings.
(b) A pulley supported by the shaft.
(c) The shaft and bearing when no couple is applied to the shaft.
(d) A couple causes the shaft to roll within the bearing.
(e) Free-body diagram of the shaft.
(f) The two forces on the shaft must be equal and opposite.

Here we analyze journal bearings consisting of brackets with holes through which the shaft passes. The radius of the shaft is slightly smaller than the radius of the holes in the bearings. Our objective is to determine the couple that must be applied to the shaft to cause it to rotate in the bearings. Let F be the total load supported by the shaft including the weight of the shaft itself. When no couple is exerted on the shaft, the force F presses it against the bearings as shown in Fig. 8.19c. When a couple M is exerted on the shaft, it rolls up the surfaces of the bearings (Fig. 8.19d). The term α is the angle from the original point of contact of the shaft to its point of contact when M is applied.

In Fig. 8.19e, we draw the free-body diagram of the shaft when M is sufficiently large that slip is impending. The force R is the total reaction exerted on the shaft by the two bearings. Since R and F are the only forces acting on the shaft, equilibrium requires that $\alpha = \theta_s$ and $R = F$ (Fig. 8.19f). The reaction exerted on the shaft by the bearings is displaced a distance $r \sin \theta_s$ from the vertical line through the center of the shaft. By summing moments about the center of the shaft, we obtain the couple M that causes the shaft to be on the verge of slipping:

$$M = rF \sin \theta_s. \tag{8.12}$$

This is the largest couple that can be exerted on the shaft without causing it to start rotating. Replacing θ_s in this expression by the angle of kinetic friction θ_k gives the couple necessary to rotate the shaft at a constant rate.

The simple type of journal bearing we have described is too primitive for most applications. The surfaces where the shaft and bearing are in contact would quickly become worn. Designers usually incorporate "ball" or "roller" bearings in journal bearings to minimize friction (Fig. 8.20).

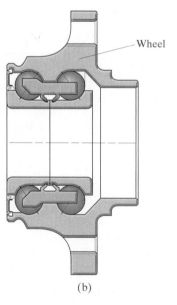

Wheel

Figure 8.20
(a) A journal bearing with one row of balls.
(b) Journal bearing assembly of the wheel of a car. There are two rows of balls between the rotating wheel and the fixed inner cylinder.

(a)   (b)

Example 8.7

Pulley Supported by Journal Bearings

The mass of the suspended load in Fig. 8.21 is 450 kg. The pulley P has a 150-mm radius and is rigidly attached to a horizontal shaft supported by journal bearings. The radius of the horizontal shaft is 12 mm and the coefficient of kinetic friction between the shaft and the bearings is 0.2. The masses of the pulley and shaft are negligible. What tension must the winch A exert on the cable to raise the load at a constant rate?

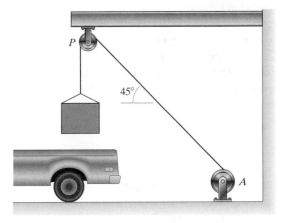

Figure 8.21

Strategy

Equation (8.12) with θ_s replaced by θ_k relates the couple M required to turn the pulley at a constant rate to the total force F on the shaft. By expressing M and F in terms of the load and the tension exerted by the winch, we can obtain an equation for the required tension.

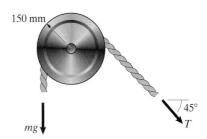

(a) Free-body diagram of the pulley.

Solution

Let T be the tension exerted by the winch (Fig. a). By calculating the magnitude of the sum of the forces exerted by the tension and the load (Fig. b), we obtain an expression for the total force F on the shaft supporting the pulley:

$$F = \sqrt{(mg + T\sin 45°)^2 + (T\cos 45°)^2}.$$

The (clockwise) couple exerted on the pulley by the tension and the load is

$$M = 0.15(T - mg).$$

The radius of the shaft is $r = 0.012$ m and the angle of kinetic friction is $\theta_k = \arctan(0.2) = 11.3°$. We substitute our expressions for F and M into Eq. (8.12).

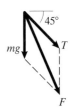

(b) The total force F on the shaft.

$$M = rF\sin\theta_k:$$

$$0.15\big[T - (450)(9.81)\big] = 0.012\sqrt{\big[(450)(9.81) + T\sin 45°\big]^2 + (T\cos 45°)^2}\,\sin(11.3°).$$

Solving for the tension, we obtain $T = 4.54$ kN.

Thrust Bearings and Clutches

A *thrust bearing* supports a rotating shaft that is subjected to an axial load. In the type shown in Figs. 8.22a and 8.22b, the conical end of the shaft is pressed against the mating conical cavity by an axial load F. Let us determine the couple M necessary to rotate the shaft.

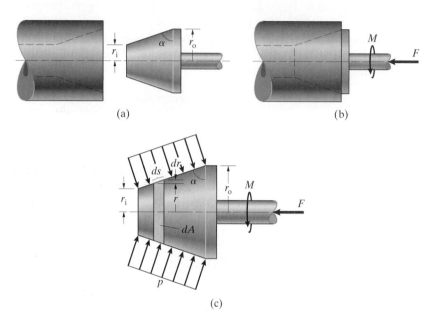

(a)   (b)

Figure 8.22
(a), (b) A thrust bearing supports a shaft subjected to an axial load.
(c) The differential element dA and the uniform pressure p exerted by the cavity.

(c)

The differential element of area dA in Fig. 8.22(c) is

$$dA = 2\pi r \, ds = 2\pi r \left(\frac{dr}{\cos \alpha}\right).$$

Integrating this expression from $r = r_i$ to $r = r_o$, we obtain the area of contact:

$$A = \frac{\pi\left(r_o^2 - r_i^2\right)}{\cos \alpha}.$$

If we assume that the mating surface exerts a uniform pressure p, the axial component of the total force due to p must equal F: $pA \cos \alpha = F$. Therefore the pressure is

$$p = \frac{F}{A \cos \alpha} = \frac{F}{\pi\left(r_o^2 - r_i^2\right)}.$$

As the shaft rotates about its axis, the moment about the axis due to the friction force on the element dA is $r\mu_k\,(p \, dA)$. The total moment equals M:

$$M = \int_A \mu_k r p \, dA = \int_{r_i}^{r_o} \mu_k r \left[\frac{F}{\pi\left(r_o^2 - r_i^2\right)}\right]\left(\frac{2\pi r \, dr}{\cos \alpha}\right).$$

Integrating, we obtain the couple M necessary to rotate the shaft at a constant rate:

$$M = \frac{2\mu_k F}{3 \cos \alpha}\left(\frac{r_o^3 - r_i^3}{r_o^2 - r_i^2}\right). \tag{8.13}$$

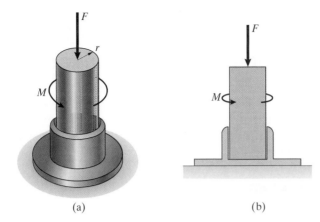

(a)   (b)

Figure 8.23
A thrust bearing that supports a flat-ended shaft.

A simpler thrust bearing is shown in Figs. 8.23a and 8.23b. The bracket supports the flat end of a shaft of radius r that is subjected to an axial load F. We can obtain the couple necessary to rotate the shaft at a constant rate from Eqs. (8.13) by setting $\alpha = 0$, $r_i = 0$, and $r_o = r$:

$$M = \frac{2}{3}\mu_k Fr. \tag{8.14}$$

Although they are good examples of the analysis of friction forces, the thrust bearings we have described would become worn too quickly to be used in most applications. The designer of the thrust bearing in Fig. 8.24 minimizes friction by incorporating "roller" bearings.

A *clutch* is a device used to connect and disconnect two coaxial rotating shafts. The type shown in Figs. 8.25a and 8.25b consists of disks of radius r attached to the ends of the shafts. When the disks are separated (Fig. 8.25a), the clutch is *disengaged*, and the shafts can rotate freely relative to each other. When the clutch is engaged by pressing the disks together with axial forces F (Fig. 8.25b), the shafts can support a couple M due to the friction forces between the disks. If the couple M becomes too large, the clutch slips.

The friction forces exerted on one face of the clutch by the other face are identical to the friction forces exerted on the flat-ended shaft by the bracket in Fig. 8.23. We can therefore determine the largest couple the clutch can support without slipping by replacing μ_k by μ_s in Eqs. (8.14):

$$M = \frac{2}{3}\mu_s Fr. \tag{8.15}$$

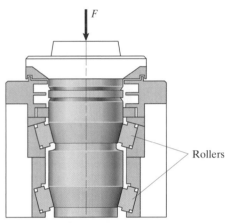

Figure 8.24
A thrust bearing with two rows of cylindrical rollers between the shaft and the fixed support.

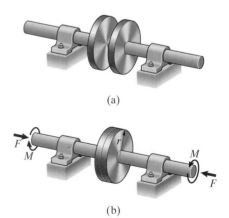

(a)

(b)

Figure 8.25
A clutch.
(a) Disengaged position.
(b) Engaged position.

Study Questions

1. What is a journal bearing?
2. If the shaft of a journal bearing is subjected to a lateral force F, how do you determine the couple M necessary to rotate the shaft at a constant rate?
3. When the axis of a clutch is subjected to an axial force F (Fig. 8.25b), how do you determine the largest couple M the clutch can support without slipping?

Example 8.8

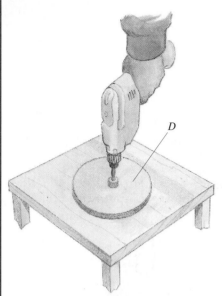

Figure 8.26

Friction on a Disk Sander

The handheld sander in Fig. 8.26 has a rotating disk D of 4-in. radius with sandpaper bonded to it. The total downward force exerted by the operator and the weight of the sander is 15 lb. The coefficient of kinetic friction between the sandpaper and the surface is $\mu_k = 0.6$. What couple (torque) M must the motor exert to turn the sander at a constant rate?

Strategy

As the disk D rotates, it is subjected to friction forces analogous to the friction forces exerted on the flat-ended shaft by the bracket in Fig. 8.23. We can determine the couple required to turn the disk D at a constant rate from Eq. (8.14).

Solution

The couple required to turn the disk at a constant rate is

$$M = \frac{2}{3}\mu_k rF = \frac{2}{3}(0.6)\left(\frac{4}{12}\right)(15) = 2 \text{ ft-lb}.$$

Problems

8.77 The horizontal shaft is supported by two journal bearings. The coefficient of kinetic friction between the shaft and the bearings is $\mu_k = 0.2$. The radius of the shaft is 20 mm, and its mass is 5 kg. Determine the couple M necessary to rotate the shaft at a constant rate.

Strategy: You can obtain the couple necessary to rotate the shaft at a constant rate by replacing θ_s by θ_k in Eq. (8.12).

P8.77

8.78 The horizontal shaft is supported by two journal bearings. The coefficient of static friction between the shaft and the bearings is $\mu_s = 0.3$. The radius of the shaft is 20 mm, and its mass

is 5 kg. Determine the largest mass m that can be suspended as shown without causing the stationary shaft to slip in the bearings.

P8.78

8.79 Suppose that in Problem 8.78 the mass $m = 8$ kg and the coefficient of kinetic friction between the shaft and the bearings is $\mu_k = 0.26$. What couple must be applied to the shaft to raise the mass at a constant rate?

8.80 The pulley is mounted on a horizontal shaft supported by journal bearings. The coefficient of kinetic friction between the shaft and the bearings is $\mu_k = 0.3$. The radius of the shaft is 20 mm, and the radius of the pulley is 150 mm. The mass $m = 10$ kg. Neglect the masses of the pulley and shaft. What force T must be applied to the cable to move the mass upward at a constant rate?

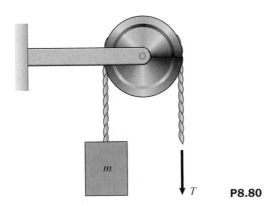

P8.80

8.81 In Problem 8.80, what force T must be applied to the cable to lower the mass at a constant rate?

8.82 The pulley of 8-in. radius is mounted on a shaft of 1-in. radius. The shaft is supported by two journal bearings. The coefficient of static friction between the bearings and the shaft is $\mu_s = 0.15$. Neglect the weights of the pulley and shaft. The 50-lb block A rests on the floor. If sand is slowly added to the bucket B, what do the bucket and sand weigh when the shaft slips in the bearings?

P8.82

8.83 The pulley of 50-mm radius is mounted on a shaft of 10-mm radius. The shaft is supported by two journal bearings. The mass of the block A is 8 kg. Neglect the weights of the pulley and shaft. If a force $T = 84$ N is necessary to raise block A at a constant rate, what is the coefficient of kinetic friction between the shaft and the bearings?

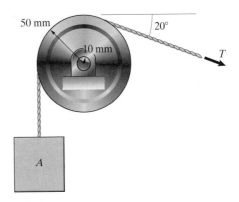

P8.83

8.84 The mass of the suspended object is 4 kg. The pulley has a 100-mm radius and is rigidly attached to a horizontal shaft supported by journal bearings. The radius of the horizontal shaft is 10 mm and the coefficient of kinetic friction between the shaft and the bearings is 0.26. What tension must the person exert on the rope to raise the load at a constant rate?

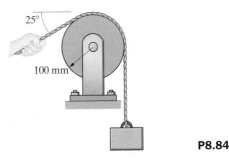

P8.84

8.85 The circular flat-ended shaft is pressed into the thrust bearing by an axial load of 100 N. Neglect the weight of the shaft. The coefficients of friction between the end of the shaft and the bearing are $\mu_s = 0.20$ and $\mu_k = 0.15$. What is the largest couple M that can be applied to the stationary shaft without causing it to rotate in the bearing?

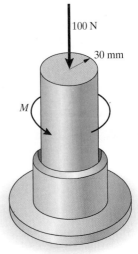

P8.85

8.86 In Problem 8.85, what couple M is required to rotate the shaft at a constant rate?

8.87 Suppose that the end of the shaft in Problem 8.85 is supported by a thrust bearing of the type shown in Fig. 8.22, where $r_o = 30$ mm, $r_i = 10$ mm, $\alpha = 30°$, and $\mu_k = 0.15$. What couple M is required to rotate the shaft at a constant rate?

8.88 The disk D is rigidly attached to the vertical shaft. The shaft has flat ends supported by thrust bearings. The disk and the shaft together have a mass of 220 kg and the diameter of the shaft is 50 mm. The vertical force exerted on the end of the shaft by the upper thrust bearing is 440 N. The coefficient of kinetic friction between the ends of the shaft and the bearings is 0.25. What couple M is required to rotate the shaft at a constant rate?

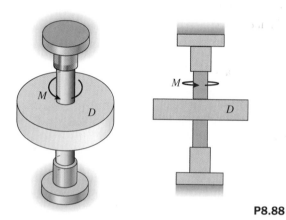

P8.88

8.89 Suppose that the ends of the shaft in Problem 8.88 are supported by thrust bearings of the type shown in Fig. 8.22, where $r_o = 25$ mm, $r_i = 6$ mm, $\alpha = 45°$, and $\mu_k = 0.25$. What couple M is required to rotate the shaft at a constant rate?

8.90 The shaft is supported by thrust bearings that subject it to an axial load of 800 N. The coefficients of kinetic friction between

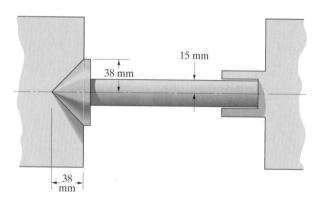

P8.90

the shaft and the left and right bearings are 0.20 and 0.26, respectively. What couple is required to rotate the shaft at a constant rate?

8.91 A motor is used to rotate a paddle for mixing chemicals. The shaft of the motor is coupled to the paddle using a friction clutch of the type shown in Fig. 8.25. The radius of the disks of the clutch is 120 mm, and the coefficient of static friction between the disks is 0.6. If the motor transmits a maximum torque of 15 N-m to the paddle, what minimum normal force between the plates of the clutch is necessary to prevent slipping?

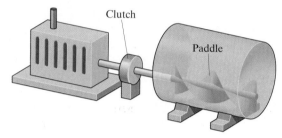

P8.91

8.92 The thrust bearing is supported by contact of the collar C with a fixed plate. The area of contact is an annulus with an inside diameter $D_1 = 40$ mm and an outside diameter $D_2 = 120$ mm. The coefficient of kinetic friction between the collar and the plate is $\mu_k = 0.3$. The force $F = 400$ N. What couple M is required to rotate the shaft at a constant rate?

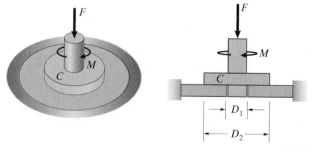

P8.92

Belt Friction

If a rope is wrapped around a fixed post as shown in Fig. 8.27, a large force T_2 exerted on one end can be supported by a relatively small force T_1 applied to the other end. In this section we analyze this familiar phenomenon. It is referred to as *belt friction* because a similar approach can be used to analyze belts used in machines, such as the belts that drive alternators and other devices in a car.

Let's consider a rope wrapped through an angle β around a fixed cylinder (Fig. 8.28a). We will assume that the tension T_1 is known. Our objective is to determine the largest force T_2 that can be applied to the other end of the rope without causing the rope to slip.

We begin by drawing the free-body diagram of an element of the rope whose boundaries are at angles α and $\alpha + \Delta\alpha$ from the point where the rope comes into contact with the cylinder (Figs. 8.28b and 8.28c). The force T is the tension in the rope at the position defined by the angle α. We know that the tension in the rope varies with position, because it increases from T_1 at $\alpha = 0$ to T_2 at $\alpha = \beta$. We therefore write the tension in the rope at the position $\alpha + \Delta\alpha$ as $T + \Delta T$. The force ΔN is the normal force exerted on the element by the cylinder. Because we want to determine the largest value of T_2 that will not cause the rope to slip, we assume that the friction force is equal to its maximum possible value $\mu_s \Delta N$, where μ_s is the coefficient of static friction between the rope and the cylinder.

The equilibrium equations in the directions tangential to and normal to the centerline of the rope are

$$\Sigma F_{\text{(tangential)}} = \mu_s \Delta N + T \cos\left(\frac{\Delta\alpha}{2}\right) - (T + \Delta T)\cos\left(\frac{\Delta\alpha}{2}\right) = 0,$$

$$\Sigma F_{\text{(normal)}} = \Delta N - (T + \Delta T)\sin\left(\frac{\Delta\alpha}{2}\right) - T\sin\left(\frac{\Delta\alpha}{2}\right) = 0. \qquad (8.16)$$

Figure 8.27
A rope wrapped around a post.

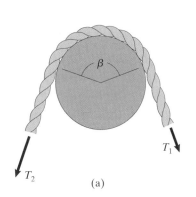

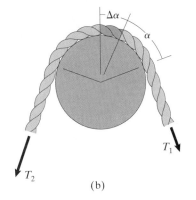

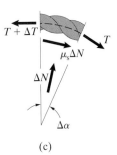

Figure 8.28
(a) A rope wrapped around a fixed cylinder.
(b) A differential element with boundaries at angles α and $\alpha + \Delta\alpha$.
(c) Free-body diagram of the element.

Eliminating ΔN, we can write the resulting equation as

$$\left[\cos\left(\frac{\Delta\alpha}{2}\right) - \mu_s \sin\left(\frac{\Delta\alpha}{2}\right)\right]\frac{\Delta T}{\Delta\alpha} - \mu_s T\frac{\sin(\Delta\alpha/2)}{(\Delta\alpha/2)} = 0.$$

Evaluating the limit of this equation as $\Delta\alpha \to 0$ and observing that

$$\frac{\sin(\Delta\alpha/2)}{(\Delta\alpha/2)} \to 1,$$

we obtain

$$\frac{dT}{d\alpha} - \mu_s T = 0.$$

This differential equation governs the variation of the tension in the rope. By separating variables,

$$\frac{dT}{T} = \mu_s\, d\alpha,$$

we can integrate to determine the tension T_2 in terms of the tension T_1 and the angle β:

$$\int_{T_1}^{T_2}\frac{dT}{T} = \int_0^\beta \mu_s\, d\alpha.$$

Thus we obtain the largest force T_2 that can be applied without causing the rope to slip when the force on the other end is T_1:

$$T_2 = T_1 e^{\mu_s\beta}. \tag{8.17}$$

The angle β in this equation must be expressed in radians. Replacing μ_s by the coefficient of kinetic friction μ_k gives the force T_2 required to cause the rope to slide at a constant rate.

Equation (8.17) explains why a large force can be supported by a relatively small force when a rope is wrapped around a fixed support. The force required to cause the rope to slip increases exponentially as a function of the angle through which the rope is wrapped. Suppose that $\mu_s = 0.3$. When the rope is wrapped one complete turn around the post ($\beta = 2\pi$), the ratio $T_2/T_1 = 6.59$. When the rope is wrapped four complete turns around the post ($\beta = 8\pi$), the ratio $T_2/T_1 = 1880$.

Study Questions

1. What is the definition of the term β in Eq. (8.17)?
2. If a rope is wrapped through a given angle around a fixed post and one end is subjected to a given tension T_1, how can you determine the tension T_2 necessary to cause the rope to be on the verge of slipping in the direction of T_2? How can you determine the smallest value of T_2 that will prevent the rope from slipping in the direction of T_1?

Example 8.9

Rope Wrapped Around Two Cylinders

The 50-kg crate in Fig. 8.29 is suspended from a rope that passes over two fixed cylinders. The coefficient of static friction is 0.2 between the rope and the left cylinder and 0.4 between the rope and the right cylinder. What is the smallest force the woman can exert and support the crate?

Figure 8.29

Strategy

She exerts the smallest possible force when slip of the rope is impending on both cylinders. Because we know the weight of the crate, we can use Eq. (8.17) to determine the tension in the rope between the two cylinders and then use Eq. (8.17) again to determine the force she exerts.

Solution

The weight of the crate is $W = (50)(9.81) = 491$ N. Let T be the tension in the rope between the two cylinders (Fig. a). The rope is wrapped around the left cylinder through an angle $\beta = \pi/2$ rad. The tension T necessary to prevent the rope from slipping on the left cylinder is related to W by

$$W = Te^{\mu_s\beta} = Te^{(0.2)(\pi/2)}.$$

Solving for T, we obtain

$$T = We^{-(0.2)(\pi/2)} = (491)e^{-(0.2)(\pi/2)} = 358 \text{ N}.$$

The rope is also wrapped around the right cylinder through an angle $\beta = \pi/2$ rad. The force F the woman must exert to prevent the rope from slipping on the right cylinder is related to T by

$$T = Fe^{\mu_s\beta} = Fe^{(0.4)(\pi/2)}.$$

The solution for F is

$$F = Te^{-(0.4)(\pi/2)} = (358)e^{-(0.4)(\pi/2)} = 191 \text{ N}.$$

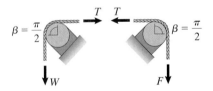

(a) The tensions in the rope.

Problems

8.93 Suppose that you want to lift a 50-lb crate off the ground by using a rope looped over a tree limb as shown. The coefficient of static friction between the rope and the limb is 0.4, and the rope is wound 120° around the limb. What force must you exert to lift the crate?

Strategy: The tension necessary to cause impending slip of the rope on the limb is given by Eq. (8.17), with $T_1 = 50$ lb, $\mu_s = 0.4$, and $\beta = (\pi/180)(120)$ rad.

P8.93

8.94 In Problem 8.93, once you have lifted the crate off the ground, what is the minimum force you must exert on the rope to keep it suspended?

8.95 *Winches* are used on sailboats to help support the forces exerted by the sails on the ropes (*sheets*) holding them in position. The winch shown is a post that will rotate in the clockwise direction (seen from above), but will not rotate in the counterclockwise direction. The sail exerts a tension $T_S = 800$ N on the sheet, which is wrapped two complete turns around the winch. The coefficient of static friction between the sheet and the winch is $\mu_s = 0.2$. What tension T_C must the crew member exert on the sheet to prevent it from slipping on the winch?

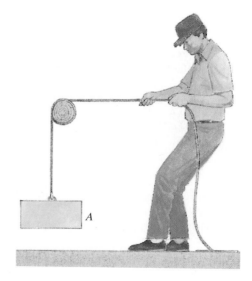

P8.95

8.96 The coefficient of kinetic friction between the sheet and the winch in Problem 8.95 is $\mu_k = 0.16$. If the crew member wants to let the sheet slip at a constant rate, releasing the sail, what initial tension T_C must he exert on the sheet as it begins slipping?

8.97 The mass of the block A is 18 kg. The rope is wrapped one and one-fourth turns around the fixed wooden post. The coefficients of friction between the rope and post are $\mu_s = 0.15$ and $\mu_k = 0.12$. What force would the person have to exert to raise the block at a constant rate?

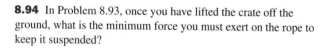

P8.97

8.98 The weight of block A is W. The disk is supported by a smooth bearing. The coefficient of kinetic friction between the disk and the belt is μ_k. What couple M is necessary to turn the disk at a constant rate?

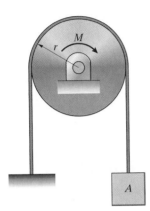

P8.98

8.99 The couple required to turn the wheel of the exercise bicycle is adjusted by changing the weight W. The coefficient of kinetic friction between the wheel and the belt is μ_k. Assume the wheel turns clockwise.
(a) Show that the couple M required to turn the wheel is
$M = WR(1 - e^{-3.4\mu_k})$.
(b) If $W = 40$ lb and $\mu_k = 0.2$, what force will the scale S indicate when the bicycle is in use?

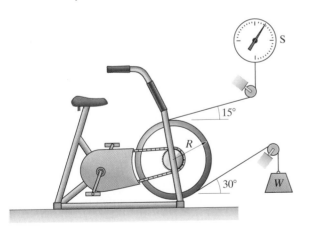

P8.99

8.100 The box B weighs 50 lb. The coefficients of friction between the cable and the fixed round supports are $\mu_s = 0.4$ and $\mu_k = 0.3$.
(a) What is the minimum force F required to support the box?
(b) What force F is required to move the box upward at a constant rate?

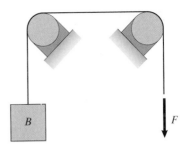

P8.100

8.101 The 20-kg box A is held in equilibrium on the inclined surface by the force T acting on the rope wrapped over the fixed cylinder. The coefficient of static friction between the box and the inclined surface is 0.1. The coefficient of static friction between the rope and the cylinder is 0.05. Determine the largest value of T that will not cause the box to slip up the inclined surface.

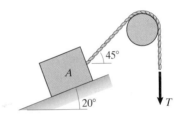

P8.101

8.102 In Problem 8.101, determine the smallest value of T necessary to hold the box in equilibrium on the inclined surface.

8.103 The mass of the block A is 14 kg. The coefficient of kinetic friction between the rope and the cylinder is 0.2. If the cylinder is rotated at a constant rate, first in the counterclockwise direction and then in the clockwise direction, the difference in the height of block A is 0.3 m. What is the spring constant k?

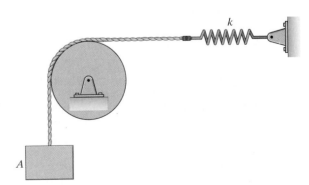

P8.103

Chapter Summary

Dry Friction

The forces resulting from the contact of two plane surfaces can be expressed in terms of the normal force N and friction force f (Fig. a) or the magnitude R and angle of friction θ (Fig. b).

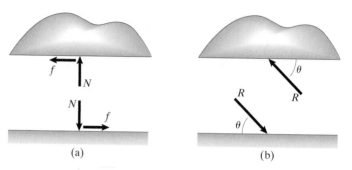

(a)   (b)

If slip is impending, the magnitude of the friction force is

$$f = \mu_s N,$$ Eq. (8.1)

and its direction opposes the impending slip. The angle of friction equals the angle of static friction $\theta_s = \arctan\left(\mu_s\right)$.

If the surfaces are sliding, the magnitude of the friction force is

$$f = \mu_k N,$$ Eq. (8.2)

and its direction opposes the relative motion. The angle of friction equals the angle of kinetic friction $\theta_k = \arctan\left(\mu_k\right)$.

Threads

The slope α of the thread (Fig. c) is related to its pitch p by

$$\tan \alpha = \frac{p}{2\pi r}.$$ Eq. (8.7)

The couple required for impending rotation and axial motion opposite to the direction of F is

$$M = rF \tan\left(\theta_s + \alpha\right),$$ Eq. (8.9)

and the couple required for impending rotation and axial motion of the shaft in the direction of F is

$$M = rF \tan\left(\theta_s - \alpha\right).$$ Eq. (8.11)

When $\theta_s < \alpha$, the shaft will rotate and move in the direction of the force F with no couple applied.

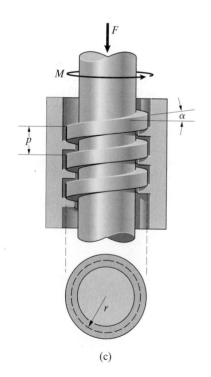

(c)

Journal Bearings

The couple required for impending slip of the circular shaft (Fig. d) is

$$M = rF \sin\theta_s, \qquad \text{Eq. (8.12)}$$

where F is the total load on the shaft.

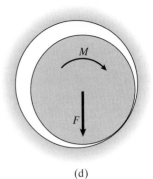

(d)

Thrust Bearings and Clutches

The couple required to rotate the shaft at a constant rate (Fig. e) is

$$M = \frac{2\mu_k F}{3\cos\alpha}\left(\frac{r_o^3 - r_i^3}{r_o^2 - r_i^2}\right). \qquad \text{Eq. (8.13)}$$

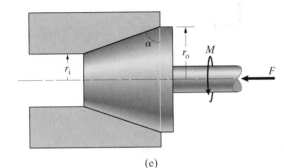

(e)

Belt Friction

The force T_2 required for impending slip in the direction of T_2 (Fig. f) is

$$T_2 = T_1 e^{\mu_s \beta}, \qquad \text{Eq. (8.17)}$$

where β is in radians.

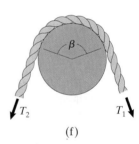

(f)

Review Problems

8.104 The weight of the box is $W = 30$ lb, and the force F is perpendicular to the inclined surface. The coefficient of static friction between the box and the inclined surface is $\mu_s = 0.2$.
(a) If $F = 30$ lb, what is the magnitude of the friction force exerted on the stationary box?
(b) If $F = 10$ lb, show that the box cannot remain at rest on the inclined surface.

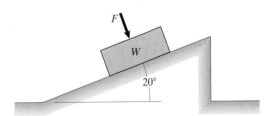

P8.104

8.105 In Problem 8.104, what is the smallest force F necessary to hold the box stationary on the inclined surface?

8.106 The mass of the van is 2250 kg, and the coefficient of static friction between its tires and the road is 0.6. If its front wheels are locked and its rear wheels can turn freely, what is the largest value of α for which it can remain in equilibrium?

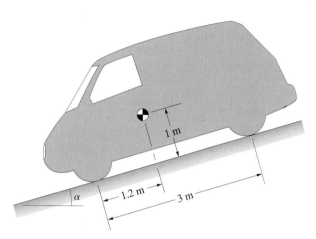

P8.106

8.107 In Problem 8.106, what is the largest value of α for which the van can remain in equilibrium if it points up the slope?

8.108 Each of the uniform 1-m bars has a mass of 4 kg. The coefficient of static friction between the bar and the surface at B is 0.2. If the system is in equilibrium, what is the magnitude of the friction force exerted on the bar at B?

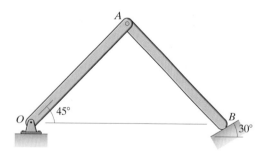

P8.108

8.109 In Problem 8.108, what is the minimum coefficient of static friction between the bar and the surface at B necessary for the system to be in equilibrium?

8.110 The clamp presses two pieces of wood together. The pitch of the threads is $p = 2$ mm, the mean radius of the thread is $r = 8$ mm, and the coefficient of kinetic friction between the thread and the mating groove is 0.24. What couple must be exerted on the threaded shaft to press the pieces of wood together with a force of 200 N?

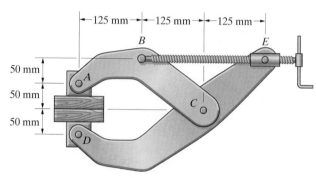

P8.110

8.111 In Problem 8.110, the coefficient of static friction between the thread and the mating groove is 0.28. After the threaded shaft is rotated sufficiently to press the pieces of wood together with a force of 200 N, what couple must be exerted on the shaft to loosen it?

8.112 The axles of the tram are supported by journal bearing. The radius of the wheels is 75 mm, the radius of the axles is 15 mm, and the coefficient of kinetic friction between the axles and the bearings is $\mu_k = 0.14$. The mass of the tram and its load is 160 kg. If the weight of the tram and its load is evenly divided between the axles, what force P is necessary to push the tram at a constant speed?

P8.112

8.113 The two pulleys have a radius of 6 in. and are mounted on shafts of 1-in. radius supported by journal bearings. Neglect the weights of the pulleys and shafts. The coefficient of kinetic friction between the shafts and the bearings is $\mu_k = 0.2$. If a force $T = 200$ lb is required to raise the man at a constant rate, what is his weight?

*𝒟*esign Experience   Design and build a device to measure the coefficient of static friction μ_s between two materials. Use it to measure μ_s for several of the materials listed in Table 8.1 and compare your results with the values in the table. Discuss possible sources of error in your device and determine how closely your values agree when you perform repeated experiments with the same two materials.

T

P8.113

This highway bridge over the Colorado River near Austin, Texas is supported by cables connected to steel arches that span the river.

Measures of Stress and Strain

In this chapter we define stresses and strains, which are measures quanti-
fying the internal forces and deformations in materials. We need these
measures to assess the capability of structural elements such as the
bridge's cables and arches to support loads and to determine the deformations
that result from those loads.

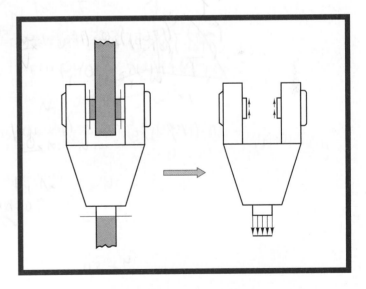

9.1 Stresses

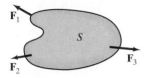

Figure 9.1
Sample of material subjected to forces.

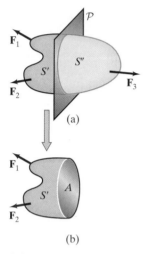

Figure 9.2
(a) The plane $\mathcal{P}$ divides the sample into parts S' and S''.
(b) Isolating part S'.

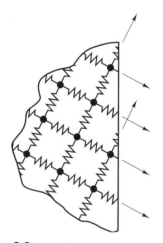

Figure 9.3
Bond forces at the plane $\mathcal{P}$.

Suppose that we subject a sample S of some solid material, such as iron, to the system of forces shown in Fig. 9.1. We assume that the sample is in equilibrium, and that its weight is negligible compared to the other external forces, so that $\mathbf{F}_1 + \mathbf{F}_2 + \mathbf{F}_3 = 0$. Let us pass an imaginary plane $\mathcal{P}$ through the sample, dividing it into parts S' and S'', and isolate part S' (Fig. 9.2). Let A be the area of intersection of the plane $\mathcal{P}$ with the sample. It is clear from Fig. 9.2b that part S' cannot be in equilibrium under the action of forces $\mathbf{F}_1$ and $\mathbf{F}_2$ alone. Part S'' must exert forces at the plane $\mathcal{P}$ that keep part S' in equilibrium.

What are those forces? They are literally the forces that hold the material together. Iron in its solid state consists of atoms "connected" to neighboring atoms by chemical bonds, electromagnetic forces that we may visualize as springs. By exerting external forces on the sample of iron, we alter the distances between atoms, changing the forces exerted by the bonds. The bond forces are internal forces within the material. But when we hypothetically separate the sample of material into two parts, the bond forces at the separating plane become external forces on the individual parts. It is these bond forces exerted at the plane $\mathcal{P}$ by part S'' which keep part S' in equilibrium (Fig. 9.3).

We could model the forces acting on part S' at the plane $\mathcal{P}$ by representing each bond force by a vector as shown in Fig. 9.3, but this approach would be impractical. Not only would there be an unwieldy number of vectors to contend with, we don't know the actual arrangement of the atoms in any given sample of material. Instead, *we will represent the forces as a distributed load.*

Traction, Normal Stress, and Shear Stress

We define a vector-valued function $\mathbf{t}$, the *traction*, such that the force exerted on each infinitesimal element dA of the area A is $\mathbf{t}\,dA$ (Fig. 9.4). Notice the similarity between the definition of $\mathbf{t}$ and the definition of the pressure p in a stationary liquid. The pressure is a scalar function, but here we must use a vector-valued function because the forces on A may act in any direction.

Since the value of $\mathbf{t}$ at a point on A is a vector, we can resolve it into components normal and tangential to A (Fig. 9.5). The scalar normal component σ is called the *normal stress* at the point, and the scalar tangential component τ is called the *shear stress* at the point. The normal stress σ is defined to be positive if it points outward, or away from the material, and is then said to be *tensile*. If σ is negative, the normal stress points toward the material and is said to be *compressive*. The force exerted on an element dA is $\sigma\,dA$ in the normal direction and $\tau\,dA$ in the tangential direction.

Figure 9.4
(a) Representing the forces on A by a distributed load.
(b) The force on an element dA is $\mathbf{t}\,dA$.

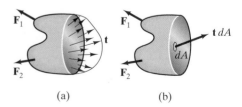

(a)　　　　　　(b)

The products $\mathbf{t}\,dA$, $\sigma\,dA$, and $\tau\,dA$ are forces, so the dimensions of $\mathbf{t}$, σ, and τ are force/area. In SI units, the traction and the normal and shear stresses are normally expressed in pascals (Pa), which are newtons per square meter, and in U.S. Customary units they are normally expressed in pounds per square inch (psi) or pounds per square foot (psf). We also use kips per square foot (ksf) and kips per square inch (ksi).

We introduced the definitions of the traction and the normal and shear stresses using a sample of iron in equilibrium as an example, but the same definitions are used to describe the internal forces in other media, and they need not be in equilibrium. The sample with which we began, Fig. 9.1, could model a different solid material, a liquid, or a gas, and it could be in an arbitrary state of motion. Depending on the medium, the internal forces represented may be quite different in nature from the bond forces between the atoms in iron. For example, the internal forces in a gas arise from impacts between the atoms or molecules of the gas resulting from their thermal motions. Nevertheless, we can represent the forces by a distributed load in exactly the same way (Fig. 9.4). The same definitions of the traction and its components are used in both solid and fluid mechanics to represent the internal forces in materials. In the special case of a liquid or gas at rest, the normal stress $\sigma = -p$, where p is the pressure, and the shear stress $\tau = 0$. Recall that the normal stress σ is defined to be positive in tension, whereas the pressure p is defined to be positive in compression.

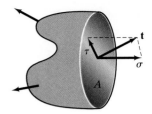

Figure 9.5
The components of $\mathbf{t}$ normal and tangential to A are the normal and shear stresses.

Average Stresses

The traction $\mathbf{t}$ or its components, the normal stress σ and shear stress τ, allow us in principle to represent the forces acting on area A in Fig. 9.5. But how can we determine their values as functions of position on A for a given sample of material and state of motion? This determination, evaluating the *stress distribution* for a given plane $\mathcal{P}$ within a material, is called *stress analysis*. In the following chapters we discuss the stress distributions in simple structural elements subjected to loads. In this section our goal is less ambitious: We want to determine the average stress at a plane when an object is in equilibrium.

The average value of the traction $\mathbf{t}$ on A is defined by

$$\mathbf{t}_{av} = \frac{1}{A}\int_A \mathbf{t}\,dA, \qquad (9.1)$$

where the subscript A on the integral sign signifies that integration is carried out over the entire area A. Therefore, we can express the total force exerted on the area A in terms of the average traction (Fig. 9.6):

$$\int_A \mathbf{t}\,dA = \mathbf{t}_{av}A.$$

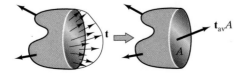

Figure 9.6
Expressing the total force exerted on A by the traction distribution in terms of the average traction.

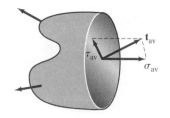

Figure 9.7
Decomposing the average traction into the average normal and shear stresses.

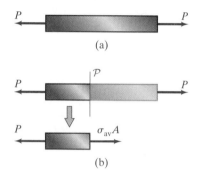

Figure 9.8
(a) Subjecting a bar to axial loads.
(b) Obtaining a free-body diagram by passing a plane through the bar perpendicular to its axis.

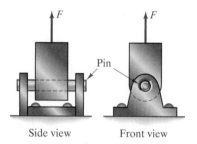

Figure 9.9
Two views of a pin support.

The components of the average traction normal and tangential to A are the average normal and shear stresses on A (Fig. 9.7). The total normal and tangential forces exerted on A are $\sigma_{av} A$ and $\tau_{av} A$, respectively.

If we know the external forces acting on an object in equilibrium, we can determine the average normal and shear stresses acting on an arbitrary plane $\mathcal{P}$. Two particular examples are important in applications.

Average Normal Stress in an Axially Loaded Bar   A straight bar whose cross section is uniform throughout its length is said to be *prismatic*. (The familiar triangular glass prism is a prismatic bar.) Figure 9.8a shows a prismatic bar subjected to axial forces parallel to its axis. In Fig. 9.8b we pass a plane $\mathcal{P}$ perpendicular to the bar's axis and isolate part of the bar. We show the force exerted by the average normal stress at $\mathcal{P}$, where A is the bar's cross-sectional area. We know that the average shear stress at $\mathcal{P}$ is zero because there is no external force tangential to $\mathcal{P}$. The sum of the forces equals zero,

$$\sigma_{av} A - P = 0,$$

so we can solve for the average normal stress:

$$\sigma_{av} = \frac{P}{A}.$$

We see that the average normal stress is proportional to the external axial load and inversely proportional to the bar's cross-sectional area. Notice that the value of the average normal stress does not depend on the location of the plane $\mathcal{P}$ along the bar's axis. We discuss the analysis of stresses in axially loaded bars fully in Chapter 10.

Average Shear Stress in a Pin   The pin support in Fig. 9.9 holds a bar subjected to an axial load F. The support consists of a bracket supporting the cylindrical pin, which passes through a hole in the bar. In Fig. 9.10 we obtain a free-body diagram by passing two planes through the pin perpendicular to its axis, one on each side of the supported bar. We show the forces resulting from the average shear stresses in the pin at the two cutting planes, where A is the pin's cross-sectional area. From the equilibrium equation

$$F - 2\tau_{av} A = 0,$$

we obtain the average shear stress:

$$\tau_{av} = \frac{F}{2A}.$$

These two examples demonstrate the ease with which we can determine the average stresses on a given plane within an object. But knowing the average

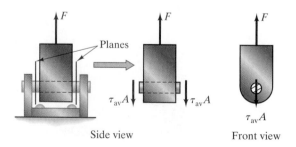

Figure 9.10
Passing two planes through the pin to determine the average shear stress.

stress on a plane is of limited usefulness in design, because it is the maximum, not the average, values of the normal and shear stresses that determine whether material failure will occur. Depending on the nature of the actual distributions of stress, the maximum stresses may be much greater than their average values. Moreover, it is not sufficient to determine the maximum stresses for a particular plane. The stresses acting on every plane must be considered. We examine the problem of determining maximum normal and shear stresses in Chapter 12.

There is one circumstance in which knowing the average stress on a given plane is useful in design. If experiments or analyses are first used to establish the relationship between the largest safe load on a given structural element and the value of the average normal or shear stress on a given plane, design can be carried out on that basis. For example, bolts made of a particular material that are to be used for a specific application can be tested to determine the largest safe value of average shear stress they will support.

Study Questions

1. If you know the traction distribution on a plane $\mathcal{P}$, how can you determine the force exerted on an element of area dA of $\mathcal{P}$?
2. What are the definitions of the normal stress σ and shear stress τ?
3. Why is knowledge of the average normal and shear stress on a plane of limited usefulness in design?
4. If a straight prismatic bar is subjected to an axial load, how can you determine the average normal stress on a plane perpendicular to the bar's axis?

Example 9.1

Determining Average Stresses

The truss in Fig. 9.11 supports a 10-kN force. Its members are solid cylindrical bars of 40-mm radius. Determine the average normal and shear stresses on the plane $\mathcal{P}$.

Strategy

We must first determine the axial force in member BC, which we can do by drawing the free-body diagram of joint C. We can then use the free-body diagram of the part of member BC on either side of the plane $\mathcal{P}$ to determine the average normal and shear stresses.

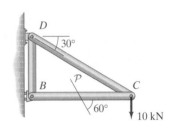

Figure 9.11

Solution

We draw the free-body diagram of joint C of the truss in Fig. (a). From the equilibrium equations

$$\Sigma F_x = -P_{BC} - P_{CD} \cos 30° = 0,$$

$$\Sigma F_y = P_{CD} \sin 30° - 10 = 0,$$

we obtain $P_{BC} = -17.3$ kN, $P_{CD} = 20$ kN. Member BC is subject to a compressive axial load of 17.3 kN.

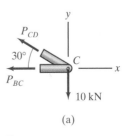

(a) Joint C.

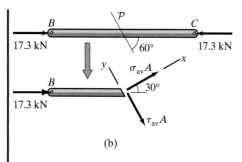

(b)

(b) Isolating the part of member BC on one side of plane $\mathcal{P}$.

In Fig. (b) we isolate the part of member BC to the left of the plane $\mathcal{P}$ and complete the free-body diagram by showing the forces exerted by the average normal and shear stresses.

By aligning a coordinate system normal and tangential to the plane $\mathcal{P}$ as shown in Fig. (b) and summing forces in the x and y directions, we obtain simple equilibrium equations for the forces exerted by the stresses:

$$\Sigma F_x = \sigma_{av} A + 17.3 \cos 30° = 0,$$

$$\Sigma F_y = -\tau_{av} A - 17.3 \sin 30° = 0.$$

We find that $\sigma_{av} A = -15$ kN and $\tau_{av} A = -8.66$ kN. The area A is the intersection of the plane $\mathcal{P}$ with the bar, the area upon which the average stresses σ_{av} and τ_{av} act. The relationship between A and the cross-sectional area of the cylindrical bar of 0.04-m radius is

$$A \cos 30° = \pi(0.04)^2,$$

so $A = \pi(0.04)^2/\cos 30° = 0.00580$ m². Solving for the average normal and shear stresses, we obtain $\sigma_{av} = -2.58$ MPa and $\tau_{av} = -1.49$ MPa.

Discussion

The negative value of σ_{av} tells us that the average normal stress is compressive. Although we have established a convention for the positive direction of the normal stress, we have not yet done so for the shear stress. In drawing the free-body diagram in Fig. (b), we chose the direction of the stress τ_{av} arbitrarily, and the negative value we obtained indicates that it acts in the opposite direction.

Example 9.2

Average Shear Stress in a Pin

In Fig. 9.12 detailed views of joint C of the truss are shown. The joint has a pin of 20-mm radius. What is the average shear stress in the pin?

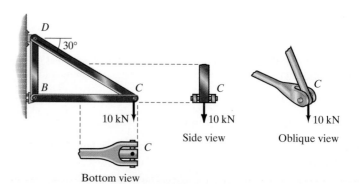

Figure 9.12

Strategy

By passing planes through the pin on both sides of member *CD*, we can isolate either member *BC* or *CD* and obtain a free-body diagram from which we can determine the average shear stress in the pin. Since the 10-kN external load acts on member *CD*, we will obtain a simpler free-body diagram by isolating member *BC*.

Solution

To draw the free-body diagram of member *BC*, we must know the axial force to which it is subjected. In the solution to Example 9.1, we found that member *BC* has a compressive axial load of 17.3 kN. In Fig. (a) we show the bottom view of member *BC* with member *CD* still attached. By passing planes through the pin on both sides of member *CD*, we obtain the free-body diagram of member *BC* in Fig. (b). The 17.3-kN compressive axial load is supported by the average shear stresses in the pin.

From the equilibrium equation

$$17.3 - 2\tau_{av} A = 0,$$

we obtain $\tau_{av} A = 8.66$ kN. The pin's cross-sectional area is $A = \pi(0.02)^2 = 0.00126$ m^2, so the average shear stress in the pin is $\tau_{av} = 6.89$ MPa.

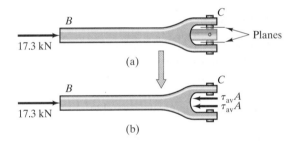

(a)

(b)

(a) Passing planes through the pin to isolate member *BC*.
(b) Free-body diagram of member *BC* showing the forces exerted on the pin.

Discussion

We should demonstrate that we can also determine the average shear stress in the pin by using the free-body diagram of member *CD*. We found in Example 9.1 that member *CD* has a tensile axial load of 20 kN. We show its free-body diagram (in side view) in Fig. (c). In this case the direction of the force exerted by the average shear stresses is not obvious, so we specify its direction by the angle β. From the equilibrium equations

$$\Sigma F_x = 2\tau_{av} A \cos\beta - 20\cos 30° = 0,$$
$$\Sigma F_y = 2\tau_{av} A \sin\beta + 20\sin 30° - 10 = 0,$$

we again obtain $\tau_{av} A = 8.66$ kN.

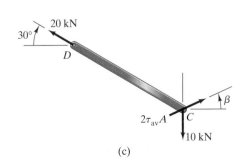

(c)

(c) Free-body diagram of member *CD*.

Example 9.3

Average Shear Stress Induced by a Punch

The arrangement in Fig. 9.13 is designed to cut a circular "blank" from the plate by exerting a sufficient force F on the punch. The plate is of thickness t and the punch and the matching hole beneath the plate are of diameter D. For a given force F on the punch, what average shear stress is induced in the plate?

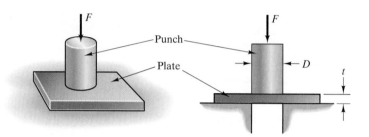

Figure 9.13

Strategy

To determine the average shear stress induced in the plate we obtain a free-body diagram by making a cylindrical "cut" of diameter D through the plate directly below the punch [Fig. (a)]. The force F on the free-body diagram is supported by vertical shear stress on the surface of the part of the plate we cut by the cylinder. We can use equilibrium to determine the average shear stress.

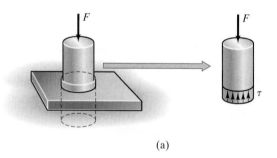

(a) Cutting the plate with a cylindrical surface.

(a)

Solution

The surface area of the part of the plate included in the free-body diagram on which the shear stress acts is equal to the product of the thickness t of the plate and the circumference πD. Summing the forces on the free-body diagram,

$$\tau_{av}(t)(\pi D) - F = 0,$$

we find that the average shear stress is $\tau_{av} = F/\pi t D$.

Example 9.4

Analyzing a Given Stress Distribution

The normal stress acting on the rectangular cross section A in Fig. 9.14 is given by the equation $\sigma = 200y^2$ lb/in^2. Determine (a) the maximum normal stress; (b) the total normal force; (c) the average normal stress.

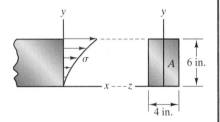

Figure 9.14

Strategy

(a) The value of σ increases monotonically with increasing y, so its maximum value occurs at $y = 6$ in.
(b) The normal stress σ depends only upon y, so we can integrate to determine the total normal force by using an element of area dA in the form of a horizontal strip [Fig. (a)].
(c) The average stress equals the total normal force divided by A.

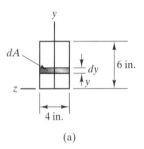

(a)

(**a**) Integration element dA.

Solution

(a) The maximum normal stress is

$$\sigma_{\max} = (200)(6)^2 = 7200 \text{ lb/in}^2.$$

(b) Using the strip element dA in Fig. (a), the total normal force is

$$\int_A \sigma \, dA = \int_0^6 (200y^2)(4 \, dy)$$

$$= 800 \left[\frac{y^3}{3} \right]_0^6$$

$$= 57{,}600 \text{ lb}.$$

(c) The average stress is

$$\sigma_{\text{av}} = \frac{1}{A} \int_A \sigma \, dA$$

$$= \frac{1}{(4)(6)} (57{,}600)$$

$$= 2400 \text{ lb/in}^2.$$

Discussion

Notice that the maximum normal stress is three times the average normal stress. This illustrates why knowledge of the average stress is usually insufficient for design. The actual distribution of the stress must be known to determine its maximum value.

Problems

9.1 The prismatic bar has a circular cross section with 50-mm radius and is subjected to 4-kN axial loads. Determine the average normal stress at the plane $\mathcal{P}$.

P9.1

9.2 In Problem 9.1, what is the average shear stress at the plane $\mathcal{P}$?

9.3 The prismatic bar has cross-sectional area $A = 30$ in^2 and is subjected to axial loads. Determine the average normal stress (a) at plane $\mathcal{P}_1$; (b) at plane $\mathcal{P}_2$.

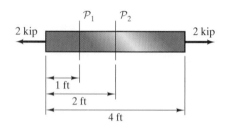

P9.3

9.4 The prismatic bar has a solid circular cross section with 2-in. radius. Determine the average normal stress (a) at plane $\mathcal{P}_1$; (b) at plane $\mathcal{P}_2$.

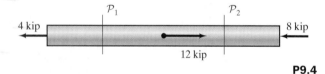

P9.4

9.5 A prismatic bar with cross-sectional area A is subjected to axial loads. Determine the average normal and shear stresses at the plane $\mathcal{P}$ if $A = 0.02$ m^2, $P = 4$ kN, and $\theta = 25°$.

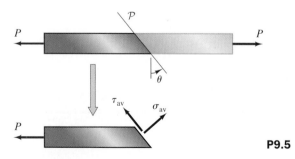

P9.5

9.6 Suppose that the prismatic bar shown in Problem 9.5 has cross-sectional area $A = 0.024$ m^2. If the angle $\theta = 35°$ and the average normal stress on the plane $\mathcal{P}$ is $\sigma_{av} = 200$ kPa, what are τ_{av} and the axial force P?

9.7 For the prismatic bar in Problem 9.5, derive equations for the average normal and shear stresses at the plane $\mathcal{P}$ as functions of θ.

9.8 The prismatic bar has a solid circular cross section with 30-mm radius. It is suspended from one end and is loaded only by its own weight. The mass density of the homogeneous material is 2800 kg/m^3. Determine the average normal stress at the plane $\mathcal{P}$, where x is the distance from the bottom of the bar in meters. (*Strategy:* Draw a free-body diagram of the part of the bar below the plane $\mathcal{P}$.)

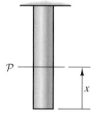

P9.8

9.9 The beam has cross-sectional area $A = 0.0625$ m^2. What are the average normal stress and the magnitude of the average shear stress at the plane $\mathcal{P}$? (*Strategy:* Draw the free-body diagram of the entire beam and determine the reactions at the pin and roller supports. Then determine the average normal and shear stresses by drawing the free-body diagram of the part of the beam to the left of the plane $\mathcal{P}$.)

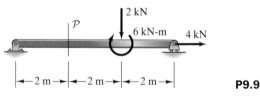

P9.9

9.10 Determine the average normal stress and the magnitude of the average shear stress at the plane $\mathcal{P}$ of the beam in Problem 9.9 by drawing the free-body diagram of the part of the beam to the right of the plane $\mathcal{P}$ and compare your answers to those of Problem 9.9.

9.11 The beams have cross-sectional area $A = 60$ in^2. What are the average normal stress and the magnitude of the average shear stress at the plane $\mathcal{P}$ in cases (a) and (b)?

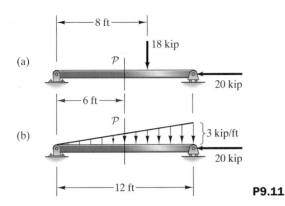

(a)

8 ft

18 kip

P

20 kip

(b)

6 ft

P

3 kip/ft

20 kip

12 ft

P9.11

9.12 Figure (a) is a diagram of the bones and biceps muscle of a person's arm supporting a mass. Figure (b) is a biomechanical model of the arm in which the biceps muscle AB is represented by a bar with pin supports. The suspended mass is $m = 2$ kg and the weight of the forearm is 9 N. If the cross-sectional area of the tendon connecting the biceps to the forearm at A is 28 mm^2, what is the average normal stress in the tendon?

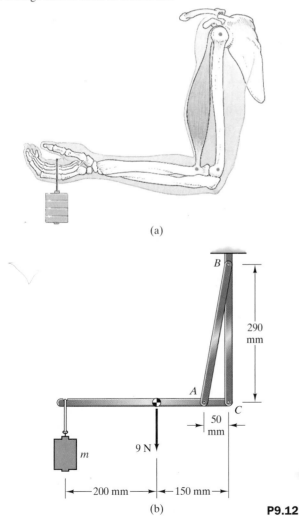

(a)

(b)

B

290 mm

A

C

50 mm

m

9 N

200 mm

150 mm

P9.12

9.13 The force **F** exerted on the bar is $20\mathbf{i} - 20\mathbf{j} - 10\mathbf{k}$ (lb). The plane $\mathcal{P}$ is parallel to the y–z plane and is 5 in. from the origin O. The bar's cross-sectional area at $\mathcal{P}$ is 0.65 in^2. What is the average normal stress in the bar at $\mathcal{P}$?

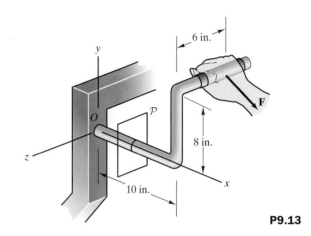

6 in.

y

O

$\mathcal{P}$

8 in.

z

10 in.

x

F

P9.13

9.14 In Problem 9.13, what is the magnitude of the average shear stress in the bar at $\mathcal{P}$?

9.15 The plane $\mathcal{P}$ is parallel to the y–z plane of the coordinate system. The cross-sectional area of the tennis racquet at $\mathcal{P}$ is 400 mm^2. Including the force exerted on the racquet by the ball and inertial effects, the total force on the racquet above the plane $\mathcal{P}$ is $35\mathbf{i} - 16\mathbf{j} - 85\mathbf{k}$ (N). What is the average normal stress on the racquet at $\mathcal{P}$?

x

y

z

$\mathcal{P}$

P9.15

9.16 The fixture shown connects a 50-mm-diameter bridge cable to a flange that is attached to the bridge. A 60-mm-diameter circular pin connects the fixture to the flange. If the average normal stress in the cable is $\sigma_{av} = 120$ MPa, what average shear stress σ_{av} must the pin support?

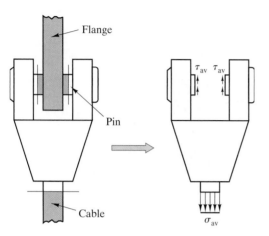

P9.16

9.17 Consider the fixture shown in Problem 9.16. The cable will safely support an average normal stress of 700 MPa and the circular pin will safely support an average shear stress of 220 MPa. Based on these criteria, what is the largest tensile load the cable will safely support?

9.18 The truss is made of prismatic bars with cross-sectional area $A = 0.25$ ft². Determine the average normal stress in member BE acting on a plane perpendicular to the axis of the member.

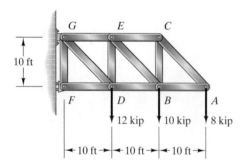

P9.18

9.19 For the truss in Problem 9.18, determine the average normal stress in member BD acting on a plane perpendicular to the axis of the member.

9.20 Three views of joint A of the truss in Problem 9.18 are shown. The joint is supported by a cylindrical pin 2 in. in diameter. What is the magnitude of the average shear stress in the pin?

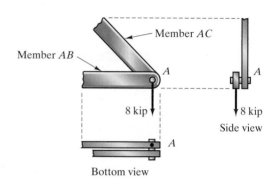

P9.20

9.21 The top view of pin A of the pliers is shown. The cross-sectional area of the pin is 4.5 mm². What is the average shear stress in the pin when 150-N forces are applied to the pliers as shown?

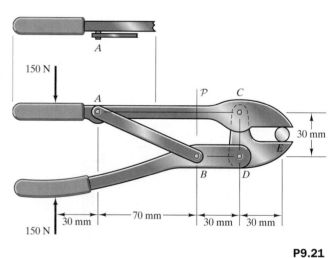

P9.21

9.22 In Problem 9.21 the vertical plane $\mathcal{P}$ is 30 mm to the left of C. The cross-sectional area of member AC of the pliers at the plane $\mathcal{P}$ is 50 mm². Determine the average normal stress and the magnitude of the average shear stress at $\mathcal{P}$ when 150-N forces are applied to the pliers as shown.

9.23 The suspended crate weighs 2000 lb and the angle $\alpha = 30°$. The top view of the pin support A of the crane's boom is shown. The cross-sectional area of the pin is 23 in². What is the average shear stress in the pin?

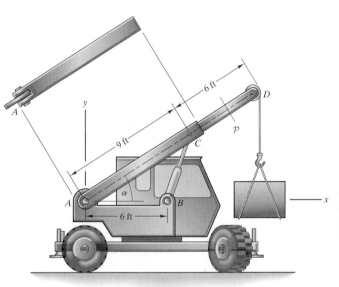

P9.23

9.24 In Problem 9.23, the plane $\mathcal{P}$ is 3 ft from end D of the crane's boom and is perpendicular to the boom. The cross-sectional area of the boom at $\mathcal{P}$ is 15 in². Determine the average normal stress and the magnitude of the average shear stress in the boom at $\mathcal{P}$.

9.25 Three rectangular boards are glued together and subjected to axial loads as shown. What is the average shear stress on each glued surface?

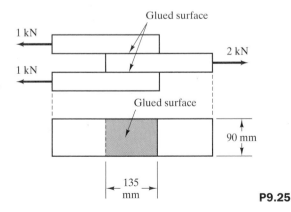

P9.25

9.26 Two boards with 4 in. × 4 in. square cross sections are mitered and glued together as shown. If the axial forces $P = 600$ lb, what average shear stress must the glue support?

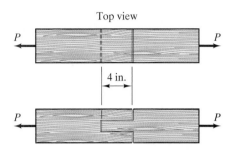

P9.26

9.27 A $\frac{1}{8}$-in.-diameter punch is used to cut blanks out of a $\frac{1}{16}$-in.-thick plate of aluminum. If an average shear stress of 20,000 psi must be induced in the plate to create a blank, what force F must be applied?

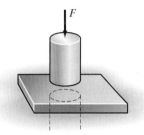

P9.27

9.28 Two pipes are connected by bolted flanges. The bolts are 20 mm in diameter. One pipe has a built-in support and the other is subjected to a torque $T = 6$ kN-m about its axis. Estimate the resulting average shear stress in each bolt. Why is this result an estimate?

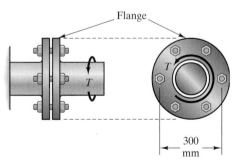

P9.28

9.29 The bolts in Problem 9.28 will each safely support an average shear stress of 130 MPa. Based on this criterion, estimate the largest safe torque T that can be applied.

9.30 "Shears" such as the familiar scissors have two blades that subject a material to shear stress. For the shearing process shown, draw a suitable free-body diagram and determine the average shear stress the blades exert on the sheet of material of thickness t and width b. (*Strategy:* Obtain a free-body diagram by passing a vertical plane through the material between the two blades.)

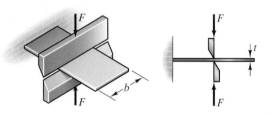

P9.30

9.31 A 2-in.-diameter cylindrical steel bar is attached to a 3-in.-thick fixed plate by a cylindrical rubber grommet. If the axial load $P = 60$ lb, what is the average shear stress on the cylindrical surface of contact between the bar and the grommet?

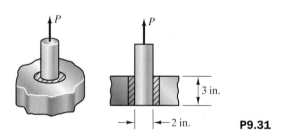

P9.31

9.32 The outer diameter of the cylindrical rubber grommet in Problem 9.31 is 3.5 in. What is the average shear stress on the cylindrical surface of contact between the grommet and the fixed plate?

9.33 The steel bar described in Problem 9.31 is subjected to a torque $T = 100$ in-lb about its axis. What is the average shear stress on the cylindrical surface of contact between the bar and the grommet?

P9.33

9.34 A traction distribution $\mathbf{t}$ acts on a plane surface A. The value of $\mathbf{t}$ at a given point on A is $\mathbf{t} = 45\mathbf{i} + 40\mathbf{j} - 30\mathbf{k}$ (kPa). The unit vector $\mathbf{i}$ is perpendicular to A and points away from the material. What is the normal stress σ at the given point?

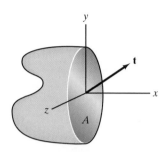

P9.34

9.35 In Problem 9.34, what is the magnitude of the shear stress τ at the given point?

9.36 A traction distribution $\mathbf{t}$ acts on a plane surface A. The value of $\mathbf{t}$ at a given point on A is $\mathbf{t} = 3000\mathbf{i} - 2000\mathbf{j} + 6000\mathbf{k}$ (psi). The unit vector $\mathbf{e} = \frac{6}{7}\mathbf{i} + \frac{3}{7}\mathbf{j} + \frac{2}{7}\mathbf{k}$ is perpendicular to A and points away from the material. What is the normal stress σ at the given point?

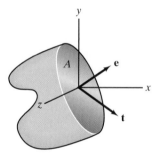

P9.36

9.2 **Strains**

In this section we consider what happens to a sample of material when it is subjected to external loads. From everyday experiments—stretching rubber bands, bending popsicle sticks—we are all familiar with the fact that objects can *deform*, or change shape, under the action of loads. Our ultimate objective is to be able to determine the deformations of simple structural elements resulting from given loads. For now, we merely want to introduce quantities that describe an object's change in shape.

If we put ourselves in the place of the pioneers of this subject, the task before us seems very difficult. How can we even describe the shape of a given object analytically, much less a change in its shape? To make the problem tractable, we don't approach it in this global way, asking "What is the object's new shape?" Instead, we begin by considering what happens to the material near a single point of the object.

Extensional Strain

Consider a sample of a solid material such as the piece of iron with which we began discussing stresses. We can imagine taking a pen and drawing an infinitesimal line dL somewhere within the material (Fig. 9.15a). Suppose that we then subject the sample to some set of loads, causing it to undergo a deformation. What happens to the imaginary line? We don't know, since we don't know what the loads are and cannot yet predict the effects those loads have on the material, but the line of original length dL may contract or stretch to some new length dL' and may also change direction (Fig. 9.15b). The *extensional strain* ε is a measure of the change in length of the line dL. It is defined to be the change in length divided by the original length:

$$\varepsilon = \frac{dL' - dL}{dL}. \tag{9.2}$$

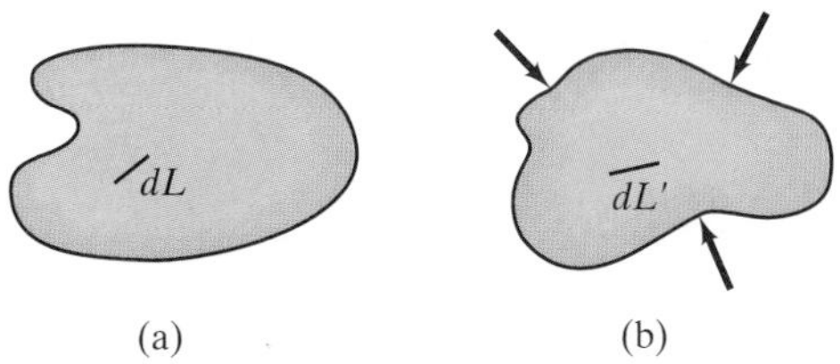

(a) (b)

Figure 9.15
(a) Infinitesimal line within a sample of material.
(b) Sample and line after a deformation.

For a given point of the material and a given direction (the direction of the line dL), ε measures how much the material contracts or stretches in the subsequent deformation. Notice that ε is negative if the material contracts and positive if it stretches. Since ε is a change in length divided by a length, it is dimensionless.

It is important to realize that the value of the extensional strain corresponds to a given point and a given direction in the material prior to the deformation. To define it, we had to choose a point in the material to place the line dL and choose a direction for dL. In general, the value of the extensional

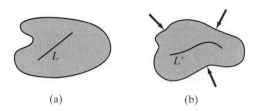

Figure 9.16
(a) Finite line within a sample of material.
(b) Sample and line after a deformation.

(a) (b)

strain may vary from point to point in a material and may be different in different directions. For example, a material may contract in one direction but stretch in a different direction. Also, notice that the extensional strain measures the contraction or stretch of the material relative to some initial state, which we call the *reference state*.

If we know the value of ε corresponding to a given point and a given direction prior to the deformation, what does that tell us? We know that if we draw an infinitesimal line of length dL in the material located at that point and in that direction, its change in length resulting from the deformation is $dL' - dL = \varepsilon\, dL$, so its new length is

$$dL' = (1 + \varepsilon)\, dL. \tag{9.3}$$

We can determine the new length of a finite line if we know the value of the extensional strain ε in the direction tangent to the line at each point of the line. Let L be the length of a finite line in a sample of material in a reference state (Fig. 9.16a). We obtain the length L' of the line in a deformed state (Fig. 9.16b) by integrating Eq. (9.3),

$$L' = \int_L (1 + \varepsilon)\, dL = L + \int_L \varepsilon\, dL, \tag{9.4}$$

where the subscript L on the integral signs means that the integrations are carried out over the entire length of the line. We denote the change in length of a finite line by δ:

$$\delta = L' - L = \int_L \varepsilon\, dL. \tag{9.5}$$

If the value of ε is uniform (constant) along the line, the change in length is

$$\delta = L' - L = \varepsilon L \qquad (\text{if } \varepsilon \text{ is constant along } L). \tag{9.6}$$

Notice that Eqs. (9.4)–(9.6) apply even if the line is not straight in the reference state as long as ε is the extensional strain in the direction tangent to the line at each point.

Shear Strain

The extensional strain tells us how much a material contracts or stretches in a given direction at a given point of a material. We will find in subsequent chapters that another type of strain is also useful for describing the deformation of a material. Let's return to our sample of material and imagine drawing two perpendicular infinitesimal lines dL_1 and dL_2 somewhere within a reference state (Fig. 9.17a). The angle between the two lines is 90°, or $\pi/2$ radi-

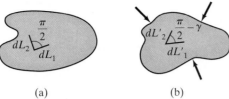

Figure 9.17
(a) Perpendicular infinitesimal lines at a point in the reference state.
(b) The lines in a deformed state.

ans. After the material undergoes a deformation, these two lines may no longer be perpendicular. Let us denote the angle in radians between the two lines after the deformation as $\pi/2 - \gamma$ (Fig. 9.17b). The angle γ will be positive, zero, or negative if the angle between the two lines decreases, remains the same, or increases, respectively.

The angle γ is called the *shear strain* referred to the directions dL_1 and dL_2. Thus the shear strain is a measure of the change in the angle between two particular lines that were perpendicular in the reference state. In general, its value may vary from point to point in a material and may be different for different directions of the lines dL_1 and dL_2. Since γ is an angle in radians, the shear strain is dimensionless.

If we consider an infinitesimal rectangle in the reference state with sides dL_1 and dL_2 (Fig. 9.18a), the shear strain referred to the directions dL_1 and dL_2 tells us the angles between the sides of the resulting parallelogram in the deformed state (Fig. 9.18b).

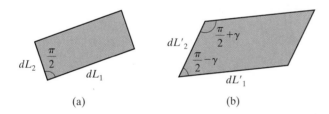

Figure 9.18
(a) Rectangle in the reference state.
(b) Shear strain causes the rectangle to become a parallelogram.

Notice that in defining the extensional and shear strains we have made no assumption about the state of motion of the material. In the examples and problems presented in this introductory treatment of mechanics of materials, we assume that the material in its deformed state is in equilibrium. But our definitions of strain are not limited to that situation. The deformed state can instead represent the state at an instant in time of a material undergoing an arbitrary motion.

Study Questions

1. What is the definition of the extensional strain ε? What are the dimensions of the extensional strain?
2. What does it tell you if the extensional strain in a particular direction at a point within a material is positive or negative?
3. Consider a line with finite length L within a reference state of a material. What do you need to know to determine the line's length in a deformed state of the material? How do you determine the new length?
4. What is the definition of the shear strain γ?

Example 9.5

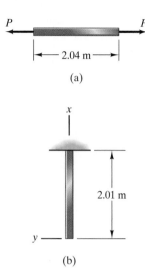

P ——— P

|← 2.04 m →|

(a)

x

2.01 m

y

(b)

Figure 9.19

Extensional Strain in a Bar

A bar has length $L = 2$ m in the unloaded state.

(a) In Fig. 9.19a, the bar is subjected to axial loads which increase its length to 2.04 m. If the extensional strain ε in the direction of the bar's axis is assumed to be uniform throughout the bar's length, what is ε?

(b) In Fig. 9.19b, the bar is suspended from one end, causing its length to increase to 2.01 m. The resulting extensional strain in the x direction is given by the equation $\varepsilon = ax/L$, where a is a constant. What is the value of a?

Strategy

In part (a) the extensional strain is uniform, so the bar's change in length δ in the deformed state is given in terms of its reference length L and the strain ε by Eq. (9.6). We know the change in length, so we can solve for ε. In part (b) the strain is a function of distance along the bar's axis, so we must use Eq. (9.5). By substituting the change in length and the given equation for ε, we can solve for the constant a.

Solution

(a) From Eq. (9.6),

$$\delta = L' - L = \varepsilon L:$$
$$2.04 - 2 = \varepsilon(2)$$
$$0.04 = \varepsilon(2).$$

We obtain $\varepsilon = 0.02$.

(b) From Eq. (9.5), the bar's change in length is

$$\delta = L' - L = \int_L \varepsilon \, dL:$$
$$2.01 - 2 = \int_0^2 \frac{ax}{2} \, dx$$
$$0.01 = \frac{a}{2} \left[\frac{x^2}{2} \right]_0^2 = a.$$

The constant $a = 0.01$.

Discussion

In case (b) we gave the form of the equation for the distribution of extensional strain along the axis of a suspended bar. In Chapter 10 we relate the extensional strain to the normal stress in a linearly elastic bar subjected to axial load. You will then be able to show that the extensional strain is a linear function of x in a suspended prismatic bar.

Example 9.6

Strain of a Rectangular Element

Consider the infinitesimal rectangle in Fig. 9.20a at a point of a material in a reference state. The diagram in a deformed state is shown in Fig. 9.20b. Let the extensional strains in the dL_1 and dL_2 directions be ε_1 and ε_2, and let the shear strain referred to the dL_1 and dL_2 directions be γ. What is the extensional strain in the direction of the diagonal dL?

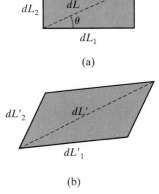

(a)

(b)

Figure 9.20

Strategy

From the extensional strains in the dL_1 and dL_2 directions we can determine the lengths dL_1' and dL_2'. From the shear strain we know the angles between the sides of the parallelogram in the deformed state (see Fig. 9.18). By applying the law of cosines to one of the oblique triangles comprising the parallelogram, we can determine the length dL'.

Solution

The lengths dL_1' and dL_2' are

$$dL_1' = (1 + \varepsilon_1)\, dL_1 = (1 + \varepsilon_1)\, dL \cos \theta$$

$$dL_2' = (1 + \varepsilon_2)\, dL_2 = (1 + \varepsilon_2)\, dL \sin \theta$$

Consider the upper triangle of the parallelogram in the deformed state [Fig. (c)]. From the law of cosines,

$$(dL')^2 = (dL_1')^2 + (dL_2')^2 - 2\, dL_1'\, dL_2' \cos\left(\frac{\pi}{2} + \gamma\right).$$

(c)

(c) Upper triangle of the element.

Substituting our expressions for dL_1' and dL_2' into this equation, we obtain

$$(dL')^2 = \left[(1 + \varepsilon_1)^2 \cos^2 \theta + (1 + \varepsilon_2)^2 \sin^2 \theta \right.$$
$$\left. - 2(1 + \varepsilon_1)(1 + \varepsilon_2) \cos \theta \sin \theta \cos\left(\frac{\pi}{2} + \gamma\right)\right](dL)^2.$$

The extensional strain in the direction of the diagonal dL is therefore

$$\varepsilon = \frac{dL' - dL}{dL}$$

$$= \sqrt{(1 + \varepsilon_1)^2 \cos^2 \theta + (1 + \varepsilon_2)^2 \sin^2 \theta - 2(1 + \varepsilon_1)(1 + \varepsilon_2) \cos \theta \sin \theta \cos\left(\frac{\pi}{2} + \gamma\right)} - 1.$$

Discussion

This example demonstrates why the shear strain is needed for determining the deformation near a point of a material. The extensional strains in the dL_1 and dL_2 directions are not sufficient to determine the extensional strain in the dL direction. We also need the value of the shear strain.

Problems

9.37 A line of length dL at a particular point of a material in a reference state has length $dL' = 1.2\,dL$ in a deformed state. What is the extensional strain corresponding to that particular point and the direction of the line dL?

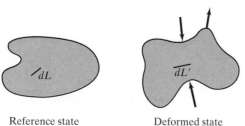

Reference state　　　　Deformed state

P9.37

9.38 The extensional strain corresponding to a point of a material and the direction of a line of length dL in the reference state is $\varepsilon = 0.15$. What is the length dL' of the line in the deformed state?

9.39 A straight line within a reference state of an object is 50 mm long. In a deformed state, the line is 54 mm long. If the extensional strain ε in the direction tangent to the line is uniform thoughout the line's length, what is ε?

9.40 The length of the curved line within the material in the reference state is $L = 0.2$ m. The material then undergoes a deformation such that the value of the extensional strain ε in the direction tangent to the curved line is $\varepsilon = 0.03$ at each point of the line. What is the length L' of the line in the deformed state?

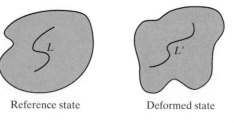

Reference state　　　　Deformed state

P9.40

9.41 The length of the curved line L in the reference state shown in Problem 9.40 is 0.5 m. Its length in the deformed state is $L' = 0.488$ m. If the value of the extensional strain ε in the direction tangent to the line is the same at each point of the line, what is ε?

9.42 The coordinate s measures distance along the curved line in the reference state. The length of the line is $L = 0.2$ m. The material then undergoes a deformation such that the value of the exten-

sional strain ε in the direction tangent to the curved line is $\varepsilon = 0.03 + 2s^2$. What is the length L' of the line in the deformed state?

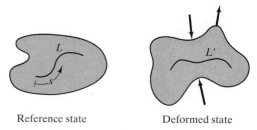

Reference state　　　　Deformed state

P9.42

9.43 In Problem 9.42, suppose that the material undergoes a deformation that induces an extensional strain tangent to the line given by the equation $\varepsilon = 0.01[1 + (s/L)^3]$. What is the length L' of the line in the deformed state?

9.44 In a shock-wave experiment, the left side of a 100-mm-thick plate of steel is subjected to a constant velocity of 1.5 km/s to the right at time $t = 0$. As a result, a shock wave travels across the plate with a constant velocity $U > 1.5$ km/s. To the right of the shock wave, the material of the plate is stationary and undeformed. To the left of the shock wave, the material is moving with a uniform velocity of 1.5 km/s and is subject to a homogeneous (uniform) extensional strain ε. If optical instrumentation indicates that the shock wave arrives at the right side of the plate at time $t = 16 \times 10^{-6}$ s, what is ε?

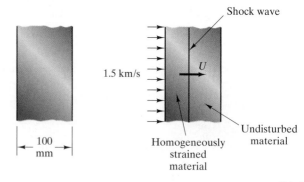

P9.44

9.45 In Problem 9.44, suppose that the time at which the shock wave arrives at the right side of the plate is unknown, but an embedded strain gauge indicates that the homogeneous extensional strain to the left of the shock wave is $\varepsilon = -0.3$. What is the velocity U of the shock wave?

9.46 When it is unloaded, the nonprismatic bar is 12 in. long. The loads cause axial strain given by the equation

$\varepsilon = 0.04/(x + 12)$, where x is the distance from the left end of the bar in inches. What is the change in length of the bar?

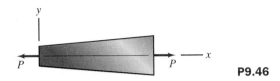

P9.46

9.47 The force F causes point B to move downward 0.002 m. If you assume the resulting extensional strain ε parallel to the axis of the bar AB is uniform along the bar's length, what is ε?

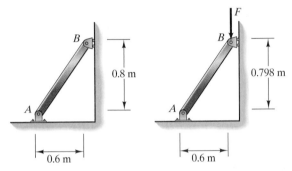

P9.47

9.48 When the truss is subjected to the vertical force F, joint A moves a distance $v = 0.3$ m vertically and a distance $u = 0.1$ m horizontally. If the extensional strain ε_{AB} in the direction parallel to member AB is uniform thoughout the length of the member, what is ε_{AB}?

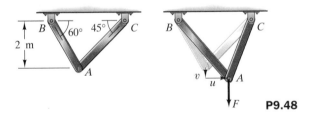

P9.48

9.49 In Problem 9.48, if the extensional strain ε_{AC} in the direction parallel to member AC is uniform thoughout the length of the member, what is ε_{AC}?

9.50 Suppose that a downward force is applied at point A of the truss, causing point A to move 0.360 in. downward and 0.220 in. to the left. If the resulting extensional strain ε_{AB} in the direction parallel to the axis of bar AB is uniform, what is ε_{AB}?

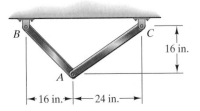

P9.50

9.51 In Problem 9.50, if the resulting extensional strain ε_{AC} in the direction parallel to the axis of bar AC is uniform, what is ε_{AC}?

9.52 A steel tube (a) has an outer radius $r = 20$ mm. The tube is then pressurized, increasing its outer radius to $r' = 20.04$ mm (b). What is the resulting extensional strain of the bar's outer circumference in the direction tangent to the circumference?

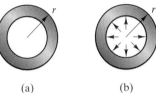

(a)   (b)   **P9.52**

9.53 The angle between two infinitesimal lines dL_1 and dL_2 that are perpendicular in a reference state is 120° in a deformed state. What is the shear strain at this point corresponding to the directions dL_1 and dL_2?

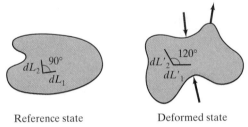

Reference state   Deformed state

P9.53

9.54 When the airplane's wing is unloaded (the reference state), the perpendicular lines L_1 and L_2 on the upper surface of the left wing are each 600 mm long. In the loaded state shown, L_1 is 600.2 mm long and L_2 is 595 mm long. If you assume that they are uniform, what are the extensional strains in the L_1 and L_2 directions?

P9.54

9.55 In Problem 9.54, the angle between the lines L_1 and L_2 at the point where they intersect is 90.2° in the loaded state. What is the shear strain referred to the directions L_1 and L_2 at that point?

9.56 Two infinitesimal lines dL_1 and dL_2 are shown in a reference state and in a deformed state. (The lines dL_1, dL_2, dL'_1, and dL'_2 are contained in the x–y plane.) What is the shear strain at this point corresponding to the directions dL_1 and dL_2?

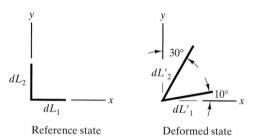

Reference state   Deformed state

P9.56

9.57 Two infinitesimal lines dL_1 and dL_2 within a material are parallel to the x and y axes in a reference state [Fig. (a)]. After a motion and deformation of the material, dL_1 points in the direction of the unit vector $e_1 = 0.667i + 0.667j + 0.333k$, and dL_2 points in the direction of the unit vector $e_2 = -0.408i + 0.816j - 0.408k$ [Fig. (b)]. What is the shear strain referred to the directions dL_1 and dL_2?

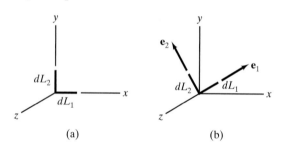

(a)   (b)

P9.57

9.58 In Problem 9.57, suppose that after the motion and deformation of the material, dL_1 points in the direction of the unit vector $e_1 = 0.667i + 0.667j + 0.333k$, and dL_2 points in the direction of the unit vector $e_2 = -0.514i + 0.686j + 0.514k$. What is the shear strain referred to the directions dL_1 and dL_2?

9.59 An infinitesimal rectangle at a point in a reference state of a material is shown. In a deformed state the extensional strains in the dL_1 and dL_2 directions are $\varepsilon_1 = 0.04$ and $\varepsilon_2 = -0.02$ and the shear strain referred to the dL_1 and dL_2 directions is $\gamma = 0.02$. What is the extensional strain in the dL direction? (See Example 9.6.)

P9.59

9.60 For the infinitesimal rectangle at a point in a reference state of a material shown in Problem 9.59, suppose that in a deformed state the extensional strains in the dL_1, dL_2, and dL directions are $\varepsilon_1 = 0.030$, $\varepsilon_2 = 0.020$, and $\varepsilon = 0.038$. What is the shear strain referred to the dL_1 and dL_2 directions?

Chapter Summary

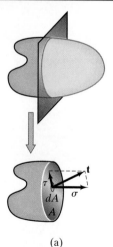

(a)

In this chapter we have introduced fundamental measures of the stresses and strains in a material. We defined average stresses on an area and showed how to evaluate them in particular cases. In Chapter 10 we apply these definitions to structural applications involving axially loaded bars.

Traction, Normal Stress, and Shear Stress

The *traction* **t** [Fig. (a)] is defined such that the force exerted on each infinitesimal element dA of the area A is **t** dA. The normal component σ is the *normal stress* and the tangential component τ is the *shear stress*. The normal stress σ is defined to be positive if it points outward, or away from the material, and is then said to be *tensile*. If σ is negative, the normal stress points toward the material and is said to be *compressive*. The dimensions of **t**, σ, and τ are force/area.

The average value of the traction **t** on the area A in Fig. (a) is

$$\mathbf{t}_{av} = \frac{1}{A} \int_A \mathbf{t}\, dA. \qquad \text{Eq. (9.1)}$$

The components of the average traction normal and tangential to A are the average normal stress σ_{av} and average shear stress τ_{av}. The total normal and tangential forces exerted on A are $\sigma_{av} A$ and $\tau_{av} A$, respectively.

Extensional and Shear Strains

Consider an infinitesimal line dL in a sample of material in a reference state [Fig. (b)]. Let the length of the line in a deformed state be dL' [Fig. (c)]. The *extensional strain* ε is defined by

$$\varepsilon = \frac{dL' - dL}{dL}. \qquad \text{Eq. (9.2)}$$

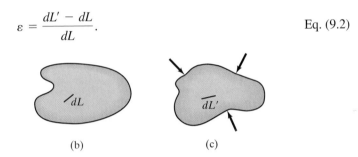

(b)   (c)

Let L be the length of a finite line within the material in a reference state. The length L' in a deformed state is

$$L' = L + \int_L \varepsilon\, dL, \qquad \text{Eq. (9.4)}$$

where ε is the extensional strain in the direction tangent to the line at each point. The change in length of the line is denoted by δ:

$$\delta = L' - L = \int_L \varepsilon\, dL. \qquad \text{Eq. (9.5)}$$

If ε is uniform along the line, the change in length is

$$\delta = L' - L = \varepsilon L. \qquad \text{Eq. (9.6)}$$

Consider two perpendicular infinitesimal lines dL_1 and dL_2 in a sample of material in a reference state [Fig. (d)]. Let the angle in radians between the two lines in a deformed state be denoted by $\pi/2 - \gamma$ [Fig. (e)]. The angle γ is the *shear strain* referred to the directions dL_1 and dL_2.

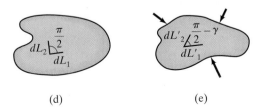

(d)   (e)

If an infinitesimal rectangle in the reference state has sides dL_1 and dL_2 [Fig. (f)], the shear strain referred to the directions dL_1 and dL_2 determines the angles between the sides of the resulting parallelogram in the deformed state [Fig. (g)].

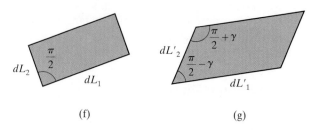

(f)

(g)

Review Problems

9.61 The bar is made of material 1 in. thick. Its width varies linearly from 2 in. at its left end to 4 in. at its right end. If the axial load $P = 200$ lb, what is the average normal stress (a) at plane $\mathcal{P}_1$; (b) at plane $\mathcal{P}_2$?

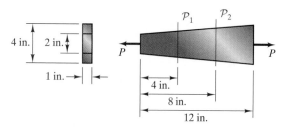

P9.61

9.62 If the average normal stress at plane $\mathcal{P}_1$ of the bar described in Problem 9.61 is $\sigma_{av} = 300$ psi, what are the axial load P and the average normal stress at plane $\mathcal{P}_2$?

9.63 The beam has cross-sectional area $A = 0.1$ m². What are the average normal stress and the magnitude of the average shear stress at the plane $\mathcal{P}$?

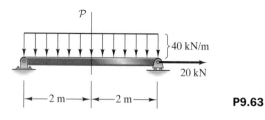

P9.63

9.64 In Problem 9.63, what are the average normal stress and the magnitude of the average shear stress at the plane $\mathcal{P}$ if the plane is 1 m from the left end of the beam?

9.65 The prismatic bar AB has cross-sectional area $A = 0.01$ m². If the force $F = 6$ kN, what is the average normal stress at the plane $\mathcal{P}$?

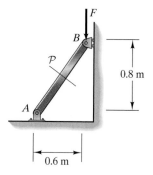

P9.65

9.66 The prismatic bar AB in Problem 9.65 will safely support an average compressive normal stress of 1.2 MPa on the plane $\mathcal{P}$. Based on this criterion, what is the largest downward force F that can safely be applied?

9.67 The jaws of the bolt cutter are connected by two links AB. The cross-sectional area of each link is 750 mm². What average normal stress is induced in each link by the 90-N forces exerted on the handles?

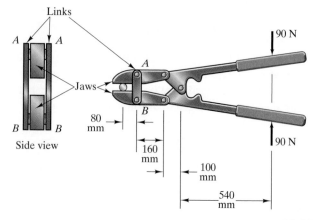

P9.67

9.68 The pins connecting the links *AB* to the jaws of the bolt cutter in Problem 9.67 are 20 mm in diameter. What average shear stress is induced in the pins by the 90-N forces exerted on the handles?

9.69 Suppose that you subject a 2-m prismatic bar to compressive axial forces that cause a uniform extensional strain $\varepsilon = -0.003$ in the axial direction. What is the bar's length in the deformed state?

9.70 A prismatic bar is subjected to loads that cause uniform axial strains $\varepsilon = 0.002$ in its left half and $\varepsilon = -0.004$ in its right half. What is the resulting change in length of the 28-in. bar?

14 in. ———— 14 in.

P9.70

9.71 The prismatic bar in Problem 9.70 is subjected to loads that cause a uniform axial strain $\varepsilon_L = 0.006$ in its left half and a uniform axial strain ε_R in its right half. As a result, the length of the 28-in. bar increases by 0.032 in. What is ε_R?

The front wheels of the Formula One car are attached to the car's frame by a light and aerodynamically efficient structure of axially loaded steel rods.

CHAPTER 10

Axially Loaded Bars

T
russ structures—widely used to support bridges, buildings, vehicles, and other mechanical devices—are frameworks of bars that support loads by subjecting their members to axial loads. Exerting axial loads on a bar and measuring the resulting change in its length is a common method of testing the response of materials to stress. Not only is this test relatively straightforward to carry out, it subjects the material to a very simple and easily determined distribution of stress. Because of this simplicity, and the great importance of axially loaded bars from the standpoint of applications, we analyze them in detail in this chapter.

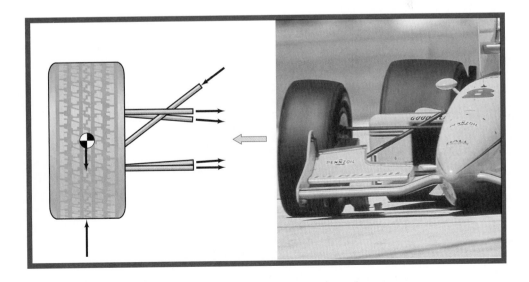

10.1 Stresses in Prismatic Bars

In Chapter 9 we determined average normal and shear stresses in axially loaded bars. We emphasized, however, that it is the actual distributions of stress, not their average values, that usually must be known for design. We now describe these distributions.

Stresses on Perpendicular Planes

We first consider the stresses acting on planes perpendicular to the axis of an axially loaded bar. Suppose that we could somehow load a prismatic bar by applying uniform normal stresses σ at its ends (Fig. 10.1). Then it can be shown that the stress distribution on *every* plane perpendicular to the axis of the bar consists of the same uniform normal stress and no shear stress (Fig. 10.2). In Fig. 10.3 we draw the free-body diagram of an element of the bar having rectangular faces, two of which are perpendicular to the bar's axis. The faces perpendicular to the axis are subjected to the uniform normal stress σ and the other faces are free of stress.

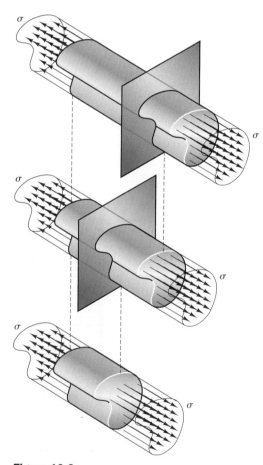

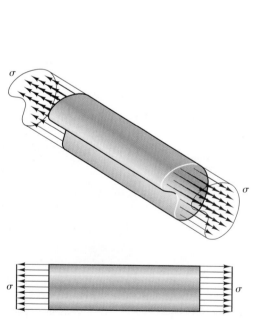

Figure 10.1
Loading a bar by normal tractions at its ends.

Figure 10.2
The same uniform normal stress acts on every plane perpendicular to the bar's axis.

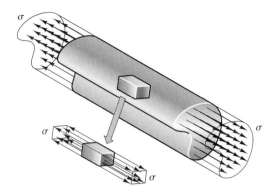

Figure 10.3
Isolating an element within the bar.

The proof of these results, beyond our scope in this elementary treatment, involves showing that no other distribution of stress can satisfy both the boundary conditions (the tractions on the ends and the traction-free lateral surfaces of the bar) and equilibrium. They also depend upon the material having a property called isotropy, which we discuss in Chapter 13.

So we know the stress distribution on planes perpendicular to the axis of an axially loaded prismatic bar, but only if the bar is loaded in a very idealized way. What does that tell us about real axially loaded bars? To answer this question we first need to demonstrate that applying a uniform normal stress to the end of a bar is equivalent to applying a single force at the centroid of its cross section. (Recall that we define two systems of forces and moments to be equivalent if the sums of the forces in the two systems are equal and the sums of the moments about any point due to the two systems are equal.) The total force exerted by a uniform normal stress σ on the bar's cross-sectional area A is σA. Let us place a coordinate system with its origin at the centroid of the cross section (Fig. 10.4a), so that the y and z coordinates of the centroid are zero:

$$\bar{y} = \frac{\int_A y \, dA}{A} = 0, \qquad \bar{z} = \frac{\int_A z \, dA}{A} = 0.$$

Then the moment about the z axis due to the uniform normal stress distribution is zero,

$$\int_A -y\sigma \, dA = -\sigma \int_A y \, dA = 0,$$

and the moment about the y axis is zero,

$$\int_A z\sigma \, dA = \sigma \int_A z \, dA = 0.$$

If we place a force $P = \sigma A$ at the centroid of the bar's cross section (Fig. 10.4b), the moments due to P about the y and z axes are, of course, zero, so P is equivalent to the uniform normal stress distribution.

Now suppose that we apply loads to the ends of a prismatic bar that are equivalent to axial forces P applied at the centroid of the cross section, but do

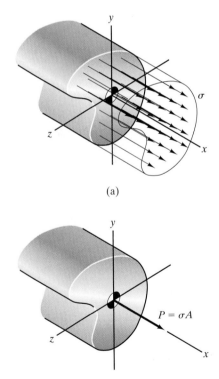

(a)

(b)

Figure 10.4
(a) Coordinate system with its origin at the centroid.
(b) The force P acting at the centroid is equivalent to the uniform stress σ.

Figure 10.5
Applying an axial load to a bar by pulling on the ends.

it in some realizable way (Fig. 10.5). On planes at increasing distances from the ends of the bar, the stress distribution approaches a uniform normal stress

$$\sigma = \frac{P}{A}. \tag{10.1}$$

No matter how we apply the loads, the stress distribution at axial distances greater than a few times the bar's width is approximately the same one obtained by subjecting the ends of the bar to uniform normal stresses (Fig. 10.6).

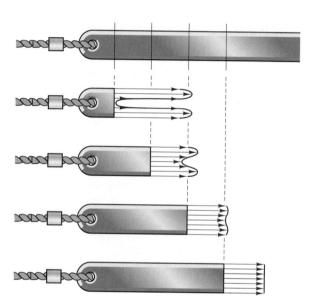

Figure 10.6
Stress distributions at increasing distances from the end of the bar.

This result was stated by Barré de Saint-Venant (1797–1886), who conducted experiments with rubber bars to demonstrate it for particular cases. In a more general form it is known as *Saint-Venant's principle*, and has been proven analytically in recent years. Based on this result, we assume in examples and problems that the stress distribution on a plane perpendicular to the axis of an axially loaded bar consists of uniform normal stress. But keep in mind that this assumption is invalid near the ends of the bar. A separate analysis, again beyond the scope of our discussion, is necessary to determine the distribution of stress near the ends.

Study Questions

1. If a prismatic bar could be loaded by applying equal uniform normal stress distributions σ at its ends, what is the resulting stress distribution on planes perpendicular to the bar's axis?

2. If a prismatic bar is loaded by arbitrary systems of forces and couples at the ends that are each equivalent to an axial tensile load P applied at the centroid of the bar's cross section, what do you know about the resulting stress distribution on planes perpendicular to the bar's axis?

Example 10.1

Normal Stresses in an Axially Loaded Bar

Part A of the bar in Fig. 10.7 has a 2-in.-diameter circular cross section and part B has a 4-in.-diameter circular cross section. Determine the normal stress on a plane perpendicular to the bar's axis (a) in part A; (b) in part B.

Figure 10.7

Strategy

By passing a plane through part A of the bar and isolating the part of the bar on one side of the plane, we can determine the axial load in part A, then use Eq. (10.1) to determine the normal stress. We can determine the normal stress in part B in the same way.

Solution

(a) In Fig. (a) we pass a plane through part A and isolate the part of the bar to the left of the plane.

From the resulting free-body diagram, the axial load in part A is $P = 4$ kip. The normal stress in part A is

$$\sigma = \frac{P}{A} = \frac{4000}{\pi(2)^2/4} = 1273 \text{ psi.}$$

(b) In Fig. (b) we pass a plane through part B and isolate the part of the bar to the left of the plane.

From the equilibrium equation $P + 12 - 4 = 0$, the axial load in part B is $P = -8$ kip. The normal stress in part B is

$$\sigma = \frac{P}{A} = \frac{-8000}{\pi(4)^2/4} = -637 \text{ psi.}$$

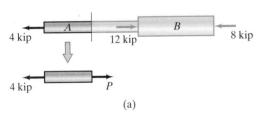

(a)

(a) Obtaining a free-body diagram by passing a plane through part A.

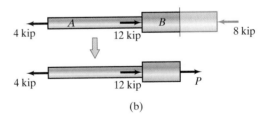

(b)

(b) Obtaining a free-body diagram by passing a plane through part B.

Discussion

In determining the axial load in part B, we could have obtained a simpler free-body diagram by isolating the part of the bar to the right of the plane [Fig. (c)]. We obtain the same result, $P = -8$ kip.

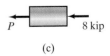

(c)

(c) Free-body diagram of the right part of the bar.

Example 10.2

Normal Stresses in Truss Members

The members of the truss in Fig. 10.8 have equal cross-sectional areas $A = 400 \text{ mm}^2$. The suspended mass is $m = 3400$ kg. What are the normal stresses in the members?

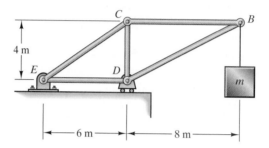

Figure 10.8

Strategy

We can use the method of joints to determine the axial force in each member and divide by A to determine the normal stress.

Solution

In Fig. (a) we draw the free-body diagram of joint B of the truss. The angle $\theta = \arctan(4/8) = 26.6°$.
From the equilibrium equations

$$\Sigma F_x = -P_{BC} - P_{BD} \cos\theta = 0,$$
$$\Sigma F_y = -P_{BD} \sin\theta - mg = 0,$$

we obtain $P_{BC} = 2mg$, $P_{BD} = -2.24mg$. Continuing in this way, we obtain the axial forces:

Member:	BC	BD	CD	CE	DE
Axial force:	$2mg$	$-2.24mg$	$-1.33mg$	$2.40mg$	$-2mg$

Substituting the values $m = 3400$ kg and $g = 9.81 \text{ m/s}^2$ and dividing by $400 \times 10^{-6} \text{ m}^2$, the stresses are

Member:	BC	BD	CD	CE	DE
Normal stress (MPa):	167	-186	-111	200	-167

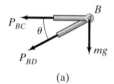

(a)

(a) Joint B.

Stresses on Oblique Planes

In design we cannot consider the stress distribution on only one particular plane or subset of planes. We can demonstrate this for an axially loaded bar. Let us determine the normal and shear stresses acting on a plane $\mathcal{P}$ oriented as shown in Fig. 10.9. The angle θ is measured from the bar's axis to a line *normal* to the plane $\mathcal{P}$. To determine the stresses on $\mathcal{P}$, we define a free-body diagram as shown in Fig. 10.10: We first isolate a rectangular element from

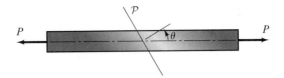

Figure 10.9
Passing a plane $\mathcal{P}$ through a bar at an arbitrary angle relative to its axis.

the bar, then isolate the part of the element to the left of plane $\mathcal{P}$. Our objective is to obtain a geometrically simple free-body diagram subjected to the normal and shear stresses on $\mathcal{P}$. We denote the normal stress by σ_θ and the shear stress by τ_θ, indicating that they act upon the plane defined by the angle θ We have arbitrarily chosen the direction of τ_θ.

We will determine σ_θ and τ_θ by writing equilibrium equations for the triangular free-body diagram we cut from the element. Letting A_θ be the area of the slanted face, we can express the areas of the horizontal and vertical faces in terms of A_θ and θ (Fig. 10.11a). Then the forces on the free-body diagram are the products of the areas and the uniform stresses (Fig. 10.11b). Summing forces in the σ_θ direction,

$$\sigma_\theta A_\theta - \left(\sigma A_\theta \cos\theta\right)\cos\theta = 0,$$

we determine the normal stress on $\mathcal{P}$:

$$\sigma_\theta = \sigma \cos^2\theta. \tag{10.2}$$

Summing forces in the τ_θ direction,

$$\tau_\theta A_\theta + \left(\sigma A_\theta \cos\theta\right)\sin\theta = 0,$$

we determine the shear stress on $\mathcal{P}$:

$$\tau_\theta = -\sigma \sin\theta \cos\theta. \tag{10.3}$$

Equations (10.2) and (10.3) give the normal and shear stresses acting on a plane oriented at an arbitrary angle relative to the axis of a prismatic bar. In Fig. 10.12 we plot the ratios σ_θ/σ and τ_θ/σ from $\theta = 0$ to $\theta = 180°$. We also

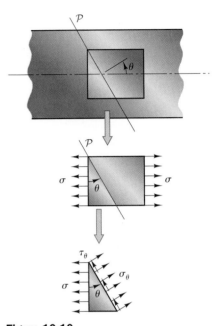

Figure 10.10
Obtaining a free-body diagram for determining the stresses on $\mathcal{P}$.

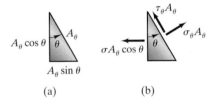

(a) (b)

Figure 10.11
(a) Areas of the faces.
(b) Forces on the free-body diagram.

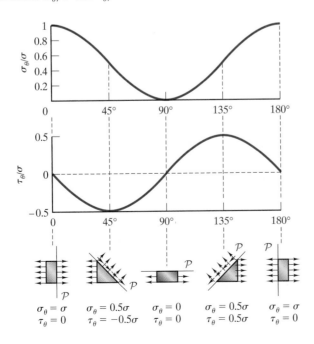

Figure 10.12
Normal and shear stresses as functions of θ.

show the elements and the values of σ_θ and τ_θ corresponding to particular values of θ.

Figure 10.12 permits us to examine the normal and shear stresses acting on all planes through an axially loaded prismatic bar. Notice that there is no plane for which the normal stress is larger in magnitude than the normal stress on the plane perpendicular to the bar's axis ($\theta = 0$). When we use the equation $\sigma = P/A$ to determine the normal stress on the plane perpendicular to a bar's axis, we determine the largest tensile or compressive normal stress acting on any plane within the bar. On the other hand, there is no shear stress on the plane perpendicular to the bar's axis, but there are shear stresses on oblique planes. On planes oriented at 45° relative to the bar's axis, the magnitude of the shear stress is one-half the magnitude of the maximum normal stress: $|\tau_\theta| = |\sigma/2|$. We can confirm that this is the maximum magnitude of the shear stress from Eq. (10.3). The derivative of τ_θ with respect to θ is

$$\frac{d\tau_\theta}{d\theta} = \sigma\left(\sin^2\theta - \cos^2\theta.\right)$$

Equating this expression to zero to determine the angles at which τ_θ has a maximum or minimum, we obtain $\theta = 45°$ and $\theta = 135°$. At $\theta = 45°$, $\tau_\theta = -\sigma/2$, and at $\theta = 135°$, $\tau_\theta = \sigma/2$.

We have shown that the maximum tensile or compressive normal stress on any plane through an axially loaded prismatic bar is $\sigma = P/A$, and the magnitude of the maximum shear stress is $|\sigma/2|$.

As an example, suppose that a column of concrete with cross sectional area A supports a weight W (Fig. 10.13). Drawing a free-body diagram by passing a plane through the column perpendicular to its axis (Fig. 10.14), we see that the compressive stress is $\sigma = -W/A$. This is the maximum compressive stress on any plane through the column. The magnitude of the maximum shear stress, which occurs on planes oriented at 45° relative to the column's axis, is $|\sigma/2| = W/2A$. If W is increased sufficiently, a column of brittle material such as concrete tends to fail along a plane on which the magnitude of the shear stress is a maximum (Fig. 10.15).

When we refer to the normal stress in an axially loaded bar without specifying the plane, we will mean the normal stress on a plane perpendicular to the bar's axis.

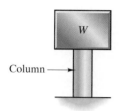

Figure 10.13
Column supporting a weight W.

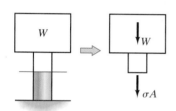

Figure 10.14
Obtaining a free-body diagram by passing a plane through the column perpendicular to its axis.

Figure 10.15
Failure of the column.

Study Questions

1. If a bar is subjected to axial loads, how can you determine the normal and shear stresses acting on a plane that is not perpendicular to the bar's axis?
2. If a bar with cross sectional area A is subjected to axial tensile loads P, what is the maximum tensile stress to which the material is subjected? What is the orientation of the planes on which the maximum tensile stress occurs?
3. If a bar with cross sectional area A is subjected to axial tensile loads P, what is the magnitude of the maximum shear stress to which the material is subjected? What is the orientation of the planes on which the magnitude of the shear stress is a maximum?

Example 10.3

Stresses on an Oblique Plane

To test a glue, two plates are glued together as shown in Fig. 10.16. The bar formed by the joined plates is then subjected to a tensile axial load of 200 N. What normal and shear stresses act on the plane where the plates are glued together? (In other words, what stresses must the glue support?)

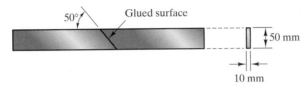

Figure 10.16

Strategy

We know the axial load and the bar's cross-sectional area, so we can determine the normal stress $\sigma = P/A$ on a plane perpendicular to the bar's axis. Then we can use Eqs. (10.2) and (10.3) to determine the normal and shear stresses on the glued plane.

Solution

The normal stress on a plane perpendicular to the bar's axis is

$$\sigma = \frac{P}{A} = \frac{200}{(0.01)(0.05)} = 400{,}000 \text{ Pa.}$$

Our objective is to determine the normal stress σ_θ and shear stress τ_θ on the plane where the plates are glued [Fig. (a)].

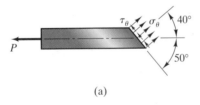

(a)

(a) Stresses on the glued plane.

The angle between the normal to the glued plane and the bar's axis is 40°. From Eq. (10.2),

$$\sigma_\theta = \sigma \cos^2 \theta = (400{,}000) \cos^2 40° = 235{,}000 \text{ Pa,}$$

and from Eq. (10.3),

$$\tau_\theta = -\sigma \sin\theta \cos\theta = -(400{,}000) \sin 40° \cos 40°$$

$$= -197{,}000 \text{ Pa.}$$

The glued surface is subjected to a tensile normal stress of 235 kPa and a shear stress of magnitude 197 kPa. The negative value of the shear stress indicates that the shear stress acts in the direction opposite to the defined direction of τ_θ.

Problems

10.1 A prismatic bar with cross-sectional area $A = 0.1 \text{ m}^2$ is loaded at the ends in two ways: (a) by 100-Pa uniform normal tractions; (b) by 10-N axial forces acting at the centroid of the bar's cross section. What are the normal and shear stress distributions at the plane $\mathcal{P}$ in the two cases?

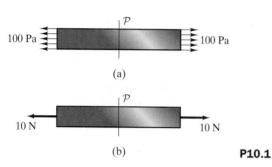

(a)

(b)

P10.1

10.2 A prismatic bar with cross-sectional area $A = 4 \text{ in}^2$ is subjected to tensile axial loads P. It consists of a material that will safely support a tensile normal stress of 60 ksi. Based on this criterion, what is the largest safe value of P?

10.3 A prismatic bar has a solid circular cross section with 20-mm diameter. It consists of a material that will safely support a tensile normal stress of 300 MPa. Based on this criterion, what is the largest tensile load P to which the bar can be subjected?

10.4 The cross-sectional area of bar AB is 0.5 in². If the force $F = 3 \text{ kip}$, what is the normal stress on a plane perpendicular to the axis of bar AB? (*Strategy:* You can determine the axial force in bar AB by drawing a free-body diagram of the horizontal bar and summing moments about point C.)

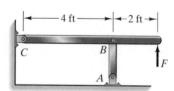

P10.4

10.5 Bar AB of the frame in Problem 10.4 consists of a material that will safely support a tensile normal stress of 20 ksi. If you want to design the frame to support forces F as large as 8 kip, what is the minimum required cross-sectional area of bar AB?

10.6 The mass of the suspended box is 800 kg. The mass of the crane's arm (not including the hydraulic actuator BC) is 200 kg, and its center of mass is 2 m to the right of A. The cross-sectional area of the upper part of the hydraulic actuator is 0.004 m². What

is the normal stress on a plane perpendicular to the axis of the upper part of the actuator?

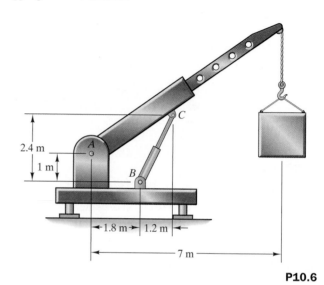

P10.6

10.7 The cross-sectional area of the lower part of the hydraulic actuator BC in Problem 10.6 is 0.010 m². What is the normal stress on a plane perpendicular to the axis of the lower part of the actuator?

10.8 The cross-sectional area of each bar is A. What is the normal stress on a plane perpendicular to the axis of one of the bars?

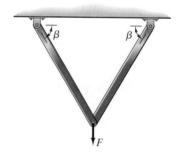

P10.8

10.9 The angle β of the system in Problem 10.8 is 60°. The bars are made of a material that will safely support a tensile normal stress of 8 ksi. Based on this criterion, if you want to design the system so that it will support a force $F = 3 \text{ kip}$, what is the minimum necessary value of the cross-sectional area A?

10.10 Suppose that the horizontal distance between the supports of the system in Problem 10.8 and the load F are specified, and the prismatic bars are made of a material that will safely support a tensile normal stress σ_0. You want to choose the angle β and the cross-sectional area A of the bars so that the total volume of material used is a minimum. What are β and A?

10.11 The cross-sectional area of each bar is 60 mm². If $F = 40$ kN, what are the normal stresses on planes perpendicular to the axes of the bars?

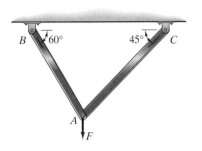

P10.11

10.12 The bars of the truss in Problem 10.11 are made of material that will safely support a tensile normal stress of 600 MPa. Based on this criterion, what is the largest safe value of the force F?

10.13 The cross-sectional area of each bar of the truss is 400 mm². If $F = 30$ kN, what is the normal stress on a plane perpendicular to the axis of member BE?

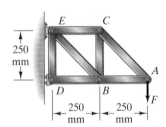

P10.13

10.14 In Problem 10.13, what is the normal stress on a plane perpendicular to the axis of member BC?

10.15 The truss in Problem 10.13 is made of a material that will safely support a normal stress (tension or compression) of 340 MPa. Based on this criterion, what is the largest safe value of the force F?

10.16 The cross-sectional area of the prismatic bar is $A = 2$ in² and the axial force $P = 20$ kip. Determine the normal and shear stresses on the plane $\mathcal{P}$. Draw a diagram isolating the part of the bar to the left of plane $\mathcal{P}$ and show the stresses.

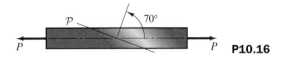

P10.16

10.17 If the normal stress on the plane $\mathcal{P}$ in Problem 10.16 is 6000 psi, what is the axial force P?

10.18 The cross-sectional area of the prismatic bar is 0.02 m². If the normal and shear stresses on the plane $\mathcal{P}$ are $\sigma_\theta = 1.25$ MPa and $\tau_\theta = 1.50$ MPa, what are the angle θ and the axial force P?

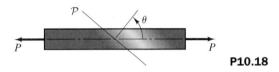

P10.18

10.19 The cross-sectional area of the bars is $A = 0.5$ in² and the force $F = 3000$ lb. Determine the normal stresses and the magnitudes of the shear stresses on the planes (a) and (b).

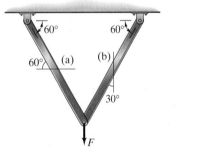

P10.19

10.20 The truss in Problem 10.19 is constructed of a material that will safely support a normal stress of 8 ksi and a shear tress of 3 ksi. Based on these criteria, what is the largest force F that can safely be applied?

10.2 Strains in Prismatic Bars

Pulling on a rubber band stretches it. If you were to pull on a bar of tungsten steel, it would also stretch. You wouldn't be able to see it stretch—in fact, the deformations of many structural elements under their operating loads are so small they can't be seen—but we show in this section that you can calculate how much the bar would stretch. We also show how you can determine the change in the bar's dimensions in the directions perpendicular to its axis.

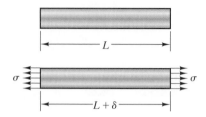

Figure 10.17
Loading a bar by normal stresses at its ends.

Axial Strain and Modulus of Elasticity

Our objective is to determine how much a bar stretches or contracts due to an axial load. Let's return for a moment to our hypothetical example of a prismatic bar loaded at its ends by uniform normal stresses σ (Fig. 10.17). Every plane perpendicular to the bar's axis is subjected to exactly the same normal stress, so it is reasonable to assume that the extensional strain ε in the direction parallel to the bar's axis, the axial strain, is the same at each point along the bar's axis. (We verify this assumption in Chapter 13.) We can therefore use Eq. (9.6) to determine the bar's change in length in terms of its reference length L:

$$\delta = \varepsilon L. \tag{10.4}$$

But what is the value of ε for a given stress σ?

Suppose that we could perform an experiment in which we apply uniform stresses to the ends of a bar of iron and measure the resulting change in its length. We would find that as long as the stresses are not too large, there is an approximately linear relationship between σ and the axial strain ε. This relationship is written as

$$\sigma = E\varepsilon, \tag{10.5}$$

where E is a constant called the *modulus of elasticity* or *Young's modulus*. Since ε is dimensionless, the dimensions of E are the same as the dimensions of σ, force/area. The value of E we would measure for a steel bar would be different from the value we would measure for a bar of a different material. For example, a given strain ε would cause a much larger stress σ in a bar of steel than in a bar of rubber. The modulus of elasticity E is a property of solid materials. Typical values are given in Appendix D.

We have pointed out that if we apply loads to the ends of a prismatic bar of cross-sectional area A that are equivalent to axial forces P applied at the centroid of its cross section, the stress distribution on planes perpendicular to its axis approximates a uniform normal stress $\sigma = P/A$ except near the ends of the bar. As a result, Eq. (10.5) relates the normal stress and axial strain in a prismatic bar except near the ends, and we can use Eq. (10.4) to approximate the change in the bar's length:

$$\delta = \frac{PL}{EA}. \tag{10.6}$$

If you know a prismatic bar's length and cross-sectional area and the elastic modulus of the material comprising it, you can use this equation to determine the change in the bar's length resulting from a given axial load. This result is based upon the assumption that the material obeys Eq. (10.5). Many solid materials used in engineering applications do satisfy this relation within some range of values of ε and some range of temperatures. We discuss relationships between stress and strain further in Section 10.6 and in Chapter 13.

In many engineering applications the strains that occur in axially loaded bars are small, and as a result their changes in length are small in comparison to their lengths. For example, the modulus of elasticity of pure aluminum is

70 GPa, and the largest normal stress a bar of pure aluminum can support without breaking is approximately 70 MPa. If we subject a 1-m-long bar to this normal stress, the resulting change in length is $\delta = (70 \times 10^6)(1)/(70 \times 10^9) = 0.001$ m, or 1 mm. This often simplifies the analysis of structures. For example, when you draw free-body diagrams of a truss to determine the axial forces in its members, you can often ignore changes in the geometry of the truss due to the changes in length of the members. In Example 10.6 we demonstrate this and also show how small changes in the lengths of its members simplify determination of the deformation of a truss.

Lateral Strain and Poisson's Ratio

When you stretch a rubber band, you can see that it becomes thinner. A bar subjected to tensile axial stresses contracts in the lateral direction (Fig. 10.18). Imagine drawing an infinitesimal line dL on an arbitrary cross section of the bar in its undeformed state (Fig. 10.19a). After tensile axial forces are applied to the bar, the line contracts to a length dL' (Fig. 10.19b). The extensional strain of the material in the lateral direction,

$$\varepsilon_{\text{lat}} = \frac{dL' - dL}{dL},$$

is negative. Of course, if the bar is subjected to compressive axial forces, it expands in the lateral direction and ε_{lat} is positive. The strain ε_{lat} has the same value in every direction perpendicular to the bar's cross section and is uniform over the cross section. (Here we assume that the material has the property of isotropy, which we discuss in Chapter 13.) Therefore, if we know the lateral strain we can calculate the length of a finite lateral line L in the deformed state,

$$L' = (1 + \varepsilon_{\text{lat}})L,$$

and its change in length,

$$\delta_{\text{lat}} = L' - L = \varepsilon_{\text{lat}}L.$$

For example, if a bar has a circular cross section of diameter D in the unloaded state, its diameter in the deformed state is $D' = (1 + \varepsilon_{\text{lat}})D$ (Fig.10.20). But how can we determine the lateral strain?

The negative of the ratio of the lateral strain of a prismatic bar to its axial strain when it is subjected to axial forces is called *Poisson's ratio:*

$$\nu = -\frac{\varepsilon_{\text{lat}}}{\varepsilon}. \tag{10.7}$$

Since ε_{lat} is negative if the bar is in tension, the minus sign is included in the definition so that the ratio is a positive number. Like the modulus of elasticity, Poisson's ratio is a property of solid materials, and typical values are given in Appendix D. Being a ratio, it is dimensionless.

If we know the modulus of elasticity and Poisson's ratio of a material, we can determine the lateral strain in a prismatic bar of cross-sectional area A

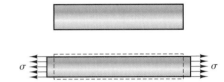

Figure 10.18
A bar lengthens in the axial direction and contracts in the lateral direction when subjected to tensile stress.

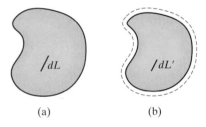

(a)   (b)

Figure 10.19
(a) Infinitesimal line on the unloaded bar's cross section.
(b) Cross section and line after the bar is subjected to tensile loads.

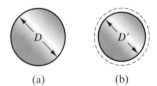

(a)   (b)

Figure 10.20
Circular cross section:
(a) in the undeformed state;
(b) after the bar is subjected to tensile forces.

subjected to an axial load P. The normal stress is $\sigma = P/A$, and the axial strain is $\varepsilon = \sigma/E = P/EA$. Then from Eq. (10.7), the lateral strain is

$$\varepsilon_{\text{lat}} = -\frac{\nu P}{EA}. \tag{10.8}$$

Study Questions

1. A prismatic bar with length L is subjected to axial load, and as a result its length increases by an amount δ. What is the extensional strain ε parallel to the bar's axis?
2. If the modulus of elasticity of the bar in question 1 is E, what is the normal stress δ on a plane perpendicular to the bar's axis?
3. If a prismatic bar with length L and cross-sectional area A is subjected to tensile axial loads P at its ends, what additional information do you need to determine the resulting change in the bar's length?
4. A prismatic bar has length L and a circular cross section with diameter D. Suppose that you subject the bar to axial loads and measure the resulting changes in its length and diameter. How can you determine the Poisson's ratio ν of the material?

Example 10.4

Determining Material Properties by Stretching a Bar

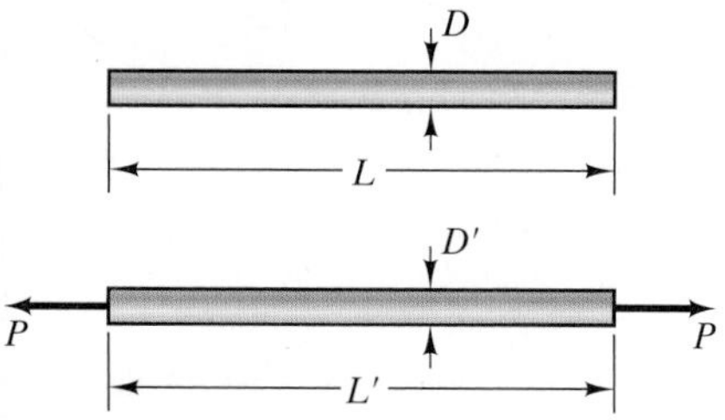

Figure 10.21

A prismatic bar with length $L = 200$ mm and a circular cross section with diameter $D = 10$ mm is subjected to a tensile load $P = 16$ kN (Fig. 10.21). The length and diameter of the deformed bar are measured and determined to be $L' = 200.60$ mm and $D' = 9.99$ mm. What are the modulus of elasticity and Poisson's ratio of the material?

Strategy

We can calculate the normal stress $\sigma = P/A$. Since we know L, L', D, and D', we can also calculate the axial and lateral strains. Then we can use Eqs. (10.5) and (10.7) to determine E and ν.

Solution

The normal stress is

$$\sigma = \frac{P}{A} = \frac{16000}{\pi(0.01)^2/4} = 203.7 \text{ MPa.}$$

The axial strain is

$$\varepsilon = \frac{L' - L}{L} = \frac{200.60 - 200}{200} = 0.003,$$

and the lateral strain is

$$\varepsilon_{\text{lat}} = \frac{D' - D}{D} = \frac{9.99 - 10}{10} = -0.001.$$

The modulus of elasticity is the ratio of the normal stress to the axial strain,

$$E = \frac{\sigma}{\varepsilon} = \frac{203.7 \times 10^6}{0.003} = 67.9 \text{ GPa,}$$

and Poisson's ratio is

$$\nu = -\frac{\varepsilon_{\text{lat}}}{\varepsilon} = -\frac{-0.001}{0.003} = 0.333.$$

Discussion

In Chapter 13 we show that the elastic properties of an important class of materials called isotropic linearly elastic materials are completely characterized by two parameters, the modulus of elasticity and Poisson's ratio. In this example we demonstrate that a simple test, subjecting a bar to axial load and measuring its axial and lateral strains, is sufficient to determine both of these parameters. We discuss this test further in Section 10.6.

Example 10.5

Deflection of One End of a Constrained Bar

The bar in Fig. 10.22 has cross-sectional area $A = 0.4$ in^2 and modulus of elasticity $E = 12 \times 10^6$ psi. If a 10-kip downward force is applied at B, how far down does point B move?

Strategy

By drawing the free-body diagram of joint B, we can determine the axial force in the bar. We can then use Eq. (10.6) to determine the bar's change in length. We must then use geometry to determine how far down point B moves. To simplify the analysis we will take advantage of the fact that the change in length of the bar is small in comparison to its length.

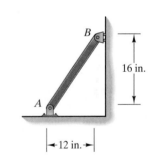

Figure 10.22

Solution

We draw the free-body diagram of joint B showing the 10-kip force in Fig. (a). The angle $\theta = \arctan(16/12) = 53.1°$.

From the equilibrium equation in the vertical direction,

$$\Sigma F_y = -10 - P \sin \theta = 0,$$

we obtain $P = -12.5$ kip. (The bar is in compression.) The bar's length is $L = 20$ in., so from Eq. (10.6), its change in length is

$$\delta = \frac{PL}{EA} = \frac{(-12,500)(20)}{(12 \times 10^6)(0.4)} = -0.0521 \text{ in.}$$

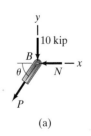

(a)

(a) Joint B.

Let v be the distance point B moves downward when the force is applied [Fig. (b)]. The bar's deformed length is $(20 + \delta)$ in. Applying the Pythagorean theorem to the dashed triangle,

$$(12)^2 + (16 - v)^2 = (20 + \delta)^2,$$

we obtain an equation relating v to δ:

$$-32v + v^2 = 40\delta + \delta^2. \tag{10.9}$$

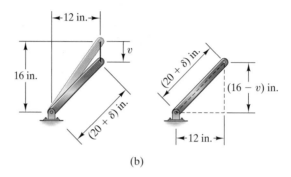

(b)

(b) Analyzing the bar's geometry.

Since the change in length of the bar is small in comparison to its length, we can neglect the second-order terms v^2 and δ^2, obtaining the linear equation

$$-32v = 40\delta. \tag{10.10}$$

Solving, we obtain $v = (-40/32)(-0.0521) = 0.0651$ in.

We can also derive Eq. (10.10) using a geometric approach that is more intuitive and direct but also obscures the approximation being made. Consider the right triangle in Fig. (c). Since v is small, the angle of the triangle labeled θ is approximately equal to θ and the side of the triangle opposite θ is approximately equal to $|\delta|$. From this triangle we obtain the relation

$$\sin \theta = \frac{|\delta|}{v}.$$

Since $\sin \theta = 16/20 = 32/40$ and δ is negative, we obtain Eq. (10.10).

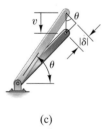

(c)

(c) Triangle for obtaining an approximate relation between v and δ.

Discussion

In this example there doesn't appear to be any motivation to neglect the second-order terms in the quadratic equation (10.9). Why didn't we simply solve it for v and obtain a more accurate answer? The reason is that we had already introduced an approximation based on the assumption of a small change in length—we neglected the bar's change in length when we determined P—so retaining the second-order terms in Eq. (10.9) would not be justified.

Example 10.6

Deflection of a Joint of a Truss

Bars *AB* and *AC* in Fig. 10.23 each have cross-sectional area $A = 60$ mm^2 and modulus of elasticity $E = 200$ GPa. The dimension $h = 200$ mm. If a downward force $F = 40$ kN is applied at *A*, what are the resulting horizontal and vertical displacements of point *A*?

Strategy

By drawing the free-body diagram of joint *A*, we can determine the axial forces in the two bars. Knowing the axial forces, we can use Eq. (10.6) to determine the change in length of each bar. We must then use geometry to determine the horizontal and vertical displacements of point *A*. We simplify the analysis by taking advantage of the fact that the changes in length of the bars are small in comparison to their lengths. (See Example 10.5.)

Solution

In Fig. (a) we draw the free-body diagram of joint *A* showing the force *F*.

The equilibrium equations are

$$\Sigma F_x = -P_{AB} \cos 60° + P_{AC} \cos 45° = 0,$$
$$\Sigma F_y = P_{AB} \sin 60° + P_{AC} \sin 45° - F = 0.$$

Solving these equations, the axial loads in the bars are

$$P_{AB} = \frac{F \cos 45°}{D}, \qquad P_{AC} = \frac{F \cos 60°}{D}, \qquad (10.11)$$

where $D = \sin 45° \cos 60° + \cos 45° \sin 60°$.

The lengths of the bars are $L_{AB} = h/\sin 60°$ and $L_{AC} = h/\sin 45°$. Using Eqs. (10.6) and (10.11), the changes in length of the bars are

$$\delta_{AB} = \frac{P_{AB} L_{AB}}{EA} = \frac{F L_{AB} \cos 45°}{EAD},$$

$$\delta_{AC} = \frac{P_{AC} L_{AC}}{EA} = \frac{F L_{AC} \cos 60°}{EAD}. \qquad (10.12)$$

To determine the displacement of point *A*, we first consider bar *AB*, denoting the horizontal and vertical displacements of point *A* by *u* and *v* [Fig. (b)]. The direction of *u* is chosen arbitrarily; we don't know beforehand whether point *A* will move to the left or right.

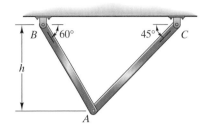

Figure 10.23

(a) Joint *A*.

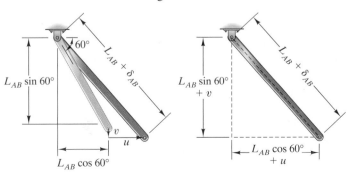

(b)

(b) Analyzing the geometry of bar *AB*.

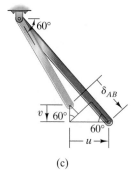

(c)

(c) Obtaining the linear equation relating u, v, and δ_{AB}.

The deformed length of the bar is $L_{AB} + \delta_{AB}$. From the dashed right triangle in Fig. (b), we obtain an equation relating δ_{AB}, u, and v:

$$\left(L_{AB}\sin 60° + v\right)^2 + \left(L_{AB}\cos 60° + u\right)^2 = \left(L_{AB} + \delta_{AB}\right)^2.$$

Squaring these expressions and using the identity $\sin^2 60° + \cos^2 60° = 1$, we obtain

$$2vL_{AB}\sin 60° + v^2 + 2uL_{AB}\cos 60° + u^2 = 2\delta_{AB}L_{AB} + \delta_{AB}^2.$$

Because u, v, and δ_{AB} are small in comparison to L_{AB}, we can neglect the second-order terms v^2, u^2, and δ_{AB}^2, obtaining a linear equation relating δ_{AB}, u, and v:

$$v\sin 60° + u\cos 60° = \delta_{AB}. \tag{10.13}$$

We can also derive this result geometrically as shown in Fig. (c). Because u and v are small, the angles of the right triangles labeled 60° are approximately 60°, and the distance labeled δ_{AB} is approximately δ_{AB}. From these triangles, we obtain Eq. (10.13).

We next consider bar AC [Fig. (d)].

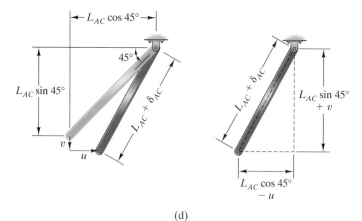

(d)

(d) Analyzing the geometry of bar AC.

From the dashed right triangle, we obtain an equation relating δ_{AC}, u, and v:

$$\left(L_{AC}\sin 45° + v\right)^2 + \left(L_{AC}\cos 45° - u\right)^2 = \left(L_{AC} + \delta_{AC}\right)^2.$$

Squaring these expressions, we obtain

$$2vL_{AC}\sin 45° + v^2 - 2uL_{AC}\cos 45° + u^2 = 2\delta_{AC}L_{AC} + \delta_{AC}^2.$$

Neglecting the second-order terms v^2, u^2, and δ_{AC}^2 gives a linear equation relating δ_{AC}, u, and v:

$$v\sin 45° - u\cos 45° = \delta_{AC}. \tag{10.14}$$

The geometrical derivation of this result is shown in Fig. (e).

Solving Eqs. (10.13) and (10.14) for u and v and substituting Eqs. (10.12), we obtain

$$u = \frac{F\left(L_{AB}\sin 45°\cos 45° - L_{AC}\sin 60°\cos 60°\right)}{EAD^2},$$

$$v = \frac{F\left(L_{AB}\cos^2 45° + L_{AC}\cos^2 60°\right)}{EAD^2}.$$

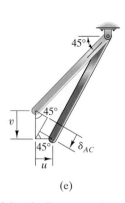

(e)

(e) Obtaining the linear equation relating u, v, and δ_{AC}.

Substituting the numerical values, the displacements are

$$u = -0.0250 \text{ mm} \qquad v = 0.6652 \text{ mm}.$$

Point A moves 0.0250 mm to the left and 0.6652 mm downward.

Discussion

Notice that in drawing the free-body diagram of joint A in Fig. (a), we disregarded the changes in the 45° and 60° angles due to the changes in length of the bars. This approximation requires that the changes in length of the bars be small compared to their lengths.

Problems

10.21 Two marks are made 2 in. apart on an unloaded bar. When the bar is subjected to axial forces P, the marks are 2.004 in. apart. What is the axial strain of the loaded bar?

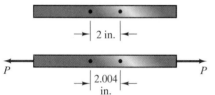

P10.21

10.22 The total length of the unloaded bar in Problem 10.21 is 10 in. Use the result of Problem 10.21 to determine the total length of the loaded bar. What assumption are you making when you do so?

10.23 If the forces exerted on the bar in Problem 10.21 are $P = 20$ kip and the bar's cross-sectional area is $A = 1.5 \text{ in}^2$, what is the modulus of elasticity of the material?

10.24 A prismatic bar with length $L = 6$ m and a circular cross section with diameter $D = 0.02$ m is subjected to 20-kN compressive forces at its ends. The length and diameter of the deformed bar are measured and determined to be $L' = 5.940$ m and $D' = 0.02006$ m. What are the modulus of elasticity and Poisson's ratio of the material?

10.25 The bar has modulus of elasticity $E = 30 \times 10^6$ psi and Poisson's ratio $v = 0.32$. It has a circular cross section with diameter $D = 0.75$ in. What compressive force would have to be exerted on the right end of the bar to increase its diameter to 0.752 in.?

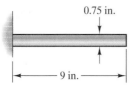

P10.25

10.26 What tensile force would have to be exerted on the right end of the bar in Problem 10.25 to increase its length to 9.02 in.? What is the bar's diameter after this load is applied?

10.27 A prismatic bar is 300 mm long and has a circular cross section with 20-mm diameter. Its modulus of elasticity is 120 GPa and its Poisson's ratio is 0.33. Axial forces P are applied to the ends of the bar that cause its diameter to decrease to 19.948 mm.
(a) What is the length of the loaded bar?
(b) What is the value of P?

10.28 When unloaded, bars AB and AC are each 36 in. in length and have a cross-sectional area of 2 in². Their modulus of elasticity is $E = 1.6 \times 10^6$ psi. When the weight W is suspended at A, bar AB increases in length by 0.1 in. What is the change in length of bar AC?

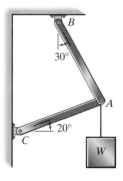

P10.28

10.29 If a weight $W = 12,000$ lb is suspended from the truss in Problem 10.28, what are the changes in length of the two bars?

10.30 Bars AB and AC are each 300 mm in length, have a cross-sectional area of 500 mm², and have modulus of elasticity $E = 72$ GPa. If a 24-kN downward force is applied at A, what is the resulting displacement of point A?

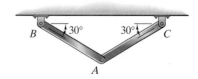

P10.30

10.31 Bars *AB* and *AC* of the truss shown in Problem 10.30 are each 300 mm in length, have a cross-sectional area of 500 mm², and are made of the same material. When a 30-kN downward force is applied at point *A*, it deflects downward 0.4 mm. What is the modulus of elasticity of the material?

10.32 Bar *AB* has cross-sectional area $A = 100$ mm² and modulus of elasticity $E = 102$ GPa. The distance $H = 400$ mm. If a 200-kN downward force is applied to bar *CD* at *D*, through what angle in degrees does bar *CD* rotate? (You can neglect the deformation of bar *CD*.) [*Strategy:* Because the change in length of bar *AB* is small, you can assume that the downward displacement *v* of point *B* is vertical and that the angle (in radians) through which bar *CD* rotates is v/H.]

10.33 Bar *AB* in Problem 10.32 is made of a material that will safely support a normal stress (in tension or compression) of 5 GPa. Based on this criterion, through what angle in degrees can bar *CD* safely be rotated relative to the position shown? (You can neglect the deformation of bar *CD*.)

10.34 If an upward force is applied at *H* that causes bar *GH* to rotate 0.02° in the counterclockwise direction, what are the axial strains in bars *AB*, *CD*, and *EF*? (You can neglect the deformation of bar *GH*.)

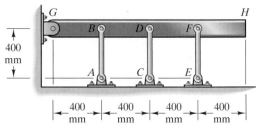

P10.34

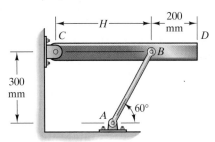

P10.32

10.3 Statically Indeterminate Problems

You have already encountered statics problems in which the number of unknown reactions exceeded the number of independent equilibrium equations. Such problems are said to be *statically indeterminate*. They are common in engineering, because safe and conservative design frequently requires the use of redundant supports—supports in addition to the minimum number necessary for equilibrium. We can now solve statically indeterminate problems involving axially loaded bars by supplementing the equilibrium equations with the relationships between the axial loads in bars and their changes in length.

Example

The bar in Fig. 10.24a consists of two segments *A* and *B* with different lengths and cross-sectional areas. It is fixed at both ends and subjected to an

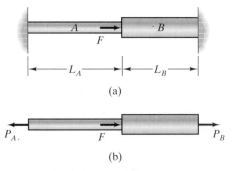

(a)

(b)

Figure 10.24
(a) Axially loaded bar fixed at both ends.
(b) Free-body diagram of the bar.

axial force. We draw its free-body diagram in Fig. 10.24b. The equilibrium equation is

$$F - P_A + P_B = 0. \qquad (10.15)$$

We cannot determine the two reactions P_A and P_B from this equation. This simple problem is statically indeterminate.

Let us disregard the fact that we don't know the reactions P_A and P_B, and determine the changes in length, or deformations, of the two parts of the bar as if we did know the reactions. The axial force in part A is P_A (Fig. 10.25a). Its change in length is therefore

$$\delta_A = \frac{P_A L_A}{EA_A}. \qquad (10.16)$$

The axial force in part B is P_B (Fig. 10.25b), so its change in length is

$$\delta_B = \frac{P_B L_B}{EA_B}. \qquad (10.17)$$

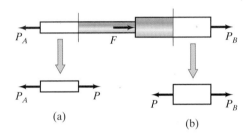

(a)

(b)

Figure 10.25
Determining the internal axial forces in parts A and B.

But we know that the change in length of the entire bar is zero, because it is fixed at both ends. Therefore, the changes in length of the two parts must satisfy the equation

$$\delta_A + \delta_B = 0. \qquad (10.18)$$

This is called a *compatibility condition*. It ensures that the changes in length of the two parts are compatible with the constraint that the bar's total length cannot change. We substitute Eqs. (10.16) and (10.17) into Eq. (10.18), obtaining

$$\frac{P_A L_A}{EA_A} + \frac{P_B L_B}{EA_B} = 0. \qquad (10.19)$$

We can solve this equation together with Eq. (10.15) to determine P_A and P_B, obtaining

$$P_A = \frac{F}{1 + L_A A_B / L_B A_A}, \qquad P_B = \frac{-F}{1 + L_B A_A / L_A A_B}.$$

Now that we know the reactions, we can determine the change in length of each part of the bar from Eqs. (10.16) and (10.17). We can also determine the normal stress in each part of the bar: $\sigma_A = P_A / A_A$ and $\sigma_B = P_B / A_B$.

Notice that our solution was based on three elements: (1) relations between the axial forces in the parts of the bar and their changes in length, or deformations; (2) compatibility; and (3) equilibrium. Other statically indeterminate problems involving axially loaded bars, as well as more elaborate problems in structural analysis, can be solved using these three elements.

Flexibility and Stiffness Methods

In this section we describe two approaches that are commonly used to analyze statically indeterminate problems. By applying them to the example discussed in the preceding section, we compare the two methods in a simple context and explain their terminology. They are both based on the same three elements that we applied in the preceding section: relations between forces and deformations, compatibility, and equilibrium. For the bar in Fig. 10.24, the equations expressing these elements are:

$$\left. \begin{aligned} \delta_A &= \frac{P_A L_A}{EA_A} \\ \delta_B &= \frac{P_B L_B}{EA_B} \end{aligned} \right\} \qquad \textbf{Force-Deformation Relations}$$

$$\delta_A + \delta_B = 0 \qquad \textbf{Compatibility}$$

$$F - P_A + P_B = 0 \qquad \textbf{Equilibrium}$$

Flexibility Method   We write the force-deformation relations as

$$\delta_A = \frac{L_A}{EA_A} P_A = f_A P_A,$$

$$\delta_B = \frac{L_B}{EA_B} P_B = f_B P_B,$$

where the constants $f_A = L_A/EA_A$ and $f_B = L_B/EA_B$ are called the *flexibilities* of parts A and B of the bar. We substitute these equations into the compatibility equation, obtaining

$$f_A P_A + f_B P_B = 0.$$

This equation can be solved together with the equilibrium equation for the reactions P_A and P_B. Once P_A and P_B are known, the changes in length can be determined from the force-deformation relations.

Except for some added terminology, this is the method of solution we used in the preceding section. It is also called the *force method* because it results in a system of equations in terms of forces.

Stiffness Method   In this method we write the force-deformation relations as

$$P_A = \frac{EA_A}{L_A} \delta_A = k_A \delta_A,$$

$$P_B = \frac{EA_B}{L_B} \delta_B = k_B \delta_B,$$

where the constants $k_A = EA_A/L_A$ and $k_B = EA_B/L_B$ are called the *stiffnesses* of parts A and B of the bar. We substitute these equations into the equilibrium equation, obtaining

$$F - k_A \delta_A + k_B \delta_B = 0.$$

This equation can be solved together with the compatibility equation for the changes in length δ_A and δ_B. Once they are known, the forces P_A and P_B can be determined from the force-deformation relations. This approach is also called the *displacement method* because it results in a system of equations in terms of deformations, which can be expressed in terms of displacements.

From these examples you may wonder why the flexibility and stiffness methods are given distinguishing names, since they apparently differ only in terminology and the order in which equations are solved. The reason is that the differences between the methods become significant when they are applied to more elaborate problems in structural analysis. For the statically indeterminate problems in this book it is not necessary (unless you are asked to do so) to apply one of these specific approaches and use its associated terminology. You need only apply their essential elements—force-deformation relations, compatibility, and equilibrium—and you can choose how and in what order they are applied.

Study Questions

1. What is a statically indeterminate problem?
2. For the axially loaded bar shown in Fig. 10.24a, what is the compatibility condition?
3. What are the three essential elements used in solving statically indeterminate problems?

Example 10.7

Statically Indeterminate Axially Bar

The bar in Fig. 10.26 has modulus of elasticity $E = 12 \times 10^6$ psi. Part A has length $L_A = 10$ in. and a 2-in.-diameter circular cross section. Part B has length $L_B = 8$ in. and a 4-in.-diameter circular cross section. Part A is fixed at its left end, and there is a gap $b = 0.02$ in. between the right end of part B and the rigid wall. If a 160-kip axial force pointing to the right is applied at the joint between parts A and B, what are the normal stresses in parts A and B?

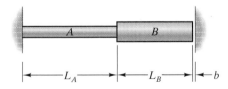

$\leftarrow L_A \rightarrow$ $\leftarrow L_B \rightarrow$ $\leftarrow b$ **Figure 10.26**

Strategy

We will first consider the possibility that the 160-kip force doesn't cause the right end of the bar to come into contact with the wall. If that is the case, the normal stress in part B is zero and the normal stress in part A is $160{,}000/A_A$. If the bar does contact the right wall, the problem is statically indeterminate and we can apply force-deformation relations, compatibility, and equilibrium to determine the axial forces in parts A and B. Here the compatibility condition is that the total change in length of the two parts of the bar must equal b.

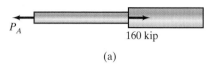

(a)

(a) Assuming that the bar doesn't contact the wall.

(b) Free-body diagram showing the reactions at both walls.

Solution

In Fig. (a) we draw the free-body diagram of the bar under the assumption that it doesn't contact the right wall.

The axial force in part A is $P_A = 160{,}000$ lb and there is no axial force in part B. The change in length of part A is

$$\delta_A = \frac{P_A L_A}{EA_A} = \frac{(160{,}000)(10)}{(12 \times 10^6)\left[\pi(2)^2/4\right]} = 0.0424 \text{ in.}$$

Since $\delta_A > b$, the bar does come into contact with the right wall and the problem is statically indeterminate.

In Fig. (b) we draw the free-body diagram of the bar, including the reaction exerted on part B by the wall.

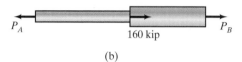

(b)

Equilibrium The equilibrium equation is

$$-P_A + P_B + 160{,}000 = 0.$$

Force-deformation relations The changes in length of the two parts of the bar are

$$\delta_A = \frac{P_A L_A}{EA_A}, \qquad \delta_B = \frac{P_B L_B}{EA_B}.$$

Compatibility The total change in length of the bar equals b:

$$\delta_A + \delta_B = b.$$

We substitute the force-deformation relations into this equation, obtaining

$$\frac{P_A L_A}{EA_A} + \frac{P_B L_B}{EA_B} = b:$$

$$\frac{P_A(10)}{(12 \times 10^6)\left[\pi(2)^2/4\right]} + \frac{P_B(8)}{(12 \times 10^6)\left[\pi(4)^2/4\right]} = 0.02.$$

We can solve this equation together with the equilibrium equation for the axial forces P_A and P_B. The results are $P_A = 89{,}500$ lb, $P_B = -70{,}500$ lb. The normal stresses are

$$\sigma_A = \frac{P_A}{A_A} = \frac{89{,}500}{\pi(2)^2/4} = 28{,}490 \text{ psi},$$

$$\sigma_B = \frac{P_B}{A_B} = \frac{-70{,}500}{\pi(4)^2/4} = -5610 \text{ psi}.$$

Example 10.8

Statically Indeterminate Structure

In Fig. 10.27 two aluminum bars $(E_{Al} = 10.0 \times 10^6 \, \text{psi})$ are attached to a rigid support at the left and a cross-bar at the right. An iron bar $(E_{Fe} = 28.5 \times 10^6 \, \text{psi})$ is attached to the rigid support at the left and there is a gap $b = 0.02$ in. between the right end of the iron bar and the cross-bar. The cross-sectional area of each bar is $A = 0.5 \, \text{in}^2$ and $L = 10$ in. If the iron bar is stretched until it contacts the cross-bar and welded to it, what are the normal stresses in the bars afterward?

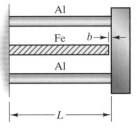

Figure 10.27

Strategy

By drawing a free-body diagram of the cross-bar we can obtain one equilibrium equation, but the problem is statically indeterminate because the axial forces in the aluminum and iron bars may be different. (Notice, however, that the axial forces in the two aluminum bars are equal because of the symmetry of the system.) The compatibility condition is that the change in length of the iron bar must equal the change in length of the aluminum bars plus the gap b.

Solution

Equilibrium   In Fig. (a) we assume the iron bar has been welded to the cross-bar and obtain a free-body diagram of the cross-bar by passing a plane through the right ends of the three bars. The equilibrium equation for the cross-bar is

$$2P_{Al} + P_{Fe} = 0.$$

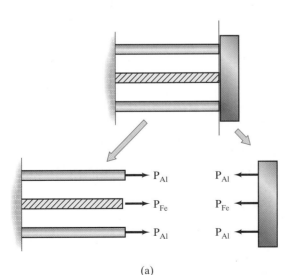

(a)

(a) Isolating the cross-bar.

Force-deformation relations The change in length of one of the aluminum bars due to its axial force is

$$\delta_{Al} = \frac{P_{Al}L}{E_{Al}A},$$

and the change in length of the iron bar due to its axial force is

$$\delta_{Fe} = \frac{P_{Fe}(L - b)}{E_{Fe}A}.$$

Compatibility The compatibility condition is

$$\delta_{Fe} = \delta_{Al} + b.$$

We substitute the force-deformation relations into this equation, obtaining

$$\frac{P_{Fe}(L - b)}{E_{Fe}A} = \frac{P_{Al}L}{E_{Al}A} + b.$$

We can solve this equation together with the equilibrium equation for the axial forces P_{Al} and P_{Fe}. The results are $P_{Al} = -5880$ lb, $P_{Fe} = 11,760$ lb. The normal stresses in the bars are

$$\sigma_{Al} = \frac{P_{Al}}{A} = \frac{-5880}{0.5} = -11,760 \text{ psi},$$

$$\sigma_{Fe} = \frac{P_{Fe}}{A} = \frac{11,760}{0.5} = 23,520 \text{ psi}.$$

Discussion

When the stretched iron bar is welded to the cross-bar and then released, it contracts, compressing the aluminum bars. The forces exerted on the cross-bar by the compressed aluminum bars prevent the iron bar from returning to its original length, so an equilibrium state is reached in which the aluminum bars are in compression and the iron bar is in tension.

Example 10.9

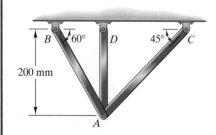

Figure 10.28

Statically Indeterminate Truss

The bars in Fig. 10.28 each have cross-sectional area $A = 60 \text{ mm}^2$ and modulus of elasticity $E = 200$ GPa. If a 40-kN downward force is applied at A, what are the resulting horizontal and vertical displacements of point A?

Strategy

In Example 10.6 we solved this same problem except that the vertical bar AD was absent. (You should review that solution.) In this case the problem is statically indeterminate because we cannot determine the axial forces in the bars from equilibrium alone. We can obtain two equilibrium equations from the free-body diagram of joint A, but there are three unknown axial forces.

However, we can approach the problem in exactly the same way that we approached Example 10.6. We can express the change in length of each bar in terms of its unknown axial force. Compatibility conditions arise from the fact that the bars are pinned together: The horizontal and vertical displacements of point A must be the same for each bar. The equilibrium equations, force-deformation equations, and compatibility conditions will provide a complete system of equations for the axial forces, the changes in length of the bars, and the horizontal and vertical displacements of point A.

Solution

Equilibrium In Fig. (a) we draw the free-body diagram of joint A showing the 40-kN force.
  The equilibrium equations are

$$\Sigma F_x = -P_{AB}\cos 60° + P_{AC}\cos 45° = 0,$$
$$\Sigma F_y = P_{AB}\sin 60° + P_{AD} + P_{AC}\sin 45° - 40,000 = 0.$$

Force-deformation relations The lengths of the bars are $L_{AB} = 0.2/\sin 60°$ m, $L_{AC} = 0.2/\sin 45°$ m, and $L_{AD} = 0.2$ m. We can express the changes in length of the bars in terms of their lengths and the unknown axial forces:

$$\delta_{AB} = \frac{P_{AB}L_{AB}}{EA}, \quad \delta_{AC} = \frac{P_{AC}L_{AC}}{EA}, \quad \delta_{AD} = \frac{P_{AD}L_{AD}}{EA}. \quad (10.20)$$

Compatibility We now relate the changes in length of the bars to the horizontal and vertical displacements of point A. Consider bar AD, denoting the horizontal and vertical displacements of point A by u and v [Fig. (b)].
  From the figure we obtain the relationship

$$\left(L_{AD} + v\right)^2 + u^2 = \left(L_{AD} + \delta_{AD}\right)^2,$$

which we can write as

$$2vL_{AD} + v^2 + u^2 = 2\delta_{AD}L_{AD} + \delta_{AD}^2.$$

Neglecting the second-order terms v^2, u^2, and δ_{AD}^2 yields the equation

$$v = \delta_{AD}. \quad (10.21)$$

Because u and v are small, the change in length of the vertical bar AD is approximately equal to the vertical displacement of point A. In Example 10.6 we derived the relationships between the changes in length of bars AB and AC and the displacements u and v, obtaining

$$v\sin 60° + u\cos 60° = \delta_{AB}, \quad (10.22)$$
$$v\sin 45° - u\cos 45° = \delta_{AC}. \quad (10.23)$$

Equations (10.21)–(10.23) are the compatibility conditions for this problem. The changes in length of the bars are constrained because they are pinned together at A. These constraints are enforced by these equations, which require the end of each bar to undergo the same horizontal and vertical displacement.
  The equilibrium equations, Eqs. (10.20), and Eqs. (10.21)–(10.23) provide eight equations in the eight unknowns P_{AB}, P_{AC}, P_{AD}, δ_{AB}, δ_{AC}, δ_{AD}, u, and v. Solving them, we obtain

$$u = -0.013 \text{ mm}, \qquad v = 0.333 \text{ mm}.$$

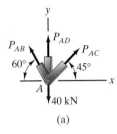

(a)

(a) Joint A.

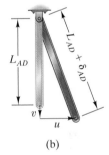

(b)

(b) Analyzing the geometry of bar AD.

Problems

10.35 The bar has cross-sectional area A and modulus of elasticity E. The left end of the bar is fixed. There is initially a gap b between the right end of the bar and the rigid wall (Fig. 1). The bar is stretched until it comes into contact with the rigid wall and is welded to it (Fig. 2). Notice that this problem is statically indeterminate because the axial force in the bar after it is welded to the wall cannot be determined from statics alone.
(a) What is the compatibility condition in this problem?
(b) What is the axial force in the bar after it is welded to the wall?

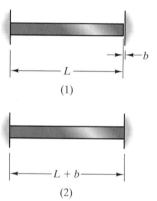

(1)

(2)   **P10.35**

10.36 The bar has cross-sectional area A and modulus of elasticity E. If an axial force F directed toward the right is applied at C, what is the normal stress in the part of the bar to the left of C? (*Strategy:* Draw the free-body diagram of the entire bar and write the equilibrium equation. Then apply the compatibility condition that the increase in length of the part of the bar to the left of C must equal the decrease in length of the part to the right of C.)

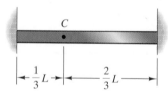

P10.36

10.37 In Problem 10.36, what is the resulting displacement of point C?

10.38 The bar in Problem 10.36 has cross-sectional area $A = 0.005$ m², modulus of elasticity $E = 72$ GPa, and $L = 1$ m. It is made of a material that will safely support a normal stress (in tension and compression) of 120 MPa. Based on this criterion, what is the largest axial force that can be applied at C?

10.39 In Example 10.7, determine the normal stresses in parts A and B of the bar if the force applied at the joint between parts A and B is (a) 40 kip; (b) 200 kip.

10.40 The bar has a circular cross section and modulus of elasticity $E = 70$ GPa. Parts A and C are 40 mm in diameter and part B is 80 mm in diameter. If $F_1 = 60$ kN and $F_2 = 30$ kN, what is the normal stress in part B?

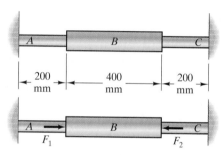

P10.40

10.41 In Problem 10.40, if $F_1 = 60$ kN, what force F_2 will cause the normal stress in part C to be zero?

10.42 The bar in Problem 10.40 consists of a material that will safely support a normal stress of 40 MPa. If $F_2 = 20$ kN, what is the largest safe value of F_1?

10.43 Two aluminum bars $(E_{Al} = 10.0 \times 10^6 \text{ psi})$ are attached to a rigid support at the left and a cross-bar at the right. An iron bar $(E_{Fe} = 28.5 \times 10^6 \text{ psi})$ is attached to the rigid support at the left and there is a gap b between the right end of the iron bar and the cross-bar. The cross-sectional area of each bar is $A = 0.5$ in² and $L = 10$ in. The iron bar is stretched until it contacts the cross-bar and welded to it. Afterward, the axial strain of the iron bar is measured and determined to be $\varepsilon_{Fe} = 0.002$. What was the size of the gap b?

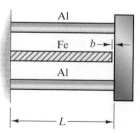

P10.43

10.44 In Problem 10.43, the iron will safely support a tensile stress of 100 ksi and the aluminum will safely support a compressive stress of 40 ksi. What is the largest safe value of the gap b?

10.45 Bars AB and AC each have cross-sectional area A and modulus of elasticity E. If a downward force F is applied at A, show that the resulting downward displacement of point A is

$$\frac{Fh}{EA}\left(\frac{1}{1 + \cos^3\theta}\right).$$

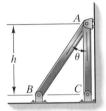

P10.45

10.46 If a downward force F is applied at point A of the system shown in Problem 10.45, what are the resulting normal stresses in bars AB and AC?

10.47 Each bar has a 500-mm^2 cross-sectional area and modulus of elasticity $E = 72$ GPa. If a 160-kN downward force is applied at A, what is the resulting displacement of point A?

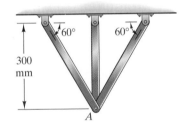

P10.47

10.48 The bars in Problem 10.47 are made of material that will safely support a tensile stress of 270 MPa. Based on this criterion, what is the largest downward force that can safely be applied at A?

10.49 Each bar has a 500-mm^2 cross-sectional area and modulus of elasticity $E = 72$ GPa. If there is a gap $h = 2$ mm between the hole in the vertical bar and the pin A connecting bars AB and AD, what are the normal stresses in the three bars after the vertical bar is connected to the pin at A?

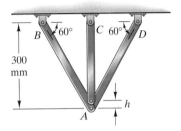

P10.49

10.50 The bars in Problem 10.49 are made of material that will safely support a normal stress (tension or compression) of 400 MPa. Based on this criterion, what is the largest safe value of the gap h?

10.4 Nonprismatic Bars and Distributed Loads

In this section we extend our analysis of prismatic bars to additional important applications by considering bars whose cross sections vary with distance along their axes and also bars subjected to axial loads that are distributed along their axes.

Bars with Gradually Varying Cross Sections

Let us consider a bar whose cross-sectional area varies with distance along its axis as shown in Fig. 10.29a. The cross-sectional area is a function of x, which we indicate by writing it as $A(x)$. If we subject the bar to axial loads

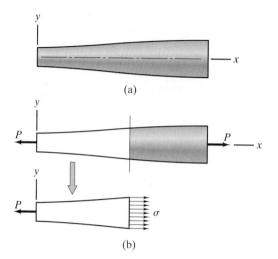

Figure 10.29
(a) Bar with a varying cross-sectional area.
(b) Approximate stress distribution.

and the change in $A(x)$ with x is gradual [$dA(x)/dx$ is small], the stress distribution on a plane perpendicular to the bar's axis can be approximated by a uniform normal stress (Fig. 10.29b):

$$\sigma = \frac{P}{A(x)}. \tag{10.24}$$

In other words, for a given perpendicular plane the stress distribution is approximately the same as for the case of a prismatic bar. But since $A(x)$ varies with distance along the bar's axis, the normal stress does also.

Since the cross-sectional area and normal stress vary along the length of the bar, how can we determine the change in the bar's length? We begin by considering an infinitesimal element of the bar whose length is dx in the unloaded state (Fig. 10.30a). We can use Eq. (10.6) to determine the element's change in length (Fig. 10.30b):

$$dx' - dx = \frac{P\,dx}{EA(x)}.$$

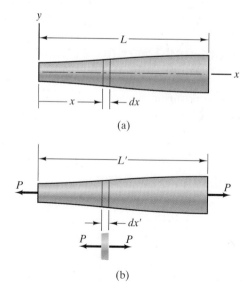

Figure 10.30
(a) Element of the unloaded bar.
(b) Length of the element in the loaded state.

We obtain the bar's change in length by integrating this expression from $x = 0$ to $x = L$:

$$\delta = L' - L = \int_0^L \frac{P \, dx}{EA(x)}. \tag{10.25}$$

The change in length of each element depends on its cross-sectional area, and we must add up, or integrate, the changes in length of the elements to determine the total change in length.

Example 10.10

Nonprismatic Axially Loaded Bar

The bar in Fig. 10.31 consists of material with modulus of elasticity $E = 120$ GPa. The area of the bar's circular cross section is given by the equation $A(x) = 0.03 + 0.008x^2$ m². If the bar is subjected to axial forces $P = 20$ MN at the ends, determine (a) the normal stress in the bar at $x = 1$ m; (b) the bar's change in length.

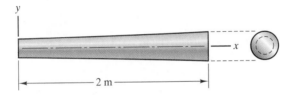

Figure 10.31

Strategy

(a) From the given equation for $A(x)$ we can determine the cross-sectional area at $x = 1$ m. Then the normal stress is $P/A(x)$.
(b) Since the bar's cross-sectional area is a function of x, we must determine the bar's change in length from Eq. (10.25).

Solution

(a) The cross-sectional area at $x = 1$ m is

$$A(1) = 0.03 + 0.008(1)^2 = 0.038 \text{ m}^2,$$

so the normal stress at $x = 1$ m is

$$\sigma = \frac{P}{A(1)} = \frac{20 \times 10^6}{0.038} = 526 \text{ MN}.$$

(b) From Eq. (10.25), the change in length of the bar is

$$\delta = \int_0^L \frac{P \, dx}{EA(x)} = \int_0^2 \frac{(20 \times 10^6) \, dx}{(120 \times 10^9)(0.03 + 0.008x^2)}$$

$$= 8.62 \text{ mm}.$$

Figure 10.32
A driven pile is subjected to a distributed axial force.

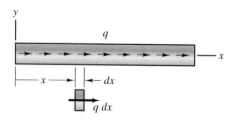

Figure 10.33
Describing a distributed axial force by a function. The force on an element of length dx is $q\, dx$.

Distributed Axial Loads

In some situations bars are subjected to axial forces that are continuously distributed along some part of the bar's axis, or to axial forces that can be modeled as continuous distributions. For example, a pile driven into the ground is subjected to a resisting friction force that is distributed along the pile's length (Fig. 10.32).

To describe a distributed axial force, we introduce a function q defined such that the force on each element dx of the bar is $q\, dx$ (Fig. 10.33). Since the product of q and dx is a force, the dimensions of q are force/length.

For example, the bar in Fig. 10.34a is fixed at the left end and subjected to a distributed axial force throughout its length. In Fig. 10.34b we pass a plane through the bar at an arbitrary position x and draw the free-body diagram of the part of the bar to the right of the plane. From the equilibrium equation

$$-P + \int_x^L q_0\left(\frac{x}{L}\right)^2 dx = 0,$$

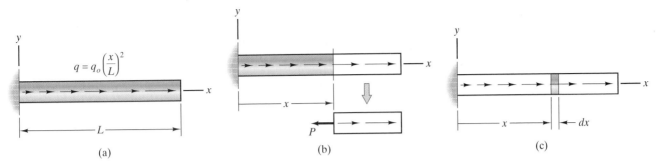

Figure 10.34
(a) Bar subjected to a distributed load.
(b) Determining the internal axial force at x.
(c) Element of length dx.

we determine the axial load in the bar at the position x:

$$P = \frac{q_0}{3}\left(L - \frac{x^3}{L^2}\right).$$

The distributed axial force causes the internal axial force in the bar to vary with axial position. To determine the bar's change in length, we consider an element of the bar of length dx (Fig. 10.34c). From Eq. (10.6), its change in length is

$$\delta_{\text{element}} = \frac{P\,dx}{EA} = \frac{q_0}{3EA}\left(L - \frac{x^3}{L^2}\right)dx.$$

Integrating this result from $x = 0$ to $x = L$, we obtain the change in length of the bar:

$$\delta = \int_0^L \frac{q_0}{3EA}\left(L - \frac{x^3}{L^2}\right)dx = \frac{q_0 L^2}{4EA}.$$

Notice the parallel between the definition of the distributed axial force q and the definitions of other distributed loads with which you are familiar: We have represented distributed lateral loads on beams by a function w defined such that the lateral force on an element dx of the beam's axis is $w\,dx$, and we represented the stress distribution on a surface by introducing a function $\mathbf{t}$ defined such that the force on an element dA of the surface is $\mathbf{t}\,dA$.

Example 10.11

Distributed Axial Load

The pile in Fig. 10.35 has a circular cross section 1 ft in diameter and its modulus of elasticity is $E = 2 \times 10^6$ psi. A downward force with initial value $F = 480$ kip is applied to the top of the pile, driving it into the ground at a constant rate. The resisting force on the bottom of the pile is negligible in comparison to the axial frictional force on its lateral surface. Assume that the frictional force is uniformly distributed. As the driving process begins, determine (a) the maximum compressive normal stress; (b) the change in length of the pile.

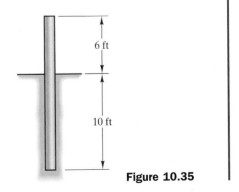

6 ft

10 ft

Figure 10.35

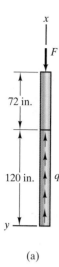

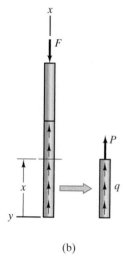

(a)

(a) Free-body diagram of the entire pile.

(b)

(b) Passing a plane at an arbitrary position x.

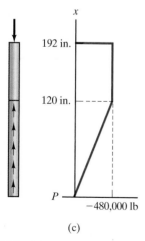

(c)

(c) Axial load P as a function of x.

Strategy

The part of the pile below the ground is subjected to a uniform distributed force q [Fig. (a)]. From the equilibrium equation for the entire pile we can determine the value of q. Then, by passing a plane through the pile at an arbitrary position x, we can determine the internal axial force in the pile [Fig. (b)]. Once we know the axial force, we can determine the maximum compressive stress and the change in the pile's length.

Solution

(a) From the equilibrium equation for the entire pile [Fig. (a)],

$$\int_0^{120} q\,dx - F = 0,$$

we determine that $q = F/120 = 4000$ lb/in. Then from the free-body diagram in Fig. (b), we obtain an equation for the internal axial force at an arbitrary position x:

$$\int_0^x q\,dx + P = 0.$$

Solving this equation, the internal axial load from $x = 0$ to $x = 120$ in. is $P = -qx = -4000x$ lb. We show the distribution of the internal axial load in the pile in Fig. (c). The maximum compressive load is $P = -480,000$ lb, so the maximum compressive stress in the pile is

$$\sigma = \frac{P}{A} = \frac{-480,000}{\pi(12)^2/4} = -4240 \text{ psi.}$$

(b) The internal axial force is given by the equation $P = -4000x$ lb from $x = 0$ to $x = 120$ in. (the part of the pile below the ground) and has the constant value $P = -480,000$ lb from $x = 120$ in. to $x = 192$ in. (the part above the ground). To determine the pile's change in length, we need to analyze the parts above and below the ground separately.

Above the ground   The change in length is

$$\delta_{\text{above}} = \frac{PL}{EA} = \frac{(-480,000)(72)}{(2 \times 10^6)\left[\pi(12)^2/4\right]} = -0.153 \text{ in.}$$

Below the ground   Below the ground the internal axial load varies with x. The change in length of an element of the beam of length dx is

$$\delta_{\text{element}} = \frac{P\,dx}{EA} = \frac{-4000x\,dx}{EA}.$$

Integrating this expression from $x = 0$ to $x = 120$ in., the change in length of the part of the pile below the ground is

$$\delta_{\text{below}} = \int_0^{120} \frac{-4000x\,dx}{EA} = \frac{(-4000)(120)^2}{(2)(2 \times 10^6)\left[\pi(12)^2/4\right]}$$

$$= -0.127 \text{ in.}$$

The change in length of the pile is $\delta_{\text{above}} + \delta_{\text{below}} = -0.280$ in.

Example 10.12

Suspended Bar

The prismatic bar in Fig. 10.36 is suspended from the ceiling and loaded only by its own weight. If the unloaded bar has length L, cross-sectional area A, weight density γ, and modulus of elasticity E, what is its length when suspended?

Strategy

We can treat the weight as an axial force distributed along the bar's axis. The force exerted on an element of the bar of length dx by its weight is equal to the product of the weight density and the volume of the element [Fig. (a)].

Figure 10.36

Solution

To determine the internal axial force, we obtain a free-body diagram by passing a plane through the bar at an arbitrary distance x from the bottom [Fig. (b)].

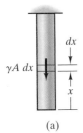

(a)

(a) Axial force on an element of length dx.

(b)

(b) Determining the internal axial force.

From the equilibrium equation

$$P - \int_0^x \gamma A\, dx = 0,$$

we determine that $P = \gamma A x$. The internal axial force increases linearly from zero at the bottom of the bar to $\gamma A L = W$, the bar's weight, at the top. The change in length of the element dx in Fig. (a) is

$$\delta_{element} = \frac{P\, dx}{EA} = \frac{\gamma A x\, dx}{EA}.$$

We integrate this expression from $x = 0$ to $x = L$ to determine the bar's change in length:

$$\delta = \int_0^L \frac{\gamma A x\, dx}{EA} = \frac{\gamma A L^2}{2EA}.$$

In terms of the bar's weight $W = \gamma A L$, the change in length is

$$\delta = \frac{WL}{2EA}.$$

The length of the suspended bar is $L + \delta = L + WL/2EA$.

Problems

10.51 The bar's cross-sectional area is $A = (1 + 0.1x)$ in^2 and the modulus of elasticity of the material is $E = 12 \times 10^6$ psi. If the bar is subjected to tensile axial forces $P = 20$ kip at its ends, what is the normal stress at $x = 6$ in.?

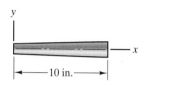

P10.51

10.52 What is the change in length of the bar in Problem 10.51?

10.53 The cross-sectional area of the bar in Problem 10.51 is $A = (1 + ax)$ in^2, where a is a constant, and the modulus of elasticity of the material is $E = 8 \times 10^6$ psi. When the bar is subjected to tensile axial forces $P = 14$ kip at its ends, its change in length is $\delta = 0.01$ in. What is the value of the constant a? (*Strategy*: Estimate the value of a by drawing a graph of δ as a function of a.)

10.54 From $x = 0$ to $x = 100$ mm, the bar's height is 20 mm. From $x = 100$ mm to $x = 200$ mm, its height varies linearly from 20 to 40 mm. From $x = 200$ mm to $x = 300$ mm, its height is 40 mm. The flat bar's thickness is 20 mm. The modulus of elasticity of the material is $E = 70$ GPa. If the bar is subjected to tensile axial forces $P = 50$ kN at its ends, what is its change in length?

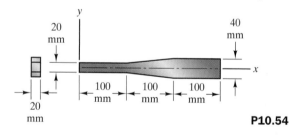

P10.54

10.55 From $x = 0$ to $x = 10$ in., the bar's cross-sectional area is $A = 1$ in^2. From $x = 10$ in. to $x = 20$ in., $A = (0.1x)$ in^2. The modulus of elasticity of the material is $E = 12 \times 10^6$ psi. There is a gap $b = 0.02$ in. between the right end of the bar and the rigid wall. If the bar is stretched so that it contacts the rigid wall and is welded to it, what is the axial force in the bar afterward?

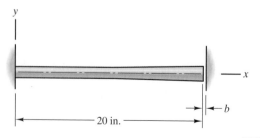

P10.55

10.56 From $x = 0$ to $x = 10$ in., the cross-sectional area of the bar shown in Problem 10.55 is $A = 1$ in^2. From $x = 10$ in. to $x = 20$ in., $A = (0.1x)$ in^2. The modulus of elasticity of the material is $E = 12 \times 10^6$ psi. There is a gap $b = 0.02$ in. between the right end of the bar and the rigid wall. If a 40-kip axial force toward the right is applied to the bar at $x = 10$ in., what is the resulting normal stress in the left half of the bar?

10.57 The diameter of the bar's circular cross section varies linearly from 10 mm at its left end to 20 mm at its right end. The modulus of elasticity of the material is $E = 45$ GPa. If the bar is subjected to tensile axial forces $P = 6$ kN at its ends, what is the normal stress at $x = 80$ mm?

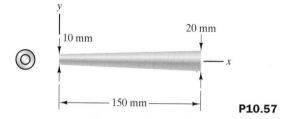

P10.57

10.58 What is the change in length of the bar in Problem 10.57?

10.59 The bar is fixed at the left end and subjected to a uniformly distributed axial force. It has cross-sectional area A and modulus of elasticity E.
(a) Determine the internal axial force P in the bar as a function of x.
(b) What is the bar's change in length?

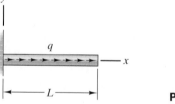

P10.59

10.60 The bar shown in Problem 10.59 has length $L = 2$ m, cross-sectional area $A = 0.03$ m^2, and modulus of elasticity $E = 200$ GPa. It is subjected to a distributed axial force $q = 12(1 + 0.4x)$ MN/m. What is the bar's change in length?

10.61 A cylindrical bar with 1-in. diameter fits tightly in a circular hole in a 5-in. thick plate. The modulus of elasticity of the material is $E = 14 \times 10^6$ psi. A 1000-lb tensile force is applied at the left end of the bar, causing it to begin slipping out of the hole. At the instant slipping begins, determine (a) the magnitude of the uniformly distributed axial force exerted on the bar by the plate; (b) the total change in the bar's length.

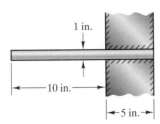

1 in.

10 in.

5 in.

P10.61

10.62 The bar has a circular cross section with 0.002-m diameter and its modulus of elasticity is $E = 86.6$ GPa. It is subjected to a uniformly distributed axial force $q = 75$ kN/m and an axial force $F = 15$ kN. What is its change in length?

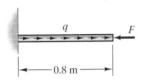

q

F

0.8 m

P10.62

10.63 In Problem 10.62, what axial force F would cause the bar's change in length to be zero?

10.64 If the bar in Problem 10.63 is subjected to a distributed force $q = 75(1 + 0.2x)$ kN/m and an axial force $F = 15$ kN, what is its change in length?

10.65 The bar is fixed at A and B and is subjected to a uniformly distributed axial force. It has cross-sectional area A and modulus of elasticity E. What are the reactions at A and B?

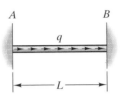

A B

q

L

P10.65

10.66 What point of the bar in Problem 10.65 undergoes the largest displacement, and what is the displacement?

All structures that are not provided with a controlled environment are subject to changes in temperature. Most materials tend to expand when their temperatures rise and contract when they fall, and this can have important effects on their use in structures and mechanisms. You are familiar with such dramatic effects as heated glass dishes breaking when suddenly cooled and concrete sidewalks developing cracks due to environmental temperature changes. Changes in temperature can also affect the mechanical properties of materials. Structural engineers must be aware of thermal effects and account for them in design. In this section we introduce the concept of thermal strain and demonstrate its effects in axially loaded bars.

Consider a sample of solid material that is at a uniform temperature T and is unconstrained—not subject to any forces or couples. Imagine drawing an infinitesimal line dL within the material (Fig. 10.37a). If the temperature of the sample changes by an amount ΔT, the material will expand or contract and the length of the line will change to a value dL' (Fig. 10.37b). The *thermal strain* of the material at the location and in the direction of the line dL is

$$\varepsilon_T = \frac{dL' - dL}{dL}.$$

If ΔT is positive, most materials will expand and ε_T will be positive, and if ΔT is negative, the material will contract and ε_T will be negative. For many materials used in engineering, the relationship between the thermal strain and the change in temperature can be approximated by a linear equation over some range of temperature:

$$\varepsilon_T = \alpha \, \Delta T. \tag{10.26}$$

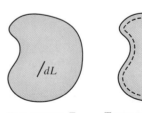

Temperature T Temperature $T + \Delta T$

(a)   (b)

Figure 10.37
(a) Infinitesimal line within a sample of material.
(b) Sample and line after a temperature change.

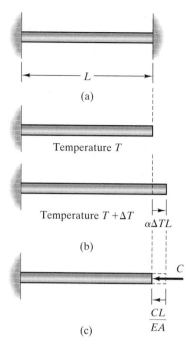

Figure 10.38
(a) Bar fixed at both ends.
(b) "Releasing" the right end and increasing the temperature.
(c) Compressive force exerted by the support.

This equation gives the thermal strain relative to the unconstrained material at temperature T. The constant α is a material property called the *coefficient of thermal expansion*. Since strain is dimensionless, the dimensions of α are (temperature)$^{-1}$. Typical values are given in Appendix D. Some materials exhibit different thermal strains in different directions (the value of α depends on direction), but we consider only materials in which the thermal strain in an unconstrained sample is the same in every direction.

The bar in Fig. 10.38a is fixed at both ends. What happens if we raise its temperature by an amount ΔT? To answer this question we first imagine that the right support is removed and raise the temperature by the amount ΔT. The temperature increase causes the bar's length to increase an amount $\varepsilon_T L = \alpha \, \Delta T \, L$ (Fig. 10.38b). But the length of the actual bar cannot increase, so the right support must exert a compressive force sufficient to compress the released bar by the amount the temperature increase caused it to expand (Fig. 10.38c):

$$\frac{CL}{EA} = \alpha \, \Delta T \, L.$$

(Notice that we calculated the change in length of the bar due to the compressive force C in terms of the original length L of the bar, neglecting the increase in the bar's length due to the change in temperature. We can do so because the change in length is small compared to L.) From this expression we see that the supports subject the bar to compressive axial forces $C = \alpha \, \Delta T \, EA$, and the resulting compressive normal stress in the bar is $\sigma = -C/A = -\alpha \, \Delta T \, E$.

This example explains one of the reasons that concrete sidewalks sometimes develop cracks. Concrete is weak in tension, and if the sidewalk is not designed so that it can contract freely when its temperature decreases, it is subjected to tensile stresses and cracks form.

Study Questions

1. What is the definition of the coefficient of thermal expansion?
2. Consider an infinitesimal line dL within an unconstrained sample of material with temperature T. If the temperature of the material is increased an amount ΔT, how can you determine the resulting length of the infinitesimal line?
3. Consider an unconstrained bar of length L and temperature T. If the temperature of the bar is increased an amount ΔT, how can you determine the resulting change in length of the bar?
4. After the temperature of the unconstrained bar in question 3 is increased, what is the normal stress on a plane perpendicular to the bar's axis? (Draw a free-body diagram of the part of the bar on one side of the plane.)

Example 10.13

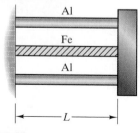

Figure 10.39

Thermal Strains in a Structure

In Fig. 10.39 two aluminum bars $\left(E_{\text{Al}} = 10.0 \times 10^6 \text{ psi}, \alpha_{\text{Al}} = 13.3 \times 10^{-6} \text{ °F}^{-1}\right)$ and an iron bar $\left(E_{\text{Fe}} = 28.5 \times 10^6 \text{ psi}, \alpha_{\text{Fe}} = 6.5 \times 10^{-6} \text{ °F}^{-1}\right)$ are attached to a rigid support at the left and a cross-bar at the right. The cross-sectional area of each bar is $A = 0.5 \text{ in}^2$ and $L = 10 \text{ in}$. The normal stresses in the bars are initially zero. If the temperature is increased by $\Delta T = 100°\text{F}$, what are the resulting normal stresses in the three bars? The deformation of the cross-bar can be neglected.

Strategy

Because the two materials have different coefficients of thermal expansion, the aluminum and iron bars tend to lengthen by different amounts. As a result, they will develop internal axial forces when the temperature increases. By drawing a free-body diagram of the cross-bar we can obtain one equilibrium equation, but the problem is statically indeterminate because the axial forces in the aluminum and iron bars may be different. (Notice, however, that the axial forces in the two aluminum bars are equal because of the symmetry of the system.) We can express the changes in length of the bars as the sum of their changes in length due to their internal axial forces and their changes in length due to thermal strain. We can then apply the compatibility condition that the changes in length of the bars must be equal because they are attached to the rigid cross-bar.

Solution

Equilibrium   In Fig. (a) we obtain a free-body diagram of the cross-bar by passing a plane through the right ends of the three bars.

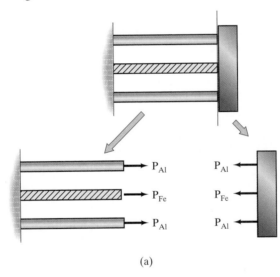

(a)

(a) Isolating the cross-bar.

The equilibrium equation for the cross-bar is

$$2P_{Al} + P_{Fe} = 0.$$

Deformation relations   The change in length of one of the aluminum bars due to its axial force and the change in temperature is

$$\delta_{Al} = \frac{P_{Al}L}{E_{Al}A} + \alpha_{Al}\Delta T\,L,$$

and the change in length of the iron bar is

$$\delta_{Fe} = \frac{P_{Fe}L}{E_{Fe}A} + \alpha_{Fe}\Delta T\,L.$$

Compatibility   The compatibility condition is that the changes in length of the bars are equal:

$$\delta_{Al} = \delta_{Fe}.$$

Substituting the deformation relations into this equation gives

$$\frac{P_{Al}L}{E_{Al}A} + \alpha_{Al}\Delta T\, L = \frac{P_{Fe}L}{E_{Fe}A} + \alpha_{Fe}\Delta T\, L.$$

This equation can be solved together with the equilibrium equation for the axial forces P_{Al} and P_{Fe}. Solving them, we obtain $P_{Al} = -2000$ lb, $P_{Fe} = 4000$ lb. The normal stresses in the bars are

$$\sigma_{Al} = \frac{P_{Al}}{A} = \frac{-2000}{0.5} = -4000 \text{ psi,}$$

$$\sigma_{Fe} = \frac{P_{Fe}}{A} = \frac{4000}{0.5} = 8000 \text{ psi.}$$

Discussion

Beyond being an exercise to illustrate thermal strains, this example demonstrates an important problem in design. The elements of structures and machines constructed of various materials are subject to different thermal strains due to their different coefficients of thermal expansion. As a result, temperature changes can cause both stresses and deformations that must be considered in design.

Example 10.14

Thermal Strains in a Truss

The bars in Fig. 10.40 each have cross-sectional area $A = 60$ mm^2, modulus of elasticity $E = 200$ GPa, and coefficient of thermal expansion $\alpha = 12 \times 10^{-6}$ °C^{-1}. If a 40-kN downward force is applied at A and the temperature of the bars is raised 30°C, what are the resulting horizontal and vertical displacements of point A?

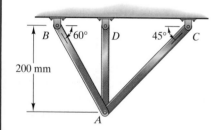

Figure 10.40

Strategy

In this problem we extend Example 10.9 by introducing a change in temperature. The only change in the method of solution we used in that example is that we must express the change in length of each bar as a sum of the change due to its unknown axial force and the change due to its thermal strain.

Solution

Equilibrium In Fig. (a) we draw the free-body diagram of joint A showing the 40-kN force.

The equilibrium equations are

$$\Sigma F_x = -P_{AB}\cos 60° + P_{AC}\cos 45° = 0,$$

$$\Sigma F_y = P_{AB}\sin 60° + P_{AD} + P_{AC}\sin 45° - 40{,}000 = 0.$$

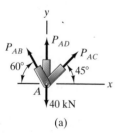

(a)

(a) Joint A.

Deformation relations The lengths of the bars are $L_{AB} = 0.2/\sin 60°$ m, $L_{AC} = 0.2/\sin 45°$ m, and $L_{AD} = 0.2$ m. The change in length of each bar will be the sum of its change in length due to its axial force and the change in length due to thermal strain:

$$\delta_{AB} = \frac{P_{AB}L_{AB}}{EA} + \alpha \, \Delta T \, L_{AB},$$

$$\delta_{AC} = \frac{P_{AC}L_{AC}}{EA} + \alpha \, \Delta T \, L_{AC}, \qquad (10.27)$$

$$\delta_{AD} = \frac{P_{AD}L_{AD}}{EA} + \alpha \, \Delta T \, L_{AD}.$$

Compatibility We denote the horizontal and vertical displacements of joint A by u and v [Fig. (b)]. From Example 10.9, the relationships between the changes in length of the bars and the displacements u and v are

$$v \sin 60° + u \cos 60° = \delta_{AB},$$

$$v \sin 45° - u \cos 45° = \delta_{AC}, \qquad (10.28)$$

$$v = \delta_{AD}.$$

These equations are the compatibility conditions. The equilibrium equations, Eqs. (10.27), and Eqs. (10.28) provide eight equations in the unknowns P_{AB}, P_{AC}, P_{AD}, δ_{AB}, δ_{AC}, δ_{AD}, u, and v. Solving them, we obtain

$$u = -0.042 \text{ mm}, \qquad v = 0.426 \text{ mm}.$$

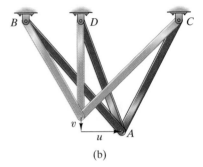

(b) Displacements of joint A.

Problems

10.67 A line L within an unconstrained sample of material is 200 mm long. The coefficient of thermal expansion of the material is $\alpha = 22 \times 10^{-6}$ °C^{-1}. If the temperature of the material is increased by 30°C, what is the length of the line?

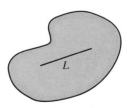

P10.67

10.68 The length of the line L within the unconstrained sample of material shown in Problem 10.67 is 2 in. The coefficient of thermal expansion of the material is $\alpha = 8 \times 10^{-6}$ °F^{-1}. After the temperature of the material is increased, the length of the line is 2.002 in. How much was the temperature increased?

10.69 Consider a 1 in. $\times$ 1 in. $\times$ 1 in. cube within an unconstrained sample of material. The coefficient of thermal expansion of the material is $\alpha = 14 \times 10^{-6}$ °F^{-1}. If the temperature of the material is decreased by 40°F, what is the volume of the cube?

10.70 A prismatic bar is 200 mm long and has a circular cross section with 30-mm diameter. After the temperature of the unconstrained bar is increased, its length is measured and determined to be 200.160 mm. What is the bar's diameter after the increase in temperature?

10.71 If the increase in temperature in Problem 10.70 is 20°C, what is the coefficient of thermal expansion of the bar?

10.72 If the increase in temperature in Problem 10.70 is 20°C and the modulus of elasticity of the material is $E = 72$ GPa, what is the normal stress on a plane perpendicular to the bar's axis after the increase in temperature? (*Strategy:* Obtain a free-body diagram by passing a plane through the bar.)

10.73 The prismatic bar is made of material with modulus of elasticity $E = 28 \times 10^6$ psi and coefficient of thermal expansion $\alpha = 8 \times 10^{-6}$ °F^{-1}. The temperature of the unconstrained bar is increased by 50°F above its initial temperature T.
(a) What is the change in the bar's length?
(b) What is the change in the bar's diameter?
(c) What is the normal stress on a plane perpendicular to the bar's axis after the increase in temperature?

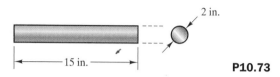

2 in.

15 in.

P10.73

10.74 Suppose that the temperature of the unconstrained bar in Problem 10.73 is increased by 50°F above its initial temperature T and the bar is also subjected to 30,000-lb tensile axial forces at the ends. What is the resulting change in the bar's length? Determine the change in length assuming that (a) the temperature is first increased and then the axial forces are applied; (b) the axial forces are first applied and then the temperature is increased.

10.75 The prismatic bar is made of material with modulus of elasticity $E = 28 \times 10^6$ psi and coefficient of thermal expansion $\alpha = 8 \times 10^{-6}$ °F^{-1}. It is constrained between rigid walls. If the temperature is increased by 50°F above the bar's initial temperature T, what is the normal stress on a plane perpendicular to the bar's axis?

15 in.

P10.75

10.76 The walls between which the prismatic bar in Problem 10.75 is constrained will safely support a compressive normal stress of 30,000 psi. Based on this criterion, what is the largest safe temperature increase to which the bar can be subjected?

10.77 The prismatic bar in Problem 10.75 has a cross-sectional area $A = 3$ in^2 and is made of material with modulus of elasticity $E = 28 \times 10^6$ psi and coefficient of thermal expansion $\alpha = 8 \times 10^{-6}$ °F^{-1}. It is constrained between rigid walls. The temperature is increased by 50°F above the bar's initial temperature T and a 20,000-lb axial force to the right is applied midway between the two walls. What is the normal stress on a plane perpendicular to the bar's axis to the right of the point where the force is applied?

10.78 The prismatic bar is made of material with modulus of elasticity $E = 28 \times 10^6$ psi and coefficient of thermal expansion $\alpha = 8 \times 10^{-6}$ °F^{-1}. It is fixed to a rigid wall at the left. There is a

gap $b = 0.002$ in. between the bar's right end and the rigid wall. If the temperature is increased by 50°F above the bar's initial temperature T, what is the normal stress on a plane perpendicular to the bar's axis?

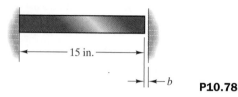

15 in.

b

P10.78

10.79 Bar A has a cross-sectional area of 0.04 m^2, modulus of elasticity $E = 70$ GPa, and coefficient of thermal expansion $\alpha = 14 \times 10^{-6}$ °C^{-1}. Bar B has a cross-sectional area of 0.01 m^2, modulus of elasticity $E = 120$ GPa, and coefficient of thermal expansion $\alpha = 16 \times 10^{-6}$ °C^{-1}. There is a gap $b = 0.4$ mm between the ends of the bars. What minimum increase in the temperature of the bars above their initial temperature T is necessary to cause them to come into contact?

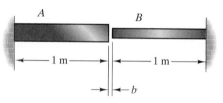

A

B

1 m

1 m

b

P10.79

10.80 If the temperature of the bars in Problem 10.79 is increased by 40°C above their initial temperature T, what are the normal stresses in the bars?

10.81 Each bar has a 2-in^2 cross-sectional area, modulus of elasticity $E = 14 \times 10^6$ psi, and coefficient of thermal expansion $\alpha = 11 \times 10^{-6}$ °F^{-1}. If their temperature is increased by 40°F from their initial temperature T, what is the resulting displacement of point A?

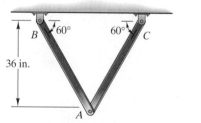

B 60°

60° C

36 in.

A

P10.81

10.82 If the temperature of the bars in Problem 10.81 is decreased by 30°F from their initial temperature T, what force would need to be applied at A so that the total displacement of point A caused by the temperature change and the force is zero?

10.6 Material Behavior

Until now we have assumed that the normal stress and axial strain in an axially loaded bar are related by the linear equation $\sigma = E\varepsilon$, where E is the modulus of elasticity of the material. In this section we discuss this assumption and introduce the richly varied subject of material behavior.

Axial Force Tests

Leonardo da Vinci (1452–1519) and Galileo (1564–1642) investigated the strengths of wires and bars by subjecting them to axial tension. In a modern *tension test*, still the most common test of the mechanical behavior of materials, a bar of material is mounted in a machine that subjects it to a tensile axial force (Fig. 10.41). For a given value of the tensile force P, the normal stress $\sigma = P/A$ can be determined. (The stress is normally defined in terms of the cross-sectional area A of the undeformed bar. The term *true stress* denotes the stress calculated using the cross-sectional area of the deformed bar. In most engineering applications the difference between these definitions of stress is negligible because the change in the cross-sectional area is small.) The bar's axial strain ε is determined by measuring the change in the distance between two axial positions relative to the distance in the unloaded state, which is called the *gauge length*. The axial positions are not chosen near the ends of the bar so that ε will be approximately uniform throughout the gauge length. When the test proceeds until the bar fractures, or breaks, the *elongation*, the final change in the gauge length expressed as a percentage of the gauge length, can also be measured. Because the axial strain becomes increasingly nonuniform as fracture approaches, the gauge length used in determining the elongation must be specified.

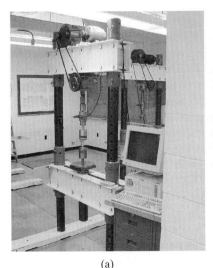
(a)

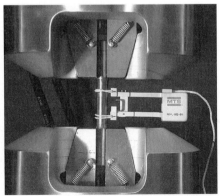
(b)

Figure 10.41
(a) Machine for subjecting a bar of material to axial force.
(b) Aluminum specimen mounted in the machine. The extensometer measures the specimen's change in length.

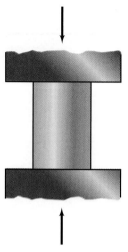

Figure 10.42
Typical compression specimen.
Compressive axial load is applied to the
ends by flat platens.

A *compression test* is carried out in much the same way as a tension test, applying compressive axial force to a specimen and measuring the strain. A much shorter square or cylindrical specimen is used and the axial forces are usually applied by compressing the material between flat platens (Fig. 10.42).

Ductile Materials We first describe some phenomena that occur in a tension test of a low-carbon steel. Progressively increasing the axial strain and recording the corresponding stress, the graph of the stress as a function of the strain, or *stress-strain diagram,* appears qualitatively as shown in Fig. 10.43. For small values of the strain, the relationship between the stress and strain is linear. The slope of this linear portion of the graph is the modulus of elasticity E. At a point called the *proportional limit*, the graph deviates from a straight line. (The stress is no longer proportional to the strain.) At a point called the *yield point*, the strain begins increasing with no change in stress. The corresponding stress is called the *yield stress* σ_Y. Eventually, the stress again begins increasing with increasing strain, a phenomenon called *strain hardening*, and reaches a maximum value, the *ultimate stress* σ_U.

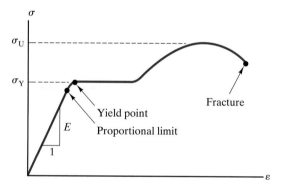

Figure 10.43
Stress-strain diagram for a tension test of
low-carbon steel.

As the strain continues to increase beyond the occurrence of the ultimate stress, the stress decreases until the bar fractures, or breaks. The decreasing stress is associated with the formation of a region of decreased cross-sectional area in the bar, a phenomenon called *necking* (Fig. 10.44). The apparently decreasing stress is an artifact caused by defining σ in terms of the cross-sectional area of the undeformed bar. The true stress in the necked portion of the bar continues to increase with strain until fracture occurs.

If the strain to which the steel is subjected remains below the proportional limit, so that $\sigma = E\varepsilon$, we say that its stress-strain relationship is *linearly elastic*, and if the stress is removed, the bar returns to its original length. If the bar is strained beyond the yield point and the stress is then decreased, the resulting path in the σ–ε plane does not return to the origin along the original curve. Instead, it returns to zero stress along a straight line with slope E, and a residual strain remains (Fig. 10.45a). If the bar is then reloaded, it follows the new straight path until it reaches the stress-strain curve obtained with monotonically increasing strain (Fig. 10.45b). Strain beyond the value of strain corresponding to the yield point, which remains as residual strain if the bar is unloaded, is called *plastic strain*. Thus the yield stress σ_Y is the stress at which plastic strain begins. Materials with stress-strain relationships of this type are sometimes modeled by representing their stress-strain relationship as shown in Fig. 10.46, which is referred to as an *elastic-perfectly plastic material*. ("Perfectly" plastic means it is assumed that there is no strain hardening.)

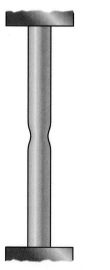

Figure 10.44
Necking in a tension test specimen.

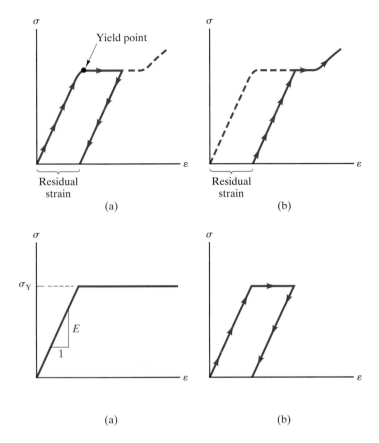

Figure 10.45
(a) Straining the bar beyond the yield point and then unloading it.
(b) Reloading the bar.

Figure 10.46
(a) Model of an elastic-perfectly plastic material.
(b) Loading-unloading behavior.

The explanation for this somewhat bizarre behavior is that plastic strain in a crystalline material like steel occurs in part by mechanisms that result in rearrangements of its lattices of atoms. Figure 10.47 illustrates how one such mechanism, called *slip*, results in plastic strain. When increasing stress is applied to the unloaded material, strain first occurs due to changes in the

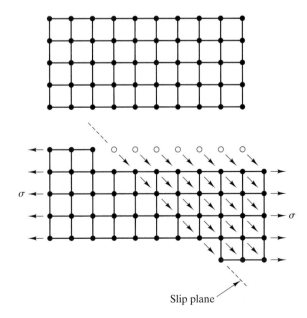

Figure 10.47
Plastic strain of a crystal lattice by the mechanism of slip.

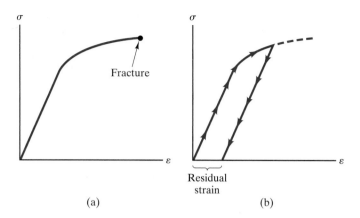

(a)   (b)

Figure 10.48
(a) Stress-strain relationship for a ductile
material that does not exhibit an
identifiable yield stress.
(b) Loading-unloading behavior.

distances between atoms, and the stress-strain relationship is linearly elastic.
Beyond the value of strain corresponding to the yield stress, strain occurs due
to rearrangements of the crystal lattices. But when unloading occurs, strain
decreases due to changes in the distances between atoms, and the stress-strain
relationship is again linearly elastic.

Materials that undergo significant plastic strain before fracture are said to
be *ductile*. Our discussion of tensile behavior has focused on low-carbon steel
in part due to its importance in structural applications, but also because describing the stress-strain behavior of steel permitted us to introduce concepts
and terminology—linearly elastic, yield stress, ultimate stress, plastic strain,
fracture—that apply to many other ductile materials. Some ductile materials,
such as aluminum and copper, do not exhibit an easily identifiable yield stress
but undergo a smooth transition from linearly elastic to plastic behavior with
strain hardening (Fig. 10.48a). However, the yield stress is a very useful parameter in design, so it has become traditional to artificially designate a yield
point and yield stress for such materials. This is done by drawing a line parallel to the linear part of the stress-strain diagram which intersects the strain
axis at some arbitrarily chosen value, often $\varepsilon = 0.002$. The point at which
this line intersects the stress-strain diagram is defined to be the yield point,
and the corresponding stress is defined to be the yield stress (Fig. 10.49). The
arbitrary strain chosen is called the *offset*, and this technique is called the offset method of determining the yield point.

When a typical ductile material is subjected to a compression test
(Fig. 10.50), it initially exhibits linearly elastic behavior governed by $\sigma = E\varepsilon$.

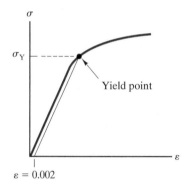

Figure 10.49
Designating a yield point and yield stress
by the offset method.

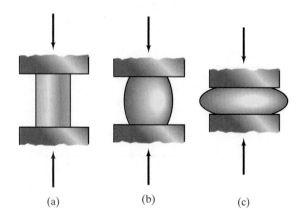

(a)   (b)   (c)

Figure 10.50
(a) Beginning a compression test of a
ductile material.
(b) Soon after yielding occurs, the
specimen becomes barrel shaped.
(c) The specimen eventually becomes
flattened.

After yielding occurs, the stress-strain behavior is different in character from that observed in the tension test (Fig. 10.51). Although the slope of the stress as a function of strain initially decreases as the material begins yielding and the specimen becomes barrel shaped, the slope increases again as the specimen begins to flatten out.

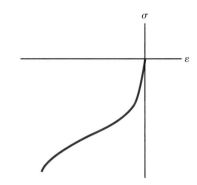

Figure 10.51
Stress as a function of strain for a compression test of a typical ductile material.

Brittle Materials   Materials such as cast iron, high-carbon steel, and masonry that exhibit relatively little plastic strain prior to fracture are said to be *brittle*. In a tension test the stress increases monotonically, so that the ultimate stress occurs at fracture. A notable feature of brittle materials is that their ultimate stress in compression is considerably greater in magnitude than their ultimate stress in tension. This is illustrated in Fig. 10.52, in which we show the stress as a function of strain for a typical brittle material in both tension and compression. Unlike ductile materials, brittle materials subjected to a compression test undergo fracture. In different materials the sample may crush or may fracture along a plane of maximum shear stress as shown in Fig. 10.15.

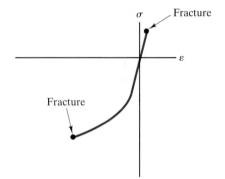

Figure 10.52
Stress as a function of strain for tension and compression tests of a typical brittle material.

Study Questions

1. What is the true stress?
2. What is the slope of the stress-strain diagram of a ductile material?
3. What is the definition of the yield stress σ_Y?
4. What is an elastic-perfectly plastic material?

10.7 Design Issues

In this chapter we have provided some of the essential background for designing bars to support constant axial loads and structures made up of such bars. If there is a dictum for the structural designer equivalent to the physician's "First, do no harm," it is "First, prevent failure." Although *failure* must be defined in different ways for different applications, at a minimum the normal and shear stresses in a structural member must not be allowed to exceed the values the material will support.

Allowable Stress

In a prismatic bar subjected to a specified axial load P, we have shown in Section 10.1 that there is no plane on which the normal stress is larger in magnitude than the stress $\sigma = P/A$ on a plane perpendicular to the bar's axis. The magnitude of the maximum shear stress, which occurs on planes oriented at 45° relative to the bar's axis, is $|\sigma/2| = |P/2A|$. A criterion frequently used in design is to try to ensure that σ does not exceed the yield stress σ_Y of the material, or a specified fraction of the yield stress called the *allowable stress* σ_{allow}. The ratio of the yield stress to the allowable stress is called the *factor of safety:*

$$S = \frac{\sigma_Y}{\sigma_{\text{allow}}}.$$

In choosing the factor of safety, the design engineer must consider how accurately the loads to which the bar will be subjected in normal use can be estimated. Whenever possible, a conservatively large factor of safety would be desirable, but compromises are usually necessary. For example, cost may require compromise in the properties of the materials used. Excessive structural weight can discourage the use of high factors of safety, especially in vehicle applications. Contingencies beyond normal use must also be considered, such as potential earthquake loads in the structure of a building, or loads exerted on the members of a car's suspension during emergency maneuvers. If the item being designed will be mass produced, the factor of safety must be chosen to account for anticipated variations in dimensions and material properties. The engineer must balance these considerations within the essential constraint of arriving at a reliable and safe design.

Other Design Considerations

A bar subjected to compression can fail by buckling (geometric instability) at a much smaller axial load than is necessary to cause yielding of the material. We analyze buckling in Chapter 17. Even when the stresses to which a structural member is subjected are small compared to the yield stress, failure can occur if they are applied repeatedly. Also, our analysis of the stress distribution in an axially loaded bar does not apply near the ends where the loads are applied. Those regions normally require a detailed stress analysis that is beyond our scope.

Our examples and problems are limited primarily to designing axially loaded members and structures to meet the objective of supporting given loads. But in addition to the overriding concern of preventing failure, the structural designer is usually confronted with a broad array of decisions relevant to a particular application, and the finest designs are achieved by successfully meeting a spectrum of requirements. Material cost and availability, cost and feasibility of processing and manufacture, resistance to corrosion in the expected environment, compatibility with other materials, and the effect of aging on the material's properties may be important considerations. Decisions on a given design can also be influenced to a greater or lesser extent by concern for safety, ease of maintenance, and aesthetics.

Example 10.15

Design of Truss Members

The truss in Fig. 10.53 is to be constructed of members with yield stress $\sigma_Y = 700$ MPa and equal cross-sectional areas. If the structure must support a mass m as large as 3400 kg with factor of safety $S = 3$, what should the cross-sectional area of the members be?

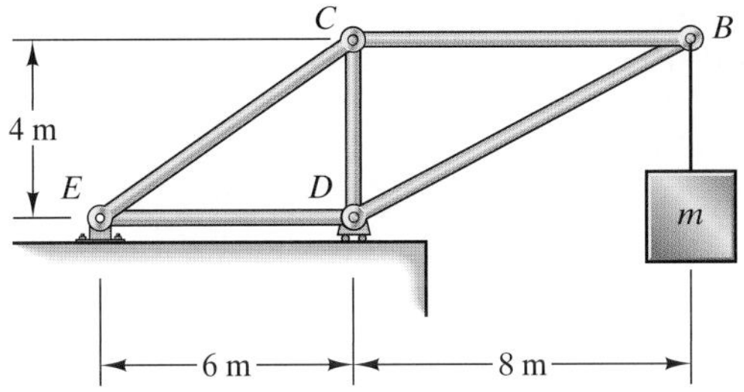

Figure 10.53

Strategy

We can use the method of joints to determine the axial forces in the members when $m = 3400$ kg. We must then choose the cross-sectional area so that the member subjected to the largest axial load has a factor of safety of 3.

Solution

With $m = 3400$ kg, the axial forces in the members are (see Example 10.2):

Member:	BC	BD	CD	CE	DE
Axial force (N):	66,700	−74,600	−44,500	80,200	−66,700

The largest axial force, 80,200 N, occurs in member CE. With a factor of safety of 3, the allowable stress is

$$\sigma_{\text{allow}} = \frac{\sigma_Y}{S} = 233 \text{ MPa.}$$

Equating the normal stress in member CE to the allowable stress,

$$\frac{80,200}{A} = 233 \times 10^6,$$

we obtain $A = 0.000344$ m^2.

Problems

10.83 You are designing a bar with a solid circular cross section that is to support a 4-kN tensile axial load. You have decided to use 6061-T6 aluminum alloy (see Appendix D), and you want the factor of safety to be $S = 2$. Based on this criterion, what should the bar's diameter be?

4 kN   4 kN

P10.83

10.84 You are designing a bar with a solid circular cross section with 5-mm diameter that is to support a 4-kN tensile axial load, and you want the factor of safety to be at least $S = 2$. Choose an aluminum alloy from Appendix D that satisfies this requirement.

10.85 You are designing a bar with a solid circular cross section that is to support a 4000-lb tensile axial load. You have decided to use ASTM-A572 structural steel (see Appendix D), and you want the factor of safety to be $S = 1.5$. Based on this criterion, what should the bar's diameter be?

4000 lb   4000 lb

P10.85

10.86 You are designing a bar with a solid circular cross section with $\frac{1}{2}$-in. diameter that is to support a 4000-lb tensile axial load, and you want the factor of safety to be at least $S = 3$. Choose a structural steel from Appendix D that satisfies this requirement.

10.87 The horizontal beam of length $L = 2$ m supports a load $F = 30$ kN. The beam is supported by a pin support and the brace BC. The dimension $h = 0.54$ m. Suppose that you want to make the brace out of existing stock that has cross-sectional area $A = 0.0016$ m^2 and yield stress $\sigma_Y = 400$ MPa. If you want the brace to have a factor of safety $S = 1.5$, what should the angle θ be?

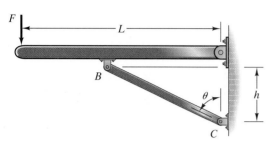

P10.87

10.88 Consider the system shown in Problem 10.87. The horizontal beam of length $L = 4$ ft supports a load $F = 20$ kip. The beam is supported by a pin support and the brace BC. The dimension $h = 1$ ft and the angle $\theta = 60°$. Suppose that you want to make the brace out of existing stock that has yield stress $\sigma_Y = 50$ ksi. If you want to design the brace BC to have a factor of safety $S = 2$, what should its cross-sectional area be?

10.89 The horizontal beam shown in Problem 10.87 is of length L and supports a load F. The beam is supported by a pin support and the brace BC. Suppose that the brace is to consist of a specified material for which you have chosen an allowable stress σ_{allow}, and you want to design the brace so that its weight is a minimum. You can do this by assuming that the brace is subjected to the allowable stress and choosing the angle θ so that the volume of the brace is a minimum. What is the necessary angle θ?

10.90 In Problem 10.89, draw a graph showing the dependence of the volume of the brace on the angle θ for $5° \le \theta \le 85°$. Notice that the graph is relatively flat near the optimum angle, meaning that the designer can choose θ within a range of angles near the optimum value and still obtain a nearly optimum design.

10.91 The truss is a preliminary design for a structure to attach one end of a stretcher to a rescue helicopter. Based on dynamic simulations, the design engineer estimates that the downward forces the stretcher will exert will be no greater than 360 lb at A and at B. Assume that the members of the truss have the same cross-sectional area. Choose a material from Appendix D and determine the cross-sectional area so that the structure has a factor of safety $S = 2.5$.

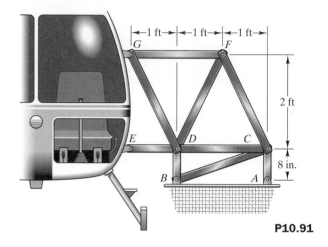

P10.91

10.92 Upon learning of an upgrade in the helicopter's engine, the engineer designing the truss shown in Problem 10.91 does new simulations and concludes that the downward forces the stretcher will exert at A and at B may be as large as 400 lb. He also decides the truss will be made of existing stock with cross-sectional area $A = 0.1$ in^2. Choose an aluminum alloy from Appendix D so that the structure will have a factor of safety of at least $S = 5$.

10.93 Two candidate truss designs to support the load F are shown. Members of a given cross-sectional area A and yield stress σ_Y are to be used. Compare the factors of safety and weights of the two designs and discuss reasons that might lead you to choose one design over the other. (The weights can be compared by calculating the total lengths of their members.)

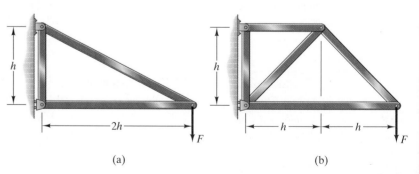

(a)   (b)

P10.93

Chapter Summary

Stresses in Prismatic Bars

If axial forces P are applied at the centroid of the cross section of a prismatic bar, the stress distribution on any plane perpendicular to the bar's axis that is not near the ends of the bar [Fig. (a)] can be approximated by a uniform normal stress

$$\sigma = \frac{P}{A}.$$

Eq. (10.1)

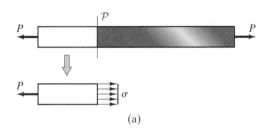

(a)

The normal and shear stresses on a plane $\mathcal{P}$ oriented as shown in Fig. (b) are

$$\sigma_\theta = \sigma \cos^2 \theta,$$ Eq. (10.2)
$$\tau_\theta = -\sigma \sin\theta \cos\theta.$$ Eq. (10.3)

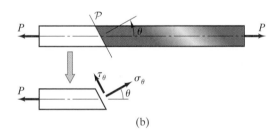

(b)

There is no plane on which the normal stress is larger in magnitude than the stress $\sigma = P/A$ on the plane perpendicular to the bar's axis. The magnitude of the maximum shear stress, which occurs on planes oriented at $45°$ relative to the bar's axis, is $|\sigma/2|$.

Strains in Prismatic Bars

The change in length δ of a prismatic bar subjected to axial load [Fig. (c)] is related to the axial strain ε by

$$\delta = \varepsilon L,$$ Eq. (10.4)

where L is the length of the undeformed bar. The normal stress is related to the axial strain by

$$\sigma = E\varepsilon,$$ Eq. (10.5)

where E is the *modulus of elasticity*. Typical values of E are given in Appendix D. The change in the bar's length resulting from a given axial load P is

$$\delta = \frac{PL}{EA},$$ Eq. (10.6)

where A is the cross-sectional area.

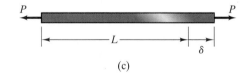

(c)

The negative of the ratio of the lateral strain of a prismatic bar to its axial strain when it is subjected to axial forces is called *Poisson's ratio*

$$\nu = -\frac{\varepsilon_{\text{lat}}}{\varepsilon}.$$
Eq. (10.7)

Typical values are given in Appendix D. The lateral strain resulting from an axial load P is

$$\varepsilon_{\text{lat}} = -\frac{\nu P}{EA}.$$
Eq. (10.8)

Statically Indeterminate Problems

Solutions to problems involving axially loaded bars in which the number of unknown reactions exceeds the number of independent equilibrium equations involve three elements: (1) relations between the axial forces in bars and their changes in length; (2) compatibility relations imposed on the changes in length by the geometry of the problem; and (3) equilibrium.

Bars with Gradually Varying Cross Sections

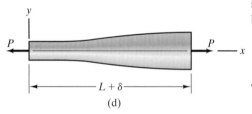

If the cross-sectional area of a bar is a gradually varying function of axial position $A(x)$ [Fig. (d)], the stress distribution on a plane perpendicular to the bar's axis can be approximated by a uniform normal stress:

$$\sigma = \frac{P}{A(x)}.$$
Eq. (10.24)

(d)

The bar's change in length is

$$\delta = \int_0^L \frac{P\,dx}{EA(x)}.$$
Eq. (10.25)

Distributed Axial Loads

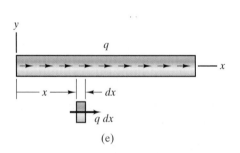

A distributed axial force on a bar can be described by a function q defined such that the force on each element dx of the bar is $q\,dx$ [Fig. (e)].

(e)

Thermal Strains

Consider an infinitesimal line dL within an unconstrained material at a uniform temperature T. If the temperature of the sample changes by an amount ΔT, the material will expand or contract and the length of the line will change to a value dL' [Fig. (f)]. The *thermal strain* of the material at the location and in the direction of the line dL is

$$\varepsilon_T = \frac{dL' - dL}{dL}.$$

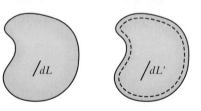

Temperature T Temperature $T + \Delta T$

(f)

The thermal strain is related to the change in temperature by

$$\varepsilon_T = \alpha\,\Delta T,$$
Eq. (10.26)

where α is the *coefficient of thermal expansion*. Typical values of α are given in Appendix D.

Review **Problems**

10.94 The cross-sectional area of bar AB is 0.015 m². If the force $F = 20$ kN, what is the normal stress on a plane perpendicular to the axis of bar AB?

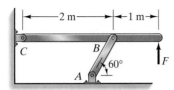

P10.94

10.95 Bar AB of the frame in Problem 10.94 consists of a material that will safely support a tensile normal stress of 20 MPa. Based on this criterion, what is the largest safe value of the force F?

10.96 The system shown supports half of the weight of the 680-kg excavator. The cross-sectional area of member AB is 0.0012 m². If the system is stationary, what normal stress acts on a plane perpendicular to the axis of member AB?

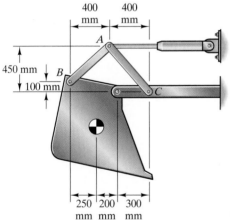

P10.96

10.97 Member AC in Problem 10.96 has a cross-sectional area of 0.0014 m². If the system is stationary, what normal stress acts on a plane perpendicular to the axis of member AC?

10.98 The bar has modulus of elasticity $E = 30 \times 10^6$ psi, Poisson's ratio $\nu = 0.32$, and a circular cross section with diameter $D = 0.75$ in. There is a gap $b = 0.02$ in. between the right end of the bar and the rigid wall. If the bar is stretched so that it contacts the rigid wall and is welded to it, what is the bar's diameter afterward?

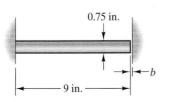

P10.98

10.99 After the bar in Problem 10.98 is welded to the rigid wall, what is the normal stress on a plane perpendicular to the bar's axis?

10.100 The link AB of the pliers has a cross-sectional area of 40 mm² and elastic modulus $E = 210$ GPa. If forces $F = 150$ N are applied to the pliers, what is the change in length of link AB?

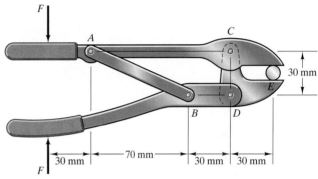

P10.100

10.101 Suppose that you want to design the pliers in Problem 10.100 so that forces F as large as 450 N can be applied. The link AB is to be made of a material that will support a compressive normal stress of 200 MPa. Based on this criterion, what minimum cross-sectional area must link AB have?

10.102 Each bar has a cross-sectional area of 3 in² and modulus of elasticity $E = 12 \times 10^6$ lb/in². If a 40-kip horizontal force directed toward the right is applied at A, what are the normal stresses in the bars?

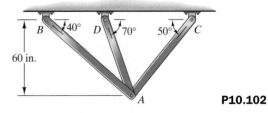

P10.102

10.103 The bars of the system in Problem 10.102 consist of a material that will safely support a tensile normal stress of 20 ksi. Based on this criterion, what is the largest *downward* force that can safely be applied at A?

This vintage Corvette has many structural elements designed to support and transmit torsional loads, including the drive shaft, axles, and crankshaft.

Torsion

A car's drive shaft transmits power from the engine to the rear wheels. The power produced by turbine engines and electrical generators is also transmitted by shafts subjected to torques about their longitudinal axes. Bars are subjected to axial torques in many engineering applications, and we analyze the resulting stresses and deformations in this chapter. To do so, we must first introduce the concept of a state of pure shear stress and the shear modulus.

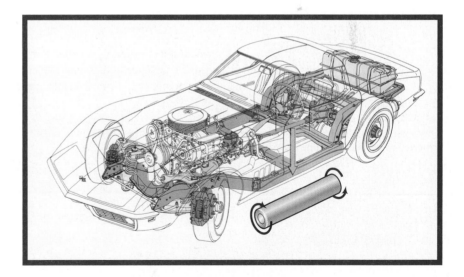

11.1 Pure Shear Stress

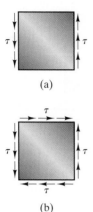

(a)

(b)

Figure 11.1
(a) A cube is not in equilibrium if only two faces are subjected to shear stress.
(b) A cube subjected to shear stresses in this way is in equilibrium.

Figure 11.2
Planes parallel to the faces of the cube are subjected to the same uniform shear stress.

We have seen in Chapter 10 that subjecting a bar to axial forces results in a state of uniform normal stress and no shear stress on planes perpendicular to the bar's axis. We begin this chapter by describing how, at least in principle, a uniform distribution of shear stress and no normal stress can be achieved on particular planes within a material.

State of Stress

If we were to subject opposite faces of a cube of material to uniform shear stresses τ as shown in Fig. 11.1a the cube would not be in equilibrium because the stresses exert a couple on it. However, if we also apply uniform shear stresses to the top and bottom faces as shown in Fig. 11.1b, the cube is in equilibrium. If a cube could be loaded in this way, the stress on any plane parallel to the loaded faces consists of uniform shear stress and no normal stress, as shown in Fig. 11.2. For this reason, we say that the cube is in a state of *pure shear stress*. An element from within the cube having rectangular faces parallel to the faces of the cube is also subjected to the same state of pure shear stress (Fig. 11.3).

Shear Modulus

The stresses on a cube subjected to pure shear stress cause a shear strain γ referred to the directions parallel to the loaded faces as shown in Fig. 11.4. If we were to perform an experiment in which we apply a pure shear stress τ to a cube and measure the resulting shear strain γ, for many materials we would find that as long as the stress is not too large, there is an approximately linear relationship between τ and γ. This relationship is written

$$\tau = G\gamma, \tag{11.1}$$

where the constant G is called the *shear modulus* or *modulus of rigidity*. Because γ is dimensionless, the dimensions of G are the same as the dimensions of τ, force/area. Typical values are given in Appendix D. In Chapter 13 we

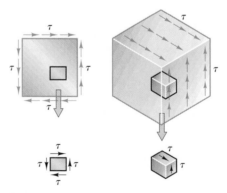

Figure 11.3
An element with faces parallel to the faces of the cube is subjected to the same state of pure shear stress.

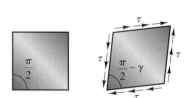

Figure 11.4
Shear strain resulting from the shear stresses on the cube.

show that the shear modulus is not independent of the modulus of elasticity E and Poisson's ratio ν, but is related to them by the equation

$$G = \frac{E}{2(1 + \nu)}. \qquad (11.2)$$

Stresses on Oblique Planes

We have described the stresses acting on the faces of a cube subjected to pure shear stress, but in design you cannot consider only the stresses on particular planes. In our discussion of bars subjected to axial forces in Chapter 10, we found that the values of the normal and shear stresses were different for planes having different orientations. Let us consider a cube of material subjected to pure shear stress and obtain a free-body diagram as shown in Fig. (11.5). The normal and shear stresses on the plane $\mathcal{P}$ are denoted by σ_θ and τ_θ Letting A_θ be the area of the slanted face of the free-body diagram, the areas of the horizontal and vertical faces in terms of A_θ and θ are shown in Fig. 11.6a. The forces on the free-body diagram are the products of the areas and the uniform stresses (Fig. 11.6b). Summing forces in the σ_θ direction,

$$\sigma_\theta A_\theta - \left(\tau A_\theta \cos\theta\right)\sin\theta - \left(\tau A_\theta \sin\theta\right)\cos\theta = 0,$$

we determine the normal stress on $\mathcal{P}$:

$$\sigma_\theta = 2\tau \sin\theta \cos\theta. \qquad (11.3)$$

Then by summing forces in the τ_θ direction,

$$\tau_\theta A_\theta - \left(\tau A_\theta \cos\theta\right)\cos\theta + \left(\tau A_\theta \sin\theta\right)\sin\theta = 0,$$

we determine the shear stress on $\mathcal{P}$:

$$\tau_\theta = \tau\left(\cos^2\theta - \sin^2\theta\right). \qquad (11.4)$$

In Fig. 11.7 we plot the ratios σ_θ/τ and τ_θ/τ obtained from Eqs. (11-3) and (11-4). We also show the free-body diagrams and values of σ_θ and τ_θ corresponding to particular values of θ. There is no value of θ for which the shear

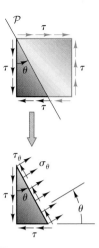

Figure 11.5
Obtaining a free-body diagram by passing a plane $\mathcal{P}$ through the cube.

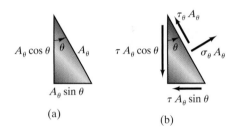

Figure 11.6
(a) Areas of the faces.
(b) Forces on the free-body diagram.

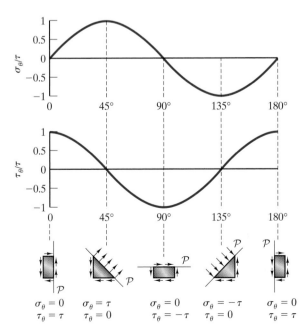

Figure 11.7
Normal and shear stresses as functions of θ.

stress is larger in magnitude than τ. Notice that although there are no normal stresses on the faces of the element in Fig. 11.3, normal stresses do occur on oblique planes. On the plane oriented at $\theta = 45°$, $\sigma_\theta = \tau$ and at $\theta = 135°$, $\sigma_\theta = -\tau$. These are the maximum tensile and compressive stresses for any value of θ. Although we have considered only a subset of planes, it can be shown that these are the maximum shear, tensile, and compressive stresses.

In summary, we have shown that if a material is subjected to a state of pure shear stress of magnitude τ, the maximum shear stress, maximum tensile stress, and maximum compressive stress each equal τ.

Study Questions

1. What is the definition of pure shear stress?
2. What is the definition of the shear modulus G of a material? If you know the modulus of elasticity E and the Poisson's ratio ν of a material, how can you determine its shear modulus?
3. If a material is subjected to a state of pure shear stress τ, what is the magnitude of the maximum shear stress to which the material is subjected? What are the values of the maximum tensile and compressive stresses to which the material is subjected?

Problems

11.1 The cube of material is subjected to a pure shear stress $\tau = 9$ MPa. The angle β is measured and determined to be 89.98°. What is the shear modulus G of the material?

P11.1

11.2 If the cube in Problem 11.1 consists of material with shear modulus $G = 4.6 \times 10^6$ psi and the shear stress $\tau = 8000$ psi, what is the angle β in degrees?

11.3 If the cube in Problem 11.1 consists of aluminum alloy that will safely support a pure shear stress of 270 MPa and $G = 26.3$ GPa, what is the largest shear strain to which the cube can safely be subjected?

11.4 The cube of material is subjected to a pure shear stress $\tau = 12$ MPa. What are the normal stress and the magnitude of the shear stress on the plane $\mathcal{P}$?

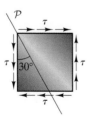

P11.4

11.5 In Problem 11.4, what are the magnitudes of the maximum tensile, compressive, and shear stresses to which the material is subjected?

11.6 The cube of material shown in Problem 11.4 is subjected to a pure shear stress τ. If the normal stress on the plane $\mathcal{P}$ is 14 MPa, what is τ?

11.7 The cube of material shown in Problem 11.4 is subjected to a pure shear stress τ. The shear modulus of the material is $G = 28$ GPa. If the normal stress on the plane $\mathcal{P}$ is 80 MPa, what is the shear strain of the cube?

11.8 The cube of material is subjected to a pure shear stress $\tau = 20$ ksi. (a) What are the normal stress and the magnitude of the shear stress on the plane $\mathcal{P}$? (b) What are the magnitudes of the maximum tensile, compressive, and shear stresses to which the material is subjected?

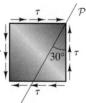

P11.8

11.9 The cube of material shown in Problem 11.8 is subjected to a pure shear stress τ. If the normal stress on the plane $\mathcal{P}$ is -20 ksi, what is τ?

11.10 The cube of material shown in Problem 11.8 is subjected to a pure shear stress τ. The shear modulus of the material is $G = 4 \times 10^6$ psi. If the normal stress on the plane $\mathcal{P}$ is -12 ksi, what is the shear strain of the cube?

11.2 Torsion of Prismatic Circular Bars

Our analysis of prismatic bars subjected to axial forces in Chapter 10 applied
to bars of arbitrary cross section. In contrast, simple analytical solutions for
the deformation and stresses in a bar subjected to axial torsion exist only for
bars with circular cross sections. We analyze such bars in this section.

Stresses and Strains

Our objective is to describe the deformation and stresses that result when a
cylindrical bar fixed at one end is subjected to an axial torque T at the free
end (Fig. 11.8). To describe the bar's deformation, we begin with a cylindrical
shell within the bar that has inner radius r and infinitesimal thickness dr
(Fig. 11.9a). We draw a radial line from the center of the end of the bar to the
cylindrical shell and extend it along the length of the shell parallel to the bar's
axis. When the torque is applied (Fig. 11.9b), *we assume that each cross section of the bar undergoes a rigid rotation.* The angle ϕ through which the end
of the bar rotates is the *angle of twist.* If ϕ is expressed in radians, the circumferential distance $c = r\phi$.

Now imagine slicing the deformed cylindrical shell along its length and
laying it out flat (Fig. 11.10). We assume that the longitudinal line viewed in
this way remains straight after the torque is applied. If the angle γ expressed

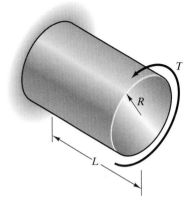

Figure 11.8
Subjecting a bar to an axial torque.

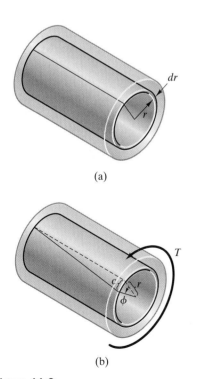

(a)

(b)

Figure 11.9
(a) Cylindrical shell within the bar.
(b) Rotations of the radial and longitudinal
lines when torque is applied.

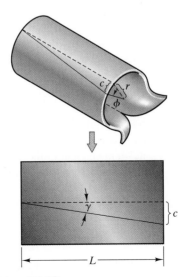

Figure 11.10
Laying the shell out flat to visualize its
deformation.

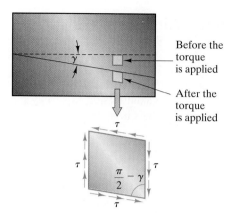

Figure 11.11
Element of the shell before and after the torque is applied.

in radians is small, we can approximate it by $\gamma = c/L$. Using the relation $c = r\phi$, we obtain

$$\gamma = \frac{r\phi}{L}. \tag{11.5}$$

Let us consider an infinitesimal rectangular element of the shell with sides parallel and perpendicular to the bar's axis before the torque is applied (Fig. 11.11). If we assume there are no extensional strains in the axial and circumferential directions, the torque subjects the element to a shear strain γ, which implies that the element is in a state of pure shear stress

$$\tau = G\gamma = \frac{Gr\phi}{L}. \tag{11.6}$$

We have given a suggestive argument (not a proof) that the applied torque subjects infinitesimal rectangular elements of the cylindrical shell oriented as shown in Fig. 11.12 to pure shear stress given by Eq. (11.6). If we pass a plane perpendicular to the shell's axis (Fig. 11.13), every element around the circumference is acted upon by the same state of shear stress, so the exposed surfaces are subjected to a uniform circumferential stress τ. By using Fig. 11.13 and Eq. (11.6), we can describe the stress distribution on a plane perpendicular to the axis of the cylindrical bar. It is subjected to a circumferential shear stress distribution τ whose magnitude is proportional to the distance from the center of the bar (Fig. 11.14).

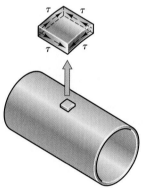

Figure 11.12
Stress distribution on an element of the shell.

We have now described the stress distribution in the bar, but Eq. (11.6) gives the shear stress in terms of the bar's angle of twist. Our next objective is to determine the shear stress in terms of the applied torque T. From the free-body diagram in Fig. 11.14, equilibrium requires that the moment about the bar's axis due to the shear stress distribution must equal T. Let dA be an infinitesimal element of the bar's cross-sectional area at a distance r from the center. The moment about the bar's axis due to the shear stress acting on dA is $r\tau\, dA$ (Fig. 11.15). Integrating to determine the total moment due to the shear stress, we obtain the equilibrium equation

$$\int_A r\tau\, dA = T.$$

Substituting Eq. (11.6) into this equation, we can solve for the bar's angle of twist ϕ. The result is

$$\phi = \frac{TL}{GJ}, \tag{11.7}$$

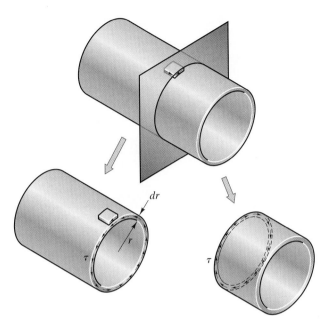

Figure 11.13
Passing a plane perpendicular to the shell's axis.

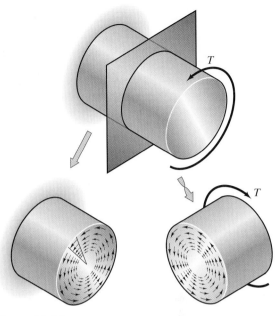

Figure 11.14
Stress distribution on a plane perpendicular to the axis of the bar.

where

$$J = \int_A r^2 dA \qquad (11.8)$$

is the polar moment of inertia of the bar's cross-sectional area about its axis. [Notice the similarity between Eq. (11.7) and the equation $\delta = PL/EA$ for the change in length of a bar subjected to axial forces.] We can now substitute Eq. (11.7) back into Eq. (11.6), obtaining an expression for the shear stress in terms of the applied torque T and the distance r from the bar's axis:

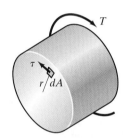

Figure 11.15
Calculating the moment about the bar's axis due to the shear stress distribution.

$$\tau = \frac{Tr}{J}. \qquad (11.9)$$

We have attempted to make Eqs. (11.7) and (11.9) plausible, but we have not proven that they determine the angle of twist and distribution of stress in a solid or hollow circular bar. Although the proof is beyond our scope, it can be shown that if such a bar could be loaded at the ends by shear stress distributions satisfying Eq. (11.9), our expressions for the angle of twist and the shear stress distribution within the bar are the exact and unique solutions. Furthermore, no matter how the torques are applied, the stress distribution at axial distances from the ends greater than a few times the bar's width is given approximately by Eq. (11.9). This is another example of Saint-Venant's principle (see Section 10.1). As a consequence, Eq. (11.7) approximates the angle of twist of a slender bar no matter how the torques are applied at the ends. These results require that the material have the property of isotropy, which we discuss in Chapter 13.

You can perform a simple experiment that dramatically demonstrates the effect of the state of stress in a bar subjected to torsion. Apply torsion to a

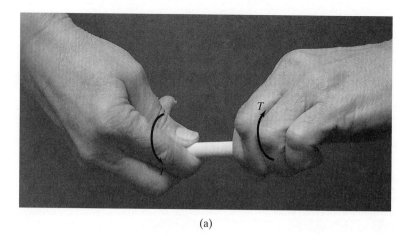

(a)

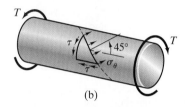

(b)

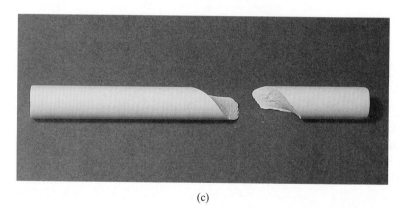

(c)

Figure 11.16
(a) Applying axial torques to a piece of chalk.
(b) The maximum tensile stress occurs at 45° relative to the longitudinal axis.
(c) Failure occurs along the plane subjected to the maximum tensile stress.

piece of blackboard chalk as shown in Fig. 11.16a. The maximum tensile stress resulting from the state of pure shear stress you apply occurs on planes at 45° relative to the chalk's axis (see Figs. 11.7 and 11.16b). Brittle materials such as chalk are weak in tension, and you will observe that the chalk fails along a clearly defined 45° spiral line (Fig. 11.16c).

Polar Moment of Inertia

To evaluate J, we can use an annular element of area with radius r and thickness dr (Fig. 11.17). The area of the element is the product of its circumference and thickness, $dA = 2\pi r\, dr$, so the polar moment of intertia is

$$J = \int_A r^2\, dA = \int_0^R r^2 (2\pi r\, dr).$$

Integrating, we obtain

Figure 11.17
Element of area for calculating J.

$$J = \frac{\pi}{2} R^4. \qquad \text{solid circular cross section} \qquad (11.10)$$

Equations (11.7) and (11.9) also apply to a bar with a hollow circular cross section. In that case the polar moment of inertia is given by

$$J = \frac{\pi}{2}\left(R_o^4 - R_i^4\right), \qquad \text{hollow circular cross section} \qquad (11.11)$$

where R_i are R_o the bar's inner and outer radii.

Positive Directions of the Torque and Angle of Twist

Equations (11.7) and (11.9), with the polar moment of inertia J evaluated using Eq. (11.10) or (11.11), determine the angle of twist and distribution of shear stress for a circular bar subjected to a torque T. For some applications it will be convenient to have sign conventions for the torque and angle of twist.

In Fig. 11.18a we isolate a segment of a bar and indicate the positive directions of the torque acting on the segment: If you point the thumb of your right hand outward from the cross section under consideration, your fingers point in the direction of positive torque. We define the angle of twist of a segment of a bar to be positive if it is in the direction resulting from a positive torque (Fig. 11.18b).

Study Questions

A cylindrical bar of length L, radius R, and shear modulus G is fixed at one end and subjected to an axial torque T at the other end.

1. How can you determine the resulting angle of twist of the end of the bar?
2. What is the maximum shear stress in the bar, and where does it occur?
3. What is the maximum tensile stress in the bar, and where does it occur?
4. What is the shear stress at a point on the bar's axis?

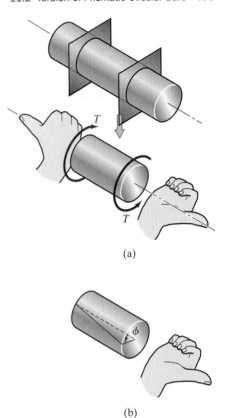

(a)

(b)

Figure 11.18
(a) Positive directions of the torque.
(b) Segment with a positive angle of twist ϕ.

Example 11.1

Bar Subjected to Torsion

The bar in Fig. 11.19 consists of material with shear modulus $G = 28$ GPa and has a solid circular cross section. Part A is 40 mm in diameter and part B is 20 mm in diameter.
(a) Determine the magnitudes of the maximum shear stresses in parts A and B.
(b) Determine the angle of twist of the right end of the bar relative to the wall.

Strategy

By passing a plane through part A of the bar and isolating the part of the bar on one side of the plane, we can determine the internal torque in part A. We can then use Eq. (11.9) to determine the maximum shear stress in part A and use Eq. (11.7) to determine the angle of twist of part A. We can determine the maximum shear stress in part B and the angle of twist of part B in the same way.

Solution

(a) In Fig. (a) we draw the free-body diagram of the entire bar.
From the equilibrium equation $400 - 1200 - T = 0$, we find that the torque exerted on the bar by the wall is $T = -800$ N-m. In Fig. (b) we pass a plane through part A and isolate the part of the bar to the left of the plane.

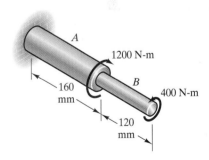

Figure 11.19

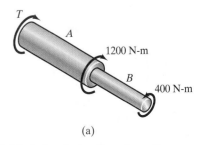

(a)

(a) Isolating the bar from the wall.

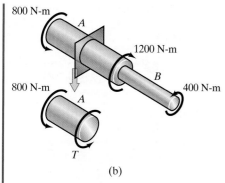

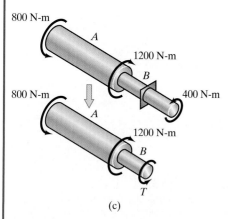

(b) Obtaining a free-body diagram by passing a plane through part A.

(c) Obtaining a free-body diagram by passing a plane through part B.

From the equilibrium equation $T + 800 = 0$, the torque in part A of the bar is $T = -800$ N-m. The polar moment of inertia of the cross section in part A is

$$J = \frac{\pi}{2} R^4 = \frac{\pi}{2} (0.02)^4 = 2.51 \times 10^{-7} \text{ m}^4.$$

The maximum shear stress in part A occurs at $r = 0.02$ m. Therefore, the magnitude of the maximum shear stress is

$$\tau = \frac{|T|r}{J} = \frac{(800)(0.02)}{2.51 \times 10^{-7}} = 63.7 \text{ MPa}.$$

In Fig. (c) we pass a plane through part B and isolate the part of the bar to the left of the plane.

From the equilibrium equation $T - 1200 + 800 = 0$, the torsion in part B is $T = 400$ N-m The polar moment of inertia of the cross section in part B is

$$J = \frac{\pi}{2} R^4 = \frac{\pi}{2} (0.01)^4 = 1.57 \times 10^{-8} \text{ m}^4.$$

The maximum shear stress in part B occurs at $r = 0.01$ m. The magnitude of the maximum shear stress in part B is

$$\tau = \frac{|T|r}{J} = \frac{(400)(0.01)}{1.57 \times 10^{-8}} = 255 \text{ MPa}.$$

(b) The torque in part A of the bar is $T = -800$ N-m, so the angle of twist of part A is

$$\phi_{\text{part } A} = \frac{TL}{GJ} = \frac{(-800)(0.16)}{(28 \times 10^9)(2.51 \times 10^{-7})} = -0.0182 \text{ rad} = -1.04°.$$

This is the angle of twist of the right end of part A relative to the wall [Fig. (d)]. The torque in part B is $T = 400$ N-m, so the angle of twist of part B is

$$\phi_{\text{part } B} = \frac{TL}{GJ} = \frac{(400)(0.12)}{(28 \times 10^9)(1.57 \times 10^{-8})} = 0.1091 \text{ rad} = 6.25°.$$

This is the angle of twist of the right end of part B relative to the end attached to part A [Fig. (e)]. The angle of twist of the right end of part B relative to the wall is

$$\phi = \phi_{\text{part } A} + \phi_{\text{part } B} = -1.04 + 6.25 = 5.21°.$$

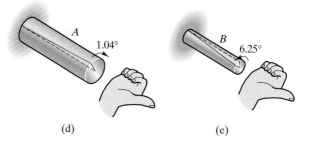

(d) Angle of twist of part A.
(e) Angle of twist of part B.

Problems

11.11 If a bar has a solid circular cross section with 15-mm diameter, what is the polar moment of inertia of its cross section in m^4?

11.12 If a bar has a hollow circular cross section with 2-in. outer radius and 1-in. inner radius, what is the polar moment of inertia of its cross section?

11.13 The bar has a circular cross section with 15-mm diameter and the shear modulus of the material is $G = 26$ GPa. If the torque $T = 10$ N-m, determine (a) the magnitude of the maximum shear stress in the bar; (b) the angle of twist of the end of the bar in degrees.

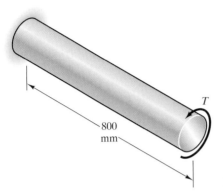

P11.13

11.14 If the bar in Problem 11.13 is subjected to a torque T that causes the end of the bar to rotate $4°$, what is the magnitude of the maximum shear stress in the bar?

11.15 The bar in Problem 11.13 is to be used in an application that requires that it be subjected to an angle of twist no greater than $1°$. What is the maximum allowable value of the torque T?

11.16 The solid circular shaft that connects the turbine blades of the hydroelectric power unit to the generator has a 0.4-m radius and supports a torque $T = 2$ MN-m. What is the maximum shear stress in the shaft?

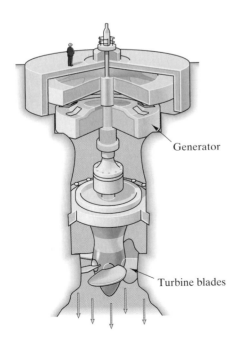

Generator

Turbine blades

P11.16

11.17 Consider the solid circular shaft in Problem 11.16. The shear modulus of the material is $G = 80$ GPa. What angle of twist per unit meter of length is caused by the 2-MN-m torque?

11.18 If the shaft in Problem 11.16 has a hollow circular cross section with 0.5-m outer radius and 0.3-m inner radius, what is the maximum shear stress?

11.19 The propeller of a wind generator is supported by a hollow circular shaft with 0.4-m outer radius and 0.3-m inner radius. The shear modulus of the material is $G = 80$ GPa. If the propeller exerts an 840 kN-m torque on the shaft, what is the resulting maximum shear stress?

P11.19

11.20 In Problem 11.19, what is the angle of twist of the propeller shaft per meter of length?

11.21 In designing a new shaft for the wind generator in Problem 11.19 the engineer wants to limit the maximum shear stress in the shaft to 10 MPa, but design constraints require retaining the 0.4-m outer radius. What new inner radius should she use?

11.22 The bar has a circular cross section with 1-in. diameter and the shear modulus of the material is $G = 5.8 \times 10^6$ psi. If the torque $T = 1000$ in-lb, determine (a) the magnitude of the maximum shear stress in the bar; (b) the magnitude of the angle of twist of the right end of the bar relative to the wall in degrees.

500 in-lb

T

8 in.

6 in.

P11.22

11.23 For the bar in Problem 11.22, what value of the torque T would cause the angle of twist of the end of the bar to be zero?

11.24 Part A of the bar has a solid circular cross section and part B has a hollow circular cross section. The shear modulus of the material is $G = 3.8 \times 10^6$ psi. Determine the magnitudes of the maximum shear stresses in parts A and B of the bar.

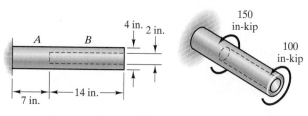

A B 4 in. 2 in.

150 in-kip

100 in-kip

7 in.   14 in.

P11.24

11.25 For the bar in Problem 11.24, determine the magnitude of the angle of twist of the end of the bar relative to the wall in degrees.

11.26 For the bar in Problem 11.24, determine the magnitudes of the maximum shear stresses in parts A and B of the bar and the magnitude of the angle of twist of the end of the bar in degrees if the 150 in-kip couple acts in the opposite direction.

11.27 The lengths $L_A = L_B = 200$ mm and $L_C = 240$ mm. The diameter of parts A and C of the bar is 25 mm and the diameter of part B is 50 mm. The shear modulus of the material is $G = 80$ GPa. If the torque $T = 2.2$ kN-m, determine the magnitude of the angle of twist of the right end of the bar relative to the wall in degrees.

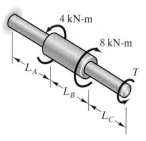

4 kN-m

8 kN-m

L_A

L_B

T

L_C

P11.27

11.28 For the bar in Problem 11.27, what value of the torque T would cause the angle of twist of the right end of the bar relative to the wall to be zero?

11.29 The bar in Problem 11.27 is made of a material that can safely support a pure shear stress of 1.1 GPa. Based on this criterion, what is the range of positive values of the torque T that can safely be applied?

11.30 The bars AB and CD each have a solid circular cross section with 30-mm diameter, are each 1 m in length, and consist of a material with shear modulus $G = 28$ GPa. The radii of the gears are $r_B = 120$ mm and $r_C = 90$ mm. If the torque $T_A = 200$ N-m, what are the maximum shear stresses in the bars?

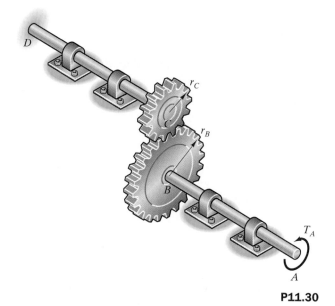

D

r_C

r_B

B

T_A

A

P11.30

11.31 In Problem 11.30 what is the angle of rotation at A? (Assume that the deformations of the gears are negligible.)

11.32 Consider the system shown in Problem 11.30. The bars AB and CD each have a solid circular cross section with 30-mm diameter. The radii of the gears must satisfy the relation $r_B + r_C = 210$ mm. If the torque $T_A = 200$ N-m and the bars are made of a material that will safely support a pure shear stress of 40 MPa, what is the largest safe value of the radius r_C?

11.3 Statically Indeterminate Problems

The approach we used in Chapter 10 for solving statically indeterminate problems involving axially loaded bars—applying equilibrium, force–deformation relations, and compatibility—applies to virtually all statically indeterminate problems. In this section we demonstrate the solution of statically indeterminate problems involving bars subjected to torsion. The force–deformation relations will now be relations between torques and angles of twist, and the compatibility conditions will be constraints imposed on the angles of twist of torsionally loaded bars.

To emphasize that you can use the same procedure you applied to axially loaded bars, we present an example equivalent to the one discussed in Section 10.3 and solve it by the same steps. (You should compare the two solutions.) The bar in Fig. 11.20a consists of two segments A and B with different lengths and diameters. It is fixed at both ends and subjected to an axial torsion. We draw its free-body diagram in Fig. 11.20b. The equilibrium equation is

$$T_0 - T_A + T_B = 0 \tag{11.12}$$

We cannot determine the two reactions T_A and T_B from this equation. The problem is statically indeterminate.

The torque in part A of the bar is T_A (Fig. 11.21a), so the angle of twist of part A is

$$\phi_A = \frac{T_A L_A}{GJ_A}. \tag{11.13}$$

The torque in part B is T_B (Fig. 11.21b). Its angle of twist is

$$\phi_B = \frac{T_B L_B}{GJ_B}. \tag{11.14}$$

Since the bar is fixed at both ends, the compatibility condition is that the angle of twist of the entire bar equals zero:

$$\phi_A + \phi_B = 0. \tag{11.15}$$

We substitute Eqs. (11.13) and (11.14) into this equation, obtaining

$$\frac{T_A L_A}{GJ_A} + \frac{T_B L_B}{GJ_B} = 0. \tag{11.16}$$

We can solve this equation simultaneously with the equilibrium equation to determine the reactions T_A and T_B. The results are

$$T_A = \frac{T_0}{1 + L_A J_B / L_B J_A}, \qquad T_B = \frac{-T_0}{1 + L_B J_A / L_A J_B}.$$

Now that we know the reactions, we can determine the angle of twist of each part of the bar from Eqs. (11.13) and (11.14). We can also determine the shear stress distribution in each part of the bar: $\tau_A = T_A r / J_A$ and $\tau_B = T_B r / J_B$.

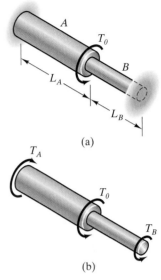

(a)

(b)

Figure 11.20
(a) Torsionally loaded bar fixed at both ends.
(b) Free-body diagram of the bar.

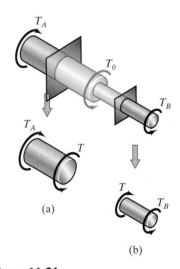

(a)

(b)

Figure 11.21
Determining the internal torques in parts A and B.

Problems

11.33 The bar has a circular cross section with 1-in. diameter. If the torque $T_O = 1000$ in-lb, determine the magnitudes of the maximum shear stresses in parts A and B of the bar.

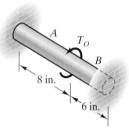

P11.33

11.34 Suppose that the bar in Problem 11.33 consists of a material that will safely support a maximum shear stress of 40 ksi. Based on this criterion, what is the maximum safe magnitude of the torque T_O?

11.35 Suppose that the bar in Problem 11.33 is subjected to a torque $T_O = 10,000$ in-lb and consists of a material that will safely support a maximum shear stress of 40 ksi. Based on this criterion, what is the largest distance from the left end of the bar at which the torque can safely be applied?

11.36 The bar is fixed at both ends. It consists of material with shear modulus $G = 28$ GPa and has a solid circular cross section. Part A is 40 mm in diameter and part B is 20 mm in diameter. Determine the torques exerted on the bar by the walls.

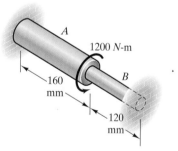

P11.36

11.37 Determine the magnitudes of the maximum shear stresses in parts A and B of the bar in Problem 11.36.

11.38 Each bar is 10 in. long and has a solid circular cross section. Bar A has a diameter of 1 in. and its shear modulus is 6×10^6 psi. Bar B has a diameter of 2 in. and its shear modulus is 3.8×10^6 psi. The ends of the bars are separated by a small gap. The free end of bar A is rotated 2° about the bar's axis and the bars are welded together. What are the magnitudes of the angles of twist (in degrees) of the two bars afterward?

P11.38

11.39 In Problem 11.38, the ends of the bars are separated by a small gap. Suppose that the free end of bar A is rotated 2° about its axis, the free end of bar B is rotated 2° about its axis in the opposite direction, and the bars are welded together. What are the magnitudes of the maximum shear stresses in the two bars afterward?

11.40 The lengths $L_A = L_B = 200$ mm and $L_C = 240$ mm. The diameter of parts A and C of the bar is 25 mm and the diameter of part B is 50 mm. The shear modulus of the material is $G = 80$ GPa. What is the magnitude of the maximum shear stress in the bar?

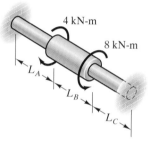

P11.40

11.41 In Problem 11.40, through what angle does the bar rotate at the position where the 8-kN-m couple is applied?

11.42 A collar is rigidly attached to bar A. The cylindrical bar A is 80 mm in diameter and its shear modulus is $G = 66$ GPa. There are gaps $b = 2$ mm between the arms of the collar and the ends of the identical bars B and C. Bars B and C are 30 mm in diameter and their modulus of elasticity is $E = 170$ GPa. If the bars B and C are stretched until they come into contact with the arms of the collar and are welded to them, what is the magnitude of the maximum shear stress in bar A afterward?

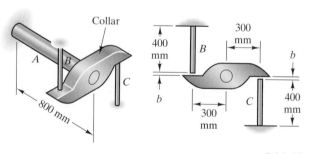

P11.42

11.4 Nonprismatic Bars and Distributed Loads

In this section we discuss torsion of circular bars whose diameters vary with distance along their axes and also bars subjected to torsional loads that are distributed along their axes. The derivations closely follow our treatment in Section 10.4 of analogous applications involving axially loaded bars.

Bars with Gradually Varying Cross Sections

Figure 11.22a shows a bar with a hollow circular cross section whose inner and outer radii vary with distance along the bar's axis. The polar moment of inertia of the cross section depends upon x, which we indicate by writing it as $J(x)$. If we subject the bar to a torsional load (Fig. 11.22b) and the change in $J(x)$ with x is gradual, we can approximate the stress distribution on a plane at axial position x by using Eq. (11.9):

$$\tau = \frac{Tr}{J(x)}. \tag{11.17}$$

For a given perpendicular plane, the stress distribution is approximately the same as for the case of a prismatic bar. But since $J(x)$ varies with distance along the bar's axis, the shear stress distribution does also.

To determine the bar's angle of twist, we begin by considering an infinitesimal element of the bar of length dx (Fig. 11.23). We can use Eq. (11.7) to determine the element's angle of twist:

$$\phi_{\text{element}} = \frac{T\,dx}{GJ(x)}.$$

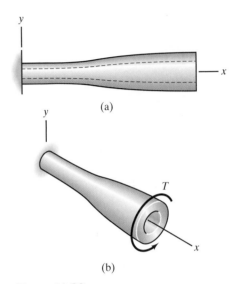

Figure 11.22
(a) Bar with a varying cross section.
(b) Subjecting the bar to a torque T.

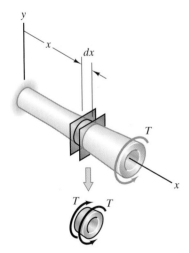

Figure 11.23
Determining the angle of twist of an infinitesimal element of the bar.

We obtain the angle of twist of the entire bar by integrating this expression from $x = 0$ to $x = L$:

$$\phi = \int_0^L \frac{T\,dx}{GJ(x)}. \tag{11.18}$$

The angle of twist of each element depends on its polar moment of inertia, and we must add up the angles of twist of the elements to determine the angle of twist of the bar.

Example 11.2

Torsion of a Nonprismatic Bar

The bar in Fig. 11.24 has a solid circular cross section and consists of material with shear modulus $G = 47$ GPa. The bar's polar moment of inertia is given by the equation $J(x) = 0.00016 + 0.0006x^2$ m^4. If the bar is subjected to a torque $T = 200$ kN-m at its free end, determine (a) the magnitude of the maximum shear stress in the bar at $x = 1$ m; (b) the magnitude of the angle of twist of the entire bar in degrees.

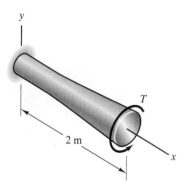

Figure 11.24

Strategy

(a) From the given equation for $J(x)$ we can determine the polar moment of inertia and the bar's radius R at $x = 1$ m. Then the maximum shear stress is $TR/J(x)$.
(b) Since the bar's polar moment of inertia is a function of x, we must determine the bar's angle of twist from Eq. (11.18).

Solution

(a) The polar moment of inertia at $x = 1$ m is

$$J(1) = 0.00016 + 0.0006(1)^2 = 0.00076 \text{ m}^4.$$

Solving the equation $J = (\pi/2)R^4$ for the bar's radius at $x = 1$ m, we obtain $R = 0.148$ m. Therefore, the magnitude of the maximum shear stress at $x = 1$ m is

$$|\tau| = \frac{TR}{J(1)} = \frac{(200,000)(0.148)}{0.00076} = 39.0 \text{ MPa}.$$

(b) From Eq. (11.18) the magnitude of the bar's angle of twist is

$$|\phi| = \int_0^L \frac{T\,dx}{GJ(x)} = \int_0^2 \frac{(200,000)\,dx}{(47 \times 10^9)(0.00016 + 0.0006x^2)} = 0.0181 \text{ rad},$$

which is 1.04°.

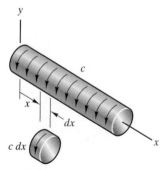

Figure 11.25
Describing a distributed torque by a function. The torque on an element of length dx is $c\,dx$.

Distributed Torsional Loads

We can describe a distributed torsional load on a bar by introducing a function c defined such that the axial torque on each element of the bar is $c\,dx$ (Fig. 11.25). Since the product of c and dx is a moment, the dimensions c are moment/length. For example, the bar in Fig. 11.26a is fixed at the left end and subjected to a distributed axial torque throughout its length. In Fig. 11.26b we pass a plane through the bar at an arbitrary position x and draw the free-body diagram of the part of the bar to the right of the plane. From the equilibrium equation

$$-T + \int_x^L c_0\left(\frac{x}{L}\right)^2 dx = 0,$$

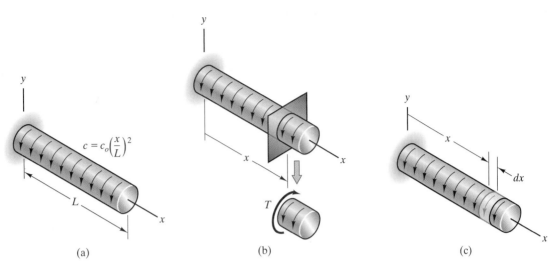

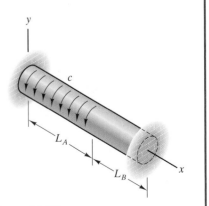

Figure 11.26
(a) Bar subjected to a distributed torque.
(b) Determining the internal torque at x.
(c) Element of length dx.

we determine the internal torque in the bar at the position x:

$$T = \frac{c_0}{3}\left(L - \frac{x^3}{L^2}\right).$$

The distributed torque causes the internal torque in the bar to vary with axial position. To determine the angle of twist of the right end of the bar relative to the wall, we consider an element of the bar of length dx (Fig. 11.26c). From Eq. (11.7), its angle of twist is

$$\phi_{\text{element}} = \frac{T\,dx}{GJ} = \frac{c_0}{3GJ}\left(L - \frac{x^3}{L^2}\right)dx.$$

Integrating this result from $x = 0$ to $x = L$, we obtain the angle of twist of the right end:

$$\phi = \int_0^L \frac{c_0}{3GJ}\left(L - \frac{x^3}{L^2}\right)dx = \frac{c_0 L^2}{4GJ}.$$

Example 11.3

Bar with Distributed Torsional Load

The cylindrical bar in Fig. 11.27 is fixed at both ends and is subjected to a uniform distributed torque c from $x = 0$ to $x = L_A$. What torques are exerted on the bar by the walls?

Strategy

By using the equilibrium equation for the entire bar and the compatibility condition that the angle of twist of the right end of the bar relative to the left end is zero, we can solve for the two torques exerted on the bar by the walls.

Figure 11.27

Solution

In Fig. (a) we draw the free-body diagram of the entire bar.

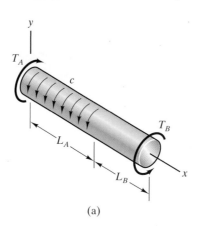

(a)

(a) Free-body diagram of the bar.

Because c is constant, the total moment exerted by the distributed torque is the product of c and L_A. The equilibrium equation is

$$-T_A + T_B + cL_A = 0.$$

We cannot solve this equation for T_A and T_B; the problem is statically indeterminate. In Fig. (b) we pass a plane through part A of the bar at an arbitrary position x. From the equilibrium equation $-T_A + T + cx = 0$, the internal torque in part A of the bar is

$$T = T_A - cx.$$

Consider an element of part A of the bar that has length dx and is located at the position x. The angle of twist of this element is

$$d\phi_A = \frac{T \, dx}{GJ} = \frac{(T_A - cx) \, dx}{GJ}.$$

Integrating this expression to determine the angle of twist of the right end of part A relative to the left end, we obtain

$$\phi_A = \int_0^{L_A} \frac{T \, dx}{GJ} = \int_0^{L_A} \frac{(T_A - cx) \, dx}{GJ} = \frac{T_A L_A}{GJ} - \frac{cL_A^2}{2GJ}.$$

The internal torque in part B of the bar is $T = T_B$ [Fig. (c)], so the angle of twist of the right end of part B relative to the left end of part B is

$$\phi_B = \frac{T_B L_B}{GJ}.$$

The compatibility condition is

$$\phi_A + \phi_B = \frac{T_A L_A}{GJ} - \frac{cL_A^2}{2GJ} + \frac{T_B L_B}{GJ} = 0.$$

Solving this equation simultaneously with the equilibrium equation, we obtain the torques exerted by the walls:

$$T_A = \frac{1/2 + L_B/L_A}{1 + L_B/L_A} cL_A, \qquad T_B = -\frac{1/2}{1 + L_B/L_A} cL_A.$$

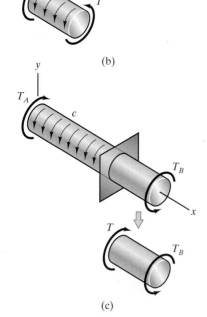

(b)

(c)

(b) Determining the internal torque in part A.
(c) Determining the internal torque in part B.

Problems

11.43 In Example 11.2, what is the magnitude of the maximum shear stress in the bar?

11.44 In Example 11.2, suppose that the torque T is applied to the bar at $x = 1$ m. What is the magnitude of the angle of twist of the entire bar?

11.45 The bar has a solid circular cross section. Its polar moment of inertia is given by $J = (0.1 + 0.15x)$ in^4, where x is the axial position in inches, and the shear modulus of the material is $G = 4.6 \times 10^6$ psi. If the bar is subjected to an axial torque $T = 20$ in-kip, what is the magnitude of the maximum shear stress at $x = 6$ in.?

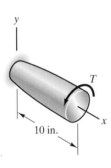

P11.45

11.46 What is the angle of twist (in degrees) of the entire bar in Problem 11.45?

11.47 Suppose that an axial hole is drilled through the bar in Problem 11.45 so that it has a hollow circular cross section with inner radius $r_i = 0.3$ in. What is the angle of twist (in degrees) of the entire bar due to the 20-in-kip torque?

11.48 The radius of the bar's circular cross section varies linearly from 10 mm at $x = 0$ to 5 mm at $x = 150$ mm. The shear modulus of the material is $G = 17$ GPa. What torque T would cause a maximum shear stress of 10 MPa at $x = 80$ mm?

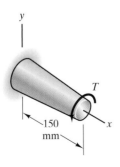

P11.48

11.49 In Problem 11.48, what torque T would cause the end of the bar to rotate $1°$?

11.50 In Problem 11.48, suppose that the torque T at the end of the bar is 20 N-m and you want to apply a torque in the opposite direction at $x = 75$ mm so that the angle through which the end of the bar rotates is zero. What is the magnitude of the torque you must apply?

11.51 Bars A and B have solid circular cross sections and consist of material with shear modulus $G = 17$ GPa. Bar A is 150 mm long and its radius varies linearly from 10 mm at its left end to 5 mm at its right end. The prismatic bar B is 100 mm long and its radius is 5 mm. There is a small gap between the bars. The end of bar A is given an axial rotation of $1°$ and the bars are welded together. What is the torque in the bars afterward?

P11.51

11.52 The aluminum alloy bar has a circular cross section with 20-mm diameter, length $L = 120$ mm, and a shear modulus of 28 GPa. If the distributed torque is uniform and causes the end of the bar to rotate $0.5°$, what is the magnitude of the maximum shear stress in the bar?

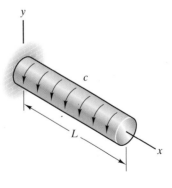

P11.52

11.53 If the distributed torque in Problem 11.52 is given by the equation $c = c_0(x/L)^3$ and causes the end of the bar to rotate $0.5°$, what is the magnitude of the maximum shear stress in the bar?

11.54 A cylindrical bar with 1-in. diameter fits tightly in a circular hole in a 5-in.-thick plate. The shear modulus of the material is

$G = 5.6 \times 10^6$ psi. A 12,000-in-lb axial torque is applied at the left end of the bar. The distributed torque exerted on the bar by the plate is given by the equation

$$c = c_0\left[1 - \left(\frac{x}{5}\right)^{1/2}\right] \text{in-lb/in.,}$$

where c_0 is a constant and x is the axial position in inches measured from the left side of the plate. Determine the constant c_0 and the magnitude of the maximum shear stress in the bar at $x = 2$ in.

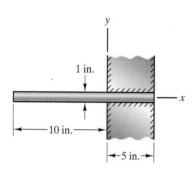

1 in.

10 in.

5 in.

P11.54

11.55 In Problem 11.54, what is the magnitude of the angle of twist of the left end of the bar relative to its right end?

11.56 The aluminum alloy bar has a circular cross section with 20-mm diameter and a shear modulus of 28 GPa. What is the magnitude of the maximum shear stress in the bar due to the uniformly distributed torque?

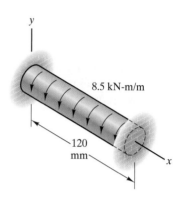

8.5 kN-m/m

120 mm

P11.56

Chapter Summary

Pure Shear Stress

In Fig. (a) a cube is subjected to a state of pure shear stress. If the stress is not too large, for many materials it is related to the shear strain by

$$\tau = G\gamma, \qquad \text{Eq. (11.1)}$$

where the constant G is the *shear modulus*. Typical values of G are given in Appendix D. The shear modulus is related to the modulus of elasticity and Poisson's ratio by

$$G = \frac{E}{2(1 + v)}. \qquad \text{Eq. (11.2)}$$

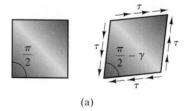

(a)

Stresses on Oblique Planes

The normal and shear stresses on a plane $\mathcal{P}$ oriented as shown in Fig. (b) are

$$\sigma_\theta = 2\tau \sin\theta \cos\theta, \qquad \text{Eq. (11.3)}$$
$$\tau_\theta = \tau(\cos^2\theta - \sin^2\theta). \qquad \text{Eq. (11.4)}$$

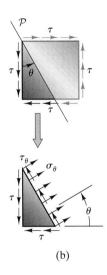

(b)

There is no value of θ for which the shear stress is larger in magnitude than τ. The maximum tensile and compressive stresses occur at $\theta = 45°$ where $\sigma_\theta = \tau$ and at $\theta = 135°$ where $\sigma_\theta = -\tau$.

Torsion of Prismatic Circular Bars

Consider a cylindrical bar of length L and radius R subjected to an axial torque T [Fig. (c)]. The resulting angle of twist of the end of the bar is

$$\phi = \frac{TL}{GJ},$$ Eq. (11.7)

where

$$J = \frac{\pi}{2} R^4$$ Eq. (11.10)

is the polar moment of inertia of the bar's cross-sectional area about its axis. The shear stress on the plane shown at a distance r from the bar's axis is

$$\tau = \frac{Tr}{J}.$$ Eq. (11.9)

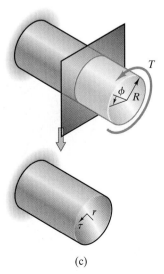

(c)

Equations (11.7) and (11.9) also apply to a bar with a hollow circular cross section. In that case the polar moment of inertia is given by

$$J = \frac{\pi}{2}\left(R_o^4 - R_i^4\right),$$ Eq. (11.11)

where R_i and R_o are the bar's inner and outer radii.

Statically Indeterminate Problems

Solutions to problems involving torsionally loaded bars in which the number of unknown reactions exceeds the number of independent equilibrium equations involve three elements: (1) relations between the torques in bars and their angles of twist; (2) compatibility relations imposed on the angles of twist by the geometry of the problem; and (3) equilibrium.

Bars with Gradually Varying Cross Sections

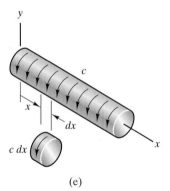

If the polar moment of inertia is a gradually varying function of axial position $J(x)$ [Fig. (d)], the stress distribution on a plane perpendicular to the bar's axis at axial position x can be approximated by

$$\tau = \frac{Tr}{J(x)}.$$ Eq. (11.17)

The angle of twist of the bar is

$$\phi = \int_0^L \frac{T\,dx}{GJ(x)}.$$ Eq. (11.18)

Distributed Torsional Loads

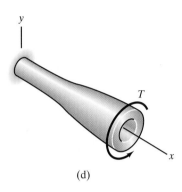

A distributed torsional load on a bar can be described by a function c defined such that the axial torque on each element dx of the bar is $c\,dx$ [Fig. (e)].

(d)

(e)

Review Problems

11.57 One type of high-strength steel drill pipe used in drilling oil wells has a 5-in. outside diameter and 4.28-in. inside diameter. If the steel will safely support a shear stress of 95 ksi, what is the largest torque to which the pipe can safely be subjected?

11.58 The drill pipe described in Problem 11.57 has a shear modulus $G = 12 \times 10^6$ psi. If it is used to drill an oil well 20,000 ft deep and the drilling operation subjects the bottom of the pipe to a torque $T = 7500$ in-lb, what is the resulting angle of twist (in degrees) of the 20,000-ft pipe?

11.59 The radius $R = 200$ mm. The infinitesimal element is at the surface of the bar. What are the normal stress and the magnitude of the shear stress on the plane $\mathcal{P}$?

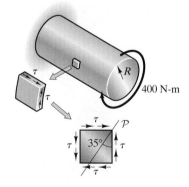

P11.59

11.60 For the element in Problem 11.59, determine the normal stress and the magnitude of the shear stress on the plane $\mathcal{P}$ shown.

P11.60

11.61 Part A of the bar has a solid circular cross section and part B has a hollow circular cross section. The bar is fixed at both ends and the shear modulus of the material is $G = 3.8 \times 10^6$ psi. Determine the torques exerted on the bar by the walls.

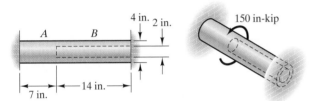

P11.61

11.62 Determine the magnitudes of the maximum shear stresses in parts A and B of the bar in Problem 11.61.

11.63 Suppose that you want to decrease the weight of the bar in Problem 11.61 by increasing the inside diameter of part B. The bar is made of material that will safely support a pure shear stress of 10 ksi. Based on this criterion, what is the largest safe value of the inside diameter?

11.64 The bar has a circular cross section with polar moment of inertia J and shear modulus G. The distributed torque $c = c_0(x/L)^2$. What are the magnitudes of the torques exerted on the bar by the left and right walls?

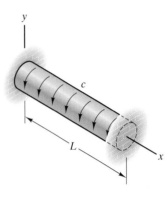

P11.64

11.65 In Problem 11.64, at what axial position x is the bar's angle of twist the greatest, and what is its magnitude?

11.66 If the bar in Problem 11.64 is acted upon by the distributed load $c = c_0(x/L)^2$ from $x = 0$ to $x = L/2$ and is free of external torque from $x = L/2$ to $x = L$, what are the magnitudes of the torques exerted on the bar by the left and right walls?

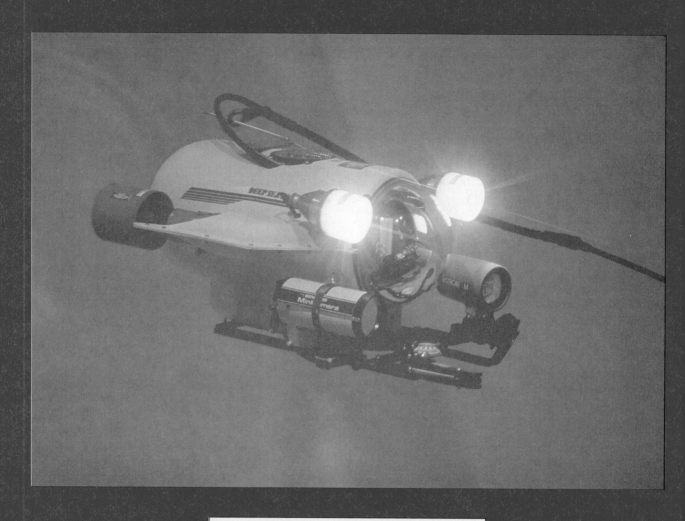

The hull of the deep submersible vehicle must be designed to support the large stresses resulting from the pressure of the surrounding water.

CHAPTER 12

States of Stress

I n this chapter we explain how to completely describe the state of stress at a point of an object such as a submersible vehicle's hull, and show that the state of stress can be used to determine the normal and shear stresses acting on an arbitrary plane through that point. Because of the importance of the maximum values of the normal and shear stresses in design, we describe how to determine maximum stresses and the orientations of the planes on which they act.

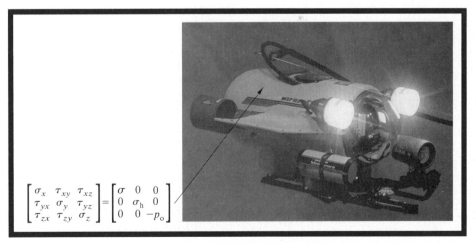

$$\begin{bmatrix} \sigma_x & \tau_{xy} & \tau_{xz} \\ \tau_{yx} & \sigma_y & \tau_{yz} \\ \tau_{zx} & \tau_{zy} & \sigma_z \end{bmatrix} = \begin{bmatrix} \sigma & 0 & 0 \\ 0 & \sigma_h & 0 \\ 0 & 0 & -p_o \end{bmatrix}$$

12.1 Components of Stress

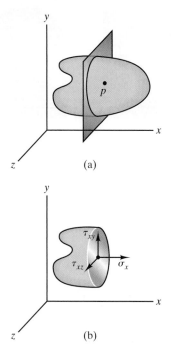

Figure 12.1
(a) Passing a plane perpendicular to the x axis.
(b) Normal stress and components of the shear stress at p.

Suppose that we are interested in the stresses acting at a point p. Let us introduce a coordinate system and pass a plane through p that is perpendicular to the x axis (Fig. 12.1a). The normal stress acting on this plane at p is denoted by σ_x (Fig. 12.1b). The shear stress may act at any direction parallel to the y–z plane and therefore may have components in both the y and z directions. These components are denoted by τ_{xy} (the shear stress on the plane perpendicular to the x axis that acts in the y direction) and τ_{xz} (the shear stress on the plane perpendicular to the x axis that acts in the z direction). Next, we pass a plane through p that is perpendicular to the y axis (Fig. 12.2a). The normal stress on this plane is σ_y and the components of the shear stress are τ_{yx} and τ_{yz} (Fig. 12.2b). Finally, we pass a plane through p perpendicular to the z axis (Fig. 12.3a). The normal stress is σ_z and the components of the shear stress are τ_{zx} and τ_{zy} (Fig. 12.3b).

At this point you may object that we have not done anything new. We have simply passed three different planes through a point and named the normal and shear stresses on those planes. If a different plane is passed through the point, the normal and shear stresses will generally be different from those acting on the planes perpendicular to the coordinate axes, so what have we achieved? The answer is that if the normal and shear stresses on these three planes are known at a point, the normal and shear stresses on any plane through the point can be determined. For this reason, the stresses on these planes are called the *state of stress* at the point. We can

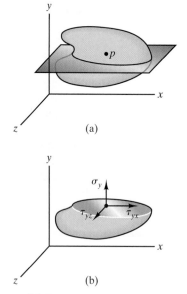

Figure 12.2
(a) Passing a plane perpendicular to the y axis.
(b) Normal stress and components of the shear stress at p.

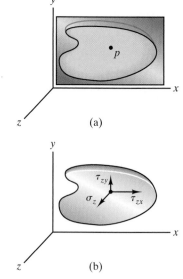

Figure 12.3
(a) Passing a plane perpendicular to the z axis.
(b) Normal stress and components of the shear stress at p.

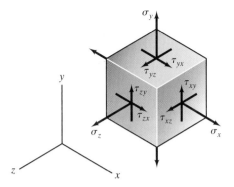

Figure 12.4
Components of stress on an element containing p.

represent the state of stress compactly as a matrix in terms of the *components of stress*:

$$\begin{bmatrix} \sigma_x & \tau_{xy} & \tau_{xz} \\ \tau_{yx} & \sigma_y & \tau_{yz} \\ \tau_{zx} & \tau_{zy} & \sigma_z \end{bmatrix}.$$

(12.1)

Figure 12.4 shows the components of stress on an element whose faces are perpendicular to the coordinate axes. If the state of stress is uniform, or *homogeneous*, in a finite neighborhood surrounding the point, you can regard the stresses in Fig. 12.4 as the stresses on an element of finite size containing p. Otherwise, you must interpret them as the average values of the stress components on an element containing p. These average values approach the state of stress at p in the limit as the element shrinks.

The shear stresses $\tau_{xy} = \tau_{yx}, \tau_{yz} = \tau_{zy}$, and $\tau_{xz} = \tau_{zx}$, so that the stress matrix (5-1) is symmetric. To show this, let the element in Fig. 12.5 be a cube with dimension b. The sum of the moments about the z axis is

$$(\sigma_x b^2)(b/2) - (\sigma_x b^2)(b/2) + (\sigma_y b^2)(b/2) - (\sigma_y b^2)(b/2) + (\tau_{xy} b^2)(b)$$
$$- (\tau_{yx} b^2)(b) + (\tau_{zy} b^2)(b/2) - (\tau_{zy} b^2)(b/2)$$
$$- (\tau_{zx} b^2)(b/2) + (\tau_{zx} b^2)(b/2) = 0.$$

Therefore, $\tau_{xy} = \tau_{yx}$, and by summing moments about the x and y axes it can be shown that $\tau_{yz} = \tau_{zy}$ and $\tau_{xz} = \tau_{zx}$. Although we have assumed the material to be in equilibrium, these results hold even when the material is not in equilibrium.

As a simple example of a state of stress, consider the pressure in a fluid (a liquid or gas) at rest. The stresses exerted on the faces of an element are the pressures exerted by the surrounding fluid (Fig. 12.6). The normal stresses

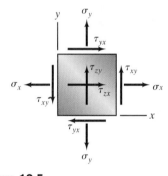

Figure 12.5
Summing moments about the z axis to show that $\tau_{xy} = \tau_{yx}$.

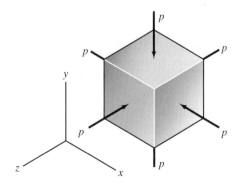

Figure 12.6
Stresses on an element of a fluid at rest.

are $\sigma_x = \sigma_y = \sigma_z = -p$ and the shear stresses are zero. (The shear stresses on an element of a flowing fluid are not generally zero, although in some situations they can be neglected.) Therefore, the state of stress at a point in a fluid at rest is

$$\begin{bmatrix} \sigma_x & \tau_{xy} & \tau_{xz} \\ \tau_{yx} & \sigma_y & \tau_{yz} \\ \tau_{zx} & \tau_{zy} & \sigma_z \end{bmatrix} = \begin{bmatrix} -p & 0 & 0 \\ 0 & -p & 0 \\ 0 & 0 & -p \end{bmatrix}. \tag{12.2}$$

Applying axial loads to the ends of a prismatic bar results in another simple state of stress. If we orient the coordinate system with its x axis parallel to the axis of the bar (Fig. 12.7), the only nonzero stress component is $\sigma_x = P/A$, where A is the bar's cross-sectional area. The state of stress is

$$\begin{bmatrix} \sigma_x & \tau_{xy} & \tau_{xz} \\ \tau_{yx} & \sigma_y & \tau_{yz} \\ \tau_{zx} & \tau_{zy} & \sigma_z \end{bmatrix} = \begin{bmatrix} \sigma_x & 0 & 0 \\ 0 & 0 & 0 \\ 0 & 0 & 0 \end{bmatrix}.$$

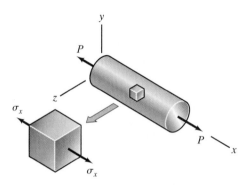

Figure 12.7
Stresses on an element of a bar subjected to axial forces.

In general, this state of stress applies only to elements that are not near the ends of the bar where the forces are applied.

In Fig. 12.8 we isolate an infinitesimal element of a cylindrical bar subjected to axial torsion. If we orient the coordinate system so that its x axis is parallel to the axis of the bar and the element lies in the $x - y$ plane, the element is in a state of pure shear stress:

$$\begin{bmatrix} \sigma_x & \tau_{xy} & \tau_{xz} \\ \tau_{yx} & \sigma_y & \tau_{yz} \\ \tau_{zx} & \tau_{zy} & \sigma_z \end{bmatrix} = \begin{bmatrix} 0 & \tau_{xy} & 0 \\ \tau_{yx} & 0 & 0 \\ 0 & 0 & 0 \end{bmatrix}.$$

This state of stress also generally applies only to elements that are not near the ends of the bar.

In citing these examples, we have emphasized that *the values of the components of normal and shear stress that define the state of stress at a point depend on the orientation of the coordinate system*. If the coordinate system is rotated, the orientations of the planes perpendicular to the axes change, and so in general the normal and shear stresses acting on them change. By doing so, we can determine the normal and shear stresses acting on different planes through a point. We undertake this task in the following section.

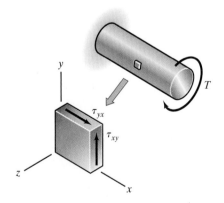

Figure 12.8
Stresses on an element of a bar subjected to torsion.

Study Questions

1. What are the definitions of the components of stress at a point?
2. How many components of stress must be known to define the state of stress at a point?
3. Suppose that you know the components of stress at a point in terms of a coordinate system xyz. Define a new coordinate system $x'y'z'$ with the x' axis coincident with the y axis, the y' axis coincident with the z axis, and the z' axis coincident with the x axis. What are the components of stress at the point in terms of the $x'y'z'$ coordinate system?

12.2 | Transformations of Plane Stress

The stress at a point is said to be a state of *plane stress* if it is of the form

$$\begin{bmatrix} \sigma_x & \tau_{xy} & 0 \\ \tau_{yx} & \sigma_y & 0 \\ 0 & 0 & 0 \end{bmatrix}. \tag{12.3}$$

That is, the stress components σ_z, τ_{xz}, and τ_{yz} are zero (Fig. 12.9). This does not imply that the three stress components σ_x, σ_y, and τ_{xy} are each necessarily nonzero, but they are the only stress components that may be nonzero in plane stress. The state of stress shown in Fig. 12.7 for a bar subjected to axial forces is plane stress, and the state of stress shown in Fig. 12.8 for a bar subjected to axial torsion is plane stress. But the state of stress shown in Fig. 12.6 for a fluid at rest is not plane stress, because $\sigma_z \neq 0$.

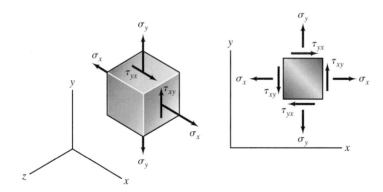

Figure 12.9
Plane stress.

Many important applications in stress analysis besides bars subjected to axial and torsional loads result in states of plane stress. In this section we assume that the state of plane stress at a point of a material is known and we wish to know the states of stress on planes other than the three planes perpendicular to the coordinate axes. We also address the most crucial questions from the standpoint of design: What are the maximum normal and shear stresses, and what are the orientations of the planes on which they act?

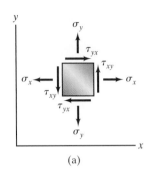

(a)

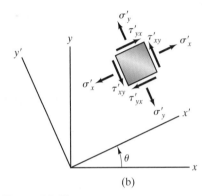

(b)

Figure 12.10

Stress components at p:
(a) in terms of the xyz coordinate system;
(b) in terms of the $x'y'z'$ coordinate system.

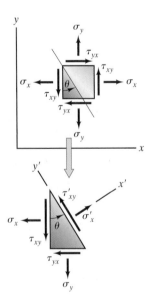

Figure 12.11

Free-body diagram for determining σ'_x and τ'_{xy}.

Coordinate Transformations

Suppose that we know the state of plane stress at a point p of a material in terms of the coordinate system shown in Fig. 12.10a, and we want to know the state of stress at p in terms of a coordinate system $x'y'z'$ oriented as shown in Fig. 12.10b. (The z and z' axes are coincident.) We begin with the element in Fig. 12.10a and pass a plane through it as shown in Fig. 12.11. The oblique surface of the resulting free-body diagram is perpendicular to the x' axis, so the normal and shear stresses acting on it are σ'_x and τ'_{xy}. We can determine σ'_x and τ'_{xy} by writing the equilibrium equations for this free-body diagram.

Let the area of the oblique surface of the free-body diagram be ΔA. The sum of the forces in the x' direction is

$$\sigma'_x \Delta A - (\sigma_x \Delta A \cos\theta)\cos\theta - (\sigma_y \Delta A \sin\theta)\sin\theta$$
$$- (\tau_{xy}\Delta A \cos\theta)\sin\theta - (\tau_{yx}\Delta A \sin\theta)\cos\theta = 0.$$

Solving for σ'_x, we obtain

$$\sigma'_x = \sigma_x \cos^2\theta + \sigma_y \sin^2\theta + 2\tau_{xy}\sin\theta\cos\theta. \tag{12.4}$$

The sum of the forces in the y' direction is

$$\tau'_{xy}\Delta A + (\sigma_x \Delta A \cos\theta)\sin\theta - (\sigma_y \Delta A \sin\theta)\cos\theta$$
$$- (\tau_{xy}\Delta A \cos\theta)\cos\theta + (\tau_{yx}\Delta A \sin\theta)\sin\theta = 0.$$

The solution for τ'_{xy} is

$$\tau'_{xy} = -(\sigma_x - \sigma_y)\sin\theta\cos\theta + \tau_{xy}(\cos^2\theta - \sin^2\theta). \tag{12.5}$$

By using the trigonometric identities

$$2\cos^2\theta = 1 + \cos 2\theta,$$
$$2\sin^2\theta = 1 - \cos 2\theta,$$
$$2\sin\theta\cos\theta = \sin 2\theta,$$
$$\cos^2\theta - \sin^2\theta = \cos 2\theta, \tag{12.6}$$

we can write Eqs. (12.4) and (12.5) in alternative forms that will be useful:

$$\sigma'_x = \frac{\sigma_x + \sigma_y}{2} + \frac{\sigma_x - \sigma_y}{2}\cos 2\theta + \tau_{xy}\sin 2\theta, \tag{12.7}$$

$$\tau'_{xy} = -\frac{\sigma_x - \sigma_y}{2}\sin 2\theta + \tau_{xy}\cos 2\theta. \tag{12.8}$$

We can obtain an equation for σ'_y by setting θ equal to $\theta + 90°$ in the expression for σ'_x. The result is

$$\sigma'_y = \frac{\sigma_x + \sigma_y}{2} - \frac{\sigma_x - \sigma_y}{2}\cos 2\theta - \tau_{xy}\sin 2\theta. \tag{12.9}$$

Given the state of plane stress shown in Fig. 12.10a, Eqs. (12.7)–(12.9) determine the state of plane stress shown in Fig. 12.10b for any value of θ. This means that when we know a state of plane stress at a point p, we can determine the normal and shear stresses on planes through p other than the three planes perpendicular to the coordinate axes. Notice, however, that we can do so only for planes that are parallel to the z axis.

Study Questions

1. What is a state of plane stress?
2. How many components of stress must be known to define a state of plane stress at a point?
3. Is the state of stress for a fluid at rest, given by Eq. (12.2), a state of plane stress? Explain.
4. What is the definition of the angle θ in Eqs. (12.7)–(12.9)?

Example 12.1

Stresses on a Specified Plane

The components of plane stress at point p of the material in Fig. 12.12 are $\sigma_x = 4$ ksi, $\sigma_y = -2$ ksi, and $\tau_{xy} = 2$ ksi. What are the normal stress and the magnitude of the shear stress on the plane $\mathcal{P}$ at point p?

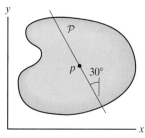

Figure 12.12

Strategy

If we orient the $x'y'$ coordinate system so that the x' axis is perpendicular to the plane $\mathcal{P}$, then σ_x' and τ_{xy}' are the normal and shear stresses on $\mathcal{P}$ (Fig. 12.11). We can use Eqs. (12.7) and (12.8) to determine σ_x' and τ_{xy}'.

Solution

The x' axis is perpendicular to the plane $\mathcal{P}$ if $\theta = 30°$ [Fig. (a)].
From Eq. (12.7),

$$\sigma_x' = \frac{4 + (-2)}{2} + \frac{4 - (-2)}{2}\cos 60° + 2\sin 60° = 4.23 \text{ ksi},$$

and from Eq. (12.8),

$$\tau_{xy}' = -\frac{4 - (-2)}{2}\sin 60° + 2\cos 60° = -1.60 \text{ ksi}.$$

The normal stress on $\mathcal{P}$ at point p is 4.23 ksi and the magnitude of the shear stress is 1.60 ksi.

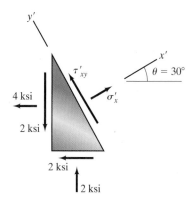

(a) Orienting the $x'y'$ coordinate system so that the x' axis is perpendicular to $\mathcal{P}$.

Example 12.2

Stresses as Functions of θ

The state of plane stress at a point p is shown on the left element in Fig. 12.13.

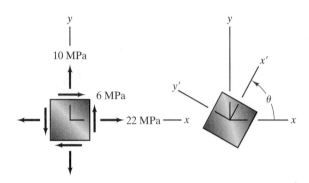

Figure 12.13

(a) Determine the state of plane stress at p acting on the right element in Fig. 12.13 if $\theta = 60°$. Draw a sketch of the element showing the stresses acting on it.

(b) Draw graphs of σ'_x and τ'_{xy} as functions of θ for values of θ from zero to 360°.

Strategy

The components of plane stress on the left element in Fig. 12.13 are $\sigma_x = 22$ MPa, $\sigma_y = 10$ MPa, and $\tau_{xy} = 6$ MPa. The components of plane stress on the right element in Fig. 5.13 are given by Eqs. (12.7)–(12.9).

Solution

(a) For $\theta = 60°$, the components of stress are

$$\sigma'_x = \frac{22 + 10}{2} + \frac{22 - 10}{2}\cos 120° + 6 \sin 120° = 18.20 \text{ MPa},$$

$$\sigma'_y = \frac{22 + 10}{2} - \frac{22 - 10}{2}\cos 120° - 6 \sin 120° = 13.80 \text{ MPa},$$

$$\tau'_{xy} = -\frac{22 - 10}{2}\sin 120° + 6 \cos 120° = -8.20 \text{ MPa}.$$

We show the components of stress acting on the element in Fig. (a). Notice the directions of the shear stresses due to the negative value of τ'_{xy}.

(b) The stresses σ'_x and τ'_{xy} as functions of θ are

$$\sigma'_x = \frac{22 + 10}{2} + \frac{22 - 10}{2}\cos 2\theta + 6 \sin 2\theta$$

$$= 16 + 6 \cos 2\theta + 6 \sin 2\theta,$$

$$\tau'_{xy} = -\frac{22 - 10}{2}\sin 2\theta + 6 \cos 2\theta$$

$$= -6 \sin 2\theta + 6 \cos 2\theta.$$

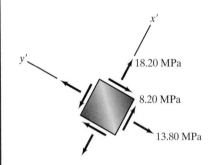

(a) State of stress for $\theta = 60°$.

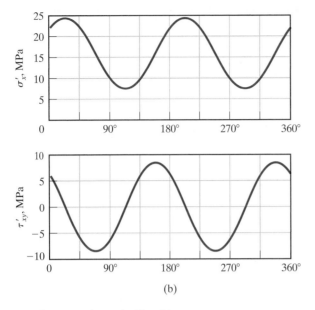

(b) Graphs of σ'_x and τ'_{xy} as functions of θ.

(b)

The graphs of these expressions are shown in Fig. (b).

Discussion

The graphs of σ'_x and τ'_{xy} show that the normal stress and shear stress attain maximum and minimum values at particular values of θ. (Not, however, at the same values of θ. Notice that at angles for which the normal stress is a maximum or minimum, the shear stress is zero.) Since the maximum values of the stresses are so important with regard to design, we could use graphs such as these to determine the maximum stresses and the orientations of the planes on which they occur. But in the following sections we introduce more efficient ways to obtain this information.

Example 12.3

Determining the Orientation of an Element

The state of plane stress at a point p is shown on the left element in Fig. 12.14, and the values of the stresses σ'_x and τ'_{xy} are shown on a rotated element. What are the normal stress σ'_y and the angle θ?

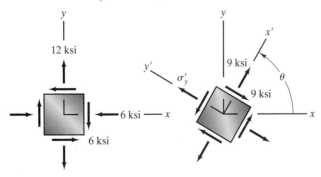

Figure 12.14

Strategy

Equations (12.7)–(12.9) give the components of stress on the rotated element in terms of θ. Since σ'_x and τ'_{xy} are known, we can solve Eqs. (12.7) and (12.8) for θ, then determine σ'_y from Eq. (12.9).

Solution

The components of stress on the left element are $\sigma_x = -6$ ksi, $\tau_{xy} = -6$ ksi, and $\sigma_y = 12$ ksi. On the rotated element, $\sigma'_x = 9$ ksi and $\tau'_{xy} = -9$ ksi. Equation (12.7) is

$$\sigma'_x = \frac{\sigma_x + \sigma_y}{2} + \frac{\sigma_x - \sigma_y}{2}\cos 2\theta + \tau_{xy}\sin 2\theta:$$

$$9 = \frac{(-6) + 12}{2} + \frac{(-6) - 12}{2}\cos 2\theta + (-6)\sin 2\theta$$

$$= 3 - 9\cos 2\theta - 6\sin 2\theta,$$

and Eq. (12.8) is

$$\tau'_{xy} = -\frac{\sigma_x - \sigma_y}{2}\sin 2\theta + \tau_{xy}\cos 2\theta:$$

$$-9 = -\frac{(-6) - 12}{2}\sin 2\theta + (-6)\cos 2\theta$$

$$= 9\sin 2\theta - 6\cos 2\theta.$$

We can solve these two equations for $\sin 2\theta$ and $\cos 2\theta$. The results are $\sin 2\theta = -1$ and $\cos 2\theta = 0$, from which we obtain $\theta = 135°$. Substituting this result into Eq. (12.9), the stress σ'_y is

$$\sigma'_y = \frac{\sigma_x + \sigma_y}{2} - \frac{\sigma_x - \sigma_y}{2}\cos 2\theta - \tau_{xy}\sin 2\theta$$

$$= \frac{(-6) + 12}{2} - \frac{(-6) - 12}{2}\cos 2(135°) - (-6)\sin 2(135°)$$

$$= -3 \text{ ksi.}$$

The stresses are shown on the rotated element in Fig. (a).

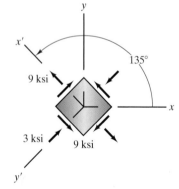

(a) Stresses on the properly oriented element.

Maximum and Minimum Stresses

Given the state of plane stress at a point p, we have seen that the normal and shear stresses on the plane shown in Fig. 12.15 are given by Eqs. (12.7) and (12.8):

$$\sigma'_x = \frac{\sigma_x + \sigma_y}{2} + \frac{\sigma_x - \sigma_y}{2}\cos 2\theta + \tau_{xy}\sin 2\theta, \qquad (12.10)$$

$$\tau'_{xy} = -\frac{\sigma_x - \sigma_y}{2}\sin 2\theta + \tau_{xy}\cos 2\theta. \qquad (12.11)$$

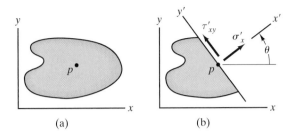

Figure 12.15
(a) Point p of a material.
(b) Normal and shear stresses at p.

Thus we can determine the normal and shear stresses at p for any value of θ. But it is the maximum values of the stresses that determine whether a material will fail. How can we determine them and the orientations of the planes through the point p on which they act?

Principal Stresses Let a value of θ for which the normal stress σ_x' is a maximum or minimum be denoted by θ_p. By evaluating the derivative of Eq. (5.10) with respect to 2θ and setting it equal to zero, we obtain the equation

$$\tan 2\theta_p = \frac{2\tau_{xy}}{\sigma_x - \sigma_y}. \tag{12.12}$$

When σ_x, σ_y, and τ_{xy} are known, we can solve this equation for $\tan 2\theta_p$, which allows us to determine θ_p. Then we can substitute θ_p into Eq. (12.10) to determine the maximum or minimum value of the normal stress.

Equation (12.12) yields more than one solution for θ_p, because of the periodic nature of the tangent. Observe in Fig. 12.16 that if $2\theta_p$ is a solution of Eq. (12.12), so are $2\theta_p + 180°, 2\theta_p + 2(180°), \ldots$ This means that the normal stress is a maximum or minimum at $\theta_p, \theta_p + 90°, \theta_p + 2(90°), \ldots$ (Fig. 12.17a). *The planes on which the maximum and minimum normal stresses act correspond to the faces of a rectangular element* (Fig. 12.17b).

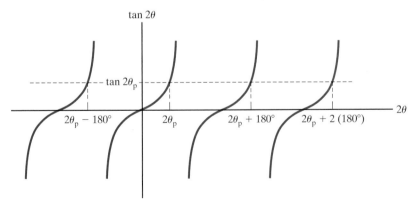

Figure 12.16
The periodic nature of the tangent gives rise to multiple roots.

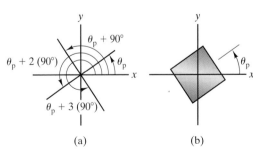

Figure 12.17
(a) Angles at which maximum or minimum normal stresses occur.
(b) The planes correspond to the faces of a rectangular element.

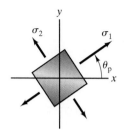

Figure 12.18

Principal stresses and planes on which they act.

Recognizing that the maximum and minimum normal stresses act on the faces of a particular rectangular element makes it easy to determine and visualize the planes on which these stresses act. We simply determine one value of θ_p from Eq. (12.12), establishing the orientation of the element. We can then determine the values of the maximum and minimum stresses, called the principal stresses and denoted σ_1 and σ_2 (Fig. 12.18), from Eq. (12.10).

We can obtain analytical expressions for the values of the principal stresses that are useful when we are not concerned with the planes on which they act. By solving the equations

$$\frac{\sin 2\theta_p}{\cos 2\theta_p} = \tan 2\theta_p = \frac{2\tau_{xy}}{\sigma_x - \sigma_y} \tag{12.13}$$

and

$$\sin^2 2\theta_p + \cos^2 2\theta_p = 1 \tag{12.14}$$

for $\sin 2\theta_p$ and $\cos 2\theta_p$ and substituting the results into Eq. (12.10), we obtain

$$\sigma_1, \sigma_2 = \frac{\sigma_x + \sigma_y}{2} \pm \sqrt{\left(\frac{\sigma_x - \sigma_y}{2}\right)^2 + \tau_{xy}^2}. \tag{12.15}$$

By substituting the expressions for $\sin 2\theta_p$ and $\cos 2\theta_p$ into Eq. (12.11), we obtain $\tau'_{xy} = 0$. *On the planes on which the principal stresses act, the shear stresses are zero.*

The principal stresses are the maximum and minimum normal stresses acting on planes through point p that are parallel to the z axis. There are no normal stresses of greater magnitude on any plane through p. We will see that the situation is more complicated in the case of the maximum shear stress.

Maximum Shear Stresses   We approach the determination of maximum or minimum shear stresses in the same way that we did normal stresses. Let a value of θ for which the shear stress is a maximum or minimum be denoted by θ_s. Evaluating the derivative of Eq. (12.11) with respect to 2θ and setting it equal to zero, we obtain the equation

$$\tan 2\theta_s = -\frac{\sigma_x - \sigma_y}{2\tau_{xy}}. \tag{12.16}$$

With this equation we can determine θ_s and substitute it into Eq. (12.11) to determine the maximum or minimum value of the shear stress.

As in the case of the normal stress, if the shear stress is a maximum or minimum at θ_s, it is also a maximum or minimum at $\theta_s + 90°$, $\theta_s + (2)(90°), \ldots$. *The planes on which the maximum and minimum shear stresses act also correspond to the faces of a rectangular element.* Since the shear stresses on the faces of a rectangular element are equal in magnitude, the magnitudes of the maximum and minimum shear stresses are equal. Furthermore, the orientation of this element is related in a simple way to the orientation of the element on which the principal stresses act. Notice from Eqs. (12.12) and (12.16) that $\tan 2\theta_s$ is the negative inverse of $\tan 2\theta_p$. This implies that the directions defined by the angles $2\theta_s$ and $2\theta_p$ are perpendicular (Fig. 12.19), which means that *the element on which the maximum and minimum shear stresses act is rotated 45° relative to the element on which the principal stresses act.* Once we have determined the orientation of the element on

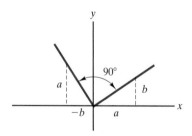

Figure 12.19

The tangents of the angles defining these two perpendicular directions relative to the x axis are b/a and $-a/b$. One tangent is the negative inverse of the other.

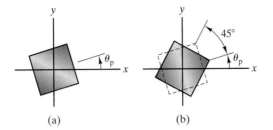

Figure 12.20
Relationship between the orientation of the elements
(a) on which the principal stresses act and
(b) on which the maximum and minimum shear stresses act.

(a)   (b)

which the principal stresses act, we also know the orientation of the element on which the maximum and minimum shear stresses act (Fig. 12.20).

To obtain an analytical expression for the magnitude of the maximum shear stress, we solve the equations

$$\frac{\sin 2\theta_s}{\cos 2\theta_s} = \tan 2\theta_s = -\frac{\sigma_x - \sigma_y}{2\tau_{xy}} \tag{12.17}$$

and

$$\sin^2 2\theta_s + \cos^2 2\theta_s = 1 \tag{12.18}$$

for $\sin 2\theta_s$ and $\cos 2\theta_s$ and substitute the results into Eq. (12.11). The result is

$$\tau_{max} = \sqrt{\left(\frac{\sigma_x - \sigma_y}{2}\right)^2 + \tau_{xy}^2}. \tag{12.19}$$

This stress is called the *maximum in-plane shear stress*, because it is the greatest shear stress that occurs on any plane parallel to the z axis. However, we will see that greater shear stresses may occur on other planes through p.

To determine the complete state of plane stress on the element on which the maximum in-plane shear stresses act, we must evaluate the normal stresses. Substituting our results for $\sin 2\theta_s$ and $\cos 2\theta_s$ into Eqs. (12.7) and (12.9), we obtain

$$\sigma_x' = \sigma_y' = \frac{\sigma_x + \sigma_y}{2}.$$

The normal stresses on the element on which the maximum in-plane shear stresses act are equal. We denote this normal stress by σ_s:

$$\sigma_s = \frac{\sigma_x + \sigma_y}{2}. \tag{12.20}$$

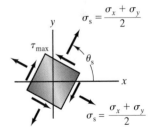

Figure 12.21
Element on which the maximum and minimum in-plane shear stresses act.

Figure 12.21 shows the complete state of stress.

Now let us consider whether shear stresses greater in magnitude than the value given by Eq. (12.19) occur on other planes through p. We begin with the element on which the principal stresses occur, and realign the coordinate system with the faces of the element (Fig. 12.22a). In terms of this new coordinate system, the components of plane stress are $\sigma_x = \sigma_1, \sigma_y = \sigma_2$, and $\tau_{xy} = 0$. Substituting these components into Eq. (12.19), we obtain a different expression for the magnitude of the maximum in-plane shear stress:

$$\sqrt{\left(\frac{\sigma_x - \sigma_y}{2}\right)^2 + \tau_{xy}^2} = \sqrt{\left(\frac{\sigma_1 - \sigma_2}{2}\right)^2 + 0} = \left|\frac{\sigma_1 - \sigma_2}{2}\right|. \tag{12.21}$$

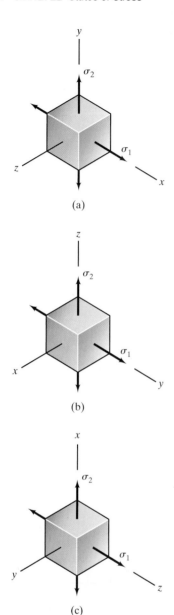

Figure 12.22
Element on which the principal stresses act.
(a) Realigned coordinate system.
(b), (c) Other orientations of the coordinate
system.

This equation is expressed in terms of the principal stresses, but it gives the same result as Eq. (12.19). We have still considered only planes parallel to the original z axis. Now, however, let's consider the element on which the principal stresses occur and reorient the coordinate system as shown in Fig. 12.22b. In terms of this coordinate system, $\sigma_x = 0$, $\sigma_y = \sigma_1$, and $\tau_{xy} = 0$. Substituting these components into Eq. (12.19), we obtain the maximum shear stress on planes parallel to the new z axis:

$$\sqrt{\left(\frac{\sigma_x - \sigma_y}{2}\right)^2 + \tau_{xy}^2} = \sqrt{\left(\frac{0 - \sigma_1}{2}\right)^2 + 0} = \left|\frac{\sigma_1}{2}\right|. \quad (12.22)$$

Next, we reorient the coordinate system as shown in Fig. 12.22c. In terms of this coordinate system, $\sigma_x = \sigma_2$, $\sigma_y = 0$, and $\tau_{xy} = 0$. Substituting these components into Eq. (12.19), we obtain the maximum shear stress on planes parallel to this z axis:

$$\sqrt{\left(\frac{\sigma_x - \sigma_y}{2}\right)^2 + \tau_{xy}^2} = \sqrt{\left(\frac{\sigma_2 - 0}{2}\right)^2 + 0} = \left|\frac{\sigma_2}{2}\right|. \quad (12.23)$$

Depending on the values of the principal stresses, Eq. (12.22) and/or Eq. (12.23) can result in larger values of the magnitude of the maximum shear stress than the maximum in-plane shear stress. Although we have still considered only a subset of the possible planes through point p, there are no shear stresses of greater magnitude on any plane through p than the largest value given by Eq. (12.21), (12.22), or (12.23), which is called the *absolute maximum shear stress*.

Summary: Determining the Principal Stresses and the Maximum Shear Stress Here we give a sequence of steps to determine the maximum and minimum stresses at a point p subjected to a known state of plane stress and the orientations of the planes on which they act.

1. Use Eq. (12.12) to determine θ_p, establishing the orientation of the element on which the principal stresses act.

2. Determine the two principal stresses by substituting first θ_p and then $\theta_p + 90°$ into Eq. (12.10). Since no shear stresses act on the element on which the principal stresses act, this determines the complete state of stress on the element. Notice that the values of the principal stresses are given by Eq. (12.15), but this does not indicate which principal stress acts on which faces of the element.

3. Use Eq. (12.16) to determine θ_s, establishing the orientation of the element on which the maximum and minimum in-plane shear stresses act. Alternatively, since this element is rotated 45° relative to the element on which the principal stresses act, simply use $\theta_s = \theta_p + 45°$.

4. Determine the shear stress on the element by substituting θ_s into Eq. (12.11). The magnitude of this shear stress is the maximum in-plane shear stress. The normal stress on each face of this element is $(\sigma_x + \sigma_y)/2$, which completes the determination of the state of stress on the element.

5. The absolute maximum shear stress (the maximum shear stress on any plane through p) is the largest of the three values

$$\left|\frac{\sigma_1 - \sigma_2}{2}\right|, \qquad \left|\frac{\sigma_1}{2}\right|, \qquad \left|\frac{\sigma_2}{2}\right|. \quad (12.24)$$

Study Questions

1. What are the principal stresses?
2. If you know the orientation of the plane on which one of the principal stresses acts, what do you know about the orientation of the plane on which the other principal stress acts?
3. What shear stresses act on the planes on which the principal stresses act?
4. If you know the orientation of the element on which the principal stresses act, what is the orientation of the element on which the maximum and minimum in-plane shear stresses act?

Example 12.4

Determining Principal Stresses

The state of plane stress at a point p is shown on the element in Fig. 12.23. Determine the principal stresses and the maximum in-plane shear stress and show them acting on properly oriented elements. Also determine the absolute maximum shear stress.

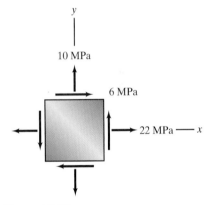

Figure 12.23

Strategy

The components of plane stress on the element are $\sigma_x = 22$ MPa, $\sigma_y = 10$ MPa, and $\tau_{xy} = 6$ MPa. We can follow the steps given in the preceding summary for determining the principal stresses and the maximum shear stress.

Solution

Step 1 From Eq. (12.12),

$$\tan 2\theta_p = \frac{2\tau_{xy}}{\sigma_x - \sigma_y} = \frac{2(6)}{22 - 10} = 1.$$

Solving this equation, we obtain $\theta_p = 22.5°$. This angle tells us the orientation of the element on which the principal stresses act.

Step 2 We substitute θ_p into Eq. (12.10) to determine the first principal stress.

$$\sigma_1 = \frac{\sigma_x + \sigma_y}{2} + \frac{\sigma_x - \sigma_y}{2} \cos 2\theta_p + \tau_{xy} \sin 2\theta_p$$

$$= \frac{22 + 10}{2} + \frac{22 - 10}{2} \cos 45° + 6 \sin 45°$$

$$= 24.49 \text{ MPa.}$$

We then substitute $\theta_p + 90°$ into Eq. (12.10) to determine the second principal stress.

$$\sigma_2 = \frac{\sigma_x + \sigma_y}{2} + \frac{\sigma_x - \sigma_y}{2} \cos 2(\theta_p + 90°) + \tau_{xy} \sin 2(\theta_p + 90°)$$

$$= \frac{22 + 10}{2} + \frac{22 - 10}{2} \cos 225° + 6 \sin 225°$$

$$= 7.51 \text{ MPa.}$$

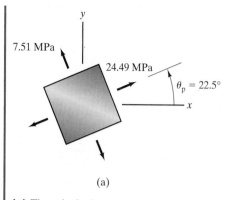

(a)

(a) The principal stresses.

The principal stresses are shown on the properly oriented element in Fig. (a). No shear stresses act on this element.

We can also determine the values of the principal stresses from Eq. (12.15),

$$\sigma_1, \sigma_2 = \frac{\sigma_x + \sigma_y}{2} \pm \sqrt{\left(\frac{\sigma_x - \sigma_y}{2}\right)^2 + \tau_{xy}^2}$$

$$= \frac{22 + 10}{2} \pm \sqrt{\left(\frac{22 - 10}{2}\right)^2 + (6)^2}$$

$$= 24.49, 7.51 \text{ MPa},$$

but this procedure does not tell us which faces of the element the stresses act on.

Step 3 From Eq. (12.16),

$$\tan 2\theta_s = -\frac{\sigma_x - \sigma_y}{2\tau_{xy}} = -\frac{22 - 10}{2(6)} = -1,$$

from which we obtain $\theta_s = -22.5°$. [We could choose instead to determine θ_s by using the fact that the element on which the maximum in-plane shear stresses act is rotated $45°$ relative to the element on which the principal stresses act. In this way we obtain $\theta_s = \theta_p + 45° = 67.5°$. This angle differs from our previous result by $90°$, so the resulting orientation of the element is the same. This emphasizes that the angle θ_s can only be determined from Eq. (12.16) within a multiple of $90°$.]

Step 4 We substitute θ_s into Eq. (12.11) to determine the maximum in-plane shear stress.

$$\tau_{\max} = -\frac{\sigma_x - \sigma_y}{2} \sin 2\theta_s + \tau_{xy} \cos 2\theta_s$$

$$= -\frac{22 - 10}{2} \sin(-45°) + 6\cos(-45°)$$

$$= 8.49 \text{ MPa}.$$

The normal stress on each face of this element is

$$\sigma_s = \frac{\sigma_x + \sigma_y}{2} = \frac{22 + 10}{2} = 16 \text{ MPa}.$$

The maximum in-plane shear stresses and associated normal stresses are shown on the properly oriented element in Fig. (b).

Step 5 The absolute maximum shear stress is given by the largest of the three values

$$\left|\frac{\sigma_x - \sigma_y}{2}\right| = \left|\frac{24.49 - 7.51}{2}\right| = 8.49 \text{ MPa},$$

$$\left|\frac{\sigma_1}{2}\right| = \left|\frac{24.49}{2}\right| = 12.24 \text{ MPa},$$

$$\left|\frac{\sigma_2}{2}\right| = \left|\frac{7.51}{2}\right| = 3.76 \text{ MPa}.$$

The absolute maximum shear stress is 12.24 MPa.

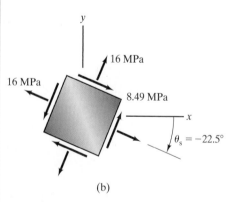

(b)

(b) Maximum in-plane shear stresses and associated normal stresses.

Problems

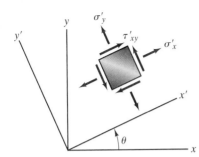

P12.1–P12.7

P12.8

12.1 The components of plane stress at a point p of a material are $\sigma_x = 20$ MPa, $\sigma_y = 0$, and $\tau_{xy} = 0$. If $\theta = 45°$, what are the stresses σ_x', σ_y', and τ_{xy}' at point p?

12.2 The components of plane stress at a point p of a material are $\sigma_x = 0$, $\sigma_y = 0$, and $\tau_{xy} = 25$ ksi. If $\theta = 45°$, what are the stresses σ_x', σ_y', and τ_{xy}' at point p?

12.3 The components of plane stress at a point p of a material are $\sigma_x = -8$ ksi, $\sigma_y = 6$ ksi, and $\tau_{xy} = -6$ ksi. If $\theta = 30°$, what are the stresses σ_x', σ_y', and τ_{xy}' at point p?

12.4 During liftoff, strain gauges attached to one of the Space Shuttle main engine nozzles determine that the components of plane stress $\sigma_x' = 66.46$ MPa, $\sigma_y' = 82.54$ MPa, and $\tau_{xy}' = 6.75$ MPa at $\theta = 20°$. What are the stresses σ_x, σ_y, and τ_{xy} at that point?

12.5 The components of plane stress at a point p of a material are $\sigma_x = 240$ MPa, $\sigma_y = -120$ MPa, and $\tau_{xy} = 240$ MPa, and the components referred to the $x'y'z'$ coordinate system are $\sigma_x' = 347$ MPa, $\sigma_y' = -227$ MPa, and $\tau_{xy}' = -87$ MPa. What is the angle θ? [*Strategy:* Solve Eqs. (12.7) and (12.8) for $\sin 2\theta$ and $\cos 2\theta$. Knowing these two quantities, you can determine θ. (Why are the values of both $\sin 2\theta$ and $\cos 2\theta$ needed to uniquely determine θ?)]

12.6 The components of plane stress at a point p of a bit during a drilling operation are $\sigma_x = 40$ ksi, $\sigma_y = -30$ ksi, and $\tau_{xy} = 30$ ksi, and the components referred to the $x'y'z'$ coordinate system are $\sigma_x' = 12.5$ ksi, $\sigma_y' = -2.5$ ksi, and $\tau_{xy}' = 45.5$ ksi. What is the angle θ?

12.7 The components of plane stress at a point p of a material referred to the $x'y'z'$ coordinate system are $\sigma_x' = -8$ MPa, $\sigma_y' = 6$ MPa, and $\tau_{xy}' = -16$ MPa. If $\theta = 20°$, what are the stresses σ_x, σ_y, and τ_{xy} at point p?

12.8 A point p of the car's frame is subjected to the components of plane stress $\sigma_x' = 32$ MPa, $\sigma_y' = -16$ MPa, and $\tau_{xy}' = -24$ MPa. If $\theta = 20°$, what are the stresses σ_x, σ_y, and τ_{xy} at p?

12.9 In Problem 12.8, what are the stresses σ_x, σ_y, and τ_{xy} at point p if $\theta = -40°$?

12.10 The components of plane stress at point p of the material shown are $\sigma_x = 4$ ksi, $\sigma_y = -2$ *ksi, and* $\tau_{xy} = 2$ ksi. What are the normal stress and the magnitude of the shear stress on the plane $\mathcal{P}$ at point p?

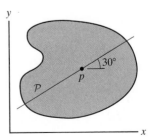

P12.10

12.11 The components of plane stress at point p of the material shown in Problem 12.10 are $\sigma_x = -10.5$ MPa, $\sigma_y = 6.0$ MPa, and $\tau_{xy} = -4.5$ MPa. What are the normal stress and the magnitude of the shear stress on the plane $\mathcal{P}$ at point p?

12.12 Determine the stresses σ and τ (a) by writing equilibrium equations for the element shown; (b) by using Eqs. (12.7) and (12.8).

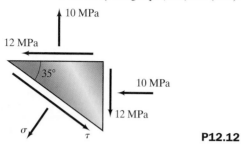

P12.12

12.13 The stress $\tau_{xy} = 14$ MPa and the angle $\theta = 25°$. Determine the components of stress on the right element.

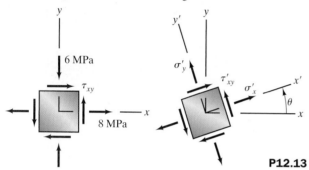

P12.13

12.14 On the elements shown in Problem 12.13, the stresses $\tau_{xy} = 12$ MPa, $\sigma'_x = 14$ MPa, and $\sigma'_y = -12$ MPa. Determine the stress τ'_{xy} and the angle θ.

12.15 On the elements shown in Problem 12.13, the stress $\tau'_{xy} = 12$ MPa and the angle $\theta = 35°$. Determine σ'_x, σ'_y, and τ_{xy}.

12.16 A point p of the airplane's wing is subjected to plane stress. When $\theta = 55°$, $\sigma_x = 100$ psi, $\sigma_y = -200$ psi, and $\sigma'_x = -175$ psi. Determine the stresses τ_{xy} and τ'_{xy} at p.

P12.16

12.17 Under a different flight condition of the airplane in Problem 12.16, the stress components $\sigma_x = 80$ psi, $\sigma_y = -120$ psi, $\tau_{xy} = -100$ psi, $\sigma'_x = -80$ psi, and $\sigma'_y = 40$ psi. Determine the stress τ'_{xy} and the angle θ.

12.18 Equations (12.7)–(12.9) apply to plane stress, but they also apply to states of stress of the form

$$\begin{bmatrix} \sigma_x & \tau_{xy} & 0 \\ \tau_{yx} & \sigma_y & 0 \\ 0 & 0 & \sigma_z \end{bmatrix}.$$

Show that for a fluid at rest, $\sigma'_x = -p$ and $\tau'_{xy} = 0$ for any value of θ (see Eq. 12.2). That is, the normal stress at a point is the negative of the pressure and the shear stress is zero for any plane through the point.

12.19 By substituting the trigonometric identities (12.6) into Eqs. (12.4) and (12.5), derive Eqs. (12.7) and (12.8).

12.20 The components of plane stress acting on an element of a bar subjected to axial loads are shown in Fig. 12.7. Assuming the stress σ_x to be known, determine the principal stresses and the maximum in-plane shear stress and show them acting on properly oriented elements.

12.21 The components of plane stress acting on an element of a bar subjected to torsion are shown in Fig. 12.8. Assuming the stress τ_{xy} to be known, determine the principal stresses and the maximum in-plane shear stress and show them acting on properly oriented elements.

For the states of plane stress given in Problems 12.22–12.25, determine the principal stresses and the maximum in-plane shear stress and show them acting on properly oriented elements.

12.22 $\sigma_x = 20$ MPa, $\sigma_y = 10$ MPa, and $\tau_{xy} = 0$.

12.23 $\sigma_x = 25$ ksi, $\sigma_y = 0$, and $\tau_{xy} = -25$ ksi.

12.24 $\sigma_x = -8$ ksi, $\sigma_y = 6$ ksi, and $\tau_{xy} = -6$ ksi.

12.25 $\sigma_x = 240$ MPa, $\sigma_y = -120$ MPa, and $\tau_{xy} = 240$ MPa.

12.26 For the state of plane stress $\sigma_x = 20$ MPa, $\sigma_y = 10$ MPa, and $\tau_{xy} = 0$, what is the absolute maximum shear stress?

12.27 For the state of plane stress $\sigma_x = 25$ ksi, $\sigma_y = 0$, and $\tau_{xy} = -25$ ksi, what is the absolute maximum shear stress?

12.28 For the state of plane stress $\sigma_x = 8$ ksi, $\sigma_y = 6$ ksi, and $\tau_{xy} = -6$ ksi, what is the absolute maximum shear stress?

12.3 Mohr's Circle for Plane Stress

Mohr's circle is a graphical method for solving Eqs. (12.7), (12.8), and (12.9). Given a state of plane stress (Fig. 12.24a), Mohr's circle allows you to determine the components of stress in terms of a coordinate system rotated through a specified angle θ (Fig. 12.24b). Why do we discuss this method in an age when computers have made most graphical methods obsolete? The reason is that Mohr's circle allows us to visualize the solutions to Eqs. (12.7)–(12.9), and understand their properties, to an extent not possible with other approaches. We first explain how to apply Mohr's circle and then show why it works.

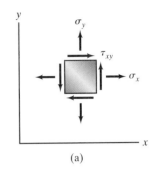

(a)

Constructing the Circle

Suppose that we know the components σ_x, τ_{xy}, and σ_y, and we want to determine the components σ'_x, τ'_{xy}, and σ'_y for a given angle θ. Determining this information with Mohr's circle involves four steps:

1. Establish a set of horizontal and vertical axes with normal stress measured along the horizontal axis and shear stress measured along the vertical axis (Fig. 12.25a). Positive normal stress is measured to the right and positive shear stress is measured *downward*.

2. Plot two points, point P with coordinates (σ_x, τ_{xy}) and point Q with coordinates $(\sigma_y, -\tau_{xy})$, as shown in Fig. 12.25b.

3. Draw a straight line connecting points P and Q. Using the intersection of the straight line with the horizontal axis as the center, draw a circle that passes through the two points (Fig. 12.25c).

4. Draw a straight line through the center of the circle at an angle 2θ measured counterclockwise from point P (Fig. 12.25d). The point P' at which this line intersects the circle has coordinates (σ'_x, τ'_{xy}), and the point Q' has coordinates $(\sigma'_y, -\tau'_{xy})$.

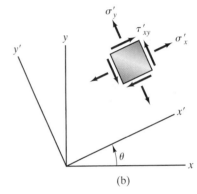

(b)

Figure 12.24
(a) State of plane stress.
(b) Components in terms of a rotated coordinate system.

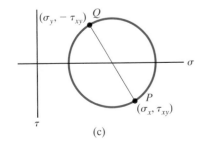

(a)

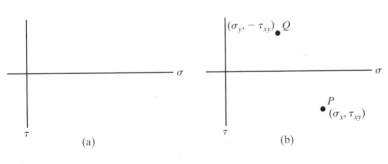

(b)

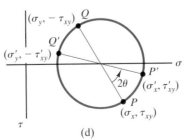

(c)

(d)

Figure 12.25
(a) Establishing the axes. Shear stress is positive downward.
(b) Plotting points P and Q.
(c) Drawing Mohr's circle. The center of the circle is the intersection of the line between points P and Q with the horizontal axis.
(d) Determining the stresses.

This construction indicates that for any value of the angle θ, we can determine the stress components σ'_x, τ'_{xy}, and σ'_y from the coordinates of two points on Mohr's circle. But we must prove this result.

Why Mohr's Circle Works

We will now prove that Mohr's circle solves Eqs. (12.7)–(12.9):

$$\sigma'_x = \frac{\sigma_x + \sigma_y}{2} + \frac{\sigma_x - \sigma_y}{2} \cos 2\theta + \tau_{xy} \sin 2\theta,$$

$$\tau'_{xy} = -\frac{\sigma_x - \sigma_y}{2} \sin 2\theta + \tau_{xy} \cos 2\theta,$$

$$\sigma'_y = \frac{\sigma_x + \sigma_y}{2} - \frac{\sigma_x - \sigma_y}{2} \cos 2\theta - \tau_{xy} \sin 2\theta.$$

In Fig. 12.26a we show the points P and Q and Mohr's circle. Notice that the horizontal coordinate of the center of the circle is $(\sigma_x + \sigma_y)/2$ and R, the radius of the circle, is given by

$$R = \sqrt{\left(\frac{\sigma_x - \sigma_y}{2}\right)^2 + \left(\tau_{xy}\right)^2}.$$

The sine and cosine of the angle β are

$$\sin \beta = \frac{\tau_{xy}}{R}, \qquad \cos \beta = \frac{\sigma_x - \sigma_y}{2R}.$$

From Fig. 12.26b, the horizontal coordinate of point P' is

$$\frac{\sigma_x + \sigma_y}{2} + R\cos(\beta - 2\theta) = \frac{\sigma_x + \sigma_y}{2} + R(\cos \beta \cos 2\theta + \sin \beta \sin 2\theta)$$

$$= \frac{\sigma_x + \sigma_y}{2} + \frac{\sigma_x - \sigma_y}{2} \cos 2\theta + \tau_{xy} \sin 2\theta$$

$$= \sigma'_x,$$

and the horizontal coordinate of point Q' is

$$\frac{\sigma_x + \sigma_y}{2} - R\cos(\beta - 2\theta) = \frac{\sigma_x + \sigma_y}{2} - R(\cos \beta \cos 2\theta + \sin \beta \sin 2\theta)$$

$$= \frac{\sigma_x + \sigma_y}{2} - \frac{\sigma_x - \sigma_y}{2} \cos 2\theta - \tau_{xy} \sin 2\theta$$

$$= \sigma'_y.$$

The vertical coordinate of point P' is

$$R\sin(\beta - 2\theta) = R(-\cos \beta \sin 2\theta + \sin \beta \cos 2\theta)$$

$$= -\frac{\sigma_x - \sigma_y}{2} \sin 2\theta + \tau_{xy} \cos 2\theta$$

$$= \tau'_{xy},$$

which implies that the vertical coordinate of point Q' is $-\tau'_{xy}$. Thus we have shown that the coordinates of point P' are (σ'_x, τ'_{xy}) and the coordinates of point Q' are $(\sigma'_y, -\tau'_{xy})$.

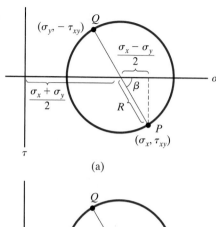

(a)

(b)

Figure 12.26
(a) Dimensions of Mohr's circle.
(b) Dimensions including the points P' and Q'.

Determining Principal Stresses and the Maximum In-Plane Shear Stress

Mohr's circle is a map of the stresses σ_x', τ_{xy}', and σ_y' for all values of θ, so once we have constructed the circle we can immediately see the values of the maximum and minimum normal stresses and the magnitude of the maximum in-plane shear stress. The coordinates of the points where the circle intersects the horizontal axis determine the two principal stresses (Fig. 12.27). The coordinates of the points at the bottom and top of the circle determine the maximum and minimum in-plane shear stresses, so the radius of the circle equals the magnitude of the maximum in-plane shear stress. Notice that Mohr's circle demonstrates very clearly that the shear stresses are zero on the planes on which the principal stresses act, and also that the normal stresses are equal on the planes on which the maximum and minimum in-plane shear stresses act.

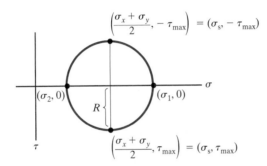

Figure 12.27
Mohr's circle indicates the values of the principal stresses and the magnitude of the maximum in-plane shear stress.

We can also use Mohr's circle to determine the orientations of the elements on which the principal stresses and maximum in-plane shear stresses act. If we let the point P' coincide with either principal stress (Fig. 12.28a), we can measure the angle $2\theta_p$ and thereby determine the orientation of the plane on which that principal stress acts (Fig. 12.28b).

We can then let P' coincide with either the maximum or minimum shear stress (Fig. 12.29a) and measure the angle $2\theta_s$, determining the orientation of

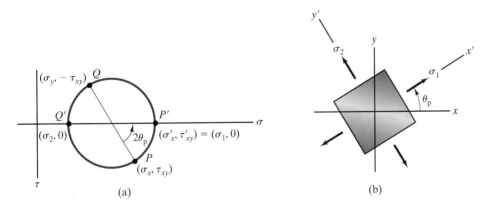

Figure 12.28
Using Mohr's circle to determine the orientation of the element on which the principal stresses act.

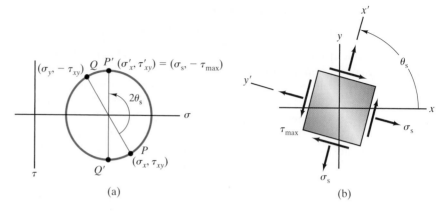

Figure 12.29
Using Mohr's circle to determine the orientation of the element on which the maximum in-plane shear stresses act.

the plane on which that shear stress acts (Fig. 12.29b). Notice in the example illustrated in Fig. 12.29 that point P' coincides with the minimum shear stress, so $\tau'_{xy} = -\tau_{max}$.

Example 12.5

Determining Stresses by Mohr's Circle

The state of plane stress at a point p is shown on the left element in Fig. 12.30. Use Mohr's circle to determine the state of plane stress at p acting on the right element in Fig. 12.30. Draw a sketch of the element showing the stresses acting on it.

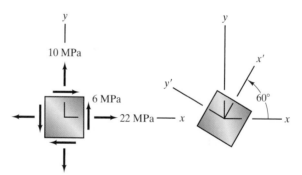

Figure 12.30

Strategy

The components of plane stress on the left element in Fig. 12.30 are $\sigma_x = 22$ MPa, $\sigma_y = 10$ MPa, and $\tau_{xy} = 6$ MPa. We can follow the steps given earlier for constructing Mohr's circle and determining the stress components σ'_x, τ'_{xy}, and σ'_y.

Solution

Step 1 Establish a set of horizontal and vertical axes with normal stress measured along the horizontal axis and shear stress measured along the vertical axis [Fig. (a)]. Positive shear stress is measured downward. The scales of normal stress and shear stress must be equal and chosen so that the circle will fit on the page but be large enough for reasonable accuracy. (This may require some trial and error.)

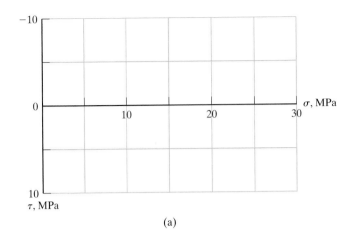

(a)

(a) Establishing the axes.

Step 2 Plot point P with coordinates $(\sigma_x, \tau_{xy}) = (22, 6)$ and point Q with coordinates $(\sigma_y, -\tau_{xy}) = (10, -6)$, as shown in Fig. (b).

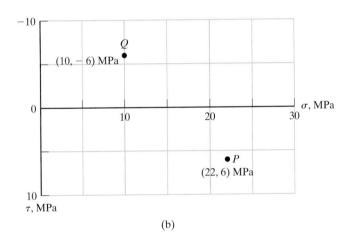

(b)

(b) Plotting points P and Q.

Step 3 Draw a straight line connecting points P and Q. Using the intersection of the straight line with the horizontal axis as the center, draw a circle that passes through the two points [Fig. (c)].

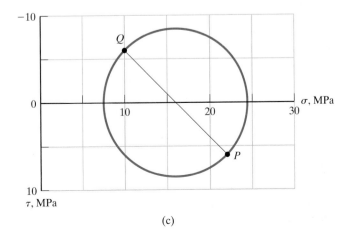

(c)

(c) Drawing the circle.

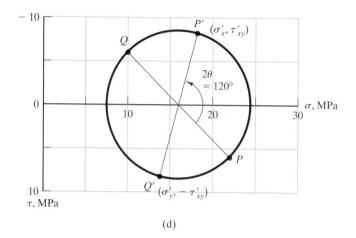

(d) Locating points P' and Q'.

(d)

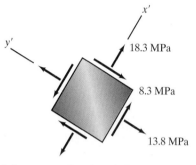

(e) Stresses on the rotated element.

Step 4 Draw a straight line through the center of the circle at an angle $2\theta = 120°$ measured counterclockwise from point P [Fig. (d)]. The point P' at which this line intersects the circle has coordinates (σ'_x, τ'_{xy}). The values we estimate from the graph are $\sigma'_x = 18.3$ MPa and $\tau'_{xy} = -8.3$ MPa. The point Q' has coordinates $(\sigma'_y, -\tau'_{xy})$, from which we estimate that $\sigma'_y = 13.8$ MPa. The stresses are shown in Fig. (e).

Discussion

Compare this application of Mohr's circle to the analytical solution of the same problem in Example 12.2.

Example 12.6

Principal Stresses by Mohr's Circle

The state of plane stress at a point p is shown on the element in Fig. 12.31. Use Mohr's circle to determine the principal stresses and the maximum in-plane shear stress and show them acting on properly oriented elements.

Strategy

The components of plane stress on the element are $\sigma_x = 22$ MPa, $\sigma_y = 10$ MPa, and $\tau_{xy} = 6$ MPa. Mohr's circle determines the principal stresses and the maximum and minimum in-plane shear stresses. By letting the point P' coincide first with one of the principal stresses and then with the maximum or minimum in-plane shear stress, we can determine the orientations of the elements on which these stresses act.

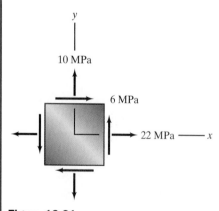

Figure 12.31

Solution

We first plot points P and Q and draw Mohr's circle [Fig. (a)]. Then we let the point P' coincide with one of the principal stresses [Fig. (b)]. From the

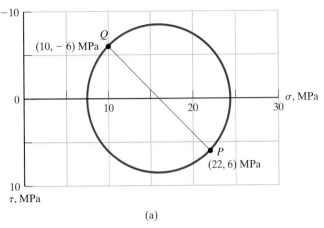

(a)

(a) Mohr's circle.

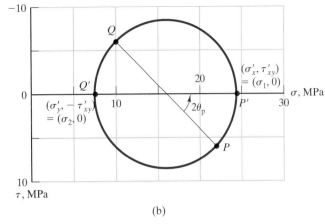

(b)

(b) Letting P' coincide with the principal stress σ_1.

circle we estimate that $\sigma_1 = 24.5$ MPa and $\sigma_2 = 7.5$ MPa. Measuring the angle $2\theta_p$, we estimate that $\theta_p = 22.5°$, which determines the orientation of the plane on which the principal stress σ_1 acts [Fig. (c)].

We next let the point P' coincide with either the minimum or maximum in-plane shear stress. In this case we choose the minimum stress [Fig. (d)]. We estimate that $\tau_{max} = 8.5$ MPa and the normal stress $\sigma_s = 16$ MPa. Measuring the angle $2\theta_s$, we estimate that $\theta_s = 67.5°$, which determines the orientation of the plane on which the minimum in-plane shear stress acts [Fig. (e)].

Discussion

Compare this solution to Example 12.4, in which we begin with the same state of stress and analytically determine the principal stresses and maximum in-plane shear stress.

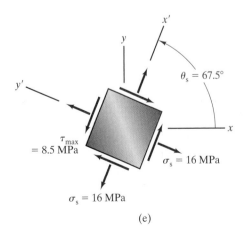

(c)

(c) Element on which the principal stresses act.

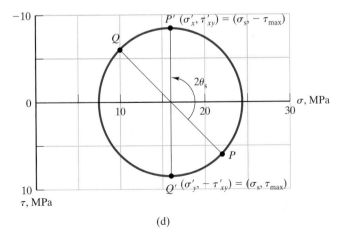

(d)

(d) Letting P' coincide with the minimum in-plane shear stress.

(e) Element on which the maximum and minimum in-plane shear stresses act. Notice that $\tau'_{xy} = -\tau_{max}$.

Example 12.7

Determining Orientation of an Element by Mohr's Circle

The state of plane stress at a point p is shown on the left element in Fig. 12.32, and the values of the stresses σ'_x and τ'_{xy} are shown on the right element. What are the normal stress σ'_y and the angle θ?

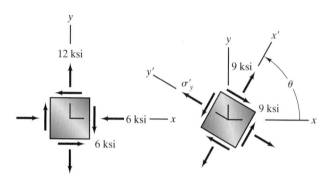

Figure 12.32

Strategy

The components of stress on the left element are $\sigma_x = -6$ ksi, $\tau_{xy} = -6$ ksi, and $\sigma_y = 12$ ksi. With this information we can plot the points P and Q and draw Mohr's circle. On the rotated element, the stresses $\sigma'_x = 9$ ksi and $\tau'_{xy} = -9$ ksi, which permits us to locate the point P'. Then we can measure the angle 2θ and determine the normal stresss σ'_y from the coordinates of the point Q'.

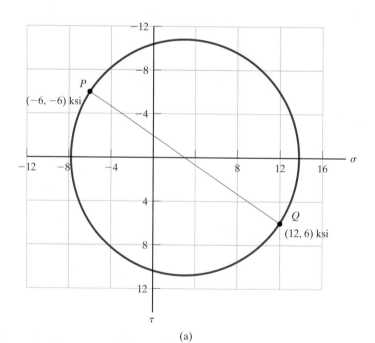

(a) Plotting the points P and Q and drawing Mohr's circle.

(a)

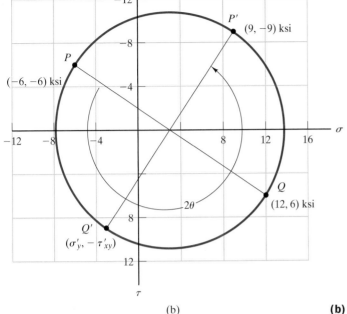

(b)

(b) Plotting P'.

Solution

Figure (a) shows the points P and Q and Mohr's circle. In Fig. (b) we plot the point P' and draw a straight line from P' through the center of the circle. From the coordinates of point Q', we estimate that $\sigma'_y = -3$ ksi. Measuring the angle 2θ, we estimate that $\theta = 135°$. The stresses are shown on the properly oriented element in Fig. (c).

Discussion

Compare this solution with Example 12.3, in which we solve the same problem analytically.

(c) Stresses on the properly oriented element.

Problems

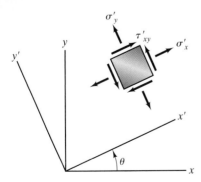

P12.29–P12.34

12.29 The components of plane stress at a point p of a material are $\sigma_x = 20$ MPa, $\sigma_y = 0$, and $\tau_{xy} = 0$, and the angle $\theta = 45°$. Use Mohr's circle to determine the stresses σ'_x, σ'_y, and τ'_{xy} at point p.

12.30 The components of plane stress at a point p of a material are $\sigma_x = 0$, $\sigma_y = 0$, and $\tau_{xy} = 25$ ksi, and the angle $\theta = 45°$. Use Mohr's circle to determine the stresses σ'_x, and σ'_y, and τ'_{xy} at point p.

12.31 The components of plane stress at a point p of a material are $\sigma_x = 240$ MPa, $\sigma_y = -120$ MPa, and $\tau_{xy} = 240$ MPa, and the components referred to the $x'y'z'$ coordinate system are

$\sigma'_x = 347$ MPa, $\sigma'_y = -227$ MPa, and $\tau'_{xy} = -87$ MPa. Use Mohr's circle to determine the angle θ.

12.32 Use Mohr's circle to determine the components of stress at a point of one of the Space Shuttle's main engine nozzles in Problem 12.4.

12.33 The components of plane stress at a point p of a material referred to the $x'y'z'$ coordinate system are $\sigma'_x = -8$ MPa, $\sigma'_y = 6$ MPa, and $\tau'_{xy} = -16$ MPa, and the angle $\theta = 20°$. Use Mohr's circle to determine the stresses σ_x, σ_y, and τ_{xy} at point p.

12.34 The components of plane stress at a point p of a bit during a drilling operation are $\sigma_x = 40$ ksi, $\sigma_y = -30$ ksi, and $\tau_{xy} = 30$ ksi, and the components referred to the $x'y'z'$ coordinate system are $\sigma'_x = 12.5$ ksi, $\sigma'_y = -2.5$ ksi, and $\tau'_{xy} = 45.5$ ksi. Use Mohr's circle to estimate the angle θ.

12.35 The components of plane stress at point p of the material shown are $\sigma_x = 4$ ksi, $\sigma_y = -2$ ksi, and $\tau_{xy} = 2$ ksi. Use Mohr's circle to determine the normal stress and the magnitude of the shear stress on the plane $\mathcal{P}$ at point p.

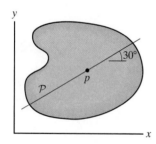

P12.35

12.36 The components of plane stress at point p of the material shown in Problem 12.35 are $\sigma_x = -10.5$ MPa, $\sigma_y = 6.0$ MPa, and $\tau_{xy} = -4.5$ MPa. Use Mohr's circle to determine the normal stress and the magnitude of the shear stress on the plane $\mathcal{P}$ at point p.

12.37 Determine the stresses σ and τ (a) by using Mohr's circle; (b) by using Eqs. (12.7) and (12.8).

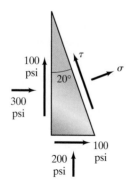

P12.37

12.38 Solve Problem 12.37 if the 300-psi stress on the element is in tension instead of compression.

12.39 Determine the stresses σ and τ (a) by using Mohr's circle; (b) by using Eqs. (12.7) and (12.8).

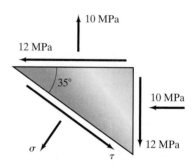

P12.39

For the states of plane stress given in Problems 12.40–12.43, use Mohr's circle to determine the principal stresses and the maximum in-plane shear stress and show them acting on properly oriented elements.

12.40 $\sigma_x = 20$ MPa, $\sigma_y = 10$ MPa, and $\tau_{xy} = 0$.

12.41 $\sigma_x = 25$ ksi, $\sigma_y = 0$, and $\tau_{xy} = -25$ ksi.

12.42 $\sigma_x = -8$ ksi, $\sigma_y = 6$ ksi, and $\tau_{xy} = -6$ ksi.

12.43 At touchdown, a point p of the Space Shuttle's landing gear is subjected to the state of plane stress $\sigma_x = -120$ MPa, $\sigma_y = 80$ MPa, and $\tau_{xy} = -50$ MPa. Use Mohr's circle to determine the principal stresses and the maximum in-plane shear stress.

P12.43

12.44 For the state of plane stress $\sigma_x = 8$ ksi, $\sigma_y = 6$ ksi, and $\tau_{xy} = -6$ ksi, use Mohr's circle to determine the absolute maximum shear stress. [*Strategy:* Use Mohr's circle to determine the principal stresses and then determine the absolute maximum shear stress from the expressions (12.24).]

12.4 Principal Stresses in Three Dimensions

Many important applications involve states of stress more general than plane stress. Because of the crucial importance of maximum stresses in design, in this section we explain how to determine the principal stresses and absolute maximum shear stress for a general state of stress and for a particular three-dimensional state of stress called *triaxial stress*.

General State of Stress

We have seen in our discussion of plane stress that the components of the state of stress depend on the orientation of the coordinate system in which they are expressed. Suppose that we know the components of stress at a point p in terms of a particular coordinate system xyz:

$$[I] = \begin{bmatrix} \sigma_x & \tau_{xy} & \tau_{xz} \\ \tau_{yx} & \sigma_y & \tau_{yz} \\ \tau_{zx} & \tau_{zy} & \sigma_z \end{bmatrix}.$$

These components will generally have different values when expressed in terms of a coordinate system $x'y'z'$ having a different orientation (Fig. 12.33). For *any* state of stress, at least one coordinate system $x'y'z'$ exists for which the state of stress is of the form

$$\begin{bmatrix} \sigma'_x & \tau'_{xy} & \tau'_{xz} \\ \tau'_{yx} & \sigma'_y & \tau'_{yz} \\ \tau'_{zx} & \tau'_{zy} & \sigma'_z \end{bmatrix} = \begin{bmatrix} \sigma_1 & 0 & 0 \\ 0 & \sigma_2 & 0 \\ 0 & 0 & \sigma_3 \end{bmatrix}.$$

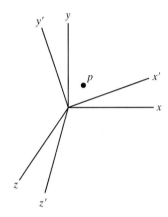

Figure 12.33
The xyz coordinate system and a system $x'y'z'$ with a different orientation.

The axes x', y', z' are called *principal axes* and σ_1, σ_2, and σ_3 are the principal stresses. An infinitesimal element at p that is oriented with the principal axes is subject to the principal stresses and no shear stress (Fig. 12.34). It can be shown that the principal stresses are the roots of the cubic equation

$$\sigma^3 - I_1\sigma^2 + I_2\sigma - I_3 = 0, \tag{12.25}$$

where

$$I_1 = \sigma_x + \sigma_y + \sigma_z,$$
$$I_2 = \sigma_x\sigma_y + \sigma_y\sigma_z + \sigma_z\sigma_x - \tau_{xy}^2 - \tau_{yz}^2 - \tau_{zx}^2, \tag{12.26}$$
$$I_3 = \sigma_x\sigma_y\sigma_z - \sigma_x\tau_{yz}^2 - \sigma_y\tau_{xz}^2 - \sigma_z\tau_{xy}^2 + 2\tau_{xy}\tau_{yz}\tau_{zx}.$$

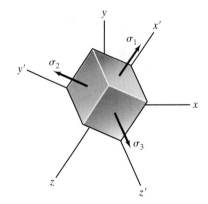

Figure 12.34
Stresses on an element oriented with the principal axes.

[Although the components of the state of stress depend on the orientation of the coordinate system in which they are expressed, the values of these three coefficients do not. This can be deduced from the fact that the principal stresses, the roots of Eq. (12.25), cannot depend on the orientation of the co-ordinate system used to evaluate them. For this reason I_1, I_2, and I_3 are called *stress invariants*.] Thus for a given state of stress, the principal stresses can be determined by evaluating the coefficients I_1, I_2, and I_3 and solving Eq. (12.25).

The absolute maximum shear stress can be determined by the same approach as that applied to plane stress in Section 12.2. We begin with the element on which the principal stresses occur, and align the coordinate system

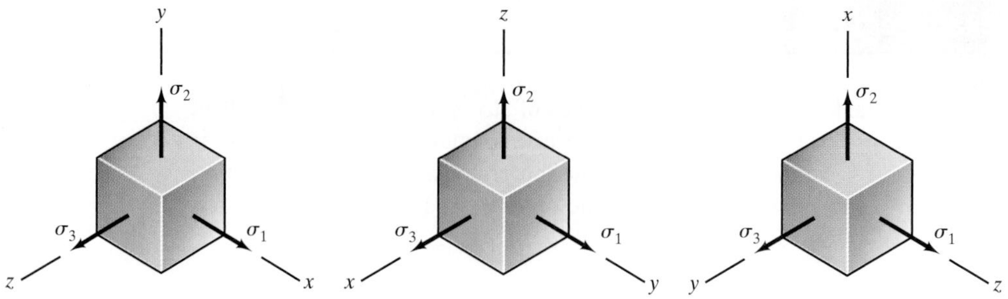

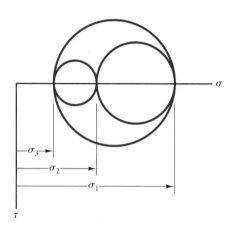

Figure 12.35
Different orientations of the coordinate system relative to the element on which the principal stresses act.

with the faces of the element in the three ways shown in Fig. 12.35. Applying Eq. (12.19) to each of these orientations, we find that the absolute maximum shear stress is the largest of the three values

$$\left|\frac{\sigma_1 - \sigma_2}{2}\right|, \quad \left|\frac{\sigma_1 - \sigma_3}{2}\right|, \quad \left|\frac{\sigma_2 - \sigma_3}{2}\right|. \tag{12.27}$$

Although this analysis considers only a subset of the possible planes through point p, no shear stresses of greater magnitude act on any plane.

In Fig. 12.36 we show that the absolute maximum shear stress determined in this way can be visualized very clearly by superimposing the Mohr's circles obtained from the three orientations of the coordinate system in Fig. 12.35. Notice that if $\sigma_1 > \sigma_2 > \sigma_3$, the absolute maximum shear stress is $(\sigma_1 - \sigma_3)/2$.

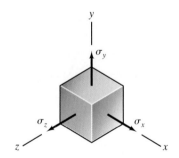

Figure 12.36
Superimposing the Mohr's circles graphically demonstrates the absolute maximum shear stress.

Triaxial Stress

The state of stress at a point is said to be *triaxial* if it is of the form

$$\begin{bmatrix} \sigma_x & 0 & 0 \\ 0 & \sigma_y & 0 \\ 0 & 0 & \sigma_z \end{bmatrix}. \tag{12.28}$$

The shear stress components τ_{xy}, τ_{xz}, and τ_{yz} are zero (Fig. 12.37). For example, the state of stress at a point in a fluid at rest [Eq. 12.2] is triaxial. In triaxial stress, x, y, and z are principal axes and σ_x, σ_y, and σ_z are the principal stresses. From Eq. (12.27), the absolute maximum shear stress in triaxial stress is the largest of the three values

$$\left|\frac{\sigma_x - \sigma_y}{2}\right|, \quad \left|\frac{\sigma_x - \sigma_z}{2}\right|, \quad \left|\frac{\sigma_y - \sigma_z}{2}\right|. \tag{12.29}$$

Figure 12.37
Triaxial stress.

Study Questions

1. How can you determine the principal stresses for a general state of stress?
2. What are the principal axes? How can you determine their directions?
3. Why are the coefficients I_1, I_2, and I_3 in Eq. (12.25) called stress invariants?
4. What is triaxal stress? What are the principal stresses for a state of triaxal stress?

Example 12.8

Principal Stresses in Three Dimensions

A point p in the frame of the racing motorcycle in Fig. 12.38 is subjected to the state of stress (in MPa)

$$
\begin{bmatrix}
\sigma_x & \tau_{xy} & \tau_{xz} \\
\tau_{yx} & \sigma_y & \tau_{yz} \\
\tau_{zx} & \tau_{zy} & \sigma_z
\end{bmatrix}
=
\begin{bmatrix}
4 & 2 & 1 \\
2 & 2 & 1 \\
1 & 1 & 3
\end{bmatrix}.
$$

Determine the principal stresses and the absolute maximum shear stress at p.

Figure 12.38

Strategy

Since we know the state of stress, we can use Eqs. (12.26) to evaluate the co-efficients I_1, I_2, and I_3, then solve Eq. (12.25) to determine the principal stresses. The absolute maximum shear stress is the largest of the three values in Eq. (12.27).

Solution

Principal Stresses   Substituting the components of stress in MPa into Eqs. (12.26), we obtain $I_1 = 9$, $I_2 = 20$, and $I_3 = 10$, so Eq. (12.25) is

$$\sigma^3 - 9\sigma^2 + 20\sigma - 10 = 0.$$

We can estimate the roots of this cubic equation by drawing a graph of the left side as a function of σ [Fig. (a)]. By using software designed to obtain roots of nonlinear algebraic equations, we obtain $\sigma_1 = 5.895$ MPa, $\sigma_2 = 2.397$ MPa, and $\sigma_3 = 0.708$ MPa.

Absolute Maximum Shear Stress   The three values in Eq. (12.27) are

$$\left| \frac{\sigma_1 - \sigma_2}{2} \right| = \left| \frac{5.895 - 2.397}{2} \right| = 1.749 \text{ MPa},$$

$$\left| \frac{\sigma_1 - \sigma_3}{2} \right| = \left| \frac{5.895 - 0.708}{2} \right| = 2.594 \text{ MPa},$$

$$\left| \frac{\sigma_2 - \sigma_3}{2} \right| = \left| \frac{2.397 - 0.708}{2} \right| = 0.845 \text{ MPa}.$$

The absolute maximum shear stress is 2.594 MPa.

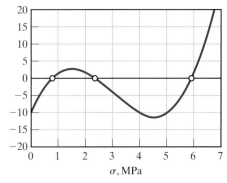

(a) Graph of $\sigma^3 - 9\sigma^2 + 20\sigma - 10$.

Problems

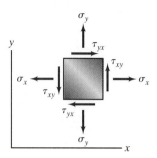

P12.45–P12.47

12.45 At a point p a material is subjected to the state of *plane* stress $\sigma_x = 20$ MPa, $\sigma_y = 10$ MPa, $\tau_{xy} = 0$. Use Eq. (12.25) to determine the principal stresses and use Eq. (12.27) to determine the absolute maximum shear stress. Confirm the absolute maximum shear stress by drawing the superimposed Mohr's circle as shown in Fig. 12.36.

12.46 At a point p a material is subjected to the state of *plane* stress $\sigma_x = 25$ ksi, $\sigma_y = 0$, $\tau_{xy} = -25$ ksi. Use Eq. (12.25) to determine the principal stresses and use Eq. (12.27) to determine the absolute maximum shear stress.

12.47 At a point p a material is subjected to the state of *plane* stress $\sigma_x = 240$ MPa, $\sigma_y = -120$ MPa, $\tau_{xy} = 240$ MPa. Use Eq. (12.25) to determine the principal stresses and use Eq. (12.27) to determine the absolute maximum shear stress. Confirm the absolute maximum shear stress by drawing the superimposed Mohr's circle as shown in Fig. 12.36.

12.48 Strain gauges attached to one of the Space Shuttle main engine nozzles determine that the components of *plane* stress are $\sigma_x = 67.34$ MPa, $\sigma_y = 82.66$ MPa, and $\tau_{xy} = 6.43$ MPa. Use Eq. (12.25) to determine the principal stresses and use Eq. (12.27) to determine the absolute maximum shear stress.

P12.48

12.49 Use Eq. (12.25) to determine the principal stresses for an arbitrary state of plane stress $\sigma_x, \sigma_y, \tau_{xy}$ and confirm Eq. (12.15).

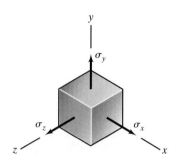

P12.50, P12.51

12.50 At a point p a material is subjected to the state of *triaxial* stress $\sigma_x = 240$ MPa, $\sigma_y = -120$ MPa, $\sigma_z = 240$ MPa. Determine the principal stresses and the absolute maximum shear stress.

12.51 At a point p a material is subjected to the state of *triaxial* stress $\sigma_x = 40$ ksi, $\sigma_y = 80$ ksi, $\sigma_z = -20$ ksi. Determine the principal stresses and the absolute maximum shear stress.

12.52 At a point p a material is subjected to the state of stress (in ksi)

$$\begin{bmatrix} \sigma_x & \tau_{xy} & \tau_{xz} \\ \tau_{yx} & \sigma_y & \tau_{yz} \\ \tau_{zx} & \tau_{zy} & \sigma_z \end{bmatrix} = \begin{bmatrix} 300 & 150 & -100 \\ 150 & 200 & 100 \\ -100 & 100 & -200 \end{bmatrix}.$$

Determine the principal stresses and the absolute maximum shear stress. Confirm the absolute maximum shear stress by drawing the superimposed Mohr's circle as shown in Fig. 12.36.

12.53 A finite element analysis of a bearing housing indicates that at a point p the material is subjected to the state of stress (in MPa)

$$\begin{bmatrix} \sigma_x & \tau_{xy} & \tau_{xz} \\ \tau_{yx} & \sigma_y & \tau_{yz} \\ \tau_{zx} & \tau_{zy} & \sigma_z \end{bmatrix} = \begin{bmatrix} 20 & 20 & 0 \\ 20 & -30 & -10 \\ 0 & -10 & 40 \end{bmatrix}.$$

Determine the principal stresses and the absolute maximum shear stress.

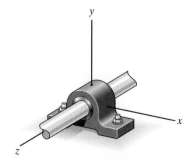

P12.53

12.54 The components of *plane* stress at a point p of a material are $\sigma_x = -8$ ksi, $\sigma_y = 6$ ksi, and $\tau_{xy} = -6$ ksi. Use Eqs. (12.7)–(12.9) to determine the components of stress σ'_x, σ'_y, and τ'_{xy} corresponding to a coordinate system $x'y'z'$ oriented at $\theta = 30°$. (a) Determine the principal stresses from Eq. (12.25) using the components of stress σ_x, σ_y, and τ_{xy}. (b) Determine the principal stresses from Eq. (12.25) using the components of stress σ'_x, σ'_y, and τ'_{xy}.

12.5 Design Issues: Pressure Vessels

A faculty member in our college once proposed that the first course in mechanics of materials no longer be part of the mechanical engineering curriculum, to which a colleague responded "Well, if that happens I'm going to stop standing near boilers!" Underlying his facetious remark on the need for mechanical engineers to understand solid mechanics (the proposal was soundly rejected) was a serious concern: the proper design of pressure vessels. The critical importance of this problem in engineering has been emphasized by pressure vessel accidents from the early days of steam power to *Apollo 13*. We discuss this subject in this chapter because pressure vessels provide interesting examples of triaxial states of stress (see Section 12.4).

Spherical Vessels

Consider a spherical pressure vessel with radius R and wall thickness t, where $t \ll R$ (Fig. 12.39). We assume that the vessel contains a gas with uniform pressure p_i and that the outer wall is subjected to a uniform pressure p_o, and we denote the difference between the inner and outer pressures by $p = p_i - p_o$. In applications, p_o is often atmospheric pressure (approximately 10^5 Pa or 14.7 psi).

We can approximate the state of stress in a thin-walled spherical pressure vessel if we assume that the effects of loads other than those exerted by the internal and external pressures are negligible. In Fig. 12.40 we bisect the vessel by a plane and draw a free-body diagram of one half, *including the gas it contains*. The normal stress in the wall, which can be approximated as being uniformly distributed for a thin-walled vessel, is denoted by σ. We want to determine the stress σ by summing the horizontal forces on the free-body diagram, but what horizontal force is exerted by the pressure p_o? Suppose that a solid hemisphere of radius R is subjected to a uniform pressure p_o (Fig. 12.41). No net force is exerted on an object by a uniform distribution of pressure, so the horizontal force to the right exerted on the hemispherical surface must equal the force to the left exerted on the plane circular face: $p_o(\pi R^2)$. Therefore, the sum of the horizontal forces on the free-body diagram in Fig. 12.40 is

$$p_o(\pi R^2) + \sigma(2\pi Rt) - p_i(\pi R^2) = 0.$$

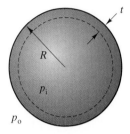

Figure 12.39
Spherical pressure vessel. The wall thickness is exaggerated.

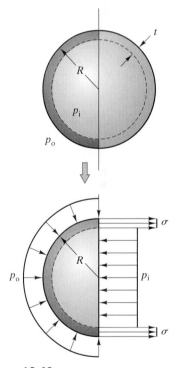

Figure 12.40
Free-body diagram of half of the pressure vessel, including the enclosed gas.

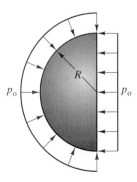

Figure 12.41
Hemisphere subjected to uniform pressure p_o.

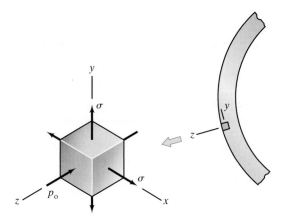

Figure 12.42
Stresses on an element at the outer surface.

Notice that the assumption $t \ll R$ is used in writing this equation. From this equation we obtain the normal stress σ in terms of the dimensions of the vessel and the pressure difference p:

$$\sigma = \frac{(p_i - p_o)R}{2t} = \frac{pR}{2t}. \tag{12.30}$$

An element isolated from the vessel's outer surface is subjected to the stresses shown in Fig. 12.42. In terms of the coordinate system shown (the z axis is perpendicular to the wall), the triaxial state of stress is

$$\begin{bmatrix} \sigma_x & \tau_{xy} & \tau_{xz} \\ \tau_{yx} & \sigma_y & \tau_{yz} \\ \tau_{zx} & \tau_{zy} & \sigma_z \end{bmatrix} = \begin{bmatrix} \sigma & 0 & 0 \\ 0 & \sigma & 0 \\ 0 & 0 & -p_o \end{bmatrix}. \tag{12.31}$$

As we discussed in Section 12.4, the principal stresses for this state of stress are σ, σ, $-p_o$, and from Eq. (12.29) the absolute maximum shear stress is $|(\sigma + p_o)/2|$. The triaxial state of stress on an element isolated from the vessel's inner surface (Fig. 12.43) is

$$\begin{bmatrix} \sigma_x & \tau_{xy} & \tau_{xz} \\ \tau_{yx} & \sigma_y & \tau_{yz} \\ \tau_{zx} & \tau_{zy} & \sigma_z \end{bmatrix} = \begin{bmatrix} \sigma & 0 & 0 \\ 0 & \sigma & 0 \\ 0 & 0 & -p_i \end{bmatrix}. \tag{12.32}$$

In this case the principal stresses are σ, σ, $-p_i$ and the absolute maximum shear stress is $|(\sigma + p_i)/2|$.

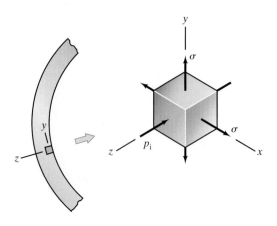

Figure 12.43
Stresses on an element at the inner surface.

Cylindrical Vessels

We now consider a cylindrical vessel with radius R and wall thickness t (Fig. 12.44). The vessel shown has hemispherical ends, but the results we will obtain for the state of stress in the cylindrical wall do not depend on the shapes of the ends. We again assume that the vessel contains a gas with uniform pressure p_i and that the outer wall is subjected to a uniform pressure p_o.

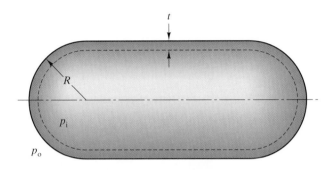

Figure 12.44
Cylindrical pressure vessel.

In Fig. 12.45 we obtain a free-body diagram by passing a plane through the cylindrical wall perpendicular to its axis. The normal stress in the wall is denoted by σ. The horizontal forces on this free-body diagram are identical to those on the free-body diagram of the spherical vessel in Fig. 12.40, so the normal stress σ is given by Eq. (12.30):

$$\sigma = \frac{(p_i - p_o)R}{2t} = \frac{pR}{2t}. \tag{12.33}$$

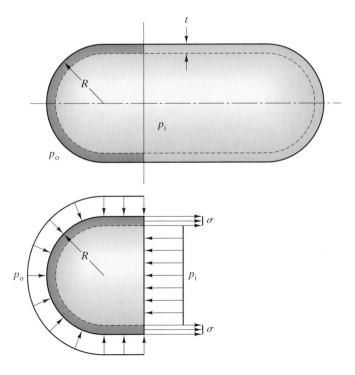

Figure 12.45
Free-body diagram obtained by passing a plane perpendicular to the cylinder axis.

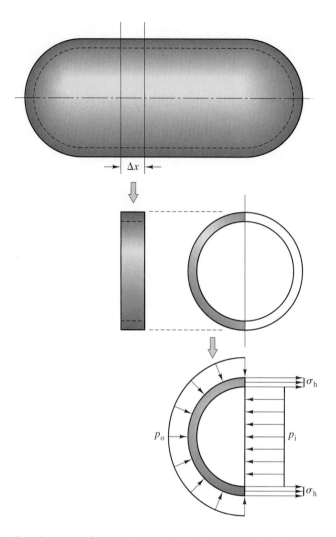

Figure 12.46
Free-body diagram for determining the hoop stress σ_h.

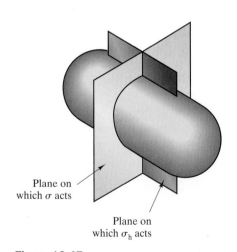

Plane on
which σ acts

Plane on
which σ_h acts

Figure 12.47
Planes on which the normal stresses σ and σ_h act.

But the state of stress in a cylindrical vessel is slightly more complicated than that in a spherical vessel. In Fig. 12.46 we first isolate a "slice" of the cylindrical wall of length Δx, including the gas it contains, then obtain a free-body diagram by bisecting the slice by a plane parallel to the cylinder axis. The normal stress σ_h on the resulting free-body diagram is called the *hoop stress* (so named because it plays the same role as the tensile stresses in the metal hoops used to reinforce wooden barrels). Planes on which σ_h acts are perpendicular to planes on which the normal stress σ acts (Fig. 12.47).

The sum of the horizontal forces on the free-body diagram in Fig. 12.46 is

$$p_o(2R\,\Delta x) + \sigma_h(2t\,\Delta x) - p_i(2R\,\Delta x) = 0.$$

Solving for the hoop stress, we obtain

$$\sigma_h = \frac{(p_i - p_o)R}{t} = \frac{pR}{t}. \tag{12.34}$$

Observe that $\sigma_h = 2\sigma$, which helps explain the popularity of spherical pressure vessels.

An element isolated from the outer surface of the cylindrical wall is subjected to the stresses shown in Fig. 12.48. In terms of the coordinate system

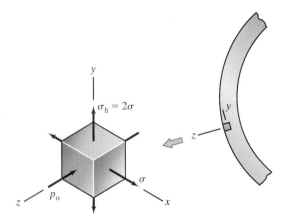

Figure 12.48
Stresses on an element at the outer surface.

shown (the z axis is perpendicular to the wall and the x axis is parallel to the axis of the cylinder), the triaxial state of stress is

$$
\begin{bmatrix}
\sigma_x & \tau_{xy} & \tau_{xz} \\
\tau_{yx} & \sigma_y & \tau_{yz} \\
\tau_{zx} & \tau_{zy} & \sigma_z
\end{bmatrix}
=
\begin{bmatrix}
\sigma & 0 & 0 \\
0 & \sigma_h & 0 \\
0 & 0 & -p_o
\end{bmatrix}.
\tag{12.35}
$$

The principal stresses are σ_h, σ, $-p_o$ and from Eq. (12.29) the absolute maximum shear stress is the largest of the three values

$$
\left|\frac{\sigma}{2}\right|, \qquad \left|\frac{\sigma + p_o}{2}\right|, \qquad \left|\frac{\sigma_h + p_o}{2}\right|.
\tag{12.36}
$$

The triaxial state of stress on an element isolated from the vessel's inner surface (Fig. 12.49) is

$$
\begin{bmatrix}
\sigma_x & \tau_{xy} & \tau_{xz} \\
\tau_{yx} & \sigma_y & \tau_{yz} \\
\tau_{zx} & \tau_{zy} & \sigma_z
\end{bmatrix}
=
\begin{bmatrix}
\sigma & 0 & 0 \\
0 & \sigma_h & 0 \\
0 & 0 & -p_i
\end{bmatrix}.
\tag{12.37}
$$

In this case the principal stresses are σ_h, σ, $-p_i$ and the absolute maximum shear stress is the largest of the three values

$$
\left|\frac{\sigma}{2}\right|, \qquad \left|\frac{\sigma + p_i}{2}\right|, \qquad \left|\frac{\sigma_h + p_i}{2}\right|.
\tag{12.38}
$$

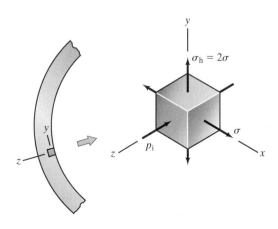

Figure 12.49
Stresses on an element at the inner surface.

Allowable Stress

The one aspect of pressure vessel design we consider is the most important: making sure that the stresses in the vessel walls do not exceed those the material will support. We have determined the principal stresses and absolute maximum shear stresses in the walls of spherical and cylindrical vessels. The criterion for design we use in our example and problems is to ensure that the absolute maximum shear stress in the material does not exceed a shear yield stress τ_Y, or some specified allowable shear stress τ_{allow}. This is called the *Tresca criterion*, and is discussed further in Chapter 12. The factor of safety is now defined by

$$S = \frac{\tau_Y}{\tau_{\text{allow}}}. \tag{12.39}$$

Since the maximum shear stress in a tensile test is one-half the applied normal stress, we will assume that the shear yield stress τ_Y is given in terms of the normal yield stress σ_Y tabulated in Appendix D by the relation

$$\tau_Y = \tfrac{1}{2}\sigma_Y. \tag{12.40}$$

Of course, the values in Appendix D are merely representative, and in actual design the yield stress must be determined for the specific material being used. We also emphasize that our analyses apply only to thin-walled vessels made of homogeneous and isotropic materials.

Study Questions

1. What are the principal stresses at a point of the inner surface of a spherical pressure vessel?
2. What is the hoop stress?
3. From the standpoint of stresses, why is a spherical pressure vessel superior to a cylindrical one?
4. What is the Tresca criterion? How can it be used in pressure vessel design?

Example 12.9

Design of a Cylindrical Pressure Vessel

A cylindrical pressure vessel with 2-m radius and hemispherical ends is to be designed to support an internal pressure as large as $p_i = 8 \times 10^5$ Pa (8 atmospheres) with the outer pressure equal to atmospheric pressure $p_o = 1 \times 10^5$ Pa. It is to be constructed of ASTM-A514 steel. Determine the vessel's wall thickness so that it has a factor of safety $S = 4$.

Strategy

We can determine the normal yield stress σ_Y for ASTM-A514 steel from Appendix D, then use Eqs. (12.40) and (12.39) to determine the allowable value of the absolute maximum shear stress τ_{allow}. Comparing the terms (12.36) and (12.38), and remembering that $\sigma_h = 2\sigma$, it is clear that the absolute maximum shear stress is $(\sigma_h + p_i)/2$. (The absolute value signs are unnecessary because σ_h is positive.) By equating this expression to τ_{allow} and using Eq. (12.34), we can determine the necessary wall thickness.

Solution

From Appendix D the normal yield stress σ_Y for ASTM-A514 steel is 700 MPa, so from Eq. (12.40), $\tau_Y = 350$ MPa. Then from Eq. (12.39), the allowable value of the absolute maximum shear stress is

$$\tau_{allow} = \frac{\tau_Y}{S} = \frac{350}{4} = 87.5 \text{ MPa.}$$

We equate the absolute maximum shear stress to the allowable value:

$$\frac{\sigma_h + p_i}{2} = \tau_{allow}.$$

Substituting Eq. (12.34) into this expression and solving for the wall thickness, we obtain

$$t = \frac{(p_i - p_o)R}{2\tau_{allow} - p_i}$$

$$= \frac{(7 \times 10^5)(2)}{(2)(87.5 \times 10^6) - 8 \times 10^5}$$

$$= 0.00804 \text{ m.}$$

The necessary wall thickness is 8.04 mm.

Problems

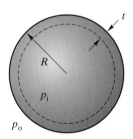

P12.55–P12.57

12.55 A spherical pressure vessel has a 2.5-m radius and a 5-mm wall thickness. It contains a gas with pressure $p_i = 6 \times 10^5$ Pa and the outer wall is subjected to atmospheric pressure $p_o = 1 \times 10^5$ Pa. Determine the maximum normal stress in the vessel wall.

12.56 A spherical pressure vessel has a 24-in. radius and a $\frac{1}{64}$-in. wall thickness. It contains a gas with pressure $p_i = 200$ psi and the outer wall is subjected to atmospheric pressure $p_o = 14.7$ psi. Determine the maximum normal stress and the absolute maximum shear stress at the vessel's inner surface.

12.57 Suppose that the spherical pressure vessel described in Problem 12.56 is made of material with a yield shear stress

$\tau_Y = 100$ ksi. If the vessel is designed to contain gas with a maximum pressure $p_i = 150$ psi and the outer wall is subjected to atmospheric pressure $p_o = 14.7$ psi, what is the factor of safety?

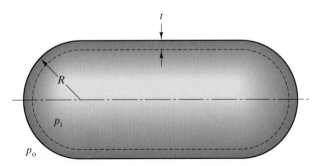

P12.58–P12.61

12.58 A cylindrical pressure vessel with hemispherical ends has a 2.5-m radius and a 5-mm wall thickness. It contains a gas with pressure $p_i = 6 \times 10^5$ Pa, and the outer wall is subjected to atmospheric pressure $p_o = 1 \times 10^5$ Pa. Determine the maximum normal stress in the vessel wall. Compare your answer to the answer to Problem 12.55.

12.59 In Example 12.9, the wall thickness of the cylindrical vessel is determined to be 8.04 mm. What is the resulting maximum normal stress in the vessel wall?

12.60 A cylindrical pressure vessel has a 600-mm radius and an 8-mm wall thickness. It contains a gas with pressure $p_i = 3 \times 10^5$ Pa and the outer wall is subjected to atmospheric pressure $p_o = 1 \times 10^5$ Pa. Determine the maximum normal stress

and the absolute maximum shear stress at the inner surface of the vessel's cylindrical wall.

12.61 A cylindrical pressure vessel used for natural gas storage has a 6-ft radius and a $\frac{1}{2}$-in. wall thickness. It contains gas with pressure $p_i = 80$ psi and the outer wall is subjected to atmospheric pressure $p_o = 14.7$ psi. Determine the maximum normal stress and the absolute maximum shear stress at the inner surface of the vessel's cylindrical wall.

Chapter Summary

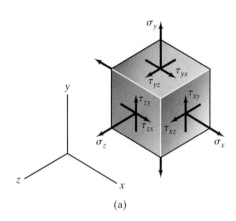

(a)

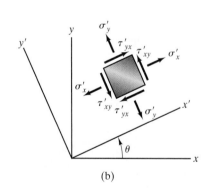

(b)

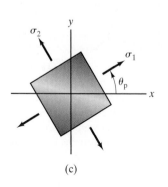

(c)

Components of Stress

In terms of a given coordinate system, the *state of stress* at a point p of a material is defined by the *components of stress*

$$\begin{bmatrix} \sigma_x & \tau_{xy} & \tau_{xz} \\ \tau_{yx} & \sigma_y & \tau_{yz} \\ \tau_{zx} & \tau_{zy} & \sigma_z \end{bmatrix}.$$
Eq. (12.1)

The directions of these stresses, and the orientations of the planes on which they act, are shown in Fig. (a). The shear stresses $\tau_{xy} = \tau_{yx}, \tau_{yz} = \tau_{zy}$, and $\tau_{xz} = \tau_{zx}$.

Transformations of Plane Stress

The stress at a point is said to be a state of *plane stress* if it is of the form

$$\begin{bmatrix} \sigma_x & \tau_{xy} & 0 \\ \tau_{yx} & \sigma_y & 0 \\ 0 & 0 & 0 \end{bmatrix}.$$
Eq. (12.3)

In terms of a coordinate system $x'y'z'$ oriented as shown in Fig. (b), the components of stress are

$$\sigma_x' = \frac{\sigma_x + \sigma_y}{2} + \frac{\sigma_x - \sigma_y}{2} \cos 2\theta + \tau_{xy} \sin 2\theta, \qquad \text{Eq. (12.7)}$$

$$\tau_{xy}' = -\frac{\sigma_x - \sigma_y}{2} \sin 2\theta + \tau_{xy} \cos 2\theta, \qquad \text{Eq. (12.8)}$$

$$\sigma_y' = \frac{\sigma_x + \sigma_y}{2} - \frac{\sigma_x - \sigma_y}{2} \cos 2\theta - \tau_{xy} \sin 2\theta. \qquad \text{Eq. (12.9)}$$

Maximum and Minimum Stresses in Plane Stress

The orientation of the element on which the principal stresses act [Fig. (c)] is determined from the equation

$$\tan 2\theta_p = \frac{2\tau_{xy}}{\sigma_x - \sigma_y}.$$
Eq. (12.12)

The values of the principal stresses can be obtained by substituting θ_p into Eqs. (12.7) and (12.9). Their values can also be determined from the equation

$$\sigma_1, \sigma_2 = \frac{\sigma_x + \sigma_y}{2} \pm \sqrt{\left(\frac{\sigma_x - \sigma_y}{2}\right)^2 + \tau_{xy}^2}, \qquad \text{Eq. (12.15)}$$

although this equation does not indicate the planes on which they act.

The orientation of the element on which the maximum in-plane shear stresses act [Fig. (d)] can be determined from the equation

$$\tan 2\theta_s = -\frac{\sigma_x - \sigma_y}{2\tau_{xy}}, \qquad \text{Eq. (12.16)}$$

or by using the relation $\theta_s = \theta_p + 45°$. The normal stresses σ_s acting on this element are shown. The value of the maximum shear stress can be obtained by substituting θ_s into Eq. (12.8). Its value can also be determined from the equation

$$\tau_{max} = \sqrt{\left(\frac{\sigma_x - \sigma_y}{2}\right)^2 + \tau_{xy}^2}, \qquad \text{Eq. (12.19)}$$

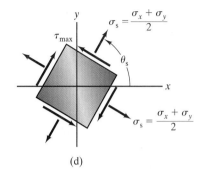

(d)

although this equation does not indicate the direction of the stress on the element.

The *absolute maximum shear stress* (the maximum shear stress on any plane through p) is the largest of the three values

$$\left|\frac{\sigma_1 - \sigma_2}{2}\right|, \qquad \left|\frac{\sigma_1}{2}\right|, \qquad \left|\frac{\sigma_2}{2}\right|. \qquad \text{Eq. (12.24)}$$

Mohr's Circle for Plane Stress

Given a state of plane stress σ_x, τ_{xy}, and σ_y, establish a set of horizontal and vertical axes with normal stress measured to the right along the horizontal axis and shear stress measured downward along the vertical axis. Plot two points, point P with coordinates (σ_x, τ_{xy}) and point Q with coordinates $(\sigma_y, -\tau_{xy})$. Draw a straight line connecting points P and Q. Using the intersection of the straight line with the horizontal axis as the center, draw a circle that passes through the two points [Fig. (e)]. Draw a straight line through the center of the circle at an 2θ measured counterclockwise from point P. The point P' at which this line intersects the circle has coordinates (σ_x', τ_{xy}'), and the point Q' has coordinates $(\sigma_y', -\tau_{xy}')$, as shown in Fig. (f).

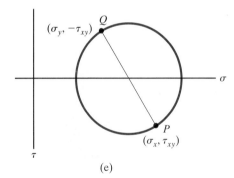

(e)

Principal Stresses in Three Dimensions

The principal stresses for a general state of stress are the roots of the cubic equation

$$\sigma^3 - I_1\sigma^2 + I_2\sigma - I_3 = 0, \qquad \text{Eq. (12.25)}$$

where

$$I_1 = \sigma_x + \sigma_y + \sigma_z,$$
$$I_2 = \sigma_x\sigma_y + \sigma_y\sigma_z + \sigma_z\sigma_x - \tau_{xy}^2 - \tau_{yz}^2 - \tau_{zx}^2, \qquad \text{Eq. (12.26)}$$
$$I_3 = \sigma_x\sigma_y\sigma_z - \sigma_x\tau_{yz}^2 - \sigma_y\tau_{xz}^2 - \sigma_z\tau_{xy}^2 + 2\tau_{xy}\tau_{yz}\tau_{zx}.$$

The absolute maximum shear stress is the largest of the three values

$$\left|\frac{\sigma_1 - \sigma_2}{2}\right|, \qquad \left|\frac{\sigma_1 - \sigma_3}{2}\right|, \qquad \left|\frac{\sigma_2 - \sigma_3}{2}\right|. \qquad \text{Eq. (12.27)}$$

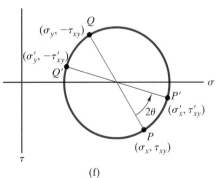

(f)

The state of stress at a point is said to be *triaxial* if it is of the form

$$
\begin{bmatrix}
\sigma_x & 0 & 0 \\
0 & \sigma_y & 0 \\
0 & 0 & \sigma_z
\end{bmatrix}.
$$

Eq. (12.28)

In triaxial stress, x, y, and z are principal axes and σ_x, σ_y, and σ_z are the principal stresses. The absolute maximum shear stress is the largest of the three values

$$
\left| \frac{\sigma_x - \sigma_y}{2} \right|, \qquad \left| \frac{\sigma_x - \sigma_z}{2} \right|, \qquad \left| \frac{\sigma_y - \sigma_z}{2} \right|.
$$

Eq. (12.29)

Review Problems

P12.62

12.62 A point p of the antenna's supporting structure is subjected to the state of plane stress $\sigma_x = 40$ MPa, $\sigma_y = -20$ MPa, and $\tau_{xy} = 30$ MPa. Determine the principal stresses and the absolute maximum shear stress.

12.63 A point p of the supporting structure of the antenna shown in Problem 12.62 is subjected to plane stress. The components $\sigma_y = -20$ MPa and $\tau_{xy} = 30$ MPa. The allowable normal stress of the material (in tension and compression) is 80 MPa. Based on this criterion, what is the allowable range of values of the stress component σ_x?

12.64 The components of plane stress acting on an element of a bar subjected to axial loads are shown in Fig. 12.7. Assuming the stress σ_x to be known, use Mohr's circle to determine the principal stresses and the maximum in-plane shear stress and show them acting on properly oriented elements.

12.65 The components of plane stress acting on an element of a bar subjected to torsion are shown in Fig. 12.8. Assuming the stress τ_{xy} to be known, use Mohr's circle to determine the principal stresses and the maximum in-plane shear stress and show them acting on properly oriented elements.

12.66 A machine element is subjected to the state of *triaxial* stress $\sigma_x = 300$ MPa, $\sigma = -200$ MPa, σ_z. If the material is not to be subjected to a shear stress greater than 400 MPa, what is the acceptable range of the stress σ_z?

12.67 At a point p a material is subjected to the state of stress (in MPa)

$$
\begin{bmatrix}
\sigma_x & \tau_{xy} & \tau_{xz} \\
\tau_{yx} & \sigma_y & \tau_{yz} \\
\tau_{zx} & \tau_{zy} & \sigma_z
\end{bmatrix}
=
\begin{bmatrix}
4 & 2 & 1 \\
2 & -2 & 1 \\
1 & 1 & -3
\end{bmatrix}.
$$

Determine the principal stresses and the absolute maximum shear stress. Confirm the absolute maximum shear stress by drawing the superimposed Mohr's circle as shown in Fig. 12.36.

12.68 A spherical pressure vessel has a 1-m radius and a 0.002-m wall thickness. It contains a gas with pressure $p_i = 1.8 \times 10^5$ Pa, and the outer wall is subjected to atmospheric pressure $p_o = 1 \times 10^5$ Pa. Determine the maximum normal stress and the absolute maximum shear stress at the vessel's inner surface.

12.69 For the spherical pressure vessel described in Problem 12.70, suppose that the allowable value of the absolute maximum shear stress is $\tau_{allow} = 14$ MPa. If the outer wall is subjected to atmospheric pressure $p_o = 1 \times 10^5$ Pa, what is the maximum allowable internal pressure?

Aerodynamic forces cause the wings of the Concorde to bend and warp, subjecting the material to extensional and shear strains.

States of Strain and Stress-Strain Relations

I n this chapter we define the state of strain at a point of an object and show that it specifies the deformation of the material in the neighborhood of the point. We then develop the relationship between the state of stress and the state of strain for an elastic material subject to small strains.

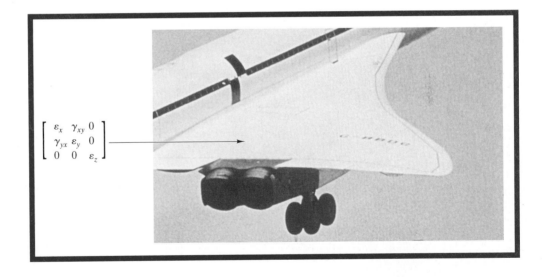

$$\begin{bmatrix} \varepsilon_x & \gamma_{xy} & 0 \\ \gamma_{yx} & \varepsilon_y & 0 \\ 0 & 0 & \varepsilon_z \end{bmatrix}$$

13.1 Components of Strain

We have introduced two types of strain, extensional strain and shear strain. The extensional strain ε is a measure of the change in length of a material line element whose length is dL in a reference state. If the length of the line element in a deformed state is dL' (Fig. 13.1), the extensional strain is defined to be the change in length divided by its original length:

$$\varepsilon = \frac{dL' - dL}{dL}.$$

Figure 13.1
Material line element in the reference and deformed states.

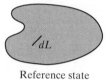

Reference state   Deformed state

The value of the extensional strain at a point depends in general on the direction of the line element. We say that ε is the extensional strain in the direction of dL.

The shear strain γ is a measure of the change in the angle between two line elements dL_1 and dL_2 that are perpendicular in a reference state. The angle between the elements in a deformed state is defined to be $(\pi/2) - \gamma$ (Fig. 13.2). The value of γ at a point depends in general on the directions of the two elements. We say that γ is the shear strain referred to the directions of the elements dL_1 and dL_2.

Figure 13.2
Line elements that are perpendicular in the reference state. The decrease in the angle between them in the deformed state is the shear strain.

Reference state   Deformed state

To define the state of strain at a point p of a material, we introduce a coordinate system (Fig. 13.3a) and introduce six components of strain:

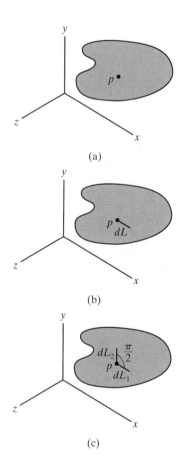

(a)

(b)

(c)

Figure 13.3
(a) Introducing a coordinate system.
(b) Line element for determining the value of ε_x at p.
(c) Line elements for determining the value of γ_{xy} at p.

ε_x The extensional strain determined with the element dL parallel to the x axis (Fig. 13.3b)

ε_y The extensional strain determined with the element dL parallel to the y axis

ε_z The extensional strain determined with the element dL parallel to the z axis

γ_{xy} The shear strain determined with the elements dL_1 and dL_2 in the positive x and y directions (Fig. 13.3c)

γ_{yz} The shear strain determined with the elements dL_1 and dL_2 in the positive y and z directions

γ_{xz} The shear strain determined with the elements dL_1 and dL_2 in the positive x and z directions

If these six components of strain are known at a point, the extensional strain in any direction and the shear strain referred to any two perpendicular directions can be determined. (These results require that the components of strain be sufficiently small that products of the components are negligible in comparison to the components themselves. In this introductory treatment we consider only small strains.) For this reason these components are called the *state of strain* at the point. We say that the strain is *homogeneous* in a region if the state of strain is the same at each point. In analogy to the state of stress, we can represent the state of strain as the matrix

$$
\begin{bmatrix}
\varepsilon_x & \gamma_{xy} & \gamma_{xz} \\
\gamma_{yx} & \varepsilon_y & \gamma_{yz} \\
\gamma_{zx} & \gamma_{zy} & \varepsilon_z
\end{bmatrix},
\tag{13.1}
$$

where $\gamma_{yx} = \gamma_{xy}$, $\gamma_{zx} = \gamma_{xz}$, and $\gamma_{zy} = \gamma_{yz}$.

The state of strain determines the change in volume of a material due to a deformation. Consider an element of material in the reference state with dimension dx_0, dy_0, dz_0, (Fig. 13.4). The volume of the element is $V_0 = dx_0\, dy_0\, dz_0$. In the deformed state, the lengths of the edges of the element are $dx = (1 + \varepsilon_x)dx_0$, $dy = (1 + \varepsilon_y)dy_0$, and $dz = (1 + \varepsilon_z)dz_0$ (Fig. 13.4). Its volume in the deformed state is

$$
dV = dx\, dy\, dz = (1 + \varepsilon_x)(1 + \varepsilon_y)(1 + \varepsilon_z)dx_0\, dy_0\, dz_0.
$$

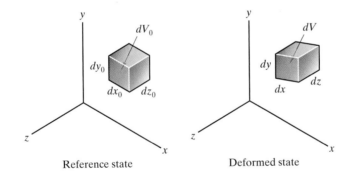

Reference state   Deformed state

Figure 13.4
Element of material in the reference and deformed states.

Neglecting products of the strains, we can write this result as

$$
dV = (1 + \varepsilon_x + \varepsilon_y + \varepsilon_z)\, dV_0.
\tag{13.2}
$$

(When products of strains are negligible, the components of shear strain do not affect the change in volume of the material.) The change in volume of the material per unit volume, denoted by e, is called the *dilatation:*

$$
e = \frac{dV - dV_0}{dV_0} = \varepsilon_x + \varepsilon_y + \varepsilon_z.
\tag{13.3}
$$

Study Questions

1. What are the definitions of the components of strain at a point?
2. What does it mean when the strain is said to be homogeneous in a region?
3. Consider an element of material that has volume dV_0 in a reference state. If the material is subjected to a known state of strain, how can you determine the change in volume of the element?

13.2 Transformations of Plane Strain

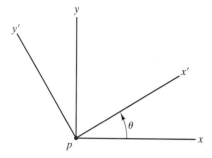

Figure 13.5
Coordinate systems xyz and $x'y'z'$. The z and z' axes are coincident.

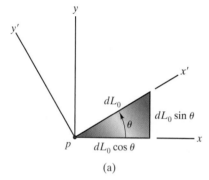

(a)

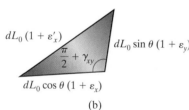

(b)

Figure 13.6
(a) Triangular element in the reference state.
(b) Element in the deformed state.

For a state of *plane strain,* defined by

$$\begin{bmatrix} \varepsilon_x & \gamma_{xy} & 0 \\ \gamma_{yx} & \varepsilon_y & 0 \\ 0 & 0 & 0 \end{bmatrix}, \tag{13.4}$$

we can derive transformation equations equivalent to the equations for plane stress developed in Section 12.2. Suppose that we know the state of plane strain at a point p of a material in terms of the xyz coordinate system shown in Fig. 13.5 and we want to know the components of strain ε_x', γ_{xy}', and ε_y'.

To determine the extensional strain ε_x', we begin with an infinitesimal element of material at p which in the reference state has the triangular shape shown in Fig. 13.6a. We denote the infinitesimal length of the hypotenuse by dL_0. In the deformed state (Fig. 13.6b), we know the lengths of the sides in terms of the extensional strains ε_x, ε_y, and ε_x', and we can express the angle between the sides that were perpendicular in the reference state in terms of γ_{xy}. By analyzing this triangle we can determine ε_x', in terms of the strains ε_x, ε_y, and γ_{xy}.

Applying the law of cosines to the deformed element gives

$$dL_0^2(1 + \varepsilon_x')^2 = dL_0^2 \sin^2 \theta (1 + \varepsilon_y)^2 + dL_0^2 \cos^2 \theta (1 + \varepsilon_x)^2$$
$$- 2\, dL_0^2 \sin \theta \cos \theta (1 + \varepsilon_x)(1 + \varepsilon_y) \cos(\pi/2 + \gamma_{xy}). \tag{13.5}$$

We apply the identity

$$\cos(\pi/2 + \gamma_{xy}) = \cos(\pi/2) \cos \gamma_{xy} - \sin(\pi/2) \sin \gamma_{xy}$$
$$= -\sin \gamma_{xy},$$

and since strains are assumed to be small, we can approximate $\sin \gamma_{xy}$ by γ_{xy}, obtaining

$$\cos(\pi/2 + \gamma_{xy}) = -\gamma_{xy}.$$

Substituting this result into Eq. (13.5) and neglecting products of strains, it becomes

$$\varepsilon_x' = \varepsilon_x \cos^2 \theta + \varepsilon_y \sin^2 \theta + \gamma_{xy} \sin \theta \cos \theta. \tag{13.6}$$

By using the trigonometric identities

$$2 \cos^2 \theta = 1 + \cos 2\theta,$$
$$2 \sin^2 \theta = 1 - \cos 2\theta,$$
$$2 \sin \theta \cos \theta = \sin 2\theta,$$

we can write Eq. (13.6) in the form

$$\varepsilon_x' = \frac{\varepsilon_x + \varepsilon_y}{2} + \frac{\varepsilon_x - \varepsilon_y}{2} \cos 2\theta + \frac{\gamma_{xy}}{2} \sin 2\theta. \tag{13.7}$$

We can obtain an equation for ε_y' by setting θ equal to $\theta + 90°$ in this expression. The result is

$$\varepsilon_y' = \frac{\varepsilon_x + \varepsilon_y}{2} - \frac{\varepsilon_x - \varepsilon_y}{2}\cos 2\theta - \frac{\gamma_{xy}}{2}\sin 2\theta. \qquad (13.8)$$

We determine the shear strain γ_{xy}' by using Eq. (13.7). Instead of using this equation to express the extensional strain ε_x' in terms of the strains ε_x, ε_y, and γ_{xy}, we can reverse its role and use it to express the extensional strain ε_x in terms of the strains ε_x', ε_y' and γ_{xy}' simply by replacing θ by $-\theta$. The result is

$$\varepsilon_x = \frac{\varepsilon_x' + \varepsilon_y'}{2} + \frac{\varepsilon_x' - \varepsilon_y'}{2}\cos 2\theta - \frac{\gamma_{xy}'}{2}\sin 2\theta.$$

Substituting Eqs. (13.7) and (13.8) into this equation, we can solve γ_{xy}' in terms of ε_x, ε_y, and γ_{xy}:

$$\frac{\gamma_{xy}'}{2} = -\frac{\varepsilon_x - \varepsilon_y}{2}\sin 2\theta + \frac{\gamma_{xy}}{2}\cos 2\theta. \qquad (13.9)$$

Compare Eqs. (13.7), (13.8), and (13.9) to the transformation equations for plane stress, Eqs. (12.7), (12.9), and (12.8). They are identical in form, with the normal stress replaced by the extensional strain and the shear stress replaced by one-half the shear strain. The state of stress and the state of strain, with the shear strains γ_{xy}, γ_{yz}, and γ_{xz} replaced by $\gamma_{xy}/2$, $\gamma_{yz}/2$, and $\gamma_{xz}/2$, are both quantities called *tensors*. Although a complete discussion of tensors is beyond our scope, these examples demonstrate how components of tensors transform between coordinate systems. From our present point of view, the similarities of these equations means that the analysis of strains follows the same path used for stresses, and you will therefore find the results quite familiar.

Although we have derived the strain transformation equations under the assumption of a state of plane strain, we will show in Section 13.4 that for many materials they also apply to the components of strain resulting from a state of plane stress. The strain component ε_z is not generally zero in plane stress, but that does not affect the derivations of Eqs. (13.7), (13.8), and (13.9).

Study Questions

1. What is a state of plane strain?
2. Do Eqs. (13.7)–(13.9) hold for any values of the components of plane strain ε_x, ε_y, and γ_{xy}? Explain.
3. What is the definition of the term γ_{xy}' in Eq. (13.9)?
4. Equations (13.7)–(13.9) hold for a state of plain strain. Do they also apply when ε_x, ε_y, and γ_{xy} are components of strain resulting from a state of plane stress?

Example 13.1

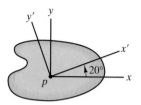

Figure 13.7

Transformations of Strain Components

The components of plane strain at point p of the material shown in Fig. 13.7 are $\varepsilon_x = 0.003$, $\varepsilon_y = 0.001$, and $\gamma_{xy} = -0.006$. Determine the components of plane strain in terms of the $x'y'$ coordinate system.

Strategy

We can use Eqs. (13.7)–(13.9) to determine the strain components ε_x', ε_y', and γ_{xy}'.

Solution

The components of plane strain in terms of the $x'y'$ coordinate system are

$$\varepsilon_x' = \frac{\varepsilon_x + \varepsilon_y}{2} + \frac{\varepsilon_x - \varepsilon_y}{2} \cos 2\theta + \frac{\gamma_{xy}}{2} \sin 2\theta$$

$$= \frac{0.003 + 0.001}{2} + \frac{0.003 - 0.001}{2} \cos 2(20°) + \frac{-0.006}{2} \sin 2(20°)$$

$$= 0.00084,$$

$$\varepsilon_y' = \frac{\varepsilon_x + \varepsilon_y}{2} - \frac{\varepsilon_x - \varepsilon_y}{2} \cos 2\theta - \frac{\gamma_{xy}}{2} \sin 2\theta$$

$$= \frac{0.003 + 0.001}{2} - \frac{0.003 - 0.001}{2} \cos 2(20°) - \frac{-0.006}{2} \sin 2(20°)$$

$$= 0.00316,$$

$$\gamma_{xy}' = 2\left(-\frac{\varepsilon_x - \varepsilon_y}{2} \sin 2\theta + \frac{\gamma_{xy}}{2} \cos 2\theta\right)$$

$$= -(0.003 - 0.001) \sin 2(20°) + (-0.006) \cos 2(20°)$$

$$= -0.00588.$$

Strain Gauge Rosette

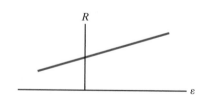

Figure 13.8
The electrical resistance R of a wire is a function of its axial strain ε.

Before continuing our discussion of the analysis of strains, we will describe an interesting and important application of the strain transformation equations. The term *strain gauge* refers to an instrument for measuring strains. The type we are concerned with here, called a *resistance strain gauge*, is based on the observation that the electrical resistance of a wire varies when the wire is subjected to axial strain (Fig. 13.8). Once the relationship between the strain of a given wire and its electrical resistance has been established experimentally (a procedure called *calibration*), the axial strain of the wire can be determined by measuring its resistance, which can be done very accurately.

To use the calibrated wire as a strain gauge, it is bonded to the surface of an unloaded specimen. When the specimen is loaded, the strain of the speci-

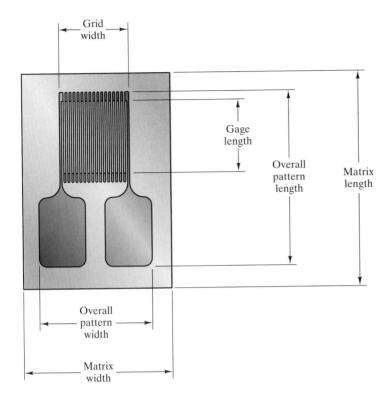

Figure 13.9
Typical resistance strain gauge.

men in the direction of the wire is determined by measuring the wire's resistance. (The wire must be sufficiently thin that the force the strained wire exerts on the specimen is negligible.) The wire is typically arranged in the pattern shown in Fig. 13.9 to minimize the size of the gauge and so measure the strain within a relatively small neighborhood of a point.

Extensional strain can be measured by the type of strain gauge we have described, but how can shear strain be measured? This can be done in a clever way by a *strain gauge rosette*, which consists of three superimposed strain gauges measuring extensional strain in three directions, as shown in Fig. 13.11a. [The term *rosette* is said to derive from the resemblance between strain gauge rosettes mounted on colored felt and a small cloth ornament called a rosette that was worn on hats around the time of the French Revolution (Fig. 13.10).] We introduce a coordinate system and denote the directions of the strain gauges by θ_a, θ_b, and θ_c (Fig. 13.11b). Using

Figure 13.10
The original rosette.

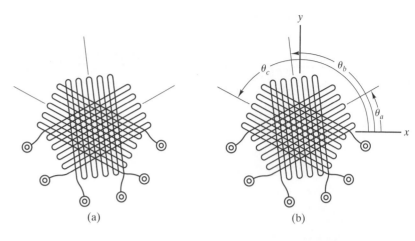

Figure 13.11

(a) Strain gauge rosette.

(b) Introducing a coordinate system. The angles θ_a, θ_b, and θ_c specify the directions of the three strain gauges.

Eq. (13.6) to express the extensional strains in the directions of the three strain gauges in terms of the strain components ε_x, ε_y, and γ_{xy}, we obtain

$$\varepsilon_a = \varepsilon_x \cos^2 \theta_a + \varepsilon_y \sin^2 \theta_a + \gamma_{xy} \sin \theta_a \cos \theta_a,$$
$$\varepsilon_b = \varepsilon_x \cos^2 \theta_b + \varepsilon_y \sin^2 \theta_b + \gamma_{xy} \sin \theta_b \cos \theta_b, \tag{13.10}$$
$$\varepsilon_c = \varepsilon_x \cos^2 \theta_c + \varepsilon_y \sin^2 \theta_c + \gamma_{xy} \sin \theta_c \cos \theta_c.$$

By measuring the extensional strains ε_a, ε_b, and ε_c, this system of equations can be solved for the strain components ε_x, ε_y, and γ_{xy}, thus determining the shear strain.

Example 13.2

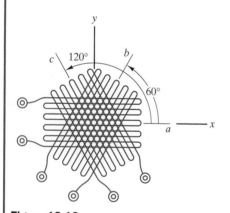

Figure 13.12

Strain Gauge Rosette

The strains measured by a strain gauge rosette oriented as shown in Fig. 13.12 are $\varepsilon_a = 0.004$, $\varepsilon_b = -0.003$, and $\varepsilon_c = 0.002$. Determine the components of strain ε_x, ε_y, and γ_{xy}.

Strategy

The angles specifying the directions of the strain gauges relative to the x axis are $\theta_a = 0$, $\theta_b = 60°$, and $\theta_c = 120°$. By substituting these values and the values of the strains ε_a, ε_b, and ε_c into Eqs. (13.10), we can solve for ε_x, ε_y, and γ_{xy}.

Solution

Equations (13.10) are

$$0.004 = \varepsilon_x,$$
$$-0.003 = \varepsilon_x \cos^2 60° + \varepsilon_y \sin^2 60° + \gamma_{xy} \sin 60° \cos 60°,$$
$$0.002 = \varepsilon_x \cos^2 120° + \varepsilon_y \sin^2 120° + \gamma_{xy} \sin 120° \cos 120°.$$

Solving, we obtain $\varepsilon_x = 0.00400$, $\varepsilon_y = -0.00200$, and $\gamma_{xy} = -0.00577$.

Maximum and Minimum Strains

Given a state of plane strain at a point p, Eqs. (13.7) (13.8) and (13.9) determine the strain components ε_x', ε_y', and γ_{xy}' corresponding to the coordinate system shown in Fig. 13.5 for any value of θ. In this section we consider the following questions: For what values of θ are the extensional strain ε_x' and shear strain γ_{xy}' a maximum or minimum, and what are their values? In other words, what are the maximum and minimum extensional and shear strains in the x–y plane?

Principal Strains   Let a value of θ for which ε_x' is a maximum or minimum be denoted by θ_p. By evaluating the derivative of Eq. (13.7) with respect to 2θ and setting it equal to zero, we obtain the equation

$$\tan 2\theta_p = \frac{\gamma_{xy}}{\varepsilon_x - \varepsilon_y}. \tag{13.11}$$

When ε_x, ε_y, and γ_{xy} are known, you can solve this equation for θ_p and substitute it into Eq. (13.7) to determine the maximum or minimum value of ε_x'. Just as in the case of the maximum and minimum normal stresses, if $2\theta_p$ is a solution of Eq. (13.11), then so is $2\theta_p + 180°$. This means that the extensional strain is a maximum or minimum in the direction θ_p and also in the direction $\theta_p + 90°$, which means that *the maximum and minimum extensional strains occur in the x' and y' axis directions*. Once we have determined θ_p, we can determine the maximum and minimum extensional strains in the x–y plane, called the *principal strains* and denoted ε_1 and ε_2, from Eqs. (13.7) and (13.8). There are no extensional strains of greater magnitude in any direction at the point p.

To obtain analytical expressions for the values of the principal strains, we solve the equations

$$\frac{\sin 2\theta_p}{\cos 2\theta_p} = \tan 2\theta_p = \frac{\gamma_{xy}}{\varepsilon_x - \varepsilon_y} \tag{13.12}$$

and

$$\sin^2 2\theta_p + \cos^2 2\theta_p = 1 \tag{13.13}$$

for $\sin 2\theta_p$ and $\cos 2\theta_p$ and substitute the results into Eq. (13.7), obtaining

$$\varepsilon_1, \varepsilon_2 = \frac{\varepsilon_x + \varepsilon_y}{2} \pm \sqrt{\left(\frac{\varepsilon_x - \varepsilon_y}{2}\right)^2 + \left(\frac{\gamma_{xy}}{2}\right)^2}. \tag{13.14}$$

By substituting the expressions for $\sin 2\theta_p$ and $\cos 2\theta_p$ into Eq. (13.9), we find that the value of the shear γ_{xy}' at $\theta = \theta_p$ is zero. The physical interpretation of this result is interesting: An infinitesimal square element oriented as shown in Fig. 13.13a is subjected to the principal strains in the x' and y' directions and undergoes no shear strain. The element is rectangular in the deformed state (Fig. 13.13b).

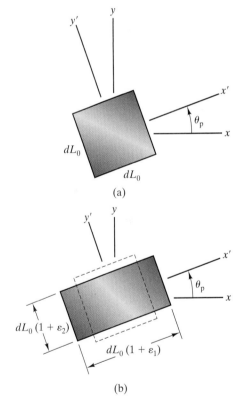

Figure 13.13
(a) Reference state of a square element aligned with the directions of the principal strains.
(b) The deformed element undergoes no shear strain. (The strains are exaggerated. The largest principal strain may occur in either the x' or the y' direction.)

Maximum Shear Strains   Let a value of θ for which γ'_{xy} is a maximum or minimum be denoted by θ_s. Evaluating the derivative of Eq. (13.9) with respect to 2θ and setting it equal to zero, we obtain the equation

$$\tan 2\theta_s = -\frac{\varepsilon_x - \varepsilon_y}{\gamma_{xy}}. \tag{13.15}$$

With this equation we can determine θ_s and substitute it into Eq. (13.9) to obtain the maximum or minimum value of γ'_{xy}. If θ_s is a solution of Eq. (13.15), then so are $\theta_s + 90°$, $\theta_s + 2(90°)$, ... As we illustrate in Fig. 13.14, *the maximum and minimum shear strains describe the shear strain of a rectangular element.* Furthermore, because $\tan 2\theta_s$ is the negative inverse of $\tan 2\theta_p$, this element is rotated 45° relative to the element which is subjected to the maximum and minimum extensional strains. We can demonstrate why this is the case by superimposing these two elements (Fig. 13.15).

To obtain an analytical expression for the maximum magnitude of the shear strain, we solve the equations

$$\frac{\sin 2\theta_s}{\cos 2\theta_s} = \tan 2\theta_s = -\frac{\varepsilon_x - \varepsilon_y}{\gamma_{xy}} \tag{13.16}$$

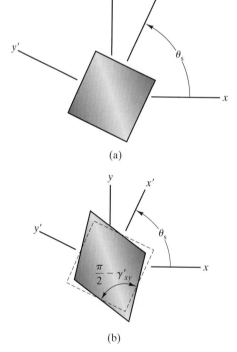

(a)

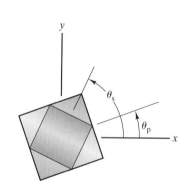

Reference state

(b)

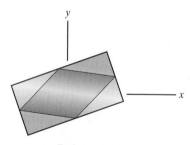

Deformed state

Figure 13.14
(a) Reference state of the square element that is subjected to the maximum and minimum shear strains.
(b) Deformed element. (The shear strain is exaggerated. The strain γ'_{xy} may be either positive or negative.)

Figure 13.15
Element subjected to the greatest shear strain superimposed onto the element subjected to the principal strains.

and

$$\sin^2 2\theta_s + \cos^2 2\theta_s = 1 \tag{13.17}$$

for $\sin 2\theta_s$ and $\cos 2\theta_s$ and substitute the results into Eq. (13.9). The result is

$$\gamma_{\max} = \sqrt{(\varepsilon_x - \varepsilon_y)^2 + \gamma_{xy}^2}. \tag{13.18}$$

This is called the *maximum in-plane shear strain*, because it is the maximum shear strain in the x–y plane.

By substituting our results for $\sin 2\theta_s$ and $\cos 2\theta_s$ into Eqs. (13.7) and (13.8), we obtain

$$\varepsilon_x' = \varepsilon_y' = \frac{\varepsilon_x + \varepsilon_y}{2}.$$

The element that is subjected to the maximum and minimum shear strains (Fig. 13.14) is subjected to equal extensional strains in the x' and y' directions.

We can obtain expressions for the absolute maximum shear strain the same way that we determined the absolute maximum shear stress. We begin with the element that is subjected to the principal strains (Fig. 13.13) and realign the coordinate system with the faces of the element (Fig. 13.16a). In terms of this new coordinate system, the components of plane strain are $\varepsilon_x = \varepsilon_1$, $\varepsilon_y = \varepsilon_2$, and $\gamma_{xy} = 0$. Substituting these components into Eq. (13.18), we obtain a different expression for the magnitude of the maximum in-plane shear strain:

$$\sqrt{(\varepsilon_x - \varepsilon_y)^2 + \gamma_{xy}^2} = \sqrt{(\varepsilon_1 - \varepsilon_2)^2 + 0}$$

$$= |\varepsilon_1 - \varepsilon_2|. \tag{13.19}$$

This equation is expressed in terms of the principal strains, but it gives the same result as Eq. (13.18). Now we reorient the coordinate system as shown in Fig. 13.16b. In terms of this coordinate system, $\varepsilon_x = 0$, $\varepsilon_y = \varepsilon_1$, and $\gamma_{xy} = 0$. Substituting these components into Eq. (13.18), we obtain

$$\sqrt{(\varepsilon_x - \varepsilon_y)^2 + \gamma_{xy}^2} = \sqrt{(0 - \varepsilon_1)^2 + 0} = |\varepsilon_1|. \tag{13.20}$$

Next, we reorient the coordinate system as shown in Fig. 13.16c. In terms of this coordinate system, $\varepsilon_x = 0$, $\varepsilon_y = \varepsilon_2$, and $\gamma_{xy} = 0$. Substituting these components into Eq. (13.18), we obtain

$$\sqrt{(\varepsilon_x - \varepsilon_y)^2 + \gamma_{xy}^2} = \sqrt{(0 - \varepsilon_2)^2 + 0} = |\varepsilon_2|. \tag{13.21}$$

Depending on the values of the principal strains, Eq. (13.20) and/or Eq. (13.21) can result in larger values of the magnitude of the maximum shear strain than the maximum in-plane shear strain. The absolute maximum shear strain is the largest value given by Eq. (13.19), (13.20), or (13.21). No greater shear strain occurs for any orientation of the coordinate system.

Summary: Determining the Principal Strains and the Maximum Shear Strain

Here we give a sequence of steps you can use to determine the principal strains and the maximum shear strain for a given state of plane strain.

1. Use Eq. (13.11) to determine θ_p, establishing the orientation $x'y'$ coordinate system that is aligned with the directions of the principal strains.

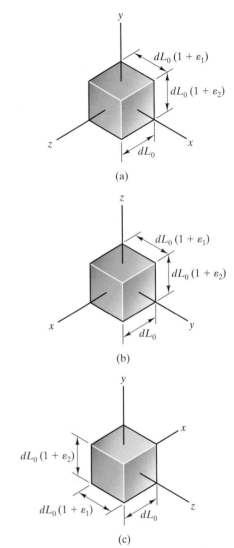

Figure 13.16
Element subjected to the principal strains.
(a) Realigned coordinate system.
(a), (b), (c) Other orientations of the coordinate system.

2. Determine ε_1 and ε_2 and the directions in which they act by substituting θ_p into Eqs. (13.7) and (13.8). You can also determine the values of ε_1 and ε_2 from Eq. (13.14), but this equation does not tell you which principal strain acts in the x' direction and which one acts in the y' direction.

3. Use Eq. (13.15) to determine θ_s, establishing directions of the x' and y' axes for which γ'_{xy} is a maximum or minimum.

4. Determine γ'_{xy} by substituting θ_s into Eq. (13.9). The magnitude of this shear strain is the maximum in-plane shear strain. You can also determine the maximum in-plane shear strain from Eq. (13.18).

5. The absolute maximum shear strain is the largest of the three values

$$|\varepsilon_1 - \varepsilon_2|, \qquad |\varepsilon_1|, \qquad |\varepsilon_2|. \tag{13.22}$$

Study Questions

1. What are the principal strains?
2. If you know the direction of one of the principal strains, what do you know about the direction of the other principal strain?
3. What is the shear strain of the element that is subjected to the principal strains?
4. If you know the orientation of the element that is subjected to the principal strains, what do you know about the orientation of the element that is subjected to the maximum in-plane shear strain?

Example 13.3

Determining Principal Strains

The state of plane strain at a point p is $\varepsilon_x = 0.003$, $\varepsilon_y = 0.001$, and $\gamma_{xy} = -0.006$. Determine the principal strains and the maximum in-plane shear strain and show the orientations of the elements subjected to these strains. Also determine the absolute maximum shear strain.

Strategy

We can follow the steps given in the preceeding summary to determine the principal strains and the maximum shear strain.

Solution

Step 1 From Eq. (13.11),

$$\tan 2\theta_p = \frac{\gamma_{xy}}{\varepsilon_x - \varepsilon_y} = \frac{-0.006}{0.003 - 0.001} = -3.0.$$

Solving this equation, we obtain $\theta_p = -35.78°$. This angle tells us the orientation of the $x'y'$ coordinate system aligned with the principal strains.

Step 2 We substitute θ_p into Eqs. (13.7) and (13.8) to determine the principal strains.

$$\varepsilon_x' = \frac{\varepsilon_x + \varepsilon_y}{2} + \frac{\varepsilon_x - \varepsilon_y}{2}\cos 2\theta_p + \frac{\gamma_{xy}}{2}\sin 2\theta_p$$

$$= \frac{0.003 + 0.001}{2} + \frac{0.003 - 0.001}{2}\cos 2(-35.78°)$$

$$+ \frac{-0.006}{2}\sin 2(-35.78°)$$

$$= 0.00516,$$

$$\varepsilon_y' = \frac{\varepsilon_x + \varepsilon_y}{2} - \frac{\varepsilon_x - \varepsilon_y}{2}\cos 2\theta_p - \frac{\gamma_{xy}}{2}\sin 2\theta_p$$

$$= \frac{0.003 + 0.001}{2} - \frac{0.003 - 0.001}{2}\cos 2(-35.78°)$$

$$- \frac{-0.006}{2}\sin 2(-35.78°)$$

$$= -0.00116.$$

The principal strains are $\varepsilon_1 = 0.00516$ and $\varepsilon_2 = -0.00116$. They are shown on the properly oriented element in Fig. (a). This element is subjected to no shear strain.

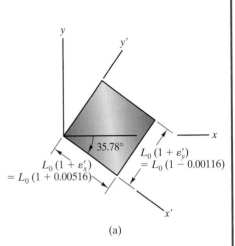

(a) The principal strains. L_0 is the dimension of the square element in the reference state.

Step 3 From Eq. (13.15),

$$\tan 2\theta_s = -\frac{\varepsilon_x - \varepsilon_y}{\gamma_{xy}} = -\frac{0.003 - 0.001}{-0.006} = 0.333,$$

from which we obtain $\theta_s = 9.22°$.

Step 4 We substitute θ_s into Eq. (13.9) to determine the maximum in-plane shear strain.

$$\frac{\gamma_{xy}'}{2} = -\frac{\varepsilon_x - \varepsilon_y}{2}\sin 2\theta_s + \frac{\gamma_{xy}}{2}\cos 2\theta_s$$

$$= -\frac{0.003 - 0.001}{2}\sin 2(9.22°) + \frac{-0.006}{2}\cos 2(9.22°)$$

$$= -0.00316.$$

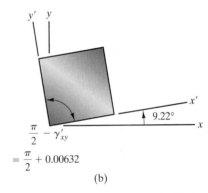

(b) Maximum in-plane shear strain. Notice that γ_{xy}' is negative.

We see that $\gamma_{xy}' = -0.00632$. The maximum in-plane shear strain is shown on the properly oriented element in Fig. (b).

Step 5 The absolute maximum shear strain is given by the largest of the three values

$$|\varepsilon_1 - \varepsilon_2| = |0.00516 - (-0.00116)| = 0.00632,$$

$$|\varepsilon_1| = |0.00516| = 0.00516,$$

$$|\varepsilon_2| = |-0.00116| = 0.00116.$$

In this example the absolute maximum shear strain equals the magnitude of the maximum in-plane shear strain, 0.00632.

Discussion

Observe that the element subjected to the maximum in-plain shear strain, Fig. (b), is rotated 45° relative to the element that is subjected to the principal strains, Fig. (a).

Problems

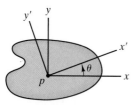

P13.1–P13.9

13.1 The components of plane strain at point p are $\varepsilon_x = 0.003$, $\varepsilon_y = 0$, and $\gamma_{xy} = 0$. If $\theta = 45°$, what are the strains ε'_x, ε'_y, and γ'_{xy} at point p?

13.2 The components of plane strain at point p are $\varepsilon_x = 0$, $\varepsilon_y = 0$, and $\gamma_{xy} = 0.004$. If $\theta = 45°$, what are the strains ε'_x, ε'_y, and γ'_{xy} at point p?

13.3 The components of plane strain at point p are $\varepsilon_x = -0.0024$, $\varepsilon_y = 0.0012$, and $\gamma_{xy} = -0.0012$. If $\theta = 25°$, what are the strains ε'_x, ε'_y, and γ'_{xy} at point p?

13.4 The components of plane strain at a point p of a bit during a drilling operation are $\varepsilon_x = 0.00400$, $\varepsilon_y = -0.00300$, and $\gamma_{xy} = 0.00600$, and the components referred to the $x'y'z'$ coordinate system are $\varepsilon'_x = 0.00125$, $\varepsilon'_y = -0.00025$, and $\gamma'_{xy} = 0.00910$. What is the angle θ?

13.5 The components of plane strain at point p are $\varepsilon_x = 0.0024$, $\varepsilon_y = -0.0012$, and $\gamma_{xy} = 0.0048$. The extensional strains $\varepsilon'_x = 0.00347$ and $\varepsilon'_y = -0.00227$. Determine γ'_{xy} and the angle θ.

13.6 During liftoff, strain gauges attached to one of the Space Shuttle main engine nozzles determine that the components of plane strain at point p are $\varepsilon'_x = 0.00665$, $\varepsilon'_y = 0.00825$, and $\gamma'_{xy} = 0.00135$ for a coordinate system oriented at $\theta = 20°$. What are the strains ε_x, ε_y, and γ_{xy} at that point?

13.7 The components of plane strain at point p referred to the $x'y'$ coordinate system are $\varepsilon'_x = 0.0066$, $\varepsilon'_y = -0.0086$, and $\gamma'_{xy} = 0.0028$. If $\theta = 20°$, what are the strains ε_x, ε_y, and γ_{xy} at point p?

13.8 The strains $\varepsilon_x = 0.008$, $\varepsilon_y = -0.006$, $\gamma_{xy} = 0.024$, $\varepsilon'_x = 0.014$, and $\varepsilon'_y = -0.012$. Determine the strain γ'_{xy} and the angle θ.

13.9 The strains $\varepsilon_x = 0.008$, $\varepsilon_y = -0.006$, and $\gamma'_{xy} = 0.024$, and the angle $\theta = 35°$. Determine ε'_x, ε'_y, and γ_{xy}.

13.10 Points P and Q are 1 mm apart in the reference state of a material. If the material is subjected to the homogeneous state of plane strain $\varepsilon_x = 0.003$, $\varepsilon_y = -0.002$, and $\gamma_{xy} = -0.006$, what is the distance between points P and Q in the deformed material?

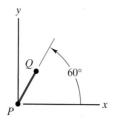

P13.10

13.11 Two points P and Q of the Concorde's wing are 2 mm apart when the wing is unstressed. In a particular flight condition, the material containing these points is subjected to a homogeneous state of plane stress and the strain components $\varepsilon_x = 0.008$, $\varepsilon_y = 0.002$, and $\gamma_{xy} = -0.003$. What is the distance between points P and Q?

P13.11

13.12 The points P and Q of the Concorde's wing shown in Problem 13.11 are 2 mm apart when the wing is unstressed. In a particular flight condition, they are 1.992 mm apart. The material is in plane stress. If $\varepsilon_x = -0.0088$ and $\varepsilon_y = 0.0024$, what is γ_{xy}?

13.13 Point O and Q are 1 mm apart and points O and P are 2 mm apart in the reference state of a material. If the material is subjected to a homogeneous state of plane strain $\varepsilon_x = 0.006$, $\varepsilon_y = 0.002$, and $\gamma_{xy} = 0.004$, what is the distance between points P and Q in the deformed material?

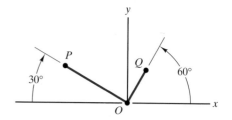

P13.13

13.14 In Problem 13.13, what is the angle between the lines OQ and OP in the deformed material?

13.15 Points O and Q shown in Problem 13.13 are 1 mm apart and points O and P are 2 mm apart in the reference state of a material. After the material is subjected to a homogeneous state of strain, points O and Q are 1.002 mm apart, points O and P are 1.998 mm apart, and points P and Q are 2.242 mm apart. What are the strain components ε_x, ε_y, and γ_{xy}?

13.16 A bar is subjected to axial forces. The strains measured by a strain gauge rosette oriented as shown are $\varepsilon_a = 0.003$, $\varepsilon_b = 0.001$, and $\varepsilon_c = -0.001$. Determine the shear strain γ_{xy}.

For the states of plane strain given in Problems 13.18–13.21, determine the principal strains and the maximum in-plane shear strain and show the orientations of the elements subjected to these strains.

13.18 $\varepsilon_x = 0.002$, $\varepsilon_y = 0.001$, and $\gamma_{xy} = 0$.

13.19 $\varepsilon_x = 0.0025$, $\varepsilon_y = 0$, and $\gamma_{xy} = -0.0050$.

13.20 $\varepsilon_x = -0.008$, $\varepsilon_y = 0.006$, and $\gamma_{xy} = -0.012$.

13.21 $\varepsilon_x = 0.0024$, $\varepsilon_y = -0.0012$, and $\gamma_{xy} = 0.0024$.

13.22 For the state of plane strain $\varepsilon_x = 0.0024$, $\varepsilon_y = 0.0012$, and $\gamma_{xy} = 0.0024$, what is the absolute maximum shear strain?

13.23 A point p of the MacPherson strut suspension is subjected to the state of plane strain $\varepsilon_x = -0.0088$, $\varepsilon_y = 0.0024$, $\gamma_{xy} = -0.0036$. Determine the principal strains and the absolute maximum shear strain.

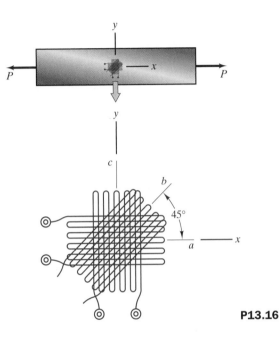

P13.16

13.17 The strains measured by a strain gauge rosette mounted on the bicycle brake are $\varepsilon_a = 0.00220$, $\varepsilon_b = -0.00100$, and $\varepsilon_c = -0.00360$. Determine the strains ε_x, ε_y, and γ_{xy}.

P13.23

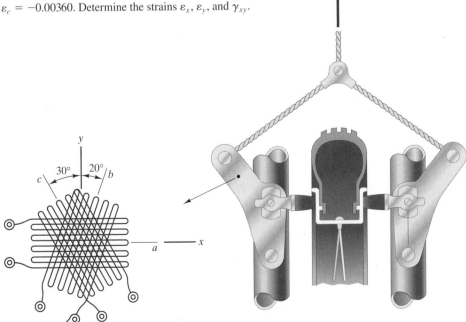

P13.17

13.24 A circular line in the x–y plane of circumference $C = 2\pi R$ is drawn in the reference state of a material. If the material is then subjected to a homogeneous strain ε_x and other components of strain are zero, show that the length of the deformed line is $C(1 + 0.5\varepsilon_x)$.

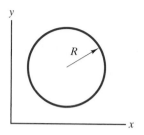

P13.24

<hr>

13.3 Stress-Strain Relations

We have discussed the state of stress, which is related to the internal forces at a material point, and the state of strain, related to the deformation in the neighborhood of a material point. Is there a relationship between them? Consider the example of a crystalline material, such as iron, that can be modeled as a lattice of atoms connected by bonds that behave like springs. When such a material is deformed, the distances between atoms change and the "springs" become stretched or compressed, altering the forces they exert. Since the internal forces near a point will depend only on the changes in the distances between atoms near that point, it is reasonable to postulate that the state of stress depends only on the state of strain. This is a very simple conceptual model of material behavior, and in general the state of stress at a point of a material can depend on the temperature as well as the history of the state of strain. A material for which the state of stress at a point is a single-valued function of the current state of strain at that point is said to be *elastic*. Some materials, including most metals subjected to limited ranges of temperature and stress, can be modeled adequately by assuming them to be elastic. In this section we derive the equations relating the state of stress in an elastic material to its state of strain for linearly elastic materials having the property of isotropy.

Linearly Elastic Materials

An elastic material is one for which the state of stress at a point is a function only of the current state of strain at that point. We can express each component of stress as a function of the components of strain:

$$
\begin{aligned}
\sigma_x &= \sigma_x(\varepsilon_x, \varepsilon_y, \varepsilon_z, \gamma_{xy}, \gamma_{yz}, \gamma_{xz}),\\
\sigma_y &= \sigma_y(\varepsilon_x, \varepsilon_y, \varepsilon_z, \gamma_{xy}, \gamma_{yz}, \gamma_{xz}),\\
\sigma_z &= \sigma_z(\varepsilon_x, \varepsilon_y, \varepsilon_z, \gamma_{xy}, \gamma_{yz}, \gamma_{xz}),\\
\tau_{xy} &= \tau_{xy}(\varepsilon_x, \varepsilon_y, \varepsilon_z, \gamma_{xy}, \gamma_{yz}, \gamma_{xz}),\\
\tau_{yz} &= \tau_{yz}(\varepsilon_x, \varepsilon_y, \varepsilon_z, \gamma_{xy}, \gamma_{yz}, \gamma_{xz}),\\
\tau_{xz} &= \tau_{xz}(\varepsilon_x, \varepsilon_y, \varepsilon_z, \gamma_{xy}, \gamma_{yz}, \gamma_{xz}).
\end{aligned}
\tag{13.23}
$$

These *stress-strain relations*, which determine the state of stress in a given material in terms of its state of strain, are examples of what are called *constitutive equations*. They are functions that depend on the constitution, or physical structure, of a material. Let us express the equation for σ_x as a power series in terms of the components of strain,

$$
\begin{aligned}
\sigma_x &= a_{10} + a_{11}\varepsilon_x + a_{12}\varepsilon_y + a_{13}\varepsilon_z + a_{14}\gamma_{xy} + a_{15}\gamma_{yz}\\
&\quad + a_{16}\gamma_{xz} + a_{17}\varepsilon_x^2 + a_{18}\varepsilon_x\varepsilon_y + \cdots,
\end{aligned}
$$

where the coefficients $a_{10}, a_{11}, \ldots$ are constants. If we assume that the stress σ_x is zero when the components of strain are zero, the coefficient $a_{10} = 0$. If we also assume that the components of strain are sufficiently small that products of the components are negligible in comparison to the components themselves, we obtain

$$\sigma_x = a_{11}\varepsilon_x + a_{12}\varepsilon_y + a_{13}\varepsilon_z + a_{14}\gamma_{xy} + a_{15}\gamma_{yz} + a_{16}\gamma_{xz}.$$

Expressing each component of stress in this way, Eqs. (13.23) become

$$\sigma_x = a_{11}\varepsilon_x + a_{12}\varepsilon_y + a_{13}\varepsilon_z + a_{14}\gamma_{xy} + a_{15}\gamma_{yz} + a_{16}\gamma_{xz}.$$
$$\sigma_y = a_{21}\varepsilon_x + a_{22}\varepsilon_y + a_{23}\varepsilon_z + a_{24}\gamma_{xy} + a_{25}\gamma_{yz} + a_{26}\gamma_{xz}.$$
$$\sigma_z = a_{31}\varepsilon_x + a_{32}\varepsilon_y + a_{33}\varepsilon_z + a_{34}\gamma_{xy} + a_{35}\gamma_{yz} + a_{36}\gamma_{xz}. \qquad (13.24)$$
$$\tau_{xy} = a_{41}\varepsilon_x + a_{42}\varepsilon_y + a_{43}\varepsilon_z + a_{44}\gamma_{xy} + a_{45}\gamma_{yz} + a_{46}\gamma_{xz}.$$
$$\tau_{yz} = a_{51}\varepsilon_x + a_{52}\varepsilon_y + a_{53}\varepsilon_z + a_{54}\gamma_{xy} + a_{55}\gamma_{yz} + a_{56}\gamma_{xz}.$$
$$\tau_{xz} = a_{61}\varepsilon_x + a_{62}\varepsilon_y + a_{63}\varepsilon_z + a_{64}\gamma_{xy} + a_{65}\gamma_{yz} + a_{66}\gamma_{xz}.$$

An elastic material that can be modeled by these stress-strain relations is said to be *linearly elastic*. The components of stress are linear functions of the components of strain. To model a given material, the 36 constants $a_{11}, a_{12}, \ldots$ must be known. At this juncture we are apparently faced with the daunting prospect of performing 36 independent experiments to determine the stress-strain relations of a single material. But we will show in the following section that far fewer than 36 constants need to be determined to characterize most linearly elastic materials.

Isotropic Materials

The simplest way to explain what is meant by an isotropic material is to consider a familiar material that is not isotropic—wood. If we apply a normal stress σ to opposite faces of a cube of wood and measure the resulting extensional strain ε, it is clear that we obtain one result if the grain of the wood is parallel to the direction of the strain ε (Fig. 13.17a) and a different result if the grain of the wood is perpendicular to the direction of the strain ε (Fig. 13.17b). The behavior of the wood depends on the direction of its grain. The behavior of a material that is not isotropic, or *anisotropic*, depends on the orientation of the material. The behavior of a material that is *isotropic* (which, roughly translated from Latin, means "the same in all directions") does not depend on the orientation of the material. Its stress-strain relations are the same for any orientation of the material. Many materials used in engineering are approximately isotropic, although the use of intentionally created anisotropic materials, such as fiber-reinforced and layered composite materials, is increasing.

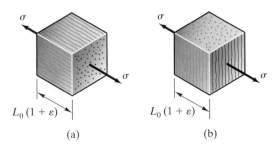

(a)   (b)

Figure 13.17
Stretching a cube of wood:
(a) parallel to the grain;
(b) perpendicular to the grain.

Another way to state the definition of an isotropic material is that *the stress-strain relations are the same for any orientation of the coordinate system relative to the material*. In other words, instead of requiring the material properties to be the same for different orientations of the material, we require the material properties to be the same for different orientations of the frame of reference relative to the material. Using this definition, we now investigate how material isotropy affects the stress-strain relations of a linearly elastic material.

Isotropic Stress-Strain Relations   If we regard the stress-strain equations (13.24) as six linear algebraic equations for the components of strain in terms of the components of stress, we can in principle invert them to obtain linear equations for the components of strain in terms of the components of stress. We write the resulting equations as

$$\varepsilon_x = b_{11}\sigma_x + b_{12}\sigma_y + b_{13}\sigma_z + b_{14}\tau_{xy} + b_{15}\tau_{yz} + b_{16}\tau_{xz},$$
$$\varepsilon_y = b_{21}\sigma_x + b_{22}\sigma_y + b_{23}\sigma_z + b_{24}\tau_{xy} + b_{25}\tau_{yz} + b_{26}\tau_{xz},$$
$$\varepsilon_z = b_{31}\sigma_x + b_{32}\sigma_y + b_{33}\sigma_z + b_{34}\tau_{xy} + b_{35}\tau_{yz} + b_{36}\tau_{xz},$$
$$\gamma_{xy} = b_{41}\sigma_x + b_{42}\sigma_y + b_{43}\sigma_z + b_{44}\tau_{xy} + b_{45}\tau_{yz} + b_{46}\tau_{xz},$$
$$\gamma_{yz} = b_{51}\sigma_x + b_{52}\sigma_y + b_{53}\sigma_z + b_{54}\tau_{xy} + b_{55}\tau_{yz} + b_{56}\tau_{xz},$$
$$\gamma_{xz} = b_{61}\sigma_x + b_{62}\sigma_y + b_{63}\sigma_z + b_{64}\tau_{xy} + b_{65}\tau_{yz} + b_{66}\tau_{xz}. \tag{13.25}$$

Isotropy places severe restrictions on the possible values of the constants $b_{11}, b_{12}, \ldots$. For example, suppose that we subject an isotropic material to the normal stress σ shown in Fig. 13.18. In terms of the coordinate system shown, the only nonzero stress component is $\sigma_x = \sigma$. From Eqs. (13.25), the strain components ε_x and ε_y are

$$\varepsilon_x = b_{11}\sigma,$$
$$\varepsilon_y = b_{21}\sigma.$$

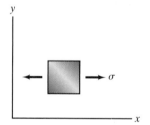

Figure 13.18
Subjecting an isotropic material to a normal stress.

We now introduce a different coordinate system obtained by rotating the coordinate system in Fig. 13.18 90° about its z axis (Fig. 13.19). Because the material is isotropic, the stress-strain relations are given by Eqs. (13.25) expressed in terms of the $x'y'z'$ coordinate system with the same coefficients $b_{11}, b_{12}, \ldots$. The only nonzero stress component is $\sigma'_y = \sigma$, so we obtain

$$\varepsilon'_x = b_{12}\sigma,$$
$$\varepsilon'_y = b_{22}\sigma.$$

Because the x and y' axes of the coordinate systems in Figs. 13.18 and 13.19 are parallel, the extensional strains ε_x and ε'_y are equal.

$$\varepsilon_x = \varepsilon'_y:$$
$$b_{11}\sigma = b_{22}\sigma. \tag{13.26}$$

The y and x' axes are parallel, so ε_y and ε'_x are also equal.

$$\varepsilon_y = \varepsilon'_x:$$
$$b_{21}\sigma = b_{12}\sigma. \tag{13.27}$$

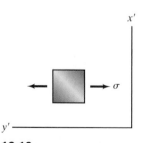

Figure 13.19
The same state of stress with a new coordinate system.

From Eqs. (13.26) and (13.27) we see that the constants $b_{11} = b_{22}$ and $b_{12} = b_{21}$ for an isotropic material.

Next, we subject the cube to the shear stress τ shown in Fig. 13.20a. In terms of the coordinate system shown, the only nonzero stress component is $\tau_{xy} = \tau$. From Eqs. (13.25), the strain component ε_x is

$$\varepsilon_x = b_{14}\tau.$$

We now introduce an $x'y'z'$ coordinate system obtained by rotating the coordinate system in Fig. 13.20a $180°$ about its y axis (Fig. 13.20b). The only nonzero stress component is $\tau'_{xy} = -\tau$, so from Eqs. (13.25) expressed in terms of the $x'y'z'$ system, we obtain

$$\varepsilon'_x = -b_{14}\tau.$$

Because the x and x' axes of the coordinate systems in Figs. 13.20a and b are parallel, the extensional strains ε_x and ε'_x are equal.

$$\varepsilon_x = \varepsilon'_x:$$
$$b_{14}\tau = -b_{14}\tau.$$

This equation shows that the constant $b_{14} = 0$ for an isotropic material.

Isotropy—the requirement that the stress-strain relations for a material must be the same for any orientation of the coordinate system—has forced us to conclude that the coefficients in Eqs. (13.25) cannot have any values, but must satisfy the restrictions $b_{11} = b_{22}$, $b_{12} = b_{21}$, and $b_{14} = 0$. The number of independent constants in Eqs. (13.25) is therefore reduced from 36 to 33. By considering additional changes in orientation of the coordinate system, it can be shown that for an isotropic material, Eqs. (13.25) must be of the forms

$$\varepsilon_x = b_{11}\sigma_x + b_{12}\sigma_y + b_{12}\sigma_z, \tag{13.28}$$
$$\varepsilon_y = b_{12}\sigma_x + b_{11}\sigma_y + b_{12}\sigma_z, \tag{13.29}$$
$$\varepsilon_z = b_{12}\sigma_x + b_{12}\sigma_y + b_{11}\sigma_z, \tag{13.30}$$
$$\gamma_{xy} = b_{44}\tau_{xy}, \tag{13.31}$$
$$\gamma_{yz} = b_{44}\tau_{yz}, \tag{13.32}$$
$$\gamma_{xz} = b_{44}\tau_{xz}. \tag{13.33}$$

The number of coefficients is reduced from 36 to 3. Furthermore, we can express these coefficients in familiar terms. If we subject an isotropic material to a normal stress σ_x and the other components of stress are zero (Fig. 13.21), the ratio of σ_x to the resulting extensional strain ε_x is the modulus of elasticity E of the material:

$$\frac{\sigma_x}{\varepsilon_x} = E. \tag{13.34}$$

Appying Eq. (13.28) to this state of stress, we obtain

$$\varepsilon_x = b_{11}\sigma_x. \tag{13.35}$$

By comparing Eqs. (13.34) and (13.35), we determine the constant b_{11} in terms of E:

$$b_{11} = \frac{1}{E}. \tag{13.36}$$

For the state of stress shown in Fig. 13.21, the negative of the ratio of the lateral strain $\left(\text{either } \varepsilon_y \text{ or } \varepsilon_z\right)$ to the axial strain ε_x is Poisson's ratio ν of the material:

$$-\frac{\varepsilon_y}{\varepsilon_x} = \nu. \tag{13.37}$$

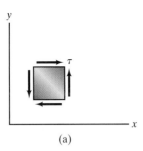

(a)

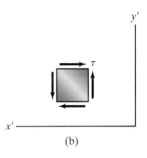

(b)

Figure 13.20
Subjecting an isotropic material to a shear stress.

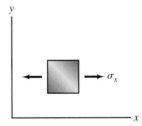

Figure 13.21
Applying a normal stress σ_x.

Applying Eq. (13.29) to this state of stress gives

$$\varepsilon_y = b_{12}\sigma_x. \tag{13.38}$$

Dividing this equation by Eq. (13.35), we obtain

$$\frac{\varepsilon_y}{\varepsilon_x} = \frac{b_{12}}{b_{11}},$$

and by comparing this equation to Eq. (13.37), we determine the constant b_{12} in terms of E and ν:

$$b_{12} = -\nu b_{11} = -\frac{\nu}{E}. \tag{13.39}$$

Now we subject the isotropic material to a shear stress τ_{xy} with the other components of stress equal to zero (Fig. 13.22). The ratio of τ_{xy} to the resulting shear strain γ_{xy} is the shear modulus G of the material:

$$\frac{\tau_{xy}}{\gamma_{xy}} = G. \tag{13.40}$$

From Eq. (13.31),

$$\gamma_{xy} = b_{44}\tau_{xy}, \tag{13.41}$$

and by comparing Eqs. (13.40) and (13.41), we determine the constant b_{44} in terms of G:

$$b_{44} = \frac{1}{G}. \tag{13.42}$$

Substituting the expressions (13.36), (13.39), and (13.42) into Eqs. (13.28)–(13.33), we obtain the stress-strain relations for an isotropic linearly elastic material in forms in which they are commonly presented:

$$\varepsilon_x = \frac{1}{E}\sigma_x - \frac{\nu}{E}(\sigma_y + \sigma_z), \tag{13.43}$$

$$\varepsilon_y = \frac{1}{E}\sigma_y - \frac{\nu}{E}(\sigma_x + \sigma_z), \tag{13.44}$$

$$\varepsilon_z = \frac{1}{E}\sigma_z - \frac{\nu}{E}(\sigma_x + \sigma_y), \tag{13.45}$$

$$\gamma_{xy} = \frac{1}{G}\tau_{xy}, \tag{13.46}$$

$$\gamma_{yz} = \frac{1}{G}\tau_{yz}, \tag{13.47}$$

$$\gamma_{xz} = \frac{1}{G}\tau_{xz}. \tag{13.48}$$

Relating E, ν, and G Equations (13.43)–(13.48) express the stress-strain relations for an isotropic linearly elastic material in terms of three coefficients, the modulus of elasticity E, the Poisson's ratio ν, and the shear modulus G. Such a material is actually characterized by only two independent constants, because G can be expressed in terms of E and ν.

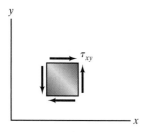

Figure 13.22
Applying a shear stress τ_{xy}.

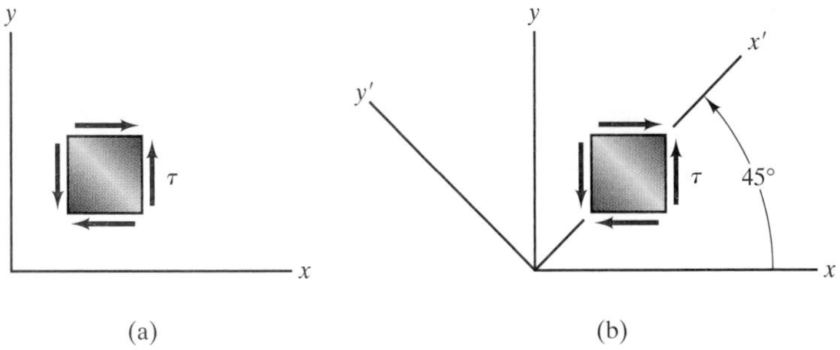

Figure 13.23
(a) Applying a shear stress $\tau_{xy} = \tau$.
(b) Introducing a rotated coordinate system $x'y'z'$.

To derive this result, we subject an isotropic material to a shear stress $\tau_{xy} = \tau$ and let other components of stress be zero (Fig. 13.23a). From Eqs. (13.43)–(13.48), the only nonzero component of strain is $\gamma_{xy} = \tau/G$. Now let us consider an $x'y'z'$ coordinate system obtained by rotating the xyz coordinate system 45° about the z axis (Fig. 13.23b). Since we know the state of stress, we can use Eqs. (12.7) and (12.9) to determine the stress components σ'_x and σ'_y:

$$\sigma'_x = \tau \sin 2(45°) = \tau,$$
$$\sigma'_y = -\tau \sin 2(45°) = -\tau. \tag{13.49}$$

Notice that the stress components $\sigma'_z = \sigma_z = 0$. We also know the state of strain, so we can use Eq. (13.7) to determine the strain component ε'_x:

$$\varepsilon'_x = \frac{\gamma_{xy}}{2} \sin 2(45°) = \frac{1}{2G} \tau.$$

Because the material is isotropic, the strain component ε'_x and the stress components σ'_x, σ'_y, and σ'_z must satisfy Eq. (13.43).

$$\varepsilon'_x = \frac{1}{E} \sigma'_x - \frac{\nu}{E} \left(\sigma'_y + \sigma'_z \right):$$

$$\frac{1}{2G} \tau = \frac{1}{E} \tau + \frac{\nu}{E} \tau.$$

Solving this equation for G, we determine the shear modulus in terms of the modulus of elasticity and Poisson's ratio:

$$G = \frac{E}{2(1 + \nu)}. \tag{13.50}$$

Lamé Constants and Bulk Modulus Equations (13.43)–(13.48) give the components of strain in terms of the components of stress for an isotropic linearly elastic material. When the components of stress are expressed in terms of the components of strain, they can conveniently be written in the forms

$$\sigma_x = (\lambda + 2\mu)\varepsilon_x + \lambda(\varepsilon_y + \varepsilon_z), \tag{13.51}$$
$$\sigma_y = (\lambda + 2\mu)\varepsilon_y + \lambda(\varepsilon_x + \varepsilon_z), \tag{13.52}$$
$$\sigma_z = (\lambda + 2\mu)\varepsilon_z + \lambda(\varepsilon_x + \varepsilon_y), \tag{13.53}$$
$$\tau_{xy} = \mu\gamma_{xy}, \tag{13.54}$$
$$\tau_{yz} = \mu\gamma_{yz}, \tag{13.55}$$
$$\tau_{xz} = \mu\gamma_{xz}, \tag{13.56}$$

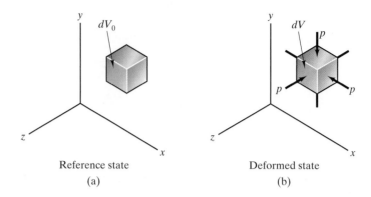

Figure 13.24
Element of material in the reference state
and when subjected to a pressure p.

Reference state
(a)

Deformed state
(b)

where λ and μ are called the *Lamé constants*. The constant $\mu = G$, and the constant λ is given in terms of the modulus of elasticity and Poisson's ratio by

$$\lambda = \frac{\nu E}{(1 + \nu)(1 - 2\nu)}. \tag{13.57}$$

Consider an element of material with volume dV_0 in the reference state (Fig. 13.24a). We subject the element to a pressure p, so that $\sigma_x = -p$, $\sigma_y = -p$, and $\sigma_z = -p$ (Fig. 13.24b). Let the volume of the deformed element be dV. The *bulk modulus K* of the material is defined to be the ratio of $-p$ to the dilatation $e = (dV - dV_0)/dV_0$:

$$K = \frac{-p}{e}. \tag{13.58}$$

From Eq. (13.3), the dilatation of a material subjected to small strains is

$$e = \varepsilon_x + \varepsilon_y + \varepsilon_z,$$

and from Eqs. (13.43)–(13.45), the extensional strains of the element are

$$\varepsilon_x = \varepsilon_y = \varepsilon_z = \frac{1 - 2\nu}{E}(-p).$$

Using these relationships, we can express the bulk modulus in the form

$$K = \frac{E}{3(1 - 2\nu)}. \tag{13.59}$$

The bulk modulus is sometimes used instead of the modulus of elasticity E or the Lamé constant λ in expressing the stress-strain relations of an isotropic elastic material.

Study Questions

1. What is an elastic material? What is a linearly elastic material?
2. We proved that for an *isotropic* linearly elastic material, the coefficients in Eqs. (13.25) must satisfy the restrictions $b_{11} = b_{22}$, $b_{12} = b_{21}$, and $b_{14} = 0$. Why doesn't the proof we used show that these restrictions must be satisfied by any linearly elastic material?
3. How many independent coefficients appear in the stress-strain relations for an isotropic linearly elastic material?
4. What is the definition of the bulk modulus?

Example 13.4

Elastic Constants

The cylindrical bar in Fig. 13.25 consists of an isotropic linearly elastic material and is subjected to axial loads. As a result, the bar is subjected to a normal stress $\sigma_x = 420$ MPa and the other components of stress are zero. By measuring the changes in the bar's length and diameter, it is determined that the axial strain is $\varepsilon_x = 0.006$ and the lateral strain is $\varepsilon_y = \varepsilon_z = -0.002$. Determine (a) the modulus of elasticity, Poisson's ratio, and shear modulus of the material; (b) the Lamé constants of the material; (c) the bulk modulus of the material.

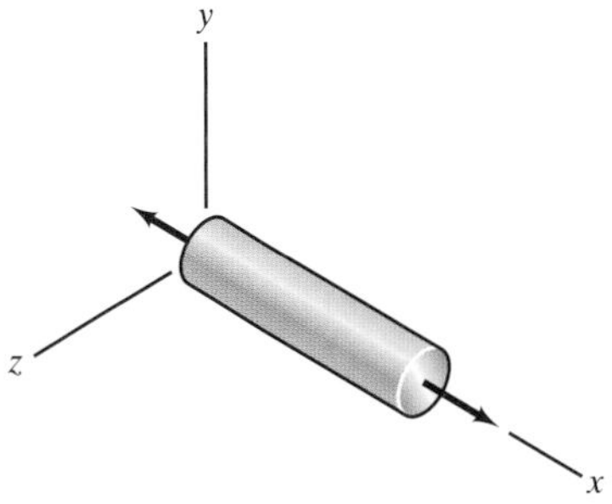

Figure 13.25

Strategy

(a) We know the state of stress and the strain components ε_x and ε_y, so we can solve Eqs. (13.43) and (13.44) for the modulus of elasticity E and the Poisson's ratio ν. We can then use Eq. (13.50) to determine the shear modulus G. (b) The Lamé constant $\mu = G$, and λ is given in terms of E and ν by Eq. (13.57). (c) The bulk modulus is given in terms of E and ν by Eq. (13.59).

Solution

(a) From Eq. (13.43),

$$\varepsilon_x = \frac{1}{E}\sigma_x - \frac{\nu}{E}(\sigma_x + \sigma_z):$$

$$0.006 = \frac{1}{E}\left(420 \times 10^6\right),$$

we obtain $E = 70.0$ GPa. Then, from Eq. (13.44),

$$\varepsilon_y = \frac{1}{E}\sigma_y - \frac{\nu}{E}(\sigma_x + \sigma_z):$$

$$-0.002 = \frac{-\nu}{70.0 \times 10^9}\left(420 \times 10^6\right),$$

we obtain $\nu = 0.333$. From Eq. (13.50), the shear modulus is

$$G = \frac{E}{2(1 + \nu)} = \frac{70.0 \times 10^9}{(2)(1 + 0.333)} = 26.3 \text{ GPa.}$$

(b) The Lamé constant $\mu = G = 26.3$ GPa. From Eq. (13.57),

$$\lambda = \frac{\nu E}{(1 + \nu)(1 - 2\nu)} = \frac{(0.333)\left(70.0 \times 10^9\right)}{(1 + 0.333)\left[1 - (2)(0.333)\right]}$$

$$= 52.5 \text{ GPa.}$$

(c) From Eq. (13.59), the bulk modulus is

$$K = \frac{E}{3(1 - 2\nu)} = \frac{70.0 \times 10^9}{3\left[1 - (2)0.333\right]} = 70.0 \text{ GPa.}$$

Problems

13.25 The state of stress at a point p in a material with modulus of elasticity $E = 28$ GPa and Poisson's ratio $\nu = 0.3$ is

$$\begin{bmatrix} 250 & -20 & 0 \\ -20 & 250 & 40 \\ 0 & 40 & 200 \end{bmatrix} \text{MPa.}$$

What is the state of strain at p?

13.26 The state of strain at a point p of the material described in Problem 13.25 is

$$\begin{bmatrix} -250 & 125 & 0 \\ 125 & 500 & -125 \\ 0 & -125 & 250 \end{bmatrix} \times 10^{-5}.$$

What is the state of stress at p?

13.27 The state of stress at a point p of a nickel pipe in a gaseous diffusion plant is

$$\begin{bmatrix} 35 & -20 & 25 \\ -20 & 45 & 32 \\ 25 & 32 & 40 \end{bmatrix} \text{ksi.}$$

What is the state of strain at p?

13.28 Show that a state of plane stress at a point p of an isotropic, linearly elastic material does not necessarily result in a state of plane strain at p. What condition must the state of plane stress satisfy to result in a state of plane strain?

13.29 The state of plane stress at a point p in a machine part made of 2014-T6 aluminum alloy is $\sigma_x = 40$ MPa, $\sigma_y = -30$ MPa, and $\tau_{xy} = 30$ MPa. What is the state of strain at p?

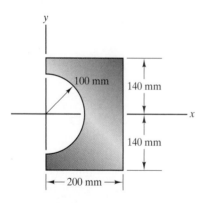

P13.29

13.30 An arm of a robotic actuator made of 7075-T6 aluminum alloy is subjected to a state of plane stress $\sigma_x, \sigma_y, \tau_{xy}$. Using a strain gauge rosette, it is determined experimentally that $\varepsilon_x = 0.00350$, $\varepsilon_y = 0.00600$, and $\gamma_{xy} = -0.02400$. What are the components of stress (in ksi)?

13.31 A material is subjected to a state of plane stress $\sigma_x = 400$ MPa, $\sigma_y = -200$ MPa, $\tau_{xy} = 300$ MPa. Using a strain gauge rosette, it is determined experimentally that $\varepsilon_x = 0.00239$, $\varepsilon_y = -0.00162$, and $\gamma_{xy} = 0.00401$. What are the modulus of elasticity and Poisson's ratio of the material?

13.32 For the material described in Problem 13.25, determine (a) the Lamé constants λ and μ; (b) the bulk modulus K.

13.33 The state of stress at a point p in a material with modulus of elasticity $E = 15 \times 10^6$ psi and Poisson's ratio $\nu = 0.33$ is

$$\begin{bmatrix} 50 & -60 & -40 \\ -60 & 40 & 40 \\ -40 & 40 & -40 \end{bmatrix} \text{ksi.}$$

What is the state of strain at p?

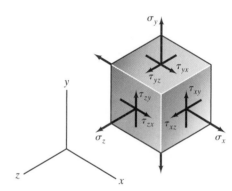

P13.33

13.34 The state of strain at a point p of the material described in Problem 13.33 is

$$\begin{bmatrix} 0.004 & 0.010 & -0.010 \\ 0.010 & -0.002 & 0 \\ -0.010 & 0 & -0.003 \end{bmatrix}$$

What is the state of stress at p?

13.35 For the material described in Problem 13.33, determine (a) the Lamé constants λ and μ; (b) the bulk modulus K.

13.36 The cylindrical bar consists of a material with modulus of elasticity E and Poisson's ratio ν and is subjected to axial loads. The resulting state of stress is

$$\begin{bmatrix} \sigma_x & 0 & 0 \\ 0 & 0 & 0 \\ 0 & 0 & 0 \end{bmatrix}.$$

(a) Determine the state of strain in terms of σ_x, E, and ν.
(b) The length of the unloaded bar is L and its cross-sectional area is A. Determine the volume of the loaded bar in terms of L, A, σ_x, E, and ν.

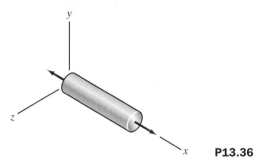

P13.36

13.37 The bar described in Problem 13.36 is subjected to axial loads that cause a normal stress $\sigma_x = 380$ MPa. The axial and lateral extensional strains are measured and determined to be $\varepsilon_x = 0.0020$, $\varepsilon_y = \varepsilon_z = -0.0007$. Determine (a) the modulus of elasticity E, Poisson's ratio ν, and shear modulus G of the material; (b) the Lamé constants λ and μ of the material.

13.38 At a point of a material subjected to a state of plane stress, the strains measured by the strain gauge rosette are $\varepsilon_a = 0.006$, $\varepsilon_b = -0.003$, and $\varepsilon_c = -0.002$. The modulus of elasticity and Poisson's ratio of the material are $E = 30$ GPa and $\nu = 0.33$.. What is the state of stress at the point?

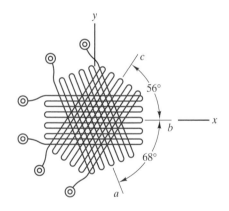

P13.38

13.39 At a point of a steel hydraulic piston that is subjected to a state of plane stress, the strains measured by the strain gauge rosette shown in Problem 13.38 are $\varepsilon_a = -0.00150$, $\varepsilon_b = 0.00140$, and $\varepsilon_c = 0.00086$. The modulus of elasticity and Poisson's ratio of the material are $E = 200$ GPa and $\nu = 0.33$. What are the principal stresses and the absolute maximum shear stress at the point?

13.40 The strain gauge rosette shown in Problem 13.38 is mounted on the outer wall of one of the Atlas launch vehicle's nozzles, where the material is in a homogeneous state of plane stress. The strains measured by the rosette are $\varepsilon_a = 0.0053$, $\varepsilon_b = 0.0038$, and $\varepsilon_c = 0.0029$. The modulus of elasticity and Poisson's ratio of the material are $E = 70$ GPa and $\nu = 0.33$. What are the components of plane stress?

P13.40

13.41 An isotropic material is subjected to a normal stress $\sigma_x = \sigma$ and other components of stress are zero. By writing Eqs. (13.43)–(13.48) and Eqs. (13.51)–(13.56) for this state of stress, derive the relation

$$\lambda = \frac{\nu E}{(1 + \nu)(1 - 2\nu)}.$$

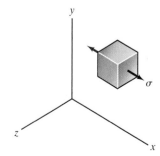

P13.41

Chapter Summary

Components of Strain

In terms of a given coordinate system, the *state of strain* at a point p of a material is defined by the *components of strain*

$$\begin{bmatrix} \varepsilon_x & \gamma_{xy} & \gamma_{xz} \\ \gamma_{yx} & \varepsilon_y & \gamma_{yz} \\ \gamma_{zx} & \gamma_{zy} & \varepsilon_z \end{bmatrix}.$$

Eq. (13.1)

The components ε_x, ε_y, and ε_z are the extensional strains in the x, y, and z directions. The component $\gamma_{xy} = \gamma_{yx}$ is the shear strain referred to the directions of the x and y axes, and the components $\gamma_{yz} = \gamma_{zy}$ and $\gamma_{xz} = \gamma_{zx}$ are defined similarly.

Consider a volume dV_0 of material in a reference state. Its volume in the deformed state is

$$dV = \left(1 + \varepsilon_x + \varepsilon_y + \varepsilon_z\right) dV_0.$$

Eq. (13.2)

The *dilatation* is the change in volume of the material per unit volume:

$$e = \frac{dV - dV_0}{dV_0} = \varepsilon_x + \varepsilon_y + \varepsilon_z.$$

Eq. (13.3)

Transformations of Plane Strain

The strain at a point p is said to be a state of *plane strain* if it is of the form

$$\begin{bmatrix} \varepsilon_x & \gamma_{xy} & 0 \\ \gamma_{yx} & \varepsilon_y & 0 \\ 0 & 0 & 0 \end{bmatrix}.$$

Eq. (13.4)

In terms of a coordinate system $x'y'z'$ oriented as shown in Fig. (a), the components of strain are

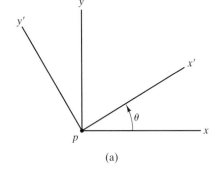

(a)

$$\varepsilon_x' = \frac{\varepsilon_x + \varepsilon_y}{2} + \frac{\varepsilon_x - \varepsilon_y}{2}\cos 2\theta + \frac{\gamma_{xy}}{2}\sin 2\theta,$$

Eq. (13.7)

$$\varepsilon_y' = \frac{\varepsilon_x + \varepsilon_y}{2} - \frac{\varepsilon_x - \varepsilon_y}{2}\cos 2\theta - \frac{\gamma_{xy}}{2}\sin 2\theta,$$

Eq. (13.8)

$$\frac{\gamma_{xy}'}{2} = -\frac{\varepsilon_x - \varepsilon_y}{2}\sin 2\theta + \frac{\gamma_{xy}}{2}\cos 2\theta.$$

Eq. (13.9)

For an isotropic linearly elastic material, Eqs. (13.7)–(13.9) also apply to the components of strain resulting from a state of plane stress.

Strain Gauge Rosette

Suppose that a material is subject to an unknown state of plane strain relative to the $x - y$ coordinate system in Fig. (a). A *strain gauge rosette* measures the extensional strain ε_x' in three different directions: θ_a, θ_b, and θ_c. Let the measured strains be ε_a, ε_b, and ε_c. Using Eq. (13.6) to express these strains in terms of the strain components ε_x, ε_y, and γ_{xy}, gives

$$\varepsilon_a = \varepsilon_x \cos^2\theta_a + \varepsilon_y \sin^2\theta_a + \gamma_{xy}\sin\theta_a\cos\theta_a,$$
$$\varepsilon_b = \varepsilon_x \cos^2\theta_b + \varepsilon_y \sin^2\theta_b + \gamma_{xy}\sin\theta_b\cos\theta_b, \qquad \text{Eq. (13.10)}$$
$$\varepsilon_c = \varepsilon_x \cos^2\theta_c + \varepsilon_y \sin^2\theta_c + \gamma_{xy}\sin\theta_c\cos\theta_c.$$

This system of equations can be solved for the strain components ε_x, ε_y, and γ_{xy}.

Maximum and Minimum Strains in Plane Strain

A value of θ for which the extensional strain is a maximum or minimum is determined from the equation

$$\tan 2\theta_p = \frac{\gamma_{xy}}{\varepsilon_x - \varepsilon_y}. \qquad \text{Eq. (13.11)}$$

The values of the principal strains can be obtained by substituting θ_p into Eqs. (13.7) and (13.8). Their values can also be determined from the equation

$$\varepsilon_1, \varepsilon_2 = \frac{\varepsilon_x + \varepsilon_y}{2} \pm \sqrt{\left(\frac{\varepsilon_x - \varepsilon_y}{2}\right)^2 + \left(\frac{\gamma_{xy}}{2}\right)^2}, \qquad \text{Eq. (13.14)}$$

although this equation does not indicate their directions. An infinitesimal square element oriented as shown in Fig. (b) is subjected to the principal strains in the x' and y' directions and undergoes no shear strain.

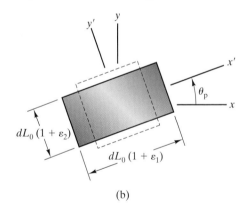

(b)

A value of θ for which the in-plane shear strain is a maximum or minimum is determined from the equation

$$\tan 2\theta_s = -\frac{\varepsilon_x - \varepsilon_y}{\gamma_{xy}}. \qquad \text{Eq. (13.15)}$$

The corresponding shear strain can be obtained by substituting θ_s into Eq. (13.9). The magnitude of the maximum in-plane shear strain can be determined from the equation

$$\gamma_{max} = \sqrt{(\varepsilon_x - \varepsilon_y)^2 + \gamma_{xy}^2}. \qquad \text{Eq. (13.18)}$$

The *absolute maximum shear strain* is the largest of the three values

$$|\varepsilon_1 - \varepsilon_2|, \quad |\varepsilon_1|, \quad |\varepsilon_2|. \qquad \text{Eq. (13.22)}$$

Stress-Strain Relations

A material for which the state of stress at a point is a single-valued function of the current state of strain at that point is said to be *elastic*. If the components of strain are linear functions of the components of stress, the material is *linearly elastic*. If those functions are the same for any orientation of the coordinate system relative to the material, the material is *isotropic*. The stress-strain relations for an isotropic linearly elastic material are

$$\varepsilon_x = \frac{1}{E}\sigma_x - \frac{\nu}{E}(\sigma_y + \sigma_z), \qquad\qquad \text{Eq. (13.43)}$$

$$\varepsilon_y = \frac{1}{E}\sigma_y - \frac{\nu}{E}(\sigma_x + \sigma_z), \qquad\qquad \text{Eq. (13.44)}$$

$$\varepsilon_z = \frac{1}{E}\sigma_z - \frac{\nu}{E}(\sigma_x + \sigma_y), \qquad\qquad \text{Eq. (13.45)}$$

$$\gamma_{xy} = \frac{1}{G}\tau_{xy}, \qquad\qquad \text{Eq. (13.46)}$$

$$\gamma_{yz} = \frac{1}{G}\tau_{yz}, \qquad\qquad \text{Eq. (13.47)}$$

$$\gamma_{xz} = \frac{1}{G}\tau_{xz}. \qquad\qquad \text{Eq. (13.48)}$$

The shear modulus is related to the modulus of elasticity and Poisson's ratio by

$$G = \frac{E}{2(1 + \nu)}. \qquad\qquad \text{Eq. (13.50)}$$

The stress-strain relations can also be expressed as

$$\sigma_x = (\lambda + 2\mu)\varepsilon_x + \lambda(\varepsilon_y + \varepsilon_z), \qquad\qquad \text{Eq. (13.51)}$$
$$\sigma_y = (\lambda + 2\mu)\varepsilon_y + \lambda(\varepsilon_x + \varepsilon_z), \qquad\qquad \text{Eq. (13.52)}$$
$$\sigma_z = (\lambda + 2\mu)\varepsilon_z + \lambda(\varepsilon_x + \varepsilon_y), \qquad\qquad \text{Eq. (13.53)}$$
$$\tau_{xy} = \mu\gamma_{xy}, \qquad\qquad \text{Eq. (13.54)}$$
$$\tau_{yz} = \mu\gamma_{yz}, \qquad\qquad \text{Eq. (13.55)}$$
$$\tau_{xz} = \mu\gamma_{xz}, \qquad\qquad \text{Eq. (13.56)}$$

where λ and μ are the *Lamé constants*. The constant $\mu = G$, and the constant λ is given in terms of the modulus of elasticity and Poisson's ratio by

$$\lambda = \frac{\nu E}{(1 + \nu)(1 - 2\nu)}. \qquad\qquad \text{Eq. (13.57)}$$

Let an isotropic linearly elastic material be subjected to a pressure p so that $\sigma_x = -p$, $\sigma_y = -p$, and $\sigma_z = -p$. The *bulk modulus K* of the material is the ratio of $-p$ to the dilatation:

$$K = \frac{-p}{e}. \qquad\qquad \text{Eq. (13.58)}$$

In terms of the modulus of elasticity and Poisson's ratio, the bulk modulus is

$$K = \frac{E}{3(1 - 2\nu)}. \qquad\qquad \text{Eq. (13.59)}$$

Review Problems

13.42 A point p of the bearing's housing is subjected to a state of plane stress, and the strain components $\varepsilon_x = -0.0024$, $\varepsilon_y = 0.0044$, and $\gamma_{xy} = -0.0030$. If $\theta = 20°$, what are the strains ε'_x, ε'_y, and γ'_{xy}, at p?

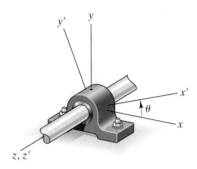

P13.42

13.43 A point p of the housing of the bearing shown in Problem 13.42 is subjected to the state of plane strain $\varepsilon_x = 0.0032$, $\varepsilon_y = -0.0026$, and $\gamma_{xy} = 0.0044$. If $\varepsilon'_x = 0.0037$ and $\varepsilon'_y = -0.0031$, determine the angle θ and the strain γ'_{xy} at p.

13.44 The strains measured by a strain gauge rosette oriented as shown are $\varepsilon_a = -0.00116$, $\varepsilon_b = -0.00065$, and $\varepsilon_c = 0.00130$. Determine the strains ε_x, ε_y, and γ_{xy}.

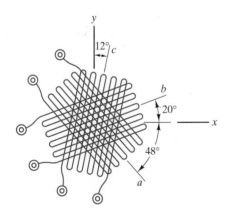

P13.44

13.45 In Example 13.3, use Eq. (13.18) to calculate the magnitude of the maximum in-plane shear strain.

13.46 For the state of plane strains $\varepsilon_x = 0.0025$, $\varepsilon_y = 0$, and $\gamma_{xy} = -0.005$, what is the absolute maximum shear strain?

13.47 For the state of plane strains $\varepsilon_x = -0.008$, $\varepsilon_y = -0.006$, and $\gamma_{xy} = -0.012$, what is the absolute maximum shear strain?

13.48 If the bearing's housing in Problem 13.42 is made of steel with elastic modulus $E = 200$ GPa and Poisson's ratio $\nu = 0.28$, what is the state of plane stress at p?

13.49 If the bearing's housing in Problem 13.42 is made of steel with elastic modulus $E = 200$ GPa and Poisson's ratio $\nu = 0.28$, what is the strain component ε_z at p?

The frame of Case Study House #22 in Los Angeles, California (Pierre Koenig, 1960) consists of steel beams supported by slender steel columns.

Internal Forces and Moments in Beams

A *beam* is a slender structural member. (The word originally meant either a structural member or a *tree* in the Germanic language that became modern English, because the beams used in constructing buildings and ships were hewn from trees. The word for a tree in modern German is still *baum*.) Beams are the most common structural elements and make up the supporting structures of cars, aircraft, and buildings. In this chapter we begin the task of determining the states of stress and strain in beams by analyzing their internal forces and moments.

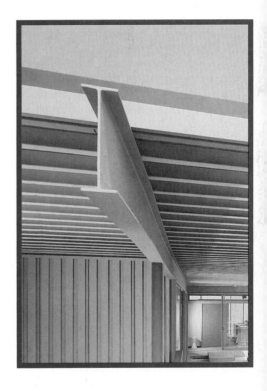

14.1 Axial Force, Shear Force, and Bending Moment

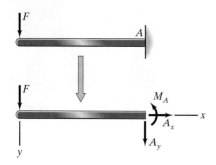

Figure 14.1
Beam subjected to a load and reactions.

Consider the beam subjected to an external load and reactions in Fig. 14.1. Notice that we orient the coordinate system with the y axis downward, which is the traditional orientation for analyzing stresses in beams. To determine the internal forces and moments within the beam, in Fig. 14.2a we cut the beam by a plane perpendicular to the beam's axis and isolate part of it. The isolated part cannot be in equilibrium unless it is subjected to some system of forces and moments at the plane where it joins the other part of the beam. We know from statics that any system of forces and moments can be represented by an equivalent system consisting of a force acting at a given point and a couple. If the system of external loads and reactions on a beam is two-dimensional, we can represent the internal forces and moments by an equivalent system consisting of two components of force and a couple as shown in Fig. 14.2b. The *axial force P* is parallel to the beam's axis. The force component V normal to the beam's axis is called the *shear force,* and the couple M is called the *bending moment.*

The directions of the axial force, shear force, and bending moment in Fig. 14.2b are the established definitions of the positive directions of these quantities. A positive axial force P subjects the beam to tension. A positive shear force V tends to rotate the axis of the beam clockwise (Fig. 14.3a). Bending moments are defined to be positive when they tend to bend the axis of the beam in the negative y-axis direction (Fig. 14.3b).

In Chapters 15 and 16 we show that knowledge of the internal forces and moment in a beam is essential for evaluating the states of stress and deformations resulting from a given system of loads. Determining the internal forces and moment at a particular cross section of a beam typically involves three steps.

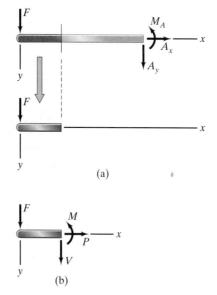

Figure 14.2
(a) Isolating part of the beam.
(b) Axial force, shear force, and bending moment.

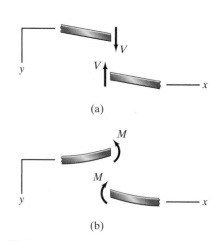

Figure 14.3
(a) Positive shear forces tend to rotate the axis of the beam clockwise.
(b) Positive bending moments tend to bend the axis of the beam in the negative y-axis direction.

1. Draw the free-body diagram of the entire beam and determine the reactions at its supports.

2. Cut the beam where you wish to determine the internal forces and moment and draw the free-body diagram of one of the resulting parts. You can choose the part with the simplest free-body diagram. If your cut divides a distributed load, don't represent the distributed load by an equivalent force until after you have obtained your free-body diagram.

3. Use the equilibrium equations to determine P, V, and M.

Study Questions

1. What are the axial force, shear force, and bending moment?
2. What are the established definitions of the positive directions of the axial force, shear force, and bending moment?

Example 14.1

Determining P, V, and M

For the beam in Fig. 14.4, determine the internal forces and moment at C.

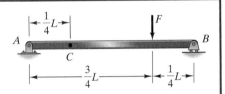

Figure 14.4

Solution

Determine the External Forces and Moments   The first step is to draw the free-body diagram of the entire beam and determine the reactions at its supports. We simply show the results of this step in Fig. (a).

Draw the Free-body Diagram of Part of the Beam   We cut the beam at C [Fig. (a)] and draw the free-body diagram of the left part, including the internal forces and moment P_C, V_C, and M_C in their defined positive directions [Fig. (b)].

Apply the Equilibrium Equations   From the equilibrium equations

$$\Sigma F_x = P_C = 0,$$

$$\Sigma F_y = V_C - \tfrac{1}{4}F = 0,$$

$$\Sigma M_{\text{point } C} = M_C - \left(\tfrac{1}{4}L\right)\left(\tfrac{1}{4}F\right) = 0,$$

we obtain $P_C = 0$, $V_C = F/4$, and $M_C = LF/16$.

Discussion

We should check our results with the free-body diagram of the other part of the beam [Fig. (c)]. The equilibrium equations are

$$\Sigma F_x = -P_C = 0,$$

$$\Sigma F_y = -V_C + F - \tfrac{3}{4}F = 0,$$

$$\Sigma M_{\text{point } C} = -M_C - \left(\tfrac{1}{2}L\right)F + \left(\tfrac{3}{4}L\right)\left(\tfrac{3}{4}F\right) = 0,$$

which confirm that $P_C = 0$, $V_C = F/4$, and $M_C = LF/16$.

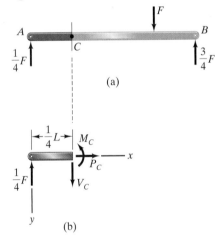

(a) Free-body diagram of the beam and a plane through point C.
(b) Free-body diagram of the part of the beam to the left of the plane through point C.

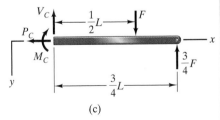

(c) Free-body diagram of the part of the beam to the right of the plane through point C.

Example 14.2

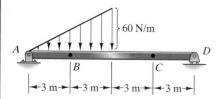

Figure 14.5

Determining P, V, and M

For the beam in Fig. 14.5, determine the internal forces and moment at B and at C.

Solution

Determine the External Forces and Moments We draw the free-body diagram of the beam and represent the distributed load by an equivalent force in Fig. (a).

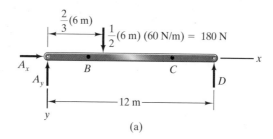

(a)

(a) Free-body diagram of the entire beam with the distributed load represented by an equivalent force.

The equilibrium equations are

$$\Sigma F_x = A_x = 0,$$

$$\Sigma F_y = 180 - A_y - D = 0,$$

$$\Sigma M_{\text{point } A} = 12D - (4)(180) = 0.$$

Solving them, we obtain $A_x = 0, A_y = 120$ N and $D = 60$ N.

Draw the Free-body Diagram of Part of the Beam We cut the beam at B, obtaining the free-body diagram in Fig. (b). Because point B is at the midpoint of the triangular distributed load, the value of the distributed load at B is 30 N/m. By representing the distributed load in Fig. (b) by an equivalent force, we obtain the free-body diagram in Fig. (c). From the equilibrium equations

$$\Sigma F_x = P_B = 0,$$

$$\Sigma F_y = V_B + 45 - 120 = 0,$$

$$\Sigma M_{\text{point } B} = M_B + (1)(45) - (3)(120) = 0,$$

we obtain $P_B = 0$, $V_B = 75$ N, and $M_B = 315$ N-m.

To determine the internal forces and moment at C, we obtain the simplest free-body diagram by isolating the part of the beam to the right of C [Fig.(d)]. From the equilibrium equations

$$\Sigma F_x = -P_C = 0,$$

$$\Sigma F_y = -V_C - 60 = 0,$$

$$\Sigma M_{\text{point } C} = -M_C + (3)(60) = 0,$$

we obtain $P_C = 0$, $V_C = -60$ N, and $M_C = 180$ N-m.

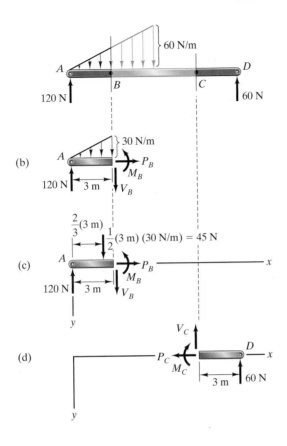

(b), (c) Free-body diagrams of the part of the beam to the left of point B.

(d) Free-body diagram of the part of the beam to the right of point C.

Discussion

If you attempt to determine the internal forces and moment at B by cutting the free-body diagram in Fig. (a) at B, you do *not* obtain correct results. (You can confirm that the resulting free-body diagram of the part of the beam to the left of B gives $P_B = 0$, $V_B = 120$ N, and $M_B = 360$ N-m.) The reason is that you do not account properly for the effect of the distributed load on your free-body diagram. You must wait until *after* you have obtained the free-body diagram of part of the beam before representing distributed loads by equivalent forces.

Problems

14.1 In Example 14.1, determine the internal forces and moment at C if the distance from A to C is $L/2$.

14.2 Determine the internal forces and moment at A, B, and C. (*Strategy:* In this case you don't need to determine the reactions at the built-in support. Cut the beam at the point where you want to determine the internal forces and moment and draw the free-body diagram of the part of the beam to the left of your cut. Remember that P, V, and M must be in their defined positive directions in your free-body diagrams.)

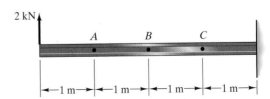

P14.2

14.3 Determine the internal forces and moment at A, B, and C.

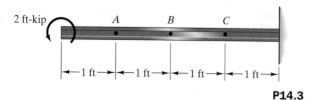

P14.3

14.4 Determine the internal forces and moment at A.

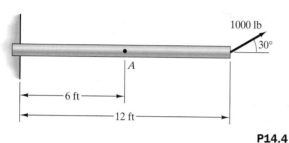

P14.4

14.5 Determine the internal forces and moment (a) at B; (b) at C.

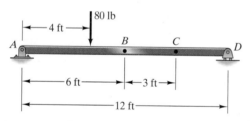

P14.5

14.6 Determine the internal forces and moment at B (a) if $x = 250$ mm; (b) if $x = 750$ mm.

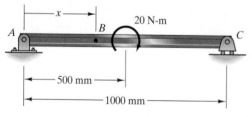

P14.6

14.7 In Example 14.2, determine the internal forces and moment at B if the distance from A to B is 4 m.

14.8 Determine the internal forces and moment at A for each loading.

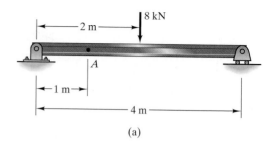

(a)

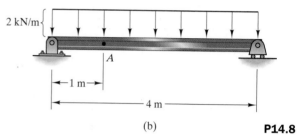

(b)

P14.8

14.9 Model the ladder rung as a simply supported (pin-supported) beam and assume that the 200-lb load exerted by the person's shoe is uniformly distributed. Determine the internal forces and moment at A.

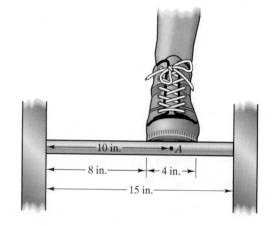

P14.9

14.10 In Problem 14.9, determine the internal forces and moment at A if the distance from A to the left edge of the rung is 9 in.

14.11 Determine the internal forces and moment at A.

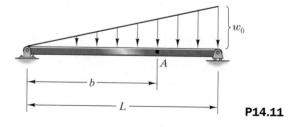

P14.11

14.12 If $x = 3$ ft, what are the internal forces and moment at A?

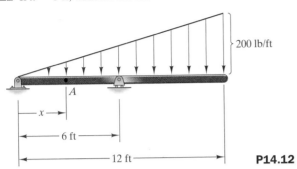

200 lb/ft

A

x

6 ft

12 ft

P14.12

14.13 If $x = 9$ ft in Problem 14.12, what are the internal forces and moment at A?

14.14 Determine the internal forces and moment at A.

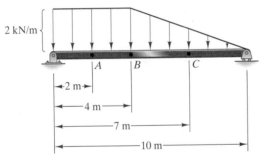

2 kN/m

A B C

2 m

4 m

7 m

10 m

P14.14

14.15 Determine the internal forces and moment at point B of the beam in Problem 14.14.

14.16 Determine the internal forces and moment at point C of the beam in Problem 14.14.

14.17 The lift force on the airplane's wing is given by the distributed load $w = -15(1 - 0.04x^2)$ kN/m and the weight of the wing is given by the distributed load $w = 5 - 0.5x$ kN/m. Determine the internal forces and moment at the wing root R.

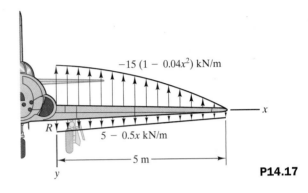

$-15(1 - 0.04x^2)$ kN/m

x

R

$5 - 0.5x$ kN/m

5 m

y

P14.17

14.18 In Problem 14.17, determine the internal forces and moment at the wing's midpoint $x = 2.5$ m.

14.2 Shear Force and Bending Moment Diagrams

To determine whether a beam will support a given set of loads, the structural designer must know the state of stress throughout the beam. To evaluate the state of stress, the internal forces and moment must be determined throughout the beam's length. In this section we show how the values of P, V, and M can be determined as functions of x and introduce shear force and bending moment diagrams.

Let's consider a simply supported beam loaded by a force (Fig. 14.6a). Instead of cutting the beam at a specific cross section to determine the internal forces and moment, we cut it at an arbitrary position x between the left end of the beam and the load F (Fig. 14.6b). Applying the equilibrium equations to this free-body diagram, we obtain

$$\left. \begin{array}{l} P = 0 \\ V = \frac{1}{3}F \\ M = \frac{1}{3}Fx \end{array} \right\} \quad 0 < x < \frac{2}{3}L.$$

To determine the internal forces and moment for values of x greater than $\frac{2}{3}L$, we obtain a free-body diagram by cutting the beam at an arbitrary

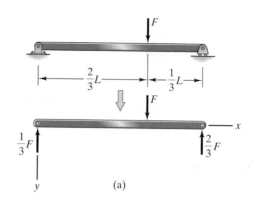

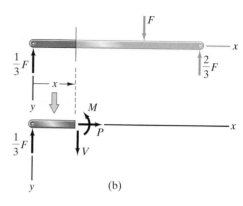

Figure 14.6
(a) Beam loaded by a force F and its free-body diagram.
(b) Cutting the beam at an arbitrary position x to the left of F.
(c) Cutting the beam at an arbitrary position x to the right of F.

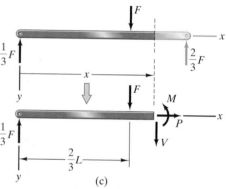

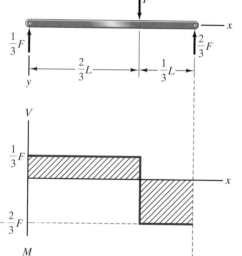

Figure 14.7
Shear force and bending moment diagrams indicating the maximum positive and negative values of V and M.

position x between the load F and the right end of the beam (Fig. 14.6c). The results are

$$\left. \begin{array}{l} P = 0 \\ V = -\tfrac{2}{3}F \\ M = \tfrac{2}{3}F(L - x) \end{array} \right\} \quad \tfrac{2}{3}L < x < L.$$

Shear force and bending moment diagrams are simply graphs of V and M, respectively, as functions of x (Fig. 14.7). They permit you to see the changes in the shear force and bending moment that occur along the beam's length as well as their maximum positive and negative values.

Thus you can determine the distributions of the internal forces and moment in a beam by considering a plane at an arbitrary distance x from the end of the beam and solving for P, V, and M as functions of x. Depending on the complexity of the loading of the beam, you may have to draw several free-body diagrams to determine the distributions over the entire length of the beam. The resulting equations allow you to determine the maximum positive and negative values of the shear force and bending moment and also draw the shear force and bending moment diagrams.

Example 14.3

Shear and Bending Moment Diagrams

Draw the shear force and bending moment diagrams for the beam in Fig. 14.8.

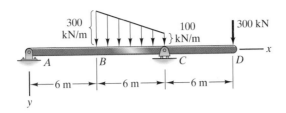

Figure 14.8

Strategy

To determine the internal forces and moment as functions of x for the entire beam, we must use three free-body diagrams: one for the range $0 < x < 6$ m, one for $6 < x < 12$ m, and one for $12 < x < 18$ m.

Solution

We begin by drawing the free-body diagram of the entire beam [Fig. (a)]. We treat the distributed load as the sum of uniform and triangular distributed loads and represent these distributed loads by equivalent forces.

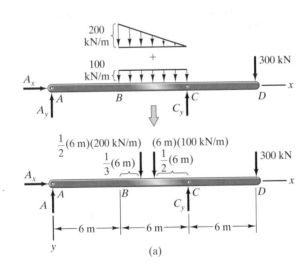

(a)

(a) Free-body diagram of the beam representing the distributed load by two equivalent forces.

From the equilibrium equations

$$\Sigma F_x = A_x = 0,$$

$$\Sigma F_y = -A_y - C_y + 600 + 600 + 300 = 0,$$

$$\Sigma M_{\text{point } A} = 12C_y - (8)(600) - (9)(600) - (18)(300) = 0,$$

we obtain the reactions $A_x = 0$, $A_y = 200$ kN, and $C_y = 1300$ kN.

We draw the free-body diagram for the range $0 < x < 6$ m in Fig. (b). From the equilibrium equations

$$\Sigma F_x = P = 0,$$

$$\Sigma F_y = -200 + V = 0,$$

$$\Sigma M_{\text{right end}} = M - 200x = 0,$$

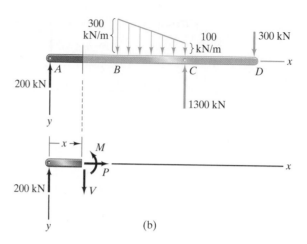

(b) Free-body diagram for $0 < x < 6$ m.

(b)

we obtain

$$\left.\begin{array}{l} P = 0 \\ V = 200 \text{ kN} \\ M = 200x \text{ kN-m} \end{array}\right\} \quad 0 < x < 6 \text{ m.}$$

We draw the free-body diagram for the range $6 < x < 12$ m in Fig. (c). To obtain the equilibrium equations, we determine the distributed load w as a function of x and integrate to determine the force and moment exerted by the distributed load. Since w is a linear function in the interval from $x = 6$ m to $x = 12$ m, we can express it in the form $w = cx + d$, where c and d are constants. Solving for c and d using the two conditions that $w = 300$ kN/m at $x = 6$ m and $w = 100$ kN/m at $x = 12$ m, we obtain

$$w = -\tfrac{100}{3}x + 500 \text{ kN/m.}$$

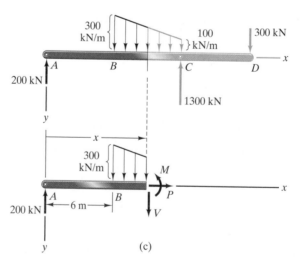

(c) Free-body diagram for $6 < x < 12$ m.

(c)

The downward force on the free body in Fig. (c) due to the distributed load is

$$F = \int_L w \, dx = \int_6^x \left(-\tfrac{100}{3}x + 500\right) dx$$

$$= -\tfrac{50}{3}x^2 + 500x - 2400 \text{ kN.}$$

The clockwise moment about the origin (point A) due to the distributed load is

$$\int_L xw \, dx = \int_6^x \left(-\tfrac{100}{3} x^2 + 500x\right) dx$$

$$= -\tfrac{100}{9} x^3 + 250x^2 - 6600 \text{ kN-m}.$$

The equilibrium equations are

$$\Sigma F_x = P = 0,$$

$$\Sigma F_y = -200 + V - \tfrac{50}{3} x^2 + 500x - 2400 = 0,$$

$$\Sigma M_{\text{point } A} = M - Vx + \tfrac{100}{9} x^3 - 250x^2 + 6600 = 0.$$

Solving them, we obtain

$$\left.\begin{array}{l} P = 0 \\ V = \tfrac{50}{3} x^2 - 500x + 2600 \text{ kN} \\ M = \tfrac{50}{9} x^3 - 250x^2 + 2600x - 6600 \text{ kN-m} \end{array}\right\} \quad 6 < x < 12 \text{ m.}$$

For the range $12 < x < 18$ m, we obtain a very simple free-body diagram by using the part of the beam on the right of the cut [Fig. (d)]. From the equilibrium equations

$$\Sigma F_x = P = 0,$$

$$\Sigma F_y = -V + 300 = 0,$$

$$\Sigma M_{\text{left end}} = -M - 300(18 - x) = 0,$$

we obtain

$$\left.\begin{array}{l} P = 0 \\ V = 300 \text{ kN} \\ M = 300x - 5400 \text{ kN-m} \end{array}\right\} \quad 12 < X < 18 \text{ m.}$$

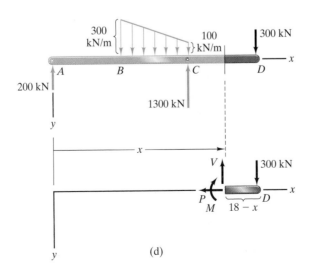

(d)

(d) Free-body diagram for $12 < x < 18$ m.

The shear force and bending moment diagrams, obtained by plotting the equations for V and M for the three ranges of x, are shown in Fig. (e).

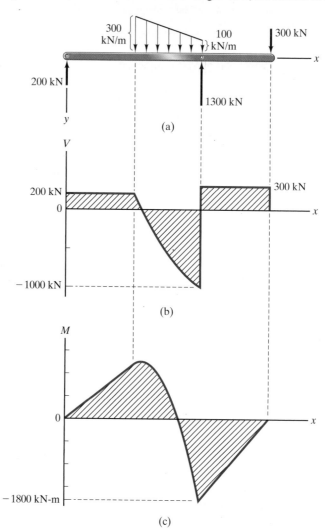

(e) Shear force and bending moment diagrams.

Problems

14.19 (a) Determine the internal forces and moment as functions of x. (b) Draw the shear force and bending moment diagrams.

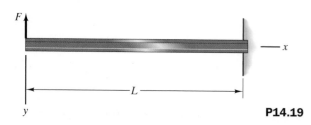

P14.19

14.20 (a) Determine the internal forces and moment as functions of x. (b) Show that the equations for V and M as functions of x satisfy the equation $V = dM/dx$. (c) Draw the shear force and bending moment diagrams.

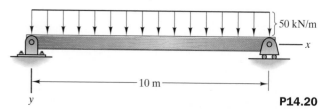

P14.20

14.21 The beam in Problem 14.20 will safely support a bending moment of 1 MN-m (meganewton-meter) at any cross section. Based on this criterion, what is the maximum safe value of the uniformly distributed load?

14.22 (a) Determine the internal forces and moment as functions of x. (b) Show that the equations for V and M as functions of x satisfy the equation $V = dM/dx$. (c) Determine the maximum bending moment in the beam and the value of x where it occurs.

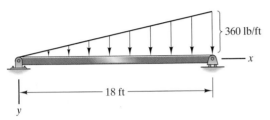

360 lb/ft

18 ft

P14.22

14.23 Draw the shear and bending moment diagrams for the beam in Problem 14.22.

14.24 Consider the beam in Problem 14.6. Determine the internal forces and moment as functions of x for $0 < x < 0.5$ m.

14.25 Consider the beam in Problem 14.6. Determine the internal forces and moment as functions of x for $0.5 < x < 1$ m.

14.26 Consider the beam in Problem 14.12. Determine the internal forces and moment as functions of x for $0 < x < 6$ ft.

14.27 Consider the beam in Problem 14.12. Determine the internal forces and moment as functions of x for $6 < x < 12$ ft.

14.28 (a) Determine the internal forces and moment as functions of x. (b) Draw the shear force and bending moment diagrams.

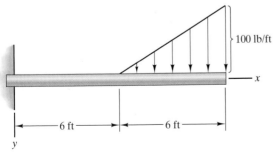

100 lb/ft

6 ft 6 ft

P14.28

14.29 Model the ladder rung as a simply supported (pin-supported) beam and assume that the 200-lb load exerted by the person's shoe is uniformly distributed. Draw the shear force and bending moment diagrams for the rung.

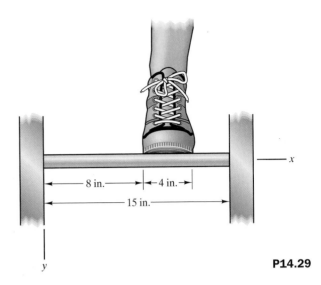

8 in. 4 in.

15 in.

P14.29

14.30 What is the maximum bending moment in the ladder rung in Problem 14.29, and where does it occur?

14.31 Assume that the surface on which the beam rests exerts a uniformly distributed load on the beam. Draw the shear force and bending moment diagrams.

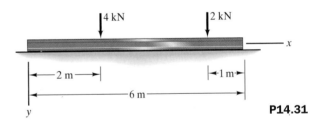

4 kN 2 kN

2 m 1 m

6 m

P14.31

14.32 The homogeneous beams AB and CD weigh 600 lb and 500 lb, respectively. Draw the shear force and bending moment diagrams for beam CD. (Remember that the beam's weight is a distributed load.)

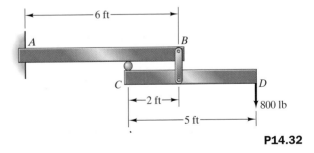

6 ft

A B

C D

2 ft 800 lb

5 ft

P14.32

14.33 Draw the shear force and bending moment diagrams for beam AB in Problem 14.32, including the beam's weight.

14.34 The load $F = 4650$ lb. Draw the shear force and bending moment diagrams for the beam.

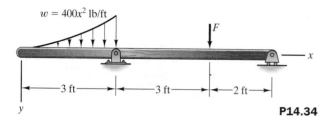

$w = 400x^2$ lb/ft

|←—3 ft—→|←—3 ft—→|←2 ft→|

y

P14.34

14.35 If the load $F = 2150$ lb in Problem 14.34, what are the maximum positive and negative values of the shear force and bending moment, and at what values of x do they occur?

14.36 Draw the shear force and bending moment diagrams.

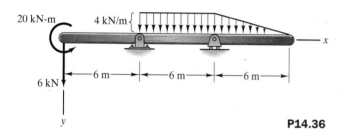

20 kN-m   4 kN/m

|←—6 m—→|←—6 m—→|←—6 m—→|

6 kN

y

P14.36

14.37 In Problem 14.36, what are the maximum positive and negative values of the shear force and bending moment, and at what values of x do they occur?

14.38 The lift force on the airplane's wing is given by the distributed load $w = -15(1 - 0.04x^2)$ kN/m and the weight of the wing is given by the distributed load $w = 5 - 0.5x$ kN/m. Determine the shear force as a function of x.

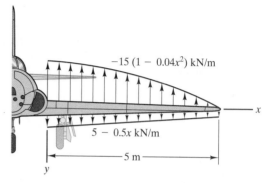

$-15 (1 - 0.04x^2)$ kN/m

$5 - 0.5x$ kN/m

|←————— 5 m —————→|

y

P14.38

14.3 Equations Relating Distributed Load, Shear Force, and Bending Moment

The shear force and bending moment in a beam subjected to a distributed load are governed by simple differential equations. In this section we derive these equations and show that they provide an interesting and enlightening way to obtain the shear force and bending moment diagrams. In Chapters 15 and 16 we show that these equations are also needed for determining the states of stress and the deflections of beams.

Suppose that a portion of a beam is subjected to a distributed load w (Fig. 14.9a). In Fig. 14.9b we obtain a free-body diagram by cutting the beam at x and at $x + \Delta x$. The terms ΔP, ΔV, and ΔM are the changes in the axial force, shear force, and bending moment, respectively, from x to $x + \Delta x$. The sum of the forces in the x direction is

$$\Sigma F_x = P + \Delta P - P = 0.$$

Dividing this equation by Δx and taking the limit as $\Delta x \rightarrow 0$, we obtain

$$\frac{dP}{dx} = 0.$$

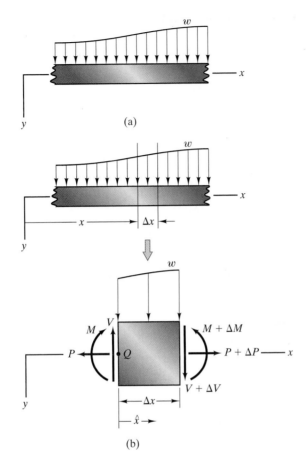

Figure 14.9
(a) Portion of a beam subjected to a distributed force w.
(b) Obtaining the free-body diagram of an element of the beam.

To sum the forces on the free-body diagram in the y direction, we must determine the downward force exerted by the distributed load. In Fig. 14.9b we introduce a coordinate $\hat{x}$ that measures distance from the left edge of the free-body diagram. In terms of this coordinate, the downward force exerted on the free-body diagram by the distributed load is

$$\int_0^{\Delta x} w(x + \hat{x}) \, d\hat{x}, \tag{14.1}$$

where $w(x + \hat{x})$ denotes the value of w at $x + \hat{x}$. To evaluate this integral, we express $w(x + \hat{x})$ as a Taylor series in terms of $\hat{x}$:

$$w(x + \hat{x}) = w(x) + \frac{dw(x)}{dx} \hat{x} + \frac{1}{2} \frac{d^2 w(x)}{dx^2} \hat{x}^2 + \cdots. \tag{14.2}$$

Substituting this expression into Eq. (14.1) and integrating term by term, the downward force exerted by the distributed load is

$$w(x)\Delta x + \frac{1}{2} \frac{dw(x)}{dx} (\Delta x)^2 + \cdots.$$

The sum of the forces on the free-body diagram in the y direction is therefore

$$\Sigma F_y = V + \Delta V - V + w(x)\Delta x + \frac{1}{2} \frac{dw(x)}{dx} (\Delta x)^2 + \cdots = 0.$$

Dividing by Δx and taking the limit as $\Delta x \rightarrow 0$, we obtain

$$\frac{dV}{dx} = -w,$$

where $w = w(x)$.

Our next step is to sum the moments on the free-body diagram in Fig. 14.9b about point Q. The clockwise moment about Q due to the distributed load is

$$\int_0^{\Delta x} \hat{x} w(x + \hat{x}) d\hat{x}.$$

Substituting Eq. (14.2) into this expression and integrating term by term, the moment is

$$\frac{1}{2} w(x)(\Delta x)^2 + \frac{1}{3} \frac{dw(x)}{dx}(\Delta x)^3 + \cdots.$$

The sum of the moments on the free-body diagram about Q is therefore

$$\Sigma M_{\text{point } Q} = M + \Delta M - M - (V + \Delta V)\Delta x$$
$$- \frac{1}{2} w(x)(\Delta x)^2 - \frac{1}{3} \frac{dw(x)}{dx}(\Delta x)^3 + \cdots = 0.$$

Dividing this equation by Δx and taking the limit as $\Delta x \rightarrow 0$ gives

$$\frac{dM}{dx} = V.$$

In summary, we have obtained three differential equations:

$$\frac{dP}{dx} = 0, \tag{14.3}$$

$$\frac{dV}{dx} = -w, \tag{14.4}$$

$$\frac{dM}{dx} = V. \tag{14.5}$$

Equation (14.3) states that the axial force does not depend on x in a portion of a beam subjected only to a lateral distributed load. Equation (14.4) relates the rate of change of the shear force to the distributed load, and Eq. (14.5) relates the rate of change of the bending moment to the shear force. In principle, these equations can be used to determine the distributions of the shear force and bending moment in a beam: We can integrate Eq. (14.4) to determine V as a function of x, then integrate Eq. (14.5) to determine M as a function of x.

However, we derived Eqs. (14.4) and (14.5) for a segment of beam subjected only to a distributed load. To apply them for a more general loading, we must also account for the effects of any point forces and couples acting on the beam. Let us determine what happens to the shear force and bending moment where a beam is subjected to a force F in the positive y direction (Fig. 14.10a). By cutting the beam just to the left and just to the right of the force, we obtain the free-body diagram in Fig. 14.10b, where the subscripts −

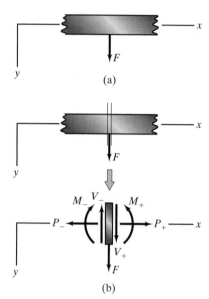

Figure 14.10
(a) Portion of a beam subjected to a force F in the positive y direction.
(b) Obtaining a free-body diagram by cutting the beam to the left and right of F.
(c) The shear force diagram undergoes a decrease of magnitude F.
(d) The bending moment diagram is continuous.

and + denote values to the left and right of the force. Equilibrium requires that

$$V_+ - V_- = -F, \tag{14.6}$$

$$M_+ - M_- = 0. \tag{14.7}$$

We see that the shear force diagram undergoes a decrease of magnitude F (Fig. 14.10c), but the bending moment diagram is continuous (Fig. 14.10d). The change in the shear force is *negative* if the force F is in the positive y direction.

Now we consider what happens to the shear force and bending moment diagrams where a beam is subjected to a counterclockwise couple C (Fig. 14.11a). Cutting the beam just to the left and just to the right of the couple (Fig. 14.11b), we determine that

$$V_+ - V_- = 0, \tag{14.8}$$

$$M_+ - M_- = -C. \tag{14.9}$$

The shear force diagram is continuous (Fig. 14.11c), but the bending moment diagram undergoes a decrease of magnitude C (Fig. 14.11d) where a beam is subjected to a couple. The change in the bending moment is *negative* if the couple is in the counterclockwise direction.

Summarizing these results:

1. A point force results in a jump discontinuity in the shear force but no discontinuity in the bending moment. A force F in the positive y direction causes a decrease in the shear force of magnitude F. Observe in Fig. 14.10 that the shear force distribution changes in the same direction as the force.

2. A couple results in a jump discontinuity in the bending moment but no discontinuity in the shear force. A counterclockwise couple C causes a decrease in the bending moment of magnitude C.

We can demonstrate these results with the cantilever beam in Fig. 14.12a. To determine the shear force diagram, we first observe that the force F at $x = 0$ results in a decrease in V of magnitude F (Fig. 14.12b). Since there is no distributed load on the beam, Eq. (14.4) states that $dV/dx = 0$. The couple at $x = L/2$ does not affect the shear force, so the shear force remains constant, $V = -F$, in the interval $0 < x < L$ (Fig. 14.12c).

To determine the bending moment diagram, we begin again at $x = 0$. There is no couple at $x = 0$, so the value of the bending moment there is zero. In the interval $0 < x < L/2$, the shear force $V = -F$. Integrating Eq. (14.5) from $x = 0$ to an arbitrary value of x within the interval $0 < x < L/2$,

$$\int_0^M dM = \int_0^x V\,dx = \int_0^x -F\,dx,$$

we determine M as a function of x in this interval:

$$M = -Fx, \qquad 0 < x < L/2.$$

The bending moment diagram in this interval is shown in Fig. 14.12d. The value of the bending moment just to the left of the couple at $x = L/2$ is $M = -FL/2$. The counterclockwise couple C causes a decrease in the value

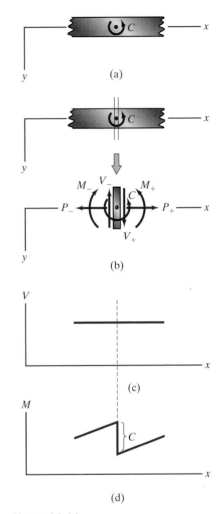

Figure 14.11
(a) Portion of a beam subjected to a counterclockwise couple C.
(b) Obtaining a free-body diagram by cutting the beam to the left and right of C.
(c) The shear force diagram is continuous.
(d) The bending moment diagram undergoes a decrease of magnitude C.

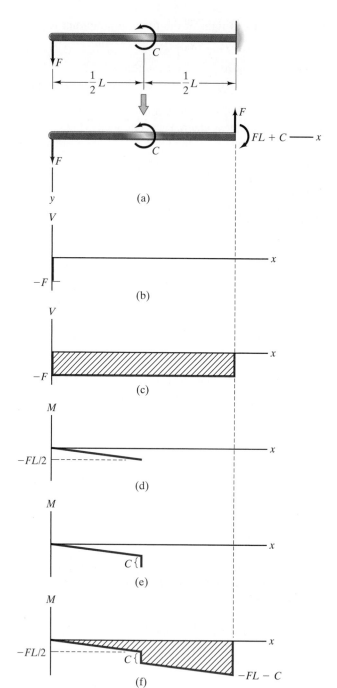

Figure 14.12
(a) Beam loaded by a force and a couple.
(b) The shear force undergoes a decrease of magnitude F at $x = 0$.
(c) The shear force is constant in the interval $0 < x < L$.
(d) Bending moment diagram from $x = 0$ to $x = L/2$.
(e) The bending moment undergoes a decrease of magnitude C at $x = L/2$.
(f) Complete bending moment diagram.

of the bending moment of magnitude C (Fig. 14.12e), so the bending moment just to the right of the couple C is $M = -(FL/2) - C$.

In the interval $L/2 < x < L$, $V = -F$. Integrating Eq. (14.5) from $x = L/2$ to an arbitrary value of x within the interval $L/2 < x < L$,

$$\int_{-(FL/2)-C}^{M} dM = \int_{L/2}^{x} V \, dx = \int_{L/2}^{x} - F \, dx,$$

we obtain M as a function of x in this interval:

$$M = -Fx - C, \qquad L/2 < x < L.$$

The completed bending moment diagram is shown in Fig. 14.12f.

Example 14.4

Shear Force and Bending Moment by Integration

Use Eqs. (14.4) and (14.5) to determine the shear force and bending moment diagrams for the beam in Fig. 14.13.

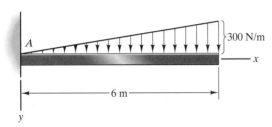

Figure 14.13

Strategy

We must first draw the free-body diagram of the entire beam and determine the reactions at the built-in support A. The result of this step is shown in Fig. (a). We can then use Eq. (14.4) to determine the shear force as a function of x, accounting for the effect of the 900-N force at A. Once the shear force is known, we can use Eq. (14.5) to determine the bending moment as a function of x, accounting for the effect of the 3600 N-m couple at A.

Solution

Shear force diagram   The force at A causes an increase in the shear force of 900-N magnitude at $x = 0$. Expressing the linearly distributed load as a function of x, we obtain $w = (x/6)300 = 50x$ N/m. Integrating Eq. (14.4) from $x = 0$ to an arbitrary value of x,

$$\int_{900}^{V} dV = \int_{0}^{x} - w\, dx = \int_{0}^{x} - 50x\, dx,$$

we obtain V as a function of x:

$$V = 900 - 25x^2 \text{N}.$$

The shear force diagram is shown in Fig. (b).

Bending moment diagram   The counterclockwise couple at A causes a decrease in the bending moment of 3600 N-m magnitude at $x = 0$. Integrating Eq. (14.5) from $x = 0$ to an arbitrary value of x,

$$\int_{-3600}^{M} dM = \int_{0}^{x} V\, dx = \int_{0}^{x} (900 - 25x^2)\, dx,$$

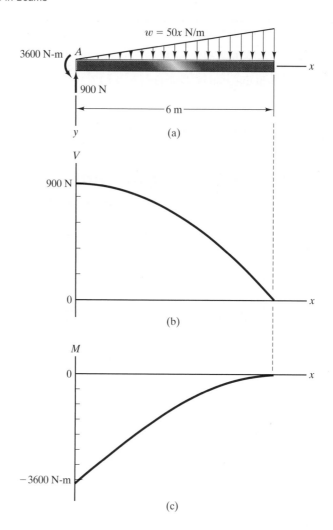

(a) Free-body diagram of the beam.
(b) Shear force diagram.
(c) Bending moment diagram.

we obtain M as a function of x:

$$M = -3600 + 900x - \tfrac{25}{3}x^3 N - m.$$

The bending moment diagram is shown in Fig. (c).

Discussion

Once you have determined V and M as functions of x, a useful check is to confirm that your results satisfy the equation $dM/dx = V$. In this example,

$$\frac{dM}{dx} = \frac{d}{dx}\left(-3600 + 900x - \tfrac{25}{3}x^3\right) = 900 - 25x^2 = V.$$

Example 14.5

Shear Force and Bending Moment by Integration

Use Eqs. (14.4) and (14.5) to determine the shear force and bending moment diagrams for the beam in Fig. 14.14.

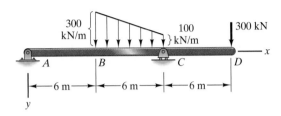

Figure 14.14

Solution

The first step, determining the reactions at the supports, was carried out for this beam and loading in Example 14.3. The results are shown in Fig. (a).

Shear Force Diagram *From A to B* There is no distributed load between A and B, so the shear force increases by 200 kN at A and then remains constant from A to B:

$$V = 200 \text{ kN}, \qquad 0 < x < 6 \text{ m}.$$

From B to C We can express the linearly distributed load between B and C in the form $w = cx + d$, where c and d are constants. Using the conditions $w = 300$ kN/m at $x = 6$ m and $w = 100$ kN/m at $x = 12$ m, we obtain the equation

$$w = -\tfrac{100}{3}x + 500 \text{ kN/m}.$$

From our solution between A and B, $V = 200$ kN at $x = 6$ m. Integrating Eq. (14.4) from $x = 6$ m to an arbitrary value of x between B and C,

$$\int_{200}^{V} dV = \int_{6}^{x} -w dx = \int_{6}^{x} \left(\tfrac{100}{3}x - 500 \right) dx,$$

we obtain an equation for V between B and C:

$$V = \tfrac{100}{6}x^2 - 500x + 2600 \text{ kN}, \qquad 6 < x < 12 \text{ m}.$$

From C to D At C, V undergoes an increase of 1300-N magnitude due to the force exerted by the pin support. Adding this change to the value of V at C obtained from our solution from B to C, the value of V just to the right of C is

$$1300 + \tfrac{100}{6}(12)^2 - 500(12) + 2600 = 300 \text{ kN}.$$

There is no loading between C and D, so V remains constant from C to D:

$$V = 300 \text{ kN}, \qquad 12 < x < 18 \text{ m}.$$

The shear force diagram is shown in Fig. (b).

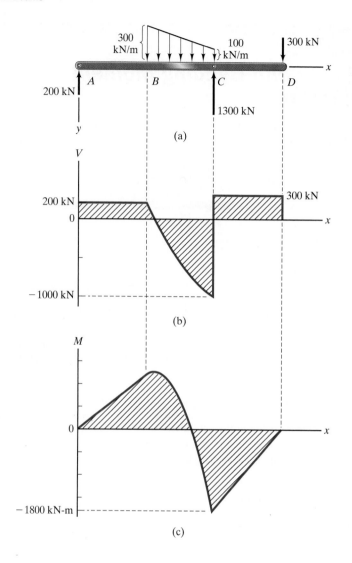

(a) Free-body diagram of the beam.
(b) Shear force diagram.
(c) Bending moment diagram.

Bending Moment Diagram *From A to B* Integrating Eq. (14.3) from $x = 0$ to an arbitrary value of x between A and B,

$$\int_0^M dM = \int_0^x V \, dx = \int_0^x 200 \, dx,$$

we obtain

$$M = 200x \text{ kN-m}, \qquad 0 < x < 6 \text{ m}.$$

At $x = 6$ m, $M = 1200$ kN-m.

From B to C Integrating Eq. (14.5) from $x = 6$ m to an arbitrary value of x between B and C,

$$\int_{1200}^M dM = \int_6^x V \, dx = \int_6^x \left(\tfrac{100}{6} x^2 - 500x + 2600 \right) dx,$$

we obtain

$$M = \tfrac{50}{9} x^3 - 250x^2 + 2600x - 6600 \text{ kN-m}, \qquad 6 < x < 12 \text{ m.}$$

At $x = 12$ m, $M = -1800$ kN-m.

 From C to D Integrating Eq. (14.5) from $x = 12$ m to an arbitrary value of x between C and D,

$$\int_{-1800}^{M} dM = \int_{12}^{x} V \, dx = \int_{12}^{x} 300 \, dx,$$

we obtain

$$M = 300x - 5400 \text{ kN-m}, \qquad 12 < x < 18 \text{ m.}$$

The bending moment diagram is shown in Fig. (c).

Discussion

Compare this example with Example 14.3, in which we use free-body diagrams to determine the shear force and bending moment as functions of x for this beam and loading.

Problems

14.39 Determine V and M as functions of x (a) by drawing free-body diagrams and using the equilibrium equations; (b) by using Eqs. (14.4) and (14.5).

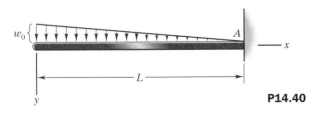

P14.39

14.40 Determine V and M as functions of x (a) by drawing free-body diagrams and using the equilibrium equations; (b) by using Eqs. (14.4) and (14.5).

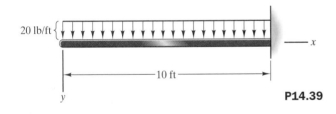

P14.40

14.41 Determine V and M as functions of x by using Eqs. (14.4) and (14.5).

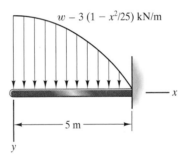

$w - 3 (1 - x^2/25)$ kN/m

P14.41

14.42 Use Eqs. (14.4) and (14.5) to determine the internal forces and moment as functions of x for the beam in Problem 14.19.

14.43 Use Eqs. (14.4) and (14.5) to determine the internal forces and moment as functions of x for the beam in Problem 14.20.

14.44 Determine V and M as functions of x by using Eqs. (14.4) and (14.5).

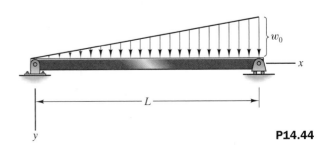

P14.44

14.45 (a) Determine V and M as functions of x by using Eqs. (14.4) and (14.5). (b) Draw the shear force and bending moment diagrams.

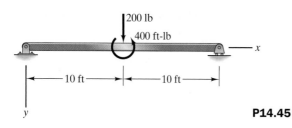

P14.45

14.46 Use Eqs. (14.4) and (14.5) to solve Problem 14.28.

14.47 Use Eqs. (14.4) and (14.5) to solve Problem 14.29.

14.48 Use Eqs. (14.4) and (14.5) to determine the internal forces and moment as functions of x for the beam in Problem 14.31.

14.49 Use Eqs. (14.4) and (14.5) to solve Problem 14.34.

14.50 Use Eqs. (14.4) and (14.5) to solve Problem 14.36.

Chapter Summary

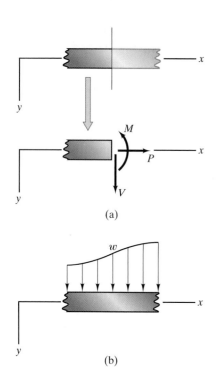

(a)

(b)

Axial Force, Shear Force, and Bending Moment

If the system of external loads and reactions on a beam is two-dimensional, the internal forces and moments can be represented by an equivalent system consisting of the *axial force P*, the *shear force V*, and the *bending moment M* [Fig. (a)]. The directions of P, V, and M in Fig. (a) are the established positive directions of these quantities. The *shear force and bending moment diagrams* are simply the graphs of V and M, respectively, as functions of x.

Equations Relating Distributed Load, Shear Force, and Bending Moment

In a portion of a beam subjected to a distributed load w [Fig. (b)], the shear force and bending moment satisfy the differential equations

$$\frac{dV}{dx} = -w, \tag{14.4}$$

$$\frac{dM}{dx} = V. \tag{14.5}$$

A point force results in a jump discontinuity in the shear force but no discontinuity in the bending moment. A force F in the positive y direction causes a decrease in the shear force of magnitude F [Fig. (c)]. A couple results in a jump discontinuity in the bending moment but no discontinuity in the shear force. A counterclockwise couple C causes a decrease in the bending moment of magnitude C [Fig. (d)].

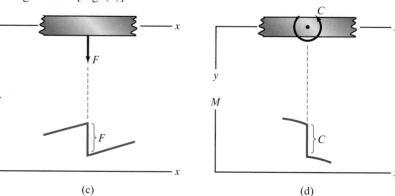

(c)

(d)

Review Problems

14.51 Determine the internal forces and moment at A.

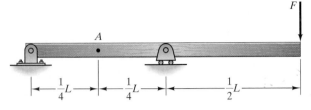

P14.51

14.52 In Problem 14.51, determine the internal forces and moment at A if point A is (a) just to the left of the roller support; (b) just to the right of the roller support.

14.53 If $x = 3$ m, what are the internal forces and moment at A?

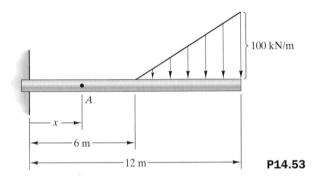

P14.53

14.54 If $x = 9$ m in Problem 14.53, what are the internal forces and moment at A?

14.55 The loads $F = 200$ N and $C = 800$ N-m. (a) Determine the internal forces and moment as functions of x. (b) Draw the shear force and bending moment diagrams.

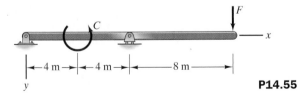

P14.55

14.56 The beam in Problem 14.55 will safely support shear forces and bending moments of magnitudes 2 kN and 6.5 kN-m, respectively. Based on this criterion, can it safely be subjected to the loads $F = 1$ kN, $C = 1.6$ kN-m?

14.57 The bar BD is rigidly fixed to the beam ABC at B. Draw the shear force and bending moment diagrams for the beam ABC

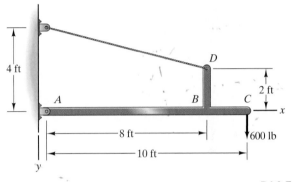

P14.57

14.58 Draw the shear force and bending moment diagrams for beam ABC.

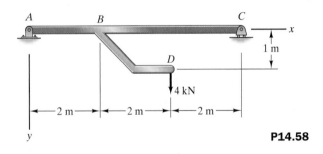

P14.58

14.59 Use Eqs. (14.4) and (14.5) to solve Problem 14.55.

14.60 Use Eqs. (14.4) and (14.5) to solve Problem 14.58.

Gymnast performing on a balance beam made of laminated wood.

Stresses in Beams

Because beams are utilized in so many ways and in so many types of structures, determining their states of stress is an important part of structural analysis and design. In this chapter we show that the stresses at a given cross section of a beam can be expressed in terms of the values of the axial force, shear force, and bending moment at that cross section.

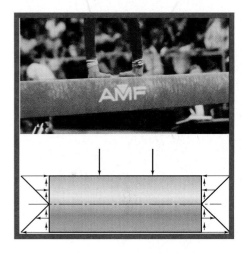

15.1 Normal Stress

The easiest way to break a small piece of firewood is to bend it (Fig. 15.1a), causing the wood to fracture (Fig. 15.1b). The stick is subjected to couples at the ends, inducing stresses within the wood that cause it to fail. In the same way, stresses induced by bending moments in the members of a structure can cause the members to become permanently deformed or even lead to collapse of the structure. In this section we analyze the stresses induced in beams by bending moments.

Geometry of Deformation

Let us consider a prismatic beam of isotropic elastic material (Fig. 15.2a). The x axis of the coordinate system is parallel to the beam's longitudinal axis, and we assume that the beam's cross section is symmetric about the y

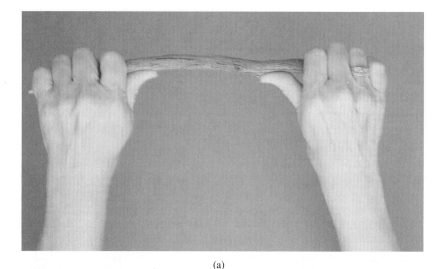

(a)

(b)

Figure 15.1
(a) Bending a stick.
(b) The resulting stresses can cause the stick to break.

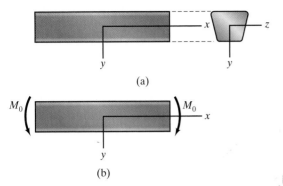

(a)

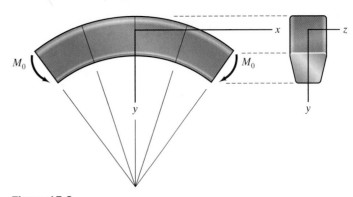

Figure 15.2
(a) Prismatic beam with a symmetrical cross section.
(b) Subjecting the beam to couples at its ends.

Figure 15.3
Deformation resulting from the applied couples. Each cross section of
the beam remains plane.

axis. Suppose that we were to carry out an experiment in which we subject
this beam to couples M_0 at its ends as shown in Fig. 15.2b. The resulting deformation of the beam is shown in Fig. 15.3. Each line of the beam that is
parallel to the longitudinal axis before the couples are applied deforms into a
circular arc parallel to the x–y plane. Each cross section of the beam remains
plane and perpendicular to the beam's curved longitudinal axis. (The latter result is traditionally expressed by the Henry Higgins–sounding statement
"Plane sections remain plane.")

When the beam bends, longitudinal lines toward the top of its cross section become longer and those near the bottom become shorter (Fig. 15.4).
The longitudinal line in the x–y plane that does not change length when the
beam bends is called the *neutral axis*. (We will demonstrate presently that the
neutral axis is coincident with the centroid of the beam's cross section.) Let
us assume that the origin of the coordinate system lies on the neutral axis, and
consider an element of the beam that is of width dx before the couples are applied (Fig. 15.5). We show this element after the couples are applied in
Fig. 15.6, isolating it so that we can analyze its geometry. The radius of the
beam's neutral axis is denoted by ρ. Since we assume that the origin coincides with the neutral axis, the width of the element at $y = 0$ equals dx after

Longitudinal lines
near the top increase
in length

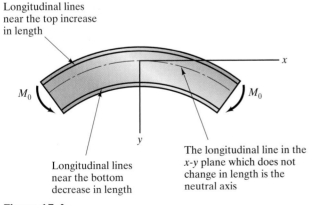

Longitudinal lines
near the bottom
decrease in length

The longitudinal line in the
x-y plane which does not
change in length is the
neutral axis

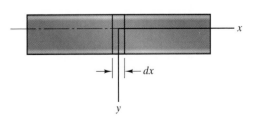

Figure 15.4
Changes in the lengths of longitudinal lines.

Figure 15.5
Element of the beam of width dx.

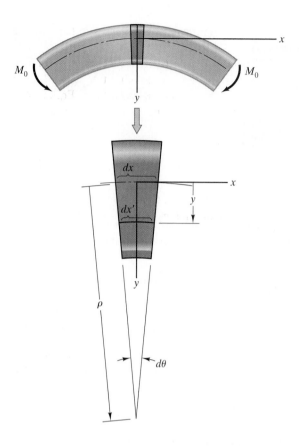

Figure 15.6
Element after the couples are applied.

the couples are applied. The width of the element decreases below the neutral axis and increases above it. Let dx' be the width of the element at a distance y from the neutral axis (Fig. 15.6). In terms of the width before and after the deformation, the extensional strain in the x direction is

$$\varepsilon_x = \frac{dx' - dx}{dx} = \frac{dx'}{dx} - 1. \tag{15.1}$$

We can express dx and dx' in terms of the radius ρ and the angle $d\theta$ shown in Fig. 15.6:

$$dx = \rho\, d\theta,$$

$$dx' = (\rho - y)\, d\theta.$$

Dividing the second equation by the first,

$$\frac{dx'}{dx} = 1 - \frac{y}{\rho},$$

and substituting this result into Eq. (15.1), we obtain

$$\varepsilon_x = -\frac{y}{\rho}. \tag{15.2}$$

The extensional strain in the direction parallel to the beam's axis is a linear function of y, which can also be seen from the shape of the element in Fig. 15.6. The negative sign confirms that the width of the beam decreases in the positive y direction (below the neutral axis) and increases in the negative y direction (above the neutral axis).

The deformation of the element in Fig. 15.6 implies the presence of normal stresses on the vertical faces of the element, causing the material to be stretched above the neutral axis and compressed below it. From Eq. (13.43), the extensional strain ε_x in an isotropic elastic material is given in terms of the components of stress by

$$\varepsilon_x = \frac{1}{E}\sigma_x - \frac{\nu}{E}(\sigma_y + \sigma_z). \tag{15.3}$$

Since the beam we are considering is subjected to no loads perpendicular to its axis, let us assume that the normal stresses σ_y and σ_z are zero. Then Eq. (15.3) states that

$$\sigma_x = E\varepsilon_x. \tag{15.4}$$

Substituting Eq. (15.2) into this expression, we obtain

$$\sigma_x = -\frac{Ey}{\rho}. \tag{15.5}$$

We see that the material is subjected to a normal stress σ_x which is a linear function of y. The result of our "effect–cause" analysis is shown in Fig. 15.7. The normal stress is negative (compressive) for positive values of y, causing the width of the element to decrease, and positive (tensile) for negative values of y, causing the width of the element to increase. Since the same analysis can be applied to any element of the beam, the normal stress at every cross section is described by Eq. (15.5).

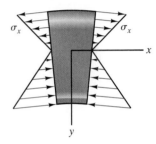

Figure 15.7
Normal stresses on the vertical faces of the element.

Relation between Normal Stress and Bending Moment

Equations (15.2) and (15.5) indicate that the extensional strain ε_x and normal stress σ_x are linear functions of the distance y from the neutral axis, and that they are inversely proportional to the radius of curvature ρ. But we have not determined the location of the neutral axis, and we don't know the relationship between ρ and the applied couple M_0.

In Fig. 15.8 we obtain a free-body diagram by passing a plane perpendicular to the beam's axis. Since the couple M_0 exerts no net force, the horizontal force exerted on the free-body diagram by the distribution of stress σ_x must be zero if the beam is in equilibrium. Letting dA be an element of the beam's cross-sectional area, the horizontal force is

$$\int_A \sigma_x \, dA = 0. \tag{15.6}$$

Figure 15.8
Free-body diagram obtained by passing a plane perpendicular to the beam's axis.

Substituting Eq. (15.5) into this equation, we find that the horizontal force exerted on the free-body diagram is zero only if

$$\int_A y \, dA = 0. \tag{15.7}$$

From the equation for the y coordinate of the centroid of the beam's cross section,

$$\bar{y} = \frac{\displaystyle\int_A y\,dA}{\displaystyle\int_A dA},$$

we see that Eq. (15.7) implies that the origin of the coordinate system coincides with the centroid. Since we had assumed the origin of the coordinate system to be at the neutral axis, *the neutral axis coincides with the centroid of the beam's cross section*.

We have determined the location of the neutral axis from the condition that the sum of the forces on the free-body diagram in Fig. 15.8 must equal zero. We can relate the radius of curvature ρ of the neutral axis to the couple M_0, and thus determine the distribution of the normal stress in terms of M_0, from the condition that the sum of the moments on the free-body diagram must equal zero. From Fig. 15.9, the moment about the z axis due to the normal stress acting on an element dA of the beam's cross section is $-y\sigma_x\,dA$. The total moment about the z axis due to the stress distribution and the couple M_0 in Fig. 15.8 is therefore

$$\int_A -y\sigma_x\,dA - M_0 = 0.$$

By substituting Eq. (15.5) into this expression, we can obtain a relation between M_0 and ρ. We write the resulting equation as

$$\frac{1}{\rho} = \frac{M_0}{EI}, \tag{15.8}$$

where

$$I = \int_A y^2\,dA$$

is the moment of inertia of the beam's cross-sectional area about the z axis. We now substitute Eq. (15.8) into Eq. (15.5) to obtain the normal stress distribution in terms of M_0:

$$\sigma_x = -\frac{M_0 y}{I}. \tag{15.9}$$

Although it has been convenient to derive these results for a beam subjected to couples M_0 in the directions shown in Fig. 15.2, those couples do not conform to our convention for the positive direction of the bending moment. We can apply our results to a beam subjected to positive couples M (Fig. 15.10) simply by making the substitution $M_0 = -M$.

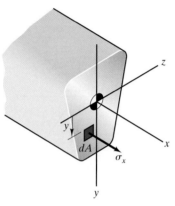

Figure 15.9
Determining the moment due to the normal stress acting on an element dA.

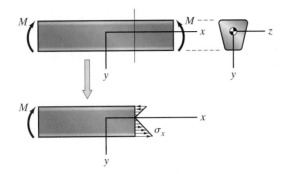

Figure 15.10
Beam subjected to positive couples M.

We summarize the results for a beam subjected to the couples shown in Fig. 15.10:

1. Radius of curvature:

$$\frac{1}{\rho} = -\frac{M}{EI}.$$ (15.10)

The sign of ρ indicates the direction of the curvature of the neutral axis. If ρ is positive, the positive y axis is on the concave side of the neutral axis. The product EI is called the flexural rigidity *of the beam.*

2. Distribution of extensional strain:

$$\varepsilon_x = -\frac{y}{\rho} = \frac{My}{EI}.$$ (15.11)

3. Distribution of normal stress:

$$\sigma_x = \frac{My}{I}.$$ (15.12)

The internal bending moment in a beam loaded as shown in Fig. 15.10 has the same value M at every cross section. We have seen that the internal bending moment in a beam subjected to arbitrary loading is a function of position along the beam's axis. If such a beam is slender (its cross-sectional dimensions are small in comparison with its length), the radius of curvature and distributions of extensional strain and normal stress *at a given cross section* can be approximated using Eqs. (15.10)–(15.12). When this is done, M is the value of the bending moment at the given cross section and ρ is the radius of curvature of the beam's neutral axis in the neighborhood of the cross section.

Study Questions

1. What is the neutral axis of a beam? Where is it located?
2. What is the term I in Eqs. (15.10)–(15.12)?
3. What is the flexural rigidity?
4. If you know the bending moment M at a given cross section of a beam, how can you use Eq. (15.12) to determine the maximum tensile and compressive stresses at that cross section?

Example 15.1

Determining Normal Stresses

The beam in Fig. 15.11 is subjected to couples $M = 4$ kN-m. It consists of aluminum alloy with modulus of elasticity $E = 70$ GPa. The shape and dimensions of the cross section are shown. Determine the resulting radius of curvature of the beam's neutral axis. What are the maximum tensile and compressive normal stresses, and where do they occur?

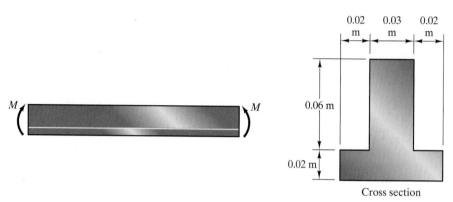

Figure 15.11

Strategy

The radius of curvature is given by Eq. (15.10). To apply that equation, we must determine the position of the neutral axis (the centroid of the beam's cross section), then use the parallel axis theorem to determine the moment of inertia I. The distribution of normal stress is given by Eq. (15.12), from which we can determine the maximum tensile and compressive stresses.

Solution

We can use any convenient coordinate system to locate the centroid of the cross section [Fig. (a)].

Dividing the cross section into the rectangles 1 and 2 shown in Fig. (a), the y coordinate of its centroid is

$$\bar{y} = \frac{\bar{y}_1 A_1 + \bar{y}_2 A_2}{A_1 + A_2} = \frac{(0.03)(0.03)(0.06) + (0.07)(0.07)(0.02)}{(0.03)(0.06) + (0.07)(0.02)}$$

$$= 0.0475 \text{ m}.$$

Placing the origin of the coordinate system at the neutral axis [Fig. (b)], we apply the parallel axis theorem to rectangles 1 and 2 to determine the moment of inertia of the cross section about the z axis:

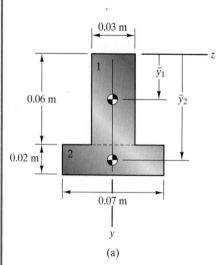

(a)

(a) Coordinate system for determining the position of the neutral axis.

$$I = I_1 + I_2$$

$$= \left(\tfrac{1}{12} b_1 h_1^3 + d_1^2 A_1\right) + \left(\tfrac{1}{12} b_2 h_2^3 + d_2^2 A_2\right)$$

$$= \left[\tfrac{1}{12}(0.03)(0.06)^3 + (0.0175)^2(0.03)(0.06)\right]$$

$$\quad + \left[\tfrac{1}{12}(0.07)(0.02)^3 + (0.0225)^2(0.07)(0.02)\right]$$

$$= 1.85 \times 10^{-6} \text{ m}^4.$$

From Eq. (15.10),

$$\frac{1}{\rho} = -\frac{M}{EI} = -\frac{4000}{(70 \times 10^9)(1.85 \times 10^{-6})},$$

we obtain $\rho = -32.3$ m.

From Eq. (15.12), the normal stress is

$$\sigma_x = \frac{My}{I} = \frac{4000y}{1.85 \times 10^{-6}} = 2.17 \times 10^9 y.$$

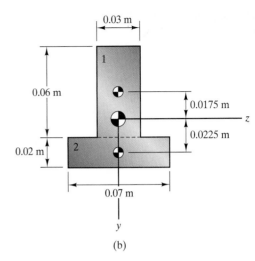

(b)

(b) Applying the parallel axis theorem.

The distribution of normal stress is shown in Fig. (c). The maximum tensile stress occurs at the bottom of the beam:

$$\sigma_x = (2.17 \times 10^9)(0.0325) = 70.4 \text{ MPa}.$$

The maximum compressive stress occurs at the top:

$$\sigma_x = (2.17 \times 10^9)(-0.0475) = -102.9 \text{ MPa}.$$

Discussion

Notice that the maximum tensile stress is smaller in magnitude than the maximum compressive stress. This occurs in this example because the shape of the cross section causes the neutral axis to be closer to the bottom of the beam, where the maximum tensile stress occurs [Fig. (c)]. For this reason, this type of cross section is often used in designing beams made of brittle materials such as concrete, which are relatively weak in tension and strong in compression.

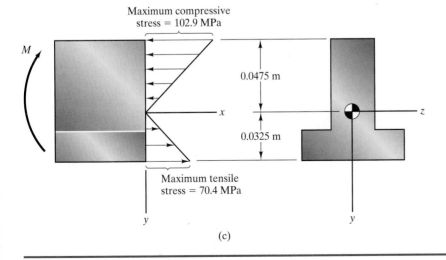

(c)

(c) Normal stress.

Example 15.2

Determining Normal Stress

For the beam in Fig. 15.12, determine the normal stress due to bending at point Q.

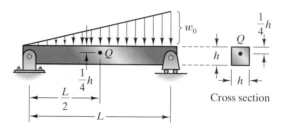

Figure 15.12

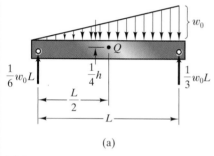

(a)

(a) Reactions at the supports.

Strategy

We must first determine the value of the bending moment M at the cross section containing point Q. Then we can obtain the normal stress from Eq. (15.12).

Solution

By applying the equilibrium equations to a free-body diagram of the entire beam, we obtain the reactions shown in Fig. (a).

We obtain a free-body diagram by passing a plane through Q [Fig. (b)] and represent the distributed load by an equivalent force [Fig. (c)].

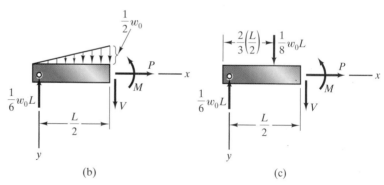

(b)   (c)

(b), (c) Free-body diagram of the part of the beam to the left of Q.

From the equilibrium equation

$$\Sigma M_{\text{right end}} = M + \left[\frac{1}{3}\left(\frac{L}{2}\right)\right]\left(\frac{1}{8}w_0 L\right) - \left(\frac{L}{2}\right)\left(\frac{1}{6}w_0 L\right) = 0,$$

we find that the bending moment at the cross section containing Q is

$$M = \tfrac{1}{16}w_0 L^2.$$

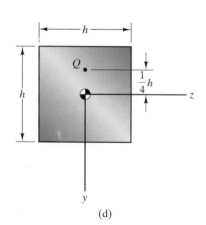

(d)

(d) Position of Q relative to the neutral axis.

Placing the origin of the coordinate system at the neutral axis [Fig. (d)], the moment of inertia of the cross section about the z axis is $I = (1/12)h^4$, and the y coordinate of point Q is $y = -h/4$.

The normal stress at Q is

$$\sigma_x = \frac{My}{I} = \frac{(\frac{1}{16}w_0L^2)(-h/4)}{\frac{1}{12}h^4} = -\frac{3w_0L^2}{16h^3}.$$

Example 15.3

Finding a Maximum Normal Stress

For the beam in Fig. 15.13, what is the maximum tensile stress due to bending, and where does it occur?

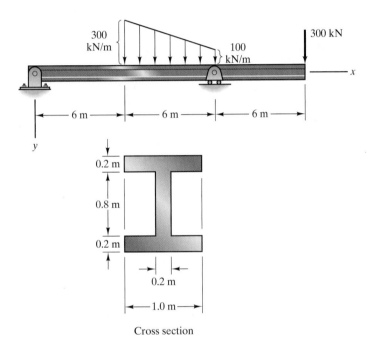

Cross section

Figure 15.13

Strategy

The maximum tensile stress will occur at the cross section where the magnitude of the bending moment M is greatest. To locate this cross section and calculate the bending moment, M must be determined as a function of x. We can then determine the maximum tensile stress from Eq. (15.12).

Solution

The distribution of the bending moment for this beam and loading were determined in Example 14.3:

$0 < x < 6$ m,	$M = 200x$ kN-m,
$6 < x < 12$ m,	$M = \frac{50}{9}x^3 - 250x^2 + 2600x - 6600$ kN-m,
$12 < x < 18$ m,	$M = 300x - 5400$ kN-m.

The bending moment diagram is shown in Fig. (a).

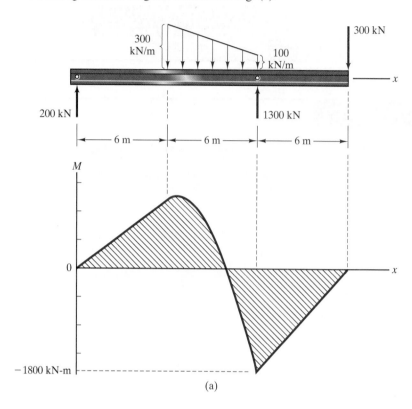

(a)

(a) Bending-moment diagram.

From the bending moment diagram we can see that the magnitude of M is greatest either where M attains its maximum positive value within the interval $6 < x < 12$ m or at $x = 12$ m. To determine the maximum value of M within the interval $6 < x < 12$ m, we equate the derivative of M with respect to x in that interval to zero:

$$\frac{dM}{dx} = \tfrac{50}{3}x^2 - 500x + 2600 = 0.$$

Solving this equation, we find that the maximum occurs at $x = 6.69$ m. Substituting this value of x into the equation for M, we obtain $M = 1270$ kN-m. Therefore, the greatest magnitude of the bending moment occurs at $x = 12$ m, where $M = -1800$ kN-m $= -1.8 \times 10^6$ N-m.

Applying the parallel axis theorem to the cross section [Fig. (b)], the moment of inertia about the z axis is

$$I = I_1 + 2I_2$$

$$= \tfrac{1}{12}(0.2)(0.8)^3 + 2\big[\tfrac{1}{12}(1.0)(0.2)^3 + (0.5)^2(1.0)(0.2)\big]$$

$$= 0.110 \text{ m}^4.$$

From Eq. (15.12), the distribution of normal stress at $x = 12$ m is

$$\sigma_x = \frac{My}{I} = \frac{-1.8 \times 10^6 y}{0.110}$$

$$= -16.4 \times 10^6 y \text{ Pa.}$$

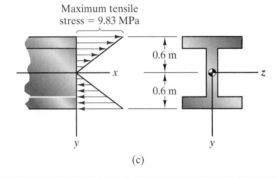

(b)

(b) Applying the parallel axis theorem.

The maximum tensile stress occurs at $y = -0.6$ m [Fig. (c)]:

$$\sigma_x = (-16.4 \times 10^6)(-0.6) = 9.83 \times 10^6 \text{ Pa}.$$

Maximum tensile
stress = 9.83 MPa

(c)

(c) Distribution of normal stress at $x = 12$ m.

Design Issues

For many applications the most essential requirement in designing a beam is to ensure that the maximum tensile and compressive stresses induced by bending moments do not exceed allowable values. This requires determining the distribution of the bending moment throughout the beam due to the maximum anticipated loads and evaluating the resulting maximum normal stresses with Eq. (15.12):

$$\sigma_x = \frac{My}{I}. \qquad (15.13)$$

The fact that the normal stress on a beam's cross section is inversely proportional to I explains in large part the designs of many of the beams in use, for example in highway overpasses and in the frames of buildings: Their cross sections are configured to increase their moments of inertia. (See the I-beam used to support the roof of Case Study House #22 at the beginning of Chapter 14.)

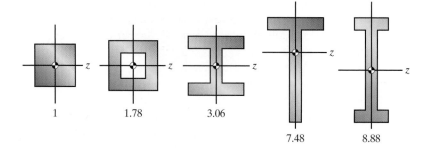

Figure 15.14

Typical beam cross sections and the ratio of I to the value for a solid square beam of equal cross-sectional area.

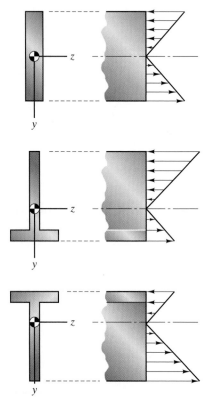

Figure 15.15

A T cross section can be used to decrease either the maximum tensile stress or the maximum compressive stress to which a beam is subjected.

The cross sections in Fig. 15.14 all have the same area. The numbers are the ratios of the value of the moment of inertia I about the z axis to the value for the solid square cross section. In some cases the cross section is also tailored to alter the position of the neutral axis, for example when the beam's material is stronger in compression than in tension (Fig. 15.15).

Using a cross section that increases I generally permits given loads to be supported with a smaller, lighter beam, or enhances the load-carrying capacity of a beam of a given weight. But configuring the cross section to increase its moment of inertia can be carried too far. Figure 15.16a shows a cantilever beam with a square cross section subjected to a moment M. In Fig. 15.16b, a beam with a "box" cross section having the same area as the beam in Fig. 15.16a is subjected to the moment M. If the two beams are made of the

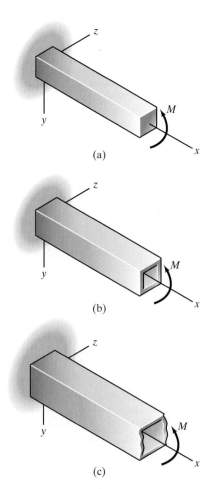

Figure 15.16

(a) Beam with a square cross section.
(b) Box beam with the same cross-sectional area as the beam in (a).
(c) A box beam with thin walls can buckle.

same material, they weigh the same. But the magnitude of the maximum normal stress in the box beam is substantially smaller, due to its larger moment of inertia. By making the walls of the box beam still thinner while holding the cross-sectional area constant, the maximum normal stress for a given moment M can be decreased still further without increasing the weight of the beam. But the walls must not be made too thin or the beam will fail, not because the yield stress of the material is exceeded, but by buckling (Fig. 15.16c).

Once the bending moment distribution due to the expected maximum loads on a beam has been determined, the beam's material and cross section must be chosen so that the maximum normal stress does not exceed the allowable stress σ_{allow}. (Depending on the properties of the material, it may be necessary to specify allowable stresses in both tension and compression.) The ratio of the material's yield stress to the allowable stress determines the factor of safety for the beam:

$$S = \frac{\sigma_Y}{\sigma_{\text{allow}}}.$$

It is essential to keep in mind that Eq. (15.13), and therefore the design procedure we describe, does not apply near a beam's supports or near locations where loads are applied. A more detailed stress analysis is necessary in those regions. Other considerations of which the structural designer must be aware in choosing the beam's material and factor of safety were discussed in Section 10.7.

Example 15.4

Factor of Safety of a Beam

The maximum anticipated magnitude of the load on the beam in Fig. 15.17 is $w_0 = 150$ kN/m. The beam's length is $L = 3$ m. The two candidate cross sections (a) and (b) have approximately the same cross-sectional area. The beam is to be made of material with yield stress $\sigma_Y = 700$ MPa. Compare the beam's factor of safety if it has cross section (a) to its factor of safety with cross section (b).

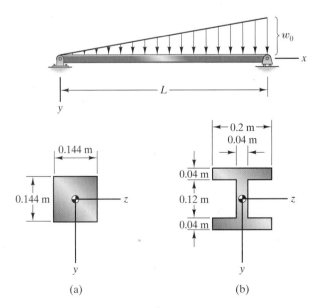

(a)

(b)

Figure 15.17

Strategy

We must determine the maximum normal stress resulting from the given load for each cross section. This requires determining the maximum bending moment in the beam and applying Eq. (15.13) for each cross section.

Solution

The distribution of the bending moment in the beam is

$$M = \frac{1}{6} w_0 \left(Lx - \frac{x^3}{L} \right).$$

To determine where the maximum value of M occurs, we equate the derivative of M with respect to x to zero,

$$\frac{dM}{dx} = \frac{1}{6} w_0 \left(L - \frac{3x^2}{L} \right) = 0,$$

and solve for x, obtaining $x = L/\sqrt{3}$. Substituting this value into the equation for M, the maximum value of the bending moment is

$$M_{max} = \frac{w_0 L^2}{9\sqrt{3}}$$

$$= \frac{(150{,}000)(3)^2}{9\sqrt{3}}$$

$$= 86{,}600 \text{ N-m.}$$

(a) Square Cross Section The moment of inertia of the square cross section about the z axis is

$$I_x = \tfrac{1}{12}(0.144)(0.144)^3 = 3.58 \times 10^{-5} \text{ m}^4.$$

The maximum normal stress is

$$\sigma_{max} = \frac{My}{I}$$

$$= \frac{(86{,}600)(0.072)}{3.58 \times 10^{-5}}$$

$$= 174 \text{ MPa,}$$

so the beam's factor of safety is

$$S = \frac{\sigma_Y}{\sigma_{allow}}$$

$$= \frac{700}{174}$$

$$= 4.02.$$

(b) I-Beam Cross Section The moment of inertia of the vertical rectangle about the z axis is

$$I_v = \tfrac{1}{12}(0.04)(0.12)^3$$

$$= 0.576 \times 10^{-5} \text{ m}^4.$$

Applying the parallel axis theorem, the moment of inertia of each horizontal rectangle about the z axis is

$$I_h = \tfrac{1}{12}(0.2)(0.04)^3 + (0.06 + 0.02)^2(0.2)(0.04)$$
$$= 5.227 \times 10^{-5}\,\text{m}^4.$$

Therefore, the moment of inertia of the cross section is

$$I = I_v + 2I_h = 11.03 \times 10^{-5}\,\text{m}^4.$$

The maximum normal stress is

$$\sigma_{\max} = \frac{My}{I}$$
$$= \frac{(86,600)(0.06 + 0.04)}{11.03 \times 10^{-5}}$$
$$= 78.5\,\text{MPa}.$$

The beam's factor of safety is

$$S = \frac{\sigma_Y}{\sigma_{\text{allow}}}$$
$$= \frac{700}{78.5}$$
$$= 8.91.$$

Discussion

The advantage of a cross section that increases the beam's moment of inertia is apparent in this example. But in particular applications, other factors, such as the availability of stock with a desired cross section, or the cost of manufacturing it, may be overriding considerations.

Problems

15.1 The beam consists of material with modulus of elasticity $E = 70$ GPa and is subjected to couples $M = 250$ kN-m at its ends. (a) What is the resulting radius of curvature of the neutral axis? (b) Determine the maximum tensile stress due to bending.

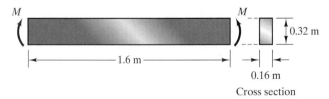

1.6 m

0.16 m

0.32 m

Cross section

P15.1

15.2 The material of the beam in Problem 15.1 will safely support a tensile stress of 180 MPa and a compressive stress of 200 MPa. Based on these criteria, what is the largest couple M to which the beam can be subjected?

15.3 The material of the beam in Problem 15.1 will safely support a tensile stress of 180 MPa and a compressive stress of 200 MPa. Suppose that the beam is rotated 90° about its axis, so that the width of its cross section is 0.32 m and its height is 0.16 m. What is the largest couple M to which the beam can be subjected? Compare your answer to the answer to Problem 15.2.

15.4 The beam consists of material with modulus of elasticity $E = 14 \times 10^6$ psi and is subjected to couples $M = 150,000$ in-lb at its ends. (a) What is the resulting radius of curvature of the neutral axis? (b) Determine the maximum tensile stress due to bending.

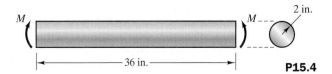

36 in.

2 in.

P15.4

15.5 The material of the beam in Problem 15.4 will safely support a tensile or compressive stress of 30,000 psi. Based on this criterion, what is the largest couple M to which the beam can be subjected?

15.6 The material of the beam in Problem 15.4 will safely support a tensile or compressive stress of 30,000 psi. If the beam has a hollow circular cross section, with 2-in. outer radius and 1-in. inner radius, what is the largest couple M to which the beam can be subjected?

15.7 Suppose that the beam in Example 15.1 is made of a brittle material that will safely support a tensile stress of 20 MPa or a compressive stress of 50 MPa. What is the largest couple M to which the beam can be subjected?

15.8 What is the maximum tensile stress due to bending in the beam in Example 15.2, and where does it occur?

15.9 The beam consists of material that will safely support a tensile or compressive stress of 350 MPa. Based on this criterion, determine the largest force F the beam will safely support if it has the cross section (a); if it has the cross section (b). (The two cross sections have approximately the same area.)

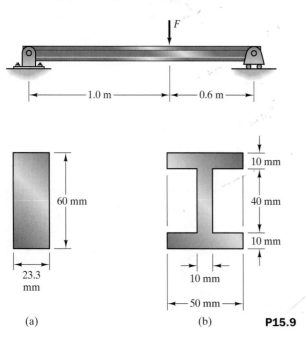

(a)   (b)   **P15.9**

15.10 If the beam in Problem 15.9 is subjected to a force $F = 6$ kN, what is the maximum tensile stress due to bending at the cross section midway between the beam's supports in cases (a) and (b)?

15.11 The beam in Problem 15.9 consists of material that will safely support a tensile or compressive stress of 350 MPa. If it has the cross section (a) and is subjected to a force $F = 17$ kN, what is the maximum distance from the ends of the beam at which F can be applied?

15.12 The beam is subjected to a uniformly distributed load $w_0 = 300$ lb/in. Determine the maximum tensile stress due to bending at $x = 20$ in. if the beam has the cross section (a); if it has the cross section (b). (The two cross sections have approximately the same area.)

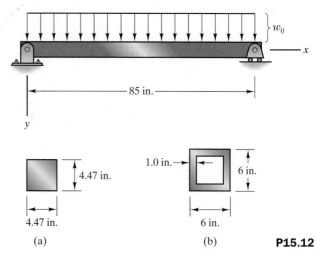

(a)   (b)   **P15.12**

15.13 The beam in Problem 15.12 consists of material that will safely support a tensile or compressive stress of 30 ksi. Based on this criterion, determine the largest distributed load w_0 (in lb/in.) the beam will safely support if it has the cross section (a); if it has the cross section (b).

15.14 A bandsaw blade with 2-mm thickness and 20-mm width is wrapped around a pulley with 160-mm radius. The blade is made of steel with modulus of elasticity $E = 200$ GPa. What maximum tensile stress is induced in the blade as a result of being wrapped around the pulley?

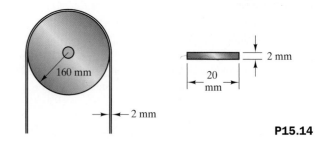

P15.14

15.15 If the beam in Example 15.1 is made of a material for which the allowable stress in tension and compression is $\sigma_{\text{allow}} = 120$ MPa, what is the largest allowable magnitude of the couple M?

15.16 Suppose that the beam in Example 15.1 is made of 7075-T6 aluminum alloy. If it will be subjected to values of M as large as 10 kN-m, what is the beam's factor of safety? (Assume that the yield stress is the same in tension and compression.)

15.17 Suppose that the beam in Example 15.1 is made of a material for which the yield stress in tension is 160 MPa and the yield stress in compression is 200 MPa. If the beam will be subjected to (positive) values of M as large as 4 kN-m, what is the beam's factor of safety?

15.18 Suppose that the length of the beam in Example 15.2 is $L = 8$ ft and it is made of ASTM-A36 structural steel. The maximum anticipated magnitude of the distributed load is $w_0 = 2400$ lb/ft. Determine the dimension h so that the beam has a factor of safety $S = 3$.

15.19 Suppose that the loads on the beam in Example 15.3 are the maximum anticipated loads and the beam is made of wood with yield stress $\sigma_Y = 40$ MPa. What is the beam's factor of safety?

15.20 A beam made of 7075-T6 aluminum alloy will be subjected to anticipated bending moments as large as 1500 N-m. Determine the beam's factor of safety for two cases:
(a) It has a solid circular cross section with 20-mm radius.
(b) It has a hollow circular cross section with 30-mm outer radius and the inner radius chosen so that the beam has the same weight as the beam in case (a).

15.21 Design a cross section for the beam in Example 15.4 so that the beam's factor of safety is $S = 2$.

15.22 The device shown is a playground seesaw. Make a conservative estimate of the maximum weight to which it will be subjected at each end when in use. (Consider contingent situations such as an adult sitting with a child.) Choose a material from Appendix D and design a cross section for the 4.8-m beam so that it has a factor of safety $S = 4$.

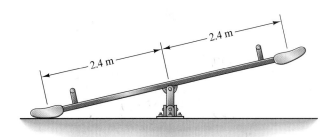

P15.22

15.2 Shear Stress

The internal forces and moments in a beam include the axial force P, shear force V, and bending moment M (Fig. 15.18a). In Chapter 10 we discussed the uniform normal stress distribution associated with the axial force (Fig. 15.18b). In the previous section we discussed the normal stress distribution associated with the bending moment (Fig. 15.18c). If the shear force is not zero at a given cross section, there must be a distribution of shear stress on the cross section that exerts a force in the y direction equal to V (Fig. 15.19).

Determining the distribution of the shear stress over a beam's cross section generally requires advanced methods of analysis or the use of a numerical solution. But we can obtain some information about the shear stress by an interesting indirect deduction that leads to a result called the shear formula.

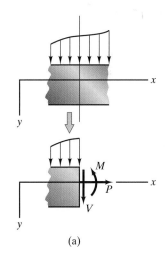

(a)

Figure 15.18
(a) Internal forces and moment.
(b) Normal stress distribution associated with the axial load.
(c) Normal stress distribution associated with the bending moment.

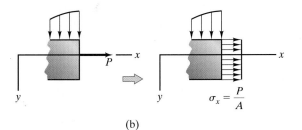

$$\sigma_x = \frac{P}{A}$$

(b)

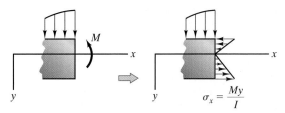

$$\sigma_x = \frac{My}{I}$$

(c)

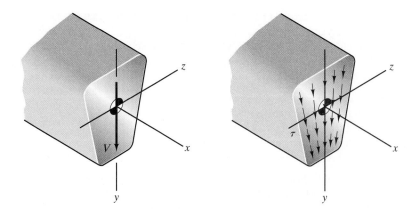

Figure 15.19
The shear load results from a distribution of shear stress.

Shear Formula

From Eq. (14.5), the shear force is related to the bending moment by

$$\frac{dM}{dx} = V, \tag{15.14}$$

which states that the shear force at a given cross section is related to the rate of change of the bending moment with respect to x. Let us consider a beam whose cross section is symmetric about the vertical (y) axis. If we isolate an element of the beam of width dx, the normal stress distributions on its faces are different if the bending moment varies with respect to x (Fig. 15.20). Let us pass a horizontal plane through this element at a position y' relative to the neutral axis and draw the free-body diagram of the part of the element below the plane (Fig. 15.21). Because of the different normal stresses on the oppo-

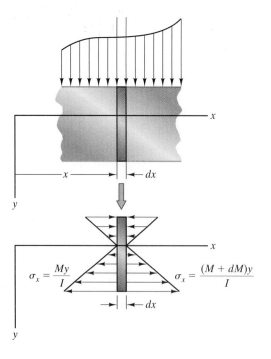

Figure 15.20
Element of a beam of width dx showing the normal stresses on the faces.

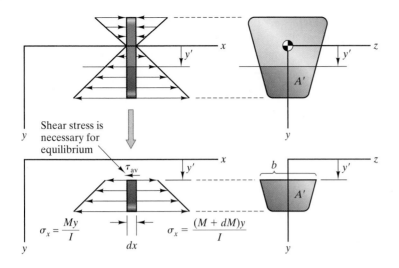

Figure 15.21
Isolating the part of the element below a horizontal plane at position y'.

site faces of the element isolated in Fig. 15.21, the element can be in equilibrium only if shear stress acts on its top surface. We denote the average value of this shear stress by τ_{av}. By passing a second horizontal plane through this element at $y = y' + dy$ and considering the shear stresses on the resulting element (Fig. 15.22), we can see that equilibrium requires that the shear stress τ_{av} also act on the vertical faces of the element. This is the shear stress on the beam's cross section whose distribution we are seeking. Notice that if the shear force V is positive, the direction of τ_{av} on the beam cross section is such that it points *into* the area A'.

From this analysis we cannot determine the distribution of the shear stress across the width b of the element. However, we can determine the dependence of τ_{av} on y' from the free-body diagram of the element in Fig. 15.21. The area acted upon by τ_{av} is $b \, dx$. Denoting the area of the part of the beam's cross section below $y = y'$ by A', the sum of the forces on the element is

$$-\tau_{av} b \, dx - \int_{A'} \frac{My}{I} \, dA + \int_{A'} \frac{(M + dM)y}{I} \, dA = 0.$$

We solve this equation for τ_{av}, obtaining

$$\tau_{av} = \frac{1}{bI} \frac{dM}{dx} \int_{A'} y \, dA.$$

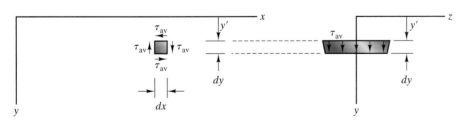

Figure 15.22
State of shear stress on an element of infinitesimal height dy obtained from the top of the element in Fig. 15.21.

From Eq. (15.14), $dM/dx = V$, so we obtain an equation for the shear stress in terms of the shear force:

$$\tau_{av} = \frac{VQ}{bI},$$ (15.15)

where

$$Q = \int_{A'} y\, dA.$$

Equation (15.15) is called the *shear formula*. It determines τ_{av} for a given cross section of a beam at a given position y' relative to the neutral axis. To apply it, we must determine the moment of inertia I of the beam's cross section and the shear force V at the cross section under consideration. We must also determine b and evaluate Q.

We can express Q in terms of the area A' and the position $\bar{y}'$ of the centroid of A' relative to the neutral axis (Fig. 15.23a). The definition of the position of the centroid of A' is

$$\bar{y}' = \frac{\int_{A'} y\, dA}{A'},$$

so Q is given by

$$Q = \bar{y}'A'.$$ (15.16)

It is sometimes convenient to express Q in terms of the area complementary to A'. We denote the complementary area and its centroid by A'' and $\bar{y}''$ in Fig. 15.23b. Because the origin of the coordinate system coincides with the centroid of the entire cross section, which we can express as

$$\bar{y} = \frac{\bar{y}'A' + \bar{y}''A''}{A} = 0,$$

we see that $\bar{y}''A'' = -\bar{y}'A'$, so

$$Q = |\bar{y}''|A''.$$

In Fig. 15.21 we passed a plane parallel to the z axis through the element shown to derive an equation for the average shear stress τ_{av}, obtaining Eq. (15.15). If we pass a plane at an arbitrary angle relative to the z axis through the element as shown in Fig. 15.24a, the derivation of the shear formula is unaltered and the average shear stress, shown on an infinitesimal element of the cross section in Fig. 15.24b, is still given by Eqs. (15.15) and (15.16). The result is the average of the component of the shear stress perpendicular to the line of length b (see Example 15.6). We again observe that if the shear force V is positive, the direction of τ_{av} on the beam cross section is such that it points into the area A'.

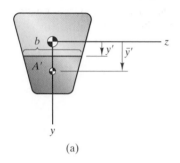

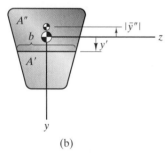

Figure 15.23
Determining Q using
(a) the area A';
(b) the complementary area A''.

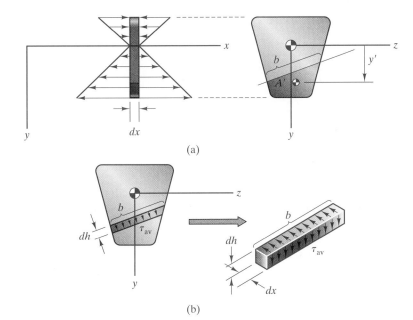

Figure 15.24
(a) Determining the average shear stress on an arbitrary plane.
(b) Average shear stress.

Rectangular Cross Section

For a beam with a rectangular cross section (Fig. 15.25a), we can obtain a simple expression for the dependence of the average shear stress on y'. From Fig. 15.25b, the area $A' = b(h/2 - y')$ and the position of the centroid of A' is $\bar{y}' = y' + \frac{1}{2}(h/2 - y')$, so

$$Q = \bar{y}'A' = \frac{b}{2}\left[\left(\frac{h}{2}\right)^2 - (y')^2\right].$$

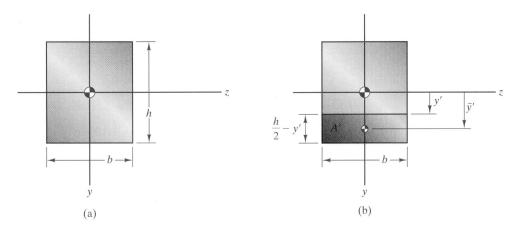

Figure 15.25
(a) Rectangular cross section.
(b) Determining Q.

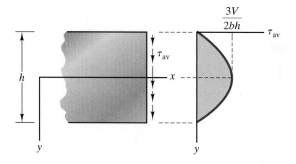

Figure 15.26
Distribution of τ_{av} on a rectangular cross section.

The moment of inertia of the rectangular cross section about the z axis is $I = \frac{1}{12}bh^3$. From Eq. (15.15), the shear stress is

$$\tau_{av} = \frac{VQ}{bI} = \frac{6V}{bh^3}\left[\left(\frac{h}{2}\right)^2 - (y')^2\right]. \tag{15.17}$$

From this equation we see that the average shear stress on a rectangular cross section is a parabolic function of y' (Fig. 15.26). Its value is zero at the top of the cross section $(y' = -h/2)$. At the neutral axis it reaches its maximum magnitude,

$$\left(\tau_{av}\right)_{y'=0} = \frac{3V}{2bh} = \frac{3V}{2A}, \tag{15.18}$$

and its value decreases to zero at the bottom of the cross section $(y' = h/2)$.

Study Questions

1. What is the shear formula?
2. What are the definitions of the terms b and I in Eq. (15.15)?
3. For a given value of y', how can you evaluate the term Q in Eq. (15.15)?
4. If you know the shear force V at a given cross section of a beam with a rectangular cross section, how can you determine the maximum magnitude of τ_{av}?

Example 15.5

Determining Shear Stresses in a Beam

The beam in Fig. 15.27 is subjected to a uniformly distributed load. For the cross section at $x = 2$ m, determine the average shear stress (a) at the neutral axis; (b) at $y' = 0.1$ m.

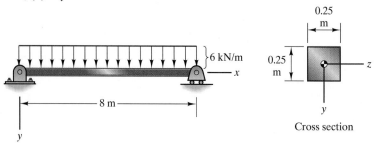

Figure 15.27

Cross section

Strategy

We must first determine the shear force V at $x = 2$ m. Then, because the beam has a square cross section, the average shear stress at the neutral axis is given by Eq. (15.18) and the shear stress is given as a function of y' by Eq. (15.17).

Solution

In Fig. (a) we draw a free-body diagram to determine the shear force at $x = 2$ m, obtaining $V = 12$ kN.

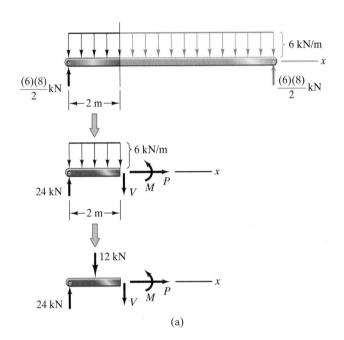

(a)

(a) Free-body diagram obtained by passing a plane through the beam at $x = 2$ m.

(a) From Eq. (15.18), the average shear stress at the neutral axis is

$$\left(\tau_{\mathrm{av}}\right)_{y'=0} = \frac{3V}{2A}$$

$$= \frac{(3)(12{,}000)}{(2)(0.25)(0.25)}$$

$$= 288 \text{ kPa.}$$

(b) From Eq. (15.17), the average shear stress at $y' = 0.1$ m is

$$\tau_{\mathrm{av}} = \frac{6V}{bh^3}\left[\left(\frac{h}{2}\right)^2 - (y')^2\right]$$

$$= \frac{6(12{,}000)}{(0.25)^4}\left[\left(\frac{0.25}{2}\right)^2 - (0.1)^2\right]$$

$$= 104 \text{ kPa.}$$

Example 15.6

Shear Stresses in a Built-up Beam

The beam whose cross section is shown in Fig. 15.28 consists of five planks of wood glued together. At a given axial position the beam is subjected to a shear force $V = 6000$ lb.
(a) What is the average shear stress at the neutral axis $y' = 0$?
(b) What are the magnitudes of the average shear stresses acting on each glued joint?

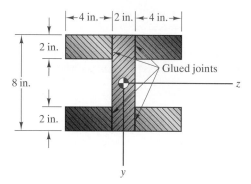

Figure 15.28

Strategy

The average shear stress is given by the shear formula, Eq. (15.15), where I is the moment of inertia of the entire cross section about the z axis. We must also determine the appropriate values of b and Q for parts (a) and (b).

Solution

We can obtain the moment of inertia of the entire cross section about the z axis by summing the moments of inertia of the planks about the z axis:

$$I = \tfrac{1}{12}(2)(8)^3 + 4\left[\tfrac{1}{12}(4)(2)^3 + (3)^2(2)(4)\right]$$

$$= 384 \text{ in}^4.$$

(a) We determine the average shear stress at the neutral axis $(y' = 0)$ by using the area A' and dimension b shown in Fig. (a). We can calculate Q by summing the contributions of the individual planks [Fig. (b)]:

$$Q = \bar{y}'A' = \bar{y}_1'A_1' + \bar{y}_2'A_2' + \bar{y}_3'A_3'$$

$$= (2)(2)(4) + (3)(4)(2) + (3)(4)(2)$$

$$= 64 \text{ in}^3.$$

The average shear stress at the neutral axis is

$$\tau_{\text{av}} = \frac{VQ}{bI}$$

$$= \frac{(6000)(64)}{(2)(384)}$$

$$= 500 \text{ psi}.$$

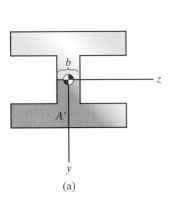

(a)

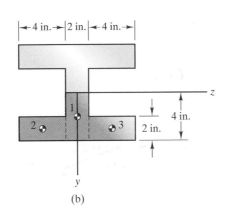

(b)

(a) Area A' for determining the average shear stress at the neutral axis.
(b) Calculating Q.

(b) We can determine the average shear stress acting on the lower-right glued joint by using the area A' and dimension b shown in Fig. (c). The value of Q is

$$Q = \bar{y}'A'$$

$$= (3)(4)(2)$$

$$= 24 \text{ in}^3.$$

The average shear stress is

$$\tau_{av} = \frac{VQ}{bI}$$

$$= \frac{(6000)(24)}{(2)(384)}$$

$$= 188 \text{ psi}.$$

We leave it as an exercise to show that each glued joint is subjected to the same average shear stress.

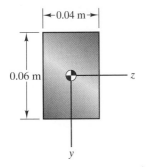

(c)

(c) Area A' for determining the average shear stress on a glued joint.

Problems

15.23 A beam with the cross section shown is subjected to a shear force $V = 8$ kN. What is the average shear stress at the neutral axis $(y' = 0)$?

15.24 In Problem 15.23, determine the average shear stress
(a) at $y' = 0.01$ m;
(b) at $y' = -0.02$ m.

15.25 In Example 15.5, consider the cross section at $x = 3$ m. What is the average shear stress at $y' = 0.05$ m?

15.26 What is the maximum magnitude of the average shear stress in the beam in Example 15.5, and where does it occur?

P15.23

15.27 The beam is subjected to a distributed load. For the cross section at $x = 40$ in., determine the average shear stress (a) at the neutral axis; (b) at $y' = 1.5$ in.

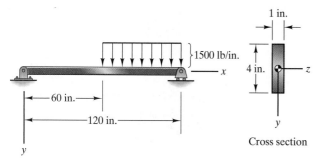

Cross section

P15.27

15.28 Solve Problem 15.27 for the cross section at $x = 80$ in.

15.29 What is the maximum magnitude of the average shear stress in the beam in Problem 15.27, and where does it occur?

15.30 By integrating the stress distribution given by Eq. (15.17), confirm that the total force exerted on the rectangular cross section by the shear stress is equal to V.

15.31 Prove that the quantity Q defined by Eq. (15.16) is a maximum at the neutral axis $(y' = 0)$.

15.32 At a particular axial position, the beam whose cross section is shown is subjected to a shear force $V = 20$ kN. Determine the average shear stress acting on the slanted infinitesimal element.

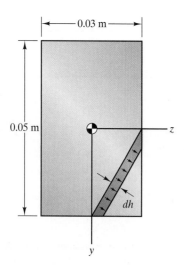

P15.32

15.33 In Example 15.6, determine the average shear stress at $y' = 1$ in.

15.34 In Example 15.6, determine the average shear stress in the upper-right glued joint.

15.35 The beam whose cross section is shown consists of three planks of wood glued together. At a given axial position it is subjected to a shear force $V = 2400$ lb. What is the average shear stress at the neutral axis $y' = 0$?

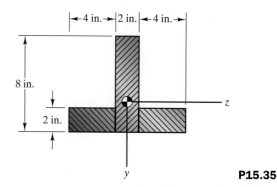

P15.35

15.36 In Problem 15.35, what are the magnitudes of the average shear stresses acting on each glued joint?

15.37 For the cross section at $x = 8$ ft, determine the average shear stress (a) at the neutral axis; (b) at $y' = 2$ in.

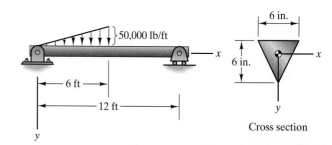

Cross section

P15.37

15.38 In Problem 15.37, determine the value of y' at which the magnitude of the average shear stress is a maximum. (Notice that the maximum magnitude does *not* occur at the neutral axis.) What is the maximum magnitude?

15.39 Solve Problem 15.37 for the cross section at $x = 4$ ft.

15.40 At a particular axial position, the beam whose cross section is shown is subjected to a shear force $V = 15$ kN. Determine the average shear stress (a) at the neutral axis $y' = 0$; (b) at $y' = 0.025$ m.

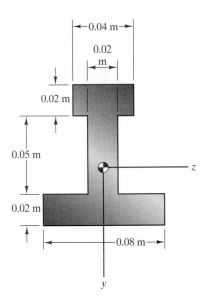

P15.40

15.41 For the beam in Problem 15.40, determine the average shear stress on the infinitesimal element shown.

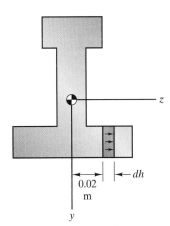

P15.41

Chapter Summary

Normal Stress

Consider a slender prismatic beam of isotropic linearly elastic material subjected to arbitrary loads. At the cross section with axial coordinate x [Fig. (a)], the radius of curvature of the beam's neutral axis is given by the equation

$$\frac{1}{\rho} = -\frac{M}{EI},$$ Eq. (15.10)

where I is the moment of inertia of the beam's cross section about the z axis and M is the value of the bending moment at x. If ρ is positive, the positive y

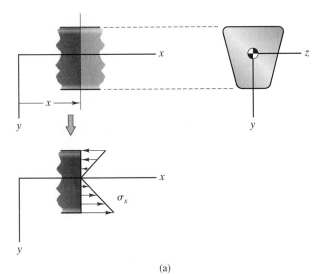

(a)

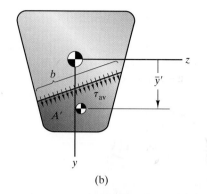

(b)

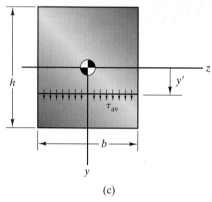

(c)

axis is on the concave side of the neutral axis. The product EI is the beam's *flexural rigidity*. The distribution of the extensional strain is

$$\varepsilon_x = -\frac{y}{\rho} = \frac{My}{EI},$$ Eq. (15.11)

and the distribution of the normal stress is

$$\sigma_x = \frac{My}{I}.$$ Eq. (15.12)

Shear Stress

Consider a slender prismatic beam of isotropic linearly elastic material subjected to arbitrary loads. At a given cross section, the average of the component of the shear stress perpendicular to the line of length b in Fig. (b) is given by the *shear formula*

$$\tau_{av} = \frac{VQ}{bI},$$ Eq. (15.15)

where I is the moment of inertia of the beam's cross section about the z axis, V is the value of the shear force at the given cross section, and

$$Q = \bar{y}'A'.$$ Eq. (15.16)

For a beam with a rectangular cross section, the average of the shear stress over the horizontal line in Fig. (c) as a function of y' is

$$\tau_{av} = \frac{VQ}{bI} = \frac{6V}{bh^3}\left[\left(\frac{h}{2}\right)^2 - (y')^2\right].$$ Eq. (15.17)

Review Problems

15.42 Assume that the surface on which the beam rests exerts a uniformly distributed load on the beam. Determine the maximum tensile and compressive stresses due to bending at $x = 3$ m.

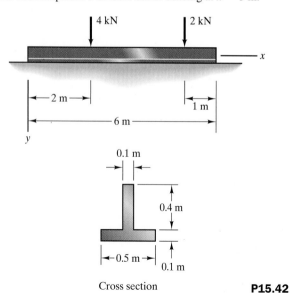

Cross section   **P15.42**

15.43 If you are selecting a material for the beam in Problem 15.42, what maximum tensile and compressive stresses must the material be able to support?

15.44 The maximum anticipated load on the beam is shown. Choose a material from Appendix D and design a cross section for the beam so that it has a factor of safety $S = 2$.

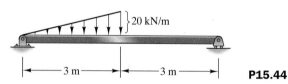

P15.44

15.45 For a preliminary design of the ladder rung, assume that it has pin supports at the ends. Make a conservative estimate of the maximum weight to which it will be subjected when in use. (Assume that the maximum weight acts as a point force at the center of the rung.) The rung is to be made of 6061-T6 aluminum alloy. Design a cross section so that it has a factor of safety $S = 4$. Consider the appropriate width the rung should have.

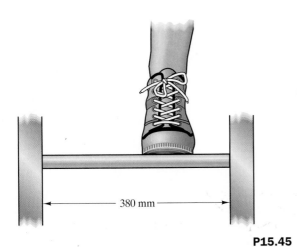

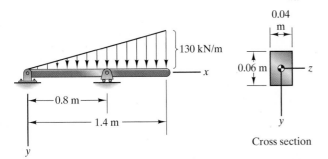

380 mm

P15.45

15.46 The beam is subjected to a distributed load. For the cross section at $x = 0.6$ m, determine the average shear stress (a) at the neutral axis; (b) at $y' = 0.02$ m.

130 kN/m

x

0.8 m

1.4 m

y

0.04 m

0.06 m

z

y

Cross section

P15.46

15.47 Solve Problem 15.46 for the cross section at $x = 1.0$ m.

15.48 At a particular axial position, the beam whose cross section is shown is subjected to a shear force $V = 40$ kN. What is the average shear stress at the neutral axis $(y' = 0)$?

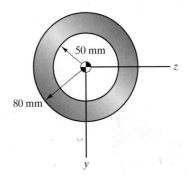

50 mm

z

80 mm

y

P15.48

15.49 For the beam in Problem 15.48, what is the average shear stress at $y' = 50$ mm?

15.50 For the beam in Problem 15.48, what is the average shear stress at $y' = 25$ mm?

Deflection of a layered beam called a leaf spring isolates the vehicle's frame from the effects of road irregularities.

CHAPTER 16

Deflections of Beams

Deflections of some beams, including leaf springs, diving boards, and vaulting poles, are central to their function. In contrast, many structural beams are designed simply to support loads; deflections are not important considerations in their design. But even in such cases, we show in this chapter that calculation of deflections can be of crucial importance for another reason: It is through calculating deflections that we can determine the reactions on statically indeterminate beams.

16.1 Determination of the Deflection

Let v be the deflection of a beam's neutral axis relative to the x axis, and let θ be the angle between the neutral axis and the x axis (Fig. 16.1). Our objective is to determine v and θ as functions of x for a beam with given loads.

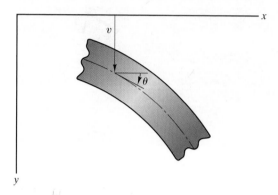

Figure 16.1
Deflection v and the angle θ between the neutral and x axes.

Differential Equation

By considering the deflection at x and at $x + dx$ (Fig. 16.2), we see that v and θ are related by

$$\frac{dv}{dx} = \tan\theta = \theta + \frac{1}{3}\theta^3 + \cdots,$$

where we express $\tan\theta$ in terms of its Taylor series. We restrict our analysis to beams and loadings for which θ is small enough to neglect terms of second and higher orders, so that

$$\frac{dv}{dx} = \theta. \tag{16.1}$$

(The magnitude of θ in the figures is greatly exaggerated.) In Fig. 16.3 we draw lines perpendicular to the neutral axis at x and at $x + dx$. The angle $d\theta$

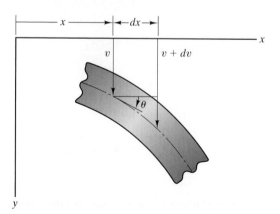

Figure 16.2
Determining the relation between v and θ.

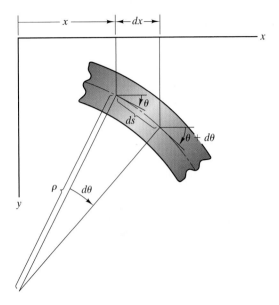

Figure 16.3
Relating θ to the radius of curvature of the neutral axis.

between these lines equals the change in θ from x to $x + dx$. In terms of the radius of curvature of the neutral axis ρ and the distance ds the angle $d\theta$ is

$$d\theta = \frac{1}{\rho}\, ds. \qquad (16.2)$$

Notice from Fig. 16.3 that

$$dx = ds\cos\theta = ds\left(1 - \frac{1}{2}\theta^2 + \cdots\right).$$

Therefore, $ds = dx$ when we neglect terms of second order and higher in θ, so we can express Eq. (16.2) as

$$\frac{d\theta}{dx} = \frac{1}{\rho}.$$

Substituting Eq. (16.1) into this expression, we obtain

$$\frac{d^2v}{dx^2} = \frac{1}{\rho}. \qquad (16.3)$$

In Chapter 15 we obtained an equation relating the beam's radius of curvature, the bending moment, and the flexural rigidity. From Eq. (15.10),

$$\frac{1}{\rho} = -\frac{M}{EI}.$$

Substituting this result into Eq. (16.3), we obtain a relationship between the beam's deflection and the bending moment:

$$v'' = -\frac{M}{EI}, \qquad (16.4)$$

where the primes denote derivatives with respect to x. With this equation we can determine deflections of beams. Although there are several details that we must discuss, the basic procedure is to determine the bending moment M in a beam as a function of x, then integrate Eq. (16.4) twice to determine the deflection v as a function of x. (If the beam is prismatic and consists of homogeneous material, the flexural rigidity EI is a constant.) Notice that once v is known as a function of x, we can use Eq. (16.1) to determine θ as a function of x.

Boundary Conditions

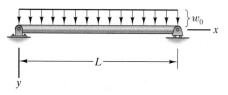

Figure 16.4

Beam subjected to a uniformly distributed load.

The beam in Fig. 16.4 has pin and roller supports and is subjected to a uniformly distributed load. The bending moment M as a function of x is

$$M = \frac{w_0}{2}\left(Lx - x^2\right).$$

We substitute this expression into Eq. (16.4):

$$v'' = \frac{w_0}{2EI}\left(-Lx + x^2\right).$$

Integrating, we obtain

$$v' = \frac{w_0}{2EI}\left(-\frac{Lx^2}{2} + \frac{x^3}{3}\right) + A,$$

where A is an integration constant. Integrating again yields

$$v = \frac{w_0}{2EI}\left(-\frac{Lx^3}{6} + \frac{x^4}{12}\right) + Ax + B, \qquad (16.5)$$

where B is a second integration constant.

To complete our determination of the deflection, we must evaluate the integration constants A and B by using the boundary conditions imposed by the beam's supports (Fig. 16.5). The deflection is zero at the pin support $(x = 0)$. We substitute this condition into Eq. (16.5), obtaining the value of B:

$$v|_{x=0} = B = 0.$$

The deflection is also zero at the roller support $(x = L)$:

$$v|_{x=L} = \frac{w_0}{2EI}\left(-\frac{L^4}{6} + \frac{L^4}{12}\right) + AL = 0.$$

From this equation we obtain $A = w_0 L^3/24EI$. Substituting the values of A and B into Eq. (16.5), we obtain the solution for the beam's deflection:

$$v = \frac{w_0 x}{24EI}\left(L^3 - 2Lx^2 + x^3\right).$$

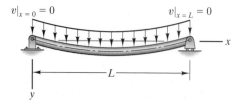

Figure 16.5

The deflection equals zero at each support.

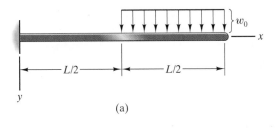

(a)

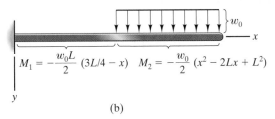

$$M_1 = -\frac{w_0 L}{2}(3L/4 - x) \quad M_2 = -\frac{w_0}{2}(x^2 - 2Lx + L^2)$$

(b)

Figure 16.6

(a) Beam subjected to a uniformly distributed load over part of its length.

(b) Functions describing the distribution of the bending moment.

In the previous example we determined the beam's deflection by substituting the expression for the bending moment as a function of x into Eq. (16.4), integrating twice, and using the boundary conditions at the supports to evaluate the integration constants. A new consideration arises when we apply this procedure to the beam in Fig. 16.6a. When we determine the bending moment as a function of x, we obtain one expression M_1 that applies for $0 \le x \le L/2$ and a different expression M_2 that applies for $L/2 \le x \le L$ (Fig. 16.6b). We need to determine the deflection independently for each of these regions. Let v_1 denote the beam's deflection from $x = 0$ to $x = L/2$. We substitute the expression for M_1 into Eq. (16.4):

$$v_1'' = -\frac{M_1}{EI} = \frac{w_0 L}{2EI}\left(\frac{3L}{4} - x\right).$$

Integrating twice gives

$$v_1' = \frac{w_0 L}{2EI}\left(\frac{3Lx}{4} - \frac{x^2}{2}\right) + A,$$

$$v_1 = \frac{w_0 L}{2EI}\left(\frac{3Lx^2}{8} - \frac{x^3}{6}\right) + Ax + B,$$

where A and B are integration constants. Letting v_2 denote the beam's deflection from $x = L/2$ to $x = L$, we substitute the expression for M_2 into Eq. (16.4) and integrate twice:

$$v_2'' = -\frac{M_2}{EI} = \frac{w_0}{2EI}\left(x^2 - 2Lx + L^2\right),$$

$$v_2' = \frac{w_0}{2EI}\left(\frac{x^3}{3} - Lx^2 + L^2 x\right) + C,$$

$$v_2 = \frac{w_0}{2EI}\left(\frac{x^4}{12} - \frac{Lx^3}{3} + \frac{L^2 x^2}{2}\right) + Cx + D,$$

where C and D are integration constants.

We must now identify four boundary conditions with which to evaluate the integration constants A, B, C, and D. Two conditions are imposed by the

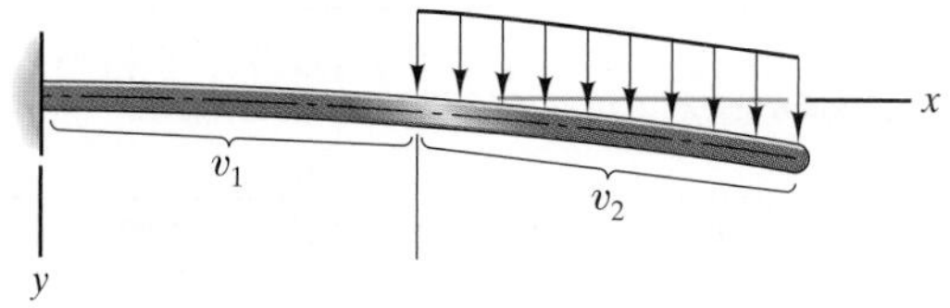

Figure 16.7
Boundary conditions at the built-in support
and where the two solutions meet.

$$v_1|_{x=0} = 0 \qquad v_1|_{x=L/2} = v_2|_{x=L/2}$$
$$v_1'|_{x=0} = 0 \qquad v_1'|_{x=L/2} = v_2'|_{x=L/2}$$

beam's built-in support. At $x = 0$, the deflection and slope equal zero
(Fig. 16.7):

$$v_1|_{x=0} = B = 0,$$

$$v_1'|_{x=0} = A = 0.$$

We also know that the deflections given by the two solutions must be equal at
$x = L/2$ (Fig. 16.7):

$$v_1|_{x=L/2} = v_2|_{x=L/2}:$$

$$\frac{14w_0L^4}{384EI} = \frac{17w_0L^4}{384EI} + \frac{CL}{2} + D.$$

The fourth condition is that the slopes given by the two solutions must be
equal at $x = L/2$.

$$v_1'|_{x=L/2} = v_2'|_{x=L/2}:$$

$$\frac{6w_0L^3}{48EI} = \frac{7w_0L^3}{48EI} + C.$$

From the four boundary conditions we find that $A = 0$, $B = 0$,
$C = -w_0L^3/48EI$, and $D = w_0L^4/384EI$. Substituting these results into our
expressions for v_1 and v_2, the beam's deflection is

$$v = \begin{cases} \dfrac{w_0L}{48EI}\left(9Lx^2 - 4x^3\right), & 0 \le x \le L/2 \qquad\qquad (16.6) \\[4mm] \dfrac{w_0}{384EI}\left(16x^4 - 64Lx^3 + 96L^2x^2 - 8L^3x + L^4\right), & L/2 \le x \le L. \quad (16.7) \end{cases}$$

The examples we have discussed indicate the steps required to determine
a beam's deflection:

1. Determine the bending moment as a function of x. As in our second
    example, this may result in two or more functions M_1, M_2 ... , each of
    which applies to a different segment of the beam's length.

2. For each segment, integrate Eq. (16.4) twice to determine v_1, v_2 ... . If
    there are N segments, this step will result in $2N$ unknown integration
    constants.

3. Use the boundary conditions to determine the integration constants. Our
    two examples illustrate the most common boundary conditions
    encountered in determining deflections of beams. These boundary
    conditions are summarized in Fig. 16.8.

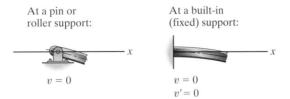

At a pin or
roll support:

$v = 0$

At a built-in
(fixed) support:

$v = 0$
$v' = 0$

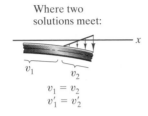

Where two
solutions meet:

$v_1 = v_2$
$v'_1 = v'_2$

Figure 16.8
Common boundary conditions.

Study Questions

1. What is the definition of the deflection v?
2. If you know the deflection v as a function of x, how can you determine the angle θ between the neutral axis and the x axis as a function of x?
3. If you know the bending moment M as a function of x, how can you use Eq. (16.4) to determine the deflection v as a function of x? Do you need any additional information to do so?

Example 16.1

Determining a Beam Deflection

The beam in Fig. 16.9 consists of material with modulus of elasticity $E = 72$ GPa, the moment of inertia of its cross section is $I = 1.6 \times 10^{-7}\,\text{m}^4$, and its length is $L = 4$ m. If $w_0 = 100$ N/m, what is the deflection at the right end of the beam?

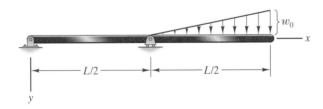

w_0

$L/2$ ← → $L/2$

y

Figure 16.9

Solution

Determine the Bending Moment as a Function of x We leave it as an exercise to show that the distributions of the bending moment in the left and right halves of the beam are given by the expressions in Fig. (a).

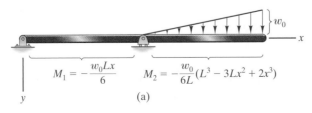

w_0

x

$M_1 = -\dfrac{w_0 L x}{6}$ $\qquad M_2 = -\dfrac{w_0}{6L}(L^3 - 3Lx^2 + 2x^3)$

y

(a)

(a) Equations for the bending moment.

For Each Segment, Integrate Eq. (16.4)   For the segment $0 \leq x \leq L/2$, we obtain

$$v_1'' = -\frac{M_1}{EI} = \frac{w_0 Lx}{6EI},$$

$$v_1' = \frac{w_0 Lx^2}{12EI} + A,$$

$$v_1 = \frac{w_0 Lx^3}{36EI} + Ax + B,$$

where A and B are integration constants, and for the segment $L/2 \leq x \leq L$ we obtain

$$v_2'' = -\frac{M_2}{EI} = \frac{w_0}{6LEI}\left(L^3 - 3Lx^2 + 2x^3\right),$$

$$v_2' = \frac{w_0}{6LEI}\left(L^3 x - Lx^3 + \frac{x^4}{2}\right) + C,$$

$$v_2 = \frac{w_0}{6LEI}\left(\frac{L^3 x^2}{2} - \frac{Lx^4}{4} + \frac{x^5}{10}\right) + Cx + D,$$

where C and D are integration constants.

Use the Boundary Conditions to Determine the Integration Constants
We can apply the four boundary conditions shown in Fig. (b).

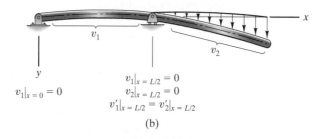

$$v_1|_{x=0} = 0$$
$$v_1|_{x=L/2} = 0$$
$$v_2|_{x=L/2} = 0$$
$$v_1'|_{x=L/2} = v_2'|_{x=L/2}$$

(b) Boundary conditions.

(b)

The deflection is zero at $x = 0$:

$$v_1|_{x=0} = B = 0.$$

The deflection is zero at $x = L/2$, a condition that applies to both v_1 and v_2:

$$v_1|_{x=L/2} = \frac{w_0 L^4}{288EI} + \frac{AL}{2} = 0,$$

$$v_2|_{x=L/2} = \frac{3w_0 L^4}{160EI} + \frac{CL}{2} + D = 0.$$

The slopes of the two solutions must be equal at $x = L/2$.

$$v_1'|_{x=L/2} = v_2'|_{x=L/2}:$$

$$\frac{w_0 L^3}{48EI} + A = \frac{13w_0 L^3}{192EI} + C.$$

Substituting the values of w_0, L, E, and I into these four equations and solving, we obtain $A = -0.00386$, $B = 0$, $C = -0.0299$, and $D = 0.0181$. Now we can use the equation for v_2 to obtain the deflection at the right end of the beam:

$$v_2 = \frac{w_0}{6LEI}\left(\frac{L^3x^2}{2} - \frac{Lx^4}{4} + \frac{x^5}{10}\right) + Cx + D$$

$$= \frac{100}{(6)(4)(72 \times 10^9)(1.6 \times 10^{-7})}\left[\frac{(4)^3(4)^2}{2} - \frac{(4)(4)^4}{4} + \frac{(4)^5}{10}\right]$$

$$+ (-0.0299)(4) + 0.0181$$

$$= 0.0282 \text{ m.}$$

Figure (c) is a graph of the beam's deflection.

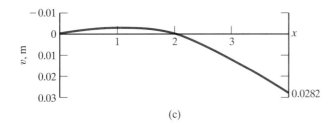

(c)

(c) Deflection as a function of x.

Discussion

From Fig. 16.8 you would conclude that the beam in this example has three boundary conditions at the roller support: $v_1|_{x=L/2} = 0$, $v_2|_{x=L/2} = 0$, and $v_1|_{x=L/2} = v_2|_{x=L/2}$. But notice that only two of these conditions are independent. Once any two of them are chosen, the third one is implied.

Problems

16.1 The beam in Fig. 16.4 has length $L = 4$ m and the magnitude of the distributed load is $w_0 = 12$ kN/m. The beam has a solid circular cross section with 80-mm radius and is made of material with modulus of elasticity $E = 200$ GPa. Determine the beam's deflection and slope at $x = 1$ m.

16.2 In Problem 16.1, what is the beam's maximum deflection? What is the slope of the beam where the maximum deflection occurs?

16.3 The beam in Fig. 16.6a has length $L = 72$ in. and the magnitude of the distributed load is $w_0 = 14$ lb/in. The beam's moment of inertia $I = 2.4$ in^4 and it is made of material with modulus of elasticity $E = 30 \times 10^6$ psi. Determine the deflection at $x = 36$ in. (a) by using Eq. (16.6); (b) by using Eq. (16.7).

16.4 In Problem 16.3, what is the beam's maximum deflection? What is the slope of the beam where the maximum deflection occurs?

16.5 Confirm that Eqs. (16.6) and (16.7) satisfy the boundary conditions for the beam in Fig. 16.6a.

16.6 In Example 16.1, what is the maximum magnitude of the beam's deflection between $x = 0$ and $x = 2$ m? At what value of x does it occur?

For the beams shown in Problems 16.7–16.17, determine the deflection v as a function of x and confirm the results in Appendix E.

16.7

P16.7

16.8

P16.8

16.9

P16.9

16.10

P16.10

16.11

P16.11

16.12

P16.12

16.13

P16.13

16.14

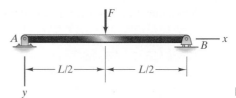

P16.14

16.15

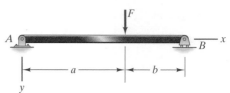

P16.15

16.16

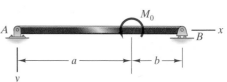

P16.16

16.17

P16.17

16.18 The beam shown in Problem 16.7 is 2 m in length. It has a solid circular cross section with 20-mm radius and is made of aluminum alloy with modulus of elasticity $E = 70$ GPa. When the force F is applied, the deflection at B is measured and determined to be 4.8 mm. What is the maximum resulting tensile stress in the beam?

16.19 The beam shown in Problem 16.10 is 96 in. in length. It has a square cross section with 4-in. width and is made of titanium alloy with modulus of elasticity $E = 15 \times 10^6$ psi. When the distributed load is applied, the deflection at B is measured and determined to be 2 in. What is the maximum resulting tensile stress in the beam?

16.20 The beam shown in Problem 16.15 is 3 m in length and $a = 2$ m. Its moment of inertia is $I = 2 \times 10^{-5}$ m^4 and it is made of material with modulus of elasticity $E = 120$ GPa. If $F = 15$ kN, what is the beam's maximum deflection, and where does it occur?

16.21 The beam consists of material with elastic modulus $E = 190$ GPa. Determine the axial position x at which the magnitude of the deflection due to the couple M_0 is a maximum.

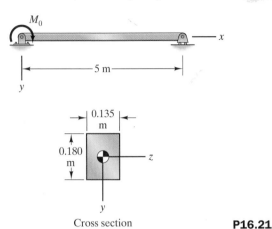

Cross section   **P16.21**

16.22 The beam in Problem 16.21 is used in a structure whose design requires that the magnitude of the beam's maximum deflection be no greater than 20 mm. Determine the maximum couple M_0 that can be applied.

16.23 A couple M_0 is applied to the beam in Problem 16.21. The deflection at $x = 2.5$ m is measured and determined to be 5 mm. What is the maximum resulting tensile stress in the beam?

16.24 The titanium beam has elastic modulus $E = 16 \times 10^6$ psi. Determine the axial position x at which the magnitude of the deflection is a maximum.

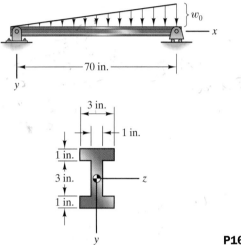

P16.24

Remarkably, the procedure we used in Section 16.1 to determine the deflections of statically determinate beams also works for statically indeterminate beams, yielding both the deflection and the unknown reactions.

Consider the beam in Fig. 16.10a. From the free-body diagram (Fig. 16.10b) we obtain the equilibrium equations

$$\Sigma F_x = A_x = 0, \tag{16.8}$$

$$\Sigma F_y = -A_y - B + w_0 L = 0, \tag{16.9}$$

$$\Sigma M_{\text{point } A} = M_A + LB - \frac{L}{2} w_0 L = 0. \tag{16.10}$$

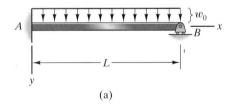

(a)

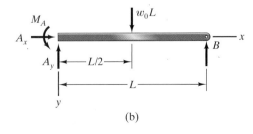

(b)

Figure 16.10
(a) Statically indeterminate beam.
(b) Free-body diagram of the entire beam.

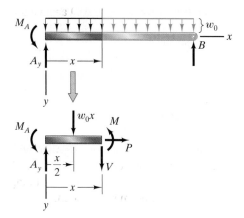

Figure 16.11
Determining the distribution of the bending moment.

We can't solve Eqs. (16.9) and (16.10) for the reactions A_y, M_A, and B. The beam is statically indeterminate.

We ignore this setback and proceed to determine the deflection in terms of the unknown reactions. From Fig. 16.11, the distribution of the bending moment is

$$M = -\frac{w_0 x^2}{2} + A_y x - M_A.$$

Substituting this expression into Eq. (16.4) and integrating gives

$$EIv'' = \frac{w_0 x^2}{2} - A_y x + M_A,$$

$$EIv' = \frac{w_0 x^3}{6} - \frac{A_y x^2}{2} + M_A x + C,$$

$$EIv = \frac{w_0 x^4}{24} - \frac{A_y x^3}{6} + \frac{M_A x^2}{2} + Cx + D,$$

where C and D are integration constants.

Our next step is to use the boundary conditions to evaluate the integration constants. But while there are two integration constants, we see from Fig. 16.12 that there are three boundary conditions. This is the key to the solution: *There are more boundary conditions than there are unknown integration constants.* (The boundary conditions are compatibility conditions imposed on the beam's deflection.) We can use the three equations obtained from the boundary conditions together with the two equilibrium equations

$$v|_{x=L} = 0$$

Figure 16.12
Boundary conditions.

$$v|_{x=0} = 0$$
$$v'|_{x=0} = 0$$

(16.9) and (16.10) to determine the integration constants C and D and the unknown reactions A_y, M_A, and B.

The boundary conditions can be written

$$EIv|_{x=0} = D = 0,$$

$$EIv'|_{x=0} = C = 0,$$

$$EIv|_{x=L} = \frac{w_0 L^4}{24} - \frac{A_y L^3}{6} + \frac{M_A L^2}{2} + CL + D = 0.$$

We see that $C = 0$ and $D = 0$. Solving the third equation together with Eqs. (16.9) and (16.10) yields the unknown reactions: $A_y = 5w_0 L/8$, $B = 3w_0 L/8$, and $M_A = w_0 L^2/8$. We complete the solution by substituting these results into the expression for v to obtain the beam's deflection:

$$v = \frac{w_0}{48EI}\left(2x^4 - 5Lx^3 + 3L^2 x^2\right).$$

This example sheds light on the analysis of statically indeterminant structures in general. An object is statically indeterminate if it has more supports than are necessary for equilibrium. If the roller support is removed from the beam in Fig. 16.10a, it remains in equilibrium and is statically determinate. The roller support introduces an additional reaction, which makes the beam statically indeterminate *but also introduces an additional boundary (compatibility) condition*. Each redundant support added to a structure introduces a new compatibility condition. As a consequence, the number of combined equilibrium equations and compatibility conditions remains sufficient to determine the reactions.

In the following example we demonstrate the steps required to analyze statically indeterminate beams:

1. Draw a free-body diagram of the entire beam and write the equilibrium equations.
2. Determine the bending moment as a function of x in terms of the unknown reactions. This may result in two or more functions $M_1, M_2 \ldots$, each of which applies to a different segment of the beam's length.
3. For each segment, integrate Eq. (16.4) twice to determine $v_1, v_2 \ldots$.
4. Use the boundary conditions together with the equilibrium equations to determine the integration constants and unknown reactions.

Example 16.2

Statically Indeterminate Beam

The beam in Fig. 16.13 has length $L = 4$ m and $w_0 = 5$ kN/m. What are the reactions at A, B, and C?

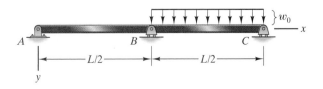

Figure 16.13

Solution

Write the Equilibrium Equations From the free-body diagram in Fig. (a), we obtain the equations

$$\Sigma F_x = A_x = 0,$$

$$\Sigma F_y = -A_y - B - C + \frac{w_0 L}{2} = 0, \tag{16.11}$$

$$\Sigma M_{\text{point } A} = \frac{L}{2}B + LC - \left(\frac{3}{4}L\right)\frac{w_0 L}{2} = 0. \tag{16.12}$$

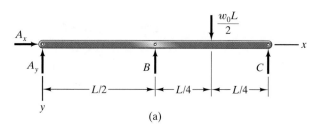

(a) Free-body diagram of the entire beam.

(a)

Determine the Bending Moment as a Function of *x* From Fig. (b), the distributions of the bending moment in the left and right halves of the beam are

$$M_1 = A_y x,$$

$$M_2 = -\frac{w_0 x^2}{2} + (w_0 L - C)x - \left(\frac{w_0 L}{2} - C\right)L.$$

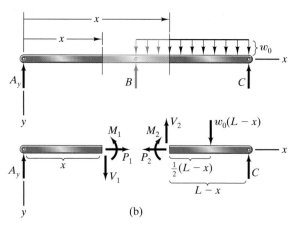

(b) Determining the distribution of the bending moment.

(b)

For Each Segment, Integrate Eq. (16.4) Substituting the expressions for M_1 and M_2 and integrating, we obtain

$$EIv_1'' = -M_1 = -A_y x,$$

$$EIv_1' = -\frac{A_y x^2}{2} + G,$$

$$EIv_1 = -\frac{A_y x^3}{6} + Gx + H,$$

$$EIv_2'' = -M_2 = \frac{w_0 x^2}{2} - (w_0 L - C)x + \left(\frac{w_0 L}{2} - C\right)L,$$

$$EIv_2' = \frac{w_0 x^3}{6} - (w_0 L - C)\frac{x^2}{2} + \left(\frac{w_0 L}{2} - C\right)Lx + J,$$

$$EIv_2 = \frac{w_0 x^4}{24} - (w_0 L - C)\frac{x^3}{6} + \left(\frac{w_0 L}{2} - C\right)\frac{Lx^2}{2} + Jx + K,$$

where G, H, J, and K are integration constants.

Use the Boundary Conditions Together with the Equilibrium Equations

We can apply the five boundary conditions shown in Fig. (c).
The deflection is zero at $x = 0$:

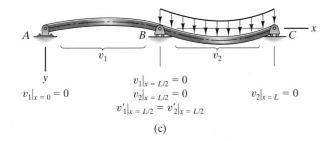

(c)

(c) Boundary conditions.

$$EIv_1|_{x=0} = H = 0. \tag{16.13}$$

Both v_1 and v_2 are zero at $x = L/2$:

$$EIv_1|_{x=L/2} = -\frac{A_y L^3}{48} + \frac{GL}{2} + H = 0, \tag{16.14}$$

$$EIv_2|_{x=L/2} = \frac{w_0 L^4}{384} - (w_0 L - C)\frac{L^3}{48} + \left(\frac{w_0 L}{2} - C\right)\frac{L^3}{8}$$

$$+ \frac{JL}{2} + K = 0. \tag{16.15}$$

The slopes of the two solutions must be equal at $x = L/2$.

$$EIv_1'|_{x=L/2} = EIv_2'|_{x=L/2}:$$

$$-\frac{A_y L^2}{8} + G = \frac{w_0 L^3}{48} - (w_0 L - C)\frac{L^2}{8} + \left(\frac{w_0 L}{2} - C\right)\frac{L^2}{2} + J. \tag{16.16}$$

The deflection is zero at $x = L$:

$$EIv_2|_{x=L} = \frac{w_0 L^4}{24} - (w_0 L - C)\frac{L^3}{6} + \left(\frac{w_0 L}{2} - C\right)\frac{L^3}{2} + JL + K$$

$$= 0. \tag{16.17}$$

We can solve the equilibrium equations (16.11) and (16.12) together with Eqs. (16.13)–(16.17) for the integration constants G, H, J, and K and the unknown reactions A_y, B, and C. Substituting the values of L and w_0 and solving, the solutions for the reactions are $A_y = -625$ N, $B = 6250$ N, and $C = 4375$ N.

Problems

16.25 The prismatic beam has elastic modulus E and moment of inertia I. Determine the reactions at A and B.

P16.25

16.26 Determine the deflection as a function of x for the beam in Problem 16.25.

16.27 The beam in Problem 16.25 has length $L = 4$ m, a solid circular cross section with 80-mm radius, and is made of material with modulus of elasticity $E = 200$ GPa. If the couple $M_0 = 2$ kN-m, what is the maximum tensile stress in the beam?

16.28 The prismatic beam has elastic modulus E and moment of inertia I. Determine the reactions at A and B.

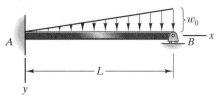

P16.28

16.29 Determine the deflection as a function of x for the beam in Problem 16.28.

16.30 The beam in Problem 16.28 has length $L = 60$ in., elastic modulus $E = 20 \times 10^6$ psi, moment of inertia $I = 12$ in^4, and $w_0 = 6000$ lb/in.
(a) Draw a graph of the beam's deflection as a function of x.
(b) Estimate the value of the maximum deflection and the axial position at which it occurs.

16.31 The prismatic beam has elastic modulus E and moment of inertia I. Determine the reactions at the left wall. (Assume that the supports exert no axial forces on the beam.)

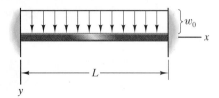

P16.31

16.32 Determine the deflection as a function of x for the beam in Problem 16.31.

16.33 The beam in Problem 16.31 has a square cross section with 4-in. width and modulus of elasticity $E = 15 \times 10^6$ psi. The beam's length is $L = 96$ in. and it is made of material that will

safely support a normal stress of 110 ksi. Based on this criterion, what is the maximum safe value of w_0?

16.34 The beam in Problem 16.31 has the cross section shown. The beam's length is $L = 4$ m and it is made of material that will safely support a normal stress of 220 MPa. What is the maximum safe value of w_0?

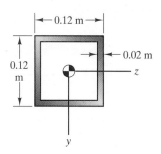

P16.34

16.35 The prismatic beam has elastic modulus E and moment of inertia I. Determine the reactions at the walls. (Assume that the supports exert no axial forces on the beam.)

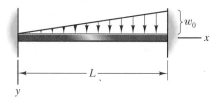

P16.35

16.36 Determine the deflection as a function of x for the beam in Problem 16.35.

16.37 The beam in Problem 16.35 has length $L = 10$ m, elastic modulus $E = 210$ GPa, moment of inertia $I = 8 \times 10^{-6}$ m^4, and $w_0 = 12$ kN/m.
(a) Draw a graph of the beam's deflection as a function of x.
(b) Estimate the value of the maximum deflection and the axial position at which it occurs.

16.38 The length of the beam in Example 16.2 is $L = 4$ m and $w_0 = 5$ kN/m. The beam consists of material with modulus of elasticity $E = 72$ GPa and the moment of inertia of its cross section is $I = 1.6 \times 10^{-7}$ m^4. Determine the beam's deflection at $x = 1$ m.

16.39 The length of the beam in Example 16.2 is $L = 4$ m and $w_0 = 5$ kN/m. The beam consists of material with modulus of elasticity $E = 72$ GPa and the moment of inertia of its cross section is $I = 1.6 \times 10^{-7}$ m^4. What is the maximum magnitude of the beam's deflection between $x = 0$ and $x = 2$ m? At what value of x does it occur?

16.40 The prismatic beam has elastic modulus E and moment of inertia I. Determine the reactions at A and B. (Assume that the supports exert no axial forces on the beam.)

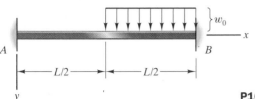

P16.40

16.41 For the beam in Problem 16.40, determine the deflection as a function of x in the region $0 \le x \le L/2$.

16.42 For the beam in Problem 16.40, determine the deflection as a function of x in the region $L/2 \le x \le L$.

16.3 Method of Superposition

Determining deflections of beams is a time-consuming process even when the loads are relatively simple. For the convenience of structural engineers, deflections of prismatic beams with typical supports and simple loads are available in tables such as the one in Appendix E. Furthermore, we will show that the solutions in such tables can be superimposed to obtain deflections of beams with more complicated loads.

Consider the beam and loading in Fig. 16.14a. Let the bending moment in the beam be denoted by M_a, and let v_a be its deflection. The bending moment and deflection satisfy Eq. (16.4)

$$v_a'' = -\frac{M_a}{EI}. \tag{16.18}$$

In Fig. 16.14b we subject the same beam to a different load. Let M_b and v_b denote the resulting bending moment and deflection. They also satisfy Eq. (16.4):

$$v_b'' = -\frac{M_b}{EI}. \tag{16.19}$$

In Fig. 16.14c we superimpose the loads in Figs. 16.14a and b. Summing Eqs. (16.18) and (16.19), we conclude that the superimposed deflections and bending moments also satisfy Eq. (16.4):

$$\left(v_a + v_b\right)'' = -\frac{M_a + M_b}{EI}.$$

This result confirms that we can obtain the deflection resulting from the loads in Fig. 16.14c by summing the deflections resulting from the loads in Fig. 16.14a and b. Using the deflections given in Appendix E, we obtain

$$v_a + v_b = \frac{-Fx^2}{6EI}(3L - x) + \frac{w_0 x^2}{24EI}(6L^2 - 4Lx + x^2).$$

The reduction in effort achieved by this approach is evident. Notice that to determine the deflection of a given beam by superposition, the loading must be matched by the superimposed loads and the boundary conditions must be satisfied by the superimposed displacements.

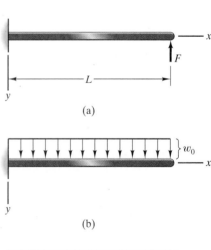

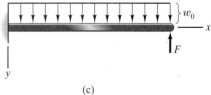

Figure 16.14
Superimposing the loads (a) and (b) results in (c).

Example 16.3

Beam Deflection by Superposition

Use superposition to determine the deflection of the beam in Fig. 16.15.

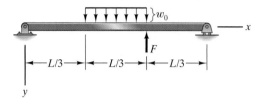

Figure 16.15

Strategy

We obtain this loading by superimposing the loads in Figs. (a), (b), and (c), so we can determine the deflection by superimposing the deflections for these three loads.

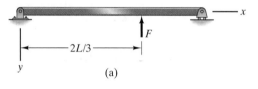

(a)

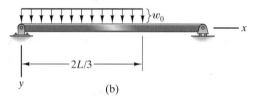

(b)

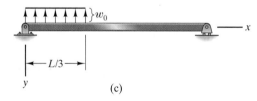

(a), (b), (c) Three loads that can be superimposed to obtain the desired load.

(c)

Solution

Using the results in Appendix E, the deflections due to the loads in Figs. (a), (b), and (c) are

$$
v_a = \begin{cases} \dfrac{-Fx}{162EI}\left(8L^2 - 9x^2\right), & 0 \le x \le 2L/3 \\[2mm] \dfrac{-F}{162EI}\left(18x^3 - 54Lx^2 + 44L^2x - 8L^3\right), & 2L/3 \le x \le L \end{cases}
$$

$$
v_b = \begin{cases} \dfrac{w_0 x}{1944EI}\left(64L^3 - 144Lx^2 + 81x^3\right), & 0 \le x \le 2L/3 \\[2mm] \dfrac{w_0 L}{1944EI}\left(72x^3 - 216Lx^2 + 160L^2x - 16L^3\right), & 2L/3 \le x \le L \end{cases}
$$

$$v_c = \begin{cases} \dfrac{-w_0 x}{1944EI}(25L^3 - 90Lx^2 + 81x^3), & 0 \le x \le L/3 \\[2mm] \dfrac{-w_0 L}{1944EI}(18x^3 - 54Lx^2 + 37L^2 x - L^3). & L/3 \le x \le L \end{cases}$$

Superimposing these results, we obtain the beam's deflection.

$0 \le x \le L/3$:

$$v = \frac{-Fx}{162EI}(8L^2 - 9x^2) + \frac{w_0 x}{1944EI}(39L^3 - 54Lx^2).$$

$L/3 \le x \le 2L/3$:

$$v = \frac{-Fx}{162EI}(8L^2 - 9x^2) + \frac{w_0}{1944EI}(81x^4 - 162Lx^3 \\ + 54L^2 x^2 + 27L^3 x + L^4).$$

$2L/3 \le x \le L$:

$$v = \frac{-F}{162EI}(18x^3 - 54Lx^2 + 44L^2 x - 8L^3) \\ + \frac{w_0 L}{1944EI}(54x^3 - 162Lx^2 + 123L^2 x - 15L^3).$$

A nondimensional measure of the beam's deflection when $F = w_0 L/3$ is shown in Fig. (d).

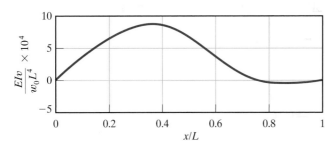

(d) Graph of the deflection.

Example 16.4

Statically Indeterminate Beam by Superposition

Use superposition to determine the deflection of the beam in Fig. 16.16.

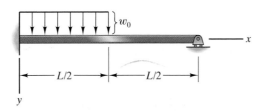

Figure 16.16

Strategy

The beam is statically indeterminate. If we superimpose the loads in Figs. (a) and (b), we can express the beam's deflection in terms of the unknown reaction B exerted by the roller support. Then we can determine B from the condition that the deflection equals zero at $x = L$.

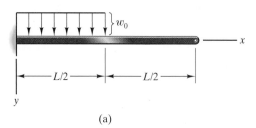

(a)

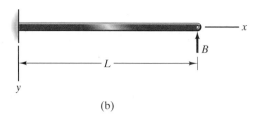

(b)

(a), (b) Superimposing two loads to determine the deflection in terms of B.

Solution

Using the results in Appendix E, the deflection due to the load in Fig. (a) is

$$v_a = \begin{cases} \dfrac{w_0 x^2}{48EI}\left(3L^2 - 4Lx + 2x^2\right), & 0 \le x \le L/2 \\[3mm] \dfrac{w_0 L^3}{384EI}\left(8x - L\right), & L/2 \le x \le L \end{cases}$$

and the deflection due to the unknown reaction in Fig. (b) is

$$v_b = \frac{-Bx^2}{6EI}\left(3L - x\right).$$

The beam's deflection in terms of B is

$$v = v_a + v_b = \begin{cases} \dfrac{w_0 x^2}{48EI}\left(3L^2 - 4Lx + 2x^2\right) - \dfrac{Bx^2}{6EI}\left(3L - x\right), & 0 \le x \le L/2 \\[3mm] \dfrac{w_0 L^3}{384EI}\left(8x - L\right) - \dfrac{Bx^2}{6EI}\left(3L - x\right). & L/2 \le x \le L \end{cases}$$

To determine B, we apply the boundary condition at $x = L$:

$$v|_{x=L} = \frac{w_0 L^3}{384EI}\left(8L - L\right) - \frac{BL^2}{6EI}\left(3L - L\right) = 0.$$

Solving, we obtain $B = 7w_0 L/128$. Substituting this result, the beam's deflection is

$$v = \begin{cases} \dfrac{w_0 x^2}{768EI}\left(27L^2 - 57Lx + 32x^2\right), & 0 \le x \le L/2 \\[3mm] \dfrac{w_0 L}{768EI}\left(7x^3 - 21Lx^2 + 16L^2x - 2L^3\right). & L/2 \le x \le L \end{cases}$$

Problems

16.43 Determine the deflection of the beam as a function of x.

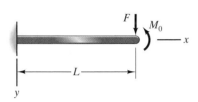

P16.43

16.44 For the beam in Problem 16.43, determine the value of the couple M_0 for which the right end of the beam has no deflection.

16.45 For the beam in Problem 16.43, determine the value of the couple M_0 for which the slope of the right end of the beam is zero.

16.46 Use the solution of Problem 16.43 to determine the bending moment M in the beam as a function of x. To confirm your answer, determine M as a function of x by drawing a free-body diagram and using equilibrium.

16.47 Determine the deflection of the beam as a function of x.

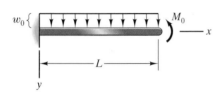

P16.47

16.48 For the beam in Problem 16.47, determine the value of the couple M_0 for which the right end of the beam has no deflection.

16.49 For the beam in Problem 16.47, determine the value of the couple M_0 for which the slope of the right end of the beam is zero.

16.50 Determine the deflection of the beam as a function of x for $0 \leq x \leq L/2$.

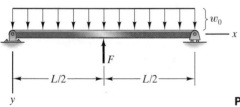

P16.50

16.51 In Problem 16.50, the force $F = w_0 L$. That is, F is equal to the total force exerted by the distributed load. What is the beam's deflection at $x = L/2$?

16.52 Determine the reactions at A, B, and C.

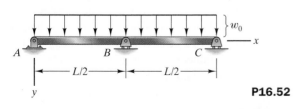

P16.52

16.53 In Problem 16.52, determine the deflection in the beam as a function of x for $0 \leq x \leq L/2$.

16.54 Determine the deflection of the beam as a function of x.

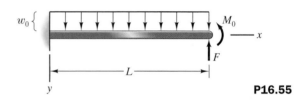

P16.54

16.55 Determine the values of F and M_0 for which both the deflection and slope of the right end of the beam are zero.

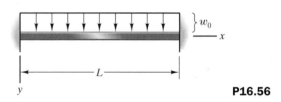

P16.55

16.56 Determine the deflection of the beam as a function of x.

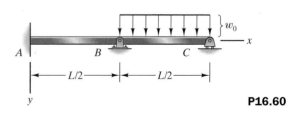

P16.56

16.57 Use the solution of Problem 16.56 to determine the bending moment M in the beam as a function of x.

16.58 Use superposition to solve Problem 16.25.

16.59 Use superposition to solve Problem 16.28.

16.60 Determine the reactions at B and C.

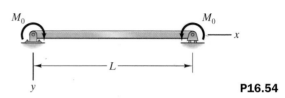

P16.60

16.61 In Problem 16.60, determine the deflection in the beam as a function of x for $0 \leq x \leq L/2$.

Chapter Summary

Determination of the Deflection

Let v be the deflection of a beam's neutral axis relative to the x axis, and let θ be the angle between the neutral axis and the x axis [Fig. (a)]. For small deflections, v and θ are related by

$$\frac{dv}{dx} = \theta. \qquad \text{Eq. (16.1)}$$

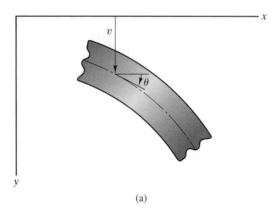

(a)

The deflection is related to the bending moment by

$$v'' = -\frac{M}{EI}, \qquad \text{Eq. (16.4)}$$

where the primes denote derivatives with respect to x. Determining a beam's deflection using Eq. (16.4) requires three steps:

1. Determine the bending moment as a function of x in terms of the loads and reactions acting on the beam. This may result in two or more functions $M_1, M_2 \ldots$, each of which applies to a different segment of the beam's length.

2. For each segment, integrate Eq. (16.4) twice to determine $v_1, v_2, \ldots$. If there are N segments, this step will result in $2N$ unknown integration constants.

3. Use the boundary conditions to determine the integration constants and, if the beam is statically indeterminate, the unknown reactions. Common boundary conditions are summarized in Fig. 16.8.

Method of Superposition

Deflections of prismatic beams with typical supports and simple loads are available in tables such as the one in Appendix E. The solutions in such tables can be superimposed to obtain deflections of beams with more complicated loads. To determine the deflection of a given beam by superposition, the loading must be matched by the superimposed loads and the boundary conditions must be satisfied by the superimposed displacements.

Review Problems

16.62 Determine the deflection v as a function of x and confirm the result in Appendix E.

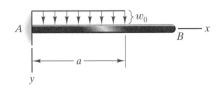

P16.62

16.63 Determine the deflection v as a function of x and confirm the result in Appendix E.

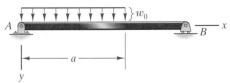

P16.63

16.64 The prismatic beam has elastic modulus E and moment of inertia I. Determine the reactions at A and B.

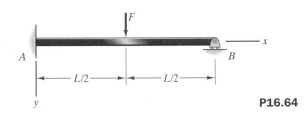

P16.64

16.65 For the beam in Problem 16.64, determine the deflection as a function of x in the region $0 \le x \le L/2$.

16.66 For the beam in Problem 16.64, determine the deflection as a function of x in the region $L/2 \le x \le L$.

16.67 The beam in Problem 16.64 has length $L = 4$ m and a solid circular cross section with 80-mm radius. If $F = 5$ kN, what is the maximum resulting tensile stress in the beam?

16.68 Determine the deflection of the beam as a function of x for $0 \le x \le L/3$.

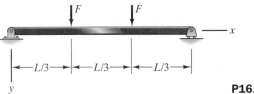

P16.68

16.69 The beam in Problem 16.68 consists of material with modulus of elasticity $E = 72$ GPa, the moment of inertia of its cross section is $I = 1.6 \times 10^{-7}$ m^4, and its length is $L = 6$ m. If $F = 100$ N, what is the deflection at $x = 3$ m?

16.70 Use superposition to solve Problem 16.64.

16.71 What force is exerted on the beam by the spring?

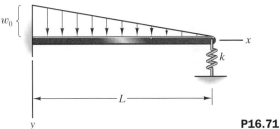

P16.71

The facade of the John F. Kennedy Center for the Performing Arts in Washington, DC (Edward Durrel Stone, 1971) is dominated by columns that are both functional and aesthetic.

Buckling
of Columns

Beams subjected to compressive loads must be designed so that they do not fail by buckling laterally. We analyze the buckling of axially loaded beams in this chapter. Because load-bearing columns of buildings must support large compressive axial loads, this subject is traditionally called *buckling of columns*.

17.1 Euler Buckling Load

Suppose that a beam with the dimensions shown in Fig. 17.1 is subjected to a compressive axial load P. If the material is steel with yield stress $\sigma_Y = 520$ MPa, the value of P that can be applied without exceeding the yield stress is

$$P = \sigma_Y A$$
$$= (520 \times 10^6)(0.012)(0.0005)$$
$$= 3120 \text{ N } (701 \text{ lb}).$$

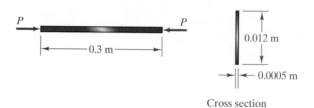

Figure 17.1
Applying compressive axial load to a beam.

Cross section

The beam we have described is a common hacksaw blade. If you hold one between your palms (Fig. 17.2a) and exert an increasing compressive force, it will quickly collapse into a bowed shape (Fig. 17.2b). Although the axial force you exert is tiny compared to the force necessary to exceed the yield stress of the material, the blade fails as a structural element; it will not support the compressive load you exert.

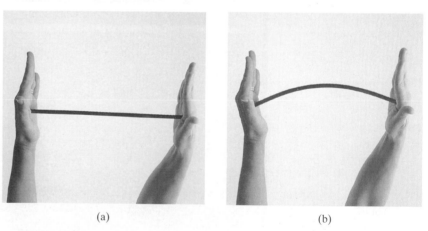

(a)   (b)

Figure 17.2
(a) Applying compressive axial load to a hacksaw blade.
(b) A small compressive load causes the blade to collapse into a bowed shape.

Clearly, the criterion for preventing failure that we have used until now—making sure that loads do not cause the yield stress of the material (or some other defined allowable stress) to be exceeded—does not apply in this situation. The hacksaw blade fails by geometric instability, or *buckling*. Buckling can occur whenever a slender structural member—a thin beam or a thin-walled plate—is subjected to compression.

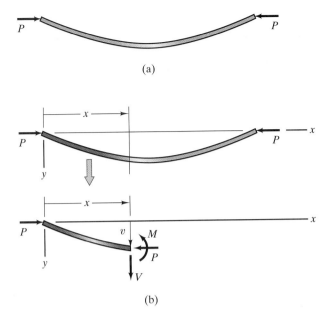

Figure 17.3
(a) Buckled beam in equilibrium.
(b) Determining the bending moment as a function of x.

In this section we derive the buckling load for a prismatic beam subjected to axial forces at the ends. We begin by assuming that the beam has already buckled (Fig. 17.3a) and seek to determine the value of P necessary to hold it in equilibrium. We can accomplish this by proceeding to determine the distribution of the beam's deflection in terms of P. In Fig. 17.3b we introduce a coordinate system and obtain a free-body diagram by passing a plane through the beam at an arbitrary position x. Solving for the bending moment yields $M = Pv$, where v is the beam's deflection at x. We substitute this expression into Eq. (16.4), $v'' = -M/EI$, and write the resulting equation as

$$v'' + \lambda^2 v = 0, \tag{17.1}$$

where

$$\lambda^2 = \frac{P}{EI}. \tag{17.2}$$

The general solution of the second-order differential equation (17.1) is

$$v = B \sin \lambda x + C \cos \lambda x, \tag{17.3}$$

where B and C are constants we must determine from the boundary conditions. [You should confirm that this expression satisfies Eq. (17.1).] From the boundary condition that the deflection equals zero at $x = 0$, $v|_{x=0} = 0$, we see that $C = 0$. The deflection is also zero at $x = L$.

$$v|_{x=L} = 0:$$

$$B \sin \lambda L = 0.$$

This boundary condition is satisfied if $B = 0$, but in that case the solution for the deflection reduces to $v = 0$. [Notice that this solution does indeed satisfy Eq. (17.1) and the boundary conditions, but we are seeking the buckled solution.] If $B \neq 0$, the second boundary condition requires that

$$\sin \lambda L = 0. \tag{17.4}$$

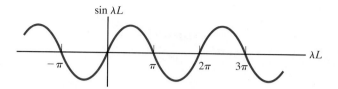

Figure 17.4
Roots of $\sin \lambda L = 0$.

The parameter λ depends on P, so it is from this condition that we can determine the axial load. Here something interesting happens. As Fig. 17.4 indicates, Eq. (17.4) has not one but an infinite number of roots for λL. It is satisfied if

$$\lambda = \frac{n\pi}{L},$$

where n is any integer. Substituting this expression into Eqs. (17.2) and (17.3), we obtain the axial load and deflection:

$$P = \frac{n^2 \pi^2 EI}{L^2}, \tag{17.5}$$

$$v = B \sin \frac{n\pi x}{L}. \tag{17.6}$$

What are all these solutions? Consider the solution for $n = 1$. As x increases from O to L, the argument of the sine in Eq. (17.6) increases from 0 to π. This is the buckled solution we have been seeking (Fig. 17.5a), and the corresponding value of P is

$$P = \frac{\pi^2 EI}{L^2}. \tag{17.7}$$

Notice that we have not determined the value of B, which is the beam's maximum deflection. The load P given by Eq. (17.7) is the force necessary to hold the beam in the buckled state for an arbitrarily small value of B, so it is interpreted as the buckling load. This result was obtained by Leonhard Euler in 1744 and is called the *Euler buckling load*.

When $n = 2$, the argument of the sine in Eq. (17.6) increases from 0 to 2π as x increases from 0 to L, resulting in the deflection in Fig. 17.5b. The deflections for $n = 3$ and $n = 4$ are shown in Figs. 17.5c and d. (Our analysis requires that the beam's slope remain small, so the deflections in Fig. 17.5 are exaggerated.) The solutions for the various values of n are all valid in the sense that they satisfy Eq. (17.1) and the boundary conditions. They are called the *buckling modes*, and are referred to as the first mode ($n = 1$), second mode ($n = 2$), and so on. Although the buckled shape of the beam could theoretically correspond to any of these modes, we know from experience that it will buckle in the first mode. The higher modes are unstable. But they are not merely of academic interest. The beam can be provided with supports so that the lowest mode in which it can buckle is the second or a higher mode (Fig. 17.6). As the figure emphasizes, this greatly increases the beam's buckling load.

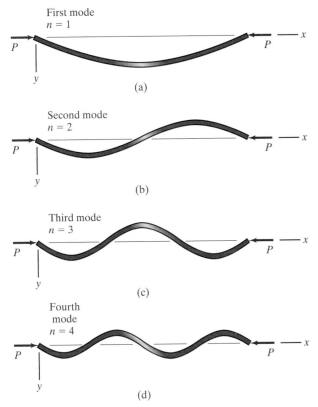

First mode
n = 1

(a)

Second mode
n = 2

(b)

Third mode
n = 3

(c)

Fourth
mode
n = 4

(d)

Figure 17.5
Deflection distributions for increasing values of n.

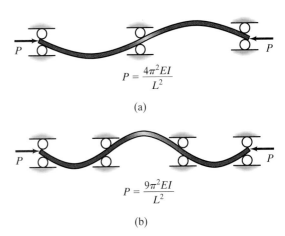

$$P = \frac{4\pi^2 EI}{L^2}$$

(a)

$$P = \frac{9\pi^2 EI}{L^2}$$

(b)

Figure 17.6
Preventing a beam from buckling in a lower mode.

We can now return to the example with which we began this discussion, the hacksaw blade in Fig. 17.1. The value of the axial load P necessary to exceed the yield stress of the material was 3120 N, or 701 lb. Let us calculate the buckling load. When the blade buckles, the axis of its cross section about which it bends is obvious (Fig. 17.7), so the moment of inertia about the z axis is

$$I = \tfrac{1}{12} bh^3$$

$$= \tfrac{1}{12}(0.012)(0.0005)^3$$

$$= 1.25 \times 10^{-13} \text{ m}^4.$$

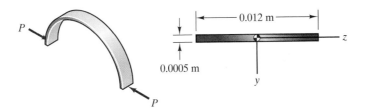

0.012 m

0.0005 m

z

y

P

P

Figure 17.7
Orientation of the cross section of the buckled blade.

(Generally, if it is free to do so, a *buckling beam will bend about the axis of its cross section which has the smallest moment of inertia*.) If the elastic modulus of the steel blade is $E = 200$ GPa, its Euler buckling load is

$$P = \frac{\pi^2 EI}{L^2}$$

$$= \frac{\pi^2 (200 \times 10^9)(1.25 \times 10^{-13})}{(0.3)^2}$$

$$= 2.74 \text{ N } (0.616 \text{ lb}).$$

The buckling load of the hacksaw blade is one-tenth of 1% of the compressive load necessary to exceed the yield stress of the material. This illustrates the relative vulnerability to buckling of beams subjected to compression.

Example 17.1

Buckling of a Truss Member

The truss in Fig. 17.8 consists of bars with the cross section shown. The material has modulus of elasticity $E = 72$ GPa and will safely support a normal stress of 270 MPa in both tension and compression. What is the largest force F that can be applied to the truss?

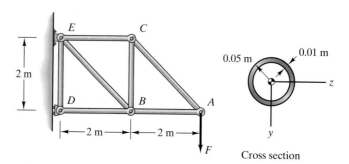

Figure 17.8

Cross section

Strategy

The truss must first be analyzed to determine the axial loads in the members in terms of F. We can then identify the member in which the magnitude of the axial load is greatest and determine the value of F that will subject that member to the allowable normal stress. We must then identify the members that are subject to compression and determine the smallest value of F that will cause a member to buckle.

Solution

We leave it as an exercise to show that the axial loads in the members are

$$\text{AB: } F \qquad (C)$$
$$\text{AC: } \sqrt{2}F \qquad (T)$$
$$\text{BC: } F \qquad (C)$$
$$\text{BD: } 2F \qquad (C)$$

BE: $\sqrt{2}F$ (T)

CE: F (T)

DE: zero

where (T) and (C) denote tension and compression. The magnitude of the axial load is greatest in member *BD*. Setting the magnitude of the normal stress in member *BD* equal to 270 MPa,

$$\frac{2F}{A} = \frac{2F}{\pi(0.05)^2 - \pi(0.04)^2} = 270 \times 10^6,$$

we obtain $F = 382$ kN. This is the largest value of F that will not exceed the normal stress the material will safely support.

We must now consider buckling. Three members, *AB*, *BC*, and *BD*, are subjected to compression. These three members are of equal length and the compressive load is greatest in member *BD*, so it is member *BD* with which we must be concerned with regard to buckling. The moment of inertia of the cross section is

$$I = \tfrac{1}{4}\pi(0.05)^4 - \tfrac{1}{4}\pi(0.04)^4$$
$$= 2.90 \times 10^{-6}\,\text{m}^4,$$

so the Euler buckling load of member *BD* is

$$P = \frac{\pi^2 EI}{L^2}$$
$$= \frac{\pi^2(72 \times 10^9)(2.90 \times 10^{-6})}{(2)^2}$$
$$= 515\,\text{kN}.$$

Equating this value to $2F$ (the compressive load in member *BD*), we determine that member *BD* buckles when $F = 257$ kN.

We have found that a force $F = 382$ kN causes the maximum normal stress in the truss to equal the allowable value, while a force $F = 257$ kN will cause member *BD* to buckle. If an increasing force F is applied, the truss fails, not by failure of the material but by geometric instability (Fig. 17.9). Thus it is the value $F = 257$ kN, with a suitable factor of safety imposed, that determines the largest force that can be applied to the truss.

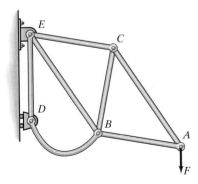

Figure 17.9
Collapse of the truss by buckling of member *BD*.

Discussion

The largest force the truss in this example could support was determined by the buckling load of the members subjected to compression. To achieve light structures while preventing buckling, trusses are sometimes designed with compression members that are relatively thick in comparison to the tension members.

This approach is strikingly evident in R. Buckminster Fuller's *tensegrity* structures (Fig. 17.10). The compression members, supported by tension members consisting of nearly invisible cords, appear to be suspended in the air.

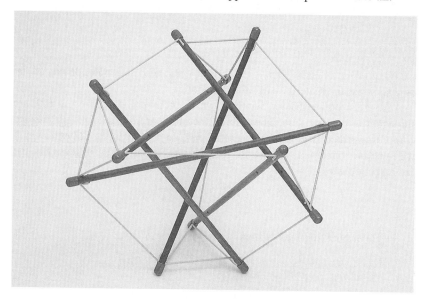

Figure 17.10
Tensegrity structure with thick compression members and slender tension members. (Model by Design Science Toys, Ltd.)

Problems

17.1 The beam is steel with elastic modulus $E = 28 \times 10^6$ psi. Determine its Euler buckling load.

2 in.

Cross section

P17.1

17.2 What is the second-mode buckling load of the beam in Problem 17.1?

17.3 If you want to design the beam in Problem 17.1 so that its length is 120 in. and its Euler buckling load is 380 kip, what should the radius of its cross section be?

17.4 The beam has a solid circular cross section with radius R. The material will safely support an allowable normal stress σ_{allow}. Suppose that you want to achieve an optimal design in the sense that the compressive axial load P that subjects the material to a normal stress of magnitude σ_{allow} is equal to the beam's Euler

buckling load. Show that this is achieved by choosing the dimensions R and L so that

$$\frac{R}{L} = \frac{2}{\pi} \sqrt{\frac{\sigma_{\text{allow}}}{E}}$$

P17.4

17.5 The bar AB has a solid circular cross section with 20-mm radius. It consists of material with elastic modulus $E = 14$ GPa. If the force F is gradually increased, at what value will bar AB buckle?

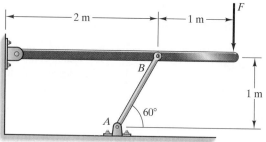

P17.5

17.6 If you want to design the bar AB in Problem 17.5 so that it doesn't buckle until the force $F = 20$ kN, what should its radius be?

17.7 In Example 17.1, suppose that the outer radius of the bars is decreased from 0.05 m to 0.04 m and their wall thickness is kept at 0.01 m. What is the largest force F that can be applied to the truss?

17.8 The bars of the truss have the cross section shown and consist of material with elastic modulus $E = 2 \times 10^6$ psi. If the force F is gradually increased, at what value will the structure fail due to buckling?

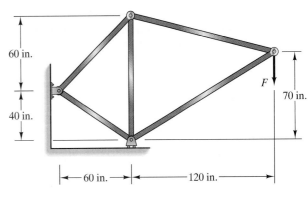

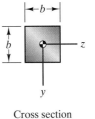

Cross section

P17.11

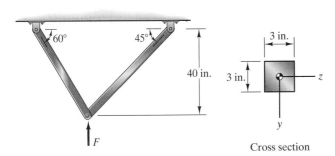

Cross section

P17.8

17.9 If the bars of the truss in Problem 17.8 consist of 2014-T6 Aluminum alloy, at what value of the force F will the structure fail due to buckling?

17.10 The bars of the truss have a solid circular cross section with 30-mm radius and consist of material with elastic modulus $E = 70$ GPa. What is the smallest value of the mass m that will cause the structure to fail due to buckling?

17.12 The truss in Problem 17.11 consists of material that will safely support an allowable normal stress $\sigma_{\text{allow}} = 1000$ psi in tension or compression. Suppose that you want to achieve an optimal design in the sense that the force F which causes the maximum normal stress in the truss to equal σ_{allow} is equal to the smallest force F that causes buckling of a member. Determine the necessary value of the dimension b and the value of F at which failure occurs.

17.13 Bars AB and CD have a solid circular cross section with 20-mm radius. They consist of material with elastic modulus $E = 14$ GPa. If the force F is gradually increased, at what value does the structure fail due to buckling?

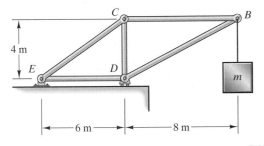

P17.10

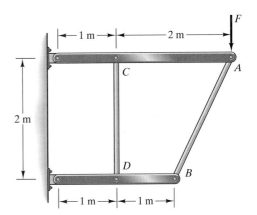

17.11 The bars of the truss have the cross section shown and consist of material with elastic modulus $E = 2 \times 10^6$ psi. If $b = 4$ in. and the force F is gradually increased, at what value of F will the structure fail due to buckling?

P17.13

17.14 The bars in Problem 17.13 will safely support a normal stress (in tension or compression) of 12 GPa. Suppose that you don't want the structure to fail until the force F reaches 5 kN, but you want the radius of each of the bars AB and CD to be as small as possible. What are the radii of the two bars?

17.15 The system supports half of the weight of the 680-kg excavator. Member AC has the cross section shown and consists of material with elastic modulus $E = 73$ GPa. What value of the dimension b would cause member AC to buckle with the stationary system in the position shown?

17.16 The identical vertical bars A, B, and C have a solid circular cross section with 5-mm radius and are made of aluminum alloy with elastic modulus $E = 70$ GPa. The horizontal bar is relatively rigid in comparison to the vertical bars. Determine the value of the force F that would cause failure of the structure by buckling.

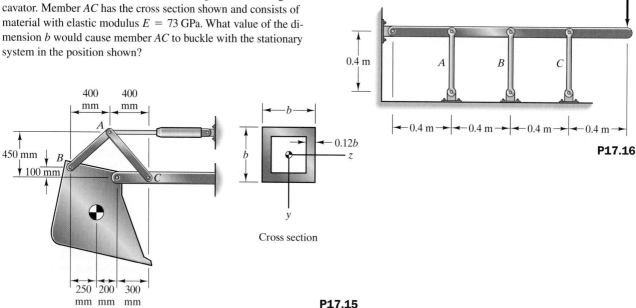

P17.16

P17.15

Other End Conditions

The Euler buckling load is derived under the assumption that the beam is free to rotate at the ends where the compressive axial loads are applied. That is, there are no reactions at the ends other than the applied axial loads. Various types of end supports are used in applications of beams subjected to compressive axial loads, and the choice of supports can greatly affect the resulting buckling load. In some of these cases the buckling load can be determined by analyzing the beam's deflection as we did in Section 17.1.

Analysis of the Deflection

The beam in Fig. 17.11a is free at the left end and has a built-in support at the right end. We begin by assuming that it is buckled (Fig. 17.11b) and seek to determine the value of axial force P necessary to hold it in equilibrium. In Fig. 17.11c we introduce a coordinate system with its origin at the beam's left end and obtain a free-body diagram by passing a plane through the beam at

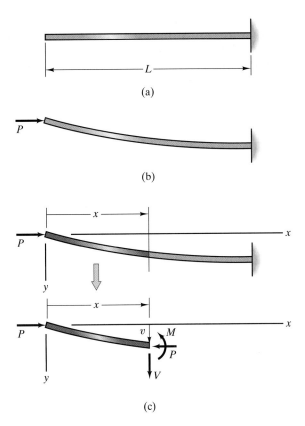

Figure 17.11
(a) Beam with built-in support.
(b) Buckled beam in equilibrium.
(c) Determining the bending moment as a function of x.

an arbitrary position x. This free-body diagram is identical to the one we obtained in deriving the Euler buckling load (Fig. 17.3b), so the steps leading to Eq. (17.3) are unchanged. The deflection is governed by

$$v = B \sin \lambda x + C \cos \lambda x, \tag{17.8}$$

where B and C are constants we must determine from the boundary conditions and

$$\lambda^2 = \frac{P}{EI}. \tag{17.9}$$

Substituting the boundary condition $v|_{x=0} = 0$ into Eq. (17.8), we find that the constant $C = 0$. Because of the built-in support at the right end, the slope is zero at $x = L$.

$$v'|_{x=L} = 0:$$
$$\lambda B \cos \lambda L = 0.$$

This boundary condition requires that

$$\cos \lambda L = 0. \tag{17.10}$$

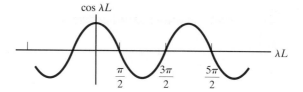

Figure 17.12
Roots of $\cos \lambda L = 0$.

From Fig. 17.12, we see that this equation is satisfied if

$$\lambda = \frac{\pi}{2L}, \frac{3\pi}{2L}, \frac{5\pi}{2L}, \ldots .$$

Substituting $\lambda = \pi/2L$ into Eqs. (17.9) and (17.8), we obtain

$$P = \frac{\pi^2 EI}{4L^2}, \tag{17.11}$$

$$v = B \sin \frac{\pi x}{2L}. \tag{17.12}$$

As x increases from O to L, the argument of the sine in Eq. (17.12) increases from 0 to $\pi/2$. This is the buckled solution shown in Fig. 17.11b, and Eq. (17.11) is the buckling load. (Notice that the buckling load is one-fourth of the Euler buckling load for a beam of equal length.) This is the first buckling mode for a prismatic beam supported in this way. We leave it as an exercise to determine the buckling loads and distributions of the deflection for higher modes.

As another example, the beam in Fig. 17.13a has a roller support that prevents lateral deflection at the left end and a built-in support at the right end. We assume that it is buckled (Fig. 17.13b) and seek to determine the value of axial force P necessary to hold it in equilibrium. In Fig. 17.13c we introduce a coordinate system and obtain a free-body diagram by passing a plane through the beam at an arbitrary position x. The force V_0 is the unknown reaction at the roller support, which equals the shear force at x. Solving for the bending moment and substituting the resulting expression into the equation $v'' = -M/EI$, we write the resulting equation as

$$v'' + \lambda^2 v = -\frac{V_0}{EI} x, \tag{17.13}$$

where

$$\lambda^2 = \frac{P}{EI}. \tag{17.14}$$

Equation (17.13) is nonhomogeneous, because the term on the right side does not contain v or one of its derivatives. Its general solution consists of the sum of the homogeneous and particular solutions:

$$v = v_h + v_p.$$

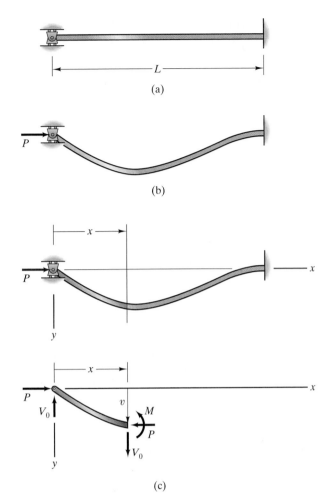

Figure 17.13
(a) Beam with roller and built-in supports.
(b) Buckled beam in equilibrium.
(c) Determining the bending moment as a function of x.

The homogeneous solution is the general solution of Eq. (17.13) with the right side set equal to zero, which we introduced in Section 17.1:

$$v_h = B \sin \lambda x + C \cos \lambda x,$$

where B and C are constants. The particular solution is one that satisfies Eq. (17.13). The nonhomogeneous term is a polynomial in x, so we seek a particular solution in the form of a polynomial of the same order: $v_p = a_0 + a_1 x$, where a_0 and a_1 are constants we must determine. Substituting this expression into Eq. (17.13), we write the resulting equation as

$$\lambda^2 a_0 + \left(\lambda^2 a_1 + \frac{V_0}{EI} \right) x = 0.$$

This equation is satisfied over an interval of x only if $a_0 = 0$ and $a_1 = -V_0/\lambda^2 EI$, yielding the particular solution

$$v_p = -\frac{V_0}{\lambda^2 EI} x.$$

[You should confirm that this is a particular solution by substituting it into Eq. (17.13).] The general solution of Eq. (17.13) is

$$v = v_h + v_p = B \sin \lambda x + C \cos \lambda x - \frac{V_0}{\lambda^2 EI} x. \tag{17.15}$$

From the boundary condition $v|_{x=0} = 0$, we obtain $C = 0$. The boundary conditions at $x = L$ are

$$v\bigg|_{x=L} = B \sin \lambda L - \frac{V_0 L}{\lambda^2 EI} = 0, \tag{17.16}$$

$$v'\bigg|_{x=L} = \lambda B \cos \lambda L - \frac{V_0}{\lambda^2 EI} = 0. \tag{17.17}$$

We solve the first of these equations for the reaction V_0:

$$V_0 = \frac{\lambda^2 EI \sin \lambda L}{L} B. \tag{17.18}$$

Substituting this result into Eq. (17.17), we find that the boundary conditions at $x = L$ are satisfied only if

$$\sin \lambda L - \lambda L \cos \lambda L = 0. \tag{17.19}$$

We also substitute Eq. (17.18) into Eq. (17.15), obtaining the distribution of the deflection in the form

$$v = B\left(\sin \lambda x - \frac{x}{L} \sin \lambda L \right). \tag{17.20}$$

Equation (17.19) is called the *characteristic equation* for this problem. For each value of λ that satisfies the characteristic equation, Eq. (17.14) determines the buckling load and Eq. (17.20) determines the shape of the buckled beam. [Equation (17.10) is the characteristic equation for a beam with free and fixed ends.] In this case we must determine the roots of the characteristic equation numerically. Figure 17.14 is a graph of the characteristic function $f(\lambda L) = \sin \lambda L - \lambda L \cos \lambda L$. The first four roots, determined numerically, are $\lambda L = 4.493, 7.725, 10.904$, and 14.066. The shapes of the resulting buckling modes and the associated buckling loads are shown in Fig. 17.15.

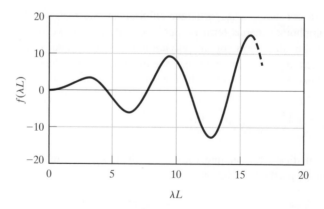

Figure 17.14
Graph of the characteristic function.

The common end conditions that can be analyzed in this way are shown in Fig. 17.16 together with their buckling loads.

First mode

$$P = \frac{(4.493)^2 EI}{L^2}$$

Second mode

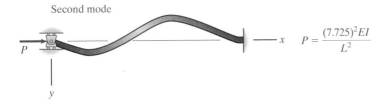

$$P = \frac{(7.725)^2 EI}{L^2}$$

Third mode

$$P = \frac{(10.904)^2 EI}{L^2}$$

Fourth mode

$$P = \frac{(14.066)^2 EI}{L^2}$$

Figure 17.15
First four buckling modes.

$$P = \frac{\pi^2 EI}{L^2}$$

(a)

$$P = \frac{\pi^2 EI}{4L^2}$$

(b)

$$P = \frac{(4.493)^2 EI}{L^2}$$

(c)

$$P = \frac{4\pi^2 EI}{L^2}$$

(d)

Figure 17.16
First-mode buckling loads of beams with
common end conditions.

Example 17.2

Buckling with Various End Conditions

The column in Fig. 17.17 supports an axial load P. The base of the column is built in. The support at the top prevents lateral deflection. The support at the top allows rotation of the column in the x–y plane but prevents rotation in the x–z plane. The moments of inertia of the cross section about the y and z axes are 5×10^{-6} m^4 and 15×10^{-6} m^4, respectively. The elastic modulus is $E = 70$ GPa. If P is gradually increased, at what value will the column buckle?

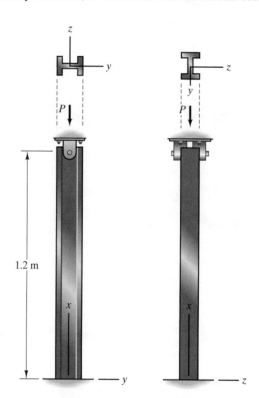

Figure 17.17

Strategy

If it is free to do so, a beam subjected to axial compression buckles by bending about the axis of the cross section that has the smallest moment of inertia. In this example the top support of the column will cause the column to buckle as shown in Fig. 17.16c if it buckles by bending in the x–y plane [Fig. (a)], but will cause it to buckle as shown in Fig. 17.16d if it buckles by bending in the x–z plane [Fig. (b)]. We must determine which of these possibilities yields the lowest buckling load.

Solution

If the column buckles by bending in the x–y plane [Fig. (a)], it fails by bending about the z axis. From Fig. 17.16c, the buckling load is

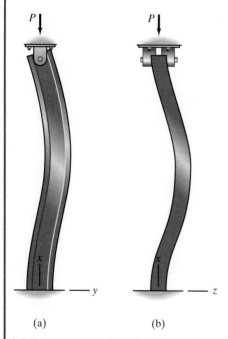

(a)
(b)

(a) Geometry of buckling in the x–y plane.
(b) Geometry of buckling in the x–z plane.

$$P = \frac{(4.493)^2 EI}{L^2}$$

$$= \frac{(4.493)^2 (70 \times 10^9)(15 \times 10^{-6})}{(1.2)^2}$$

$$= 14.72 \text{ MN}.$$

If the column buckles by bending in the x–z plane [Fig. (b)], it fails by bending about the y axis. From Fig. 17.16d, the buckling load is

$$P = \frac{4\pi^2 EI}{L^2}$$

$$= \frac{4\pi^2 (70 \times 10^9)(5 \times 10^{-6})}{(1.2)^2}$$

$$= 9.60 \text{ MN}.$$

We see that the column will buckle as shown in Fig. (b) and the buckling load is 9.60 MN. That is, the column does buckle by bending about the axis of the cross section that has the smaller moment of inertia.

Example 17.3

Buckling with Various End Conditions

The beam in Fig. 17.18 has a support that prevents lateral deflection and rotation at the left end and a built-in support at the right end. Determine its buckling load.

Figure 17.18

Solution

In Fig. (a) we introduce a coordinate system and obtain a free-body diagram by passing a plane through the buckled beam at an arbitrary position x.

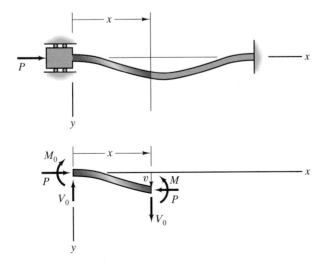

(a) Determining the bending moment.

Solving for the bending moment and substituting it into the equation $v'' = -M/EI$, we write the resulting equation as

$$v'' + \lambda^2 v = -\frac{V_0}{EI}x - \frac{M_0}{EI}, \tag{17.21}$$

where

$$\lambda^2 = \frac{P}{EI}. \tag{17.22}$$

The homogeneous solution of Eq. (17.21) is

$$v_h = B \sin \lambda x + C \cos \lambda x,$$

where B and C are constants. The nonhomogeneous term in Eq. (17.21) is a polynomial in x, so we seek a particular solution in the form of a polynomial of the same order: $v_p = a_0 + a_1 x$. Substituting this expression into Eq. (17.21) and writing the resulting equation as

$$\left(\lambda^2 a_0 + \frac{M_0}{EI}\right) + \left(\lambda^2 a_1 + \frac{V_0}{EI}\right)x = 0,$$

we find that $a_0 = -M_0/\lambda^2 EI$ and $a_1 = -V_0/\lambda^2 EI$, yielding the particular solution

$$v_p = -\frac{M_0}{\lambda^2 EI} - \frac{V_0}{\lambda^2 EI}x.$$

The general solution of Eq. (17.21) is therefore

$$v = v_h + v_p = B \sin \lambda x + C \cos \lambda x - \frac{M_0}{\lambda^2 EI} - \frac{V_0}{\lambda^2 EI}x, \tag{17.23}$$

and the slope is

$$v' = \lambda B \cos \lambda x - \lambda C \sin \lambda x - \frac{V_0}{\lambda^2 EI}. \tag{17.24}$$

The boundary conditions at $x = 0$ are

$$v\Big|_{x=0} = C - \frac{M_0}{\lambda^2 EI} = 0, \tag{17.25}$$

$$v'\Big|_{x=0} = \lambda B - \frac{V_0}{\lambda^2 EI} = 0. \tag{17.26}$$

We solve these equations for M_0 and V_0 and substitute the results into Eqs. (17.23) and (17.24), obtaining

$$v = (\sin \lambda x - \lambda x)B + (\cos \lambda x - 1)C, \tag{17.27}$$

$$v' = (\lambda \cos \lambda x - \lambda)B - \lambda C \sin \lambda x. \tag{17.28}$$

The boundary conditions at $x = L$ are

$$v|_{x=L} = (\sin \lambda L - \lambda L)B + (\cos \lambda L - 1)C = 0, \tag{17.29}$$

$$v'|_{x=L} = (\lambda \cos \lambda L - \lambda)B - \lambda C \sin \lambda L = 0. \tag{17.30}$$

Solving the first of these equations for B gives

$$B = \frac{1 - \cos \lambda L}{\sin \lambda L - \lambda L}C. \tag{17.31}$$

Substituting this result into Eq. (17.30) yields the characteristic equation:

$$\lambda L \sin \lambda L + 2 \cos \lambda L - 2 = 0. \qquad (17.32)$$

We also substitute Eq. (17.31) into Eq. (17.27), obtaining the distribution of the deflection in the form

$$v = C\left[\frac{1 - \cos \lambda L}{\sin \lambda L - \lambda L}(\sin \lambda x - \lambda x) + \cos \lambda x - 1\right]. \qquad (17.33)$$

Figure 17.19 is a graph of the characteristic function $f(\lambda L) = \lambda L \sin \lambda L + 2 \cos \lambda L - 2$. The first four roots, determined numerically, are $\lambda L = 2\pi$, 8.987, 12.566, and 15.451. The shapes of the first three buckling modes [obtained from Eq. (17.33)] and the associated buckling loads [obtained from Eq. (17.22)] are shown in Fig. 17.20.

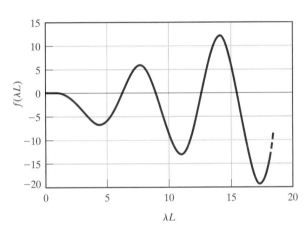

Figure 17.19
Graph of the characteristic function.

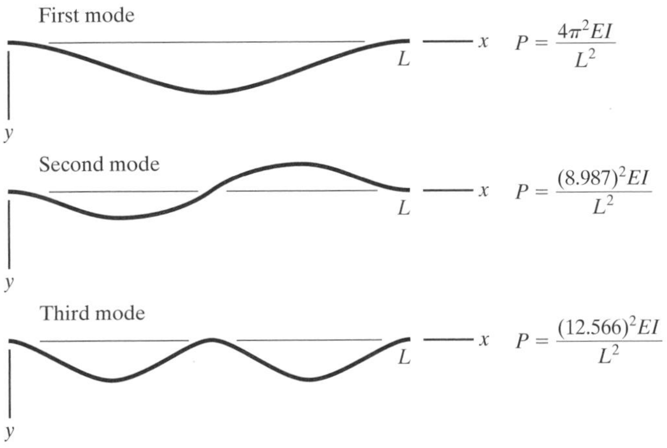

Figure 17.20
First three buckling modes.

Effective Length

In Section 17.1 we analyzed a prismatic beam that is free at the ends and subjected to compressive axial load, obtaining the Euler buckling load (Fig. 17.21a). We also obtained the buckling load of the second mode from the same analysis (Fig. 17.21b), but there is a simpler way to determine the buckling load of the second mode.

Let us assume that the beam is buckled in the second mode. In Fig. 17.22 we cut it by a plane at the midpoint and draw the resulting free-body

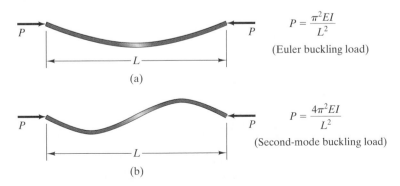

Figure 17.21
First two buckling modes of a beam that is free at the ends. The buckling load of the first mode is the Euler buckling load.

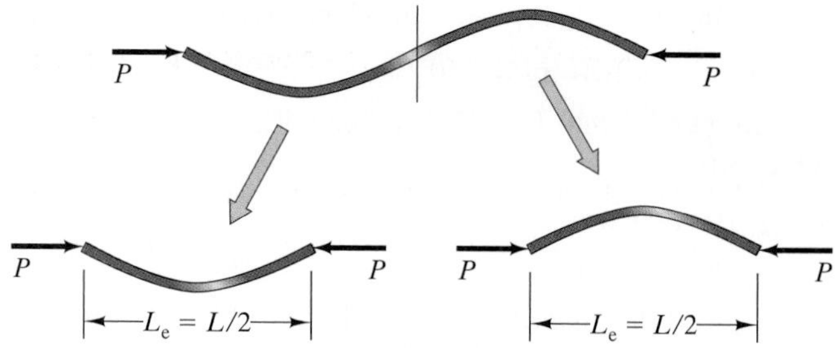

Figure 17.22
Obtaining free-body diagrams by dividing
the beam at its midpoint.

diagrams. Each of these free-body diagrams is buckled in the first mode. We
can therefore determine the second-mode buckling load by calculating the
Euler buckling load of a beam of length $L_e = L/2$:

$$P = \frac{\pi^2 EI}{L_e^2}$$

$$= \frac{\pi^2 EI}{(L/2)^2}$$

$$= \frac{4\pi^2 EI}{L^2}.$$

We see that the second-mode buckling load of the beam of length L is equal
to the Euler buckling load of a beam of length $L_e = L/2$.

This example motivates a concept called the effective length. Suppose
that a given beam buckles in a particular mode, and the buckling load is P.
The *effective length* L_e is defined to be the length of a beam of the same flexural rigidity whose Euler buckling load equals P:

$$P = \frac{\pi^2 EI}{L_e^2}. \tag{17.34}$$

If P is known for a given beam and buckling mode, this equation can be used
to determine the effective length. Alternatively, in some cases the effective
length can be determined or approximated by observation and Eq. (17.34) can
be used to determine P.

How can we determine the effective length by observation? The distribution of the deflection of a beam that is free at the ends and buckled in the first
mode, from which the Euler buckling load is derived (Fig. 17.21a), is a half-
cycle of a sine function. When such a beam buckles in the second mode
(Fig. 17.21b), its deflection forms two half-cycles, each of length $L/2$. The
buckling load of the beam in the second mode equals the Euler buckling load
of these half-cycles (Fig. 17.22), so the effective length is $L/2$. Thus we obtain the effective length by dividing L by the number of half-cycles. This approach can be applied whenever the deflection of a buckled beam consists of
an identifiable number of half-cycles (see Example 17.4).

Example 17.4

Buckling Load by Effective Length

The beam in Fig. 17.23 has a support that prevents lateral deflection and rotation at the left end and a built-in support at the right end. Figure 17.20 shows
the deflection distributions of the first three buckling modes of this beam. For

Figure 17.23

the first and third modes, use the concept of the effective length to verify the buckling loads in Fig. 17.20.

Strategy

By counting the number of half-cycles of sine functions in the deflection distribution for each mode, we can determine its effective length. The buckling load is the Euler buckling load for a beam of that length.

Solution

First mode The distribution of the deflection consists of four quarter-cycles of a sine function [Fig. (a)] or two half-cycles. The effective length is therefore $L_e = L/2$, and the buckling load is

$$
P = \frac{\pi^2 EI}{L_e^2}
$$

$$
= \frac{\pi^2 EI}{(L/2)^2}
$$

$$
= \frac{4\pi^2 EI}{L^2}.
$$

(a) First mode divided into quarter-cycles.

Third mode The distribution of the deflection consists of eight quarter-cycles of a sine function [Fig. (b)], or four half-cycles. The effective length is $L_e = L/4$, and the buckling load is

$$
P = \frac{\pi^2 EI}{L_e^2}
$$

$$
= \frac{\pi^2 EI}{(L/4)^2}
$$

$$
= \frac{(12.566)^2 EI}{L^2}.
$$

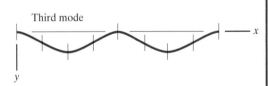

(b) Third mode divided into quarter-cycles.

Problems

17.17 The architectural column has elastic modulus $E = 1.8 \times 10^6$ psi. Its base is built in. The support at the top prevents both lateral deflection and rotation. Determine the buckling load P.

17.18 The architect's specifications for the column in Problem 17.17 would prevent both lateral deflection and rotation at the top, but the contractor installs the column in such a way that only lateral deflection is prevented and the top of the column is free to rotate. What is the actual buckling load? Compare your answer to the answer to Problem 17.17.

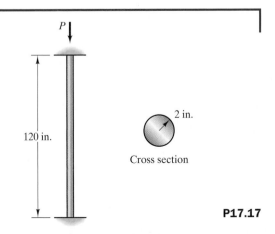

P17.17

17.19 The architect's specifications for the column in Problem 17.17 would prevent both lateral deflection and rotation at the top, but the contractor installs the column in such a way that the column is free to rotate and to deflect laterally at the top. What is the actual buckling load? Compare your answer to the answer to Problem 17.17.

17.20 The column is aluminum alloy with elastic modulus $E = 70$ GPa. Its base is built in. The support at the top prevents lateral deflection but allows rotation in the x–z plane. Determine the buckling load P.

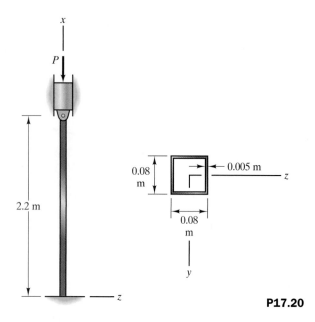

P17.20

17.21 Suppose that you want to increase the wall thickness of the cross section of the column in Problem 17.20, while keeping the 0.08-m dimension fixed, so that the column's buckling load is 700 kN. What is the necessary wall thickness?

17.22 In Example 17.2, suppose that the base of the column is supported in the same way as the top. Then the supports at the top and bottom prevent lateral deflection and prevent rotation in the x–z plane but allow rotation in the x–y plane. What is the column's buckling load? Does it buckle by bending in the x–y plane or the x–z plane?

17.23 The column supports an axial load P. The base of the column is built in. The support at the top prevents lateral deflection. The support at the top allows rotation of the column in the x–y plane but prevents rotation in the x–z plane. The dimensions $b = 0.08$ m and $h = 0.14$ m. The elastic modulus is $E = 12$ GPa. What is the column's buckling load? Does it buckle by bending in the x–y plane or the x–z plane?

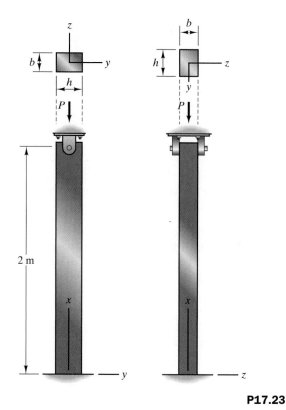

P17.23

17.24 Solve Problem 17.23 if $b = 0.12$ m and $h = 0.14$ m.

17.25 For the column shown in Problem 17.23, determine the ratio b/h for which the buckling load if the column buckles in the x–y plane is equal to the buckling load if it buckles in the x–z plane.

17.26 For the beam in Fig. 17.11, determine the buckling loads for buckling modes two, three, and four, and draw graphs of the distributions of the deflection.

17.27 For the beam in Example 17.3, determine the buckling load for the fourth buckling mode and draw a graph of the distribution of the deflection.

17.28 Figure 17.5 shows the first four buckling modes of a beam of length L that is free to bend at the ends. What are the effective lengths of the four modes?

17.29 In Fig. 17.20, the distributions of the deflection and the buckling loads are given for the first three modes of the beam in Fig. 17.18. Use the expressions for the buckling loads to calculate the effective lengths of the first three modes.

17.30 The prismatic beam of length L has a built-in support at the right end and is free at the left end. It is shown buckled in the first mode. What is its effective length? Use the effective length to determine the buckling load.

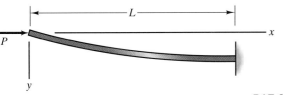

P17.30

17.31 The figure shows the fourth buckling mode of the beam in Fig. 17.18.

(a) Use the figure to determine the approximate effective length.
(b) Use the approximate effective length to determine the buckling load.

P17.31

Chapter Summary

Buckling Loads

The buckling load for a prismatic beam of length L that is free at the ends and buckles as shown in Fig. (a) is called the *Euler buckling load*:

$$P = \frac{\pi^2 EI}{L^2}.$$ Eq. (17.7)

(a)

Figure (a) is the first buckling mode for a beam that is free at the ends. The distributions of the deflection for the second, third, and fourth modes are shown in Fig. 17.5. The buckling load of the nth mode is

$$P = \frac{n^2 \pi^2 EI}{L^2}.$$ Eq. (17.5)

First-mode buckling loads of beams with other common end conditions are shown in Fig. 17.16.

Effective Length

Suppose that a given beam buckles in a particular mode, and the buckling load is P. The *effective length* L_e is the length of a beam of the same flexural rigidity whose Euler buckling load equals P:

$$P = \frac{\pi^2 EI}{L_e^2}.$$ Eq. (17.34)

For example, if a beam of length L that is free at the ends buckles in the second mode, the effective length is $L_e = L/2$ [Fig. (b)] and the buckling load is

$$P = \frac{\pi^2 EI}{L_e^2} = \frac{4\pi^2 EI}{L^2}.$$

In some cases the effective length of a buckled beam can be determined or approximated by observation and Eq. (17.34) can be used to determine the buckling load.

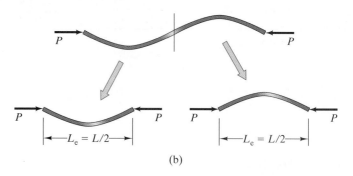

(b)

Review Problems

17.32 The beam is aluminum alloy with elastic modulus $E = 70$ GPa. Determine its Euler buckling load.

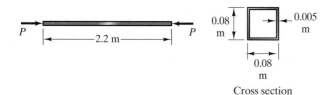

Cross section

P17.32

17.33 If you want to increase the wall thickness of the beam in Problem 17.32 so that its Euler buckling load is 300 kN, what wall thickness is required?

17.34 The bars of the truss have the cross section shown and consist of material with elastic modulus $E = 70$ GPa. If the force F is gradually increased, at what value will the structure fail due to buckling?

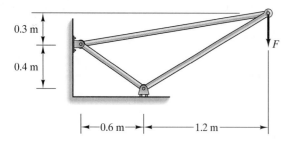

Cross section

P17.34

17.35 In Problem 17.34, suppose that you want to increase the outer radius of the cross section of the bars with the wall thickness kept at 3 mm so that the structure does not fail due to buckling until $F = 7.5$ kN. What should the outer radius be?

17.36 The link AB of the pliers has the cross section shown and is made of steel with elastic modulus $E = 190$ GPa. Determine the value of the force F that would cause failure of the link by buckling.

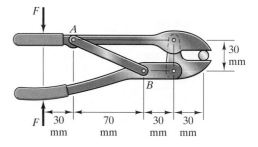

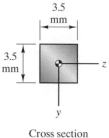

Cross section   **P17.36**

17.37 If the link AB of the pliers in Problem 17.36 is made of 7075-T6 aluminum alloy, what force F would cause failure of the link by buckling?

17.38 The rectangular platform in Fig. (a) is supported by four identical columns of length L. The platform is loaded with weights in such a way that the axial force on each column is the same until the columns buckle [Fig. (b)]. The connections of the columns to the floor and the platform behave like built-in supports. By using the same type of analysis as in Example 17.3, determine the buckling load of each column for the first buckling mode.

17.39 Suppose that the rectangular platform in Problem 17.38 is provided with supports in the horizontal plane so that the four columns buckle in the second mode. Use the same type of analysis as in Example 17.3 to determine the buckling load of each column and draw a graph of the distribution of the deflection.

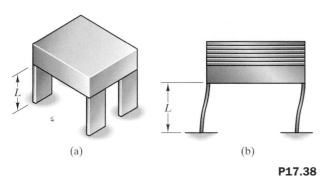

(a)   (b)

P17.38

Review of Mathematics

A.1 Algebra

Quadratic Equations

The solutions of the quadratic equation

$$ax^2 + bx + c = 0$$

are

$$x = \frac{-b \pm \sqrt{b^2 - 4ac}}{2a}.$$

Natural Logarithms

The natural logarithm of a positive real number x is denoted by $\ln x$. It is defined to be the number such that

$$e^{\ln x} = x,$$

where $e = 2.7182\ldots$ is the base of natural logarithms.
Logarithms have the following properties:

$$\ln(xy) = \ln x + \ln y,$$

$$\ln(x/y) = \ln x - \ln y,$$

$$\ln y^x = x \ln y.$$

A.2 Trigonometry

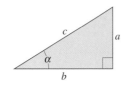

The trigonometric functions for a right triangle are

$$\sin\alpha = \frac{1}{\csc\alpha} = \frac{a}{c}, \qquad \cos\alpha = \frac{1}{\sec\alpha} = \frac{b}{c}, \qquad \tan\alpha = \frac{1}{\cot\alpha} = \frac{a}{b}.$$

The sine and cosine satisfy the relation

$$\sin^2\alpha + \cos^2\alpha = 1,$$

and the sine and cosine of the sum and difference of two angles satisfy

$$\sin(\alpha + \beta) = \sin\alpha\cos\beta + \cos\alpha\sin\beta,$$

$$\sin(\alpha - \beta) = \sin\alpha\cos\beta - \cos\alpha\sin\beta,$$

$$\cos(\alpha + \beta) = \cos\alpha\cos\beta - \sin\alpha\sin\beta,$$

$$\cos(\alpha - \beta) = \cos\alpha\cos\beta + \sin\alpha\sin\beta.$$

The **law of cosines** for an arbitrary triangle is

$$c^2 = a^2 + b^2 - 2ab\cos\alpha_c,$$

and the **law of sines** is

$$\frac{\sin\alpha_a}{a} = \frac{\sin\alpha_b}{b} = \frac{\sin\alpha_c}{c}.$$

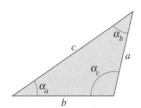

A.3 Derivatives

$$\frac{d}{dx}x^n = nx^{n-1}$$

$$\frac{d}{dx}e^x = e^x$$

$$\frac{d}{dx}\ln x = \frac{1}{x}$$

$$\frac{d}{dx}\sin x = \cos x$$

$$\frac{d}{dx}\cos x = -\sin x$$

$$\frac{d}{dx}\tan x = \frac{1}{\cos^2 x}$$

$$\frac{d}{dx}\sinh x = \cosh x$$

$$\frac{d}{dx}\cosh x = \sinh x$$

$$\frac{d}{dx}\tanh x = \frac{1}{\cosh^2 x}$$

A.4 Integrals

$$\int x^n\, dx = \frac{x^{n+1}}{n+1} \quad (n \neq -1)$$

$$\int x^{-1}\, dx = \ln x$$

$$\int (a + bx)^{1/2}\, dx = \frac{2}{3b}(a + bx)^{3/2}$$

$$\int x(a + bx)^{1/2}\, dx = -\frac{2(2a - 3bx)(a + bx)^{3/2}}{15b^2}$$

$$\int (1 + a^2x^2)^{1/2}\, dx = \frac{1}{2}\left\{x(1 + a^2x^2)^{1/2} + \frac{1}{a}\ln\left[x + \left(\frac{1}{a^2} + x^2\right)^{1/2}\right]\right\}$$

$$\int x(1 + a^2x^2)^{1/2}\, dx = \frac{a}{3}\left(\frac{1}{a^2} + x^2\right)^{3/2}$$

$$\int x^2(1 + a^2x^2)^{1/2}\, dx = \frac{1}{4}ax\left(\frac{1}{a^2} + x^2\right)^{3/2} - \frac{1}{8a^2}x(1 + a^2x^2)^{1/2}$$
$$- \frac{1}{8a^3}\ln\left[x + \left(\frac{1}{a^2} + x^2\right)^{1/2}\right]$$

$$\int (1 - a^2x^2)^{1/2}\, dx = \frac{1}{2}\left[x(1 - a^2x^2)^{1/2} + \frac{1}{a}\arcsin ax\right]$$

$$\int x(1 - a^2x^2)^{1/2}\, dx = -\frac{a}{3}\left(\frac{1}{a^2} - x^2\right)^{3/2}$$

$$\int x^2(a^2 - x^2)^{1/2}\, dx = -\frac{1}{4}x(a^2 - x^2)^{3/2}$$
$$+ \frac{1}{8}a^2\left[x(a^2 - x^2)^{1/2} + a^2\arcsin\frac{x}{a}\right]$$

$$\int \frac{dx}{(1 + a^2x^2)^{1/2}} = \frac{1}{a}\ln\left[x + \left(\frac{1}{a^2} + x^2\right)^{1/2}\right]$$

$$\int \frac{dx}{(1 - a^2x^2)^{1/2}} = \frac{1}{a}\arcsin ax, \quad \text{or} \quad -\frac{1}{a}\arccos ax$$

$$\int \sin x\, dx = -\cos x$$

$$\int \cos x\, dx = \sin x$$

$$\int \sin^2 x\, dx = -\frac{1}{2}\sin x \cos x + \frac{1}{2}x$$

$$\int \cos^2 x\, dx = \frac{1}{2}\sin x \cos x + \frac{1}{2}x$$

$$\int \sin^3 x\, dx = -\frac{1}{3}\cos x(\sin^2 x + 2)$$

$$\int \cos^3 x\, dx = \frac{1}{3}\sin x(\cos^2 x + 2)$$

$$\int \cos^4 x\, dx = \frac{3}{8}x + \frac{1}{4}\sin 2x + \frac{1}{32}\sin 4x$$

$$\int \sin^n x \cos x\, dx = \frac{(\sin x)^{n+1}}{n+1} \quad (n \neq -1)$$

$$\int \sinh x\, dx = \cosh x$$

$$\int \cosh x\, dx = \sinh x$$

$$\int \tanh x\, dx = \ln \cosh x$$

$$\int e^{ax}\, dx = \frac{e^{ax}}{a}$$

$$\int xe^{ax}\, dx = \frac{e^{ax}}{a^2}(ax - 1)$$

A.5 Taylor Series

The Taylor series of a function $f(x)$ is

$$f(a + x) = f(a) + f'(a)x + \frac{1}{2!}f''(a)x^2 + \frac{1}{3!}f'''(a)x^3 + \cdots,$$

where the primes indicate derivatives.

Some useful Taylor series are

$$e^x = 1 + x + \frac{x^2}{2!} + \frac{x^3}{3!} + \cdots,$$

$$\sin(a + x) = \sin a + (\cos a)x - \frac{1}{2}(\sin a)x^2 - \frac{1}{6}(\cos a)x^3 + \cdots,$$

$$\cos(a + x) = \cos a - (\sin a)x - \frac{1}{2}(\cos a)x^2 + \frac{1}{6}(\sin a)x^3 + \cdots,$$

$$\tan(a + x) = \tan a + \left(\frac{1}{\cos^2 a}\right)x + \left(\frac{\sin a}{\cos^3 a}\right)x^2$$

$$+ \left(\frac{\sin^2 a}{\cos^4 a} + \frac{1}{3\cos^2 a}\right)x^3 + \cdots.$$

A.6 Vector Analysis

Cartesian Coordinates

The gradient of a scalar field ψ is

$$\nabla \psi = \frac{\partial \psi}{\partial x}\mathbf{i} + \frac{\partial \psi}{\partial y}\mathbf{j} + \frac{\partial \psi}{\partial z}\mathbf{k}.$$

The divergence and curl of a vector field $\mathbf{v} = v_x\mathbf{i} + v_y\mathbf{j} + v_z\mathbf{k}$ are

$$\nabla \cdot \mathbf{v} = \frac{\partial v_x}{\partial x} + \frac{\partial v_y}{\partial y} + \frac{\partial v_z}{\partial z},$$

$$\nabla \times \mathbf{v} = \begin{vmatrix} \mathbf{i} & \mathbf{j} & \mathbf{k} \\ \dfrac{\partial}{\partial x} & \dfrac{\partial}{\partial y} & \dfrac{\partial}{\partial z} \\ v_x & v_y & v_z \end{vmatrix}.$$

Cylindrical Coordinates

The gradient of a scalar field ψ is

$$\nabla \psi = \frac{\partial \psi}{\partial r}\mathbf{e}_r + \frac{1}{r}\frac{\partial \psi}{\partial \theta}\mathbf{e}_\theta + \frac{\partial \psi}{\partial z}\mathbf{e}_z.$$

The divergence and curl of a vector field $\mathbf{v} = v_r\mathbf{e}_r + v_\theta\mathbf{e}_\theta + v_z\mathbf{e}_z$ are

$$\nabla \cdot \mathbf{v} = \frac{\partial v_r}{\partial r} + \frac{v_r}{r} + \frac{1}{r}\frac{\partial v_\theta}{\partial \theta} + \frac{\partial v_z}{\partial z},$$

$$\nabla \times \mathbf{v} = \frac{1}{r}\begin{vmatrix} \mathbf{e}_r & r\mathbf{e}_\theta & \mathbf{e}_z \\ \dfrac{\partial}{\partial r} & \dfrac{\partial}{\partial \theta} & \dfrac{\partial}{\partial z} \\ v_r & rv_\theta & v_z \end{vmatrix}.$$

Properties of Areas and Lines

B.1 Areas

The coordinates of the centroid of the area A are

$$\bar{x} = \frac{\displaystyle\int_A x \, dA}{\displaystyle\int_A dA}, \qquad \bar{y} = \frac{\displaystyle\int_A y \, dA}{\displaystyle\int_A dA}.$$

The moment of inertia about the x axis I_x, the moment of inertia about the y axis I_y, and the product of inertia I_{xy} are

$$I_x = \int_A y^2 \, dA, \qquad I_y = \int_A x^2 \, dA, \qquad I_{xy} = \int_A xy \, dA.$$

The polar moment of inertia about O is

$$J_O = \int_A r^2 \, dA = \int_A (x^2 + y^2) \, dA = I_x + I_y.$$

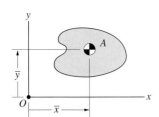

Rectangular area

Area $= bh$

$$I_x = \frac{1}{3} bh^3, \qquad I_y = \frac{1}{3} hb^3, \qquad I_{xy} = \frac{1}{4} b^2 h^2$$

$$I_{x'} = \frac{1}{12} bh^3, \qquad I_{y'} = \frac{1}{12} hb^3, \qquad I_{x'y'} = 0$$

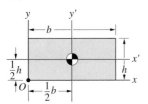

Triangular area

Area $= \frac{1}{2} bh$

$$I_x = \frac{1}{12} bh^3, \qquad I_y = \frac{1}{4} hb^3, \qquad I_{xy} = \frac{1}{8} b^2 h^2$$

$$I_{x'} = \frac{1}{36} bh^3, \qquad I_{y'} = \frac{1}{36} hb^3, \qquad I_{x'y'} = \frac{1}{72} b^2 h^2$$

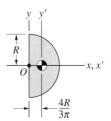

Triangular area

Area $= \frac{1}{2} bh$ $\qquad I_x = \frac{1}{12} bh^3, \qquad I_{x'} = \frac{1}{36} bh^3$

Circular area

Area $= \pi R^2$ $\qquad I_{x'} = I_{y'} = \frac{1}{4} \pi R^4, \qquad I_{x'y'} = 0$

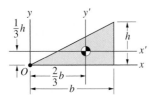

Semicircular area

Area $= \frac{1}{2} \pi R^2$ $\qquad I_x = I_y = \frac{1}{8} \pi R^4, \qquad I_{xy} = 0$

$$I_{x'} = \frac{1}{8} \pi R^4, \qquad I_{y'} = \left(\frac{\pi}{8} - \frac{8}{9\pi} \right) R^4, \qquad I_{x'y'} = 0$$

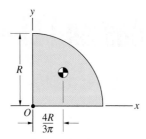

Quarter-circular area

$$\text{Area} = \frac{1}{4}\pi R^2 \qquad I_x = I_y = \frac{1}{16}\pi R^4, \qquad I_{xy} = \frac{1}{8}R^4$$

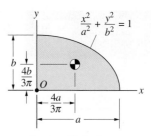

Quarter-elliptical area

$$\text{Area} = \frac{1}{4}\pi ab$$

$$I_x = \frac{1}{16}\pi ab^3, \qquad I_y = \frac{1}{16}\pi a^3 b, \qquad I_{xy} = \frac{1}{8}a^2 b^2$$

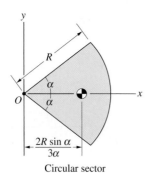

Circular sector

$$\text{Area} = \alpha R^2$$

$$I_x = \frac{1}{4}R^4\left(\alpha - \frac{1}{2}\sin 2\alpha\right), \qquad I_y = \frac{1}{4}R^4\left(\alpha + \frac{1}{2}\sin 2\alpha\right),$$

$$I_{xy} = 0$$

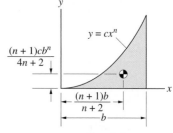

Spandrel

$$\text{Area} = \frac{cb^{n+1}}{n+1}$$

$$I_x = \frac{c^3 b^{3n+1}}{9n+3}, \qquad I_y = \frac{cb^{n+3}}{n+3}, \qquad I_{xy} = \frac{c^2 b^{2n+2}}{4n+4}$$

B.2 Lines

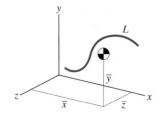

The coordinates of the centroid of the line L are

$$\bar{x} = \frac{\int_L x\, dL}{\int_L dL}, \qquad \bar{y} = \frac{\int_L y\, dL}{\int_L dL}, \qquad \bar{z} = \frac{\int_L z\, dL}{\int_L dL}.$$

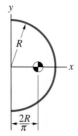

Semicircular arc

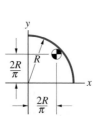

Quarter-circular arc

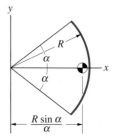

Circular arc

APPENDIX C
Properties of Volumes and Homogeneous Objects

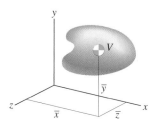

The coordinates of the centroid of the volume V are

$$\bar{x} = \frac{\int_V x \, dV}{\int_V dV}, \qquad \bar{y} = \frac{\int_V y \, dV}{\int_V dV}, \qquad \bar{z} = \frac{\int_V z \, dV}{\int_V dV}.$$

(The center of mass of a homogeneous object coincides with the centroid of its volume.)

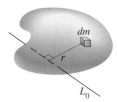

The mass moment of inertia of the object about the axis L_0 is

$$I_0 = \int_m r^2 \, dm.$$

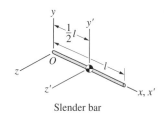

Slender bar

$$I_{(x \text{ axis})} = 0, \qquad I_{(y \text{ axis})} = I_{(z \text{ axis})} = \frac{1}{3} ml^2$$

$$I_{(x' \text{ axis})} = 0, \qquad I_{(y' \text{ axis})} = I_{(z' \text{ axis})} = \frac{1}{12} ml^2$$

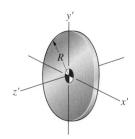

Thin circular plate

$$I_{(x' \text{ axis})} = I_{(y' \text{ axis})} = \frac{1}{4} mR^2, \qquad I_{(z' \text{ axis})} = \frac{1}{2} mR^2$$

$$I_{(x \text{ axis})} = \frac{1}{3} mh^2, \qquad I_{(y \text{ axis})} = \frac{1}{3} mb^2, \qquad I_{(z \text{ axis})} = \frac{1}{3} m(b^2 + h^2)$$

$$I_{(x' \text{ axis})} = \frac{1}{12} mh^2, \qquad I_{(y' \text{ axis})} = \frac{1}{12} mb^2, \qquad I_{(z' \text{ axis})} = \frac{1}{12} m(b^2 + h^2)$$

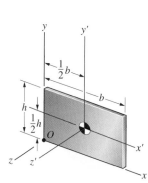

Thin rectangular plate

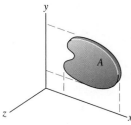

Thin plate

$$I_{(x \text{ axis})} = \frac{m}{A} I_x^A, \qquad I_{(y \text{ axis})} = \frac{m}{A} I_y^A, \qquad I_{(z \text{ axis})} = I_{(x \text{ axis})} + I_{(y \text{ axis})}$$

(The superscripts A denote moments of inertia of the plate's cross-sectional area A.)

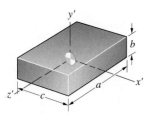

Rectangular prism

Volume $= abc$

$$I_{(x' \text{ axis})} = \frac{1}{12} m(a^2 + b^2), \qquad I_{(y' \text{ axis})} = \frac{1}{12} m(a^2 + c^2),$$

$$I_{(z' \text{ axis})} = \frac{1}{12} m(b^2 + c^2),$$

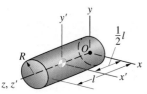

Circular cylinder

Volume $= \pi R^2 l$

$$I_{(x \text{ axis})} = I_{(y \text{ axis})} = m\left(\frac{1}{3} l^2 + \frac{1}{4} R^2\right), \qquad I_{(z \text{ axis})} = \frac{1}{2} mR^2$$

$$I_{(x' \text{ axis})} = I_{(y' \text{ axis})} = m\left(\frac{1}{12} l^2 + \frac{1}{4} R^2\right), \qquad I_{(z' \text{ axis})} = \frac{1}{2} mR^2$$

Circular cone

Volume $= \frac{1}{3} \pi R^2 h$

$$I_{(x \text{ axis})} = I_{(y \text{ axis})} = m\left(\frac{3}{5} h^2 + \frac{3}{20} R^2\right), \qquad I_{(z \text{ axis})} = \frac{3}{10} mR^2$$

$$I_{(x' \text{ axis})} = I_{(y' \text{ axis})} = m\left(\frac{3}{80} h^2 + \frac{3}{20} R^2\right), \qquad I_{(z' \text{ axis})} = \frac{3}{10} mR^2$$

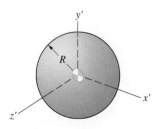

Sphere

Volume $= \frac{4}{3} \pi R^3$

$$I_{(x' \text{ axis})} = I_{(y' \text{ axis})} = I_{(z' \text{ axis})} = \frac{2}{5} mR^2$$

This appendix summarizes the properties of selected materials. These values may be used as approximations for preliminary design. However, in final design calculations you should try to use values obtained from the actual materials specified in your design. Material properties can sometimes be obtained from the sources supplying the materials, or it may be necessary to have measurements made from samples of the specified materials.

Table D-1 Elastic moduli of selected materials

Material	Modulus of elasticity E		Shear modulus G		Poisson's ratio ν
	10^6 psi	GPa	10^6 psi	GPa	
Aluminum	10	70	3.8	26	0.33
Aluminum alloys	10–12	70–80	3.8–4.4	26–30	0.33
2014-T6	10.6	73	4	28	0.33
6061-T6	10	70	3.8	26	0.33
7075-T6	10.4	72	3.9	27	0.33
Brick (compression)	1.5–3.5	10–24			
Cast iron	12–25	80–170	4.5–10	31–69	0.2–0.3
Gray cast iron	14	97	5.6	39	0.25
Concrete (compression)	2.6–4.4	18–30			0.1–0.2
Copper	17	115	6.2	43	0.35
Copper alloys	14–18	96–120	5.2–6.8	36–47	0.33–0.35
Brass	14–16	96–110	5.2–6	36–41	0.34
80% Cu, 20% Zn	15	100	5.5	38	0.33
Naval brass	15	100	5.5	38	0.33
Bronze	14–17	96–120	5.2–6.3	36–44	0.34
Manganese bronze	15	100	5.6	39	0.35
Glass	7–12	50–80	2.9–5	20–33	0.20–0.27
Magnesium	5.8	40	2.2	15	0.34
Nickel	30	210	11.4	80	0.31
Nylon	0.3–0.4	2–3			0.4
Rubber	0.0001–0.0006	0.001–0.004	0.00004–0.0002	0.0003–0.0014	0.44–0.50
Steel	28–32	190–220	10.8–12.3	75–85	0.28–0.30
Stone (compression)					
Granite	6–10	40–70			0.2–0.3
Marble	7–14	50–100			0.2–0.3
Titanium	16	110	5.8	40	0.33
Titanium alloys	15–18	100–124	5.6–6.8	39–47	0.33
Tungsten	52	360	22	150	0.2
Wood (bending)					
Ash	1.5–1.6	10–11			
Oak	1.6–1.8	11–12			
Southern pine	1.6–2	11–14			
Wrought iron	28	190	10.9	75	0.3

Table D-2 Yield and ultimate stresses of selected materials

Material	Yield stress σ_Y		Ultimate stress σ_U		Percent elongation (2-in. gauge length)
	10^3 psi	MPa	10^3 psi	MPa	
Aluminum	3	20	10	70	60
Aluminum alloys	6–75	40–520	15–80	100–560	2–45
2014-T6	60	410	70	480	13
6061-T6	40	270	45	310	17
7075-T6	70	480	80	550	11
Brick (compression)			1–10	7–70	
Cast iron (compression)			50–200	340–1400	
Cast iron (tension)	17–41	120–280	10–70	70–480	1
Gray cast iron	17	120	20–58	140–400	1
Concrete (compression)			1.5–10	10–70	
Copper					
Hard-drawn	49	340	55	380	10
Soft-annealed	8	55	33	230	50
Copper alloys					
Beryllium copper, hard	109	750	120	830	4
Brass	12–80	80–540	36–90	240–600	4–60
80% Cu, 20% Zn, hard	67	460	84	580	4
80% Cu, 20% Zn, soft	13	90	43	300	50
Naval brass, hard	58	400	84	580	15
Naval brass, soft	25	170	59	410	50
Bronze	12–100	82–700	30–120	200–830	5–60
Manganese bronze, hard	65	450	90	620	10
Manganese bronze, soft	25	170	65	450	35
Glass (plate)			9	65	
Magnesium	3–10	20–68	15–25	100–170	5–15
Nickel	20–90	140–620	45–110	310–760	2–50
Nylon			6–10	40–72	50
Rubber	0.3–1	2–7	1–3	7–20	100–800
Steel, structural	30–104	200–720	50–118	340–820	10–40
ASTM-A36	36	250	60	400	30
ASTM-A572	50	340	70	500	20
ASTM-A514	100	700	120	830	15
Stone (compression)					
Granite			10–40	70–280	
Marble			8–25	50–180	
Titanium	60	400	70	500	25
Titanium alloys	110–125	760–860	130–140	900–960	10
Tungsten			200–600	1400–4000	0–4
Wood (bending)					
Ash	6–10	40–70	8–13	50–90	
Oak	6–8	40–55	8–13	50–90	
Southern Pine	6–8	40–55	8–13	50–90	
Wood (compression parallel to grain)					
Ash	4–6	30–40	5–8	30–50	
Oak	4–6	30–40	5–8	30–50	
Southern pine	4–8	30–50	6–10	40–70	
Wrought iron	30	210	48	330	30

Table D-3 Densities and coefficients of thermal expansion of selected materials

Material	Density ρ		Coefficient of thermal expansion α	
	slug/ft^3	kg/m^3	$10^{-6}\,{}^\circ\text{F}^{-1}$	$10^{-6}\,{}^\circ\text{C}^{-1}$
Aluminum	5.2	2700	13.3	23.9
Aluminum alloys	4.9–5.4	2500–2800	13–13.4	23–24
Copper	17.4	9000	9.2	16.6
Copper alloys				
Brass	16.3–16.9	8400–8700	10.6–11.8	19–21
Bronze	14.3–17.2	7400–8900	9.9–11.6	18–21
Cast iron	13.9	7200	5.6–6.7	10–12
Gray cast iron	13.6–13.8	7000–7100	5.6	10
Concrete	2.9–4.7	1500–2400	4–8	7–14
Glass	4.6–5.8	2400–3000	3–6	5–11
Magnesium	3.4	1740	14	25
Magnesium alloys	3.4–3.5	1760–1830	14–16	26–29
Nickel	16.7	8600	7.2	13
Rubber	1.7–3.9	900–2000	70–110	130–200
Steel	14.9–15.2	7700–7830	6–10	10–18
Stone	3.9–5.1	2000–2900	3–5	5–9
Titanium	8.8	4540	4.7	8.5
Tungsten	37.4	1930	2.4	4.3
Wrought iron	14–15	7400–7800	7	12

Deflections and Slopes of Prismatic Beams

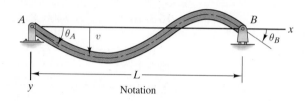

Notation

Simply Supported Beams

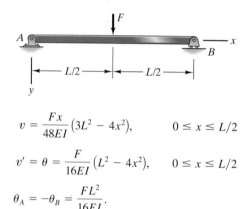

$$v = \frac{Fx}{48EI}\left(3L^2 - 4x^2\right), \qquad 0 \le x \le L/2$$

$$v' = \theta = \frac{F}{16EI}\left(L^2 - 4x^2\right), \qquad 0 \le x \le L/2$$

$$\theta_A = -\theta_B = \frac{FL^2}{16EI}.$$

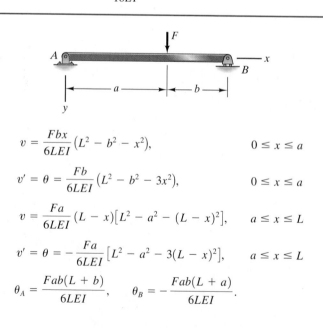

$$v = \frac{Fbx}{6LEI}\left(L^2 - b^2 - x^2\right), \qquad 0 \le x \le a$$

$$v' = \theta = \frac{Fb}{6LEI}\left(L^2 - b^2 - 3x^2\right), \qquad 0 \le x \le a$$

$$v = \frac{Fa}{6LEI}(L - x)\left[L^2 - a^2 - (L - x)^2\right], \qquad a \le x \le L$$

$$v' = \theta = -\frac{Fa}{6LEI}\left[L^2 - a^2 - 3(L - x)^2\right], \qquad a \le x \le L$$

$$\theta_A = \frac{Fab(L + b)}{6LEI}, \qquad \theta_B = -\frac{Fab(L + a)}{6LEI}.$$

$$v = \frac{M_0 x}{6LEI}\left(2L^2 - 3Lx + x^2\right), \qquad v' = \theta = \frac{M_0}{6LEI}\left(2L^2 - 6Lx + 3x^2\right),$$

$$\theta_A = \frac{M_0 L}{3EI}, \qquad \theta_B = -\frac{M_0 L}{6EI}.$$

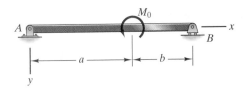

$$v = \frac{M_0 x}{6LEI}\left(6aL - 3a^2 - 2L^2 - x^2\right), \qquad\qquad 0 \le x \le a$$

$$v' = \theta = \frac{M_0}{6LEI}\left(6aL - 3a^2 - 2L^2 - 3x^2\right), \qquad\qquad 0 \le x \le a$$

$$v = \frac{M_0}{6LEI}\left(3a^2 L - 3a^2 x - 2L^2 x + 3Lx^2 - x^3\right), \qquad\qquad a \le x \le L$$

$$v' = \theta = -\frac{M_0}{6LEI}\left(3a^2 + 2L^2 - 6Lx + 3x^2\right), \qquad\qquad a \le x \le L$$

$$\theta_A = \frac{M_0}{6LEI}\left(6aL - 3a^2 - 2L^2\right), \qquad \theta_B = -\frac{M_0}{6LEI}\left(3a^2 - L^2\right).$$

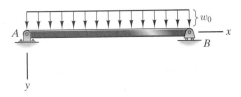

$$v = \frac{w_0 x}{24EI}\left(L^3 - 2Lx^2 + x^3\right), \qquad v' = \theta = \frac{w_0}{24EI}\left(L^3 - 6Lx^2 + 4x^3\right),$$

$$\theta_A = -\theta_B = \frac{w_0 L^3}{24EI}.$$

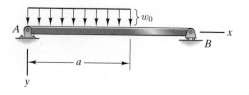

$$v = \frac{w_0 x}{24LEI}\left(a^4 - 4a^3L + 4a^2L^2 + 2a^2x^2 - 4aLx^2 + Lx^3\right), \qquad 0 \le x \le a$$

$$v' = \frac{w_0}{24LEI}\left(a^4 - 4a^3L + 4a^2L^2 + 6a^2x^2 - 12aLx^2 + 4Lx^3\right), \qquad 0 \le x \le a$$

$$v = \frac{w_0 a^2}{24LEI}\left(-a^2L + 4L^2x + a^2x - 6Lx^2 + 2x^3\right), \qquad a \le x \le L$$

$$v' = \theta = \frac{w_0 a^2}{24LEI}\left(4L^2 + a^2 - 12Lx + 6x^2\right), \qquad a \le x \le L$$

$$\theta_A = \frac{w_0 a^2}{24LEI}\left(2L - a\right)^2, \qquad \theta_B = -\frac{w_0 a^2}{24LEI}\left(2L^2 - a^2\right).$$

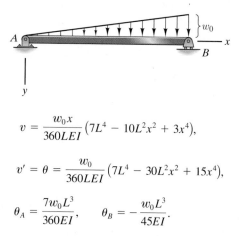

$$v = \frac{w_0 x}{360LEI}\left(7L^4 - 10L^2x^2 + 3x^4\right),$$

$$v' = \theta = \frac{w_0}{360LEI}\left(7L^4 - 30L^2x^2 + 15x^4\right),$$

$$\theta_A = \frac{7w_0 L^3}{360EI}, \qquad \theta_B = -\frac{w_0 L^3}{45EI}.$$

Cantilever Beams

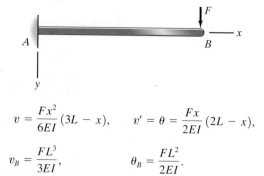

$$v = \frac{Fx^2}{6EI}\left(3L - x\right), \qquad v' = \theta = \frac{Fx}{2EI}\left(2L - x\right),$$

$$v_B = \frac{FL^3}{3EI}, \qquad \theta_B = \frac{FL^2}{2EI}.$$

$$v = \frac{Fx^2}{6EI}(3a - x), \qquad v' = \theta = \frac{Fx}{2EI}(2a - x), \qquad 0 \leq x \leq a$$

$$v = \frac{Fa^2}{6EI}(3x - a), \qquad v' = \theta = \frac{Fa^2}{2EI}, \qquad\qquad a \leq x \leq L$$

$$v_B = \frac{Fa^2}{6EI}(3L - a), \qquad \theta_B = \frac{Fa^2}{2EI}.$$

$$v = \frac{M_0 x^2}{2EI}, \qquad v' = \theta = \frac{M_0 x}{EI},$$

$$v_B = \frac{M_0 L^2}{2EI}, \qquad \theta_B = \frac{M_0 L}{EI}.$$

$$v = \frac{M_0 x^2}{2EI}, \qquad\qquad v' = \theta = \frac{M_0 x}{EI}, \qquad 0 \leq x \leq a$$

$$v = \frac{M_0 a}{2EI}(2x - a), \qquad v' = \theta = \frac{M_0 a}{EI}, \qquad a \leq x \leq L$$

$$v_B = \frac{M_0 a}{2EI}(2L - a), \qquad \theta_B = \frac{M_0 a}{EI}.$$

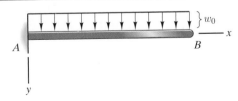

$$v = \frac{w_0 x^2}{24EI}(6L^2 - 4Lx + x^2), \qquad v' = \theta = \frac{w_0 x}{6EI}(3L^2 - 3Lx + x^2),$$

$$v_B = \frac{w_0 L^4}{8EI}, \qquad \theta_B = \frac{w_0 L^3}{6EI}.$$

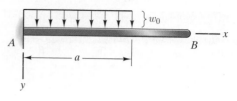

$$v = \frac{w_0 x^2}{24EI}\left(6a^2 - 4ax + x^2\right), \qquad\qquad 0 \le x \le a$$

$$v' = \theta = \frac{w_0 x}{6EI}\left(3a^2 - 3ax + x^2\right), \qquad\qquad 0 \le x \le a$$

$$v = \frac{w_0 a^3}{24EI}(4x - a), \qquad v' = \theta = \frac{w_0 a^3}{6EI}, \qquad a \le x \le L$$

$$v_B = \frac{w_0 a^3}{24EI}(4L - a), \qquad \theta_B = \frac{w_0 a^3}{6EI}.$$

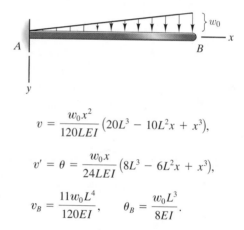

$$v = \frac{w_0 x^2}{120LEI}\left(20L^3 - 10L^2 x + x^3\right),$$

$$v' = \theta = \frac{w_0 x}{24LEI}\left(8L^3 - 6L^2 x + x^3\right),$$

$$v_B = \frac{11 w_0 L^4}{120EI}, \qquad \theta_B = \frac{w_0 L^3}{8EI}.$$

$$v = \frac{w_0 x^2}{120LEI}\left(10L^3 - 10L^2 x + 5Lx^2 - x^3\right),$$

$$v' = \theta = \frac{w_0 x}{24LEI}\left(4L^3 - 6L^2 x + 4Lx^2 - x^3\right),$$

$$v_B = \frac{w_0 L^4}{30EI}, \qquad \theta_B = \frac{w_0 L^3}{24EI}.$$

Answers to Even-Numbered Problems

Chapter 1

1.2 2.7183.
1.4 7.32 m wide, 2.44 m high.
1.6 The 1-in. wrench fits the 25-mm nut.
1.8 149 mi/hr.
1.10 (a) 5000 m/s; (b) 3.11 mi/s.
1.12 $g = 32.2$ ft/s^2.
1.14 0.310 m^2.
1.16 2.07×10^6 Pa.
1.18 $G = 3.44 \times 10^{-8}$ lb-ft^2/slug^2.
1.20 (a) The SI units of T are kg-m^2/s^2;
(b) $T = 73.8$ slug-ft^2/s^2.
1.22 (a) N/m^3. (b) $\gamma = 62.4$ lb/ft^3.
1.24 (a) 491 N; (b) 81.0 N.
1.26 163 lb.
1.28 32.1 km.
1.30 345,000 km.

Chapter 2

2.2 $|\mathbf{F}_B| = 52$ N.
2.4 $|\mathbf{F}_B| = 52.1$ N.
2.6 $|\mathbf{r}_{AC}| = 199$ mm.
2.8 $|\mathbf{F}_{AB}| = 117.0$ kN, $|\mathbf{F}_{AC}| = 62.2$ kN.
2.10 $|\mathbf{F}| = 7.02$ kN.
2.12 $AB : 1202$ lb. $AD : 559$ lb.
2.14 (a) $|\mathbf{r}_A + \mathbf{r}_B| = 70$ m; (b) $|\mathbf{r}_A + \mathbf{r}_B| = 50$ m.
2.16 $|\mathbf{F}_{BA} + \mathbf{F}_{BC}| = 918$ N.
2.18 $|\mathbf{r}| = 390$ m, $\alpha = 21.2°$.
2.22 $F_y = -102$ MN.
2.24 $|\mathbf{F}| = 447$ kip.
2.26 $V_x = 16, V_y = 12$ or $V_x = -16, V_y = -12$.
2.28 $\mathbf{F} = 56.4\mathbf{i} + 20.5\mathbf{j}$ (lb).
2.30 $\mathbf{r}_{AB} = -4\mathbf{i} - 3\mathbf{j}$ (m).
2.32 $\mathbf{r}_{AB} - \mathbf{r}_{BC} = \mathbf{i} - 1.73\mathbf{j}$ (m)
2.34 (a) $\mathbf{r}_{AB} = 48\mathbf{i} + 15\mathbf{j}$ (in.); (b) $\mathbf{r}_{BC} = -53\mathbf{i} + 5\mathbf{j}$ (in.);
(c) $|\mathbf{r}_{AB} + \mathbf{r}_{BC}| = 20.6$ in.
2.36 (a) $\mathbf{r}_{AB} = 52.0\mathbf{i} + 30\mathbf{j}$ (mm);
(b) $\mathbf{r}_{AB} = -42.4\mathbf{i} - 42.4\mathbf{j}$ (mm).
2.38 $x_B = 785$ m, $y_B = 907$ m
or $x_B = 255$ m, $y_B = 1173$ m.
2.40 $\mathbf{e}_{AC} = -0.757\mathbf{i} + 0.653\mathbf{j}$.
2.42 $\mathbf{e} = \frac{3}{5}\mathbf{i} - \frac{4}{5}\mathbf{j}$.
2.44 $\mathbf{F} = -937\mathbf{i} + 750\mathbf{j}$ (N).
2.46 $\mathbf{e}_{EM} = 0.609\mathbf{i} - 0.793\mathbf{j}$.
2.48 $|\mathbf{F}_{BA} + \mathbf{F}_{BC}| = 918$ N.
2.50 $|\mathbf{F}_A| = 1720$ lb, $\alpha = 33.3°$.
2.52 $\alpha = 36.4°$.
2.54 $|\mathbf{F}_A| = 10$ kN, $|\mathbf{F}_D| = 8.66$ kN.
2.56 $|\mathbf{L}| = 216.1$ lb, $|\mathbf{D}| = 78.7$ lb.

2.58 $|\mathbf{F}_A| = 68.2$ kN.
2.60 $|\mathbf{F}_{AC}| = 2.11$ kN, $|\mathbf{F}_{AD}| = 2.76$ kN.
2.62 $x = 75 - 0.880s, y = 12 + 0.476s$.
2.64 $\mathbf{r} = (0.814s - 6)\mathbf{i} + (0.581s + 1)\mathbf{j}$ (m).
2.66 $|\mathbf{F}| = 110$ N.
2.68 $U_x = 3.61, U_y = -7.22, U_z = -28.89$ or $U_x = -3.61$,
$U_y = 7.22, U_z = 28.89$.
2.70 (a) $|\mathbf{U}| = 7, |\mathbf{V}| = 13$; (b) $|3\mathbf{U} + 2\mathbf{V}| = 27.5$.
2.72 $\theta_x = 56.9°, \theta_y = 129.5°, \theta_z = 56.9°$.
2.74 $\mathbf{F} = -0.5\mathbf{i} + 0.2\mathbf{j} + 0.843\mathbf{k}$.
2.76 (a) 11 ft; (b) $\cos\theta_x = -0.545, \cos\theta_y = 0.818$,
$\cos\theta_z = 0.182$.
2.78 (a) 5.39 N; (b) $0.557\mathbf{i} - 0.743\mathbf{j} - 0.371\mathbf{k}$.
2.80 $\mathbf{F} = 40\mathbf{i} + 40\mathbf{j} - 70\mathbf{k}$ (kN).
2.82 (a) $|\mathbf{r}_{AB}| = 16.2$ m; (b) $\cos\theta_x = 0.615$,
$\cos\theta_y = -0.492, \cos\theta_z = -0.615$.
2.84 $\mathbf{r}_{AR}$: $\cos\theta_x = 0.667, \cos\theta_y = 0.667, \cos\theta_z = 0.333$.
$\mathbf{r}_{BR}$: $\cos\theta_x = -0.242, \cos\theta_y = 0.970, \cos\theta_z = 0$.
2.86 29,100 ft.
2.88 $\mathbf{r} = 70.7\mathbf{i} + 61.2\mathbf{j} + 35.4\mathbf{k}$ (in.).
2.90 $\mathbf{r}_{OP} = R_E(0.612\mathbf{i} + 0.707\mathbf{j} + 0.354\mathbf{k})$.
2.92 (a) $\mathbf{e}_{BC} = -0.286\mathbf{i} - 0.857\mathbf{j} + 0.429\mathbf{k}$.
(b) $\mathbf{F} = -2.29\mathbf{i} - 6.86\mathbf{j} + 3.43\mathbf{k}$ (kN).
2.94 $\cos\theta_x = -0.703, \cos\theta_y = 0.592, \cos\theta_z = 0.394$.
2.96 259 lb.
2.98 $|\mathbf{F}_{AC}| = 1116$ N, $|\mathbf{F}_{AD}| = 910$ N.
2.100 $\mathbf{T} = -15.4\mathbf{i} + 27.0\mathbf{j} + 7.7\mathbf{k}$ (lb).
2.102 $\mathbf{T} = -41.1\mathbf{i} + 28.8\mathbf{j} + 32.8\mathbf{k}$ (N).
2.104 $\mathbf{U} \cdot \mathbf{V} = -300$.
2.106 -250 ft-lb.
2.108 $U_x = 2.857, V_y = 0.857, W_z = -3.143$.
2.112 81.6°.
2.114 $\theta = 53.5°$.
2.116 Parallel component is $12\mathbf{i} - 4\mathbf{j} + 6\mathbf{k}$ (kN), normal
component is $9\mathbf{i} + 18\mathbf{j} - 6\mathbf{k}$ (kN).
2.118 (a) 42.5°; (b) $-423\mathbf{j} + 604\mathbf{k}$ (lb).
2.120 $\mathbf{F}_p = 5.54\mathbf{j} + 3.69\mathbf{k}$ (N),
$\mathbf{F}_n = 10\mathbf{i} + 6.46\mathbf{j} - 9.69\mathbf{k}$ (N).
2.122 $\mathbf{T}_n = -37.1\mathbf{i} + 31.6\mathbf{j} + 8.2\mathbf{k}$ (N).
2.124 $\mathbf{F}_p = -0.1231\mathbf{i} + 0.0304\mathbf{j} - 0.1216\mathbf{k}$ (lb).
2.126 $\mathbf{v}_p = -1.30\mathbf{i} - 1.68\mathbf{j} - 3.36\mathbf{k}$ (m/s).
2.128 $\mathbf{U} \times \mathbf{V} = -82\mathbf{i} - 60\mathbf{j} + 74\mathbf{k}$.
2.130 $\mathbf{r} \times \mathbf{F} = -80\mathbf{i} + 120\mathbf{j} - 40\mathbf{k}$ (N-m).
2.132 (a) $\mathbf{U} \times \mathbf{V} = \mathbf{0}$; (b) They are parallel.
2.134 (a), (c) $\mathbf{U} \times \mathbf{V} = -51.8\mathbf{k}$; (b), (d) $\mathbf{V} \times \mathbf{U} = 51.8\mathbf{k}$.
2.138 (a) $\mathbf{r}_{OA} \times \mathbf{r}_{OB} = -4\mathbf{i} + 36\mathbf{j} + 32\mathbf{k}$ (m^2);
(b) $-0.083\mathbf{i} + 0.745\mathbf{j} + 0.662\mathbf{k}$ or
$0.083\mathbf{i} - 0.745\mathbf{j} - 0.662\mathbf{k}$.
2.140 $\mathbf{r}_{AB} \times \mathbf{F} = -2400\mathbf{i} + 9600\mathbf{j} + 7200\mathbf{k}$ (ft-lb).
2.142 $\mathbf{r}_{CA} \times \mathbf{T} = -4.72\mathbf{i} - 3.48\mathbf{j} - 7.96\mathbf{k}$ (N-m).
2.144 $x_B = 2.81$ m, $y_B = 6.75$ m, $z_B = 3.75$ m.

2.146 $\mathbf{U} \cdot (\mathbf{V} \times \mathbf{W}) = -4$.

2.148 1.8×10^6 mm².

2.150 $U_y = -2$.

2.152 $|\mathbf{A}| = 1110$ lb, $\alpha = 29.7°$.

2.154 $\mathbf{e}_{AB} = 0.625\mathbf{i} - 0.469\mathbf{j} - 0.625\mathbf{k}$.

2.156 $\mathbf{F}_p = 8.78\mathbf{i} - 6.59\mathbf{j} - 8.78\mathbf{k}$ (lb).

2.158 $\mathbf{F}_A = 18.2\mathbf{i} + 19.9\mathbf{j} + 15.3\mathbf{k}$ (N),
$\mathbf{F}_B = -7.76\mathbf{i} + 26.9\mathbf{j} + 13.4\mathbf{k}$ (N).

2.160 $\mathbf{r}_{BC} \times \mathbf{T} = -12.0\mathbf{i} - 138.4\mathbf{j} - 117.4\mathbf{k}$ (N-m).

Chapter 3

3.2 $F_2 = 86.6$ N, $F_3 = 50$ N.

3.4 $A_y = 267$ kN, $C = 154$ kN.

3.6 $T = 763$ N, $M = 875$ N.

3.8 $k = 1960$ N/m, $m_A = 4$ kg, $m_B = 6$ kg.

3.10 Normal force = 196,907 N, friction force = 707 N.

3.12 $\alpha = 31.0°$.

3.14 (a) 254 lb; (b) 41.8°.

3.16 150 lb.

3.18 116 N.

3.20 $T_{\text{left}} = 299$ lb, $T_{\text{right}} = 300$ lb.

3.22 188 N.

3.24 (a) 56.4 lb. (b) 340.3 lb.

3.26 No. The tension in cables BD and CE would be 4.14 kN.

3.28 Upper cable tension is $0.828W$, lower cable tension is $0.132W$.

3.30 $T_{AB} = 1.21$ N, $T_{AD} = 2.76$ N.

3.32 $m = 12.2$ kg.

3.34 $k = 2250$ N/m.

3.36 $h = b$.

3.38 $T_{AB} = 688$ lb.

3.40 AB: 64.0 kN, BC: 61.0 kN.

3.44 $T = 196$ N, $\alpha = 53.1°$.

3.46 $T = 1330$ lb.

3.48 (b) Left surface: 36.6 lb; right surface: 25.9 lb.

3.50 202 N.

3.52 Normal force = 13.29 kN, friction force = 4.19 kN.

3.54 $T = m_A g/7 - (4/7) \, mg$.

3.56 $x = \frac{1}{2}(b - h \cot 30°)$, $y = -\frac{1}{2}(b \tan 30° - h)$.

3.60 $T_{AB} = 780$ N, $T_{AC} = 1976$ N, $T_{AD} = 2568$ N.

3.62 $T_{AC} = 20.6$ lb, $T_{AD} = 21.4$ lb, $T_{AE} = 11.7$ lb.

3.64 Two at B, three at C, and three at D.

3.66 $T_{AB} = 10,270$ lb, $T_{AC} = 4380$ lb, $T_{AD} = 11,010$ lb.

3.68 $D = 1176$ N, $T_{OA} = 6774$ N.

3.70 12.3 lb.

3.72 $T_{EF} = T_{EG} = 738$ kN.

3.74 (a) The tension = 2.70 kN;
(b) The force exerted by the bar is $1.31\mathbf{i} - 1.31\mathbf{k}$ (kN).

3.76 $T_{AB} = 357$ N.

3.78 $F = 36.6$ N.

3.80 $W = 25.0$ lb.

3.82 (a) 83.9 lb; (b) 230.5 lb.

3.84 $T = mg/26$.

3.86 $F = 162.0$ N.

3.88 Normal force = 12.15 kN, friction force = 4.03 kN.

Chapter 4

4.2 (a) 28 N-m. (b) −8 N-m.

4.4 Direction shown, 40.5 N-m; perpendicular, 45 N-m.

4.6 $\alpha = 61.0°$.

4.8 L = 2.4 m.

4.10 (a) 1 m; (b) 53.1° or 180°.

4.12 229 ft-lb.

4.14 $M_S = 611$ in-lb.

4.16 (a)–(c) Zero.

4.18 $G = 1400$ lb.

4.20 $F_1 = -30$ kN, $F_2 = 50$ kN.

4.22 (a) $F_A = 24.6$ N, $F_B = 55.4$ N; (b) Zero.

4.24 $T = 1.2$ kN.

4.26 $M = 2.39$ kN-m.

4.28 (a) $A_x = 18.1$ kN, $A_y = -29.8$ kN, $B = -20.4$ kN; (b) Zero.

4.30 (a) $A_x = 300$ lb, $A_y = 240$ lb, $B = 280$ lb; (b) Zero.

4.32 186 kg.

4.34 −22.3 ft-lb.

4.36 $M = -2340$ N-m.

4.38 671 lb.

4.40 617 kN-m.

4.42 1040 lb.

4.44 $M_A = -3.00$ kN-m, $M_D = 7.50$ kN-m.

4.46 (a), (b) $480\mathbf{k}$ (N-m).

4.48 (a) $800\mathbf{k}$ (kN-m); (b) $-400\mathbf{k}$ (kN-m).

4.50 $\mathbf{F} = 20\mathbf{i} + 40\mathbf{j}$ (N).

4.52 (a), (b) Zero.

4.54 (a), (b) 1270 N-m.

4.56 $|\mathbf{M}_P| = 502$ N-m, $D = 7.18$ m.

4.58 $\mathbf{F} = 40\mathbf{i} + 40\mathbf{j} + 70\mathbf{k}$ (N) or
$\mathbf{F} = -40\mathbf{i} - 40\mathbf{j} - 70\mathbf{k}$ (N).

4.60 58.0 kN.

4.62 (a) $|\mathbf{F}| = 1586$ N. (b) $|\mathbf{F}| = 1584$ N.

4.64 $-16.4\mathbf{i} - 111.9\mathbf{k}$ (N-m).

4.66 1540 ft-lb.

4.68 $\mathbf{M}_D = 1.25\mathbf{i} + 1.25\mathbf{j} - 6.25\mathbf{k}$ (kN-m).

4.70 $T_{AC} = 2.23$ kN, $T_{AD} = 2.43$ kN.

4.72 $T_{AB} = 1.60$ kN, $T_{AC} = 1.17$ kN.

4.74 $F = 2530$ lb.

4.76 $\mathbf{M} = 482\mathbf{k}$ (kN-m).

4.78 (a) $\mathbf{M}_{(x\text{axis})} = 80\mathbf{i}$ (N-m).
(b) $\mathbf{M}_{(y\text{axis})} = -140\mathbf{j}$ (N-m). (c) $\mathbf{M}_{(z\text{axis})} = \mathbf{0}$.

4.80 (a) Zero; (b) $2.7\mathbf{k}$ (kN-m).

4.82 $F_1 = 200$ lb, $F_2 = 100$ lb, $F_3 = 200$ lb.

4.84 $\mathbf{F} = 80\mathbf{i} + 80\mathbf{j} + 40\mathbf{k}$ (lb).

4.86 $-16.4\mathbf{i}$ (N-m).

4.88 (a), (b) $\mathbf{M}_{AB} = -76.1\mathbf{i} - 95.1\mathbf{j}$ (N-m).

4.90 $\mathbf{M}_{AO} = 119.1\mathbf{j} + 79.4\mathbf{k}$ (N-m).

4.92 $\mathbf{M}_{AB} = 77.1\mathbf{j} - 211.9\mathbf{k}$ (ft-lb).

4.94 $\mathbf{M}_{(y\,\text{axis})} = 215\mathbf{j}$ (N-m).

4.96 $\mathbf{M}_{(x\,\text{axis})} = 44\mathbf{i}$ (N-m).

4.98 $-338\mathbf{j}$ (ft-lb).

4.100 $|\mathbf{F}| = 13$ lb.

4.102 $\mathbf{M}_{(\text{axis})} = -478\mathbf{i} - 174\mathbf{k}$ (N-m).

4.104 1 N-m.

4.106 $124\mathbf{k}$ (ft-lb).

4.108 40 N-m counterclockwise, or $40\mathbf{k}$ (N-m).

4.110 (a) b = 3.84 m. (b) $-110\mathbf{k}$ (N-m).

4.112 (a), (b) $-400\mathbf{k}$ (N-m).

4.114 40 ft-lb clockwise, or $-40\mathbf{k}$ (ft-lb).

4.116 2200 ft-lb clockwise.

4.118 330 N-m counterclockwise, or $330\mathbf{k}$ (N-m).

4.120 (a) $\mathbf{M} = 12\mathbf{i} + 88\mathbf{j} - 216\mathbf{k}$ (N-m). (b) 4.85 m.

4.122 356 ft-lb.

4.124 $\mathbf{M}_P = 3\mathbf{i} - 2\mathbf{j} + 2\mathbf{k}$ (kN-m).

4.126 $M_{Cy} = 7$ kN-m, $M_{Cz} = -2$ kN-m.

4.128 Yes.

4.130 Systems 1, 2, and 4 are equivalent.

4.134 $F = 265$ N.

4.136 $F = 70$ lb, $M = 130$ in-lb.

4.138 (a) $\mathbf{F} = -10\mathbf{j}$ (lb), $M = -10$ ft-lb; (b) $D = 1$ ft.

4.140 (a) $A_x = 0$, $A_y = 20$ lb, $B = 80$ lb;
(b) $\mathbf{F} = 100\mathbf{j}$ (lb), $M = -1120$ in-lb.

4.142 (a) $A_x = 12$ kip, $A_y = 10$ kip, $B = -10$ kip;
(b) $\mathbf{F} = -12\mathbf{i}$ (kip), intersects at $y = 5$ ft;
(c) they are both zero.

4.144 $\mathbf{F} = 161\mathbf{i}$ (kN), $y = -0.0932$ m.

4.146 $\mathbf{F} = 100\mathbf{j}$ (lb), $\mathbf{M} = \mathbf{0}$.

4.148 (a) $\mathbf{F} = 920\mathbf{i} - 390\mathbf{j}$ (N), $M = -419$ N-m;
(b) intersects at $y = 456$ mm.

4.150 $\mathbf{F} = 800\mathbf{j}$ (lb), intersects at $x = 7.5$ in.

4.152 (a) $-360\mathbf{k}$ (in-lb); (b) $-36\mathbf{j}$ (in-lb);
(c) $\mathbf{F} = 10\mathbf{i} - 30\mathbf{j} + 3\mathbf{k}$ (lb),
$\mathbf{M} = -36\mathbf{j} - 360\mathbf{k}$ (in-lb).

4.154 (a) $\mathbf{F} = 600\mathbf{i}$ (lb), $\mathbf{M} = 1400\mathbf{j} - 1800\mathbf{k}$ (ft-lb);
(b) $\mathbf{F} = 600\mathbf{i}$ (lb), intersects at $y = 3$ ft, $z = 2.33$ ft.

4.156 $\mathbf{F} = 100\mathbf{j} + 80\mathbf{k}$ (N), $\mathbf{M} = 240\mathbf{j} - 300\mathbf{k}$ (N-m).

4.158 (a) $\mathbf{F} = \mathbf{0}$, $\mathbf{M} = rA\mathbf{i}$; (b) $\mathbf{F}' = \mathbf{0}$, $\mathbf{M}' = rA\mathbf{i}$.

4.160 (a) $\mathbf{F} = \mathbf{0}$, $\mathbf{M} = 4.60\mathbf{i} + 1.86\mathbf{j} - 3.46\mathbf{k}$ (kN-m);
(b) 6.05 kN-m.

4.162 $\mathbf{F} = -20\mathbf{i} + 20\mathbf{j} + 10\mathbf{k}$ (lb),
$\mathbf{M} = 50\mathbf{i} + 250\mathbf{j} + 100\mathbf{k}$ (in-lb).

4.164 (a) $\mathbf{F} = 28\mathbf{k}$ (kip), $\mathbf{M} = 96\mathbf{i} - 192\mathbf{j}$ (ft-kip);
(b) $x = 6.86$ ft, $y = 3.43$ ft.

4.166 $\mathbf{F} = 100\mathbf{i} + 20\mathbf{j} - 20\mathbf{k}$ (N),
$\mathbf{M} = -143\mathbf{i} + 406\mathbf{j} - 280\mathbf{k}$ (N-m).

4.168 $\mathbf{M}_t = \mathbf{0}$, line of action intersects at $y = 0$, $z = 2$ ft.

4.170 $x = 2.41$ m, $y = 3.80$ m.

4.172 $\mathbf{F} = 40.8\mathbf{i} + 40.8\mathbf{j} + 81.6\mathbf{k}$ (N),
$\mathbf{M} = -179.6\mathbf{i} + 391.9\mathbf{j} - 32.7\mathbf{k}$ (N-m).

4.174 (a) $320\mathbf{i}$ (in-lb);
(b) $\mathbf{F} = -20\mathbf{k}$ (lb), $\mathbf{M} = 320\mathbf{i} + 660\mathbf{j}$ (in-lb);
(c) $\mathbf{M}_t = \mathbf{0}$, $x = 33$ in., $y = -16$ in.

4.176 $|\mathbf{F}| = 224$ lb, $|\mathbf{M}| = 1600$ ft-lb.

4.178 671 lb.

4.180 $-228.1\mathbf{i} - 68.4\mathbf{k}$ (N-m).

4.182 $\mathbf{M}_{CD} = -173\mathbf{i} + 1038\mathbf{k}$ (ft-lb).

4.184 $\mathbf{F} = 1166\mathbf{i} + 566\mathbf{j}$ (N), $y = 13.9$ m.

Chapter 5

5.2 (b) $A_x = 0$, $A_y = 2$ kN, $M_A = -2$ kN-m.

5.4 $A_x = 0$, $A_y = -5$ kN, $B = 10$ kN.

5.6 $A_x = 0$, $A_y = -1.85$ kN, $B = 2.74$ kN.

5.8 (b) $A_x = 0$, $A_y = 5$ kN, $B = 5$ kN.

5.10 (b) $A = 100$ lb, $B = 200$ lb.

5.12 (b) $A_x = 1.15$ kN, $A_y = 0$, $B = 2.31$ kN.

5.14 $A_x = -26.7$ kN, $B_x = 26.7$ kN, $B_y = -40$ kN.

5.16 (a) 293.3 N. (b) 99.1 N.

5.18 (b) $A_x = 0$, $A_y = -1000$ lb, $M_A = -12,800$ ft-lb.

5.20 $F = 18.38$ kN.

5.22 5.93 kN.

5.24 $R = 12.5$ lb, $B_x = 11.3$ lb, $B_y = 15.3$ lb.

5.26 (a) $A = 53.8$ lb, $B = 46.2$ lb; (b) $F = 21.2$ lb.

5.28 $A = 9211$ N, $B_x = 0$, $B_y = 789$ N.

5.30 $T = 4.71$ lb.

5.34 Tension is 50 lb, $C_x = -43.3$ lb, $C_y = 25$ lb.

5.36 $A_x = 0$, $A_y = 1.5F$, $B = 2.5F$.

5.38 $A_x = -200$ lb, $A_y = -100$ lb, $M_A = 1600$ ft-lb.

5.40 $0.354W$.

5.42 $A_x = 3.46$ kN, $A_y = -2$ kN, $B_x = -3.46$ kN,
$B_y = 2$ kN.

5.44 $T = 392$ N, $A_x = 340$ N, $A_y = 196$ N.

5.46 $A_x = -1.57$ kN, $A_y = 1.57$ kN, $E = 1.57$ kN.

5.48 $A_x = 0$, $A_y = 200$ lb, $M_A = 900$ ft-lb.

5.50 $A_x = 57.7$ lb, $A_y = -13.3$ lb, $B = 15.3$ lb.

5.52 $W = 15$ kN.

5.54 $k = 13,500$ N/m.

5.56 $0.612W$.

5.58 20.3 kN.

5.60 $W_2 = 2484$ lb, $A_x = -2034$ lb, $A_y = 2425$ lb.

5.62 $W = 46.2$ N, $A_x = 22.3$ N, $A_y = 61.7$ N.

5.64 $F = 44.5$ lb, $A_x = 25.3$ lb, $A_y = -1.9$ lb.

5.72 (1) and (2) are improperly supported. For (3), reactions
are $A = F/2$, $B = F/2$, $C = F$.

5.74 (b) $A_x = -6.53$ kN, $A_y = -3.27$ kN, $A_z = 3.27$ kN,
$M_{Ax} = 0$, $M_{Ay} = -6.53$ kN-m,
$M_{Az} = -6.53$ kN-m.

5.76 $T_{BC} = 20.3$ kN.

5.78 $O_x = \pm21.6$ kN, $O_y = 0.6$ kN, $O_z = 0$,
$M_{Ox} = -4.8$ kN-m, $M_{Oy} = \pm172.5$ kN-m,
$M_{Oz} = \pm172.5$ kN-m.

5.80 (a) $-17.8\mathbf{i} - 62.8\mathbf{k}$ (N-m). (b) $A_x = 0$, $A_y = 360$ N,
$A_z = 0$, $M_{Ax} = 17.8$ N-m, $M_{Ay} = 0$, $M_{Az} = 62.8$ N-m.

5.82 $O_x = \pm900$ N, $O_y = \pm900$ N, $O_z = 0$,
$M_{Ox} = \pm135$ N-m, $M_{Oy} = \pm135$ N-m,
$M_{Oz} = \pm288$ N-m.

5.84 $|\mathbf{F}| = 10.7$ kN.

5.86 $T_{AB} = 553$ lb, $T_{AC} = 289$ lb, $O_x = 632$ lb, $O_y = 574$ lb, $O_z = 0$.

5.88 $T_A = 3.72$ kN, $T_B = 2.60$ kN, $T_C = 1.53$ kN.

5.90 $T_A = 54.7$ lb, $T_B = 22.7$ lb, $T_C = 47.7$ lb.

5.92 $\mathbf{F} = 4\mathbf{j}$ (kN) at $x = 0$, $z = 0.15$ m.

5.94 (b) $A_x = -0.74$ kN, $A_y = 1$ kN, $A_z = -0.64$ kN, $B_x = 0.74$ kN, $B_z = 0.64$ kN.

5.96 $F_y = 34.5$ lb.

5.98 $T_{BD} = 1.47$ kN, $T_{BE} = 1.87$ kN, $A_x = 0$, $A_y = 4.24$ kN, $A_z = 0$.

5.100 $F = 22.5$ kN.

5.102 Tension is 60 N, $B_x = -10$ N, $B_y = 90$ N, $B_z = 10$ N, $M_{By} = 1$ N-m, $M_{Bz} = -3$ N-m.

5.104 Tension is 60 N, $B_x = -10$ N, $B_y = 75$ N, $B_z = 15$ N, $C_y = 15$ N, $C_z = -5$ N.

5.106 $A_x = -2.86$ kip, $A_y = 17.86$ kip, $A_z = -8.10$ kip, $B_y = 3.57$ kip, $B_z = 12.38$ kip.

5.108 $E_x = 0.67$ kN, $E_y = -1.33$ kN, $E_z = 2.67$ kN, $F_x = 4.67$ kN, $F_y = 6.67$ kN.

5.110 $|\mathbf{A}| = 8.54$ kN, $|\mathbf{B}| = 10.75$ kN.

5.112 $C_y = 0$, $A_y = 66.7$ lb.

5.114 $T_{AB} = 488$ lb, $T_{CD} = 373$ lb, reaction is $31\mathbf{i} + 823\mathbf{j} - 87\mathbf{k}$ (lb).

5.116 $A_x = 474$ N, $A_y = -825$ N, $A_z = -1956$N; $B_x = 860$ N, $B_y = 2380$ N, $B_z = -44$ N.

5.118 $\alpha = 52.4°$.

5.120 Tension is 33.3 lb; magnitude of reaction is 44.1 lb.

5.122 $\alpha = 73.9°$, magnitude at A is 4.32 kN, magnitude at B is 1.66 kN.

5.124 (a) No, because of the 3 kN-m couple; (b) magnitude at A is 7.88 kN; magnitude at B is 6.66 kN; (c) no.

5.126 (b) $A_x = -8$ kN, $A_y = 2$ kN, $C = 8$ kN.

5.130 (a) There are four unknown reactions and three equilibrium equations; (b) $A_x = -50$ lb, $B_x = 50$ lb.

5.132 (b) Force on nail = 55 lb, normal force = 50.77 lb, friction force = 9.06 lb.

5.134 $A = 500$ N, $B_x = 0$, $B_y = -800$ N.

5.136 $A = 727$ lb, $H_x = 225$ lb, $H_y = 113$ lb.

5.138 $\alpha = 90°$, $T_{BC} = W/2$, $A = W/2$.

Chapter 6

6.2 (a) $A = 13.3$ kN, $B_x = -13.3$ kN, $B_y = 10$ kN; (b) AB: zero; BC: 16.7 kN (T); AC: 13.3 kN (C).

6.4 AB: 2.839 kN (T); AC: 0.926 kN (C); BC: 0.961 kN (C).

6.6 AB: 16.7 kN (T); AC: 13.3 kN (C); BC: 20 kN (C); BD: 16.7 kN (T); CD: 13.3 kN (C).

6.8 (a) Howe, $2F$ in members GH and HI; (b) they are the same: $2.12F$ in members AB and DE.

6.10 DF: 14.7 kN (C); EF: 5 kN (C); FG: zero.

6.12 AB: 13.75 kN (T); BC: zero; BD: 7.5 kN (T).

6.14 $F = 5.09$ kN.

6.16 DE: 800 lb (C); DF: 447 lb (C); DG: 894 lb (T).

6.18 1.56 kN.

6.20 AB: 375 lb (C); AC: 625 lb (T); BC: 300 lb (T).

6.22 BC: 90.1 kN (T); CD: 90.1 kN (C); CE: 300 kN (T).

6.24 BC: 1200 kN (C); BI: 300 kN (T); BJ: 636 kN (T).

6.26 AB: 2520 lb (C); BC: 2160 lb (C); CD: 1680 lb (C).

6.30 141 kN (C).

6.32 AB: 1.33F (C); BC: 1.33F (C); CE: 1.33F (T).

6.34 EG: 32 kN (T); EF: 5 kN (C); DF: 28 kN (C).

6.38 96.2 kN (T).

6.40 55.5 kN.

6.42 AC: 3.33 kN (T); BC: 1.18 kN (C); BD: 3.33 kN (C).

6.44 2.50 kN (C).

6.46 3.33 kip (C).

6.48 (a) 1160 lb (C).

6.50 IL: 16 kN (C); KM: 24 kN (T).

6.52 $A_x = 100$ N, $A_y = 100$ N.

6.54 $A_x = 57.2$ lb, $A_y = 42.8$ lb, $M_A = 257$ ft-lb, $B_x = -57.2$ lb, $B_y = -42.8$ lb.

6.56 $F = 50$ kN.

6.58 The largest lifting force is 8.94 kN. Axial force is 25.30 kN.

6.60 $D_x = -1475$ N, $D_y = -516$ N, $E_x = 0$, $E_y = -516$ N, $M_E = 619$ N-m.

6.62 $A_x = -2.35$ kN, $A_y = 2.35$ kN, $B_x = 0$, $B_y = -4.71$ kN, $C_x = 2.35$ kN, $C_y = 2.35$ kN.

6.64 $A_x = -400$ lb, $A_y = -100$ lb, tension = 361 lb, $C_x = 200$ lb, $C_y = -300$ lb, $D = 100$ lb.

6.66 $B_x = -400$ lb $B_y = -300$ lb, $C_x = 400$ lb, $C_y = 200$ lb, $D_x = 0$, $D_y = 100$ lb.

6.68 $A_x = -150$ lb, $A_y = 120$ lb, $B_x = 180$ lb, $B_y = -30$ lb, $D_x = -30$ lb, $D_y = -90$ lb.

6.70 $A_x = -310$ lb, $A_y = -35$ lb, $B_x = 80$ lb, $B_y = -80$ lb, $C_x = 310$ lb, $C_y = 195$ lb, $D_x = -80$ lb, $D_y = -80$ lb.

6.72 $|\mathbf{B}| = 1200$ N.

6.76 $A_x = -22$ lb, $A_y = 15$ lb, $C_x = -14$ lb, $C_y = 3$ lb.

6.78 300 lb (C).

6.80 110 kip.

6.82 539 N.

6.84 $A_x = 2$ kN, $A_y = -1.52$ kN, $B_x = -2$ kN, $B_y = 1.52$ kN.

6.86 Axial force is 4 kN compression, reaction at A is 4.31 kN.

6.88 BC: 1270 N (C).

6.90 $|\mathbf{B}| = 726$ N; CD: 787 N (C).

6.92 742 lb.

6.94 $A_x = -8$ kN, $A_y = 2$ kN, axial force = 8 kN.

6.96 $T_{AB} = 7.14$ kN (C), $T_{AC} = 5.71$ kN (T), $T_{BC} = 10$ kN (T).

6.98 BC: 120 kN (C); BG: 42.4 kN (T); FG: 90 kN (T).

6.100 AC: 480 N (T); CD: 240 N (C); CF: 300 N (T).

6.102 $A_x = -1.57$ kN, $A_y = 1.18$ kN, $B_x = 0$, $B_y = -2.35$ kN, $C_x = 1.57$ kN, $C_y = 1.18$ kN.

6.104 The force on the bolt is 972 N. The force at A is 576 N.

Chapter 7

7.2 $\bar{y} = 27/10$.

7.4 $\bar{y} = 4.46$.

7.6 $\bar{x} = a(n + 1)/(n + 2)$.

7.8 $\bar{x} = 0.711$ ft, $y = 0.584$ ft.

7.10 $\bar{x} = 0$, $y = 1.6$ ft.

7.12 $\bar{x} = 4$.

7.14 $\bar{x} = 0.533$.

7.16 $\bar{x} = \bar{y} = 9/20$.

7.18 $\bar{y} = -7.6$.

7.20 $\bar{x} = 2.27$.

7.22 $a = 0.656$, $b = 6.56 \times 10^{-5}$ m^{-2}.

7.24 $\bar{x} = \bar{y} = 4R/3\pi$.

7.26 $\bar{y} = \left[(2R^3/3) - R^2h + h^3/3\right]/2A$, where the area $A = (R/2)\left[(\pi R/2) - h(1 - h^2/R^2)^{1/2} - R \arcsin(h/R)\right]$.

7.28 $\bar{x} = 0$, $\bar{y} = 47.5$ mm.

7.30 $\bar{x} = 9.90$ in., $\bar{y} = 0$.

7.32 $\bar{x} = -1.12$ in., $\bar{y} = 0$.

7.34 $\bar{x} = 9$ in., $\bar{y} = 13.5$ in.

7.36 $\bar{x} = 3.67$ mm, $\bar{y} = 21.52$ mm.

7.38 $b = 39.6$ mm, $h = 18.2$ mm.

7.40 $\bar{y} = 4.60$ m.

7.42 $\bar{y} = 4.02$ in.

7.44 $\bar{x} = 6.47$ ft, $\bar{y} = 10.60$ ft.

7.46 (a) 720 N. (b) 720 N at $x = 4$ m. (c) $A_x = 0$, $A_y = 240$ N, $B = 480$ N.

7.48 $A_x = 0$, $A_y = 300$ N, $M_A = 1500$ N-m.

7.50 $A_x = 0$, $A_y = 10$ kN, $M_A = -31.3$ kN-m.

7.52 $A_x = 0$, $A_y = -25.9$ N, $B = 263.5$ N.

7.54 $A_x = 0$, $A_y = 4940$ lb, $B = 660$ lb.

7.56 $A_x = 0$, $A_y = 350$ N, $B_x = 0$, $B_y = -200$ N.

7.58 $A_x = -18$ kN, $A_y = 20$ kN, $B_x = 0$, $B_y = -4$ kN, $C_x = 18$ kN, $C_y = -16$ kN.

7.60 $V = 275$ m^3, height $= 2.33$ m.

7.62 $V = 0.032$ m^3, $\bar{x} = 0.45$ m, $\bar{y} = 0$, $\bar{z} = 0$.

7.64 $\bar{x} = 0.675R$, $\bar{y} = 0$, $\bar{z} = 0$.

7.66 $\bar{x} = h\left[(2R/3) + a/4\right]/(R + a/3)$, $\bar{y} = 0$, $\bar{z} = 0$.

7.68 $\bar{x} = 3.24$.

7.70 $\bar{x} = R\sin\alpha/\alpha$, $\bar{y} = R(1 - \cos\alpha)/\alpha$

7.72 $\bar{x} = 15.7$ in., $\bar{y} = 13.3$ in., $\bar{z} = 10$ in.

7.74 $\bar{x} = 88.4$ mm, $\bar{y} = \bar{z} = 0$.

7.76 $\bar{x} = 0$, $\bar{y} = 43.7$ mm, $\bar{z} = 38.2$ mm.

7.78 $\bar{x} = 229.5$ mm, $\bar{y} = \bar{z} = 0$.

7.80 $\bar{x} = 23.65$ mm, $\bar{y} = 36.63$ mm, $\bar{z} = 3.52$ mm.

7.82 $\bar{x} = 6$ m, $\bar{y} = 1.83$ m.

7.84 $\bar{x} = 65.9$ mm, $\bar{y} = 21.7$ mm, $\bar{z} = 68.0$ mm.

7.86 $A_x = 0$, $A_y = 294$ N, $B = 196$ N.

7.88 $A_x = 0$, $A_y = 316$ N, $B = 469$ N.

7.90 $A = 80.7$ kN, $B = 171.6$ kN.

7.92 $A_x = 0$, $A_y = 3.16$ kN, $M_A = 1.94$ kN-m.

7.94 $\bar{x} = 121$ mm, $\bar{y} = 0$, $\bar{z} = 0$.

7.96 $\bar{x}_3 = 82$ mm, $\bar{y}_3 = 122$ mm, $\bar{z}_3 = 16$ mm.

7.98 $\bar{y} = 34.05$ mm, $\bar{z} = 8.45$ mm.

7.100 Mass $= 408$ kg, $\bar{x} = 2.5$ m, $\bar{y} = -1.5$ m.

7.102 (a), (b) $I_x = \frac{1}{3}bh^3$, $k_x = \frac{1}{\sqrt{3}}h$.

7.104 $I_x = 7.20 \times 10^5$ mm^4, $k_x = 17.3$ mm, $I_y = 3.20 \times 10^5$ mm^4, $k_y = 11.5$ mm.

7.106 $I_y = \frac{1}{12}hb^3$, $k_y = \frac{1}{\sqrt{6}}b$.

7.108 $I_{xy} = \frac{1}{24}b^2h^2$.

7.110 $I_x = b^{3n+1}/(9n + 3)$.

7.112 $I_{xy} = b^{2n+2}/(4n + 4)$.

7.114 $I_x = 0.0500$, $k_x = 0.447$.

7.116 $I_{xy} = 0.0625$.

7.118 $I_x = 78.0$ ft^4, $k_x = 1.91$ ft.

7.120 (a) $I_x = \frac{1}{8}\pi R^4$, $k_x = \frac{1}{2}R$.

7.122 $I_y = 49.09$ m^4, $k_y = 2.50$ m.

7.124 $I_y = 522$, $k_y = 2.07$.

7.128 $I_y = 10.00$ m^4, $k_y = 1.29$ m.

7.130 $I_y = 1512$ m^4, $k_y = 4.58$ m.

7.132 $I_y = 5.92 \times 10^6$ mm^4, $k_y = 29.3$ mm.

7.134 $I_y = 3.6 \times 10^5$ mm^4, $J_0 = 1 \times 10^6$ mm^4.

7.136 2.65×10^8 mm^4, $k_x = 129$ mm.

7.138 $I_x = 7.79 \times 10^7$ mm^4, $k_x = 69.8$ mm.

7.140 $I_{xy} = 1.08 \times 10^7$ mm^4.

7.142 $J_0 = 363$ ft^4, $k_0 = 4.92$ ft.

7.144 $I_x = 10.7$ ft^4, $k_x = 0.843$ ft.

7.146 $I_{xy} = 7.1$ ft^4.

7.148 $J_O = 5.63 \times 10^7$ mm^4, $k_O = 82.1$ mm.

7.150 $I_x = 1.08 \times 10^7$ mm^4, $k_x = 36.0$ mm.

7.152 $J_O = 1.58 \times 10^7$ mm^4, $k_O = 43.5$ mm.

7.154 $J_0 = 2.35 \times 10^5$ in^4, $k_0 = 15.1$ in.

7.156 $I_x = 49.7$ m^4, $k_x = 2.29$ m.

7.158 $\bar{x} = 87.3$ mm, $\bar{y} = 55.3$ mm.

7.160 917 N (T).

7.162 $T_B = T_C = 15.2$ kN.

7.164 $I_y = 2.75 \times 10^7$ mm^4, $k_y = 43.7$ mm.

7.166 $I_x = 5.03 \times 10^7$ mm^4, $k_x = 59.1$ mm.

Chapter 8

8.2 (a), (b) $\mu_s = 1$.

8.4 (a) $f = 10.1$ N toward the left. (b) $F = 52.0$ N.

8.6 (a) No. (b) 46.8 N. (c) 45.1 N.

8.8 $T = 112.94$ N.

8.10 20 lb.

8.12 $\alpha = 14.0°$.

8.14 $T = 56.5$ N.

8.16 (a) Yes. The force is $\mu_s W$. (b) $3\mu_s W$.

8.18 $T = 455$ N.

8.20 $\alpha = 40.0°$.

8.22 $F = W(\sin\alpha - 5\mu_s\cos\alpha)$.

8.24 (a) $\alpha = \arctan(\mu_s)$; (b) $\mu_k W/4$.

8.26 $\mu_s = 0.529$.

8.28 (a) $x = 2.07$ m; (b) $\mu_s = 0.66$.

8.30 $M = \mu_s RW(1 + \mu_s)/(1 + \mu_s^2)$.

8.32 $\mu_k = 0.35$.

8.34 $\alpha = 2\arctan(\mu_s)$.

8.36 27.7 N-m.

8.38 240 N.

8.40 $y = 234$ mm.

8.42 $\alpha = 9.27°$.

8.44 (a) $F = 84$ lb; (b) Yes.

8.48 $\alpha = 1.54°$, $P = 202$ N.

8.50 (a) $F = \mu_s W$;
(b) $F = (W/2)(\mu_{sA} + \mu_{sB})/[1 + (h/b)(\mu_{sA} - \mu_{sB})]$.

8.52 $F = 272$ N.

8.54 Yes. It is necessary that $\mu_s \geq 0.268$.

8.56 $F = 2.30$ kN.

8.58 $F = 156$ N.

8.60 343 kg.

8.62 No. The minimum value of μ_s required is 0.176.

8.64 $F = 1160$ N.

8.66 1.84 N-m.

8.68 $\mu_s = 0.0398$.

8.70 (a) 2.39 ft-lb; (b) 1.20 ft-lb.

8.72 11.8 ft-lb.

8.74 15.1 N-m.

8.76 10.4 N-m.

8.78 2.02 kg.

8.80 106 N.

8.82 51.9 lb.

8.84 $T = 40.9$ N.

8.86 $M = 0.3$ N-m.

8.88 $M = 12.7$ N-m.

8.90 $M = 7.81$ N-m.

8.92 $M = 5.20$ N-m.

8.94 21.6 lb.

8.96 $T_C = 107$ N.

8.98 $M = rW(e^{\pi\mu_k} - 1)$.

8.100 (a) 14.2 lb; (b) 128.3 lb.

8.102 $T = 50.1$ N.

8.104 (a) $f = 10.3$ lb.

8.106 $\alpha = 24.2°$.

8.108 $f = 2.63$ N.

8.110 $M = 1.13$ N-m.

8.112 $P = 43.5$ N.

Chapter 9

9.2 $\tau_{av} = 0$.

9.4 (a) $\sigma_{av} = 318.3$ psi. (b) $\sigma_{av} = -636.6$ psi.

9.6 $\tau_{av} = -140$ kPa, $P = 7.15$ kN.

9.8 $\sigma_{av} = 27.5x$ kPa.

9.10 $\sigma_{av} = 64$ kPa, $\tau_{av} = 5.33$ kPa.

9.12 $\sigma_{av} = 5.96$ MPa (864 psi).

9.14 $\tau_{av} = 34.4$ psi.

9.16 $\tau_{av} = 41.7$ MPa.

9.18 $\sigma_{av} = 102$ ksf.

9.20 $\tau_{av} = 3601$ psi.

9.22 $\sigma_{av} = 30.3$ MPa, $\tau_{av} = 10.0$ MPa.

9.24 $\sigma_{av} = -66.7$ psi, $\tau_{av} = 115.5$ psi.

9.26 $\tau_{av} = 18.75$ psi.

9.28 $\tau_{av} = 21.2$ MPa.

9.30 $\tau_{av} = F/bt$.

9.32 $\tau_{av} = 1.82$ psi.

9.34 $\sigma = 45$ kPa.

9.36 $\sigma = 3430$ psi.

9.38 $dL' = 1.15\, dL$.

9.40 $L' = 0.206$ m.

9.42 $L' = 0.211$ m.

9.44 $\varepsilon = -0.240$.

9.46 $\delta = 0.04 \ln 2 = 0.028$ in.

9.48 $\varepsilon_{AB} = 0.134$.

9.50 $\varepsilon = 0.00454$.

9.52 $\varepsilon = 0.002$.

9.54 $\varepsilon_1 = 0.000333$, $\varepsilon_2 = -0.008333$.

9.56 $\gamma = 0.698$.

9.58 $\gamma = 0.290$.

9.60 $\gamma = 0.0242$.

9.62 $P = 800$ lb, $\sigma_{av} = 240$ psi.

9.64 $\sigma_{av} = 200$ kPa, $\tau_{av} = 400$ kPa.

9.66 $F = 9.6$ kN.

9.68 $\tau_{av} = 2.32$ MPa.

9.70 $\delta = -0.028$ in.

Chapter 10

10.2 $P = 240$ kip.

10.4 $\sigma = 9$ ksi.

10.6 $\sigma = -7.15$ MPa.

10.8 $\sigma = F/2A \sin\beta$.

10.10 $\beta = 45°$, $A = F/2\sigma_0 \sin\beta$.

10.12 $F = 49.2$ kN.

10.14 $\sigma_{BC} = -75.0$ MPa.

10.16 $\sigma_\theta = 1170$ psi, $\tau_\theta = -3214$ psi.

10.18 $\theta = 50.2°$, $P = 61.0$ kN.

10.20 $F = 5196$ lb.

10.22 10.02 in.

10.24 $E = 6.37$ GPa, $v = 0.30$.

10.26 Force = 29.5 kip, diameter = 0.749 in.

10.28 $\delta_{AC} = -0.0532$ in.

10.30 0.400 mm downward.

10.32 1.95° clockwise.

10.34 $\varepsilon_{AB} = 0.000349$, $\varepsilon_{CD} = 0.000698$, $\varepsilon_{EF} = 0.001047$.

10.36 $\sigma = 2F/3A$.

10.38 900 kN.

10.40 $\sigma_B = -7.16$ MPa.

10.42 $F_1 = 97.1$ kN.

10.44 $b = 0.0681$ in.

10.46 $\sigma_{AB} = -F\cos^2\theta/[A(1 + \cos^3\theta)]$,
$\sigma_{AC} = -F/[A(1 + \cos^3\theta)]$.

10.48 310 kN.

10.50 $h = 2.95$ mm.

10.52 $\delta = 0.0116$ in.

10.54 $\delta = 0.392$ mm.

10.56 $\sigma = 30.6$ ksi.

10.58 $\delta = 0.127$ mm.

10.60 $\delta = 6.13$ mm.

10.62 $\delta = 0.0441$ m.

10.64 $\delta = 0.0535$ m.

10.66 $x = L/2$, displacement $= qL^2/8EA$.

10.68 125°F.

10.70 30.024 mm.

10.72 $\sigma = 0$.

10.74 (a), (b) $\delta = 0.0111$ in.

10.76 134°F.

10.78 $\sigma = -7467$ psi.

10.80 $\sigma_A = -16.8$ MPa, $\sigma_B = -67.2$ MPa.

10.82 16,000 lb downward.

10.84 Either 2014-T6 or 7075-T6.

10.86 ASTM-A514.

10.88 3.69 in².

10.90

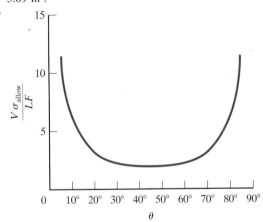

10.92 2014-T6 or 7075-T6.

10.94 $\sigma = 2.31$ MPa.

10.96 $\sigma = 1.50$ MPa.

10.98 $D' = 0.7495$ in.

10.100 -0.0150 mm.

10.102 $\sigma_{AB} = 7.45$ ksi, $\sigma_{AC} = -10.17$ ksi, $\sigma_{AD} = 3.19$ ksi.

Chapter 11

11.2 $\beta = 89.9°$.

11.4 $\sigma_\theta = 10.4$ MPa, $|\tau_\theta| = 6$ MPa.

11.6 $\tau = 16.2$ MPa.

11.8 (a) $\sigma_\theta = -17.3$ ksi, $|\tau_\theta| = 10$ ksi. (b) 20 ksi.

11.10 $\gamma = 0.00346$.

11.12 $J = 23.6$ in⁴.

11.14 $|\tau| = 17.0$ MPa.

11.16 $\tau = 19.9$ MPa.

11.18 $\tau = 11.7$ MPa.

11.20 $\phi = 0.000382$ rad (0.0219°).

11.22 (a) $|\tau| = 5093$ psi. (b) $|\phi| = 1.006°$.

11.24 $|\tau_A| = 19.89$ ksi, $|\tau_B| = 8.49$ ksi.

11.26 $|\tau_A| = 3.98$ ksi, $|\tau_B| = 8.49$ ksi, $|\phi| = 0.686°$.

11.28 $T = 1.99$ kN-m.

11.30 $\tau_{AB} = 37.7$ MPa, $\tau_{CD} = 28.3$ MPa.

11.32 $r_C = 108$ mm.

11.34 $|T_O| = 13.7$ in-kip.

11.36 $T_A = 1107.7$ N-m, $T_B = -92.3$ N-m.

11.38 $|\phi_A| = 1.820°$, $|\phi_B| = 0.180°$.

11.40 $|\tau| = 0.656$ GPa.

11.42 $|\tau| = 21.9$ MPa.

11.44 $\phi = 0.0150$ rad (0.861°).

11.46 $\phi = 4.60°$.

11.48 $T = 6.19$ N-m.

11.50 102 N-m.

11.52 $|\tau| = 40.7$ MPa.

11.54 $c_0 = 7200$ in-lb/in., $\tau = 18.7$ ksi.

11.56 $|\tau| = 325$ MPa.

11.58 $\phi = 5.28$ rad (302°).

11.60 $\sigma_\theta = 20.5$ kPa, $|\tau_\theta| = 24.4$ kPa.

11.62 $|\tau_A| = 8.13$ ksi, $|\tau_B| = 4.06$ ksi.

11.64 $T_{\text{left}} = c_0L/12$, $T_{\text{right}} = c_0L/4$.

11.66 $T_{\text{left}} = 5c_0L/192$, $T_{\text{right}} = c_0L/64$.

Chapter 12

12.2 $\sigma'_x = 25$ ksi, $\sigma'_y = -25$ ksi, $\tau'_{xy} = 0$.

12.4 $\sigma_x = 64.00$ MPa, $\sigma_y = 85.00$ MPa, $\tau_{xy} = 0.00$ MPa.

12.6 $\theta = -20.0°$ or 160°.

12.8 $\sigma_x = 41.81$ MPa, $\sigma_y = -25.81$ MPa, $\tau_{xy} = -2.96$ MPa.

12.10 $\sigma = -2.23$ ksi, $|\tau| = 1.60$ ksi.

12.12 $\sigma = -7.86$ MPa, $\tau = 13.50$ MPa.

12.14 $\tau'_{xy} = 4.90$ MPa, $\theta = 19.5°$ or $\tau'_{xy} = -4.90$ MPa, $\theta = 40.2°$.

12.16 $\tau_{xy} = -78.4$ psi, $\tau'_{xy} = -114.1$ psi.

12.20 $\sigma_1 = \sigma_x$, $\sigma_2 = 0$, $\tau_{\max} = |\sigma_x/2|$.

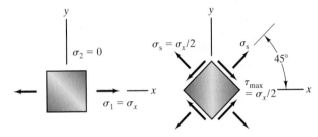

12.22 $\sigma_1 = 20$ MPa, $\sigma_2 = 10$ MPa, $\tau_{\max} = 5$ MPa.

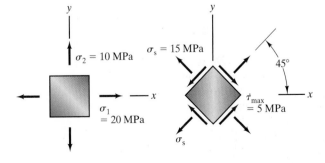

12.24 $\sigma_1 = 8.22$ ksi, $\sigma_2 = -10.22$ ksi, $\tau_{max} = 9.22$ ksi.

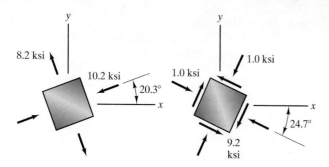

12.26 Absolute maximum shear stress = 10 MPa.
12.28 Absolute maximum shear stress = 6.54 ksi.
12.30 $\sigma_x' = 25$ ksi, $\sigma_y' = -25$ ksi, $\tau_{xy}' = 0$.
12.32 $\sigma_x = 64.00$ MPa, $\sigma_y = 85.00$ MPa, $\tau_{xy} = 0.00$ MPa.
12.34 $\theta = -20°$.
12.36 $\sigma = 5.77$ MPa, $|\tau| = 4.89$ MPa.
12.38 $\sigma = 177.2$ psi, $\tau = -237.3$ psi.
12.40 See the answer to Problem 12.22.
12.42 See the answer to Problem 12.24.
12.44 Absolute maximum shear stress = 6.54 ksi.
12.46 $\sigma_1 = 40.45$ ksi, $\sigma_2 = 0$, $\sigma_3 = -15.45$ ksi,
  $\tau_{max} = 27.95$ ksi.
12.48 $\sigma_1 = 85.00$ MPa, $\sigma_2 = 65.00$ MPa, $\sigma_3 = 0$,
  $\tau_{max} = 42.50$ MPa.
12.50 $\sigma_1 = 240$ MPa, $\sigma_2 = 240$ MPa, $\sigma_3 = -120$ MPa,
  $\tau_{max} = 180$ MPa.
12.52 $\sigma_1 = 409$ ksi, $\sigma_2 = 148$ ksi, $\sigma_3 = -257$ ksi,
  $\tau_{max} = 333$ ksi.
12.54 (a), (b) $\sigma_1 = 8.22$ ksi, $\sigma_2 = -10.22$ ksi, $\sigma_3 = 0$.
12.56 $\sigma = 142$ ksi, $\tau_{max} = 71.3$ ksi.
12.58 $\sigma_h = 250$ MPa.
12.60 $\sigma_h = 15$ MPa, $\tau_{max} = 7.65$ MPa.
12.62 $\sigma_1 = 52.4$ MPa, $\sigma_2 = -32.4$ MPa, absolute maximum
  shear stress = 42.4 MPa.
12.64 See the answer to Problem 12.20.
12.66 See the answer to Problem 12.20.
12.68 $\sigma = 20$ MPa, $\tau_{max} = 10.09$ MPa.

Chapter 13

13.2 $\varepsilon_x' = 0.002$, $\varepsilon_y' = -0.002$, $\gamma_{xy}' = 0$.
13.4 $\theta = -20.0°$.
13.6 $\varepsilon_x = 0.00640$, $\varepsilon_y = 0.00850$, $\gamma_{xy} = 0.00000$.
13.8 $\gamma_{xy}' = 0.0098$, $\theta = 19.5°$ or $\gamma_{xy}' = 0.0098$, $\theta = 40.2°$.
13.10 $PQ = 0.99665$ mm.
13.12 $\gamma_{xy} = -0.00360$.
13.14 1.57626 rad (90.313°).
13.16 $\gamma_{xy} = 0$.

13.18 $\varepsilon_1 = 0.002$, $\varepsilon_2 = 0.001$, $\gamma_{max} = 0.001$.

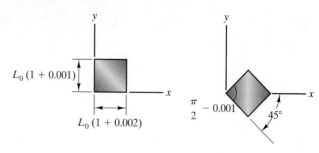

13.20 $\varepsilon_1 = 0.00822$, $\varepsilon_2 = -0.01022$, $\gamma_{max} = 0.01844$.

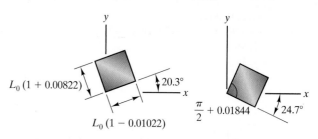

13.22 Absolute maximum shear strain = 0.00314.

13.26 $\begin{bmatrix} 26.9 & 13.5 & 0 \\ 13.5 & 188.5 & -13.5 \\ 0 & -13.5 & 134.6 \end{bmatrix}$ MPa.

13.28 The required condition is that $\sigma_x + \sigma_y = 0$.
13.30 $\sigma_x = 64.0$ ksi, $\sigma_y = 83.5$ ksi, $\tau_{xy} = -93.8$ ksi.
13.32 (a) $\lambda = 16.2$ GPa, $\mu = 10.8$ GPa. (b) $K = 23.3$ GPa.

13.34 $\begin{bmatrix} 34.2 & 56.4 & -56.4 \\ 56.4 & -33.5 & 0 \\ -56.4 & 0 & -44.8 \end{bmatrix}$ ksi.

13.36 (a) $\varepsilon_x = \sigma_x/E$, $\varepsilon_y = \varepsilon_z = -\nu\sigma_x/E$, other strain
  components equal zero.
  (b) Volume = $LA[1 + (1 - 2\nu)\sigma_x/E]$.
13.38 $\sigma_x = -55.5$ MPa, $\sigma_y = 104.4$ MPa,
  $\tau_{xy} = -94.3$ MPa.
13.40 $\sigma_x = 412.6$ MPa, $\sigma_y = 444.2$ MPa, $\tau_{xy} = -74.5$ MPa.
13.42 $\varepsilon_x' = -0.00257$, $\varepsilon_y' = 0.00457$, $\gamma_{xy}' = 0.00207$.
13.44 $\varepsilon_x = -0.00160$, $\varepsilon_y = 0.00100$, $\gamma_{xy} = 0.00201$.
13.46 Absolute maximum shear strain = 0.00559.
13.48 $\sigma_x = -253$ MPa, $\sigma_y = 809$ MPa, $\tau_{xy} = -234$ MPa.

Chapter 14

14.2 $P_A = P_B = P_C = 0$, $V_A = V_B = V_C = 2$ kN,
  $M_A = 2$ kN-m, $M_B = 4$ kN-m, $M_C = 6$ kN-m.
14.4 $P_A = 866$ lb, $V_A = -500$ lb, $M_A = 3000$ ft-lb.
14.6 (a) $P_B = 0$, $V_B = -20$ N, $M_B = -5$ N-m.
  (b) $P_B = 0$, $V_B = -20$ N, $M_B = 5$ N-m.

14.8 (a) $P_A = 0, V_A = 4$ kN, $M_A = 4$ kN-m.
(b) $P_A = 0, V_A = 2$ kN, $M_A = 3$ kN-m.

14.10 $P_A = 0, V_A = 16.7$ lb, $M_A = 575$ in-lb.

14.12 $P_A = 0, V_A = -475$ lb, $M_A = -1275$ ft-lb.

14.14 $P_A = 0, V_A = 4.8$ kN, $M_A = 13.6$ kN-m.

14.16 $P_C = 0, V_C = -3.7$ kN, $M_C = 14.1$ kN-m.

14.18 $P = 0, V = -7.81$ kN, $M = 4.56$ kN-m.

14.20 (a) $P = 0, V = -50x + 250$ kN,
$M = -25x^2 + 250x$ kN-m.

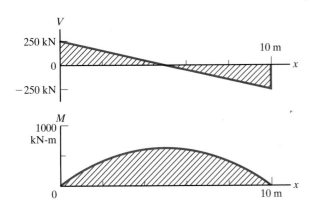

14.22 (a) $P = 0, V = 1080 - 10x^2$ lb, $M = 1080x - (10/3)x^3$ ft-lb. (b) $M = 7482$ ft-lb at $x = 10.39$ ft.

14.24 $P = 0, V = -20$ N, $M = -20x$ N-m.

14.26 $P = 0, V = -100(4 + x^2/12)$ lb,
$M = -100(4x + x^3/36)$ ft-lb.

14.28 (a) $0 < x < 6$ ft: $P = 0, V = 300$ lb,
$M = 300x - 3000$ ft-lb.
$6 < x < 12$ ft: $P = 0, V = 100(x - x^2/12)$ lb,
$M = 100(-24 + x^2/2 - x^3/36)$ ft-lb.
(b)

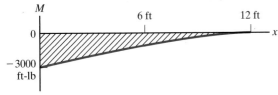

14.30 $M = 578$ in-lb at $x = 9.33$ in.

14.32

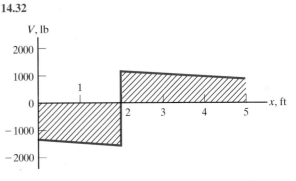

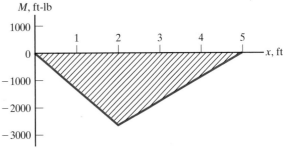

14.34

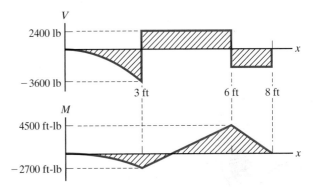

14.36

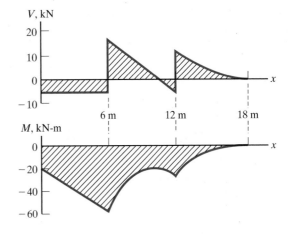

14.38 $V = -31.25 + 10.00x + 0.25x^2 - 0.20x^3$ kN.

14.40 $V = -w_0(x - x^2/2L)$, $M = -w_0(x^2/2 - x^3/6L)$.

14.42 $P = 0$, $V = F$, $M = Fx$.

14.44 $V = w_0L/6 - w_0x^2/2L$, $M = (Lx - x^3/L)w_0/6$.

14.48 $0 < x < 2$ m: $P = 0$, $V = x$ kN, $M = x^2/2$ kN-m.
$2 < x < 5$ m: $P = 0$, $V = -4 + x$ kN, $M = 8 - 4x + x^2/2$ kN-m. $5 < x < 6$ m: $P = 0$,
$V = -6 + x$ kN, $M = 18 - 6x + x^2/2$ kN-m.

14.52 (a) $P_A = 0$, $V_A = -F$, $M_A = -LF/2$. (b) $P_A = 0$,
$V_A = F$, $M_A = -LF/2$.

14.54 $P_A = 0$, $V_A = 225$ kN, $M_A = -375$ kN-m.

14.56 No. The resulting maximum bending moment
magnitude is 8 kN-m.

14.58

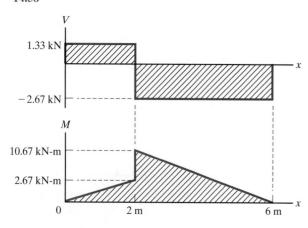

Chapter 15

15.2 $M = 492$ kN-m.

15.4 (a) $\rho = -1170$ in. (b) $\sigma_x = 23{,}900$ psi.

15.6 $M = 176{,}700$ in-lb.

15.8 $\sigma_x = 2\sqrt{3}\, w_0L^2/9h^3$ at $x = L/\sqrt{3}$, $y = h/2$.

15.10 (a) $\sigma_x = 128.8$ MPa. (b) $\sigma_x = 78.6$ MPa.

15.12 (a) $\sigma_x = 13.10$ ksi. (b) $\sigma_x = 6.75$ ksi.

15.14 $\sigma_x = 1.24$ GPa.

15.16 $S = 1.87$.

15.18 $h = 3.90$ in.

15.20 (a) $S = 2.01$. (b) $S = 4.69$.

15.24 (a) $\tau_{av} = 4.44$ MPa. (b) $\tau_{av} = 2.78$ MPa.

15.26 $|\tau_{av}| = 576$ kPa at $x = 0$, $y' = 0$ and at $x = 8$ m,
$y' = 0$.

15.28 (a) $\tau_{av} = -2810$ psi. (b) $\tau_{av} = -1230$ psi.

15.32 $\tau_{av} = 6.86$ MPa.

15.34 $\tau_{av} = 188$ psi.

15.36 $\tau_{av} = 88.5$ psi on each joint.

15.38 $y' = 1$ in., $|\tau_{av}| = 4170$ psi.

15.40 (a) $\tau_{av} = 11.42$ MPa. (b) $\tau_{av} = 1.82$ MPa.

15.42 $\sigma_x = 41.0$ kPa, $\sigma_x = -86.3$ kPa.

15.46 (a) $\tau_{av} = -19.9$ MPa. (b) $\tau_{av} = -11.1$ MPa.

15.48 $\tau_{av} = 6.31$ MPa.

15.50 $\tau_{av} = 5.35$ MPa.

Chapter 16

16.2 $v = 6.22$ mm, $v' = 0$.

16.4 $v = 0.558$ in, $v' = 0.0106$ rad.

16.6 $|v| = 2.97$ mm at $x = 1.15$ m.

16.18 $\sigma_x = 5.04$ MPa.

16.20 $v = 3.02$ mm at $x = 1.63$ m.

16.22 $M_0 = 155$ kN-m.

16.24 $x = 36.4$ in.

16.26 $v = -(M_0/4LEI)(Lx^2 - x^3)$.

16.28 $A_x = 0$, $A_y = -9w_0L/40$, $M_A = 7w_0L^2/120$
counterclockwise, $B_y = -11w_0L/40$.

16.30 (a)

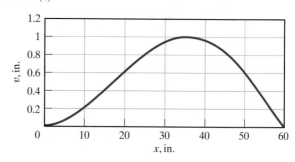

(b) $v = 0.988$ in. at $x = 35.9$ in.

16.32 $v = (w_0x^2/24EI)(L^2 - 2Lx + x^2)$.

16.34 $w_0 = 38.1$ kN/m.

16.36 $v = (w_0x^2/120LEI)(2L^3 - 3L^2x + x^3)$.

16.38 $v = -27.1$ mm.

16.40 $A_x = 0$, $A_y = -3w_0L/32$, $M_A = 5w_0L^2/192$
counterclockwise, $B_x = 0$, $B_y = -13w_0L/32$,
$M_B = 11w_0L^2/192$ clockwise.

16.42 $v = (w_0/384EI)$
$(16x^4 - 38Lx^3 + 29L^2x^2 - 8L^3x + L^4)$.

16.44 $M_0 = 2FL/3$.

16.46 $M = M_0 - F(L - x)$.

16.48 $M_0 = w_0L^2/4$.

16.50 $v = (w_0x/24EI)(L^3 - 2Lx^2 + x^3)$
$- (Fx/48EI)(3L^2 - 4x^2)$.

16.52 $A_x = 0$, $A_y = C_y = -3w_0L/16$, $B_y = -5w_0L/8$.

16.54 $v = (M_0x/2EI)(L - x)$.

16.56 $v = (w_0x^2/24EI)(L^2 - 2Lx + x^2)$.

16.58 $A_x = 0$, $A_y = 3M_0/2L$, $M_A = M_0/2$ clockwise,
$B_y = -3M_0/2L$.

16.60 $B_y = -19w_0L/56$, $C_y = -12w_0L/56$.

16.64 $A_x = 0$, $A_y = -11F/16$, $M_A = 3LF/16$
counterclockwise, $B_y = -5F/16$.

16.66 $v = (F/96EI)(5x^3 - 15Lx^2 + 12L^2x - 2L^3)$.

16.68 $v = (Fx/18EI)(2L^2 - 3x^2)$.

16.70 $A_x = 0$, $A_y = -11F/16$, $M_A = 3LF/16$
counterclockwise, $B_y = -5F/16$.

Chapter 17

17.2 $P = 965$ kip.
17.6 $R = 25.5$ mm.
17.8 $F = 80.4$ kip.
17.10 $m = 250$ kg.
17.12 $b = 3.10$ in., $F = 5.66$ kip.
17.14 $r_{AB} = 29.6$ mm, $r_{CD} = 0.892$ mm.
17.16 $F = 3.18$ kN.
17.18 $P = 31.7$ kip.
17.20 $P = 412$ kN.
17.22 $P = 7.20$ MN. It bends in the x–y plane.
17.24 $P = 1.66$ MN. It bends in the x–y plane.
17.26

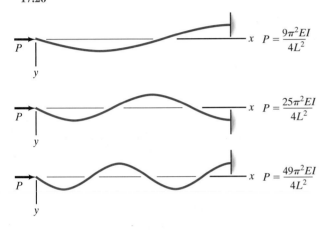

$$x \quad P = \frac{9\pi^2 EI}{4L^2}$$

$$x \quad P = \frac{25\pi^2 EI}{4L^2}$$

$$x \quad P = \frac{49\pi^2 EI}{4L^2}$$

17.28 $L_e = L$, $L/2$, $L/3$, and $L/4$.
17.30 $L_e = 2L$, $P = \pi^2 EI/4L^2$.
17.32 $P = 202$ kN.
17.34 $F = 3.02$ kN.
17.36 $F = 368$ N.
17.38 $P = \pi^2 EI/L^2$.

Index

Properties of Areas and Lines

Areas

The coordinates of the centroid of the area A are

$$\bar{x} = \frac{\displaystyle\int_A x\, dA}{\displaystyle\int_A dA}, \qquad \bar{y} = \frac{\displaystyle\int_A y\, dA}{\displaystyle\int_A dA}.$$

The moment of inertia about the x axis I_x, the moment of inertia about the y axis I_y, and the product of inertia I_{xy} are

$$I_x = \int_A y^2\, dA, \qquad I_y = \int_A x^2\, dA, \qquad I_{xy} = \int_A xy\, dA.$$

The polar moment of inertia about O is

$$J_O = \int_A r^2\, dA = \int_A (x^2 + y^2)\, dA = I_x + I_y.$$

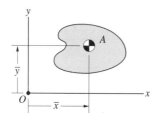

Rectangular area

Area $= bh$

$$I_x = \frac{1}{3} bh^3, \qquad I_y = \frac{1}{3} hb^3, \qquad I_{xy} = \frac{1}{4} b^2 h^2$$

$$I_{x'} = \frac{1}{12} bh^3, \qquad I_{y'} = \frac{1}{12} hb^3, \qquad I_{x'y'} = 0$$

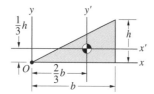

Triangular area

Area $= \dfrac{1}{2} bh$

$$I_x = \frac{1}{12} bh^3, \qquad I_y = \frac{1}{4} hb^3, \qquad I_{xy} = \frac{1}{8} b^2 h^2$$

$$I_{x'} = \frac{1}{36} bh^3, \qquad I_{y'} = \frac{1}{36} hb^3, \qquad I_{x'y'} = \frac{1}{72} b^2 h^2$$

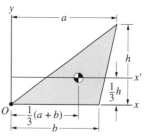

Triangular area

Area $= \dfrac{1}{2} bh$ $\qquad I_x = \dfrac{1}{12} bh^3, \qquad I_{x'} = \dfrac{1}{36} bh^3$

Circular area

Area $= \pi R^2$ $\qquad I_{x'} = I_{y'} = \dfrac{1}{4} \pi R^4, \qquad I_{x'y'} = 0$

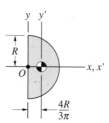

Semicircular area

Area $= \dfrac{1}{2} \pi R^2$ $\qquad I_x = I_y = \dfrac{1}{8} \pi R^4, \qquad I_{xy} = 0$

$$I_{x'} = \frac{1}{8} \pi R^4, \qquad I_{y'} = \left(\frac{\pi}{8} - \frac{8}{9\pi} \right) R^4, \qquad I_{x'y'} = 0$$